CW00351826

physical sciences data 21

radiant properties of materials

tables of radiant values for black body and real materials

physical sciences data

Other titles in this series:

1 J. Wisniak and A. Tamir, **Mixing and Excess Thermodynamic Properties**

2 J. R. Green and D. Margerison, **Statistical Treatment of Experimental Data**

3 K. Kojima and K. Tochigi, **Prediction of Vapor-Liquid Equilibria by the ASOG Method**

4 S. Fraga, J. Karwowski and K. M. S. Saxena, **Atomic Energy Levels**

5 S. Fraga, J. Karwowski and K. M. S. Saxena, **Handbook of Atomic Data**

6 M. Broul, J. Nyvlt and O. Söhnel, **Solubility in Inorganic Two-Component Systems**

7 J. Wisniak and A. Tamir, **Liquid-Liquid Equilibrium and Extraction**

8 S. Fraga and J. Muszyńska, **Atoms in External Fields**

9 A. Tslaf, **Combined Properties of Conductors**

10 J. Wisniak, **Phase Diagrams**

11 J. Wisniak and A. Tamir, **Mixing and Exces Thermodynamic Properties, Supplement 1**

12 K. Ohno and K. Morokuma, **Quantum Chemistry Literature Data Base**

13 A. Apelblat, **Table of Definite and Infinite Integrals**

14 A. Tamir, E. Tamir and K. Stephan, **Heats of Phase Change of Pure Components and Mixtures**

15 O. V. Mazurin, M. V. Streltsina and T. P. Shvaiko-Shvaikovskaya, **Handbook of Glass Data. Part A: Silica Glass and Binary Silicate Glasses**

16 S. Huzinaga (Editor), **Gaussian Basis Sets for Molecular Calculations**

17 T. Boublik, V. Fried and E. Hála, , **The Vapour Pressures of Pure Substances** (2nd revised edition)

18 J. Wisniak and M. Herskowitz, **Solubility of Gases and Solids**

19 D. Horváth and R. M. Lambrecht, **Exotic Atoms. A Bibligraphy 1939-1982**

20 R. K. Winge, V. A. Fassel, V. J. Peterson and M. A. Floyd, **Inductively Coupled Plasma-Atomic Emission Spectroscopy**

21 A. Sala, **Radiant Properties of Materials. Tables of Radiant Values for Black Body and Real Materials**

physical sciences data 21

radiant properties of materials

tables of radiant values for black body and real materials

aleksander sala

Institute of Precision Mechanics, Warsaw, Poland

ELSEVIER
Amsterdam – Oxford – New York – Tokyo
PWN–Polish Scientific Publishers
Warsaw
1986

Distribution of this book is being handled by the following publishers:
for the U. S. A. and Canada
ELSEVIER SCIENCE PUBLISHING CO., INC.
52, Vanderbilt Avenue
New York, N. Y. 10017

for Albania, Bulgaria, Cuba, Czechoslovakia, German Democratic Republic, Hungary, Korean People's Democratic Republic, Mongolia, People's Republic of China, Poland, Romania, the U. S. S. R., Vietnam and Yugoslavia
ARS POLONA
Krakowskie Przedmieście 7, 00-068 Warszawa 1, Poland

for all remaining areas
ELSEVIER SCIENCE PUBLISHERS B. V.
Molenwerf 1
P. O. Box 211, 1000 AE Amsterdam
The Netherlands

Library of Congress Cataloging in Publication Data

Sala, Aleksander.
Radiant properties of materials.

(Physical sciences data ; 21)
Bibliography: p.
Includes index.
1. Materials--Thermal properties. 2. Heat--Radiation and absorption. I. Title. II. Series.
TA418.52.S24 1984 620.1'129 84-13525
ISBN 0-444-99599-4 (U.S.)

ISBN 0-444-99599-4 (vol. 21)
ISBN 0-444-41689-7 (series)

Printed in Poland

PREFACE

This book deals with two kinds of bodies that are involved in all problems of radiative heat transfer, i.e. with an ideal body generally called the black body, and with real bodies called materials. The black body occurs in practice in the form of technical black bodies. The properties of man-made bodies only approximate to those of the theoretical black body.

The properties of a black body and the resulting radiative characteristics of this ideal body are discussed, and the radiative properties of real materials and their values are given in tables. This approach allows us to give a mathematical description of real phenomena of radiation and its magnitude without having to resort to individual experimentation in each case. This is of great importance in many branches of science and in such activities as design and construction.

The introduction to the book contains an alphabetical list of "symbols" for all the basic terms used in the book. The book itself is composed of three main parts:

Part I - Black body radiation and properties of materials;

Part II - Tables of black body radiation;

Part III - Tables of radiative properties of materials.

The first part deals with the properties of the black body and defines, on the basis of black body radiation laws, those radiation quantities whose values are included in the second part of the book. These are: hemispherical spectral emissive power of a black body for λ_{max}-$m_{\lambda max,bb}$;

hemispherical total emissive power of a black body - $M_{T,bb}$; wavelength of maximal emissive power for a given temperature T - λ_{max}; hemispherical spectral emissive power of a black body $m_{\lambda,bb}$ and hemispherical total emissive power of a black body from $\lambda_1 = 0$ to $\lambda_2 \infty$, i.e., $M_{T,o-\lambda,bb}$.

In the first part of the book the basic notion of radiative properties is defined and these properties are characterized according to the groups of materials concerned. The terms "radiative values" is taken to mean emissivity, reflectivity and transmittivity, and other physical properties of materials strictly connected with the first three properties mentioned, for ultraviolet, visible and infrared electromagnetic radiation. The radiative properties of materials involve radiation wavelength ranges of λ from 10^{-1} to 10^3 μm.

The third part of the book contains tables listing the values of the following properties: hemispherical total emissivity - ε_T; total emissivity in the normal direction - $\varepsilon_{T,n}$; spectral emissivity in the normal direction -$\varepsilon_{\lambda,n}$; spectral (red) emissivity in the normal direction -$\varepsilon_{\lambda=0.65,n}$; bidirectional spectral reflectivity in the normal direction - $r_{\lambda,n}$; spectral reflectivity in the normal direction of layer of thickness $l\tau_{\lambda,n}$. The values of the above radiative properties are listed for three groups of materials: metals and alloys, non-organic materials and biological and organic materials.

The book includes an index of materials and contains separate entries for the name of the material, its common name, its chemical formula, its standard symbol, etc. The index enables the reader to find quickly not only the page containing the required radiative value of a material but also the pages on which its remaining radiative values are quoted if these are included in the book. This permits a quick determination of the required radiative value of a given material from its other values in case the required value is not in the book.

Until now the literature on this subject, either in Poland or elsewhere, has contained no publication in the form of a collection of tables devoted exclusively to the radiation properties of materials. In most cases the tables of these values included in specialized handbooks occupy no more than a few pages.

In the present publication the author has summarised the results of more than twenty years of his investigations and compilations of all the data concerning the radiation properties of materials.

The values of the radiative properties of materials are needed in some of the very latest modern technologies. Examples include the energetics for heat transfer radiation, especially in heating elements, and in the use of solar energy; ion nondestructive testing and above all in thermovision; for military purposes, particularly in the technology of camouflaging soldiers and equipment, in criminology, for discovering forgeries of documents, banknotes, and works of art; in designing and testing heating and heat insulation systems in buildings; in the design and exploitation tests of combustion and electrical engines. The values of the radiative properties of materials are of considerable importance in scientific investigations in chemistry, biology, medicine, geology, astronautics, and astronomy.

Aleksander Sala

CONTENTS

SYMBOLS

a_T - hemispherical total absorptivity

$a_{T,\beta}$ - directional total absorptivity from direction β

a_λ - hemispherical spectral absorptivity

$a_{\lambda,\beta}$ - directional spectral absorptivity from direction β

$a_{\lambda,n}$ - directional spectral absorptivity from normal direction n

c - speed of electromagnetic radiation propagation in vacuum - $2.9979 \cdot 10^{8}$ m/s

h - Planck's constant - $6.6218 \cdot 10^{-34}$ J·s

$i_{\lambda,bb,n}$ - spectral intensity of black body in normal direction n, W/(μm²·sr)

$i_{\lambda,bb,}$ - spectral intensity of black body in direction , W/(μm·m²·sr)

$i_{\lambda,n}$ - spectral intensity of material in normal direction n, W/(μm·m²·sr)

$i_{\lambda,}$ - spectral intensity of material in direction , W/(μm·m²·sr)

k - Boltzmann constant $1.37975 \cdot 10^{-23}$ J/K

l - thickness of layer, m·μm

l_m - optical path, m·Pa

m_λ - hemispherical spectral emissive power of material, W/(μm·m²)

$m_{\lambda,bb}$ - hemispherical spectral emissive power of black body W/(μm·m²)

$m_{\lambda_{max},bb}$ - hemispherical spectral emissive power of black body, for λ_{max}, W/(μm·m²)

$n_{\lambda,bb}$ - hemispherical spectral quantity of photons of black body, photons (μm·m²·s)

p - pressure, Pa

r_T - hemispherical total reflectivity

$r_{T,\beta}$ - directional-hemispherical total reflectivity

$r_{T,\beta,}$ - bidirectional total reflectivity

$r_{T,\varphi}$ - hemispherical-directional total reflectivity

r_λ - hemispherical spectral reflectivity

$r_{\lambda,n}$ - bidirectional spectral reflectivity in normal direction n

$r_{\lambda,d}$ - spectral reflectivity of diffusely reflecting surfaces

$r_{\lambda,s}$ - spectral reflectivity of specularly reflecting surfaces

$r_{\lambda,\beta}$ - directional-hemispherical spectral reflectivity

$r_{\lambda,\beta,\varphi}$ - bidirectional spectral reflectivity

$r_{\lambda,\varphi}$ - hemispherical-directional spectral reflectivity

C_1 - first constant in Planck's spectral energy distribution - $2\pi hc^2 = 3.74\cdot10^{-16}$ W·m^2

C_2 - second constant in Planck's spectral energy distribution - $\frac{hc}{k} = 1.4388\cdot10^{-2}$ m>.< K

$I_{T,bb,n}$ - total intensity of black body in normal direction n, W/(m^2·sr)

$I_{T,bb,\varphi}$ - total intensity of black body in direction φ, W/(m^2·sr)

$I_{T,n}$ - total intensity of material in normal direction n, W/(m^2·sr)

$I_{T,\varphi}$ - total intensity of material in direction φ, W/(m^2·sr)

M_T - hemispherical total emissive power of material, W/m^2

$M_{T,bb}$ - hemispherical total emissive power of black body, W/m^2

$M_{T,0-\lambda,bb}$ - hemispherical total emissive power of black body from $\lambda_1 = 0$ to $\lambda_2=\lambda\neq\infty$, W/m^2

$N_{T,bb}$ - hemispherical total quantity of photons of black body, photons/m^2·s

$N_{T,0-\lambda,bb}$ - hemispherical total quantity of photons of black body from $\lambda_1 = 0$ to $\lambda_2=\lambda\neq\infty$, photons/$m^2\cdot s$

R_z - roughness height, μm

S_m - roughness width, μm

T - absolute temperature of emitting body, K

T_m - temperature of measurement, K

T_{ox} - temperature of oxidizing, K

α_λ - spectral absorption coefficient, 1/m

$\alpha_{\lambda,r}$ - spectral absorption coefficient of radiation media, 1/m

β - angle of incident radiation to the normal n of reflected surface, rd

γ_W - relative roughness

ΔS - elementary surface area, m^2

ε_T - hemispherical total emissivity

$\varepsilon_{T,n}$ - total emissivity in normal direction n

$\varepsilon_{T,n,o}$ - total emissivity in normal direction of smooth surface

$\varepsilon_{T,\varphi}$ - total emissivity in direction φ

ε_λ - hemispherical spectral emissivity

$\varepsilon_{\lambda=0.65,n}$ - spectral (red) emissivity in normal direction for wavelength $\lambda \cong 65$ μm

$\varepsilon_{\lambda,n}$ - spectral emissivity in normal direction n

$\varepsilon_{\lambda,n,o}$ - spectral emissivity in normal direction n of smooth surface

$\varepsilon_{\lambda,\varphi}$ - spectral emissivity in direction φ

λ - wavelength of radiation, μm

$\lambda_{max,bb}$ - wavelength of maximum radiation of black body, μm

σ - Stefan-Boltzmann constant ($5.6703\cdot10^{-8}$ W/(m^2K) - - calculated; $5.729\cdot10^{-8}$ W/(m^2K^4) - experimental)

σ_λ - spectral scattering coefficient, 1/m

$\tau_{\lambda,l}$ - spectral transmissivity along path l

$\tau_{\lambda,n}$ - spectral transmissivity in normal direction of layer of thickness l

$\tau_{\lambda,\varphi}$ - spectral transmissivity in direction φ

φ - angle between given direction and normal n to the radiation surface, rd

φ_λ	- spectral radiation to or from hemisphere, W/μm
$\varphi_{\lambda,\alpha}$	- spectral radiation absorbed from hemisphere, W/μm
$\varphi_{\lambda,a,n}$	- spectral radiation absorbed from normal direction n, W/μm
$\varphi_{\lambda,a,\beta}$	- spectral radiation absorbed from incident direction, W/μm
$\varphi_{\lambda,n}$	- spectral radiation incident from normal direction n, W/μm
$\varphi_{\lambda,r}$	- spectral radiation reflected to hemisphere, W/μm
$\varphi_{\lambda,\beta}$	- spectral radiation incident to direction β, W/μm
$\varphi_{\lambda,\beta,\varphi}$	- bidirectional spectral radiation reflected to direction φ form direction β, W/μm
$\varphi_{\lambda,\tau,n}$	- spectral radiation transmitted in normal direction n, W/μm
$\varphi_{\lambda,\varphi}$	- spectral radiation reflected in direction φ incident radiation from hemisphere, W/μm
Φ	- total radiation incident from (emitted to) hemisphere, W
Φ_a	- total radiation absorbed from hemisphere, W
$\Phi_{a,\beta}$	- total radiation absorbed from direction β, W
Φ_β	- total radiation incident from direction β, W

ABBREVIATIONS OF TERMS

G O S T	-	U S S R standard
P N	-	Polish standard
A I S I	-	American Iron and Steel Institute
S A E	-	Society of Automotive Engineers
B S	-	British standard
A S T M	-	American Society for Testing and Materials
B K	-	Bulten-Kanthal standard

Part I. BLACK BODY RADIATION AND PROPERTIES OF MATERIALS

1.1. DEFINITION OF RADIATIVE VALUES OF BODIES

The radiative value of a body is the number defining quantitatively the properties of a body in respect of its thermal radiation or thermal exchange. The chief radiative values of bodies are:

- the value of various kinds of thermal radiation of an idealized body - black body;
- the value of various radiative properties of real bodies - materials.

The radiative property of a body is its ability to emit or exchange thermal radiation. As regards radiative properties two classes of bodies can be distinquished:

- an ideal body - black body;
- a real body - material.

1.1.1. Properties of black body

A black body is an ideal body which absorbs any incident radiation. A black body has six principal properties [22], [132]:

1. maximum emissive power at a given temperature T;
2. isotropy of radiation in a space closed within black walls, irrespective of direction and position;
3. maximum emissivity for any wavelength of radiation at a given temperature T;
4. maximum emissivity in any direction of radiation at a given temperature T;
5. the total emissive power is a function of temperature T

only and spectral emissive power is a function of wavelength λ and temperature T only;

6. principal values of radiation for any wavelength, temperature and direction are in accordance with the fundamental laws of thermal radiation, called the fundamental laws of black body emission.

1.1.2. Fundamental laws of black body emission

Thermal radiation from an elementary flat surface to a hemisphere is a spectrum, which can be described by a function of temperature T and wavelength λ [8]. In the case of a black body the spectrum is the same in any direction. The hemispherical spectral emissive power of a black body $m_{\lambda,bb}$ is expressed by Planck's formula (see: [8]):

$$m_{\lambda,bb} = \pi i_{\lambda,bb,n} = \frac{2\pi h\, c^2}{\lambda^5 (e^{\frac{hc}{\lambda kT}} - 1)} = \frac{C_1}{\lambda^5 (e^{\frac{C_2}{\lambda T}} - 1)} \; [W/(\mu m \cdot m^2)] \tag{1}$$

where:

- $i_{\lambda,bb,n}$ - spectral intensity of a black body in normal direction, $W/(\mu m \cdot m^2 \cdot sr)$;
- c - velocity of electromagnetic wave propagation in a vacuum - $2.99792 \cdot 10^8 m/s \approx 3 \cdot 10^8 m/s$;
- h - Planck's constant - $6.622 \cdot 10^{-34}$ J·s;
- k - Boltzmann's constant - $1.37975 \cdot 10^{-23}$ J/K;
- λ - wavelength, µm;
- T - absolute temperature of black body, K;
- C_1 - first constant of Planck's formula - $2 \cdot \pi \cdot hc^2 = 3.74 \cdot 10^{-16}$ W·m²;
- C_2 - second constant of Planck's formula - $\frac{hc}{k} = 1.44 \cdot 10^{-2}$ mK.

The graph of $m_{\lambda,bb}$ as a function of wavelength λ for different temperatures is presented in Fig.1 [1],[9],[19]. The values of $m_{\lambda,bb}$ given on the axis of ordinates should be multiplied by the value N^5 times if the temperature range is other than N=1. The function $m_{\lambda,bb}$ has its maximum, which is denoted by $m_{\lambda_{max} bb}$ - and corresponds to the wavelength marked λ_{max}. In Fig.1 the maxima of the functions are connected by a broken line.

The product of the value λ_{max} and the value of temperature is constant, which is a form of Wien's displacement law:

$$\lambda_{max} \cdot T = const. = 2897.8 \; [\mu mK] \qquad (2)$$

The product of $m_{\lambda_{max},bb}$ value and the value of temperature raised to a minus fifth power is constant, which is another

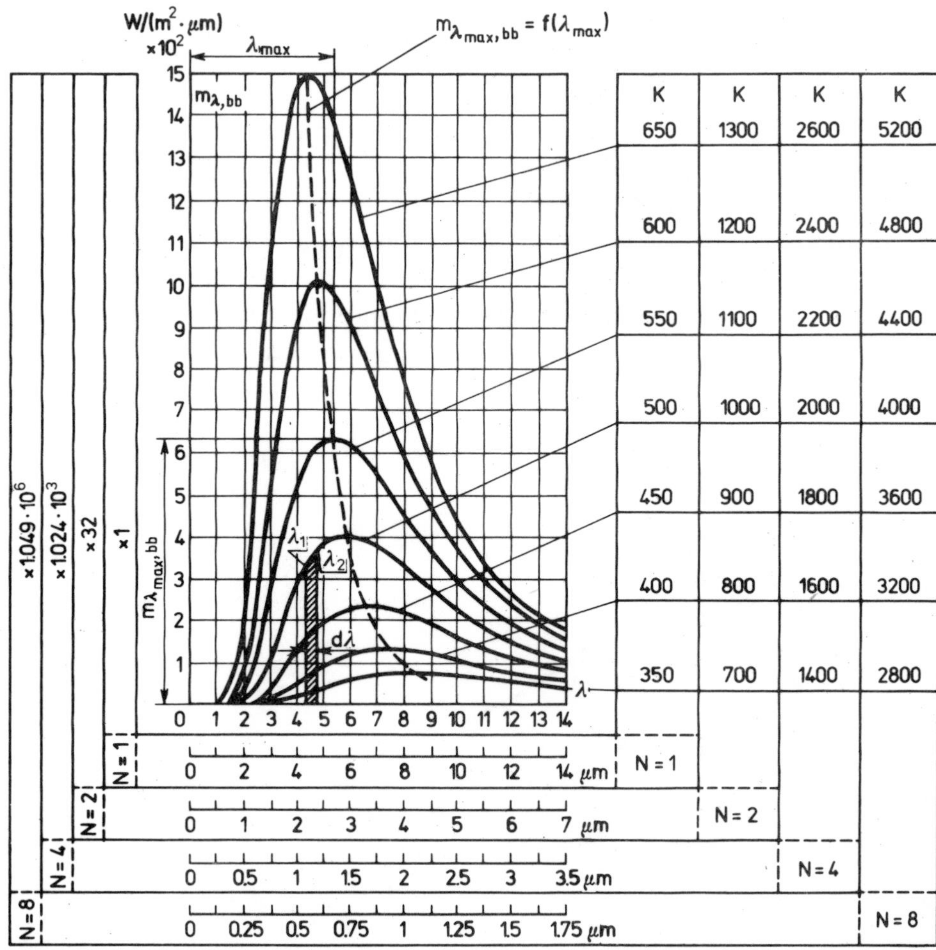

Fig.1. Functions of hemispherical spectral emissive power of black body $m_{\lambda\ bb}$ for various temperatures (---- functions of hemispherical spectral emissive power of black body for maximum wavelength $m_{\lambda_{max},bb}$) [10], [22], [38], [37], [55], [82]

form of Wien's displacement law:

$$m_{\lambda_{max},bb} \cdot T^{-5} = \text{const.} = 1.2867 \cdot 10^{-11} [W/(\mu m \cdot m^2 \cdot K)^5] \qquad (3)$$

Integrating $m_{\lambda,bb}$ on all wavelength, we obtain the equation

$$\int_{\lambda_1=0}^{\lambda_2=\infty} m_{\lambda,bb} \cdot d\lambda = \sigma T^4 = M_{T,bb} = \pi I_{T,bb,n} \; [W/m^2], \qquad (4)$$

which is known as the Stefan-Boltzmann law,
where:

- σ - Stefan constant - $5.67 \cdot 10^{-8}$ $W/(cm^2 \cdot K^4)$;
- $M_{T,bb}$ - hemispherical total emissive power of a black body, W/m^2;
- $I_{T,bb,n}$ - total intensity of a black body in normal direction, $W/(m \cdot sr)$.

The surface under the curve of function $m_{\lambda,bb}$ is the hemispherical total emissive power $M_{T,bb}$. For technical purposes it is often necessary to determine the hemispherical total emission power for the range of wavelength from $\lambda_1 = 0$ to $\lambda_2 = \lambda \neq \infty$ (Fig.2) [45]:

$$\int_{\lambda_1=0}^{\lambda_2=\lambda} m_{\lambda,bb} \cdot d\lambda = M_{T,0-\lambda,bb} \quad [W/m^2] \qquad (5)$$

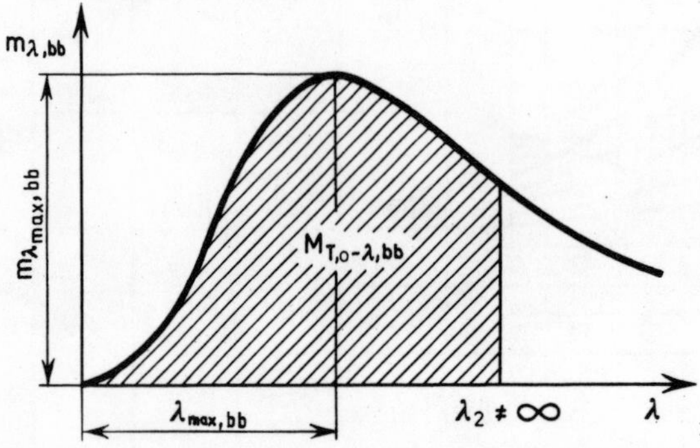

Fig.2. Function of hemispherical total emissive power of black body $M_{T,0-\lambda}$ (from $\lambda_1 = 0$ to $\lambda_2 \neq \infty$)

The geometrical distribution of radiation of the elementary surface of a black body is subject to Lambert's law, called also the cosine law. The law states that the distribution of total intensity of a black body in direction φ - $I_{T,bb,\varphi}$ is the product of the total intensity of a black body in normal direction - $I_{T,bb,n}$ and the cosine of angle φ (Fig.3):

$$I_{T,bb,\varphi} = I_{T,bb,n} \cdot \cos\varphi = \int_{\lambda_1=0}^{\lambda_2=\infty} i_{\lambda,bb,\varphi} \cdot d\lambda = \int_{\lambda_1=0}^{\lambda_2=0} i_{\lambda,bb,n} \cdot \cos\varphi d\lambda \quad [W/(m^2 sr)] \tag{6}$$

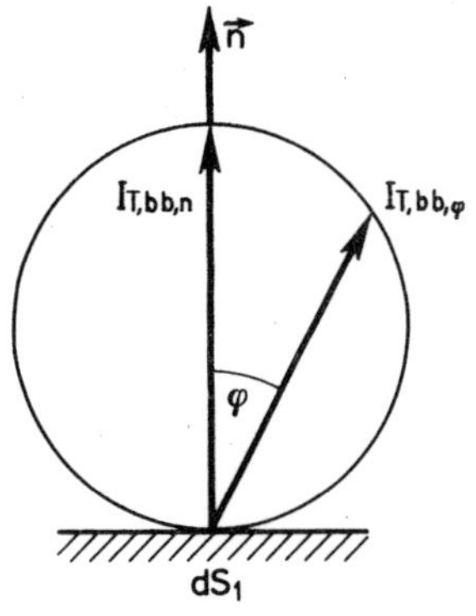

Fig.3. Graphical interpretation of Lambert's law

where:

φ - angle between the given direction and the normal to the radiation surface, rd;

$i_{\lambda,bb,\varphi}$ - spectral intensity of a black body in direction φ, $W/(\mu m \cdot m^2 \cdot sr)$

The number of photons emitted per second per unit area 1 m^2 in a unit spectral range of the wavelength (1μm) is expressed by the formula (see [10], [48]):

$$n_{\lambda,bb} = \frac{m_{\lambda,bb}}{hc/\lambda} = \frac{2\pi \cdot c}{\lambda^4 (e^{\frac{hc}{\lambda kT}} - 1)} = \frac{C_1}{hc\lambda^4 (e^{\frac{C_2}{\lambda T}} - 1)} \left[\frac{photons}{\mu m \cdot m^2 \cdot s}\right] \tag{7}$$

The number of photons emitted per second per unit area $1m^2$ in a full spectral range depends only on the temperature of a body and is expressed by the relation (see [10], [49]):

$$N_{T,bb} = 1.5213 \cdot 10^{15} T^3 \left[\frac{photons}{m^2 \cdot s}\right] \tag{8}$$

The number of photons in a spectral range is expressed by the formula (see [10], [53]):

$$N_{T,0-\lambda,bb} = \int_{\lambda_1=0}^{\lambda_2=\lambda} n_\lambda \cdot d\lambda \quad \left[\frac{photons}{m^2 \cdot s}\right] \tag{9}$$

where $\lambda \neq \infty$.

1.1.3. Definitions of radiative properties of materials

Radiative property of material is its ability to emit or exchange thermal radiation.

We can distinguish four basic kinds of radiation properties of materials:

- emissivity,
- absorptivity,
- reflectivity,
- transmissivity.

Absorptivity is a quantity which cannot be measured directly; for opaque materials it is defined by means of emissivity and reflectivity, and for, transparent materials by means of the remaining three quantities. There are several kinds of a given basic radiation property of a material [50],[52].

1.1.3.1. EMISSIVITY

Emissivity, i.e., the radiative ability of a real body - a material, is a measure of how well the material can emit thermal radiation as compared with a black body. Emissivity depends on wavelength λ and on spherical radiation distribution.

Every unit area ΔS of a black body has a spectral distribution (as a function of radiation wavelength λ) which is in accordance with Planck's formula. The radiation spectrum is shown in Fig.4. The ordinate of this function $m_{\lambda,bb}$ indicates the hemispherical spectral emissive power, i.e., the amount of radiation with a given radiation wavelength λ emitted by a unit area ΔS of a black body. The spectral distribution of the hemispherical spectral emissive power of a material m_λ has more or less the same arrangement as that of a black body; the following inequality is always satisfied:

$$m_{\lambda,bb} \geqslant m_\lambda \geqslant 0 \ . \qquad (10)$$

The area below the curve $m_{\lambda,bb}$ represents the hemispherical total emissive power $M_{T,bb}$ of a unit area of a black body ΔS. The hemispherical total emissive power of a material M_T always satisfies the inequality:

$$M_{T,bb} \geqslant M_T \geqslant 0 \ . \qquad (11)$$

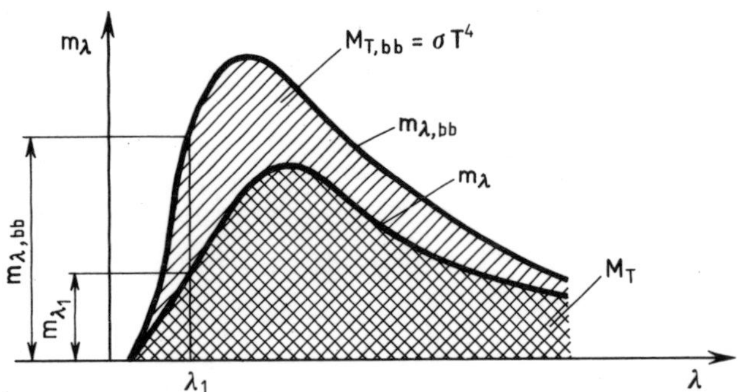

Fig.4. Function of hemispherical spectral emissive power of black body $M_{T,bb}$ and hemispherical spectral emissive power of material M_T

The radiation of a given wavelength λ from a unit surface area ΔS of a black body can be represented as a radiating sphere generating a vector of spectral intensity $i_{\lambda,bb}$, and constituting the quantity of spectral radiation in a solid angle Δω in the φ direction from a unit area of a black body (Fig.5). The

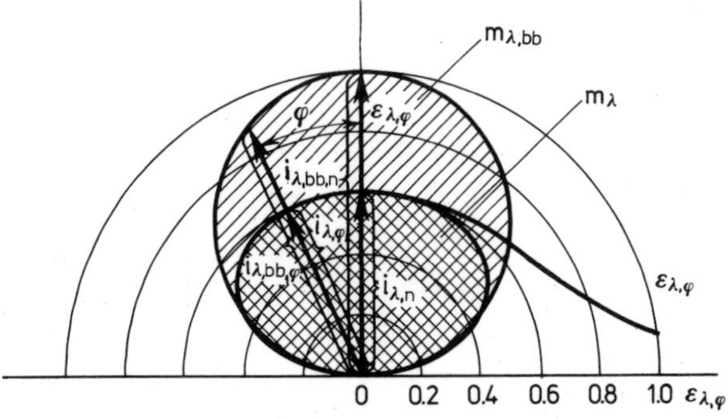

Fig.5. Function of spectral intensity of black body $i_{\lambda,bb,\varphi}$ and spectral intensity of material in hemisphere $i_{\lambda,\varphi}$

maximum intensity of radiation is in the direction normal to the unit surface area ΔS. This spectral intensity is denoted by $i_{\lambda,bb,n}$. The spectral intensity distribution of a material $i_{\lambda,\varphi}$ is more or less spherical. The total value of spherical radiation determined by the vector of spectral intensity $i_{\lambda,bb,\varphi}$ defines the hemispherical spectral emissive power of a black body $m_{\lambda,bb}$. Similarly the spectral intensity of the material $i_{\lambda,\varphi}$ defines the hemispherical spectral emissive power of the material m_{λ}.

The sum of spectral intensities of a material $i_{\lambda,\varphi}$ for all wavelengths λ can be expressed by the following formula (similar to the formula of black body radiation) which indicates the total intensity of the material in direction φ:

$$I_{T,\varphi} = \int_{\lambda_1=0}^{\lambda_2=\infty} i_{\lambda,\varphi} \, d\lambda \; [\frac{W}{m^2 \cdot sr}] \; . \tag{12}$$

The distribution of the vector of total intensity $I_{T,bb,\varphi}$ forms a regular sphere. The sum of the total intensities of a black body is the total emissive power $M_{T,bb}$. The hemispherical total emissive power of a black body can be expressed by the following integral:

$$M_{T,bb} = \int_{\varphi_1=0}^{\varphi_2=\pi/2} I_{T,bb,\varphi} \sin\varphi \, d\varphi \; [\frac{W}{m^2}] \; . \tag{13}$$

For a material:

$$M_T = \int_{\varphi_I=0}^{\varphi_2=\frac{\pi}{2}} I_{T,\varphi} \sin\varphi \, d\varphi \; [\frac{W}{m^2}] \; . \tag{14}$$

The distribution of the vector of total intensity $I_{T,\varphi}$ is not a sphere and therefore the emissivity of a non-black body depends on direction φ normal to the surface area of a material. The emissivity of materials depends also on the kind of radiation classified according to its wavelength λ.

For the purpose of simplification we can presume that:

1. the above mentioned quantities are independent of the rotation angle and are always constant in relation to this angle,

2. the spectral properties of materials are independent of temperature.

There are four kinds of emissivity **(Fig. 6)**

1. <u>Spectral emissivity in direction φ measured from the normal n</u> - $\varepsilon_{\lambda,\varphi}$. This emissivity depends on direction φ and on wavelength λ as in the formula (see [78], [132]):

$$\varepsilon_{\lambda,\varphi} = \frac{i_{\lambda,\varphi}}{i_{\lambda,bb,\varphi}} = \frac{i_{\lambda,\varphi}}{i_{\lambda,bb,n}\cos\varphi}, \tag{15}$$

where:

$i_{\lambda,\varphi}$ - spectral intensity of the material in direction φ, $W/(\mu m \cdot m^2 \cdot sr)$;

$i_{\lambda,bb,n}$ - spectral intensity of the black body in normal direction n, $W/(\mu m \cdot sr)$.

A particular kind of the spectral emissivity in φ direction occurs for $\varphi=0$, i.e., in a direction normal to the radiating surface [78], [98], [132]. This results from the fact that the value of the spectral emissivity is greatest in this direction (see Fig.5). The spectral emissivity in normal direction $\varepsilon_{\lambda,n}$ can be expressed by the formula (see [93]):

$$\varepsilon_{\lambda,n} = \frac{i_{\lambda,n}}{i_{\lambda,bb,n}}, \tag{16}$$

where $i_{\lambda,n}$ - spectral intensity of material in normal direction n, $W/(\mu m \cdot m^2 \cdot sr)$. On account of the needs of optical pyrometry, we often determine normal spectral emissivity for red light. This quantity is expressed by $\varepsilon_{\lambda=0.65,n}$, where 0.65 is the wavelength λ in μm. This quantity can be given for different wavelengths λ, e.g. 0.63; 0.67 μm and so on.

2. <u>Hemispherical spectral emissivity</u> - ε_{λ}. This emissivity is the weighted mean of $\varepsilon_{\lambda,\varphi}$ (see [109], [132]):

$$\varepsilon_{\lambda} = \frac{m_{\lambda}}{m_{\lambda,bb}} = \frac{\int_{\triangle} i_{\lambda,\varphi}\, d\omega}{m_{\lambda,bb}} = \frac{1}{\pi}\int_{\triangle} \varepsilon_{\lambda,\varphi}\cos\varphi\, d\omega, \tag{17}$$

where:

m_{λ} - hemispherical spectral emissive power of material, $W/(\mu m \cdot m^2)$,

$d\omega$ - the elementary solid angle constituting the function φ.

Because of the difficulties involved in measuring $\varepsilon_{\lambda,\varphi}$ over the whole enclosing hemisphere numerical values of ε_{λ} are very rare in the literature. Emissivity $\varepsilon_{\lambda,n}$ is very often confused with emissivity ε_λ [109],[130].

3. Total emissivity in direction φ measured from the normal n - $\varepsilon_{T,\varphi}$. This emissivity depends on the direction φ and temperature T of the emitting body. The emissivity $\varepsilon_{T,\varphi}$ is expressed by the equation (see [112],[132],[137]):

$$\varepsilon_{T,\varphi} = \frac{I_{T,\varphi}}{I_{T,bb,\varphi}} = \frac{I_{T;\varphi}}{I_{T,bb,n}\cos\varphi} = \frac{\int_{\lambda_1=0}^{\lambda_2=\infty} i_{\lambda,\varphi} d\lambda}{\frac{1}{\pi}\sigma T^4 \cos\varphi} =$$

$$= \frac{\int_{\lambda_1=0}^{\lambda_2=\infty} \varepsilon_{\lambda,\varphi} i_{\lambda,bb,\varphi} \, d\lambda}{\frac{1}{\pi}\sigma T^4 \cdot \cos\varphi} = \frac{\int_{\lambda_1=0}^{\lambda_2=\infty} \varepsilon_{\lambda,\varphi} i_{\lambda,bb,n} \, d\varphi}{\sigma T^4} \tag{18}$$

A particularly privileged directional total emissivity is the total emissivity in normal direction n - $\varepsilon_{T,n}$ (for the same reason as $\varepsilon_{\lambda,n}$). It is expressed by formula (see [69], [80],[132]:

$$\varepsilon_{T,n} = \frac{I_{T,n}}{I_{T,bb,n}} = \frac{\pi \int_{\lambda_1=0}^{\lambda_2=\infty} i_{\lambda,n} \, d\lambda}{\sigma T^4} = \frac{\int_{\lambda_1=0}^{\lambda_2=\infty} \varepsilon_{\lambda,n} i_{\lambda,bb,n} \, d\lambda}{\sigma T^4} , \tag{19}$$

where:

$I_{T,n}$ - total intensity of material in normal direction n, $W/(m^2 \cdot sr)$.

This emissivity is the most frequently measured and quoted in various publications. It is often confused with hemispherical total emissivity ε_T.

4. Hemispherical total emissivity - ε_T. It can be expressed with formula:

$$\varepsilon_T = \frac{M_T}{M_{T,bb}} = \frac{\int_{\lambda_1=0}^{\lambda_2=\infty} \varepsilon_\lambda \, m_{\lambda,bb} \, d\lambda}{\sigma T^4} = \frac{1}{\pi} \int_{\Omega} \varepsilon_{T,\varphi} \cos\varphi \, d\omega . \tag{20}$$

1.1.3.2. ABSORPTIVITY

Absorptivity is defined as that fraction of the radiation incident on a material which is absorbed by the material. The

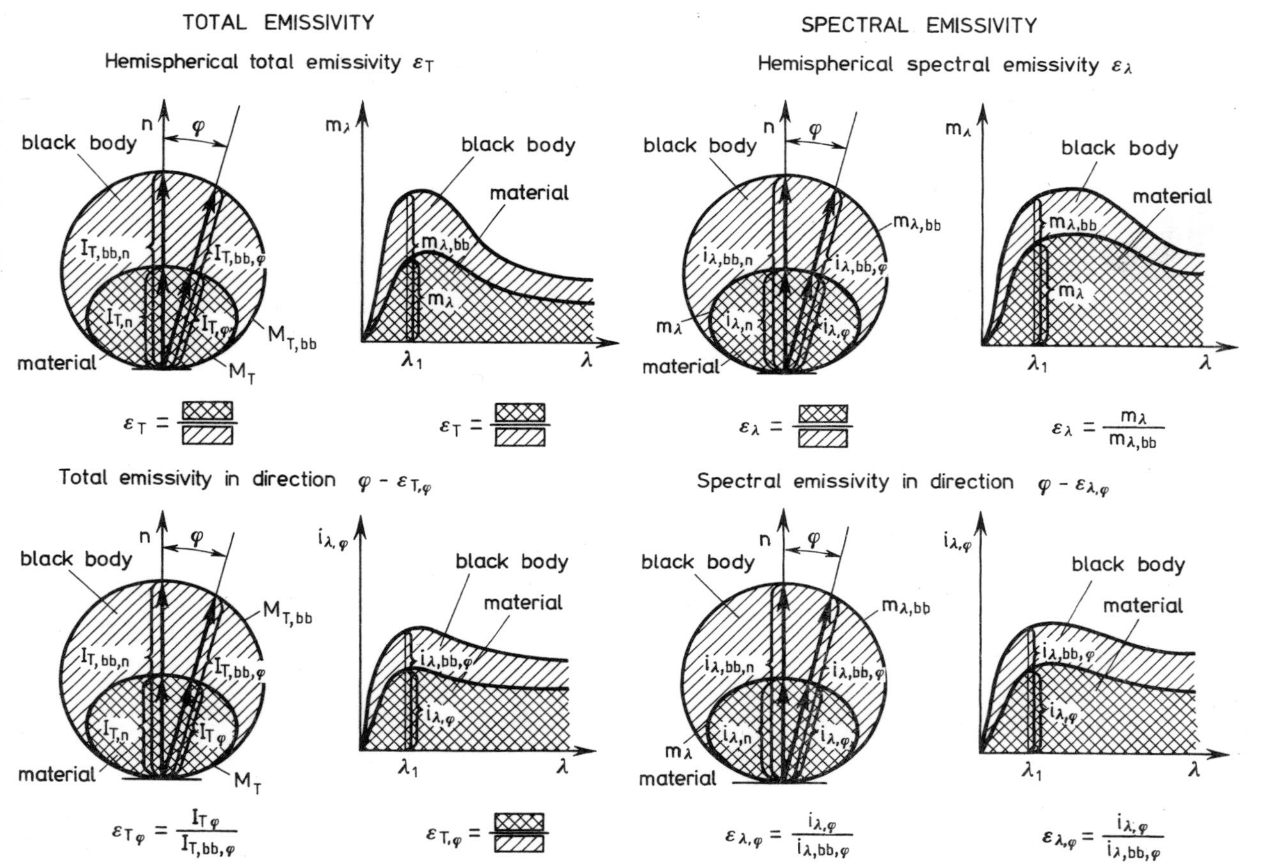

Fig.6. Kinds of emissivities

absorptivity of a black body is equal to 1 for all wavelengths λ and all directions β of incident radiation. The absorptivity of a material is dependent on wavelength λ and incident radiation angle β. Four kinds of absorptivity are as distinquished.

1. Directional spectral absorptivity from direction β - $a_{\lambda,\beta}$. Directional spectral absorptivity is the ratio of the spectral radiation absorbed from direction β - $\varphi_{\lambda,a,\beta}$ to the spectral radiation incident from direction β - $\varphi_{\lambda,\beta}$. It can be written as (see [72], [82]):

$$a_{\lambda,\beta} = \frac{\varphi_{\lambda,a,\beta}}{\varphi_{\lambda,\beta}} . \tag{21}$$

In accordance with the Kirchhoff's law the directional spectral absorptivity $a_{\lambda,\beta}$ should be equal to the spectral emissivity $\varepsilon_{\lambda,\varphi}$. This equality holds if (see [67], [81], [132]):

- the directions of incident radiation β and emissivity are the same;
- the wavelength of the incident radiation is equal to the wavelength of the emitting radiation;
- the absorbing body is placed in a space surrounded by black walls and with a temperature equal to that of the absorbing body.

We can then write:

$$a_{\lambda,\beta} = \varepsilon_{\lambda,\varphi} . \tag{22}$$

The third restriction on equation (22) is very difficult to satisfy and in assuming specified values of absorptivity it should be regarded as a certain approximation.

2. Hemispherical spectral absorptivity - a_{λ}. The hemispherical spectral absorptivity is given by the formula (see [96]):

$$a_{\lambda} = \frac{\varphi_{\lambda,a}}{\varphi_{\lambda}} \tag{23}$$

where:

$\varphi_{\lambda,a}$ - spectral radiation absorbed from the hemisphere, W/μm;

φ_{λ} - spectral radiation incident from the hemisphere on an elemental area dS, W/μm.

The general form of the Kirchhoff's law:

$$a_\lambda = \varepsilon \tag{24}$$

is valid if [76],[105],[132]:

- the conditions of the equation $a_{\lambda,\beta} = \varepsilon_{\lambda,\varphi}$ are satisfied,
- the distribution of incident radiation intensity is the same as the distribution of emitted intensity emitts in the case of ε_λ.

3. Directional total absorptivity from β-$a_{T,\beta}$. It is expressed by the formula

$$a_{T,\beta} = \frac{\Phi_{a,\beta}}{\Phi_\beta} = \frac{\int\limits_{\lambda_1=0}^{\lambda_2=\infty} a_{\lambda.\varphi}\cdot\varphi_{\lambda,\beta}d\lambda}{\int\limits_{\lambda_2=0}^{\lambda_1=\infty} \varphi_{\lambda,\beta}\, d\lambda}, \tag{25}$$

where:

$\Phi_{a,\beta}$ - total radiation absorbed from direction β, W;

Φ_β - total radiation incident from direction β, W.

In accordance with Kirchhoff's law

$$a_{T,\beta} = \varepsilon_{T,\varphi}, \tag{26}$$

if [74],[132]:

- conditions of the equation $a_{\lambda,\beta} = \varepsilon_{\lambda,\varphi}$ are satisfied,
- a radiating body is a black or grey body.

4. Hemispherical total absorptivity - a_T. It is expressed by the formuła:

$$a_T = \frac{\Phi_a}{\Phi} = \frac{\int\limits_{\lambda_1=0}^{\lambda_2=\infty} a_\lambda\cdot\varphi_\lambda\, d\lambda}{\int\limits_{\lambda_1=0}^{\lambda_2=\infty} \varphi_\lambda\, d\lambda} \tag{27}$$

where:

Φ_a - total radiation absorbed from the hemisphere, W;

Φ - total radiation incident from the hemisphere, W.

In accordance with Kirchhoff's law

$$a_T = \varepsilon_T \tag{28}$$

if (see [132],[137]):

- the conditions of equation $a_{\lambda,\beta}=\varepsilon_{\lambda,\varphi}$ are satisfied,
- the incident radiation intensity has the same distribution as the radiation emitted for the value ε_T,
- the radiating body has a spectral radiation of a black or a grey body.

In the case of $a_{T,\beta}$ and a_T the temperature subscript refers to the radiating body.

Figure 7 shows the four basic kinds of absorptivity.

When an absorbing body is diffusive and gray, its absorptivity does not depend either on direction or on incident spectral radiation. Then we can write (see [119],[132],[140]):

$$a_{\lambda,\beta} = a_\lambda=a_{T,\beta}=a_T=\varepsilon_{\lambda,\varphi}=\varepsilon_\lambda=\varepsilon_{T,\varphi}=\varepsilon_T \ . \quad (\beta=\varphi) \tag{29}$$

If the body is gray, then (see [111],[119]):

$$a_{\lambda,\beta} = a_{T,\beta} = \varepsilon_{\lambda,\varphi} = \varepsilon_{T,\varphi} , \quad (\beta=\varphi) \tag{30}$$

$$a_\lambda = a_T = \varepsilon_\lambda = \varepsilon_T \ . \tag{31}$$

If the body is diffusive, than (see [115],[140]):

$$a_{T,\beta} = a_T = \varepsilon_{T,\varphi} = \varepsilon_T , \tag{32}$$

$$a_{\lambda,\beta}= a_\lambda = \varepsilon_{\lambda,\varphi} = \varepsilon_\lambda \ . \tag{33}$$

1.1.3.3. REFLECTIVITY

Reflectivity is defined as an ability of a surface (as the boundary of two optical media) to reflect of the incident radiation. We can distinguish eight kinds of reflectivity.

1. <u>Bidirectional spectral reflectivity - $r_{\lambda,\beta,\varphi}$</u>. It can be written as: [3],[41],[64]:

$$r_{\lambda,\beta,\varphi} = \frac{\varphi_{\lambda,\beta,\varphi}}{\varphi_{\lambda,\beta}} \tag{34}$$

where: $\varphi_{\lambda,\beta,\varphi}$ - bidirectional spectral radiation reflected to direction φ from direction β, W/μm.

The spectral reflectivity to the direction φ of the radiation incident to the direction β, for $\beta=0=\varphi$, has special importance. It is a tabulated quantity denoted by the symbol $r_{\lambda,n}$.

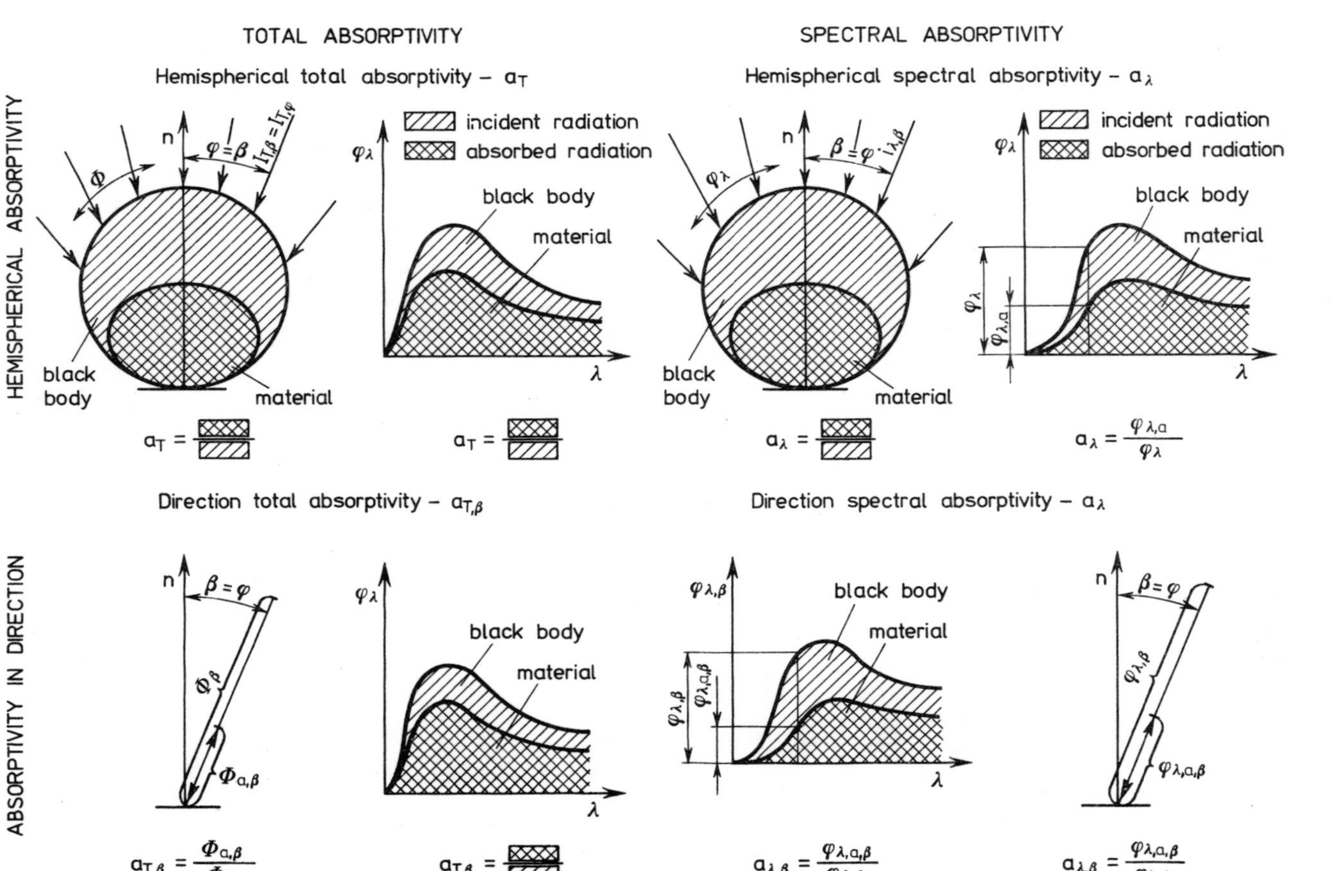

Fig.7. Kinds of absorptivities

2. Directional - hemispherical spectral reflectivity - $r_{\lambda,\beta}$. It is expressed by the formula (see [71]):

$$r_{\lambda,\beta} = \frac{\varphi_{\lambda,r}}{\varphi_{\lambda,\beta}} , \tag{35}$$

where $\varphi_{\lambda,r}$ - spectral radiation reflected to the hemisphere, W/μm.

3. Hemispherical - directional spectral reflectivity - $r_{\lambda,\varphi}$. For the uniform incident radiation this kind of reflectivity can be expressed by the formula (see [84],[132]):

$$r_{\lambda,\varphi} = \frac{\varphi_{\lambda,\varphi}}{\frac{1}{\pi}\varphi_{\lambda}} , \tag{36}$$

where $\varphi_{\lambda,\varphi}$ - spectral radiation reflected in the direction φ of the incident radiation from the hemisphere, W/μm.

4. Hemispherical spectral reflectivity - r_{λ}. It is expressed by the formula (see [87],[123]):

$$r_{\lambda} = \frac{\varphi_{\lambda,r}}{\varphi_{\lambda}} . \tag{37}$$

For a diffusive surface we have:

$$r_{\lambda,d} = r_{\lambda,\beta} = \pi r_{\lambda,\beta,\varphi} , \tag{38}$$

where $r_{\lambda,d}$ is the spectral reflectivity of surfaces diffusely reflecting to the hemisphere of the incident radiation from direction β.

For specular surfaces we have the relation see([117],[132]):

$$r_{\lambda,s} = r_{\lambda} = r_{\lambda,\beta} = r_{\lambda,\beta,\varphi} \tag{39}$$

on condition that reflectivity is independent of the angle of incident radiation β. The $r_{\lambda,s}$ in equation (39) denotes specular spectral reflectivity of the incident radiation from direction β.

Just as for the spectral refelctivities we have analogous reflectivities for total radiation:

1. Bidirectional total reflectivity in direction φ of incident radiation from direction β - $r_{T,\beta,\varphi}$.

2. Directional - hemispherical total reflectivity of incident radiation from direction β - $r_{T,\beta}$.

3. Hemispherical - directional total reflectivity of incident radiation from the hemisphere - $r_{T,\varphi}$.

4. Hemispherical total reflectivity of incident radiation from the hemisphere - r_T.

Total reflectivity is dependent on the same relations as those expressed in equations (38) and (39). The particular kinds of reflectivity can be seen in Fig. 8.

1.1.3.4. RELATIONS BETWEEN REFLECTIVITY, ABSORPTIVITY AND EMISSIVITY

From the definitions of absorptivity and reflectivity as fractions of incident energy absorbed or reflected, it is evident that for an opaque body there exist simple relations between these surface properties. Therefore we can write [12], [132],[149]:

$$a_{\lambda,\beta} + r_{\lambda,\beta} = 1 , \tag{40}$$

$$a_{\lambda} + r_{\lambda} = 1 , \tag{41}$$

$$a_{T,\beta} + r_{T,\beta} = 1 , \tag{42}$$

$$a_{T} + r_{T} = 1. \tag{43}$$

Since absorptivity cannot be measured directly, we may replace it by the appropriate emissivity on the conditions given in 1.1.3.2. Then (see [23],[132]):

$$\varepsilon_{\lambda,\varphi} + r_{r,\beta} = 1 , \tag{44}$$

if $\varphi = \beta$,

$$\varepsilon_{\lambda} + r_{\lambda} = 1 , \tag{45}$$

if incident spectral intensity has a distribution according to Lambert's law;

$$\varepsilon_{T,\varphi} + r_{T,\beta} = 1 , \tag{46}$$

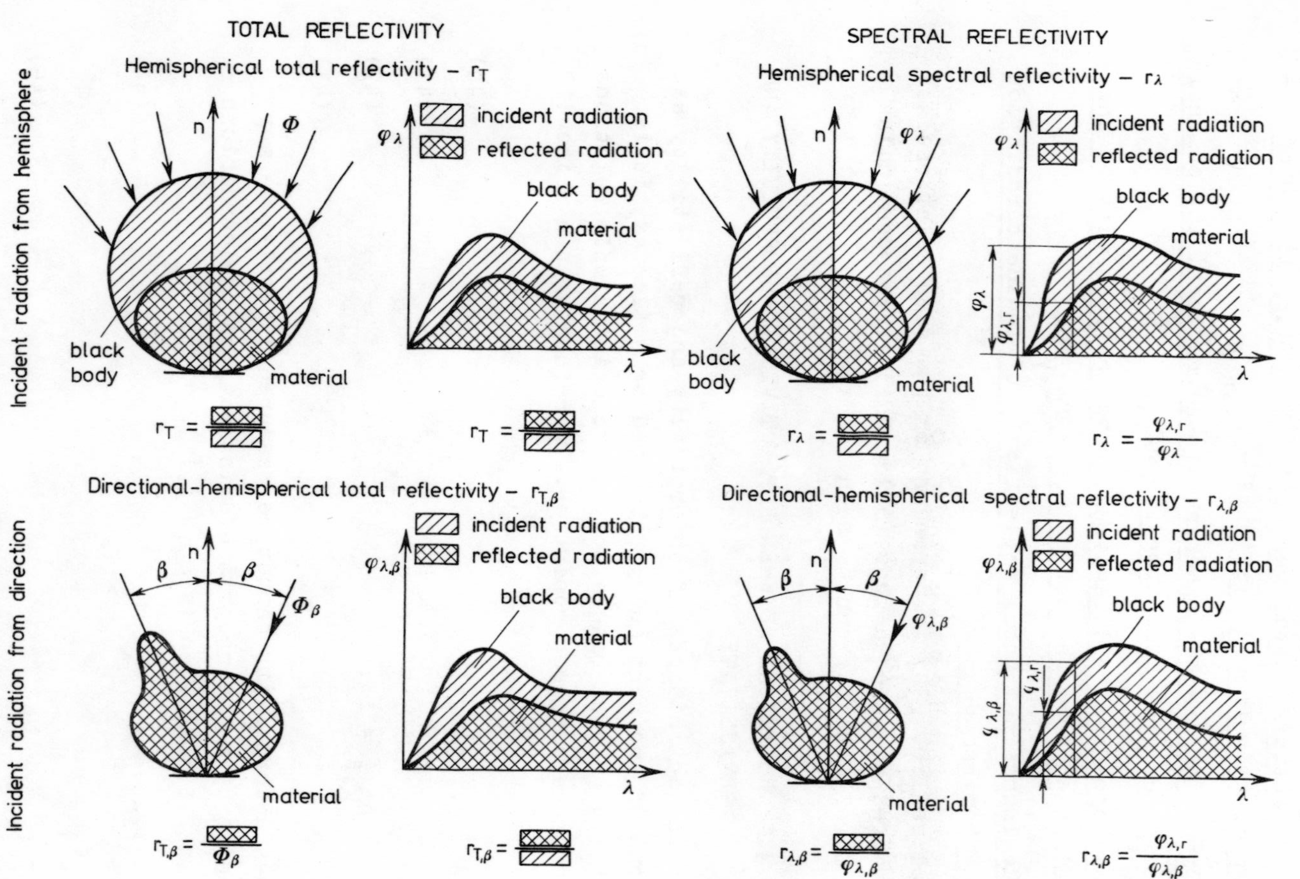
HEMISPHERICAL REFLECTIVITY
Incident radiation from hemisphere
Incident radiation from direction
TOTAL REFLECTIVITY
SPECTRAL REFLECTIVITY
Hemispherical total reflectivity – r_T
Hemispherical spectral reflectivity – r_λ
Directional-hemispherical total reflectivity – $r_{T,\beta}$
Directional-hemispherical spectral reflectivity – $r_{\lambda,\beta}$
incident radiation
reflected radiation
black body
material
n
Φ
Φ_β
β
φ_λ
$\varphi_{\lambda,r}$
$\varphi_{\lambda,\beta}$
λ
$r_T =$
$r_\lambda =$
$r_\lambda = \frac{\varphi_{\lambda,r}}{\varphi_\lambda}$
$r_{T,\beta} =$
$r_{\lambda,\beta} =$
$r_{\lambda,\beta} = \frac{\varphi_{\lambda,r}}{\varphi_{\lambda,\beta}}$

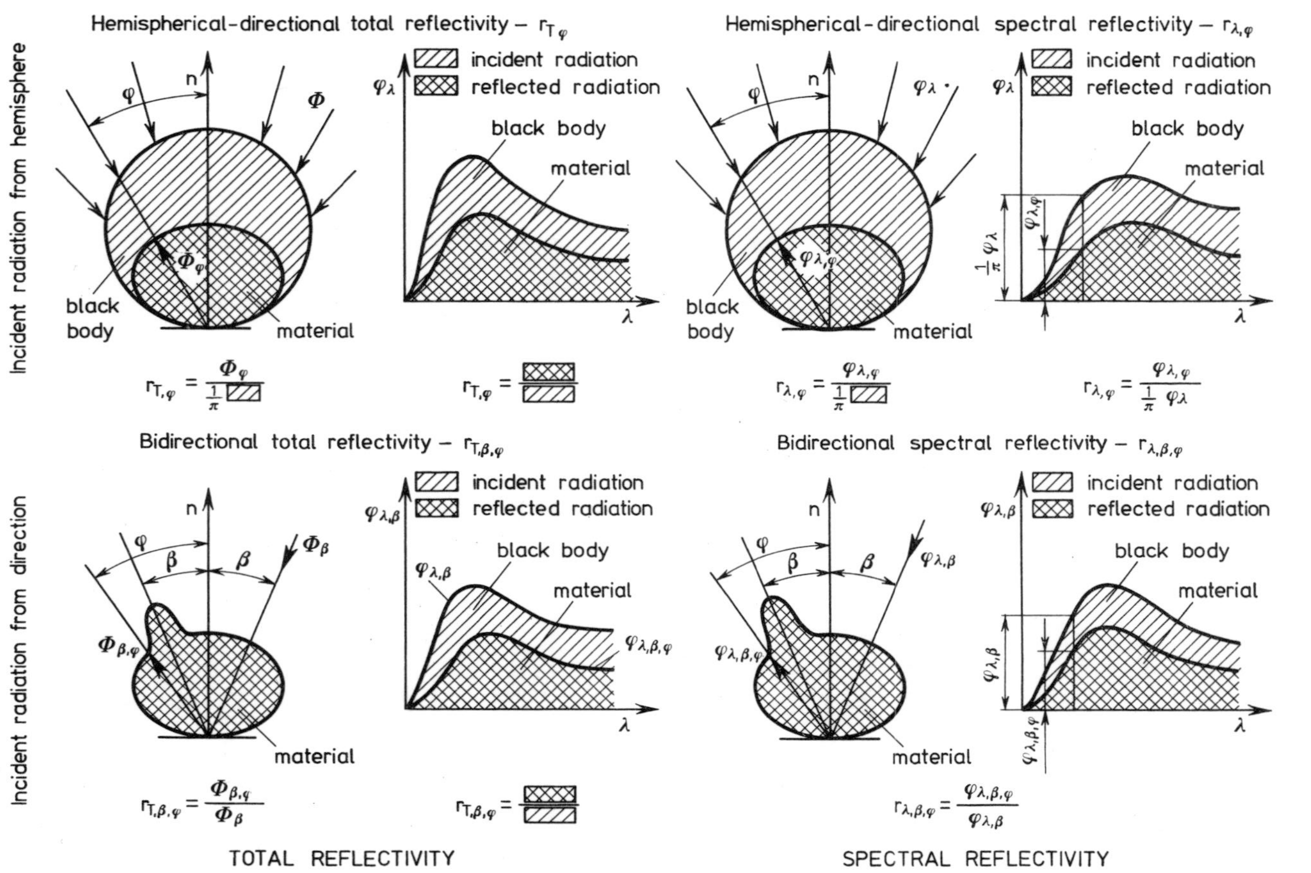

Fig. 8. Kinds of reflectivities

if $\varphi = \beta$ and incident radiation is emitted by a black or grey body

$$\varepsilon_T + r_T = 1 \;, \tag{47}$$

if incident radiation is emitted by a black body or a grey body.

1.1.3.5. TRANSMISSIVITY

Transmissivity is the ratio of the spectral radiation which passes through a given medium to the spectral radiation incident in to this medium.

Transmissivity depends on:

- the spectral quality of the incident radiation,
- the geometry of the body transmitting the radiation,
- the angle of incidence of the radiation upon the surface of the a body,
- the radiative properties of the medium transmitting the radiation.

Transmissivity should characterize only the radiative properties of the medium transmitting the radiation. To achieve maximum independence of the other relations, investigations we carried out for:

- radiation of a given wavelength λ,
- a flat plate of the material tested,
- a radiant energy flux in normal direction to the plate.

Under these restrictions, transmissivity can be expressed as (Fig. 9) [59],[71]:

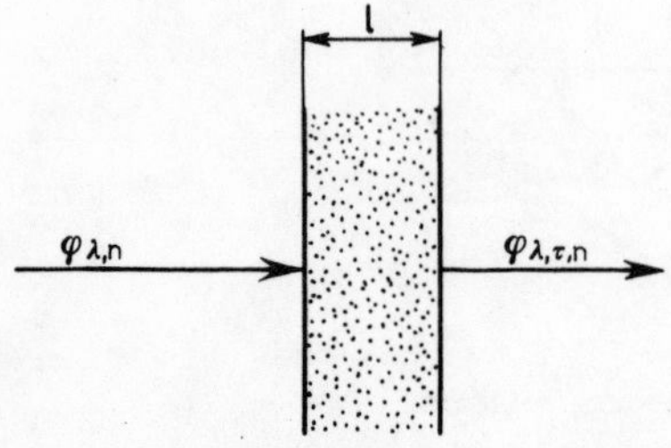

Fig.9. Scheme of transmissivity

$$\tau_{\lambda,n} = \frac{\varphi_{\lambda,\tau,n}}{\varphi_{\lambda,n}} \tag{48}$$

where:

$\varphi_{\lambda,n}$ - spectral radiation incident from normal direction, W/µm,

$\varphi_{\lambda,\tau,n}$ - spectral radiation transmitted in normal direction n, W/µm.

Defined in this way, transmissivity depends on the optical path l_m in the material tested. This parameter is explicitly defined only when the optical path of the radiation flux is given. In the case of a gas, the optical path is defined by the product of the layer thickness l of the gas and its pressure p. In the case of liquids and solids the optical path is defined by layer of plate thickness l.

1.2. RADIATIVE PROPERTIES OF MATERIALS

1.2.1. General radiative properties of materials

A real body, called a material, unlike ideal bodies (black and grey), has variable values of spectral properties ($\varepsilon_{\lambda,\varphi}$, $\varepsilon_{\lambda,n}$, ε_{λ}, $a_{\lambda,\beta}$, a_{λ}, $r_{\lambda,\beta,\varphi}$ $r_{\lambda,\beta}$, $r_{\lambda,\varphi}$, r_{λ}, $\tau_{\lambda,n}$). We often replace a variable spectral value in the range $\lambda_2-\lambda_1$ by the mean value, regarding a real body as a grey body [115],[132]:

$$\varepsilon_{\lambda_1-\lambda_2} = \frac{\int_{\lambda_1}^{\lambda_2} \varepsilon_{\lambda}\cdot m_{\lambda,bb}\cdot d\lambda}{\int_{\lambda_1}^{\lambda_2} m_{\lambda,bb}\cdot d\lambda} \cong \frac{\sum_{\lambda_1}^{\lambda_2} \varepsilon_{\lambda}\, m_{\lambda,bb}\Delta\lambda}{\sum_{\lambda_1}^{\lambda_2} m_{\lambda,bb}\Delta\lambda} . \tag{49}$$

A material, as opposed to ideal bodies, has a geometrical distribution of hemispherical emissivity and reflectivity whose value is not constant as a function of an angle to normal elemental active plane, i.e., the plane is not ideally diffusive; in other words, it is not a Lambert plane. In order to simplify calculations it is often assumed that the plane is diffusive [134]:

$$\bar{\varepsilon}_T = \frac{1}{\pi}\int_{\Omega}\varepsilon_{T,\varphi}\cdot\cos\varphi\cdot d\omega \quad \text{or} \quad \bar{\varepsilon}_T = \varepsilon_{T,\varphi}, \tag{50}$$

where $\bar{\varepsilon}_T$ is the mean total emissivity in the hemisphere (the first formula) or the mean total emissivity in part of the hemisphere (the second formula).

For dielectrics formula (50) can only be used in the range $\varphi \leqslant 45^{\circ}$. An analogical formula can be derived for spectral emissivity - $\bar{\varepsilon}_{\lambda}$. The remaining radiative properties of materials depend so much upon the angle of incidence of the radiation and on the spectral distribution that it is not possible to adopt mean formulas without experimental verification. All total radiative properties of materials can only be regarded as functions of angle φ and temperature. The properties of materials for direction n, i.e., $\varphi = 0$, and for the hemispherical quantity are given in tables.

- All materials can be divided into two groups:
  - non-transparent (in the whole spectral range),
  - transparent (in specified spectral ranges).

Spectral emissivity as a function of wavelength λ decreases for metals, increases for dielectrics and is band-like for gases, liquids and some solids [110],[130],[135],[144]. If the state of the surface does not change it is usualy assumed that function ε_λ does not depend upon temperature. More precise investigations for many metals have shown that there is a certain regularity of changes of ε_λ with the increase of body temperature [79],[93],[107].

The total emissivity of materials usually increases (for metals) or decreases (for dielectrics) with the change of ε_λ - as a function of λ [24]. The chemical and physical changes of the emitting body caused by temperature, time and pressure influence its emissivity [29],[31]. A general characteristic, independent of the kind of material is the variability of emissivity according to surface roughness. Emissivity increases with the increase of roughness [114],[116]. Assuming that roughness consists of microcavities and its measure is expressed by the ratio γ_w (relative roughness) of the area of rough surface S_r to the area of geometrical, i.e., smooth surface S_g [118],[121].

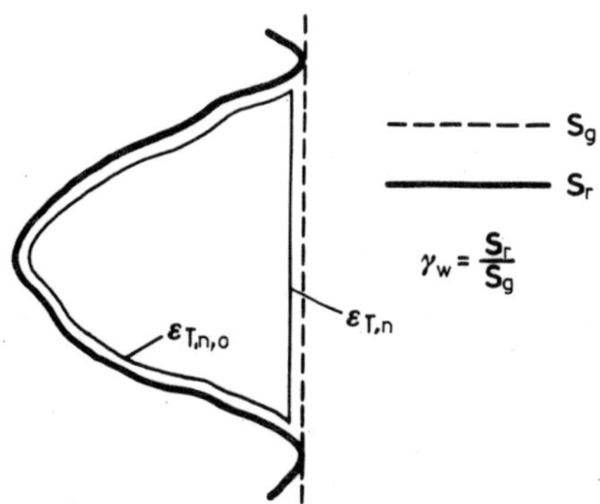

Fig.10. Scheme of relative roughness $\gamma_w = \frac{S_r}{S_g}$

Figure 10 shows γ_w for one microcavity. That is relative roughness. Relative roughness can be expressed by means of standardized parameters of roughness R_z and S_m (PN-Polish Standard):

$$\gamma_w = \frac{S_r}{S_g} = \sqrt{1+4\left(\frac{R_z}{S_m}\right)^2} \ . \tag{51}$$

R_z is the height of the roughness. It is defined as the mean distance of the highest five peaks from the lowest five cavity points of the profile under observation within the length of an elementary segment l_e, measured from the line of reference to the mean line (Fig.11a). R_z is expressed by the formula:

$$R_z = \frac{1}{5}[(R_1+R_3+R_5+R_7+R_9) - (R_2+R_4+R_6+R_8+R_{10})]. \tag{52}$$

S_m is the roughness interval. It is the length of a segment of the mean line limited by its points of intersection with two neighbouring parts of the observed profile having at those points the first derivative of the same sign (Fig.11b). S_m is expressed by the formula:

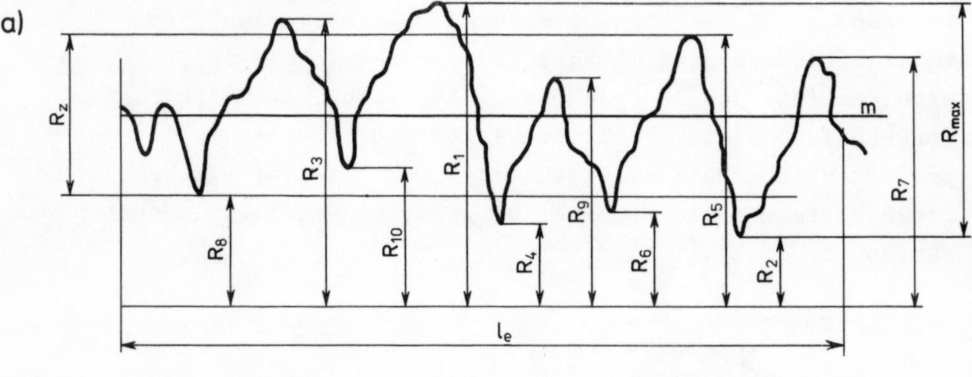

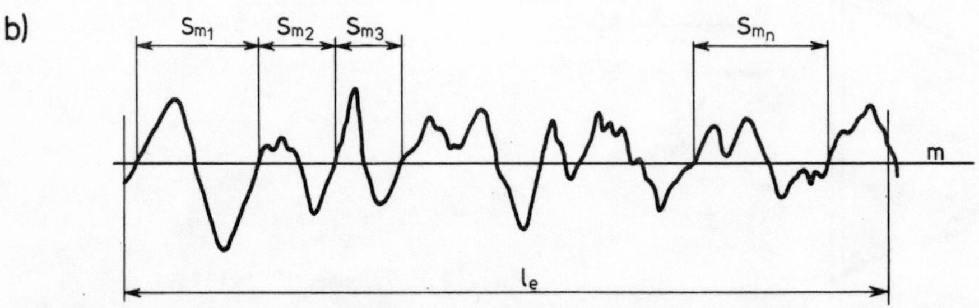

Fig.11. Parameters of roughness: a-R_z (roughness height); b-S_m - roughness width (l_e - roughness - width cutoff)

$$S_m = \frac{S_{m_1} + S_{m_2} + S_{m_3} + \dots\dots\dots + S_{m_n}}{n} \tag{53}$$

Function ε_T changes as is shown in Fig. 12. This empirical curve is described by the formula (see [122],[124]):

$$\varepsilon_{T,n} = \frac{\varepsilon_{T,n,o} \cdot \gamma_w}{1+(\gamma_w - 1)\cdot \varepsilon_{T,n,o}} = \frac{\varepsilon_{T,n,o}\sqrt{1+4(\frac{R_z}{S_m})^2}}{1+(\sqrt{1+4(\frac{R_z}{S_m})^2}-1)\varepsilon_{T,n,o}} \tag{54}$$

where:

R_z - roughness height;

S_m - roughness width;

$\varepsilon_{T,n,o}$ - total emissivity in normal direction n of a smooth surface or a surface with primary roughness.

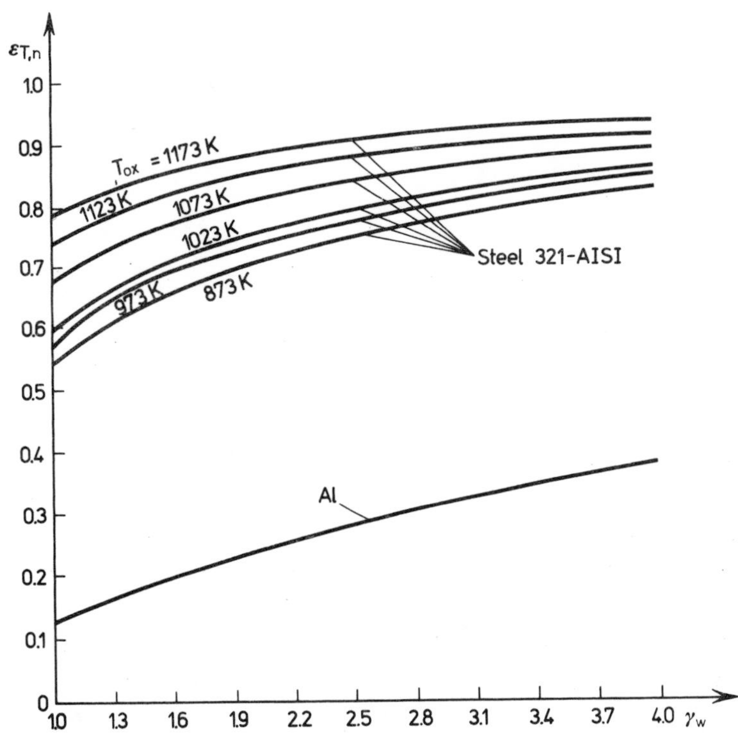

Fig.12. Emissivity $\varepsilon_{T,n}$ of materials as a function of the ratio γ_w, T_{ox} - temperature of oxidation [122]

As a smooth surface or a primary surface we should regard the smoothest surface in a given pattern, taking note of the possibility of there being a greater number of rows of roughnesses, which can essentialy influence emissivity. The resultant ratio γ_W of unevenness for the given surface is the product of several rows of roughnesses (see [20],[46],[99],[122]):

$$\gamma_W = \gamma_{W,I} \cdot \gamma_{W,II} \cdot \gamma_{W,III} \cdots \tag{55}$$

Formula (55) is also applied to the spectral emissivity $\varepsilon_{\lambda,n}$.

The smaller is the initial value of $\varepsilon_{T,n,o}$ and $\varepsilon_{\lambda,n,o}$ the greater is the increase of emissivity $\varepsilon_{T,n}$ and $\varepsilon_{\lambda,n}$, due to the increase of roughness (Fig.13).

Theoretical and experimental results can be verified at a first approximation on the basis of criterion limits (Fig.14). The values of the function $f(\varepsilon_{T,n,o}, \gamma_W)$ have to be between the limits and the function itself is increasing and tangent to the straight line $\varepsilon_{T,0} \cdot \gamma_W$ at the initial point, i.e., for $\varepsilon_{T,n} = \varepsilon_{T,n,0}$ [44],[122].

This follows from the fact that for small values of γ_W the function from formula (53) will assume the following form (see [118],[122]):

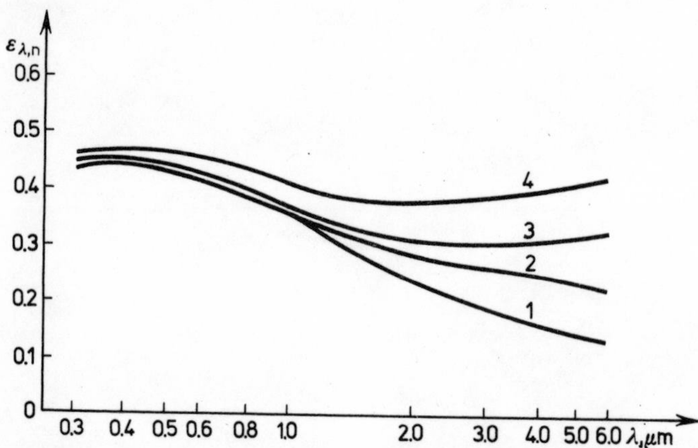

Fig.13. Emissivity $\varepsilon_{\lambda,n}$ of tungsten as a function of wavelength λ for different roughness: 1 - smooth surface, 2 - roughness rms = 0.38 μm, 3 - rms = 0.58 μm, 4 - rms = 1.14 μm (reproduced by permission of the American Institute of Physics, from "Thermal radiation from rough tungsten surface in normal and off-normal directions" by L.K. Thomas)

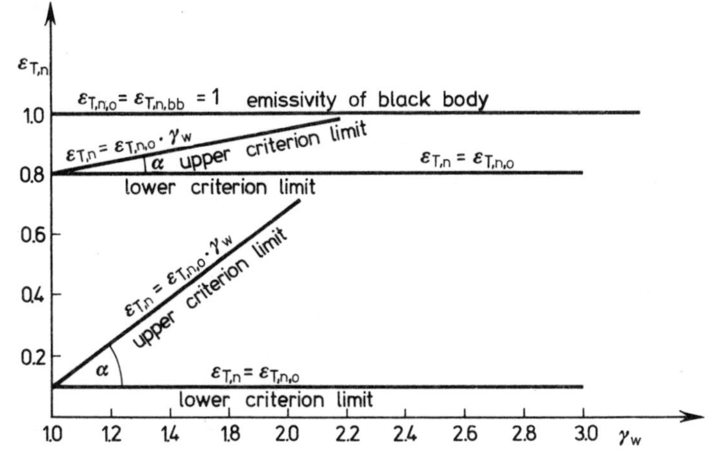

Fig.14. Criterion limits of function $f(\varepsilon_{T,0},\gamma_w)$ [118],[122]

$$\varepsilon_{T,n} = \varepsilon_{T,n,0}\cdot\gamma_w - \varepsilon^2_{T,n,0}\cdot(\gamma_w - 1) \;, \tag{56}$$

for the limit case, $\gamma_w = 1$, we have:

$$\varepsilon_T = \varepsilon_{T,n,0}\cdot\gamma_w \tag{57}$$

Formula (55) implies that for small values of $\varepsilon_{T,n,0}$ the angle between the criterion limits α will be greater than for large values. It tends to zero with the increase of $\varepsilon_{T,n,0}$ to 1, and at the limit for $\varepsilon_{T,n,0} = 1$ the derivative $\varepsilon'_{T,n}$ will be expressed by the formula:

$$\frac{d\varepsilon_{T,n}}{d\gamma_w} = \varepsilon_{T,n,0} - \varepsilon^2_{T,n,0} = 0. \tag{58}$$

This case applies to a black body, i.e., to $\varepsilon_{T,n,0} = \varepsilon_{T,n,bb} = 1$ regardless of the value of γ_w (see Fig.14).

The character of the distribution of emissivity as a function of angle φ is determined by the type of material, but the changes of that distribution due to the changes of temperature and roughness are universal. With the increase of temperature the distribution approaches the Lambert distribution ($\varepsilon_{T,\varphi}\rightarrow$constant) whence $\varepsilon_{T,\varphi}$ tends to a constant with the decrease of λ. Also $\varepsilon_{T,\varphi}$ or $\varepsilon_{\lambda,\varphi}$ tends to a constant value as the roughness γ_w of the emitting surface increases [5].

Reflectivity is much more difficult to define and to measure than emissivity, not only becauseof a greater number of types of reflectivity but also because of the character of the process. Two limiting types of reflectivity are distinguished: diffusive $r_{\lambda,d}$ and specular $r_{\lambda,s}$ [54],[90]. Actually the reflectivity of materials has an intermediate, i.e., diffusive - specular, character. This can be shown most simply on the basis of the specular distribution of the total reflectivity of incident radiation in the direction β - $r_{T,\beta,\varphi}$ (Fig.15). With the increase of the incidence angle β, the character of reflectivity becomes more and more specular one and total reflectivity increases. The directional spectral reflectivity $r_{\lambda,\beta}$ is the sum of $r_{\lambda,s}$ and $r_{\lambda,d}$ [120].

With the increase of wavelength λ for a given surface, the bidirectional spectral reflectivity $r_{\lambda,\beta,\varphi}$ approaches specular

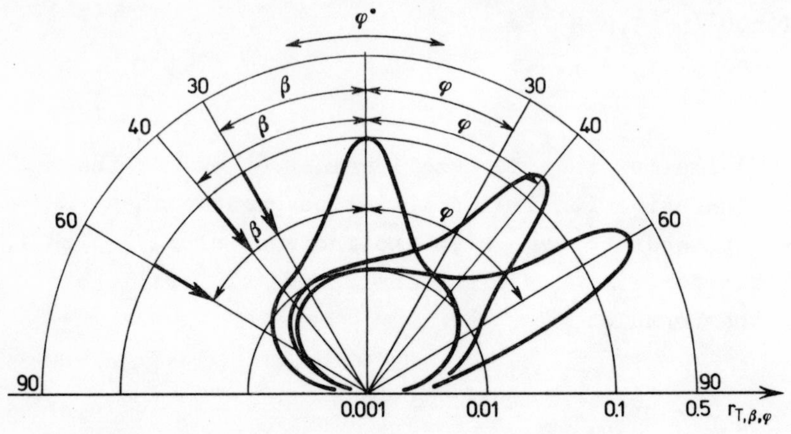

Fig.15. Reflectivity of surface material for different angles of incidence of a beam of thermal radiation β [10],[40],[37],[143]

reflectivity more and more closely regardless of the angle of incidence β. Accordingly diffusive reflectivity diminishes with the increase of wavelength λ. It is to be expected that diffusive reflectivity will increase with the increase of temperature [57],[145].

With the increase of roughness, the directional hemispherical spectral reflectivity - $r_{\lambda,\beta}$ decreases in relation to the R_z and λ [57]. Further increase does not produce change, and the value of $r_{\lambda,\beta}$ depends upon the type of material only (Fig. 16). With the increase of roughness the specular

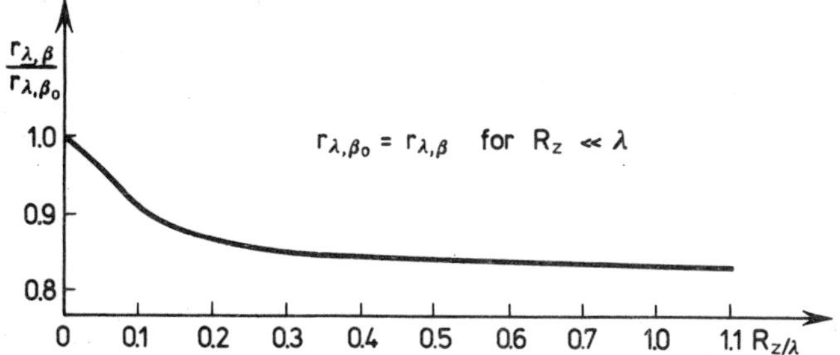

Fig.16. Empirical value $r_{\lambda,\beta}/r_{\lambda,\beta_o}$ as a function of the ratio $\sim R_z/\lambda$ for aluminium coating on glass substratum (r_{λ,β_o} - reflectivity of polished aluminium hence $r_{\lambda,\beta} = r_{\lambda.\beta_o}$ (after R.C.Birkebak and E.R.G. Eckert, 1965) [5], [6]

reflectivity $r_{T,s}$ decreases more rapidly than the directional hemispherical total reflectivity $r_{T,\beta}$ [4], [6], [148].

Only the reflectivity $r_{\lambda,n}$ is used as a standard parameter, since establishing the position of the source with respect to the reflecting surface is the simplest and independent of the other parameters of the source.

Transmissivity as a property of a material can only be considered under the assumption that the material constitutes a planar paralel layer whose roughness satisfies the condition $\lambda/R_z >> 1$. Otherwise we have to do with some additional phenomena complicating the definition of transmissivity as an unambiguous parameter [92]. The property in question exists under the assumption that incident radiation is perpendicular to the surface of the body. For homogeneous bodies the following formula can be written [13], [30], [36]:

$$\tau_{\lambda,n} = \frac{\varphi_{\lambda,\tau,n}}{\varphi_{\lambda,n}} = e^{-(K_\lambda \cdot l)} = e^{-(\alpha_\lambda + \sigma_\lambda)\cdot l}, \qquad (59)$$

where:

K_λ = spectral attenuation coefficient, 1/m,
l = thickness of transmitting layer, m,
α_λ = spectral absorption coefficient, 1/m,
σ_λ = spectral scattering coefficient, 1/m.

1.2.2. Radiative properties of opaque materials

Metals are the fundamental kind of opaque bodies. Only a thin layer of metal takes part in the process of radiative heat exchange. Radiation is caused mainly by the oscillation of atoms and molecules, according to the chemical composition of the metal and its physical state, in a layer 0 - 10^{-10} m thick.

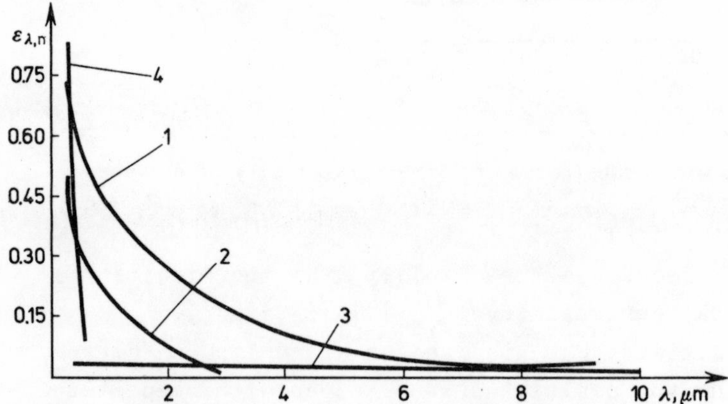

Fig.17. Emissivity ε_λ of metals with polished surface: 1 - iron, 2 - aluminium, 3 - gold, 4 - silver (reproduced by permission of the McGraw-Hill International Book Company, from "Relation and properties" by E.R.G. Eckert, eds. W.M. Rohsenow and J.P. Hartnett "Handbook of heat transfer")

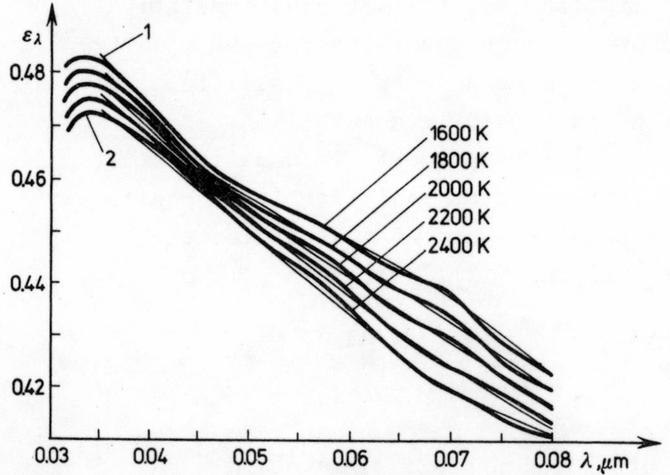

Fig.18. Emissivity $\varepsilon_{\lambda,n}$ of tungsten: 1 - computed curves, 2 - empirical curves (reproduced by permission of the Optical Society of America a from "Spectral emissivity of tungsten" by R.D.Larrabe, from the Journal of the Optical Society of America, vol. 49,1959, pp.619-625)

Spectral emissivity of metals decreases with the increase of wavelength λ, inversely to reflectivity (Fig.17); in the range of visible radiation or ultraviolet radiation, ε_λ reaches a maximum and then decreases again (Fig. 18) [16],[35].

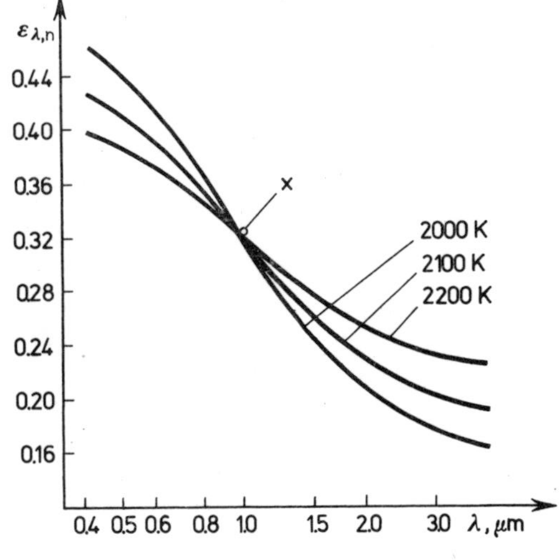

Fig.19. Emissivity $\varepsilon_{\lambda,n}$ as a function of λ for different temperature [132],[142],[143]

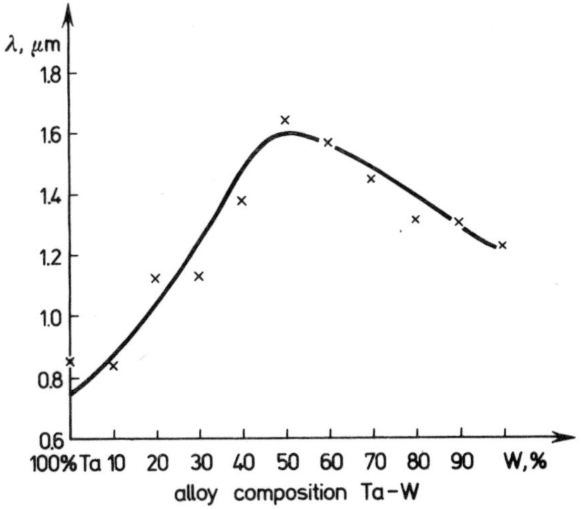

Fig.20. The possition of the point X as a function of the chemical composition of Ta - W alloy (reproduced by permission of the American Institute of Physics, from "Normal spectral emissivity of Ta-W and Nb-Mo alloys" by L.K.Thomas)

The influence of temperature on the emissivity of a clean

metal surface is negligible. With the increase of temperature the courve of emissivity as a function of wavelength becomes more and more flat (Fig.19). For λ = X (called point X) the curves $\varepsilon_\lambda = f(\lambda, T)$ intersect [65]. The position of the point X depends upon the type of the material or the chemical composition of the alloy (Fig.20) [142].

The total emissivity $\varepsilon_{T,n}$ of metals with a clean surface increases with the increase of temperature (Fig.21) [150]. This is in accordance with our former considerations concerning the increase of ε_λ with the decrease of wavelength λ [43],[101].

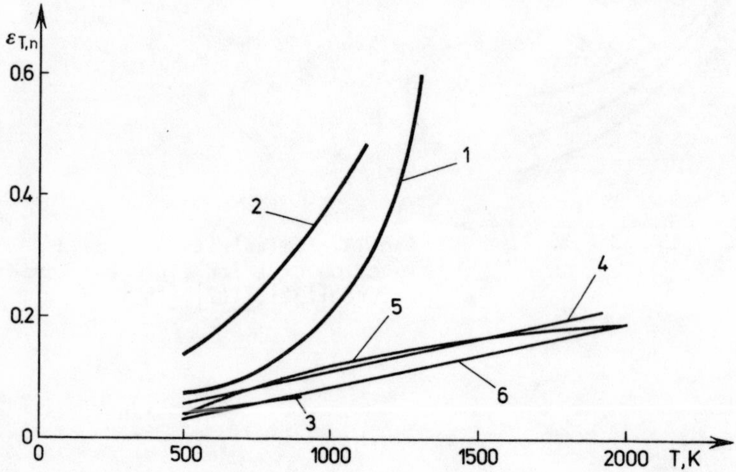

Fig.21. Emissivity $\varepsilon_{T,n}$ of different metals as a function of temperature T: 1 - chromium, 2 - iron, 3 - aluminium, 4 - tungsten, 5 - platinum, 6 - molybdenum [41],[96],[109]

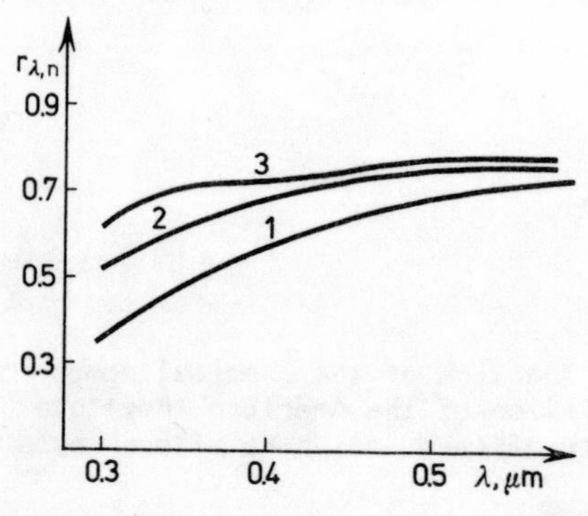

Fig.22. Reflectivity $r_{\lambda,n}$ as a function of wavelength λ: 1 - Mg, 2 - Al-Mq alloy (50%Al), 3 - Al [68],[95],[142]

As the temperature of a metal increases, its directional hemispherical total reflectivity $r_{T,\beta}$ decreases. The greater the roughness is the smaller the reflectivity. The influence of temperature is greater for specular than for diffusive reflectivity [154],[155],[158].

The radiative properties of metal alloys are intermediate in relation to the constituent pure metals (Fig. 22)[142].

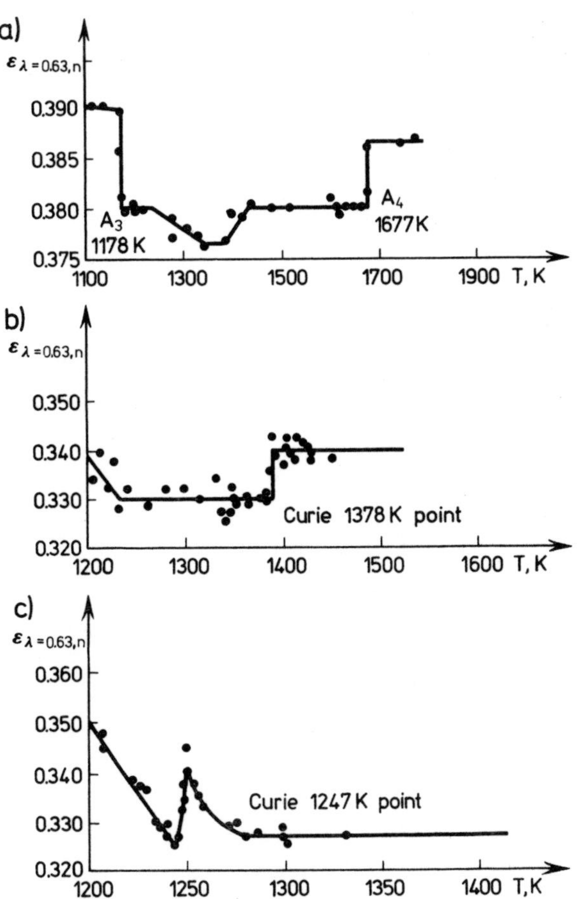

Fig.23. Jump changes of $\varepsilon_{\lambda=0.63,n}$ due to structural changes caused by temperature: a - airon, b - cobalt, c - 65% Co - 35% Ni alloy (reproduced by permission of the American Physical Society from "The spectral emissivity of iron and cobalt" by H.B. Wahlin and H.W. Jr Knop)

Structural changes of metals and alloys cause jump changes of the radiative properties [129],[152]. The changes usually amount to a few hundredths and do not exceed two tenths of the value of the given radiative property (Fig.23). The radiative properties of a metal depend upon layer thickness, which is particularly important in the case of reflecting coatings [129]. An increase of thickness of a layer causes an increase of reflectivity $r_{\lambda,\beta,\varphi}$ and when the thickness of gold or silver reaches $150 \cdot 10^{-10}$ m, the layer becomes opaque and acquires the properties of a solid material (Fig.24) [33],[136].

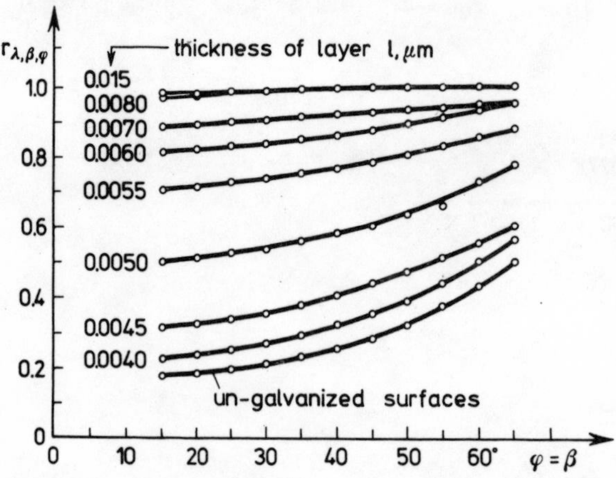

Fig.24. Reflectivity $r_{\lambda,\beta,\varphi}$ (for $\beta=\varphi$) of thin layers of gold as a function of angle β for wavelength $\lambda=10.6$ μm in perpendicular polarized light for different layers thickness (after E.M. Sparrow, R.P. Heinisch, K.K. Thin - 1973) [136]

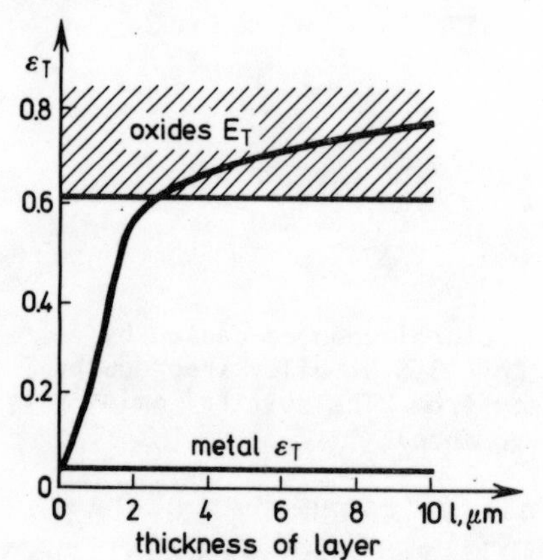

Fig.25. Emissivity ε_T as a function of thickness of layer of oxides obtained electrolytically on aluminium plate [128],[132]

Metals are often coated with oxides. Even a minimal layer of oxides changes the radiative properties of the metal in an essential way (Fig.25). In most cases it is a corrosion layer due to temperature corrosion in the atmospheric air. The state of the corrosion layer in a given atmosphere depends above all on the relation between temperature and time [17],[14],[15], [122].

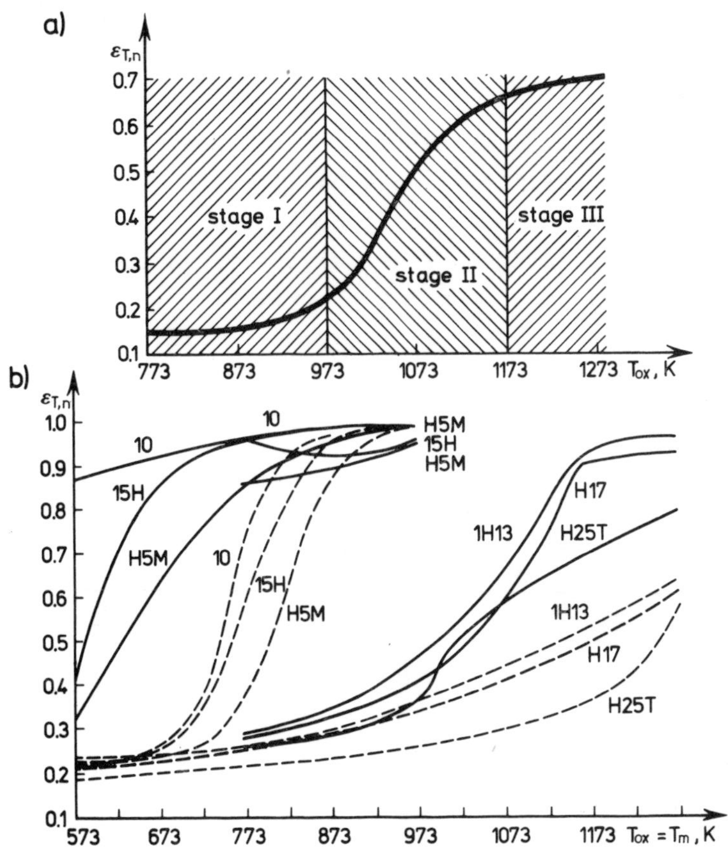

Fig.26. The increase of emissivity $\varepsilon_{T,n}$ as a function of temperature of oxidation T_{ox}. a - steel 321, t_{ox} = 8h, T_m = 773 K, b - different grades of steels $T_{ox} = T_m$, t_{ox} = 2 min (broken - line curves) and t_{ox}= 8h (full--line curves), t_{ox} - period of oxidation, T_m - temperature of measurement [14], [15],[122], [114]

Two types of temperature are distinquished: temperature of oxidation T_{ox} and temperature of measuring T_m, at which the process of oxidation can take place [15]. With the increase of the temperature of oxidation, the emissivity of the metal also increases. There are three stages of emissivity increase (Fig. 26a) [122],[140]:

- initial stage with a low increase of emissivity, there appears a layer of oxides in the form of a solid solution;
- middle stage, with a rapid increase of emissivity, the layer of oxides changes from a solid solution into a mixture;
- final stage, with a low increase of emissivity and high increase of thickness; the layer of oxides is in the form of a mixture.

The increase of emissivity is lower for metals resistant to corrosion at a higher temperature (Fig.26b) [17],[15]. For metals which are not corrosion-resistant at a high temperature - the first and the second stages of emissivity increase can be observed only for very short periods of oxidation [96]. It has been found that the emissivity increase of oxidized metal is due above all to the increase of layer thickness (stage I and III) and particularly to the structural changes in the oxidation layer (stage II) [42],[122].

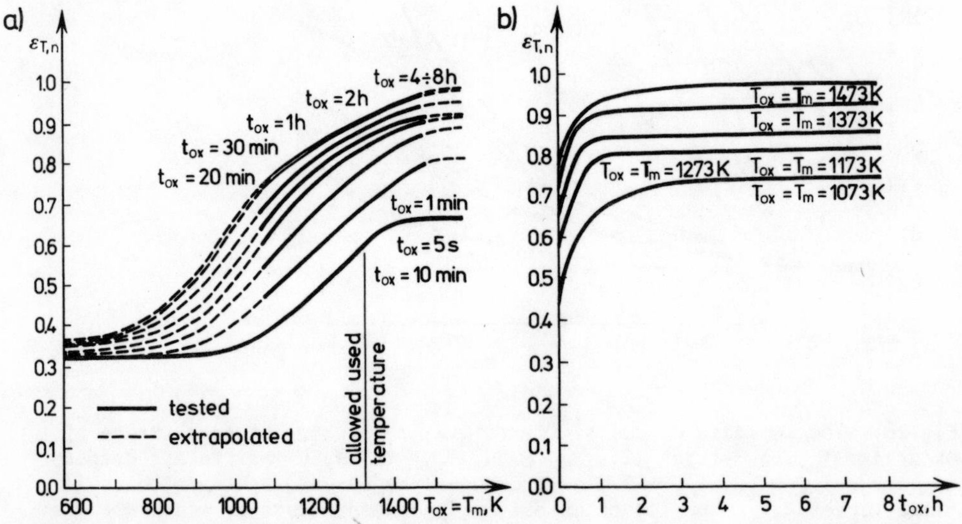

Fig.27. The increase of emissivity $\varepsilon_{T,n}$ of nikrothal 20 as a function of: a - temperature of oxidation T_{ox} (temperature of oxidation) = T_m (temperature of measurement), b - period of oxidation t_{ox} at different temperatures of oxidation T_{ox} [16],[17]

With the increase of the temperature of oxidation emissivity and absorptivity increase whereas reflectivity decreases. Reflection of incident radiation becomes more and more diffusive with the increase of oxidation temperature.

The effect of the oxidation period on the increase of emissivity of metal is considerable (Fig.27a). It is particularly evident at high oxidation temperature T_{ox}, where the initial increase is very high and then the value of emissivity is stabilized at a constant level (Fig.27b).

For polished metals the ratio $\varepsilon_T/\varepsilon_{T,n}$ varies from 1.15 to 1.20 [18],[19]. Up to an angle φ of 20-45° the difference between ε_T and $\varepsilon_{T,n}$ does not exceed 1% [19]. It is only for greater angles that the differences are considerable, reaching their maximum for $\varphi \approx 80°$ [18]. For $\varphi > 85°$ there is a continuous decrease of $\varepsilon_{T,\varphi}$ (Fig.28). The increase of angle φ induces an increase of degree of polarization of the radiation. The degree of polarization decreases with the increase of roughness.

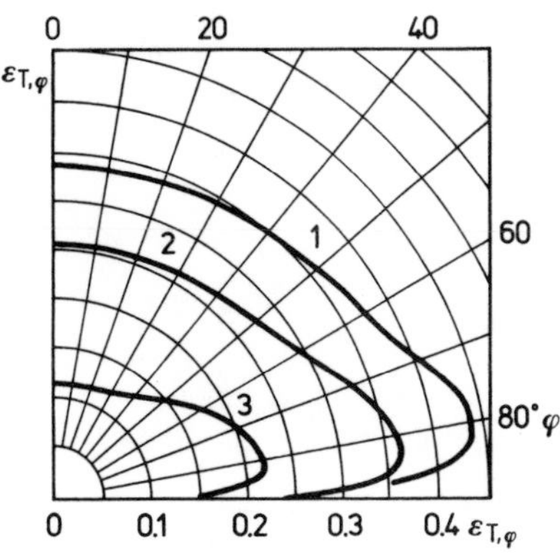

Fig.28. Geometrical distribution of $\varepsilon_{T,\varphi}$ of metals with smooth surface: 1 - bismuth, 2 - Al-Cu alloy, 3 - iron (reproduced by permission of the McGraw-Hill International Book Company, from "Relation and properties" by E.R.G. Eckert, eds W.M. Rohsenow and J.P. Hartnett, "Handbook of heat transfer")

As wavelength decreases, the ratio $\varepsilon_{\lambda,\varphi}/\varepsilon_{\lambda,n}$ for angles φ above 30° decreases for metals from values greater than one to values lower than one (Fig.29a). Together with the increase of temperature of a radiative body, for angles φ above 30° the ratio $\varepsilon_T/\varepsilon_{T,n}$ decreases as does the ratio $a_{T,\beta}/a_{T,n}$ (Fig.29b). Also the increase of the oxidized layer of the metal contributes to a greater uniformity of geometrical distribution $\varepsilon_{\lambda,\varphi}$ and $\varepsilon_{T,\varphi}$, making it more Lambert-like (Fig. 30).

Dielectrics are naturally more transparent than metals, and many of them are included among the transparent materials used

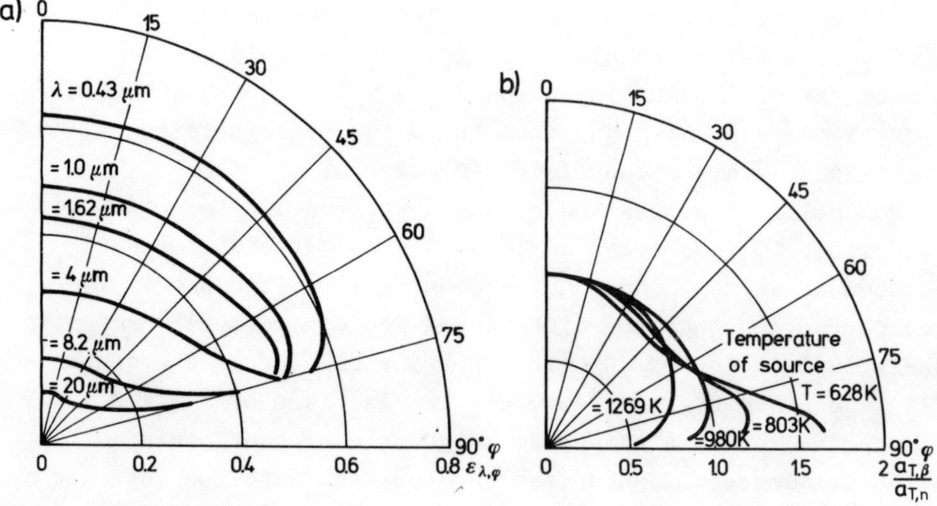

Fig.29. Geometrical distribution of radiative properties of metal: a - $\varepsilon_{\lambda,\varphi}$ of pure titanium, b - $a_{T,\beta}/a_{T,n}$ of aluminium with a thin layer of oxidization (reproduced by permission of the McGraw-Hill International Book Company, from "Thermal radiation heat transfer" by R. Siegiel and J.R.Howell)

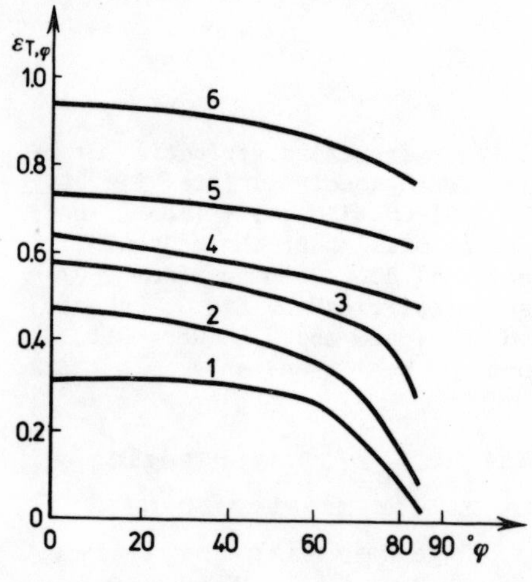

Fig.30. Geometrical distribution $\varepsilon_{T,\varphi}$ of 321 steel with polished surface after different temperatures of oxidation for a period of time t_{ox} = 8h: 1 - T_{ox}=87,3 K, 2 - T_{ox}= =973 K, 3 - T_{ox}=1023 K, 4 - T_{ox}= =1073 K, 5 - T_{ox}=1123 K, 6 - T_{ox}= =1173 K [122], [123]

in infrared engineering for an appropriate spectrum region [32],

[83]. In general, it should be stated that ε_T and a_T for dielectrics have much greater values than for metals [94]. Unlike metals, they have a spectral emissivity which increases with the increase of wavelength λ. However, this increase is not so

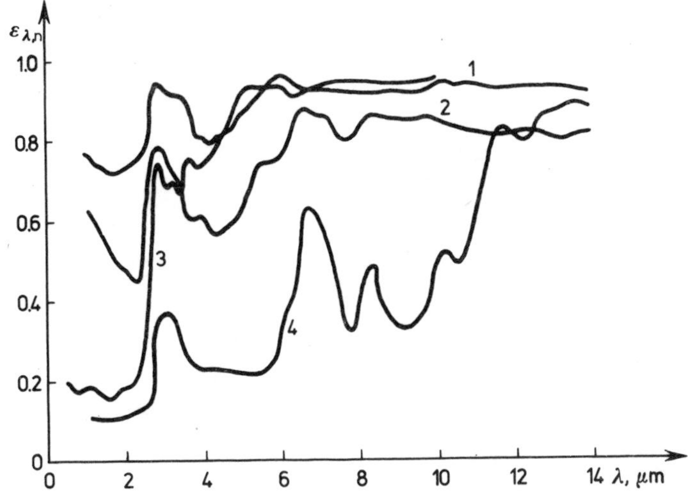

Fig.31. Change of emissivity as a function of wavelength λ: 1 - soil, 2 - Portland cement, 3 - white concrete roofing tile, 4 - magnesium oxide (reproduced by permission of the McGraw-Hill International Book Company, from "Relation and properties" by E.R.G. Eckert, eds W.M. Rohsenow and J.P. Hartnett, "Handbook of heat transfer")

uniform as for metals (Fig.31) [83]. White oxides (of thorium, magnesium, aluminium, zirconium, etc.) have low emissivity in the visible range and the near infrared range. An increase of emissivity appears at λ = 5-6 μm, and at λ ≅ 10 μm emissivity approaches black body emissivity [97],[102]. Coloured oxides (of chromium, cerium and others) have greater emissivity for near infrared ranges than white oxides [97]. Reflectivity r_λ for dielectrics decreases with wavelength λ.

In connection with the increase of ε_λ with wavelength λ, ε_T for dielectrics decreases with the increase of temperature [104]. Consequently, total reflectivity r_T increases, and absorptivity a_T decreases with the increase of temperature of the black body irradiating the surface of the dielectric (Fig. 32). Lack of appropriately clean dielectric materials, lack of a precise characterization of granularity, of the condition of the surface, and changes in spectral emissivity with wavelength λ are the cause of a considerable scatter of the results of measurement of the values of radiative properties for dielectric materials [2],[28],[66].

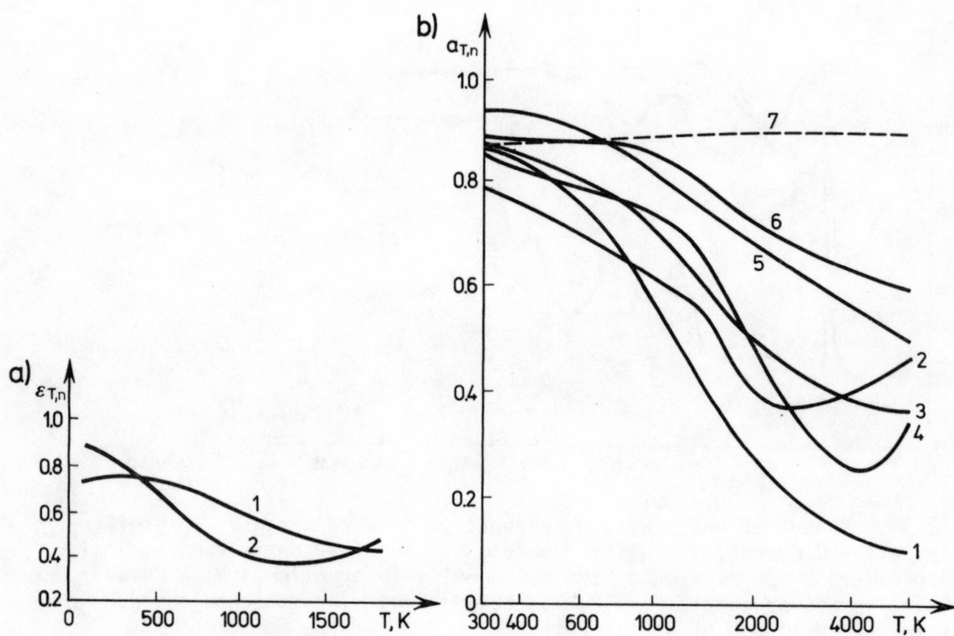

Fig.32. Radiative properties of dielectrics as a function of temperature T - a - $\varepsilon_{T,n}$: 1 - aluminium oxide LA 603, 2 - zirconium oxide, b - $a_{T,n}$: 1 - fireproof white clay, 2 - azbestos, 3 - cork, 4 - wood, 5 - porcelane, 6 - concrete, 7 - white roofing tile (reproduced by permission of the McGraw-Hill International Book Company, from "Relation and properties" by E.R.G. Eckert, eds W.M. Rohsenow and J.P. Hartnett, "Handbook of heat transfer")

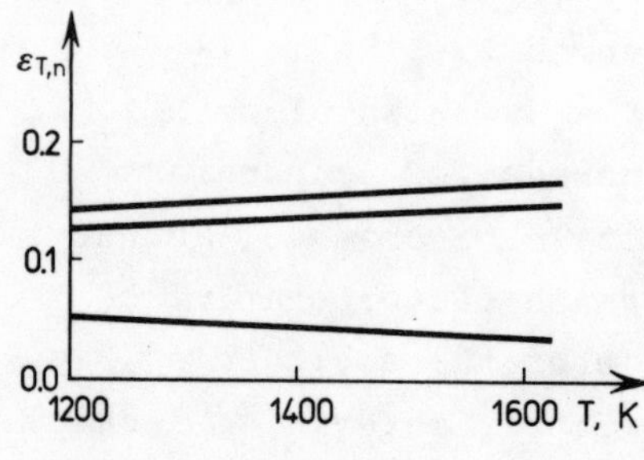

Fig.33. Emissivity $\varepsilon_{T,n}$ of Al_2O_3 as a function of T for different grain size d: 1 - d = 0.5÷1.5 μm, 2 - d = 1.5÷8 μm, 3 - d = 1.5÷6 μm [13],[20],[128],[138]

The emissivity of dielectrics increases with the increase of the grain size of the basic material, especially in the range of 1 to 30 μm of grain diameter (Fig.33). The porosity of dielectrics is conducive to an increase of emissivity [89], [98].

The radiative properties of semiconductors are between the radiative properties of metals and dielectrics [138].

1.2.3. Radiative properties of transparent materials

Transmissivity of incident radiation along a optical path l free from reflection and scattering can be expressed (see [45], [125]) for spectral transmissivity by the formula:

$$\tau_{\lambda,n} = \frac{\varphi_{\lambda,\tau,n}}{\varphi_{\lambda,n}} = 1 - a_{\lambda,n} = e^{-\alpha_{\lambda}\cdot l} \tag{60}$$

and for the total transmissivity by the formula:

$$\tau_{T,n} = 1-a_{T,n} = \frac{\int_{\lambda_1=0}^{\lambda_2=\infty} \tau_{\lambda,n}\cdot\varphi_{\lambda,n}\cdot d\lambda}{\int_{\lambda_1=0}^{\lambda_2=\infty} \varphi_{\lambda,n}\cdot d\lambda} = \frac{\int_{\lambda_1=0}^{\lambda_2=\infty} \varphi_{\lambda,n}\cdot e^{\sigma_{\lambda}\cdot l}\, d\lambda}{\int_{\lambda_1=0}^{\lambda_2=\infty} \varphi_{\lambda,n}\cdot d\lambda} \tag{61}$$

If the process of transmissivity is accompanied by reflecting (e.g., in a plane-parallel glass layer), then the measured transmissivity $\tau'_{\lambda,l}$ will be lower than the real one:

$$\tau'_{\lambda,n} = \tau_{\lambda,n} \frac{(1-r_{\lambda,n})^2}{1-r^2_{\lambda,n}\ \tau^2_{\lambda,l}} \tag{62}$$

If scattering takes place, then transmissivity $\tau_{\lambda,l}$ will be decreased by the value $e^{-\sigma_{\lambda}\cdot l}$ [132], [138].

In any medium there occur three kinds of scattering, i.e., changes of direction of the optical path [77], [106]:

- molecular, when the scattering particles are small in comparison with the wavelength λ;
- diffusive, when the scattering particle dimensions are close to the wavelength λ;
- geometrical, when the scattering particles have considerable dimensions in comparison with the wavelength λ.

Absorptivity of a transparent material can be expressed by

the following formula (see [10],[60]):
- for spectral absorptivity

$$a_{\lambda,n} = \frac{\varphi_{\lambda,a,n}}{\varphi_{\lambda,n}} = 1 - e^{-\alpha_\lambda \cdot l} , \tag{63}$$

- for total absorptivity

$$a_{T,n} = \frac{\int_{\lambda=0}^{\lambda_2=\infty} \varphi_{\lambda,n}\cdot(1-e^{-\alpha_\lambda\cdot l})d\lambda}{\int_{\lambda_1=0}^{\lambda_2=\infty} \varphi_{\lambda,n}\cdot d\lambda} = \frac{\int_{\lambda_1=0}^{\lambda_2=\infty} \varphi_{\lambda,n}\cdot a_{\lambda,n}\cdot d\lambda}{\int_{\lambda_1=0}^{\lambda_2=\infty} \varphi_{\lambda,n}\cdot d\lambda} , \tag{64}$$

where $\varphi_{\lambda,a,n}$ denotes the spectral radiation absorbed from normal direction. If the body itself radiates, the coefficient of real absorptivity $\alpha_{\lambda,r}$ is a little bigger than the one obtained by measurement. This is expressed by the following formula (see [132]):

$$\alpha_\lambda = (1 - e^{\frac{hc}{k\lambda T}}) \cdot \alpha_{\lambda,r} . \tag{65}$$

The value of the component exp(hc/kλT) is usually negligible (except when the value λT is very big). The component is 0.01 when λT = 3120 μm·K, and 0.05 when λT = 4800 μm·K [132].

The spectral emissivity of a transparent body can be expressed by the formula for $a_{\lambda,n}$ (62) and ε_T by the formula:

$$\varepsilon_T = \frac{\int_{\lambda_1=0}^{\lambda_2=\infty} m_{\lambda,bb}(1-e^{-\alpha_\lambda\cdot l})d\lambda}{\sigma T^4} . \tag{66}$$

For a transparent body $\varepsilon_{\lambda,n}$ can be expressed by means of reflectivity $r_{\lambda,n}$ and transmissivity $\tau_{\lambda,n}$ [100]:

$$\varepsilon_{\lambda,n} = \frac{(1-r_{\lambda,n})\cdot(1-\tau_{\lambda,n})}{(1-r_{\lambda,n})\cdot\tau_{\lambda,n}} \tag{67}$$

For transparent bodies the measured reflectivity $r'_{\lambda,n}$ can be higher than the true one $r_{\lambda,n}$ because reflecting takes place also in the second plane [10]:

$$r'_{\lambda,n} = r_{\lambda,n}[1 + \frac{\tau^2_{\lambda,n}(1-r_{\lambda,n})^2}{1-r^2_{\lambda,n}\cdot\tau_{\lambda,n}}] . \tag{68}$$

Transparent materials are divided according to the state of aggregation into three large groups: solid, liquid and gaseous. All of them have characteristic lines or bands of absorptivity connected with vibrational or rotational motion of the particles.

Solid-state transparent materials include: crystalline bodies, glass and plastics.

Crystalline materials form the most diversified group. Their common characteristic is absorption connected with the action of electromagnetic waves up on the crystal lattice. Crystalline bodies are further divided into mono-crystalline and polycrystalline materials. The latter can have strong bands of scattering [145]. Crystalline bodies often have wide bands of infrared transmission. There are at present more synthetic than natural crystalline materials used in infrared technology. These materials vary grealty as regards their radiative properties [113],[131],[152].

Another form of crystalline materials are semiconductors, which are characterized by additional bands of absorption, connected with internal absorption (absorption of energy by electrons passing through the barrier layer), external absorption(absorption by electrons in passing from one permitted band to another), and free electron absorption. The appropriate absorption bands can be defined by means of electric parameters. Semiconducting crystalline materials have better properties in transmitting in the infrared than in the visibile range, inversely to glass materials [62],[63].

Glass materials, unlike crystalline materials, form a close group from the point of view of radiative properties. Ordinary window panes usually have a transmissivity in the visible range up to approximately 2,8 μm (Fig. 34a). High quality glasses may have their long-wave infrared transmissivity shifted up to approximately $\lambda \cong 5$ μm. The total emissivity of glass decreases with the increase of temperature but increases with its thickness (Fig.34b). Geometrical distribution of emissivity is characteristic for dielectric distribution (Fig.34c) [12],[159].

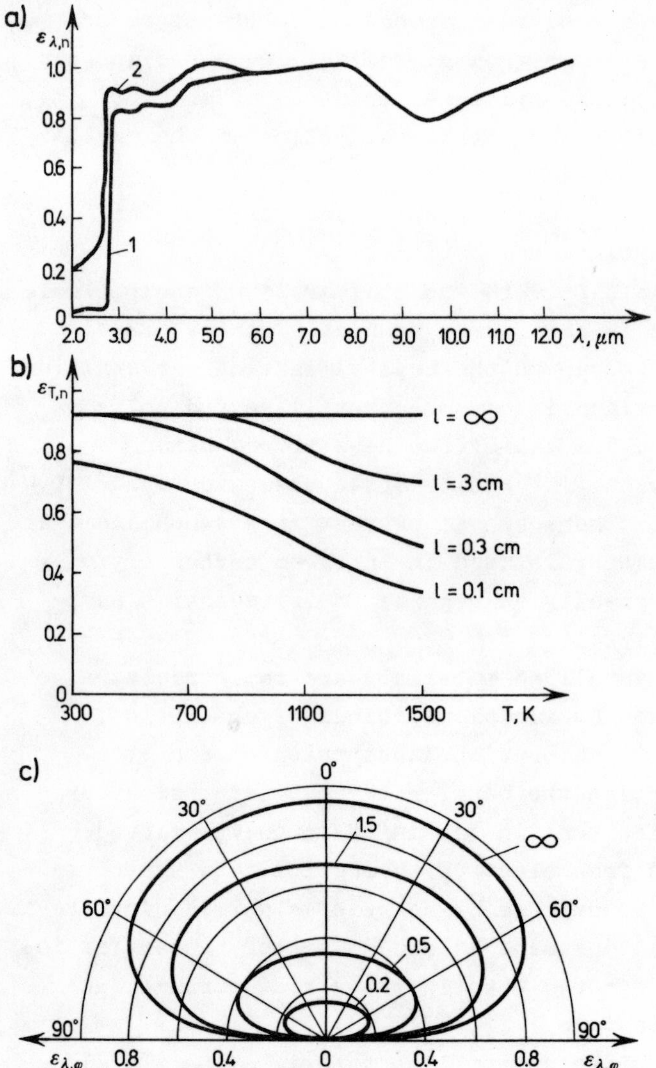

Fig.34. Radiative properties of glass - a - $\varepsilon_{\lambda,n}$ as a function wavelength λ; 1 - for thickness of plate l = 3 mm and of temperature 1273 K; 2 - for l = 6 mm and at temperature 873 K [32]; b - ε_T as a function temperature T(the numbers beside curves indicate thickness of glass); c - geometrical distribution $\varepsilon_{\lambda,\varphi}$ (λ = 2.75÷4.50 μm) for various values of $\alpha_\lambda \cdot l$ (the number at curves) [36],[41]

Plastic materials, on account of a very complex structure of chemical band chains, have many characteristic bands of absorptivity and for this reason can only be used in the form of very thin structural elements (Fig.35). In the case of thicker

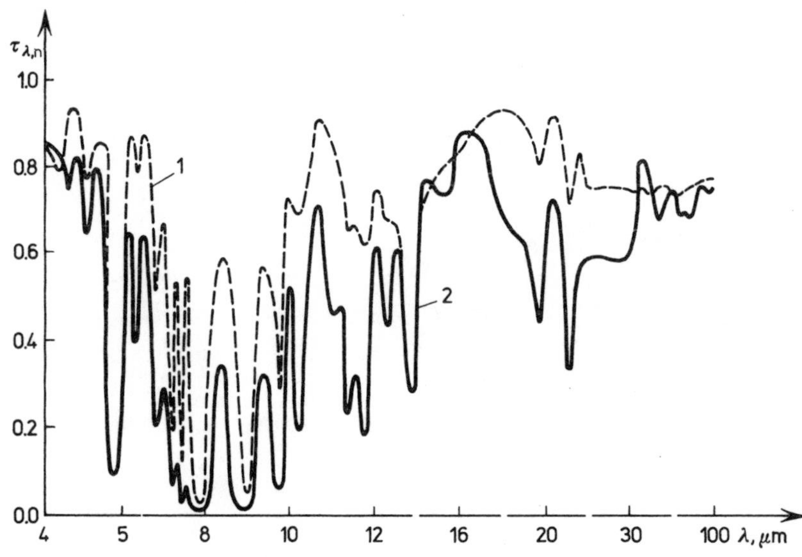

Fig.35. Transmissivity $\tau_{\lambda,n}$ of the "mylar" plastic: 1 - thickness of layer l = 6 μm; 2 - l = 25 μm (after C.L. Tien, C.K. Chan, G.R. Cunnigton - 1972) [145]

layers the absorptivity bands merge together [88],[160].

Metals, although they do not belong to transparent materials, can be used as semi-transparent films in the form of thin layers (Fig.36). Good results have been obtained with thin semi-transparent layers made of gold, silver, aluminium, thin, nickel, palladium and chromium.

Liquids are interesting first of all as materials subject to energy radiation, seldom as structural components. The most common liquid medium is water, whose radiative properties are given in Fig. 37 [15],[126].

Gases are transparent media in which radiative heat exchange takes place. Gases have exceptionally selective characteristics regarding radiative properties [21]. From the point of view of radiative heat exchange technique, all gases can be divided into two groups: gases which are non-absorbing and non-emitting media, and gases which are absorbing and emitting media. The first group includes gases with linear spectra. The most common of them are N_2, O_2, H_2, Ar. The linear spectra of these gases are so narrow that they can be disregarded. The second

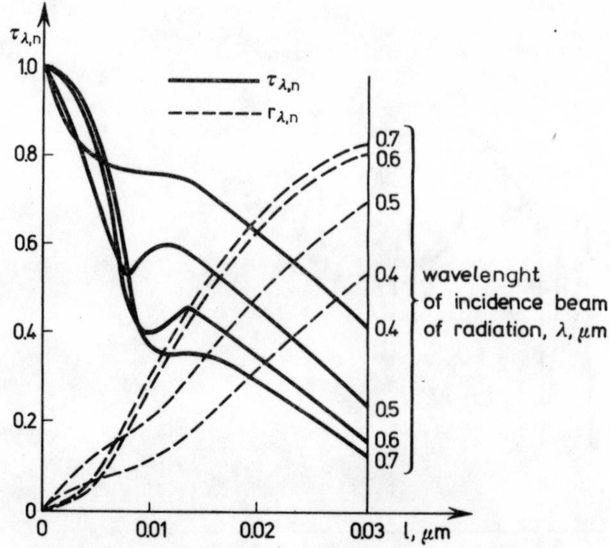

Fig.36. Transmissivity $\tau_{\lambda,n}$ and reflectivity $r_{\lambda,n}$ of silver as a function of thickness of layer l (reproduced by permission of the Optical Society of America, from "The structure of evaporated metal films and their optical properties" by R.S. Sennett and G.D. Scott, from the Journal of the Optical Society of America, vol. 40, 1950, pp. 203-211)

group includes gases with band spectra. The most common of them are CO_2, H_2O, NH_3, Co and SO_2. Their spectral characteristics are given in the tables. The optical properties of gases, contrary to those of the former materials, depend not only on the layer thickness l but also on pressure p. The product of these two quantities is called the optical path l_m (see [55],[58]):

$$l_m = l \cdot p \quad [m \cdot Pa] \tag{69}$$

On account of the selective character of optical spectral properties, their total properties are different, although it can be ascertained that $\varepsilon_{T,n}$ decreases with temperature [56]. This follows from the fact that $\varepsilon_{\lambda,n}$ decreases with temperature and the actively optical bands broaden. Although the active optical bands of gases broaden under the influence of temperature and pressure (Fig.38), the shifting of the spectrum towards shorter wavelength together with the increase of tem-

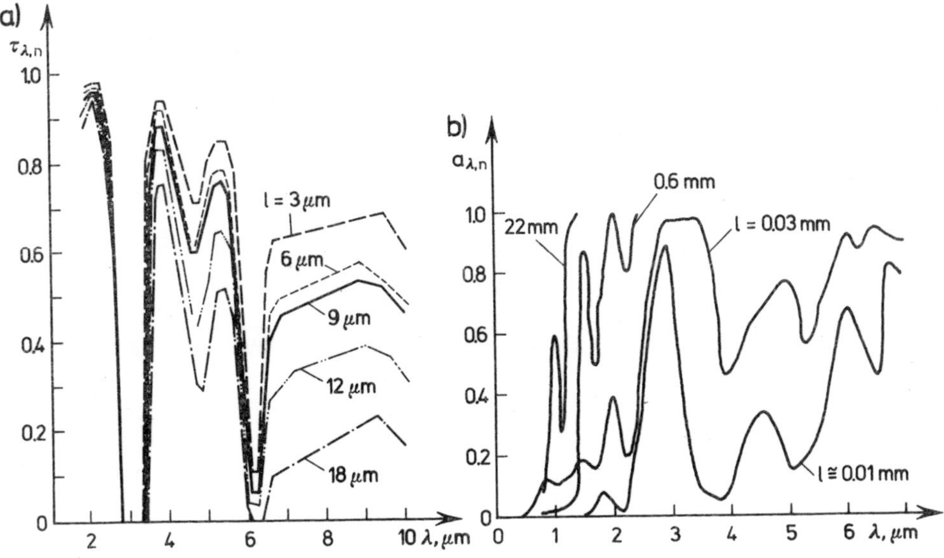

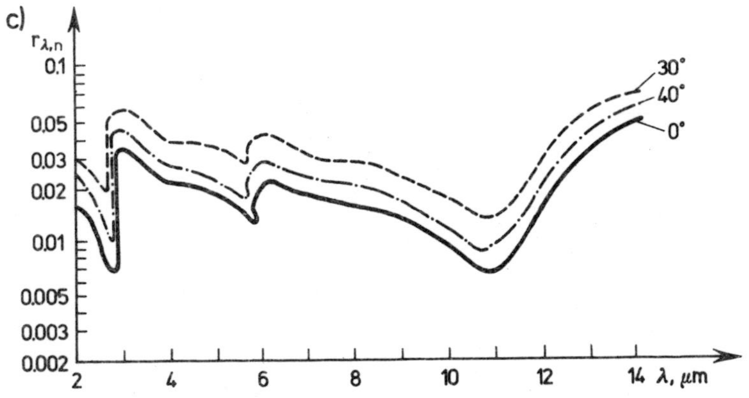

Fig.37. Radiative properties of water: a - $\tau_{\lambda,n}$ - thickness of layers l = 3÷18 μm [123]; b - $a_{\lambda,n}$ - thickness of layers l = 0.01÷22 mm [50]

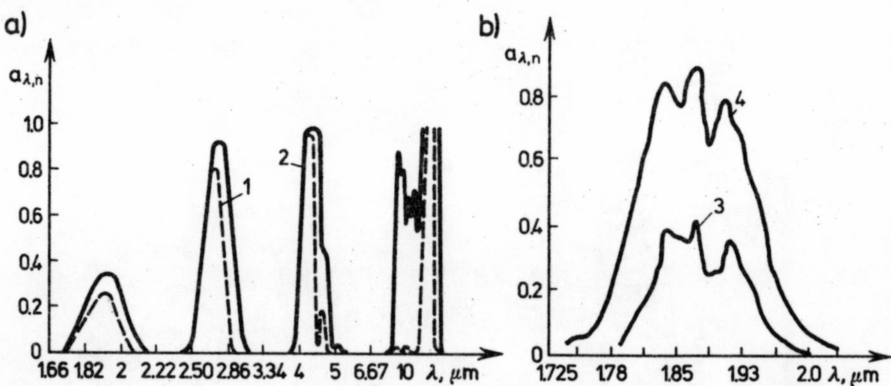

Fig.38. Absorptivity $a_{\lambda,n}$ of CO_2 as a function of λ: a - for different temperatures; 1 - temperature T = 294 K; 2 - temperature T = 833 K; b - for different pressures and for thickness of layer l = 1 mm; 3 - for pressure $p = 1.00 \cdot 10^3$Pa; 4 - for pressure $p = 10.0 \cdot 10^3$Pa (after D.K. Edwards - 1973) [29], [30]

perature is decisive for the total optical properties of gases [25], [125].

Among gases a particularly important role is played by air as a medium in which radiative heat exchange takes place. For air the most important quantity for air is transmissivity τ_λ. The transmissivity of clean atmosphere 1830 m in thickness is shown in the tables.

Part II. TABLES OF BLACK BODY RADIATION

TABLE 2.1. DEGREES OF TEMPERATURE, $m_{\lambda_{max},bb}$, $M_{T,bb}$, AND $\lambda_{max,bb}$

T [K]	$m_{\lambda_{max},bb}$ [W/μm m²]	$M_{T,bb}$ [W/m²]	$\lambda_{max,bb}$ [μm]	T [K]	$m_{\lambda_{max},bb}$ [W/μm m²]	$M_{T,bb}$ [W/m²]	$\lambda_{max,bb}$ [μm]
300	$3.1266 \cdot 10^1$	$4.5930 \cdot 10^2$	9.65930	550	$6.4756 \cdot 10^2$	$5.1887 \cdot 10^3$	5.26871
310	$3.6836 \cdot 10^1$	$5.2367 \cdot 10^2$	9.34771	560	$7.0861 \cdot 10^2$	$5.5765 \cdot 10^3$	5.17463
320	$4.3174 \cdot 10^1$	$5.9458 \cdot 10^2$	9.05559	570	$7.7418 \cdot 10^2$	$5.9856 \cdot 10^3$	5.08384
330	$5.0354 \cdot 10^1$	$6.7246 \cdot 10^2$	8.78118	580	$8.4452 \cdot 10^2$	$6.4168 \cdot 10^3$	4.99619
340	$5.8461 \cdot 10^1$	$7.5775 \cdot 10^2$	8.52291	590	$9.1987 \cdot 10^2$	$6.8709 \cdot 10^3$	4.91151
350	$6.7579 \cdot 10^1$	$8.5090 \cdot 10^2$	8.27940	600	$1.0005 \cdot 10^3$	$7.3487 \cdot 10^3$	4.82965
360	$7.7800 \cdot 10^1$	$9.5240 \cdot 10^2$	8.04942	610	$1.0867 \cdot 10^3$	$7.8510 \cdot 10^3$	4.75048
370	$8.9223 \cdot 10^1$	$1.0627 \cdot 10^3$	7.83187	620	$1.1788 \cdot 10^3$	$8.3787 \cdot 10^3$	4.67386
380	$1.0195 \cdot 10^2$	$1.1823 \cdot 10^3$	7.62576	630	$1.2769 \cdot 10^3$	$8.9324 \cdot 10^3$	4.59967
390	$1.1609 \cdot 10^2$	$1.3118 \cdot 10^3$	7.43023	640	$1.3816 \cdot 10^3$	$9.5132 \cdot 10^3$	4.52780
400	$1.3176 \cdot 10^2$	$1.4516 \cdot 10^3$	7.24448	650	$1.4929 \cdot 10^3$	$1.0122 \cdot 10^4$	4.45814
410	$1.4907 \cdot 10^2$	$1.6023 \cdot 10^3$	7.06778	660	$1.6113 \cdot 10^3$	$1.0759 \cdot 10^4$	4.39059
420	$1.6816 \cdot 10^2$	$1.7644 \cdot 10^3$	6.89950	670	$1.7372 \cdot 10^3$	$1.1426 \cdot 10^4$	4.32506
430	$1.8915 \cdot 10^2$	$1.9386 \cdot 10^3$	6.73905	680	$1.8707 \cdot 10^3$	$1.2124 \cdot 10^4$	4.26146
440	$2.1219 \cdot 10^2$	$2.1253 \cdot 10^3$	6.58589	690	$2.0124 \cdot 10^3$	$1.2853 \cdot 10^4$	4.19970
450	$2.3743 \cdot 10^2$	$2.3252 \cdot 10^3$	6.43953	700	$2.1625 \cdot 10^3$	$1.3614 \cdot 10^4$	4.13970
460	$2.6501 \cdot 10^2$	$2.5389 \cdot 10^3$	6.29954	710	$2.3215 \cdot 10^3$	$1.4409 \cdot 10^4$	4.08140
470	$2.9509 \cdot 10^2$	$2.7669 \cdot 10^3$	6.16551	720	$2.4896 \cdot 10^3$	$1.5238 \cdot 10^4$	4.02471
480	$3.2785 \cdot 10^2$	$3.0100 \cdot 10^3$	6.03706	730	$2.6674 \cdot 10^3$	$1.6103 \cdot 10^4$	3.96958
490	$3.6345 \cdot 10^2$	$3.2688 \cdot 10^3$	5.91386	740	$2.8551 \cdot 10^3$	$1.7003 \cdot 10^4$	3.91593
500	$4.0209 \cdot 10^2$	$3.5439 \cdot 10^3$	5.79558	750	$3.0533 \cdot 10^3$	$1.7941 \cdot 10^4$	3.86372
510	$4.4394 \cdot 10^2$	$3.8361 \cdot 10^3$	5.68194	760	$3.2624 \cdot 10^3$	$1.8917 \cdot 10^4$	3.81288
520	$4.8920 \cdot 10^2$	$4.1459 \cdot 10^3$	5.57267	770	$3.4827 \cdot 10^3$	$1.9033 \cdot 10^4$	3.76336
530	$5.3808 \cdot 10^2$	$4.4742 \cdot 10^3$	5.46753	780	$3.7149 \cdot 10^3$	$2.0989 \cdot 10^4$	3.71512
540	$5.9080 \cdot 10^2$	$4.8215 \cdot 10^3$	5.36628	790	$3.9592 \cdot 10^3$	$2.2086 \cdot 10^4$	3.66809

Table 2.1. (*continued*)

T [K]	$m_{\lambda_{max},bb}$ [W/µm m²]	$M_{T,bb}$ [W/m²]	$\lambda_{max,bb}$ [µm]	T [K]	$m_{\lambda_{max},bb}$ [W/µm m²]	$M_{T,bb}$ [W/m²]	$\lambda_{max,bb}$ [µm]
800	$4.2162\cdot10^3$	$2.3226\cdot10^4$	3.62224	1200	$3.2017\cdot10^4$	$1.1758\cdot10^5$	2.41483
810	$4.4864\cdot10^3$	$2.4409\cdot10^4$	3.57752	1210	$3.3373\cdot10^4$	$1.2155\cdot10^5$	2.39487
820	$4.7702\cdot10^3$	$2.5637\cdot10^4$	3.53389	1220	$3.4775\cdot10^4$	$1.2562\cdot10^5$	2.37524
830	$5.0683\cdot10^3$	$2.6910\cdot10^4$	3.49131	1230	$3.6224\cdot10^4$	$1.2979\cdot10^5$	2.35593
840	$5.3810\cdot10^3$	$2.8231\cdot10^4$	3.44975	1240	$3.7720\cdot10^4$	$1.3406\cdot10^5$	2.33693
850	$5.7090\cdot10^3$	$2.9599\cdot10^4$	3.40917	1250	$3.9266\cdot10^4$	$1.3844\cdot10^5$	2.31823
860	$6.0529\cdot10^3$	$3.1017\cdot10^4$	3.36952	1260	$4.0826\cdot10^4$	$1.4292\cdot10^5$	2.29983
870	$6.4131\cdot10^3$	$3.2485\cdot10^4$	3.33079	1270	$4.2510\cdot10^4$	$1.4751\cdot10^5$	2.28173
880	$6.7902\cdot10^3$	$3.4005\cdot10^4$	3.29294	1280	$4.4210\cdot10^4$	$1.5221\cdot10^5$	2.26390
890	$7.1849\cdot10^3$	$3.5577\ 10^4$	3.25594	1290	$4.5964\cdot10^4$	$1.5702\cdot10^5$	2.24635
900	$7.5977\cdot10^3$	$3.7203\cdot10^4$	3.21977	1300	$4.7773\cdot10^4$	$1.6195\cdot10^5$	2.22907
910	$8.0293\cdot10^3$	$3.8884\cdot10^4$	3.18439	1310	$4.9639\cdot10^4$	$1.6699\cdot10^5$	2.21205
920	$8.4802\cdot10^3$	$4.0622\cdot10^4$	3.14977	1320	$5.1563\cdot10^4$	$1.7215\cdot10^5$	2.19530
930	$8.9512\cdot10^3$	$4.2417\cdot10^4$	3.11590	1330	$5.3546\cdot10^4$	$1.7742\cdot10^5$	2.17879
940	$9.4430\cdot10^3$	$4.4271\cdot10^4$	3.08276	1340	$5.5589\cdot10^4$	$1.8282\cdot10^5$	2.16253
950	$9.9560\cdot10^3$	$4.6185\cdot10^4$	3.05031	1350	$5.7695\cdot10^4$	$1.8834\cdot10^5$	2.14651
960	$1.0491\cdot10^4$	$4.8161\cdot10^4$	3.01853	1360	$5.9864\cdot10^4$	$1.9398\cdot10^5$	2.13073
970	$1.1049\cdot10^4$	$5.0199\cdot10^4$	2.98741	1370	$6.2097\cdot10^4$	$1.9975\cdot10^5$	2.11518
980	$1.1631\cdot10^4$	$5.2301\cdot10^4$	2.95693	1380	$6.4397\cdot10^4$	$2.0565\cdot10^5$	2.09985
990	$1.2236\cdot10^4$	$5.4469\cdot10^4$	2.92706	1390	$6.6764\cdot10^4$	$2.1167\cdot10^5$	2.08474
1000	$1.2867\cdot10^4$	$5.6703\cdot10^4$	2.89779	1400	$6.9200\cdot10^4$	$2.1783\cdot10^5$	2.06985
1010	$1.3523\cdot10^4$	$5.9006\cdot10^4$	2.86910	1410	$7.1707\cdot10^4$	$2.2412\cdot10^5$	2.05517
1020	$1.4206\cdot10^4$	$6.1377\cdot10^4$	2.84097	1420	$7.4287\cdot10^4$	$2.3055\cdot10^5$	2.04070
1030	$1.4916\cdot10^4$	$6.3820\cdot10^4$	2.81339	1430	$7.6939\cdot10^4$	$2.3711\cdot10^5$	2.02643
1040	$1.5654\cdot10^4$	$6.6335\cdot10^4$	2.78634	1440	$7.9667\cdot10^4$	$2.4381\cdot10^5$	2.01236
1050	$1.6422\cdot10^4$	$6.8923\cdot10^4$	2.75980	1450	$8.2472\cdot10^4$	$2.5066\cdot10^5$	1.99848
1060	$1.7219\cdot10^4$	$7.1586\cdot10^4$	2.73377	1460	$8.5356\cdot10^4$	$2.5764\cdot10^5$	1.98479
1070	$1.8046\cdot10^4$	$7.4326\cdot10^4$	2.70822	1470	$8.8319\cdot10^4$	$2.6477\cdot10^5$	1.97129
1080	$1.8905\cdot10^4$	$7.7144\cdot10^4$	2.68314	1480	$9.1364\cdot10^4$	$2.7205\cdot10^5$	1.95797
1090	$1.9797\cdot10^4$	$8.0041\cdot10^4$	2.65852	1490	$9.4493\cdot10^4$	$2.7948\cdot10^5$	1.94483
1100	$2.0722\cdot10^4$	$8.3019\cdot10^4$	2.63436	1500	$9.7707\cdot10^4$	$2.8706\cdot10^5$	1.93186
1110	$2.1681\cdot10^4$	$8.6079\cdot10^4$	2.61062	1510	$1.0101\cdot10^5$	$2.9479\cdot10^5$	1.91907
1120	$2.2676\cdot10^4$	$8.9224\cdot10^4$	2.58731	1520	$1.0440\cdot10^5$	$3.0268\cdot10^5$	1.90644
1130	$2.3706\cdot10^4$	$9.2453\cdot10^4$	2.56442	1530	$1.0788\cdot10^5$	$3.1072\cdot10^5$	1.89398
1140	$2.4774\cdot10^4$	$9.5769\cdot10^4$	2.54192	1540	$1.1145\cdot10^5$	$3.1893\cdot10^5$	1.88168
1150	$2.5880\cdot10^4$	$9.9174\cdot10^4$	2.51982	1550	$1.1511\cdot10^5$	$3.2729\cdot10^5$	1.86954
1160	$2.7025\cdot10^4$	$1.0267\cdot10^5$	2.49810	1560	$1.1888\cdot10^5$	$3.3582\cdot10^5$	1.85756
1170	$2.8210\cdot10^4$	$1.0626\cdot10^5$	2.47674	1570	$1.2273\cdot10^5$	$3.4451\cdot10^5$	1.84573
1180	$2.9436\cdot10^4$	$1.0993\cdot10^5$	2.45576	1580	$1.2669\cdot10^5$	$3.5383\cdot10^5$	1.83405
1190	$3.0705\cdot10^4$	$1.1371\cdot10^5$	2.43512	1590	$1.3075\cdot10^5$	$3.6241\cdot10^5$	1.82251

Table 2.1. (*continued*)

T [K]	$m_{\lambda_{max},bb}$ [W/µm m²]	$M_{T,bb}$ [W/m²]	$\lambda_{max,bb}$ [µm]	T [K]	$m_{\lambda_{max},bb}$ [W/µm m²]	$M_{T,bb}$ [W/m²]	$\lambda_{max,bb}$ [µm]
1600	$1.3492 \cdot 10^5$	$3.7161 \cdot 10^5$	1.81112	2000	$4.1174 \cdot 10^5$	$9.0725 \cdot 10^5$	1.44890
1610	$1.3919 \cdot 10^5$	$3.8099 \cdot 10^5$	1.79987	2010	$4.2213 \cdot 10^5$	$9.2553 \cdot 10^5$	1.44169
1620	$1.4356 \cdot 10^5$	$3.9054 \cdot 10^5$	1.78876	2020	$4.3274 \cdot 10^5$	$9.4409 \cdot 10^5$	1.43455
1630	$1.4805 \cdot 10^5$	$4.0027 \cdot 10^5$	1.77779	2030	$4.4356 \cdot 10^5$	$9.6292 \cdot 10^5$	1.42748
1640	$1.5265 \cdot 10^5$	$4.1019 \cdot 10^5$	1.76695	2040	$4.5459 \cdot 10^5$	$9.8204 \cdot 10^5$	1.42049
1650	$1.5736 \cdot 10^5$	$4.2028 \cdot 10^5$	1.75624	2050	$4.6584 \cdot 10^5$	$1.0014 \cdot 10^6$	1.41356
1660	$1.6218 \cdot 10^5$	$4.3057 \cdot 10^5$	1.74566	2060	$4.7731 \cdot 10^5$	$1.0211 \cdot 10^6$	1.40670
1670	$1.6713 \cdot 10^5$	$4.4104 \cdot 10^5$	1.73520	2070	$4.8901 \cdot 10^5$	$1.0411 \cdot 10^6$	1.39990
1680	$1.7219 \cdot 10^5$	$4.5169 \cdot 10^5$	1.72488	2080	$5.0094 \cdot 10^5$	$1.0614 \cdot 10^6$	1.39317
1690	$1.7738 \cdot 10^5$	$4.6255 \cdot 10^5$	1.71467	2090	$5.1310 \cdot 10^5$	$1.0819 \cdot 10^6$	1.38650
1700	$1.8269 \cdot 10^5$	$4.7359 \cdot 10^5$	1.70458	2100	$5.2549 \cdot 10^5$	$1.1028 \cdot 10^6$	1.37990
1710	$1.8813 \cdot 10^5$	$4.8483 \cdot 10^5$	1.69462	2110	$5.3812 \cdot 10^5$	$1.1239 \cdot 10^6$	1.37336
1720	$1.9369 \cdot 10^5$	$4.9627 \cdot 10^5$	1.68476	2120	$5.5100 \cdot 10^5$	$1.1454 \cdot 10^6$	1.36688
1730	$1.9939 \cdot 10^5$	$5.0792 \cdot 10^5$	1.67502	2130	$5.6411 \cdot 10^5$	$1.1671 \cdot 10^6$	1.36047
1740	$2.0522 \cdot 10^5$	$5.1976 \cdot 10^5$	1.66540	2140	$5.7748 \cdot 10^5$	$1.1892 \cdot 10^6$	1.35411
1750	$2.1118 \cdot 10^5$	$5.3181 \cdot 10^5$	1.65588	2150	$5.9110 \cdot 10^5$	$1.2116 \cdot 10^6$	1.34781
1760	$2.1729 \cdot 10^5$	$5.4407 \cdot 10^5$	1.64647	2160	$6.0497 \cdot 10^5$	$1.2343 \cdot 10^6$	1.34157
1770	$2.2353 \cdot 10^5$	$5.5655 \cdot 10^5$	1.63717	2170	$6.1911 \cdot 10^5$	$1.2573 \cdot 10^6$	1.33539
1780	$2.2992 \cdot 10^5$	$5.6923 \cdot 10^5$	1.62797	2180	$6.3351 \cdot 10^5$	$1.2807 \cdot 10^6$	1.32926
1790	$2.3645 \cdot 10^5$	$5.8213 \cdot 10^5$	1.61888	2190	$6.4817 \cdot 10^5$	$1.3043 \cdot 10^6$	1.32319
1800	$2.4313 \cdot 10^5$	$5.9525 \cdot 10^5$	1.60988	2200	$6.6310 \cdot 10^5$	$1.3283 \cdot 10^6$	1.31718
1810	$2.4995 \cdot 10^5$	$6.0859 \cdot 10^5$	1.60099	2210	$6.7831 \cdot 10^5$	$1.3526 \cdot 10^6$	1.31122
1820	$2.5694 \cdot 10^5$	$6.2215 \cdot 10^5$	1.59219	2220	$6.9380 \cdot 10^5$	$1.3773 \cdot 10^6$	1.30531
1830	$2.6407 \cdot 10^5$	$6.3593 \cdot 10^5$	1.58349	2230	$7.0957 \cdot 10^5$	$1.4023 \cdot 10^6$	1.29946
1840	$2.7137 \cdot 10^5$	$6.4995 \cdot 10^5$	1.57489	2240	$7.2562 \cdot 10^5$	$1.4276 \cdot 10^6$	1.29366
1850	$2.7882 \cdot 10^5$	$6.6419 \cdot 10^5$	1.56637	2250	$7.4196 \cdot 10^5$	$1.4532 \cdot 10^6$	1.28791
1860	$2.8644 \cdot 10^5$	$6.7867 \cdot 10^5$	1.55795	2260	$7.5860 \cdot 10^5$	$1.4792 \cdot 10^6$	1.28221
1870	$2.9422 \cdot 10^5$	$6.9338 \cdot 10^5$	1.54962	2270	$7.7553 \cdot 10^5$	$1.5056 \cdot 10^6$	1.27656
1880	$3.0217 \cdot 10^5$	$7.0834 \cdot 10^5$	1.54138	2280	$7.9276 \cdot 10^5$	$1.5323 \cdot 10^6$	1.27096
1890	$3.1030 \cdot 10^5$	$7.2353 \cdot 10^5$	1.53322	2290	$8.1030 \cdot 10^5$	$1.5594 \cdot 10^6$	1.26541
1900	$3.1859 \cdot 10^5$	$7.3896 \cdot 10^5$	1.52515	2300	$8.2815 \cdot 10^5$	$1.5868 \cdot 10^6$	1.25991
1910	$3.2707 \cdot 10^5$	$7.5464 \cdot 10^5$	1.51717	2310	$8.4631 \cdot 10^5$	$1.6146 \cdot 10^6$	1.25446
1920	$3.3572 \cdot 10^5$	$7.7057 \cdot 10^5$	1.50927	2320	$8.6479 \cdot 10^5$	$1.6427 \cdot 10^6$	1.24905
1930	$3.4455 \cdot 10^5$	$7.8675 \cdot 10^5$	1.50145	2330	$8.8358 \cdot 10^5$	$1.6712 \cdot 10^6$	1.24369
1940	$3.5357 \cdot 10^5$	$8.0318 \cdot 10^5$	1.49371	2340	$9.0271 \cdot 10^5$	$1.7001 \cdot 10^6$	1.23837
1950	$3.6278 \cdot 10^5$	$8.1987 \cdot 10^5$	1.48605	2350	$9.2216 \cdot 10^5$	$1.7293 \cdot 10^6$	1.23310
1960	$3.7218 \cdot 10^5$	$8.3682 \cdot 10^5$	1.47847	2360	$9.4195 \cdot 10^5$	$1.7590 \cdot 10^6$	1.22788
1970	$3.8177 \cdot 10^5$	$8.5403 \cdot 10^5$	1.47096	2370	$9.6208 \cdot 10^5$	$1.7890 \cdot 10^6$	1.22270
1980	$3.9156 \cdot 10^5$	$8.7150 \cdot 10^5$	1.46353	2380	$9.8255 \cdot 10^5$	$1.8193 \cdot 10^6$	1.21756
1990	$4.0154 \cdot 10^5$	$8.8924 \cdot 10^5$	1.45618	2390	$1.0034 \cdot 10^6$	$1.8501 \cdot 10^6$	1.21247

Table 2.1. (*continued*)

T. [K]	$m_{\lambda_{max},bb}$ [W/µm m²]	$M_{T,bb}$ [W/m²]	$\lambda_{max,bb}$ [µm]	T [K]	$m_{\lambda_{max},bb}$ [W/µm m²]	$M_{T,bb}$ [W/m²]	$\lambda_{max,bb}$ [µm]
2400	$1.0245 \cdot 10^6$	$1.8813 \cdot 10^6$	1.20741	2800	$2.2144 \cdot 10^6$	$3.4853 \cdot 10^6$	1.03493
2410	$1.0461 \cdot 10^6$	$1.9128 \cdot 10^6$	1.20240	2810	$2.2542 \cdot 10^6$	$3.5354 \cdot 10^6$	1.03124
2420	$1.0679 \cdot 10^6$	$1.9448 \cdot 10^6$	1.19743	2820	$2.2946 \cdot 10^6$	$3.5859 \cdot 10^6$	1.02759
2430	$1.0902 \cdot 10^6$	$1.9771 \cdot 10^6$	1.19251	2830	$2.3356 \cdot 10^6$	$3.6371 \cdot 10^6$	1.02396
2440	$1.1128 \cdot 10^6$	$2.0099 \cdot 10^6$	1.18762	2840	$2.3772 \cdot 10^6$	$3.6888 \cdot 10^6$	1.02035
2450	$1.1358 \cdot 10^6$	$2.0430 \cdot 10^6$	1.18277	2850	$2.4193 \cdot 10^6$	$3.7410 \cdot 10^6$	1.01677
2460	$1.1592 \cdot 10^6$	$2.0766 \cdot 10^6$	1.17796	2860	$2.4621 \cdot 10^6$	$3.7938 \cdot 10^6$	1.01321
2470	$1.1829 \cdot 10^6$	$2.1105 \cdot 10^6$	1.17320	2870	$2.5054 \cdot 10^6$	$3.8471 \cdot 10^6$	1.00968
2480	$1.2071 \cdot 10^6$	$2.1449 \cdot 10^6$	1.16846	2880	$2.5494 \cdot 10^6$	$3.9010 \cdot 10^6$	1.00618
2490	$1.2316 \cdot 10^6$	$2.1797 \cdot 10^6$	1.16377	2890	$2.5939 \cdot 10^6$	$3.9555 \cdot 10^6$	1.00270
2500	$1.2565 \cdot 10^6$	$2.2150 \cdot 10^6$	1.15912	2900	$2.6391 \cdot 10^6$	$4.0105 \cdot 10^6$	0.99924
2510	$1.2818 \cdot 10^6$	$2.2506 \cdot 10^6$	1.15450	2910	$2.6849 \cdot 10^6$	$4.0661 \cdot 10^6$	0.99581
2520	$1.3076 \cdot 10^6$	$2.2867 \cdot 10^6$	1.14992	2920	$2.7314 \cdot 10^6$	$4.1223 \cdot 10^6$	0.99239
2530	$1.3337 \cdot 10^6$	$2.3232 \cdot 10^6$	1.14537	2930	$2.7785 \cdot 10^6$	$4.1791 \cdot 10^6$	0.98901
2540	$1.3603 \cdot 10^6$	$2.3602 \cdot 10^6$	1.14086	2940	$2.8262 \cdot 10^6$	$4.2364 \cdot 10^6$	0.98564
2550	$1.3873 \cdot 10^6$	$2.3976 \cdot 10^6$	1.13639	2950	$2.8746 \cdot 10^6$	$4.2943 \cdot 10^6$	0.98230
2560	$1.4147 \cdot 10^6$	$2.4354 \cdot 10^6$	1.13195	2960	$2.9237 \cdot 10^6$	$4.3529 \cdot 10^6$	0.97898
2570	$1.4426 \cdot 10^6$	$2.4737 \cdot 10^6$	1.12755	2970	$2.9734 \cdot 10^6$	$4.4120 \cdot 10^6$	0.97569
2580	$1.4708 \cdot 10^6$	$2.5124 \cdot 10^6$	1.12318	2980	$3.0238 \cdot 10^6$	$4.4717 \cdot 10^6$	0.97241
2590	$1.4996 \cdot 10^6$	$2.5516 \cdot 10^6$	1.11884	2990	$3.0749 \cdot 10^6$	$4.5320 \cdot 10^6$	0.96916
2600	$1.5287 \cdot 10^6$	$2.5912 \cdot 10^6$	1.11454	3000	$3.1266 \cdot 10^6$	$4.5930 \cdot 10^6$	0.96593
2610	$1.5584 \cdot 10^6$	$2.6313 \cdot 10^6$	1.11027	3010	$3.1791 \cdot 10^6$	$4.6545 \cdot 10^6$	0.96272
2620	$1.5885 \cdot 10^6$	$2.6719 \cdot 10^6$	1.10603	3020	$3.2322 \cdot 10^6$	$4.7167 \cdot 10^6$	0.95953
2630	$1.6190 \cdot 10^6$	$2.7129 \cdot 10^6$	1.10182	3030	$3.2861 \cdot 10^6$	$4.7795 \cdot 10^6$	0.95637
2640	$1.6500 \cdot 10^6$	$2.7544 \cdot 10^6$	1.09765	3040	$3.3407 \cdot 10^6$	$4.8429 \cdot 10^6$	0.95322
2650	$1.6815 \cdot 10^6$	$2.7963 \cdot 10^6$	1.09351	3050	$3.3960 \cdot 10^6$	$4.9069 \cdot 10^6$	0.95010
2660	$1.7135 \cdot 10^6$	$2.8388 \cdot 10^6$	1.08940	3060	$3.4520 \cdot 10^6$	$4.9716 \cdot 10^6$	0.94699
2670	$1.7459 \cdot 10^6$	$2.8817 \cdot 10^6$	1.08532	3070	$3.5088 \cdot 10^6$	$5.0369 \cdot 10^6$	0.94391
2680	$1.7789 \cdot 10^6$	$2.9251 \cdot 10^6$	1.08127	3080	$3.5663 \cdot 10^6$	$5.1028 \cdot 10^6$	0.94084
2690	$1.8123 \cdot 10^6$	$2.9690 \cdot 10^6$	1.07725	3090	$3.6246 \cdot 10^6$	$5.1694 \cdot 10^6$	0.93780
2700	$1.8462 \cdot 10^6$	$3.0134 \cdot 10^6$	1.07326	3100	$3.6836 \cdot 10^6$	$5.2367 \cdot 10^6$	0.93477
2710	$1.8807 \cdot 10^6$	$3.0583 \cdot 10^6$	1.06930	3110	$3.7434 \cdot 10^6$	$5.3046 \cdot 10^6$	0.93177
2720	$1.9156 \cdot 10^6$	$3.1037 \cdot 10^6$	1.06536	3120	$3.8040 \cdot 10^6$	$5.3731 \cdot 10^6$	0.92878
2730	$1.9511 \cdot 10^6$	$3.1496 \cdot 10^6$	1.06146	3130	$3.8654 \cdot 10^6$	$5.4423 \cdot 10^6$	0.92581
2740	$1.9871 \cdot 10^6$	$3.1960 \cdot 10^6$	1.05759	3140	$3.9275 \cdot 10^6$	$5.5122 \cdot 10^6$	0.92286
2750	$2.0236 \cdot 10^6$	$3.2429 \cdot 10^6$	1.05374	3150	$3.9904 \cdot 10^6$	$5.5828 \cdot 10^6$	0.91993
2760	$2.0607 \cdot 10^6$	$3.2904 \cdot 10^6$	1.04992	3160	$4.0542 \cdot 10^6$	$5.6540 \cdot 10^6$	0.91702
2770	$2.0983 \cdot 10^6$	$3.3383 \cdot 10^6$	1.04613	3170	$4.1187 \cdot 10^6$	$5.7259 \cdot 10^6$	0.91413
2780	$2.1364 \cdot 10^6$	$3.3868 \cdot 10^6$	1.04237	3180	$4.1841 \cdot 10^6$	$5.7985 \cdot 10^6$	0.91126
2780	$2.1752 \cdot 10^6$	$3.4358 \cdot 10^6$	1.03864	3190	$4.2503 \cdot 10^6$	$5.8718 \cdot 10^6$	0.90840

Table 2.1. (*continued*)

T [K]	$m_{\lambda_{max},bb}$ [W/μm m²]	$M_{T,bb}$ [W/m²]	$\lambda_{max,bb}$ [μm]	T [K]	$m_{\lambda_{max},bb}$ [W/μm m²]	$M_{T,bb}$ [W/m²]	$\lambda_{max,bb}$ [μm]
3200	$4.3174 \cdot 10^6$	$5.9458 \cdot 10^6$	0.90556	3600	$7.7800 \cdot 10^6$	$9.5240 \cdot 10^6$	0.80494
3210	$4.3852 \cdot 10^6$	$6.0204 \cdot 10^6$	0.90274	3610	$7.8887 \cdot 10^6$	$9.6302 \cdot 10^6$	0.80271
3220	$4.4540 \cdot 10^6$	$6.0958 \cdot 10^6$	0.89994	3620	$7.9986 \cdot 10^6$	$9.7374 \cdot 10^6$	0.80050
3230	$4.5236 \cdot 10^6$	$6.1719 \cdot 10^6$	0.89715	3630	$8.1096 \cdot 10^6$	$9.8454 \cdot 10^6$	0.79829
3240	$4.5940 \cdot 10^6$	$6.2487 \cdot 10^6$	0.89438	3640	$8.2220 \cdot 10^6$	$9.9544 \cdot 10^6$	0.79610
3250	$4.6654 \cdot 10^6$	$6.3262 \cdot 10^6$	0.89163	3650	$8.3355 \cdot 10^6$	$1.0064 \cdot 10^7$	0.79392
3260	$4.7376 \cdot 10^6$	$6.4044 \cdot 10^6$	0.88889	3660	$8.4503 \cdot 10^6$	$1.0175 \cdot 10^7$	0.79175
3270	$4.8107 \cdot 10^6$	$6.4833 \cdot 10^6$	0.88618	3670	$8.5664 \cdot 10^6$	$1.0287 \cdot 10^7$	0.78959
3280	$4.8847 \cdot 10^6$	$6.5630 \cdot 10^6$	0.88347	3680	$8.6838 \cdot 10^6$	$1.0399 \cdot 10^7$	0.78744
3290	$4.9596 \cdot 10^6$	$6.6434 \cdot 10^6$	0.88079	3690	$8.8024 \cdot 10^6$	$1.0513 \cdot 10^7$	0.78531
3300	$5.0354 \cdot 10^6$	$6.7246 \cdot 10^6$	0.87812	3700	$8.9223 \cdot 10^6$	$1.0627 \cdot 10^7$	0.78319
3310	$5.1122 \cdot 10^6$	$6.8064 \cdot 10^6$	0.87547	3710	$9.0435 \cdot 10^6$	$1.0742 \cdot 10^7$	0.78108
3320	$5.1899 \cdot 10^6$	$6.8891 \cdot 10^6$	0.87283	3720	$9.1661 \cdot 10^6$	$1.0859 \cdot 10^7$	0.77898
3330	$5.2685 \cdot 10^6$	$6.9724 \cdot 10^6$	0.87021	3730	$9.2899 \cdot 10^6$	$1.0976 \cdot 10^7$	0.77689
3340	$5.3481 \cdot 10^6$	$7.0566 \cdot 10^6$	0.86760	3740	$9.4151 \cdot 10^6$	$1.1094 \cdot 10^7$	0.77481
3350	$5.4287 \cdot 10^6$	$7.1415 \cdot 10^6$	0.86501	3750	$9.5417 \cdot 10^6$	$1.1213 \cdot 10^7$	0.77274
3360	$5.5102 \cdot 10^6$	$7.2271 \cdot 10^6$	0.86244	3760	$9.6696 \cdot 10^6$	$1.1333 \cdot 10^7$	0.77069
3370	$5.5927 \cdot 10^6$	$7.3135 \cdot 10^6$	0.85988	3770	$9.7989 \cdot 10^6$	$1.1454 \cdot 10^7$	0.76865
3380	$5.6761 \cdot 10^6$	$7.4007 \cdot 10^6$	0.85734	3780	$9.9295 \cdot 10^6$	$1.1576 \cdot 10^7$	0.76661
3390	$5.7606 \cdot 10^6$	$7.4887 \cdot 10^6$	0.85481	3790	$1.0062 \cdot 10^7$	$1.1699 \cdot 10^7$	0.76459
3400	$5.8461 \cdot 10^6$	$7.5775 \cdot 10^6$	0.85229	3800	$1.0195 \cdot 10^7$	$1.1823 \cdot 10^7$	0.76258
3410	$5.9325 \cdot 10^6$	$7.6670 \cdot 10^6$	0.84979	3810	$1.0330 \cdot 10^7$	$1.1948 \cdot 10^7$	0.76058
3420	$6.0200 \cdot 10^6$	$7.7573 \cdot 10^6$	0.84731	3820	$1.0466 \cdot 10^7$	$1.2074 \cdot 10^7$	0.75858
3430	$6.1086 \cdot 10^6$	$7.8485 \cdot 10^6$	0.84484	3830	$1.0604 \cdot 10^7$	$1.2201 \cdot 10^7$	0.75660
3440	$6.1981 \cdot 10^6$	$7.9404 \cdot 10^6$	0.84238	3840	$1.0743 \cdot 10^7$	$1.2329 \cdot 10^7$	0.75463
3450	$6.2887 \cdot 10^6$	$8.0331 \cdot 10^6$	0.83994	3850	$1.0884 \cdot 10^7$	$1.2458 \cdot 10^7$	0.75267
3460	$6.3804 \cdot 10^6$	$8.1267 \cdot 10^6$	0.83751	3860	$1.1026 \cdot 10^7$	$1.2588 \cdot 10^7$	0.75072
3470	$6.4732 \cdot 10^6$	$8.2210 \cdot 10^6$	0.83510	3870	$1.1169 \cdot 10^7$	$1.2719 \cdot 10^7$	0.74878
3480	$6.5670 \cdot 10^6$	$8.3162 \cdot 10^6$	0.83270	3880	$1.1314 \cdot 10^7$	$1.2851 \cdot 10^7$	0.74685
3490	$6.6619 \cdot 10^6$	$8.4122 \cdot 10^6$	0.83031	3890	$1.1461 \cdot 10^7$	$1.2984 \cdot 10^7$	0.74493
3500	$6.7579 \cdot 10^6$	$8.5090 \cdot 10^6$	0.82794	3900	$1.1609 \cdot 10^7$	$1.3118 \cdot 10^7$	0.74302
3510	$6.8549 \cdot 10^6$	$8.6067 \cdot 10^6$	0.82558	3910	$1.1759 \cdot 10^7$	$1.3253 \cdot 10^7$	0.74112
3520	$6.9532 \cdot 10^6$	$8.7052 \cdot 10^6$	0.82324	3920	$1.1910 \cdot 10^7$	$1.3389 \cdot 10^7$	0.73923
3530	$7.0525 \cdot 10^6$	$8.8045 \cdot 10^6$	0.82090	3930	$1.2062 \cdot 10^7$	$1.3526 \cdot 10^7$	0.73735
3540	$7.1529 \cdot 10^6$	$8.9047 \cdot 10^6$	0.81859	3940	$1.2217 \cdot 10^7$	$1.3664 \cdot 10^7$	0.73548
3550	$7.2545 \cdot 10^6$	$9.0058 \cdot 10^6$	0.81628	3950	$1.2372 \cdot 10^7$	$1.3804 \cdot 10^7$	0.73362
3560	$7.3573 \cdot 10^6$	$9.1077 \cdot 10^6$	0.81399	3960	$1.2530 \cdot 10^7$	$1.3944 \cdot 10^7$	0.73177
3570	$7.4612 \cdot 10^6$	$9.2104 \cdot 10^6$	0.81171	3970	$1.2689 \cdot 10^7$	$1.4085 \cdot 10^7$	0.72992
3580	$7.5663 \cdot 10^6$	$9.3141 \cdot 10^6$	0.80944	3980	$1.2849 \cdot 10^7$	$1.4228 \cdot 10^7$	0.72809
3590	$7.6726 \cdot 10^6$	$9.4186 \cdot 10^6$	0.80718	3990	$1.3012 \cdot 10^7$	$1.4371 \cdot 10^7$	0.72626

Table 2.1. (*continued*)

T [K]	$m_{\lambda_{max},bb}$ [W/μm m²]	$M_{T,bb}$ [W/m²]	$\lambda_{max,bb}$ [μm]	T [K]	$m_{\lambda_{max},bb}$ [W/μm m²]	$M_{T,bb}$ [W/m²]	$\lambda_{max,bb}$ [μm]
4000	$1.3176 \cdot 10^7$	$1.4516 \cdot 10^7$	0.72445	4400	$2.1219 \cdot 10^7$	$2.1253 \cdot 10^7$	0.65859
4010	$1.3341 \cdot 10^7$	$1.4662 \cdot 10^7$	0.72264	4410	$2.1462 \cdot 10^7$	$2.1447 \cdot 10^7$	0.65710
4020	$1.3508 \cdot 10^7$	$1.4809 \cdot 10^7$	0.72084	4420	$2.1706 \cdot 10^7$	$2.1642 \cdot 10^7$	0.65561
4030	$1.3677 \cdot 10^7$	$1.4956 \cdot 10^7$	0.71906	4430	$2.1953 \cdot 10^7$	$2.1838 \cdot 10^7$	0.65413
4040	$1.3848 \cdot 10^7$	$1.5105 \cdot 10^7$	0.71728	4440	$2.2202 \cdot 10^7$	$2.2036 \cdot 10^7$	0.65266
4050	$1.4020 \cdot 10^7$	$1.5256 \cdot 10^7$	0.71550	4450	$2.2453 \cdot 10^7$	$2.2236 \cdot 10^7$	0.65119
4060	$1.4194 \cdot 10^7$	$1.5407 \cdot 10^7$	0.71374	4460	$2.2706 \cdot 10^7$	$2.2436 \cdot 10^7$	0.64973
4070	$1.4369 \cdot 10^7$	$1.5559 \cdot 10^7$	0.71199	4470	$2.2962 \cdot 10^7$	$2.2638 \cdot 10^7$	0.64828
4080	$1.4547 \cdot 10^7$	$1.5713 \cdot 10^7$	0.71024	4480	$2.3220 \cdot 10^7$	$2.2841 \cdot 10^7$	0.64683
4090	$1.4726 \cdot 10^7$	$1.5867 \cdot 10^7$	0.70851	4490	$2.3480 \cdot 10^7$	$2.3046 \cdot 10^7$	0.64539
4100	$1.4907 \cdot 10^7$	$1.6023 \cdot 10^7$	0.70678	4500	$2.3743 \cdot 10^7$	$2.3252 \cdot 10^7$	0.64395
4110	$1.5090 \cdot 10^7$	$1.6180 \cdot 10^7$	0.70506	4510	$2.4008 \cdot 10^7$	$2.3459 \cdot 10^7$	0.64253
4120	$1.5274 \cdot 10^7$	$1.6338 \cdot 10^7$	0.70335	4520	$2.4275 \cdot 10^7$	$2.3668 \cdot 10^7$	0.64110
4130	$1.5460 \cdot 10^7$	$1.6497 \cdot 10^7$	0.70165	4530	$2.4545 \cdot 10^7$	$2.3878 \cdot 10^7$	0.63969
4140	$1.5648 \cdot 10^7$	$1.6657 \cdot 10^7$	0.69995	4540	$2.4817 \cdot 10^7$	$2.4090 \cdot 10^7$	0.63828
4150	$1.5838 \cdot 10^7$	$1.6819 \cdot 10^7$	0.69826	4550	$2.5091 \cdot 10^7$	$2.4303 \cdot 10^7$	0.63688
4160	$1.6030 \cdot 10^7$	$1.6982 \cdot 10^7$	0.69659	4560	$2.5368 \cdot 10^7$	$2.4517 \cdot 10^7$	0.63548
4170	$1.6224 \cdot 10^7$	$1.7146 \cdot 10^7$	0.69491	4570	$2.5648 \cdot 10^7$	$2.4733 \cdot 10^7$	0.63409
4180	$1.6419 \cdot 10^7$	$1.7311 \cdot 10^7$	0.69325	4580	$2.5930 \cdot 10^7$	$2.4950 \cdot 10^7$	0.63271
4190	$1.6616 \cdot 10^7$	$1.7477 \cdot 10^7$	0.69160	4590	$2.6214 \cdot 10^7$	$2.5169 \cdot 10^7$	0.63133
4200	$1.6816 \cdot 10^7$	$1.7644 \cdot 10^7$	0.68995	4600	$2.6501 \cdot 10^7$	$2.5389 \cdot 10^7$	0.62996
4210	$1.7017 \cdot 10^7$	$1.7813 \cdot 10^7$	0.68831	4610	$2.6790 \cdot 10^7$	$2.5610 \cdot 10^7$	0.62859
4220	$1.7220 \cdot 10^7$	$1.7983 \cdot 10^7$	0.68668	4620	$2.7082 \cdot 10^7$	$2.5833 \cdot 10^7$	0.62723
4230	$1.7425 \cdot 10^7$	$1.8154 \cdot 10^7$	0.68506	4630	$2.7376 \cdot 10^7$	$2.6057 \cdot 10^7$	0.62587
4240	$1.7632 \cdot 10^7$	$1.8326 \cdot 10^7$	0.68344	4640	$2.7673 \cdot 10^7$	$2.6283 \cdot 10^7$	0.62452
4250	$1.7841 \cdot 10^7$	$1.8500 \cdot 10^7$	0.68183	4650	$2.7973 \cdot 10^7$	$2.6511 \cdot 10^7$	0.62318
4260	$1.8052 \cdot 10^7$	$1.8674 \cdot 10^7$	0.68023	4660	$2.8275 \cdot 10^7$	$2.6739 \cdot 10^7$	0.62184
4270	$1.8265 \cdot 10^7$	$1.8850 \cdot 10^7$	0.67864	4670	$2.8579 \cdot 10^7$	$2.6970 \cdot 10^7$	0.62051
4280	$1.8479 \cdot 10^7$	$1.9028 \cdot 10^7$	0.67705	4680	$2.8887 \cdot 10^7$	$2.7201 \cdot 10^7$	0.61919
4290	$1.8696 \cdot 10^7$	$1.9206 \cdot 10^7$	0.67548	4690	$2.9197 \cdot 10^7$	$2.7435 \cdot 10^7$	0.61787
4300	$1.8915 \cdot 10^7$	$1.9386 \cdot 10^7$	0.67391	4700	$2.9509 \cdot 10^7$	$2.7669 \cdot 10^7$	0.61655
4310	$1.9136 \cdot 10^7$	$1.9567 \cdot 10^7$	0.67234	4710	$2.9824 \cdot 10^7$	$2.7906 \cdot 10^7$	0.61524
4320	$1.9359 \cdot 10^7$	$1.9749 \cdot 10^7$	0.67079	4720	$3.0142 \cdot 10^7$	$2.8143 \cdot 10^7$	0.61394
4330	$1.9584 \cdot 10^7$	$1.9932 \cdot 10^7$	0.66924	4730	$3.0463 \cdot 10^7$	$2.8383 \cdot 10^7$	0.61264
4340	$1.9811 \cdot 10^7$	$2.0117 \cdot 10^7$	0.66769	4740	$3.0786 \cdot 10^7$	$2.8623 \cdot 10^7$	0.61135
4350	$2.0041 \cdot 10^7$	$2.0303 \cdot 10^7$	0.66616	4750	$3.1113 \cdot 10^7$	$2.8866 \cdot 10^7$	0.61006
4360	$2.0272 \cdot 10^7$	$2.0491 \cdot 10^7$	0.66463	4760	$3.1442 \cdot 10^7$	$2.9110 \cdot 10^7$	0.60878
4370	$2.0506 \cdot 10^7$	$2.0679 \cdot 10^7$	0.66311	4770	$3.1773 \cdot 10^7$	$2.9355 \cdot 10^7$	0.60750
4380	$2.0741 \cdot 10^7$	$2.0869 \cdot 10^7$	0.66160	4780	$3.2108 \cdot 10^7$	$2.9602 \cdot 10^7$	0.60623
4390	$2.0979 \cdot 10^7$	$2.1060 \cdot 10^7$	0.66009	4790	$3.2445 \cdot 10^7$	$2.9850 \cdot 10^7$	0.60497

Table 2.1. (*continued*)

T [K]	$m_{\lambda_{max},bb}$ [W/µm m²]	$M_{T,bb}$ [W/m²]	$\lambda_{max,bb}$ [µm]
4800	$3.2785 \cdot 10^7$	$3.0100 \cdot 10^7$	0.60371
4810	$3.3128 \cdot 10^7$	$3.0352 \cdot 10^7$	0.60245
4820	$3.3474 \cdot 10^7$	$3.0605 \cdot 10^7$	0.60120
4830	$3.3822 \cdot 10^7$	$3.0860 \cdot 10^7$	0.59996
4840	$3.4174 \cdot 10^7$	$3.1116 \cdot 10^7$	0.59872
4850	$3.4528 \cdot 10^7$	$3.1374 \cdot 10^7$	0.59748
4860	$3.4886 \cdot 10^7$	$3.1634 \cdot 10^7$	0.59625
4870	$3.5246 \cdot 10^7$	$3.1895 \cdot 10^7$	0.59503
4880	$3.5610 \cdot 10^7$	$3.2158 \cdot 10^7$	0.59381
4890	$3.5976 \cdot 10^7$	$3.2422 \cdot 10^7$	0.59260
4900	$3.6345 \cdot 10^7$	$3.2688 \cdot 10^7$	0.59139
4910	$3.6718 \cdot 10^7$	$3.2956 \cdot 10^7$	0.59018
4920	$3.7093 \cdot 10^7$	$3.3225 \cdot 10^7$	0.58898
4930	$3.7472 \cdot 10^7$	$3.3496 \cdot 10^7$	0.58779
4940	$3.7853 \cdot 10^7$	$3.3769 \cdot 10^7$	0.58660
4950	$3.8238 \cdot 10^7$	$3.4043 \cdot 10^7$	0.58541
4960	$3.8626 \cdot 10^7$	$3.4319 \cdot 10^7$	0.58423
4970	$3.9017 \cdot 10^7$	$3.4597 \cdot 10^7$	0.58306
4980	$3.9411 \cdot 10^7$	$3.4876 \cdot 10^7$	0.58189
4990	$3.9808 \cdot 10^7$	$3.5157 \cdot 10^7$	0.58072
5000	$4.0209 \cdot 10^7$	$3.5439 \cdot 10^7$	0.57956
5010	$4.0612 \cdot 10^7$	$3.5724 \cdot 10^7$	0.57840
5020	$4.1019 \cdot 10^7$	$3.6010 \cdot 10^7$	0.57725
5030	$4.1429 \cdot 10^7$	$3.6298 \cdot 10^7$	0.57610
5040	$4.1843 \cdot 10^7$	$3.6587 \cdot 10^7$	0.57496
5050	$4.2260 \cdot 10^7$	$3.6878 \cdot 10^7$	0.57382
5060	$4.2680 \cdot 10^7$	$3.7171 \cdot 10^7$	0.57269
5070	$4.3103 \cdot 10^7$	$3.7466 \cdot 10^7$	0.57156
5080	$4.3530 \cdot 10^7$	$3.7763 \cdot 10^7$	0.57043
5090	$4.3960 \cdot 10^7$	$3.8061 \cdot 10^7$	0.56931
5100	$4.4394 \cdot 10^7$	$3.8361 \cdot 10^7$	0.56820
5110	$4.4830 \cdot 10^7$	$3.8663 \cdot 10^7$	0.56708
5120	$4.5271 \cdot 10^7$	$3.8966 \cdot 10^7$	0.56598
5130	$4.5715 \cdot 10^7$	$3.9271 \cdot 10^7$	0.56487
5140	$4.6162 \cdot 10^7$	$3.9579 \cdot 10^7$	0.56377
5150	$4.6613 \cdot 10^7$	$3.9887 \cdot 10^7$	0.56268
5160	$4.7067 \cdot 10^7$	$4.0198 \cdot 10^7$	0.56159
5170	$4.7525 \cdot 10^7$	$4.0511 \cdot 10^7$	0.56050
5180	$4.7986 \cdot 10^7$	$4.0825 \cdot 10^7$	0.55942
5190	$4.8451 \cdot 10^7$	$4.1141 \cdot 10^7$	0.55834
5200	$4.8920 \cdot 10^7$	$4.1459 \cdot 10^7$	0.55727
5210	$4.9392 \cdot 10^7$	$4.1779 \cdot 10^7$	0.55620
5220	$4.9868 \cdot 10^7$	$4.2101 \cdot 10^7$	0.55513
5230	$5.0347 \cdot 10^7$	$4.2424 \cdot 10^7$	0.55407
5240	$5.0831 \cdot 10^7$	$4.2750 \cdot 10^7$	0.55301
5250	$5.1317 \cdot 10^7$	$4.3077 \cdot 10^7$	0.55196
5260	$5.1808 \cdot 10^7$	$4.3406 \cdot 10^7$	0.55091
5270	$5.2302 \cdot 10^7$	$4.3737 \cdot 10^7$	0.54987
5280	$5.2801 \cdot 10^7$	$4.4070 \cdot 10^7$	0.54882
5290	$5.3302 \cdot 10^7$	$4.4405 \cdot 10^7$	0.54779

TABLE 2.2. HEMISPHERICAL EMISSIVE POWER OF BLACK BODY – SPECTRAL $m_{\lambda, bb}$ AND – TOTAL $M_{T, 0-\lambda, bb}$

temperature 250 K

λ	$m_{\lambda,bb}$	$M_{T,o-\lambda,bb}$	λ	$m_{\lambda,bb}$	$M_{T,o-\lambda,bb}$
|μm|	|W/μm m²|	|W/m²|	|μm|	|W/μm m²|	|W/m²|
0.4	$1.1943 \cdot 10^{-52}$	$3.3904 \cdot 10^{-55}$			
0.5	$1.2294 \cdot 10^{-40}$	$5.4820 \cdot 10^{-43}$	4.0	$2.0616 \cdot 10^{-1}$	$7.1041 \cdot 10^{-2}$
0.6	$1.0598 \cdot 10^{-32}$	$6.8407 \cdot 10^{-35}$	4.1	$2.5881 \cdot 10^{-1}$	$9.4217 \cdot 10^{-2}$
0.7	$4.3801 \cdot 10^{-27}$	$3.8687 \cdot 10^{-29}$	4.2	$3.2048 \cdot 10^{-1}$	$1.2310 \cdot 10^{-1}$
0.8	$6.5282 \cdot 10^{-23}$	$7.5709 \cdot 10^{-25}$	4.3	$3.9184 \cdot 10^{-1}$	$1.5864 \cdot 10^{-1}$
0.9	$1.0726 \cdot 10^{-19}$	$1.5827 \cdot 10^{-21}$	4.4	$4.7347 \cdot 10^{-1}$	$2.0181 \cdot 10^{-1}$
1.0	$3.7915 \cdot 10^{-17}$	$6.9435 \cdot 10^{-19}$	4.5	$5.6588 \cdot 10^{-1}$	$2.5369 \cdot 10^{-1}$
1.1	$4.4059 \cdot 10^{-15}$	$9.8151 \cdot 10^{-17}$	4.6	$6.6949 \cdot 10^{-1}$	$3.1536 \cdot 10^{-1}$
1.2	$2.2315 \cdot 10^{-13}$	$5.9475 \cdot 10^{-15}$	4.7	$7.8461 \cdot 10^{-1}$	$3.8797 \cdot 10^{-1}$
1.3	$5.9838 \cdot 10^{-12}$	$1.8817 \cdot 10^{-13}$	4.8	$9.1144 \cdot 10^{-1}$	$4.7267 \cdot 10^{-1}$
1.4	$9.7581 \cdot 10^{-11}$	$3.5778 \cdot 10^{-12}$	4.9	$1.0501 \cdot 10^{0}$	$5.7065 \cdot 10^{-1}$
1.5	$1.0709 \cdot 10^{-9}$	$4.5315 \cdot 10^{-11}$	5.0	$1.2005 \cdot 10^{0}$	$6.8308 \cdot 10^{-1}$
1.6	$8.5317 \cdot 10^{-9}$	$4.1296 \cdot 10^{-10}$	5.1	$1.3627 \cdot 10^{0}$	$8.1114 \cdot 10^{-1}$
1.7	$5.2275 \cdot 10^{-8}$	$2.8718 \cdot 10^{-9}$	5.2	$1.5363 \cdot 10^{0}$	$9.5599 \cdot 10^{-1}$
1.8	$2.5762 \cdot 10^{-7}$	$1.5952 \cdot 10^{-8}$	5.3	$1.7211 \cdot 10^{0}$	$1.1188 \cdot 10^{0}$
1.9	$1.0578 \cdot 10^{-6}$	$7.3371 \cdot 10^{-8}$	5.4	$1.9167 \cdot 10^{0}$	$1.3006 \cdot 10^{0}$
2.0	$3.7219 \cdot 10^{-6}$	$2.8759 \cdot 10^{-7}$	5.5	$2.1226 \cdot 10^{0}$	$1.5024 \cdot 10^{0}$
2.1	$1.1479 \cdot 10^{-5}$	$9.8321 \cdot 10^{-7}$	5.6	$2.3382 \cdot 10^{0}$	$1.7254 \cdot 10^{0}$
2.2	$3.1616 \cdot 10^{-5}$	$2.9880 \cdot 10^{-6}$	5.7	$2.5631 \cdot 10^{0}$	$1.9704 \cdot 10^{0}$
2.3	$7.8948 \cdot 10^{-5}$	$8.1990 \cdot 10^{-6}$	5.8	$2.7964 \cdot 10^{0}$	$2.2383 \cdot 10^{0}$
2.4	$1.8102 \cdot 10^{-4}$	$2.0580 \cdot 10^{-5}$	5.9	$3.0375 \cdot 10^{0}$	$2.5299 \cdot 10^{0}$
2.5	$3.8517 \cdot 10^{-4}$	$4.7773 \cdot 10^{-5}$	6.0	$3.2858 \cdot 10^{0}$	$2.8460 \cdot 10^{0}$
2.6	$7.6738 \cdot 10^{-4}$	$1.0351 \cdot 10^{-4}$	6.1	$3.5403 \cdot 10^{0}$	$3.1873 \cdot 10^{0}$
2.7	$1.4425 \cdot 10^{-3}$	$2.1096 \cdot 10^{-4}$	6.2	$3.8003 \cdot 10^{0}$	$3.5543 \cdot 10^{0}$
2.8	$2.5748 \cdot 10^{-3}$	$4.0717 \cdot 10^{-4}$	6.3	$4.0651 \cdot 10^{0}$	$3.9475 \cdot 10^{0}$
2.9	$4.3889 \cdot 10^{-3}$	$7.4856 \cdot 10^{-4}$	6.4	$4.3339 \cdot 10^{0}$	$4.3674 \cdot 10^{0}$
3.0	$7.1785 \cdot 10^{-3}$	$1.3174 \cdot 10^{-3}$	6.5	$4.6057 \cdot 10^{0}$	$4.8144 \cdot 10^{0}$
3.1	$1.1313 \cdot 10^{-2}$	$2.2290 \cdot 10^{-3}$	6.6	$4.8800 \cdot 10^{0}$	$5.2886 \cdot 10^{0}$
3.2	$1.7243 \cdot 10^{-2}$	$3.6397 \cdot 10^{-3}$	6.7	$5.1558 \cdot 10^{0}$	$5.7904 \cdot 10^{0}$
3.3	$2.5496 \cdot 10^{-2}$	$5.7549 \cdot 10^{-3}$	6.8	$5.4325 \cdot 10^{0}$	$6.3198 \cdot 10^{0}$
3.4	$3.6679 \cdot 10^{-2}$	$8.8365 \cdot 10^{-3}$	6.9	$5.7092 \cdot 10^{0}$	$6.8769 \cdot 10^{0}$
3.5	$5.1464 \cdot 10^{-2}$	$1.3211 \cdot 10^{-2}$	7.0	$5.9854 \cdot 10^{0}$	$7.4616 \cdot 10^{0}$
3.6	$7.0583 \cdot 10^{-2}$	$1.9274 \cdot 10^{-2}$	7.1	$6.2603 \cdot 10^{0}$	$8.0739 \cdot 10^{0}$
3.7	$9.4809 \cdot 10^{-2}$	$2.7497 \cdot 10^{-2}$	7.2	$6.5333 \cdot 10^{0}$	$8.7136 \cdot 10^{0}$
3.8	$1.2494 \cdot 10^{-1}$	$3.8432 \cdot 10^{-2}$	7.3	$6.8039 \cdot 10^{0}$	$9.3805 \cdot 10^{0}$
3.9	$1.6179 \cdot 10^{-1}$	$5.2709 \cdot 10^{-2}$	7.4	$7.0713 \cdot 10^{0}$	$1.0074 \cdot 10^{1}$

Table 2.2. (*continued*)

λ |μm|	$m_{\lambda,bb}$ |W/μm m²|	$M_{T,o-\lambda,bb}$ |W/m²|	λ |μm|	$m_{\lambda,bb}$ |W/μm m²|	$M_{T,o-\lambda,bb}$ |W/m²|
7.5	$7.3351\cdot10^0$	$1.0795\cdot10^1$	11.5	$1.2563\cdot10^1$	$5.4240\cdot10^1$
7.6	$7.5948\cdot10^0$	$1.1541\cdot10^1$	11.6	$1.2565\cdot10^1$	$5.5496\cdot10^1$
7.7	$7.8499\cdot10^0$	$1.2313\cdot10^1$	11.7	$1.2563\cdot10^1$	$5.6753\cdot10^1$
7.8	$8.1000\cdot10^0$	$1.3111\cdot10^1$	11.8	$1.2556\cdot10^1$	$5.8009\cdot10^1$
7.9	$8.3447\cdot10^0$	$1.3933\cdot10^1$	11.9	$1.2544\cdot10^1$	$5.9264\cdot10^1$
8.0	$8.5836\cdot10^0$	$1.4780\cdot10^1$	12.0	$1.2529\cdot10^1$	$6.0517\cdot10^1$
8.1	$8.8166\cdot10^0$	$1.5650\cdot10^1$	12.1	$1.2510\cdot10^1$	$6.1769\cdot10^1$
8.2	$9.0431\cdot10^0$	$1.6543\cdot10^1$	12.2	$1.2487\cdot10^1$	$6.3019\cdot10^1$
8.3	$9.2631\cdot10^0$	$1.7458\cdot10^1$	12.3	$1.2461\cdot10^1$	$6.4267\cdot10^1$
8.4	$9.4762\cdot10^0$	$1.8395\cdot10^1$	12.4	$1.2431\cdot10^1$	$6.5511\cdot10^1$
8.5	$9.6824\cdot10^0$	$1.9352\cdot10^1$	12.5	$1.2398\cdot10^1$	$6.6753\cdot10^1$
8.6	$9.8814\cdot10^0$	$2.0331\cdot10^1$	12.6	$1.2362\cdot10^1$	$6.7991\cdot10^1$
8.7	$1.0073\cdot10^1$	$2.1329\cdot10^1$	12.7	$1.2323\cdot10^1$	$6.9225\cdot10^1$
8.8	$1.0257\cdot10^1$	$2.2346\cdot10^1$	12.8	$1.2281\cdot10^1$	$7.0455\cdot10^1$
8.9	$1.0434\cdot10^1$	$2.3380\cdot10^1$	12.9	$1.2236\cdot10^1$	$7.1681\cdot10^1$
9.0	$1.0604\cdot10^1$	$2.4432\cdot10^1$	13.0	$1.2189\cdot10^1$	$7.2902\cdot10^1$
9.1	$1.0765\cdot10^1$	$2.5501\cdot10^1$	13.1	$1.2139\cdot10^1$	$7.4119\cdot10^1$
9.2	$1.0920\cdot10^1$	$2.6585\cdot10^1$	13.2	$1.2086\cdot10^1$	$7.5330\cdot10^1$
9.3	$1.1066\cdot10^1$	$2.7684\cdot10^1$	13.3	$1.2032\cdot10^1$	$7.6536\cdot10^1$
9.4	$1.1206\cdot10^1$	$2.8798\cdot10^1$	13.4	$1.1975\cdot10^1$	$7.7736\cdot10^1$
9.5	$1.1337\cdot10^1$	$2.9925\cdot10^1$	13.5	$1.1916\cdot10^1$	$7.8931\cdot10^1$
9.6	$1.1461\cdot10^1$	$3.1065\cdot10^1$	13.6	$1.1856\cdot10^1$	$8.0120\cdot10^1$
9.7	$1.1578\cdot10^1$	$3.2217\cdot10^1$	13.7	$1.1793\cdot10^1$	$8.1302\cdot10^1$
9.8	$1.1688\cdot10^1$	$3.3381\cdot10^1$	13.8	$1.1729\cdot10^1$	$8.2478\cdot10^1$
9.9	$1.1791\cdot10^1$	$3.4555\cdot10^1$	13.9	$1.1664\cdot10^1$	$8.3648\cdot10^1$
10.0	$1.1886\cdot10^1$	$3.5738\cdot10^1$	14.0	$1.1596\cdot10^1$	$8.4811\cdot10^1$
10.1	$1.1974\cdot10^1$	$3.6931\cdot10^1$	14.1	$1.1528\cdot10^1$	$8.5967\cdot10^1$
10.2	$1.2056\cdot10^1$	$3.8133\cdot10^1$	14.2	$1.1458\cdot10^1$	$8.7116\cdot10^1$
10.3	$1.2131\cdot10^1$	$3.9342\cdot10^1$	14.3	$1.1386\cdot10^1$	$8.8258\cdot10^1$
10.4	$1.2199\cdot10^1$	$4.0559\cdot10^1$	14.4	$1.1314\cdot10^1$	$8.9393\cdot10^1$
10.5	$1.2261\cdot10^1$	$4.1782\cdot10^1$	14.5	$1.1241\cdot10^1$	$9.0521\cdot10^1$
10.6	$1.2317\cdot10^1$	$4.3011\cdot10^1$	14.6	$1.1166\cdot10^1$	$9.1642\cdot10^1$
10.7	$1.2367\cdot10^1$	$4.4245\cdot10^1$	14.7	$1.1091\cdot10^1$	$9.2754\cdot10^1$
10.8	$1.2411\cdot10^1$	$4.5484\cdot10^1$	14.8	$1.1014\cdot10^1$	$9.3860\cdot10^1$
10.9	$1.2448\cdot10^1$	$4.6727\cdot10^1$	14.9	$1.0937\cdot10^1$	$9.4957\cdot10^1$
11.0	$1.2481\cdot10^1$	$4.7974\cdot10^1$	15.0	$1.0860\cdot10^1$	$9.6047\cdot10^1$
11.1	$1.2508\cdot10^1$	$4.9223\cdot10^1$	15.1	$1.0781\cdot10^1$	$9.7129\cdot10^1$
11.2	$1.2529\cdot10^1$	$5.0475\cdot10^1$	15.2	$1.0702\cdot10^1$	$9.8203\cdot10^1$
11.3	$1.2545\cdot10^1$	$5.1729\cdot10^1$	15.3	$1.0623\cdot10^1$	$9.9269\cdot10^1$
11.4	$1.2557\cdot10^1$	$5.2984\cdot10^1$	15.4	$1.0543\cdot10^1$	$1.0033\cdot10^1$

Table 2.2. (*continued*)

temperature 300 K

λ [μm]	$m_{\lambda,bb}$ [W/μm m^2]	$M_{T,o-\lambda,bb}$ [W/m^2]	λ [μm]	$m_{\lambda,bb}$ [W/μm m^2]	$M_{T,o-\lambda,bb}$ [W/m^2]
0.4	$3.1001\cdot10^{-42}$	$1.0605\cdot10^{-44}$			
0.5	$2.6370\cdot10^{-32}$	$1.4185\cdot10^{-34}$	4.0	$2.2679\cdot10^{0}$	$9.8013\cdot10^{-1}$
0.6	$9.2910\cdot10^{-26}$	$7.2424\cdot10^{-28}$	4.1	$2.6855\cdot10^{0}$	$1.2275\cdot10^{0}$
0.7	$3.9130\cdot10^{-21}$	$4.1781\cdot10^{-23}$	4.2	$3.1452\cdot10^{0}$	$1.5186\cdot10^{0}$
0.8	$1.0518\cdot10^{-17}$	$1.4762\cdot10^{-19}$	4.3	$3.6466\cdot10^{0}$	$1.8579\cdot10^{0}$
0.9	$4.5605\cdot10^{-15}$	$8.1525\cdot10^{-17}$	4.4	$4.1885\cdot10^{0}$	$2.2493\cdot10^{0}$
1.0	$5.5528\cdot10^{-13}$	$1.2333\cdot10^{-14}$	4.5	$4.7693\cdot10^{0}$	$2.6969\cdot10^{0}$
1.1	$2.6981\cdot10^{-11}$	$7.2974\cdot10^{-13}$	4.6	$5.3870\cdot10^{0}$	$3.2044\cdot10^{0}$
1.2	$6.6073\cdot10^{-10}$	$2.1404\cdot10^{-11}$	4.7	$6.0394\cdot10^{0}$	$3.7754\cdot10^{0}$
1.3	$9.5802\cdot10^{-9}$	$3.6656\cdot10^{-10}$	4.8	$6.7237\cdot10^{0}$	$4.4133\cdot10^{0}$
1.4	$9.2232\cdot10^{-8}$	$4.1192\cdot10^{-9}$	4.9	$7.4369\cdot10^{0}$	$5.1211\cdot10^{0}$
1.5	$6.4106\cdot10^{-7}$	$3.3079\cdot10^{-8}$	5.0	$8.1760\cdot10^{0}$	$5.9015\cdot10^{0}$
1.6	$3.4246\cdot10^{-6}$	$2.0235\cdot10^{-7}$	5.1	$8.9378\cdot10^{0}$	$6.7570\cdot10^{0}$
1.7	$1.4748\cdot10^{-5}$	$9.9011\cdot10^{-7}$	5.2	$9.7187\cdot10^{0}$	$7.6897\cdot10^{0}$
1.8	$5.3123\cdot10^{-5}$	$4.0243\cdot10^{-6}$	5.3	$1.0516\cdot10^{0}$	$8.7013\cdot10^{0}$
1.9	$1.6478\cdot10^{-4}$	$1.3998\cdot10^{-5}$	5.4	$1.1525\cdot10^{0}$	$9.7932\cdot10^{0}$
2.0	$4.5043\cdot10^{-4}$	$4.2676\cdot10^{-5}$	5.5	$1.2143\cdot10^{1}$	$1.0967\cdot10^{1}$
2.1	$1.1056\cdot10^{-3}$	$1.1624\cdot10^{-4}$	5.6	$1.2967\cdot10^{1}$	$1.2222\cdot10^{1}$
2.2	$2.4741\cdot10^{-3}$	$2.8734\cdot10^{-4}$	5.7	$1.3793\cdot10^{1}$	$1.3560\cdot10^{1}$
2.3	$5.1113\cdot10^{-3}$	$6.5304\cdot10^{-4}$	5.8	$1.4619\cdot10^{1}$	$1.4981\cdot10^{1}$
2.4	$9.8501\cdot10^{-3}$	$1.3793\cdot10^{-3}$	5.9	$1.5441\cdot10^{1}$	$1.6484\cdot10^{1}$
2.5	$1.7863\cdot10^{-2}$	$2.7317\cdot10^{-3}$	6.0	$1.6257\cdot10^{1}$	$1.8069\cdot10^{1}$
2.6	$3.0706\cdot10^{-2}$	$5.1123\cdot10^{-3}$	6.1	$1.7064\cdot10^{1}$	$1.9735\cdot10^{1}$
2.7	$5.0346\cdot10^{-2}$	$9.0989\cdot10^{-3}$	6.2	$1.7859\cdot10^{1}$	$2.1481\cdot10^{1}$
2.8	$7.9160\cdot10^{-2}$	$1.5487\cdot10^{-2}$	6.3	$1.8641\cdot10^{1}$	$2.3306\cdot10^{1}$
2.9	$1.1990\cdot10^{-1}$	$2.5328\cdot10^{-2}$	6.4	$1.9407\cdot10^{1}$	$2.5209\cdot10^{1}$
3.0	$1.7563\cdot10^{-1}$	$3.9966\cdot10^{-2}$	6.5	$2.0155\cdot10^{1}$	$2.7187\cdot10^{1}$
3.1	$2.4967\cdot10^{-1}$	$6.1064\cdot10^{-2}$	6.6	$2.0884\cdot10^{1}$	$2.9239\cdot10^{1}$
3.2	$3.4546\cdot10^{-1}$	$9.0625\cdot10^{-2}$	6.7	$2.1593\cdot10^{1}$	$3.1363\cdot10^{1}$
3.3	$4.6646\cdot10^{-1}$	$1.3100\cdot10^{-1}$	6.8	$2.2279\cdot10^{1}$	$3.3557\cdot10^{1}$
3.4	$6.1607\cdot10^{-1}$	$1.8487\cdot10^{-1}$	6.9	$2.2941\cdot10^{1}$	$3.5818\cdot10^{1}$
3.5	$7.9747\cdot10^{-1}$	$2.5527\cdot10^{-1}$	7.0	$2.3580\cdot10^{1}$	$3.8144\cdot10^{1}$
3.6	$1.0136\cdot10^{0}$	$3.4552\cdot10^{-1}$	7.1	$2.4193\cdot10^{1}$	$4.0533\cdot10^{1}$
3.7	$1.2668\cdot10^{0}$	$4.5922\cdot10^{-1}$	7.2	$2.4781\cdot10^{1}$	$4.2982\cdot10^{1}$
3.8	$1.5594\cdot10^{0}$	$6.0019\cdot10^{-1}$	7.3	$2.5342\cdot10^{1}$	$4.5488\cdot10^{1}$
3.9	$1.8927\cdot10^{0}$	$7.7245\cdot10^{-1}$	7.4	$2.5877\cdot10^{1}$	$4.8049\cdot10^{1}$

Table 2.2. (*continued*)

λ [μm]	$m_{\lambda,bb}$ [W/μm m²]	$M_{T,o-\lambda,bb}$ [W/m²]	λ [μm]	$m_{\lambda,bb}$ [W/μm m²]	$M_{T,o-\lambda,bb}$ [W/m²]
7.5	$2.6385 \cdot 10^1$	$5.0663 \cdot 10^1$	11.5	$2.9186 \cdot 10^1$	$1.7103 \cdot 10^2$
7.6	$2.6866 \cdot 10^1$	$5.3326 \cdot 10^1$	11.6	$2.8990 \cdot 10^1$	$1.7394 \cdot 10^2$
7.7	$2.7320 \cdot 10^1$	$5.6035 \cdot 10^1$	11.7	$2.8787 \cdot 10^1$	$1.7682 \cdot 10^2$
7.8	$2.7747 \cdot 10^1$	$5.8789 \cdot 10^1$	11.8	$2.8580 \cdot 10^1$	$1.7969 \cdot 10^2$
7.9	$2.8146 \cdot 10^1$	$6.1583 \cdot 10^1$	11.9	$2.8369 \cdot 10^1$	$1.8254 \cdot 10^2$
8.0	$2.8520 \cdot 10^1$	$6.4417 \cdot 10^1$	12.0	$2.8153 \cdot 10^1$	$1.8537 \cdot 10^2$
8.1	$2.8867 \cdot 10^1$	$6.7286 \cdot 10^1$	12.1	$2.7933 \cdot 10^1$	$1.8817 \cdot 10^2$
8.2	$2.9188 \cdot 10^1$	$7.0189 \cdot 10^1$	12.2	$2.7710 \cdot 10^1$	$1.9095 \cdot 10^2$
8.3	$2.9483 \cdot 10^1$	$7.3123 \cdot 10^1$	12.3	$2.7483 \cdot 10^1$	$1.9371 \cdot 10^2$
8.4	$2.9754 \cdot 10^1$	$7.6085 \cdot 10^1$	12.4	$2.7254 \cdot 10^1$	$1.9645 \cdot 10^2$
8.5	$2.9999 \cdot 10^1$	$7.9073 \cdot 10^1$	12.5	$2.7022 \cdot 10^1$	$1.9916 \cdot 10^2$
8.6	$3.0221 \cdot 10^1$	$8.2084 \cdot 10^1$	12.6	$2.6788 \cdot 10^1$	$2.0185 \cdot 10^2$
8.7	$3.0420 \cdot 10^1$	$8.5116 \cdot 10^1$	12.7	$2.6551 \cdot 10^1$	$2.0452 \cdot 10^2$
8.8	$3.0596 \cdot 10^1$	$8.8167 \cdot 10^1$	12.8	$2.6313 \cdot 10^1$	$2.0716 \cdot 10^2$
8.9	$3.0749 \cdot 10^1$	$9.1235 \cdot 10^1$	12.9	$2.6074 \cdot 10^1$	$2.0978 \cdot 10^2$
9.0	$3.0881 \cdot 10^1$	$9.4316 \cdot 10^1$	13.0	$2.5832 \cdot 10^1$	$2.1238 \cdot 10^2$
9.1	$3.0993 \cdot 10^1$	$9.7410 \cdot 10^1$	13.1	$2.5590 \cdot 10^1$	$2.1495 \cdot 10^2$
9.2	$3.1084 \cdot 10^1$	$1.0051 \cdot 10^2$	13.2	$2.5347 \cdot 10^1$	$2.1750 \cdot 10^2$
9.3	$3.1156 \cdot 10^1$	$1.0363 \cdot 10^2$	13.3	$2.5103 \cdot 10^1$	$2.2002 \cdot 10^2$
9.4	$3.1210 \cdot 10^1$	$1.0674 \cdot 10^2$	13.4	$2.4859 \cdot 10^1$	$2.2252 \cdot 10^2$
9.5	$3.1245 \cdot 10^1$	$1.0987 \cdot 10^2$	13.5	$2.4614 \cdot 10^1$	$2.2499 \cdot 10^2$
9.6	$3.1263 \cdot 10^1$	$1.1299 \cdot 10^2$	13.6	$2.4369 \cdot 10^1$	$2.2744 \cdot 10^2$
9.7	$3.1265 \cdot 10^1$	$1.1612 \cdot 10^2$	13.7	$2.4125 \cdot 10^1$	$2.2986 \cdot 10^2$
9.8	$3.1251 \cdot 10^1$	$1.1925 \cdot 10^2$	13.8	$2.3880 \cdot 10^1$	$2.3227 \cdot 10^2$
9.9	$3.1221 \cdot 10^1$	$1.2237 \cdot 10^2$	13.9	$2.3635 \cdot 10^1$	$2.3464 \cdot 10^2$
10.0	$3.1177 \cdot 10^1$	$1.2549 \cdot 10^2$	14.0	$2.3391 \cdot 10^1$	$2.3699 \cdot 10^2$
10.1	$3.1119 \cdot 10^1$	$1.2860 \cdot 10^2$	14.1	$2.3148 \cdot 10^1$	$2.3932 \cdot 10^2$
10.2	$3.1048 \cdot 10^1$	$1.3171 \cdot 10^2$	14.2	$2.2905 \cdot 10^1$	$2.4162 \cdot 10^2$
10.3	$3.0964 \cdot 10^1$	$1.3481 \cdot 10^2$	14.3	$2.2663 \cdot 10^1$	$2.4390 \cdot 10^2$
10.4	$3.0868 \cdot 10^1$	$1.3790 \cdot 10^2$	14.4	$2.2422 \cdot 10^1$	$2.4615 \cdot 10^2$
10.5	$3.0761 \cdot 10^1$	$1.4099 \cdot 10^2$	14.5	$2.2182 \cdot 10^1$	$2.4838 \cdot 10^2$
10.6	$3.0643 \cdot 10^1$	$1.4406 \cdot 10^2$	14.6	$2.1942 \cdot 10^1$	$2.5059 \cdot 10^2$
10.7	$3.0515 \cdot 10^1$	$1.4711 \cdot 10^2$	14.7	$2.1704 \cdot 10^1$	$2.5277 \cdot 10^2$
10.8	$3.0377 \cdot 10^1$	$1.5016 \cdot 10^2$	14.8	$2.1468 \cdot 10^1$	$2.5493 \cdot 10^2$
10.9	$3.0230 \cdot 10^1$	$1.5319 \cdot 10^2$	14.9	$2.1232 \cdot 10^1$	$2.5707 \cdot 10^2$
11.0	$3.0075 \cdot 10^1$	$1.5620 \cdot 10^2$	15.0	$2.0998 \cdot 10^1$	$2.5918 \cdot 10^2$
11.1	$2.9911 \cdot 10^1$	$1.5920 \cdot 10^2$	15.1	$2.0766 \cdot 10^1$	$2.6127 \cdot 10^2$
11.2	$2.9740 \cdot 10^1$	$1.6219 \cdot 10^2$	15.2	$2.0535 \cdot 10^1$	$2.6333 \cdot 10^2$
11.3	$2.9562 \cdot 10^1$	$1.6515 \cdot 10^2$	15.3	$2.0305 \cdot 10^1$	$2.6537 \cdot 10^2$
11.4	$2.9377 \cdot 10^1$	$1.6810 \cdot 10^2$	15.4	$2.0077 \cdot 10^1$	$2.6739 \cdot 10^2$

Table 2.2. (*continued*)

temperature 350 K

λ [μm]	$m_{\lambda,bb}$ [W/μm m^2]	$M_{T,o-\lambda,bb}$ [W/m^2]	λ [μm]	$m_{\lambda,bb}$ [W/μm m^2]	$M_{T,o-\lambda,bb}$ [W/m^2]
0.4	$8.5144 \cdot 10^{-35}$	$3.4126 \cdot 10^{-37}$			
0.5	$2.3558 \cdot 10^{-26}$	$1.4862 \cdot 10^{-28}$	4.0	$1.2576 \cdot 10^{1}$	$6.6283 \cdot 10^{0}$
0.6	$8.4574 \cdot 10^{-21}$	$7.7404 \cdot 10^{-23}$	4.1	$1.4281 \cdot 10^{1}$	$7.9704 \cdot 10^{0}$
0.7	$6.9700 \cdot 10^{-17}$	$8.7472 \cdot 10^{-19}$	4.2	$1.6074 \cdot 10^{1}$	$9.4875 \cdot 10^{0}$
0.8	$5.5119 \cdot 10^{-14}$	$9.1022 \cdot 10^{-16}$	4.3	$1.7943 \cdot 10^{1}$	$1.1188 \cdot 10^{1}$
0.9	$9.2287 \cdot 10^{-12}$	$1.9432 \cdot 10^{-13}$	4.4	$1.9876 \cdot 10^{1}$	$1.3078 \cdot 10^{1}$
1.0	$5.2484 \cdot 10^{-10}$	$1.3745 \cdot 10^{-11}$	4.5	$2.1863 \cdot 10^{1}$	$1.5165 \cdot 10^{1}$
1.1	$1.3679 \cdot 10^{-8}$	$4.3674 \cdot 10^{-10}$	4.6	$2.3892 \cdot 10^{1}$	$1.7452 \cdot 10^{1}$
1.2	$1.9935 \cdot 10^{-7}$	$7.6314 \cdot 10^{-9}$	4.7	$2.5950 \cdot 10^{1}$	$1.9944 \cdot 10^{1}$
1.3	$1.8631 \cdot 10^{-6}$	$8.4333 \cdot 10^{-8}$	4.8	$2.8026 \cdot 10^{1}$	$2.2643 \cdot 10^{1}$
1.4	$1.2310 \cdot 10^{-5}$	$6.5109 \cdot 10^{-7}$	4.9	$3.0110 \cdot 10^{1}$	$2.5550 \cdot 10^{1}$
1.5	$6.1741 \cdot 10^{-5}$	$3.7772 \cdot 10^{-5}$	5.0	$3.2191 \cdot 10^{1}$	$2.8665 \cdot 10^{1}$
1.6	$2.4792 \cdot 10^{-4}$	$1.7387 \cdot 10^{-5}$	5.1	$3.4258 \cdot 10^{1}$	$3.1987 \cdot 10^{1}$
1.7	$8.2990 \cdot 10^{-4}$	$6.6205 \cdot 10^{-5}$	5.2	$3.6303 \cdot 10^{1}$	$3.5515 \cdot 10^{1}$
1.8	$2.3897 \cdot 10^{-3}$	$2.1535 \cdot 10^{-4}$	5.3	$3.8316 \cdot 10^{1}$	$3.9247 \cdot 10^{1}$
1.9	$6.0667 \cdot 10^{-3}$	$6.1377 \cdot 10^{-4}$	5.4	$4.0291 \cdot 10^{1}$	$4.3177 \cdot 10^{1}$
2.0	$1.3848 \cdot 10^{-2}$	$1.5642 \cdot 10^{-3}$	5.5	$4.2219 \cdot 10^{1}$	$4.7303 \cdot 10^{1}$
2.1	$2.8875 \cdot 10^{-2}$	$3.6233 \cdot 10^{-3}$	5.6	$4.4094 \cdot 10^{1}$	$5.1619 \cdot 10^{1}$
2.2	$5.5710 \cdot 10^{-2}$	$7.7309 \cdot 10^{-3}$	5.7	$4.5911 \cdot 10^{1}$	$5.6120 \cdot 10^{1}$
2.3	$1.0052 \cdot 10^{-1}$	$1.5362 \cdot 10^{-2}$	5.8	$4.7664 \cdot 10^{1}$	$6.0799 \cdot 10^{1}$
2.4	$1.7110 \cdot 10^{-1}$	$2.8691 \cdot 10^{-2}$	5.9	$4.9350 \cdot 10^{1}$	$6.5650 \cdot 10^{1}$
2.5	$2.7679 \cdot 10^{-1}$	$5.0750 \cdot 10^{-2}$	6.0	$5.0965 \cdot 10^{1}$	$7.0667 \cdot 10^{1}$
2.6	$4.2820 \cdot 10^{-1}$	$8.5571 \cdot 10^{-2}$	6.1	$5.2507 \cdot 10^{1}$	$7.5841 \cdot 10^{1}$
2.7	$6.3681 \cdot 10^{-1}$	$1.3830 \cdot 10^{-1}$	6.2	$5.3972 \cdot 10^{1}$	$8.1166 \cdot 10^{1}$
2.8	$9.1651 \cdot 10^{-1}$	$2.1523 \cdot 10^{-1}$	6.3	$5.5360 \cdot 10^{1}$	$8.6633 \cdot 10^{1}$
2.9	$1.2731 \cdot 10^{0}$	$3.2389 \cdot 10^{-1}$	6.4	$5.6669 \cdot 10^{1}$	$9.2235 \cdot 10^{1}$
3.0	$1.7236 \cdot 10^{0}$	$4.7292 \cdot 10^{-1}$	6.5	$5.7899 \cdot 10^{1}$	$9.7964 \cdot 10^{1}$
3.1	$2.2762 \cdot 10^{0}$	$6.7202 \cdot 10^{-1}$	6.6	$5.9049 \cdot 10^{1}$	$1.0381 \cdot 10^{2}$
3.2	$2.9393 \cdot 10^{0}$	$9.3184 \cdot 10^{-1}$	6.7	$6.0119 \cdot 10^{1}$	$1.0977 \cdot 10^{2}$
3.3	$3.7195 \cdot 10^{0}$	$1.2638 \cdot 10^{0}$	6.8	$6.1111 \cdot 10^{1}$	$1.1583 \cdot 10^{2}$
3.4	$4.6215 \cdot 10^{0}$	$1.6798 \cdot 10^{0}$	6.9	$6.2024 \cdot 10^{1}$	$1.2199 \cdot 10^{2}$
3.5	$5.6476 \cdot 10^{0}$	$2.1922 \cdot 10^{0}$	7.0	$6.2861 \cdot 10^{1}$	$1.2824 \cdot 10^{2}$
3.6	$6.7980 \cdot 10^{0}$	$2.8135 \cdot 10^{0}$	7.1	$6.3622 \cdot 10^{1}$	$1.3456 \cdot 10^{2}$
3.7	$8.0709 \cdot 10^{0}$	$3.5559 \cdot 10^{0}$	7.2	$6.4309 \cdot 10^{1}$	$1.4096 \cdot 10^{2}$
3.8	$9.4622 \cdot 10^{0}$	$4.4316 \cdot 10^{0}$	7.3	$6.4924 \cdot 10^{1}$	$1.4742 \cdot 10^{2}$
3.9	$1.0966 \cdot 10^{1}$	$5.4521 \cdot 10^{0}$	7.4	$6.5469 \cdot 10^{1}$	$1.5394 \cdot 10^{2}$

Table 2.2. (*continued*)

λ [µm]	$m_{\lambda,bb}$ [W/µm m²]	$M_{T,o-\lambda,bb}$ [W/m²]	λ [µm]	$m_{\lambda,bb}$ [W/µm m²]	$M_{T,o-\lambda,bb}$ [W/m²]
7.5	$6.5945 \cdot 10^{1}$	$1.6051 \cdot 10^{2}$	11.5	$5.3642 \cdot 10^{1}$	$4.1301 \cdot 10^{2}$
7.6	$6.6356 \cdot 10^{1}$	$1.6713 \cdot 10^{2}$	11.6	$5.3025 \cdot 10^{1}$	$4.1834 \cdot 10^{2}$
7.7	$6.6703 \cdot 10^{1}$	$1.7378 \cdot 10^{2}$	11.7	$5.2407 \cdot 10^{1}$	$4.2361 \cdot 10^{2}$
7.8	$6.6988 \cdot 10^{1}$	$1.8046 \cdot 10^{2}$	11.8	$5.1790 \cdot 10^{1}$	$4.2882 \cdot 10^{2}$
7.9	$6.7214 \cdot 10^{1}$	$1.8718 \cdot 10^{2}$	11.9	$5.1173 \cdot 10^{1}$	$4.3397 \cdot 10^{2}$
8.0	$6.7384 \cdot 10^{1}$	$1.9391 \cdot 10^{2}$	12.0	$5.0558 \cdot 10^{1}$	$4.3905 \cdot 10^{2}$
8.1	$6.7500 \cdot 10^{1}$	$2.0065 \cdot 10^{2}$	12.1	$4.9944 \cdot 10^{1}$	$4.4408 \cdot 10^{2}$
8.2	$6.7563 \cdot 10^{1}$	$2.0740 \cdot 10^{2}$	12.2	$4.9332 \cdot 10^{1}$	$4.4904 \cdot 10^{2}$
8.3	$6.7578 \cdot 10^{1}$	$2.1416 \cdot 10^{2}$	12.3	$4.8723 \cdot 10^{1}$	$4.5395 \cdot 10^{2}$
8.4	$6.7545 \cdot 10^{1}$	$2.2092 \cdot 10^{2}$	12.4	$4.8116 \cdot 10^{1}$	$4.5879 \cdot 10^{2}$
8.5	$6.7467 \cdot 10^{1}$	$2.2767 \cdot 10^{2}$	12.5	$4.7512 \cdot 10^{1}$	$4.6357 \cdot 10^{2}$
8.6	$6.7347 \cdot 10^{1}$	$2.3441 \cdot 10^{2}$	12.6	$4.6912 \cdot 10^{1}$	$4.6829 \cdot 10^{2}$
8.7	$6.7187 \cdot 10^{1}$	$2.4114 \cdot 10^{2}$	12.7	$4.6315 \cdot 10^{1}$	$4.7295 \cdot 10^{2}$
8.8	$6.6988 \cdot 10^{1}$	$2.4785 \cdot 10^{2}$	12.8	$4.5722 \cdot 10^{1}$	$4.7755 \cdot 10^{2}$
8.9	$6.6754 \cdot 10^{1}$	$2.5453 \cdot 10^{2}$	12.9	$4.5134 \cdot 10^{1}$	$4.8209 \cdot 10^{2}$
9.0	$6.6486 \cdot 10^{1}$	$2.6119 \cdot 10^{2}$	13.0	$4.4550 \cdot 10^{1}$	$4.8658 \cdot 10^{2}$
9.1	$6.6187 \cdot 10^{1}$	$2.6783 \cdot 10^{2}$	13.1	$4.3970 \cdot 10^{1}$	$4.9100 \cdot 10^{2}$
9.2	$6.5858 \cdot 10^{1}$	$2.7443 \cdot 10^{2}$	13.2	$4.3395 \cdot 10^{1}$	$4.9537 \cdot 10^{2}$
9.3	$6.5501 \cdot 10^{1}$	$2.8100 \cdot 10^{2}$	13.3	$4.2826 \cdot 10^{1}$	$4.9968 \cdot 10^{2}$
9.4	$6.5117 \cdot 10^{1}$	$2.8753 \cdot 10^{2}$	13.4	$4.2261 \cdot 10^{1}$	$5.0394 \cdot 10^{2}$
9.5	$6.4710 \cdot 10^{1}$	$2.9402 \cdot 10^{2}$	13.5	$4.1701 \cdot 10^{1}$	$5.0814 \cdot 10^{2}$
9.6	$6.4280 \cdot 10^{1}$	$3.0047 \cdot 10^{2}$	13.6	$4.1147 \cdot 10^{1}$	$5.1228 \cdot 10^{2}$
9.7	$6.3830 \cdot 10^{1}$	$3.0688 \cdot 10^{2}$	13.7	$4.0598 \cdot 10^{1}$	$5.1637 \cdot 10^{2}$
9.8	$6.3360 \cdot 10^{1}$	$3.1324 \cdot 10^{2}$	13.8	$4.0055 \cdot 10^{1}$	$5.2040 \cdot 10^{2}$
9.9	$6.2872 \cdot 10^{1}$	$3.1955 \cdot 10^{2}$	13.9	$3.9518 \cdot 10^{1}$	$5.2438 \cdot 10^{2}$
10.0	$6.2368 \cdot 10^{1}$	$3.2581 \cdot 10^{2}$	14.0	$3.8986 \cdot 10^{1}$	$5.2830 \cdot 10^{2}$
10.1	$6.1849 \cdot 10^{1}$	$3.3202 \cdot 10^{2}$	14.1	$3.8460 \cdot 10^{1}$	$5.3217 \cdot 10^{2}$
10.2	$6.1316 \cdot 10^{1}$	$3.3818 \cdot 10^{2}$	14.2	$3.7940 \cdot 10^{1}$	$5.3599 \cdot 10^{2}$
10.3	$6.0771 \cdot 10^{1}$	$3.4428 \cdot 10^{2}$	14.3	$3.7426 \cdot 10^{1}$	$5.3976 \cdot 10^{2}$
10.4	$6.0215 \cdot 10^{1}$	$3.5033 \cdot 10^{2}$	14.4	$3.6918 \cdot 10^{1}$	$5.4348 \cdot 10^{2}$
10.5	$5.9648 \cdot 10^{1}$	$3.5632 \cdot 10^{2}$	14.5	$3.6415 \cdot 10^{1}$	$5.4714 \cdot 10^{2}$
10.6	$5.9073 \cdot 10^{1}$	$3.6226 \cdot 10^{2}$	14.6	$3.5919 \cdot 10^{1}$	$5.5076 \cdot 10^{2}$
10.7	$5.8489 \cdot 10^{1}$	$3.6814 \cdot 10^{2}$	14.7	$3.5429 \cdot 10^{1}$	$5.5433 \cdot 10^{2}$
10.8	$5.7899 \cdot 10^{1}$	$3.7396 \cdot 10^{2}$	14.8	$3.4944 \cdot 10^{1}$	$5.5785 \cdot 10^{2}$
10.9	$5.7303 \cdot 10^{1}$	$3.7972 \cdot 10^{2}$	14.9	$3.4466 \cdot 10^{1}$	$5.6132 \cdot 10^{2}$
11.0	$5.6701 \cdot 10^{1}$	$3.8542 \cdot 10^{2}$	15.0	$3.3994 \cdot 10^{1}$	$5.6474 \cdot 10^{2}$
11.1	$5.6095 \cdot 10^{1}$	$3.9106 \cdot 10^{2}$	15.1	$3.3528 \cdot 10^{1}$	$5.6812 \cdot 10^{2}$
11.2	$5.5485 \cdot 10^{1}$	$3.9664 \cdot 10^{2}$	15.2	$3.3068 \cdot 10^{1}$	$5.7145 \cdot 10^{2}$
11.3	$5.4873 \cdot 10^{1}$	$4.0215 \cdot 10^{2}$	15.3	$3.2613 \cdot 10^{1}$	$5.7473 \cdot 10^{2}$
11.4	$5.4258 \cdot 10^{1}$	$4.0761 \cdot 10^{2}$	15.4	$3.2165 \cdot 10^{1}$	$5.7797 \cdot 10^{2}$

Table 2.2. (*continued*)

temperature 400 K

λ [µm]	$m_{\lambda,bb}$ [W/µm m²]	$M_{T,o-\lambda,bb}$ [W/m²]	λ [µm]	$m_{\lambda,bb}$ [W/µm m²]	$M_{T,o-\lambda,bb}$ [W/m²]
0.4	$3.2303\cdot10^{-29}$	$1.4859\cdot10^{-31}$			
0.5	$6.8453\cdot10^{-22}$	$4.9617\cdot10^{-24}$	4.0	$4.5444\cdot10^{1}$	$2.8622\cdot10^{1}$
0.6	$4.4323\cdot10^{-17}$	$4.6655\cdot10^{-19}$	4.1	$5.0018\cdot10^{1}$	$3.3395\cdot10^{1}$
0.7	$1.0746\cdot10^{-13}$	$1.5528\cdot10^{-15}$	4.2	$5.4642\cdot10^{1}$	$3.8628\cdot10^{1}$
0.8	$3.3951\cdot10^{-11}$	$6.4622\cdot10^{-13}$	4.3	$5.9286\cdot10^{1}$	$4.4324\cdot10^{1}$
0.9	$2.7844\cdot10^{-9}$	$6.7650\cdot10^{-11}$	4.4	$6.3917\cdot10^{1}$	$5.0484\cdot10^{1}$
1.0	$8.9466\cdot10^{-8}$	$2.7066\cdot10^{-9}$	4.5	$6.8507\cdot10^{1}$	$5.7106\cdot10^{1}$
1.1	$1.4616\cdot10^{-6}$	$5.3961\cdot10^{-8}$	4.6	$7.3030\cdot10^{1}$	$6.4183\cdot10^{1}$
1.2	$1.4432\cdot10^{-5}$	$6.3956\cdot10^{-7}$	4.7	$7.7462\cdot10^{1}$	$7.1709\cdot10^{1}$
1.3	$9.7024\cdot10^{-5}$	$5.0898\cdot10^{-6}$	4.8	$8.1781\cdot10^{1}$	$7.9672\cdot10^{1}$
1.4	$4.8336\cdot10^{-4}$	$2.9662\cdot10^{-5}$	4.9	$8.5968\cdot10^{1}$	$8.8060\cdot10^{1}$
1.5	$1.8982\cdot10^{-3}$	$1.3488\cdot10^{-4}$	5.0	$9.0006\cdot10^{1}$	$9.6860\cdot10^{1}$
1.6	$6.1529\cdot10^{-3}$	$5.0177\cdot10^{-4}$	5.1	$9.3882\cdot10^{1}$	$1.0606\cdot10^{2}$
1.7	$1.7051\cdot10^{-2}$	$1.5854\cdot10^{-3}$	5.2	$9.7583\cdot10^{1}$	$1.1563\cdot10^{2}$
1.8	$4.1509\cdot10^{-2}$	$6.3502\cdot10^{-3}$	5.3	$1.0110\cdot10^{2}$	$1.2557\cdot10^{2}$
1.9	$9.0678\cdot10^{-2}$	$1.0703\cdot10^{-2}$	5.4	$1.0443\cdot10^{2}$	$1.3584\cdot10^{2}$
2.0	$1.8080\cdot10^{-1}$	$2.3854\cdot10^{-2}$	5.5	$1.0756\cdot10^{2}$	$1.4645\cdot10^{2}$
2.1	$3.3358\cdot10^{-1}$	$4.8946\cdot10^{-2}$	5.6	$1.1049\cdot10^{2}$	$1.5735\cdot10^{2}$
2.2	$5.7586\cdot10^{-1}$	$9.3548\cdot10^{-2}$	5.7	$1.1321\cdot10^{2}$	$1.6854\cdot10^{2}$
2.3	$9.3866\cdot10^{-1}$	$1.6813\cdot10^{-1}$	5.8	$1.1574\cdot10^{2}$	$1.7998\cdot10^{2}$
2.4	$1.4558\cdot10^{0}$	$2.8642\cdot10^{-1}$	5.9	$1.1806\cdot10^{2}$	$1.9168\cdot10^{2}$
2.5	$2.1617\cdot10^{0}$	$4.6558\cdot10^{-1}$	6.0	$1.2018\cdot10^{2}$	$2.0559\cdot10^{2}$
2.6	$3.0901\cdot10^{0}$	$7.2618\cdot10^{-1}$	6.1	$1.2211\cdot10^{2}$	$2.1571\cdot10^{2}$
2.7	$4.2711\cdot10^{0}$	$1.0920\cdot10^{0}$	6.2	$1.2384\cdot10^{2}$	$2.2800\cdot10^{2}$
2.8	$5.7307\cdot10^{0}$	$1.5897\cdot10^{0}$	6.3	$1.2538\cdot10^{2}$	$2.4047\cdot10^{2}$
2.9	$7.4884\cdot10^{0}$	$2.2481\cdot10^{0}$	6.4	$1.2674\cdot10^{2}$	$2.5307\cdot10^{2}$
3.0	$9.5572\cdot10^{0}$	$3.0977\cdot10^{0}$	6.5	$1.2792\cdot10^{2}$	$2.6581\cdot10^{2}$
3.1	$1.1943\cdot10^{1}$	$4.1701\cdot10^{0}$	6.6	$1.2893\cdot10^{2}$	$2.7865\cdot10^{2}$
3.2	$1.4643\cdot10^{1}$	$5.4967\cdot10^{0}$	6.7	$1.2977\cdot10^{2}$	$2.9159\cdot10^{2}$
3.3	$1.7650\cdot10^{1}$	$7.1089\cdot10^{0}$	6.8	$1.3046\cdot10^{2}$	$3.0460\cdot10^{2}$
3.4	$2.0949\cdot10^{1}$	$9.0365\cdot10^{0}$	6.9	$1.3099\cdot10^{2}$	$3.1767\cdot10^{2}$
3.5	$2.4518\cdot10^{1}$	$1.1308\cdot10^{1}$	7.0	$1.3138\cdot10^{2}$	$3.3079\cdot10^{2}$
3.6	$2.8333\cdot10^{1}$	$1.3948\cdot10^{1}$	7.1	$1.3163\cdot10^{2}$	$3.4395\cdot10^{2}$
3.7	$3.2366\cdot10^{1}$	$1.6982\cdot10^{1}$	7.2	$1.3174\cdot10^{2}$	$3.5711\cdot10^{2}$
3.8	$3.6584\cdot10^{1}$	$2.0428\cdot10^{1}$	7.3	$1.3174\cdot10^{2}$	$3.7029\cdot10^{2}$
3.9	$4.0955\cdot10^{1}$	$2.4303\cdot10^{1}$	7.4	$1.3161\cdot10^{2}$	$3.8346\cdot10^{2}$

λ [μm]	$m_{\lambda,bb}$ [W/μm m²]	$M_{T,o-\lambda,bb}$ [W/m²]	λ [μm]	$m_{\lambda,bb}$ [W/μm m²]	$M_{T,o-\lambda,bb}$ [W/m²]
7.5	$1.3138 \cdot 10^2$	$3.9661 \cdot 10^2$	11.5	$8.5246 \cdot 10^1$	$8.4084 \cdot 10^2$
7.6	$1.3104 \cdot 10^2$	$4.0973 \cdot 10^2$	11.6	$8.3970 \cdot 10^1$	$8.4930 \cdot 10^2$
7.7	$1.3060 \cdot 10^2$	$4.2281 \cdot 10^2$	11.7	$8.2708 \cdot 10^1$	$8.5763 \cdot 10^2$
7.8	$1.3008 \cdot 10^2$	$4.3585 \cdot 10^2$	11.8	$8.1459 \cdot 10^1$	$8.6584 \cdot 10^2$
7.9	$1.2947 \cdot 10^2$	$4.4882 \cdot 10^2$	11.9	$8.0223 \cdot 10^1$	$8.7393 \cdot 10^2$
8.0	$1.2877 \cdot 10^2$	$4.6174 \cdot 10^2$	12.0	$7.9001 \cdot 10^1$	$8.8189 \cdot 10^2$
8.1	$1.2801 \cdot 10^2$	$4.7458 \cdot 10^2$	12.1	$7.7794 \cdot 10^1$	$8.8973 \cdot 10^2$
8.2	$1.2718 \cdot 10^2$	$4.8734 \cdot 10^2$	12.2	$7.6601 \cdot 10^1$	$8.9745 \cdot 10^2$
8.3	$1.2628 \cdot 10^2$	$5.0001 \cdot 10^2$	12.3	$7.5422 \cdot 10^1$	$9.0505 \cdot 10^2$
8.4	$1.2533 \cdot 10^2$	$5.1259 \cdot 10^2$	12.4	$7.4259 \cdot 10^1$	$9.1253 \cdot 10^2$
8.5	$1.2432 \cdot 10^2$	$5.2507 \cdot 10^2$	12.5	$7.3110 \cdot 10^1$	$9.1990 \cdot 10^2$
8.6	$1.2326 \cdot 10^2$	$5.3745 \cdot 10^2$	12.6	$7.1977 \cdot 10^1$	$9.2715 \cdot 10^2$
8.7	$1.2216 \cdot 10^2$	$5.4972 \cdot 10^2$	12.7	$7.0859 \cdot 10^1$	$9.3429 \cdot 10^2$
8.8	$1.2102 \cdot 10^2$	$5.6188 \cdot 10^2$	12.8	$6.9756 \cdot 10^1$	$9.4132 \cdot 10^2$
8.9	$1.1984 \cdot 10^2$	$5.7392 \cdot 10^2$	12.9	$6.8668 \cdot 10^1$	$9.4825 \cdot 10^2$
9.0	$1.1864 \cdot 10^2$	$5.8585 \cdot 10^2$	13.0	$6.7596 \cdot 10^1$	$9.5506 \cdot 10^2$
9.1	$1.1740 \cdot 10^2$	$5.9765 \cdot 10^2$	13.1	$6.6539 \cdot 10^1$	$9.6177 \cdot 10^2$
9.2	$1.1613 \cdot 10^2$	$6.0933 \cdot 10^2$	13.2	$6.5497 \cdot 10^1$	$9.6837 \cdot 10^2$
9.3	$1.1485 \cdot 10^2$	$6.2088 \cdot 10^2$	13.3	$6.4471 \cdot 10^1$	$9.7486 \cdot 10^2$
9.4	$1.1354 \cdot 10^2$	$6.3230 \cdot 10^2$	13.4	$6.3459 \cdot 10^1$	$9.8126 \cdot 10^2$
9.5	$1.1222 \cdot 10^2$	$6.4358 \cdot 10^2$	13.5	$6.2464 \cdot 10^1$	$9.8756 \cdot 10^2$
9.6	$1.1088 \cdot 10^2$	$6.5474 \cdot 10^2$	13.6	$6.1483 \cdot 10^1$	$9.9375 \cdot 10^2$
9.7	$1.0954 \cdot 10^2$	$6.6576 \cdot 10^2$	13.7	$6.0517 \cdot 10^1$	$9.9985 \cdot 10^2$
9.8	$1.0818 \cdot 10^2$	$6.7665 \cdot 10^2$	13.8	$5.9566 \cdot 10^1$	$1.0059 \cdot 10^3$
9.9	$1.0681 \cdot 10^2$	$6.8739 \cdot 10^2$	13.9	$5.8630 \cdot 10^1$	$1.0118 \cdot 10^3$
10.0	$1.0544 \cdot 10^2$	$6.9801 \cdot 10^2$	14.0	$5.7708 \cdot 10^1$	$1.0176 \cdot 10^3$
10.1	$1.0407 \cdot 10^2$	$7.0848 \cdot 10^2$	14.1	$5.6802 \cdot 10^1$	$1.0233 \cdot 10^3$
10.2	$1.0269 \cdot 10^2$	$7.1882 \cdot 10^2$	14.2	$5.5909 \cdot 10^1$	$1.0289 \cdot 10^3$
10.3	$1.0132 \cdot 10^2$	$7.2902 \cdot 10^2$	14.3	$5.5031 \cdot 10^1$	$1.0345 \cdot 10^3$
10.4	$9.9942 \cdot 10^1$	$7.3908 \cdot 10^2$	14.4	$5.4167 \cdot 10^1$	$1.0400 \cdot 10^3$
10.5	$9.8571 \cdot 10^1$	$7.4901 \cdot 10^2$	14.5	$5.3317 \cdot 10^1$	$1.0453 \cdot 10^3$
10.6	$9.7203 \cdot 10^1$	$7.5880 \cdot 10^2$	14.6	$5.2481 \cdot 10^1$	$1.0506 \cdot 10^3$
10.7	$9.5841 \cdot 10^1$	$7.6845 \cdot 10^2$	14.7	$5.1658 \cdot 10^1$	$1.0558 \cdot 10^3$
10.8	$9.4485 \cdot 10^1$	$7.7797 \cdot 10^2$	14.8	$5.0849 \cdot 10^1$	$1.0609 \cdot 10^3$
10.9	$9.3136 \cdot 10^1$	$7.8735 \cdot 10^2$	14.9	$5.0054 \cdot 10^1$	$1.0660 \cdot 10^3$
11.0	$9.1796 \cdot 10^1$	$7.9659 \cdot 10^2$	15.0	$4.9271 \cdot 10^1$	$1.0710 \cdot 10^3$
11.1	$9.0465 \cdot 10^1$	$8.0571 \cdot 10^2$	15.1	$4.8502 \cdot 10^1$	$1.0758 \cdot 10^3$
11.2	$8.9144 \cdot 10^1$	$8.1469 \cdot 10^2$	15.2	$4.7745 \cdot 10^1$	$1.0807 \cdot 10^3$
11.3	$8.7833 \cdot 10^1$	$8.2353 \cdot 10^2$	15.3	$4.7001 \cdot 10^1$	$1.0854 \cdot 10^3$
11.4	$8.6533 \cdot 10^1$	$8.3225 \cdot 10^2$	15.4	$4.6270 \cdot 10^1$	$1.0901 \cdot 10^3$

Table 2.2. (*continued*)

temperature 450 K

λ [μm]	$m_{\lambda,bb}$ [W/μm m²]	$M_{T,o-\lambda,bb}$ [W/m²]	λ [μm]	$m_{\lambda,bb}$ [W/μm m²]	$M_{T,o-\lambda,bb}$ [W/m²]
0.4	$7.0553\cdot10^{-25}$	$3.6665\cdot10^{-27}$			
0.5	$2.0268\cdot10^{-18}$	$1.6615\cdot10^{-20}$	4.0	$1.2345\cdot10^{2}$	$9.1472\cdot10^{1}$
0.6	$3.4631\cdot10^{-14}$	$4.1272\cdot10^{-16}$	4.1	$1.3261\cdot10^{2}$	$1.0428\cdot10^{2}$
0.7	$3.2424\cdot10^{-11}$	$5.3100\cdot10^{-13}$	4.2	$1.4156\cdot10^{2}$	$1.1799\cdot10^{2}$
0.8	$5.0176\cdot10^{-9}$	$1.0836\cdot10^{-10}$	4.3	$1.5023\cdot10^{2}$	$1.3258\cdot10^{2}$
0.9	$2.3621\cdot10^{-7}$	$6.5187\cdot10^{-9}$	4.4	$1.5859\cdot10^{2}$	$1.4802\cdot10^{2}$
1.0	$4.8682\cdot10^{-6}$	$1.6747\cdot10^{-7}$	4.5	$1.6660\cdot10^{2}$	$1.6429\cdot10^{2}$
1.1	$5.5302\cdot10^{-5}$	$2.3243\cdot10^{-6}$	4.6	$1.7421\cdot10^{2}$	$1.8133\cdot10^{2}$
1.2	$4.0341\cdot10^{-4}$	$2.0374\cdot10^{-5}$	4.7	$1.8141\cdot10^{2}$	$1.9911\cdot10^{2}$
1.3	$2.0991\cdot10^{-3}$	$1.2563\cdot10^{-4}$	4.8	$1.8818\cdot10^{2}$	$2.1760\cdot10^{2}$
1.4	$8.3960\cdot10^{-3}$	$5.8847\cdot10^{-4}$	4.9	$1.9450\cdot10^{2}$	$2.3673\cdot10^{2}$
1.5	$2.7257\cdot10^{-2}$	$2.2146\cdot10^{-3}$	5.0	$2.0036\cdot10^{2}$	$2.5648\cdot10^{2}$
1.6	$7.4800\cdot10^{-2}$	$6.9826\cdot10^{-3}$	5.1	$2.0576\cdot10^{2}$	$2.7679\cdot10^{2}$
1.7	$1.7896\cdot10^{-1}$	$1.9045\cdot10^{-2}$	5.2	$2.1070\cdot10^{2}$	$2.9762\cdot10^{2}$
1.8	$3.8232\cdot10^{-1}$	$4.6064\cdot10^{-2}$	5.3	$2.1518\cdot10^{2}$	$3.4892\cdot10^{2}$
1.9	$7.4309\cdot10^{-1}$	$1.0074\cdot10^{-1}$	5.4	$2.1921\cdot10^{2}$	$3.4064\cdot10^{2}$
2.0	$1.3337\cdot10^{0}$	$2.0233\cdot10^{-1}$	5.5	$2.2279\cdot10^{2}$	$3.6274\cdot10^{2}$
2.1	$2.2373\cdot10^{0}$	$3.7791\cdot10^{-1}$	5.6	$2.2594\cdot10^{2}$	$3.8518\cdot10^{2}$
2.2	$3.5422\cdot10^{0}$	$6.6317\cdot10^{-1}$	5.7	$2.2867\cdot10^{2}$	$4.0792\cdot10^{2}$
2.3	$5.3354\cdot10^{0}$	$1.1026\cdot10^{0}$	5.8	$2.3100\cdot10^{2}$	$4.3090\cdot10^{2}$
2.4	$7.6967\cdot10^{0}$	$1.7492\cdot10^{0}$	5.9	$2.3294\cdot10^{2}$	$4.5410\cdot10^{2}$
2.5	$1.0693\cdot10^{1}$	$2.6651\cdot10^{0}$	6.0	$2.3451\cdot10^{2}$	$4.7748\cdot10^{2}$
2.6	$1.4373\cdot10^{1}$	$3.9106\cdot10^{0}$	6.1	$2.3572\cdot10^{2}$	$5.0099\cdot10^{2}$
2.7	$1.8767\cdot10^{1}$	$5.5616\cdot10^{0}$	6.2	$2.3659\cdot10^{2}$	$5.2461\cdot10^{2}$
2.8	$2.3884\cdot10^{1}$	$7.6881\cdot10^{0}$	6.3	$2.3715\cdot10^{2}$	$5.4830\cdot10^{2}$
2.9	$2.9711\cdot10^{1}$	$1.0362\cdot10^{1}$	6.4	$2.3741\cdot10^{2}$	$5.7203\cdot10^{2}$
3.0	$3.6217\cdot10^{1}$	$1.3653\cdot10^{1}$	6.5	$2.3738\cdot10^{2}$	$5.9577\cdot10^{2}$
3.1	$4.3353\cdot10^{1}$	$1.7626\cdot10^{1}$	6.6	$2.3708\cdot10^{2}$	$6.1949\cdot10^{2}$
3.2	$5.1058\cdot10^{1}$	$2.2342\cdot10^{1}$	6.7	$2.3654\cdot10^{2}$	$6.4318\cdot10^{2}$
3.3	$5.9258\cdot10^{1}$	$2.7854\cdot10^{1}$	6.8	$2.3577\cdot10^{2}$	$6.6679\cdot10^{2}$
3.4	$6.7871\cdot10^{1}$	$3.4208\cdot10^{1}$	6.9	$2.3478\cdot10^{2}$	$6.9032\cdot10^{2}$
3.5	$7.6814\cdot10^{1}$	$4.1439\cdot10^{1}$	7.0	$2.3359\cdot10^{2}$	$7.1374\cdot10^{2}$
3.6	$8.5997\cdot10^{1}$	$4.9578\cdot10^{1}$	7.1	$2.3222\cdot10^{2}$	$7.3704\cdot10^{2}$
3.7	$9.5335\cdot10^{1}$	$5.8644\cdot10^{1}$	7.2	$2.3068\cdot10^{2}$	$7.6018\cdot10^{2}$
3.8	$1.0474\cdot10^{2}$	$6.8647\cdot10^{1}$	7.3	$2.2898\cdot10^{2}$	$7.8316\cdot10^{2}$
3.9	$1.1414\cdot10^{2}$	$7.9592\cdot10^{1}$	7.4	$2.2714\cdot10^{2}$	$8.0597\cdot10^{3}$

Table 2.2. (*continued*)

λ [µm]	$m_{\lambda,bb}$ [W/µm m^2]	$M_{T,o-\lambda,bb}$ [W/m^2]	λ [µm]	$m_{\lambda,bb}$ [W/µm m^2]	$M_{T,o-\lambda,bb}$ [W/m^2]
7.5	$2.2517\cdot10^2$	$8.2859\cdot10^2$	11.5	$1.2301\cdot10^2$	$1.5230\cdot10^3$
7.6	$2.2308\cdot10^2$	$8.5100\cdot10^2$	11.6	$1.2085\cdot10^2$	$1.5352\cdot10^3$
7.7	$2.2089\cdot10^2$	$8.7320\cdot10^2$	11.7	$1.1873\cdot10^2$	$1.5472\cdot10^3$
7.8	$2.1860\cdot10^2$	$8.9518\cdot10^2$	11.8	$1.1664\cdot10^2$	$1.5590\cdot10^3$
7.9	$2.1623\cdot10^2$	$9.1692\cdot10^2$	11.9	$1.1458\cdot10^2$	$1.5705\cdot10^3$
8.0	$2.1379\cdot10^2$	$9.5842\cdot10^2$	12.0	$1.1256\cdot10^2$	$1.5819\cdot10^3$
8.1	$2.1127\cdot10^2$	$9.5967\cdot10^2$	12.1	$1.1057\cdot10^2$	$1.5930\cdot10^3$
8.2	$2.0870\cdot10^2$	$9.8067\cdot10^2$	12.2	$1.0862\cdot10^2$	$1.6040\cdot10^3$
8.3	$2.0608\cdot10^2$	$1.0014\cdot10^3$	12.3	$1.0670\cdot10^2$	$1.6148\cdot10^3$
8.4	$2.0342\cdot10^2$	$1.0219\cdot10^3$	12.4	$1.0482\cdot10^2$	$1.6253\cdot10^3$
8.5	$2.0072\cdot10^2$	$1.0421\cdot10^3$	12.5	$1.0297\cdot10^2$	$1.6357\cdot10^3$
8.6	$1.9800\cdot10^2$	$1.0620\cdot10^3$	12.6	$1.0115\cdot10^2$	$1.6459\cdot10^3$
8.7	$1.9525\cdot10^2$	$1.0817\cdot10^3$	12.7	$9.9364\cdot10^1$	$1.6560\cdot10^3$
8.8	$1.9248\cdot10^2$	$1.1011\cdot10^3$	12.8	$9.7611\cdot10^1$	$1.6658\cdot10^3$
8.9	$1.8970\cdot10^2$	$1.1202\cdot10^3$	12.9	$9.5890\cdot10^1$	$1.6755\cdot10^3$
9.0	$1.8691\cdot10^2$	$1.1390\cdot10^3$	13.0	$9.4200\cdot10^1$	$1.6850\cdot10^3$
9.1	$1.8413\cdot10^2$	$1.1576\cdot10^3$	13.1	$9.2541\cdot10^1$	$1.6943\cdot10^3$
9.2	$1.8134\cdot10^2$	$1.1758\cdot10^3$	13.2	$9.0912\cdot10^1$	$1.7035\cdot10^3$
9.3	$1.7855\cdot10^2$	$1.1938\cdot10^3$	13.3	$8.9315\cdot10^1$	$1.7125\cdot10^3$
9.4	$1.7578\cdot10^2$	$1.2116\cdot10^3$	13.4	$8.7746\cdot10^1$	$1.7214\cdot10^3$
9.5	$1.7301\cdot10^2$	$1.2290\cdot10^3$	13.5	$8.6208\cdot10^1$	$1.7301\cdot10^3$
9.6	$1.7026\cdot10^2$	$1.2462\cdot10^3$	13.6	$8.4698\cdot10^1$	$1.7386\cdot10^3$
9.7	$1.6753\cdot10^2$	$1.2630\cdot10^3$	13.7	$8.3216\cdot10^1$	$1.7470\cdot10^3$
9.8	$1.6482\cdot10^2$	$1.2797\cdot10^3$	13.8	$8.1763\cdot10^1$	$1.7552\cdot10^3$
9.9	$1.6213\cdot10^2$	$1.2960\cdot10^3$	13.9	$8.0337\cdot10^1$	$1.7633\cdot10^3$
10.0	$1.5946\cdot10^2$	$1.3121\cdot10^3$	14.0	$7.8938\cdot10^1$	$1.7713\cdot10^3$
10.1	$1.5681\cdot10^2$	$1.3279\cdot10^3$	14.1	$7.7566\cdot10^1$	$1.7791\cdot10^3$
10.2	$1.5419\cdot10^2$	$1.3434\cdot10^3$	14.2	$7.6220\cdot10^1$	$1.7868\cdot10^3$
10.3	$1.5160\cdot10^2$	$1.3587\cdot10^3$	14.3	$7.4900\cdot10^1$	$1.7944\cdot10^3$
10.4	$1.4904\cdot10^2$	$1.3738\cdot10^3$	14.4	$7.3605\cdot10^1$	$1.8018\cdot10^3$
10.5	$1.4651\cdot10^2$	$1.3885\cdot10^3$	14.5	$7.2335\cdot10^1$	$1.8091\cdot10^3$
10.6	$1.4401\cdot10^2$	$1.4031\cdot10^3$	14.6	$7.1089\cdot10^1$	$1.8163\cdot10^3$
10.7	$1.4154\cdot10^2$	$1.4173\cdot10^3$	14.7	$6.9867\cdot10^1$	$1.8233\cdot10^3$
10.8	$1.3911\cdot10^2$	$1.4314\cdot10^3$	14.8	$6.8668\cdot10^1$	$1.8302\cdot10^3$
10.9	$1.3671\cdot10^2$	$1.4452\cdot10^3$	14.9	$6.7493\cdot10^1$	$1.8371\cdot10^3$
11.0	$1.3434\cdot10^2$	$1.4587\cdot10^3$	15.0	$6.6340\cdot10^1$	$1.8437\cdot10^3$
11.1	$1.3200\cdot10^2$	$1.4720\cdot10^3$	15.1	$6.5209\cdot10^1$	$1.8503\cdot10^3$
11.2	$1.2970\cdot10^2$	$1.4851\cdot10^3$	15.2	$6.4100\cdot10^1$	$1.8568\cdot10^3$
11.3	$1.2744\cdot10^2$	$1.4980\cdot10^3$	15.3	$6.3013\cdot10^1$	$1.8631\cdot10^3$
11.4	$1.2521\cdot10^2$	$1.5106\cdot10^3$	15.4	$6.1946\cdot10^1$	$1.8694\cdot10^3$

Table 2.2. (*continued*)

temperature 500 K

λ [μm]	$m_{\lambda,bb}$ [W/μm m²]	$M_{T,o-\lambda,bb}$ [W/m²]	λ [μm]	$m_{\lambda,bb}$ [W/μm m²]	$M_{T,o-\lambda,bb}$ [W/m²]
0.4	$2.0890 \cdot 10^{-21}$	$1.2114 \cdot 10^{-23}$			
0.5	$1.2133 \cdot 10^{-15}$	$1.1110 \cdot 10^{-17}$	4.0	$2.7468 \cdot 10^{2}$	$2.3648 \cdot 10^{2}$
0.6	$7.1409 \cdot 10^{-12}$	$9.5163 \cdot 10^{-14}$	4.1	$2.8938 \cdot 10^{2}$	$2.6469 \cdot 10^{2}$
0.7	$3.1227 \cdot 10^{-9}$	$5.7249 \cdot 10^{-11}$	4.2	$3.0324 \cdot 10^{2}$	$2.9432 \cdot 10^{2}$
0.8	$2.7303 \cdot 10^{-7}$	$6.6078 \cdot 10^{-9}$	4.3	$3.1620 \cdot 10^{2}$	$3.2530 \cdot 10^{2}$
0.9	$8.2444 \cdot 10^{-6}$	$2.5525 \cdot 10^{-7}$	4.4	$3.2824 \cdot 10^{2}$	$3.5753 \cdot 10^{2}$
1.0	$1.1911 \cdot 10^{-4}$	$4.6018 \cdot 10^{-6}$	4.5	$3.3932 \cdot 10^{2}$	$3.9092 \cdot 10^{2}$
1.1	$1.0118 \cdot 10^{-3}$	$4.7810 \cdot 10^{-5}$	4.6	$3.4943 \cdot 10^{2}$	$4.2536 \cdot 10^{2}$
1.2	$5.7929 \cdot 10^{-3}$	$3.2930 \cdot 10^{-4}$	4.7	$3.5858 \cdot 10^{2}$	$4.6077 \cdot 10^{2}$
1.3	$2.4557 \cdot 10^{-2}$	$1.6562 \cdot 10^{-3}$	4.8	$3.6677 \cdot 10^{2}$	$4.9704 \cdot 10^{2}$
1.4	$8.2397 \cdot 10^{-2}$	$6.5150 \cdot 10^{-3}$	4.9	$3.7402 \cdot 10^{2}$	$5.3409 \cdot 10^{2}$
1.5	$2.2972 \cdot 10^{-1}$	$2.1079 \cdot 10^{-2}$	5.0	$3.8035 \cdot 10^{2}$	$5.7182 \cdot 10^{2}$
1.6	$5.5178 \cdot 10^{-1}$	$5.8237 \cdot 10^{-2}$	5.1	$3.8579 \cdot 10^{2}$	$6.1013 \cdot 10^{2}$
1.7	$1.1737 \cdot 10^{0}$	$1.4139 \cdot 10^{-1}$	5.2	$3.9038 \cdot 10^{2}$	$6.4895 \cdot 10^{2}$
1.8	$2.2587 \cdot 10^{0}$	$3.0838 \cdot 10^{-1}$	5.3	$3.9415 \cdot 10^{2}$	$6.8818 \cdot 10^{2}$
1.9	$3.9982 \cdot 10^{0}$	$6.1492 \cdot 10^{-1}$	5.4	$3.9714 \cdot 10^{2}$	$7.2775 \cdot 10^{2}$
2.0	$6.5971 \cdot 10^{0}$	$1.1367 \cdot 10^{0}$	5.5	$3.9938 \cdot 10^{2}$	$7.6758 \cdot 10^{2}$
2.1	$1.0256 \cdot 10^{1}$	$1.9697 \cdot 10^{0}$	5.6	$4.0093 \cdot 10^{2}$	$8.0760 \cdot 10^{2}$
2.2	$1.5151 \cdot 10^{1}$	$3.2290 \cdot 10^{0}$	5.7	$4.0182 \cdot 10^{2}$	$8.4775 \cdot 10^{2}$
2.3	$2.1424 \cdot 10^{1}$	$5.0458 \cdot 10^{0}$	5.8	$4.0209 \cdot 10^{2}$	$8.8795 \cdot 10^{2}$
2.4	$2.9166 \cdot 10^{1}$	$7.5628 \cdot 10^{0}$	5.9	$4.0178 \cdot 10^{2}$	$9.2814 \cdot 10^{2}$
2.5	$3.8417 \cdot 10^{1}$	$1.0929 \cdot 10^{1}$	6.0	$4.0094 \cdot 10^{2}$	$9.6828 \cdot 10^{2}$
2.6	$4.9161 \cdot 10^{1}$	$1.5296 \cdot 10^{1}$	6.1	$3.9960 \cdot 10^{2}$	$1.0083 \cdot 10^{3}$
2.7	$6.1334 \cdot 10^{1}$	$2.0809 \cdot 10^{1}$	6.2	$3.9780 \cdot 10^{2}$	$1.0482 \cdot 10^{3}$
2.8	$7.4823 \cdot 10^{1}$	$2.7607 \cdot 10^{1}$	6.3	$3.9559 \cdot 10^{2}$	$1.0879 \cdot 10^{3}$
2.9	$8.9485 \cdot 10^{1}$	$3.5813 \cdot 10^{1}$	6.4	$3.9299 \cdot 10^{2}$	$1.1273 \cdot 10^{3}$
3.0	$1.0514 \cdot 10^{2}$	$4.5537 \cdot 10^{1}$	6.5	$3.9003 \cdot 10^{2}$	$1.1664 \cdot 10^{3}$
3.1	$1.2161 \cdot 10^{2}$	$5.6869 \cdot 10^{1}$	6.6	$3.8676 \cdot 10^{2}$	$1.2053 \cdot 10^{3}$
3.2	$1.3868 \cdot 10^{2}$	$6.9879 \cdot 10^{1}$	6.7	$3.8321 \cdot 10^{2}$	$1.2438 \cdot 10^{3}$
3.3	$1.5616 \cdot 10^{2}$	$8.4619 \cdot 10^{1}$	6.8	$3.7939 \cdot 10^{2}$	$1.2819 \cdot 10^{3}$
3.4	$1.7384 \cdot 10^{2}$	$1.0112 \cdot 10^{2}$	6.9	$3.7534 \cdot 10^{2}$	$1.3197 \cdot 10^{3}$
3.5	$1.9153 \cdot 10^{2}$	$1.1939 \cdot 10^{2}$	7.0	$3.7108 \cdot 10^{2}$	$1.3570 \cdot 10^{3}$
3.6	$2.0907 \cdot 10^{2}$	$1.3942 \cdot 10^{2}$	7.1	$3.6664 \cdot 10^{2}$	$1.3939 \cdot 10^{3}$
3.7	$2.2628 \cdot 10^{2}$	$1.6119 \cdot 10^{2}$	7.2	$3.6205 \cdot 10^{2}$	$1.4303 \cdot 10^{3}$
3.8	$2.4303 \cdot 10^{2}$	$1.8466 \cdot 10^{2}$	7.3	$3.5731 \cdot 10^{2}$	$1.4663 \cdot 10^{3}$
3.9	$2.5920 \cdot 10^{2}$	$2.0978 \cdot 10^{2}$	7.4	$3.5246 \cdot 10^{2}$	$1.5018 \cdot 10^{3}$

Table 2.2. (*continued*)

λ [μm]	$m_{\lambda,bb}$ [W/μm m²]	$M_{T,o-\lambda,bb}$ [W/m²]	λ [μm]	$m_{\lambda,bb}$ [W/μm m²]	$M_{T,o-\lambda,bb}$ [W/m²]
7.5	$3.4751 \cdot 10^2$	$1.5368 \cdot 10^3$	11.5	$1.6596 \cdot 10^2$	$2.5356 \cdot 10^3$
7.6	$3.4247 \cdot 10^2$	$1.5713 \cdot 10^3$	11.6	$1.6271 \cdot 10^2$	$2.5521 \cdot 10^3$
7.7	$3.3737 \cdot 10^2$	$1.6052 \cdot 10^3$	11.7	$1.5953 \cdot 10^2$	$2.5682 \cdot 10^3$
7.8	$3.3221 \cdot 10^2$	$1.6387 \cdot 10^3$	11.8	$1.5641 \cdot 10^2$	$2.5840 \cdot 10^3$
7.9	$3.2701 \cdot 10^2$	$1.6717 \cdot 10^3$	11.9	$1.5336 \cdot 10^2$	$2.5995 \cdot 10^3$
8.0	$3.2178 \cdot 10^2$	$1.7041 \cdot 10^3$	12.0	$1.5036 \cdot 10^2$	$2.6147 \cdot 10^3$
8.1	$3.1654 \cdot 10^2$	$1.7360 \cdot 10^3$	12.1	$1.4744 \cdot 10^2$	$2.6295 \cdot 10^3$
8.2	$3.1129 \cdot 10^2$	$1.7674 \cdot 10^3$	12.2	$1.4457 \cdot 10^2$	$2.6441 \cdot 10^3$
8.3	$3.0605 \cdot 10^2$	$1.7983 \cdot 10^3$	12.3	$1.4176 \cdot 10^2$	$2.6585 \cdot 10^3$
8.4	$3.0081 \cdot 10^2$	$1.8286 \cdot 10^3$	12.4	$1.3901 \cdot 10^2$	$2.6725 \cdot 10^3$
8.5	$2.9560 \cdot 10^2$	$1.8585 \cdot 10^3$	12.5	$1.3632 \cdot 10^2$	$2.6863 \cdot 10^3$
8.6	$2.9041 \cdot 10^2$	$1.8878 \cdot 10^3$	12.6	$1.3368 \cdot 10^2$	$2.6998 \cdot 10^3$
8.7	$2.8526 \cdot 10^2$	$1.9165 \cdot 10^3$	12.7	$1.3110 \cdot 10^2$	$2.7130 \cdot 10^3$
8.8	$2.8014 \cdot 10^2$	$1.9448 \cdot 10^3$	12.8	$1.2858 \cdot 10^2$	$2.7260 \cdot 10^3$
8.9	$2.7507 \cdot 10^2$	$1.9726 \cdot 10^3$	12.9	$1.2611 \cdot 10^2$	$2.7387 \cdot 10^3$
9.0	$2.7004 \cdot 10^2$	$1.9998 \cdot 10^3$	13.0	$1.2369 \cdot 10^2$	$2.7512 \cdot 10^3$
9.1	$2.6507 \cdot 10^2$	$2.0266 \cdot 10^3$	13.1	$1.2132 \cdot 10^2$	$2.7635 \cdot 10^3$
9.2	$2.6015 \cdot 10^2$	$2.0528 \cdot 10^3$	13.2	$1.1901 \cdot 10^2$	$2.7755 \cdot 10^3$
9.3	$2.5529 \cdot 10^2$	$2.0786 \cdot 10^3$	13.3	$1.1674 \cdot 10^2$	$2.7873 \cdot 10^3$
9.4	$2.5049 \cdot 10^2$	$2.1039 \cdot 10^3$	13.4	$1.1452 \cdot 10^2$	$2.7988 \cdot 10^3$
9.5	$2.4576 \cdot 10^2$	$2.1287 \cdot 10^3$	13.5	$1.1235 \cdot 10^2$	$2.8102 \cdot 10^3$
9.6	$2.4109 \cdot 10^2$	$2.1531 \cdot 10^3$	13.6	$1.1022 \cdot 10^2$	$2.8213 \cdot 10^3$
9.7	$2.3649 \cdot 10^2$	$2.1769 \cdot 10^3$	13.7	$1.0814 \cdot 10^2$	$2.8322 \cdot 10^3$
9.8	$2.3196 \cdot 10^2$	$2.2004 \cdot 10^3$	13.8	$1.0611 \cdot 10^2$	$2.8429 \cdot 10^3$
9.9	$2.2750 \cdot 10^2$	$2.2233 \cdot 10^3$	13.9	$1.0411 \cdot 10^2$	$2.8534 \cdot 10^3$
10.0	$2.2311 \cdot 10^2$	$2.2659 \cdot 10^3$	14.0	$1.0216 \cdot 10^2$	$2.8637 \cdot 10^3$
10.1	$2.1880 \cdot 10^2$	$2.2680 \cdot 10^3$	14.1	$1.0026 \cdot 10^2$	$2.8739 \cdot 10^3$
10.2	$2.1455 \cdot 10^2$	$2.2896 \cdot 10^3$	14.2	$9.8390 \cdot 10^1$	$2.8838 \cdot 10^3$
10.3	$2.1038 \cdot 10^2$	$2.3109 \cdot 10^3$	14.3	$9.6562 \cdot 10^1$	$2.8935 \cdot 10^3$
10.4	$2.0629 \cdot 10^2$	$2.3317 \cdot 10^3$	14.4	$9.4774 \cdot 10^1$	$2.9031 \cdot 10^3$
10.5	$2.0226 \cdot 10^2$	$2.3521 \cdot 10^3$	14.5	$9.3023 \cdot 10^1$	$2.9125 \cdot 10^3$
10.6	$1.9831 \cdot 10^2$	$2.3722 \cdot 10^3$	14.6	$9.1310 \cdot 10^1$	$2.9217 \cdot 10^3$
10.7	$1.9443 \cdot 10^2$	$2.3918 \cdot 10^3$	14.7	$8.9633 \cdot 10^1$	$2.9308 \cdot 10^3$
10.8	$1.9062 \cdot 10^2$	$2.4110 \cdot 10^3$	14.8	$8.7992 \cdot 10^1$	$2.9396 \cdot 10^3$
10.9	$1.8689 \cdot 10^2$	$2.4299 \cdot 10^3$	14.9	$8.6386 \cdot 10^1$	$2.9484 \cdot 10^3$
11.0	$1.8323 \cdot 10^2$	$2.4484 \cdot 10^3$	15.0	$8.4813 \cdot 10^1$	$2.9569 \cdot 10^3$
11.1	$1.7963 \cdot 10^2$	$2.4666 \cdot 10^3$	15.1	$8.3274 \cdot 10^1$	$2.9653 \cdot 10^3$
11.2	$1.7611 \cdot 10^2$	$2.4844 \cdot 10^3$	15.2	$8.1767 \cdot 10^1$	$2.9736 \cdot 10^3$
11.3	$1.7266 \cdot 10^2$	$2.5018 \cdot 10^3$	15.3	$8.0292 \cdot 10^1$	$2.9817 \cdot 10^3$
11.4	$1.6928 \cdot 10^2$	$2.5189 \cdot 10^3$	15.4	$7.8848 \cdot 10^1$	$2.9896 \cdot 10^3$

Table 2.2. (*continued*)

temperature 550 K

λ [μm]	$m_{\lambda,bb}$ [W/μm m²]	$M_{T,o-\lambda,bb}$ [W/m²]	λ [μm]	$m_{\lambda,bb}$ [W/μm m²]	$M_{T,o-\lambda,bb}$ [W/m²]
0.4	$1.4461 \cdot 10^{-18}$	$9.2630 \cdot 10^{-21}$			
0.5	$2.2706 \cdot 10^{-13}$	$2.2993 \cdot 10^{-15}$	4.0	$5.2863 \cdot 10^{2}$	$5.2346 \cdot 10^{2}$
0.6	$5.5881 \cdot 10^{-10}$	$8.2440 \cdot 10^{-12}$	4.1	$5.4817 \cdot 10^{2}$	$5.7732 \cdot 10^{2}$
0.7	$1.3108 \cdot 10^{-7}$	$2.6632 \cdot 10^{-9}$	4.2	$5.6583 \cdot 10^{2}$	$6.3303 \cdot 10^{2}$
0.8	$7.1836 \cdot 10^{-6}$	$1.9289 \cdot 10^{-7}$	4.3	$5.8161 \cdot 10^{2}$	$6.9042 \cdot 10^{2}$
0.9	$1.5083 \cdot 10^{-4}$	$5.1867 \cdot 10^{-6}$	4.4	$5.9553 \cdot 10^{2}$	$7.4929 \cdot 10^{2}$
1.0	$1.6295 \cdot 10^{-3}$	$7.0000 \cdot 10^{-5}$	4.5	$6.0764 \cdot 10^{2}$	$8.0947 \cdot 10^{2}$
1.1	$1.0912 \cdot 10^{-2}$	$5.7397 \cdot 10^{-4}$	4.6	$6.1799 \cdot 10^{2}$	$8.7076 \cdot 10^{2}$
1.2	$5.1244 \cdot 10^{-2}$	$3.2462 \cdot 10^{-3}$	4.7	$6.2664 \cdot 10^{2}$	$9.3300 \cdot 10^{2}$
1.3	$1.8370 \cdot 10^{-1}$	$1.3821 \cdot 10^{-2}$	4.8	$6.3367 \cdot 10^{2}$	$9.9603 \cdot 10^{2}$
1.4	$5.3384 \cdot 10^{-1}$	$4.7143 \cdot 10^{-2}$	4.9	$6.3917 \cdot 10^{2}$	$1.0597 \cdot 10^{3}$
1.5	$1.3140 \cdot 10^{0}$	$1.3482 \cdot 10^{-1}$	5.0	$6.4321 \cdot 10^{2}$	$1.1238 \cdot 10^{3}$
1.6	$2.8303 \cdot 10^{0}$	$3.3439 \cdot 10^{-1}$	5.1	$6.4589 \cdot 10^{2}$	$1.1883 \cdot 10^{3}$
1.7	$5.4685 \cdot 10^{0}$	$7.3821 \cdot 10^{-1}$	5.2	$6.4729 \cdot 10^{2}$	$1.2530 \cdot 10^{3}$
1.8	$9.6611 \cdot 10^{0}$	$1.4799 \cdot 10^{0}$	5.3	$6.4751 \cdot 10^{2}$	$1.3177 \cdot 10^{3}$
1.9	$1.5842 \cdot 10^{1}$	$2.7367 \cdot 10^{0}$	5.4	$6.4663 \cdot 10^{2}$	$1.3824 \cdot 10^{3}$
2.0	$2.4401 \cdot 10^{1}$	$4.7277 \cdot 10^{0}$	5.5	$6.4473 \cdot 10^{2}$	$1.4470 \cdot 10^{3}$
2.1	$3.5642 \cdot 10^{1}$	$7.7064 \cdot 10^{0}$	5.6	$6.4191 \cdot 10^{2}$	$1.5113 \cdot 10^{3}$
2.2	$4.9758 \cdot 10^{1}$	$1.1952 \cdot 10^{1}$	5.7	$6.3824 \cdot 10^{2}$	$1.5753 \cdot 10^{3}$
2.3	$6.6813 \cdot 10^{1}$	$1.7756 \cdot 10^{1}$	5.8	$6.3379 \cdot 10^{2}$	$1.6390 \cdot 10^{3}$
2.4	$8.6749 \cdot 10^{1}$	$2.5411 \cdot 10^{1}$	5.9	$6.2865 \cdot 10^{2}$	$1.7021 \cdot 10^{3}$
2.5	$1.0939 \cdot 10^{2}$	$3.5196 \cdot 10^{1}$	6.0	$6.2289 \cdot 10^{2}$	$1.7647 \cdot 10^{3}$
2.6	$1.3446 \cdot 10^{2}$	$4.7369 \cdot 10^{1}$	6.1	$6.1656 \cdot 10^{2}$	$1.8266 \cdot 10^{3}$
2.7	$1.6162 \cdot 10^{2}$	$6.2158 \cdot 10^{1}$	6.2	$6.0973 \cdot 10^{2}$	$1.8880 \cdot 10^{3}$
2.8	$1.9046 \cdot 10^{2}$	$7.9749 \cdot 10^{1}$	6.3	$6.0246 \cdot 10^{2}$	$1.9486 \cdot 10^{3}$
2.9	$2.2057 \cdot 10^{2}$	$1.0029 \cdot 10^{2}$	6.4	$5.9481 \cdot 10^{2}$	$2.0084 \cdot 10^{3}$
3.0	$2.5150 \cdot 10^{2}$	$1.2389 \cdot 10^{2}$	6.5	$5.8681 \cdot 10^{2}$	$2.0675 \cdot 10^{3}$
3.1	$2.8282 \cdot 10^{2}$	$1.5060 \cdot 10^{2}$	6.6	$5.7853 \cdot 10^{2}$	$2.1258 \cdot 10^{3}$
3.2	$3.1414 \cdot 10^{2}$	$1.8045 \cdot 10^{2}$	6.7	$5.7000 \cdot 10^{2}$	$2.1832 \cdot 10^{3}$
3.3	$3.4509 \cdot 10^{2}$	$2.1342 \cdot 10^{2}$	6.8	$5.6127 \cdot 10^{2}$	$2.2398 \cdot 10^{3}$
3.4	$3.7532 \cdot 10^{2}$	$2.4945 \cdot 10^{2}$	6.9	$5.5237 \cdot 10^{2}$	$2.2955 \cdot 10^{3}$
3.5	$4.0455 \cdot 10^{2}$	$2.8845 \cdot 10^{2}$	7.0	$5.4333 \cdot 10^{2}$	$2.3502 \cdot 10^{3}$
3.6	$4.3254 \cdot 10^{2}$	$3.3032 \cdot 10^{2}$	7.1	$5.3419 \cdot 10^{2}$	$2.4041 \cdot 10^{3}$
3.7	$4.5908 \cdot 10^{2}$	$3.7491 \cdot 10^{2}$	7.2	$5.2497 \cdot 10^{2}$	$2.4571 \cdot 10^{3}$
3.8	$4.8402 \cdot 10^{2}$	$4.2208 \cdot 10^{2}$	7.3	$5.1571 \cdot 10^{2}$	$2.5091 \cdot 10^{3}$
3.9	$5.0723 \cdot 10^{2}$	$4.7166 \cdot 10^{2}$	7.4	$5.0641 \cdot 10^{2}$	$2.5602 \cdot 10^{3}$

Table 2.2. (*continued*)

λ [µm]	$m_{\lambda,bb}$ [W/µm m²]	$M_{T,o-\lambda,bb}$ [W/m²]	λ [µm]	$m_{\lambda,bb}$ [W/µm m²]	$M_{T,o-\lambda,bb}$ [W/m²]
7.5	$4.9711 \cdot 10^2$	$2.6104 \cdot 10^3$	11.5	$2.1321 \cdot 10^2$	$3.9625 \cdot 10^3$
7.6	$4.8783 \cdot 10^2$	$2.6596 \cdot 10^3$	11.6	$2.0869 \cdot 10^2$	$3.9836 \cdot 10^3$
7.7	$4.7858 \cdot 10^2$	$2.7080 \cdot 10^3$	11.7	$2.0428 \cdot 10^2$	$4.0042 \cdot 10^3$
7.8	$4.6938 \cdot 10^2$	$2.7554 \cdot 10^3$	11.8	$1.9997 \cdot 10^2$	$4.0244 \cdot 10^3$
7.9	$4.6023 \cdot 10^2$	$2.8018 \cdot 10^3$	11.9	$1.9577 \cdot 10^2$	$4.0442 \cdot 10^3$
8.0	$4.5117 \cdot 10^2$	$2.8474 \cdot 10^3$	12.0	$1.9166 \cdot 10^2$	$4.0636 \cdot 10^3$
8.1	$4.4218 \cdot 10^2$	$2.8921 \cdot 10^3$	12.1	$1.8765 \cdot 10^2$	$4.0825 \cdot 10^3$
8.2	$4.3329 \cdot 10^2$	$2.9358 \cdot 10^3$	12.2	$1.8373 \cdot 10^2$	$4.1011 \cdot 10^3$
8.3	$4.2451 \cdot 10^2$	$2.9787 \cdot 10^3$	12.3	$1.7990 \cdot 10^2$	$4.1193 \cdot 10^3$
8.4	$4.1583 \cdot 10^2$	$3.0208 \cdot 10^3$	12.4	$1.7616 \cdot 10^2$	$4.1371 \cdot 10^3$
8.5	$4.0727 \cdot 10^2$	$3.0619 \cdot 10^3$	12.5	$1.7252 \cdot 10^2$	$4.1545 \cdot 10^3$
8.6	$3.9883 \cdot 10^2$	$3.1022 \cdot 10^3$	12.6	$1.6895 \cdot 10^2$	$4.1716 \cdot 10^3$
8.7	$3.9053 \cdot 10^2$	$3.1417 \cdot 10^3$	12.7	$1.6547 \cdot 10^2$	$4.1883 \cdot 10^3$
8.8	$3.8235 \cdot 10^2$	$3.1803 \cdot 10^3$	12.8	$1.6207 \cdot 10^2$	$4.2047 \cdot 10^3$
8.9	$3.7430 \cdot 10^2$	$3.2181 \cdot 10^3$	12.9	$1.5876 \cdot 10^2$	$4.2207 \cdot 10^3$
9.0	$3.6640 \cdot 10^2$	$3.2552 \cdot 10^3$	13.0	$1.5551 \cdot 10^2$	$4.2364 \cdot 10^3$
9.1	$3.5863 \cdot 10^2$	$3.2914 \cdot 10^3$	13.1	$1.5235 \cdot 10^2$	$4.2518 \cdot 10^3$
9.2	$3.5100 \cdot 10^2$	$3.3269 \cdot 10^3$	13.2	$1.4926 \cdot 10^2$	$4.2669 \cdot 10^3$
9.3	$3.4351 \cdot 10^2$	$3.3616 \cdot 10^3$	13.3	$1.4624 \cdot 10^2$	$4.2817 \cdot 10^3$
9.4	$3.3617 \cdot 10^2$	$3.3956 \cdot 10^3$	13.4	$1.4329 \cdot 10^2$	$4.2961 \cdot 10^3$
9.5	$3.2897 \cdot 10^2$	$3.4289 \cdot 10^3$	13.5	$1.4041 \cdot 10^2$	$4.3103 \cdot 10^3$
9.6	$3.2191 \cdot 10^2$	$3.4614 \cdot 10^3$	13.6	$1.3760 \cdot 10^2$	$4.3242 \cdot 10^3$
9.7	$3.1499 \cdot 10^2$	$3.4933 \cdot 10^3$	13.7	$1.3485 \cdot 10^2$	$4.3379 \cdot 10^3$
9.8	$3.0822 \cdot 10^2$	$3.5244 \cdot 10^3$	13.8	$1.3217 \cdot 10^2$	$4.3512 \cdot 10^3$
9.9	$3.0158 \cdot 10^2$	$3.5549 \cdot 10^3$	13.9	$1.2955 \cdot 10^2$	$4.3643 \cdot 10^3$
10.0	$2.9509 \cdot 10^2$	$3.5847 \cdot 10^3$	14.0	$1.2699 \cdot 10^2$	$4.3771 \cdot 10^3$
10.1	$2.8873 \cdot 10^2$	$3.6139 \cdot 10^3$	14.1	$1.2448 \cdot 10^2$	$4.3897 \cdot 10^3$
10.2	$2.8251 \cdot 10^2$	$3.6425 \cdot 10^3$	14.2	$1.2204 \cdot 10^2$	$4.4020 \cdot 10^3$
10.3	$2.7642 \cdot 10^2$	$3.6704 \cdot 10^3$	14.3	$1.1965 \cdot 10^2$	$4.4141 \cdot 10^3$
10.4	$2.7047 \cdot 10^2$	$3.6978 \cdot 10^3$	14.4	$1.1732 \cdot 10^2$	$4.4259 \cdot 10^3$
10.5	$2.6465 \cdot 10^2$	$3.7245 \cdot 10^3$	14.5	$1.1504 \cdot 10^2$	$4.4375 \cdot 10^3$
10.6	$2.5896 \cdot 10^2$	$3.7507 \cdot 10^3$	14.6	$1.1281 \cdot 10^2$	$4.4489 \cdot 10^3$
10.7	$2.5340 \cdot 10^2$	$3.7763 \cdot 10^3$	14.7	$1.1063 \cdot 10^2$	$4.4601 \cdot 10^3$
10.8	$2.4796 \cdot 10^2$	$3.8014 \cdot 10^3$	14.8	$1.0851 \cdot 10^2$	$4.4711 \cdot 10^3$
10.9	$2.4264 \cdot 10^2$	$3.8259 \cdot 10^3$	14.9	$1.0643 \cdot 10^2$	$4.4818 \cdot 10^3$
11.0	$2.3745 \cdot 10^2$	$3.8499 \cdot 10^3$	15.0	$1.0440 \cdot 10^2$	$4.4923 \cdot 10^3$
11.1	$2.3237 \cdot 10^2$	$3.8734 \cdot 10^3$	15.1	$1.0241 \cdot 10^2$	$4.5027 \cdot 10^3$
11.2	$2.2741 \cdot 10^2$	$3.8964 \cdot 10^3$	15.2	$1.0047 \cdot 10^2$	$4.5128 \cdot 10^3$
11.3	$2.2257 \cdot 10^2$	$3.9189 \cdot 10^3$	15.3	$9.8571 \cdot 10^1$	$4.5228 \cdot 10^3$
11.4	$2.1783 \cdot 10^2$	$3.9409 \cdot 10^3$	15.4	$9.6715 \cdot 10^1$	$4.5325 \cdot 10^3$

Table 2.2. (*continued*)

temperature 600 K

λ [μm]	$m_{\lambda,bb}$ [W/μm m²]	$M_{T,o-\lambda,bb}$ [W/m²]	λ [μm]	$m_{\lambda,bb}$ [W/μm m²]	$M_{T,o-\lambda,bb}$ [W/m²]
0.4	$3.3658 \cdot 10^{-16}$	$2.3619 \cdot 10^{-18}$			
0.5	$1.7769 \cdot 10^{-11}$	$1.9733 \cdot 10^{-13}$	4.0	$9.1263 \cdot 10^{2}$	$1.0307 \cdot 10^{3}$
0.6	$2.1144 \cdot 10^{-8}$	$3.4248 \cdot 10^{-10}$	4.1	$9.3401 \cdot 10^{2}$	$1.1230 \cdot 10^{3}$
0.7	$2.9515 \cdot 10^{-6}$	$6.5910 \cdot 10^{-8}$	4.2	$9.5211 \cdot 10^{2}$	$1.2174 \cdot 10^{3}$
0.8	$1.0959 \cdot 10^{-4}$	$3.2379 \cdot 10^{-6}$	4.3	$9.6708 \cdot 10^{2}$	$1.3133 \cdot 10^{3}$
0.9	$1.7000 \cdot 10^{-3}$	$6.4392 \cdot 10^{-5}$	4.4	$9.7906 \cdot 10^{2}$	$1.4107 \cdot 10^{3}$
1.0	$1.4415 \cdot 10^{-2}$	$6.8285 \cdot 10^{-4}$	4.5	$9.8821 \cdot 10^{2}$	$1.5091 \cdot 10^{3}$
1.1	$7.9176 \cdot 10^{-2}$	$4.5977 \cdot 10^{-3}$	4.6	$9.9470 \cdot 10^{2}$	$1.6082 \cdot 10^{3}$
1.2	$3.1522 \cdot 10^{-1}$	$2.2069 \cdot 10^{-2}$	4.7	$9.9871 \cdot 10^{2}$	$1.7079 \cdot 10^{3}$
1.3	$9.8260 \cdot 10^{-1}$	$8.1799 \cdot 10^{-2}$	4.8	$1.0004 \cdot 10^{3}$	$1.8079 \cdot 10^{3}$
1.4	$2.5332 \cdot 10^{0}$	$2.4779 \cdot 10^{-1}$	4.9	$1.0000 \cdot 10^{3}$	$1.9079 \cdot 10^{3}$
1.5	$5.6204 \cdot 10^{0}$	$6.3947 \cdot 10^{-1}$	5.0	$9.9766 \cdot 10^{2}$	$2.0078 \cdot 10^{3}$
1.6	$1.1055 \cdot 10^{1}$	$1.4500 \cdot 10^{0}$	5.1	$9.9353 \cdot 10^{2}$	$2.1074 \cdot 10^{3}$
1.7	$1.9714 \cdot 10^{1}$	$2.9580 \cdot 10^{0}$	5.2	$9.8778 \cdot 10^{2}$	$2.2065 \cdot 10^{3}$
1.8	$3.2434 \cdot 10^{1}$	$5.5284 \cdot 10^{0}$	5.3	$9.8058 \cdot 10^{2}$	$2.3049 \cdot 10^{3}$
1.9	$4.9901 \cdot 10^{1}$	$9.6032 \cdot 10^{0}$	5.4	$9.7207 \cdot 10^{2}$	$2.4026 \cdot 10^{3}$
2.0	$7.2575 \cdot 10^{1}$	$1.5682 \cdot 10^{1}$	5.5	$9.6239 \cdot 10^{2}$	$2.4993 \cdot 10^{3}$
2.1	$1.0065 \cdot 10^{2}$	$2.4298 \cdot 10^{1}$	5.6	$9.5169 \cdot 10^{2}$	$2.5950 \cdot 10^{3}$
2.2	$1.3403 \cdot 10^{2}$	$3.5989 \cdot 10^{1}$	5.7	$9.4007 \cdot 10^{2}$	$2.6896 \cdot 10^{3}$
2.3	$1.7239 \cdot 10^{2}$	$5.1270 \cdot 10^{1}$	5.8	$9.2766 \cdot 10^{2}$	$2.7830 \cdot 10^{3}$
2.4	$2.1516 \cdot 10^{2}$	$7.0613 \cdot 10^{1}$	5.9	$9.1457 \cdot 10^{2}$	$2.8751 \cdot 10^{3}$
2.5	$2.6163 \cdot 10^{2}$	$9.4425 \cdot 10^{1}$	6.0	$9.0089 \cdot 10^{2}$	$2.9659 \cdot 10^{3}$
2.6	$3.1100 \cdot 10^{2}$	$1.2304 \cdot 10^{2}$	6.1	$8.8672 \cdot 10^{2}$	$3.0553 \cdot 10^{3}$
2.7	$3.6239 \cdot 10^{2}$	$1.5669 \cdot 10^{2}$	6.2	$8.7213 \cdot 10^{2}$	$3.1432 \cdot 10^{3}$
2.8	$4.1494 \cdot 10^{2}$	$1.9555 \cdot 10^{2}$	6.3	$8.5721 \cdot 10^{2}$	$3.2297 \cdot 10^{3}$
2.9	$4.6781 \cdot 10^{2}$	$2.3969 \cdot 10^{2}$	6.4	$8.4202 \cdot 10^{2}$	$3.3146 \cdot 10^{3}$
3.0	$5.2023 \cdot 10^{2}$	$2.8910 \cdot 10^{2}$	6.5	$8.2664 \cdot 10^{2}$	$3.3981 \cdot 10^{3}$
3.1	$5.7150 \cdot 10^{2}$	$3.4370 \cdot 10^{2}$	6.6	$8.1111 \cdot 10^{2}$	$3.4800 \cdot 10^{3}$
3.2	$6.2103 \cdot 10^{2}$	$4.0334 \cdot 10^{2}$	6.7	$7.9549 \cdot 10^{2}$	$3.5603 \cdot 10^{3}$
3.3	$6.6830 \cdot 10^{2}$	$4.6783 \cdot 10^{2}$	6.8	$7.7982 \cdot 10^{2}$	$3.6391 \cdot 10^{3}$
3.4	$7.1292 \cdot 10^{2}$	$5.3691 \cdot 10^{2}$	6.9	$7.6415 \cdot 10^{2}$	$3.7162 \cdot 10^{3}$
3.5	$7.5456 \cdot 10^{2}$	$6.1031 \cdot 10^{2}$	7.0	$7.4852 \cdot 10^{2}$	$3.7919 \cdot 10^{3}$
3.6	$7.9299 \cdot 10^{2}$	$6.8771 \cdot 10^{2}$	7.1	$7.3296 \cdot 10^{2}$	$3.8660 \cdot 10^{3}$
3.7	$8.2807 \cdot 10^{2}$	$7.6879 \cdot 10^{2}$	7.2	$7.1749 \cdot 10^{2}$	$3.9385 \cdot 10^{3}$
3.8	$8.5971 \cdot 10^{2}$	$8.5321 \cdot 10^{2}$	7.3	$7.0216 \cdot 10^{2}$	$4.0095 \cdot 10^{3}$
3.9	$8.8789 \cdot 10^{2}$	$9.4062 \cdot 10^{2}$	7.4	$6.8697 \cdot 10^{2}$	$4.0789 \cdot 10^{3}$

Table 2.2. (*continued*)

λ [µm]	$m_{\lambda,bb}$ [W/µm m²]	$M_{T,o-\lambda,bb}$ [W/m²]	λ [µm]	$m_{\lambda,bb}$ [W/µm m²]	$M_{T,o-\lambda,bb}$ [W/m²]
7.5	$6.7195 \cdot 10^2$	$4.1469 \cdot 10^3$	11.5	$2.6403 \cdot 10^2$	$5.8950 \cdot 10^3$
7.6	$6.5711 \cdot 10^2$	$4.2133 \cdot 10^3$	11.6	$2.5809 \cdot 10^2$	$5.9211 \cdot 10^3$
7.7	$6.4248 \cdot 10^2$	$4.2783 \cdot 10^3$	11.7	$2.5231 \cdot 10^2$	$5.9467 \cdot 10^3$
7.8	$6.2806 \cdot 10^2$	$4.3418 \cdot 10^3$	11.8	$2.4667 \cdot 10^2$	$5.9716 \cdot 10^3$
7.9	$6.1387 \cdot 10^2$	$4.4039 \cdot 10^3$	11.9	$2.4118 \cdot 10^2$	$5.9960 \cdot 10^3$
8.0	$5.9991 \cdot 10^2$	$4.4646 \cdot 10^3$	12.0	$2.3583 \cdot 10^2$	$6.0198 \cdot 10^3$
8.1	$5.8620 \cdot 10^2$	$4.5239 \cdot 10^3$	12.1	$2.3061 \cdot 10^2$	$6.0432 \cdot 10^3$
8.2	$5.7274 \cdot 10^2$	$4.5818 \cdot 10^3$	12.2	$2.2553 \cdot 10^2$	$6.0660 \cdot 10^3$
8.3	$5.5953 \cdot 10^2$	$4.6384 \cdot 10^3$	12.3	$2.2058 \cdot 10^2$	$6.0883 \cdot 10^3$
8.4	$5.4657 \cdot 10^2$	$4.6937 \cdot 10^3$	12.4	$2.1575 \cdot 10^2$	$6.1101 \cdot 10^3$
8.5	$5.3388 \cdot 10^2$	$4.7478 \cdot 10^3$	12.5	$2.1104 \cdot 10^2$	$6.1314 \cdot 10^3$
8.6	$5.2145 \cdot 10^2$	$4.8005 \cdot 10^3$	12.6	$2.0646 \cdot 10^2$	$6.1523 \cdot 10^3$
8.7	$5.0928 \cdot 10^2$	$4.8520 \cdot 10^3$	12.7	$2.0199 \cdot 10^2$	$6.1727 \cdot 10^3$
8.8	$4.9737 \cdot 10^2$	$4.9024 \cdot 10^3$	12.8	$1.9763 \cdot 10^2$	$6.1927 \cdot 10^3$
8.9	$4.8572 \cdot 10^2$	$4.9515 \cdot 10^3$	12.9	$1.9338 \cdot 10^2$	$6.2122 \cdot 10^3$
9.0	$4.7433 \cdot 10^2$	$4.9995 \cdot 10^3$	13.0	$1.8924 \cdot 10^2$	$6.2314 \cdot 10^3$
9.1	$4.6320 \cdot 10^2$	$5.0464 \cdot 10^3$	13.1	$1.8520 \cdot 10^2$	$6.2501 \cdot 10^3$
9.2	$4.5233 \cdot 10^2$	$5.0922 \cdot 10^3$	13.2	$1.8126 \cdot 10^2$	$6.2684 \cdot 10^3$
9.3	$4.4171 \cdot 10^2$	$5.1369 \cdot 10^3$	13.3	$1.7743 \cdot 10^2$	$6.2863 \cdot 10^3$
9.4	$4.3134 \cdot 10^2$	$5.1805 \cdot 10^3$	13.4	$1.7368 \cdot 10^2$	$6.3039 \cdot 10^3$
9.5	$4.2121 \cdot 10^2$	$5.2231 \cdot 10^3$	13.5	$1.7003 \cdot 10^2$	$6.3211 \cdot 10^3$
9.6	$4.1133 \cdot 10^2$	$5.2648 \cdot 10^3$	13.6	$1.6647 \cdot 10^2$	$6.3379 \cdot 10^3$
9.7	$4.0169 \cdot 10^2$	$5.3054 \cdot 10^3$	13.7	$1.6300 \cdot 10^2$	$6.3544 \cdot 10^3$
9.8	$3.9228 \cdot 10^2$	$5.3451 \cdot 10^3$	13.8	$1.5962 \cdot 10^2$	$6.3705 \cdot 10^3$
9.9	$3.8310 \cdot 10^2$	$5.3839 \cdot 10^3$	13.9	$1.5631 \cdot 10^2$	$6.3863 \cdot 10^3$
10.0	$3.7415 \cdot 10^2$	$5.4217 \cdot 10^3$	14.0	$1.5309 \cdot 10^2$	$6.4017 \cdot 10^3$
10.1	$3.6543 \cdot 10^2$	$5.4587 \cdot 10^3$	14.1	$1.4995 \cdot 10^2$	$6.4169 \cdot 10^3$
10.2	$3.5692 \cdot 10^2$	$5.4948 \cdot 10^3$	14.2	$1.4688 \cdot 10^2$	$6.4317 \cdot 10^3$
10.3	$3.4862 \cdot 10^2$	$5.5301 \cdot 10^3$	14.3	$1.4389 \cdot 10^2$	$6.4463 \cdot 10^3$
10.4	$3.4053 \cdot 10^2$	$5.5645 \cdot 10^3$	14.4	$1.4097 \cdot 10^2$	$6.4605 \cdot 10^3$
10.5	$3.3265 \cdot 10^2$	$5.5982 \cdot 10^3$	14.5	$1.3812 \cdot 10^2$	$6.4745 \cdot 10^3$
10.6	$3.2496 \cdot 10^2$	$5.6311 \cdot 10^3$	14.6	$1.3534 \cdot 10^2$	$6.4881 \cdot 10^3$
10.7	$3.1747 \cdot 10^2$	$5.6632 \cdot 10^3$	14.7	$1.3262 \cdot 10^2$	$6.5015 \cdot 10^3$
10.8	$3.1017 \cdot 10^2$	$5.6946 \cdot 10^3$	14.8	$1.2997 \cdot 10^2$	$6.5147 \cdot 10^3$
10.9	$3.0306 \cdot 10^2$	$5.7252 \cdot 10^3$	14.9	$1.2739 \cdot 10^2$	$6.5275 \cdot 10^3$
11.0	$2.9612 \cdot 10^2$	$5.7552 \cdot 10^3$	15.0	$1.2486 \cdot 10^2$	$6\ 5401 \cdot 10^3$
11.1	$2.8937 \cdot 10^2$	$5.7845 \cdot 10^3$	15.1	$1.2240 \cdot 10^2$	$6.5525 \cdot 10^3$
11.2	$2.8278 \cdot 10^2$	$5.8131 \cdot 10^3$	15.2	$1.1999 \cdot 10^2$	$6.5646 \cdot 10^3$
11.3	$2.7637 \cdot 10^2$	$5.8410 \cdot 10^3$	15.3	$1.1764 \cdot 10^2$	$6.5765 \cdot 10^3$
11.4	$2.7012 \cdot 10^2$	$5.8683 \cdot 10^3$	15.4	$1.1535 \cdot 10^2$	$6.5881 \cdot 10^3$

Table 2.2. (*continued*)

temperature 650 K

λ [µm]	$m_{\lambda,bb}$ [W/µm m²]	$M_{T,o-\lambda,bb}$ [W/m²]	λ [µm]	$m_{\lambda,bb}$ [W/µm m²]	$M_{T,o-\lambda,bb}$ [W/m²]
0.4	$3.3870 \cdot 10^{-14}$	$2.5859 \cdot 10^{-16}$			
0.5	$7.1099 \cdot 10^{-10}$	$8.5993 \cdot 10^{-12}$	4.0	$1.4495 \cdot 10^{3}$	$1.8535 \cdot 10^{3}$
0.6	$4.5746 \cdot 10^{-7}$	$8.0786 \cdot 10^{-9}$	4.1	$1.4671 \cdot 10^{3}$	$1.9993 \cdot 10^{3}$
0.7	$4.1160 \cdot 10^{-5}$	$1.0032 \cdot 10^{-6}$	4.2	$1.4799 \cdot 10^{3}$	$2.1467 \cdot 10^{3}$
0.8	$1.0994 \cdot 10^{-3}$	$3.5492 \cdot 10^{-5}$	4.3	$1.4882 \cdot 10^{3}$	$2.2951 \cdot 10^{3}$
0.9	$1.3200 \cdot 10^{-2}$	$5.4692 \cdot 10^{-4}$	4.4	$1.4923 \cdot 10^{3}$	$2.4442 \cdot 10^{3}$
1.0	$9.1181 \cdot 10^{-2}$	$4.7303 \cdot 10^{-3}$	4.5	$1.4926 \cdot 10^{3}$	$2.5935 \cdot 10^{3}$
1.1	$4.2351 \cdot 10^{-1}$	$2.6962 \cdot 10^{-2}$	4.6	$1.4894 \cdot 10^{3}$	$2.7426 \cdot 10^{3}$
1.2	$1.4662 \cdot 10^{0}$	$1.1267 \cdot 10^{-1}$	4.7	$1.4831 \cdot 10^{3}$	$2.8913 \cdot 10^{3}$
1.3	$4.0608 \cdot 10^{0}$	$3.7145 \cdot 10^{-1}$	4.8	$1.4739 \cdot 10^{3}$	$3.0391 \cdot 10^{3}$
1.4	$9.4597 \cdot 10^{0}$	$1.0179 \cdot 10^{0}$	4.9	$1.4622 \cdot 10^{3}$	$3.1860 \cdot 10^{3}$
1.5	$1.9223 \cdot 10^{1}$	$2.4088 \cdot 10^{0}$	5.0	$1.4482 \cdot 10^{3}$	$3.3315 \cdot 10^{3}$
1.6	$3.5014 \cdot 10^{1}$	$5.0636 \cdot 10^{0}$	5.1	$1.4322 \cdot 10^{3}$	$3.4755 \cdot 10^{3}$
1.7	$5.8347 \cdot 10^{1}$	$9.6633 \cdot 10^{0}$	5.2	$1.4144 \cdot 10^{3}$	$3.6179 \cdot 10^{3}$
1.8	$9.0377 \cdot 10^{1}$	$1.7023 \cdot 10^{1}$	5.3	$1.3951 \cdot 10^{3}$	$3.7583 \cdot 10^{3}$
1.9	$1.3175 \cdot 10^{2}$	$2.8051 \cdot 10^{1}$	5.4	$1.3746 \cdot 10^{3}$	$3.8968 \cdot 10^{3}$
2.0	$1.8253 \cdot 10^{2}$	$4.3687 \cdot 10^{1}$	5.5	$1.3529 \cdot 10^{3}$	$4.0332 \cdot 10^{3}$
2.1	$2.4226 \cdot 10^{2}$	$6.4856 \cdot 10^{1}$	5.6	$1.3302 \cdot 10^{3}$	$4.1674 \cdot 10^{3}$
2.2	$3.1000 \cdot 10^{2}$	$9.2407 \cdot 10^{1}$	5.7	$1.3068 \cdot 10^{3}$	$4.2992 \cdot 10^{3}$
2.3	$3.8444 \cdot 10^{2}$	$1.2708 \cdot 10^{2}$	5.8	$1.2828 \cdot 10^{3}$	$4.4287 \cdot 10^{3}$
2.4	$4.6406 \cdot 10^{2}$	$1.6947 \cdot 10^{2}$	5.9	$1.2583 \cdot 10^{3}$	$4.5558 \cdot 10^{3}$
2.5	$5.4723 \cdot 10^{2}$	$2.2001 \cdot 10^{2}$	6.0	$1.2335 \cdot 10^{3}$	$4.6804 \cdot 10^{3}$
2.6	$6.3230 \cdot 10^{2}$	$2.7897 \cdot 10^{2}$	6.1	$1.2084 \cdot 10^{3}$	$4.8025 \cdot 10^{3}$
2.7	$7.1770 \cdot 10^{2}$	$3.4648 \cdot 10^{2}$	6.2	$1.1831 \cdot 10^{3}$	$4.9220 \cdot 10^{3}$
2.8	$8.0199 \cdot 10^{2}$	$4.2248 \cdot 10^{2}$	6.3	$1.1578 \cdot 10^{3}$	$5.0391 \cdot 10^{3}$
2.9	$8.8391 \cdot 10^{2}$	$5.0679 \cdot 10^{2}$	6.4	$1.1324 \cdot 10^{3}$	$5.1536 \cdot 10^{3}$
3.0	$9.6239 \cdot 10^{2}$	$5.9914 \cdot 10^{2}$	6.5	$1.1072 \cdot 10^{3}$	$5.2656 \cdot 10^{3}$
3.1	$1.0366 \cdot 10^{3}$	$6.9913 \cdot 10^{2}$	6.6	$1.0821 \cdot 10^{3}$	$5.3750 \cdot 10^{3}$
3.2	$1.1057 \cdot 10^{3}$	$8.0628 \cdot 10^{2}$	6.7	$1.0572 \cdot 10^{3}$	$5.4820 \cdot 10^{3}$
3.3	$1.1694 \cdot 10^{3}$	$9.2008 \cdot 10^{2}$	6.8	$1.0326 \cdot 10^{3}$	$5.5865 \cdot 10^{3}$
3.4	$1.2272 \cdot 10^{3}$	$1.0400 \cdot 10^{3}$	6.9	$1.0082 \cdot 10^{3}$	$5.6885 \cdot 10^{3}$
3.5	$1.2791 \cdot 10^{3}$	$1.1653 \cdot 10^{3}$	7.0	$9.8417 \cdot 10^{2}$	$5.7881 \cdot 10^{3}$
3.6	$1.3249 \cdot 10^{3}$	$1.2956 \cdot 10^{3}$	7.1	$9.6048 \cdot 10^{2}$	$5.8853 \cdot 10^{3}$
3.7	$1.3646 \cdot 10^{3}$	$1.4301 \cdot 10^{3}$	7.2	$9.3716 \cdot 10^{2}$	$5.9802 \cdot 10^{3}$
3.8	$1.3985 \cdot 10^{3}$	$1.5683 \cdot 10^{3}$	7.3	$9.1424 \cdot 10^{2}$	$6.0728 \cdot 10^{3}$
3.9	$1.4267 \cdot 10^{3}$	$1.7096 \cdot 10^{3}$	7.4	$8.9172 \cdot 10^{2}$	$6.1631 \cdot 10^{3}$

λ [μm]	$m_{\lambda,bb}$ [W/μm m²]	$M_{T,o-\lambda,bb}$ [W/m²]	λ [μm]	$m_{\lambda,bb}$ [W/μm m²]	$M_{T,o-\lambda,bb}$ [W/m²]
7.5	$8.6964 \cdot 10^2$	$6.2511 \cdot 10^3$	11.5	$3.1781 \cdot 10^2$	$8.4330 \cdot 10^3$
7.6	$8.4799 \cdot 10^2$	$6.3370 \cdot 10^3$	11.6	$3.1032 \cdot 10^2$	$8.4644 \cdot 10^3$
7.7	$8.2679 \cdot 10^2$	$6.4207 \cdot 10^3$	11.7	$3.0304 \cdot 10^2$	$8.4951 \cdot 10^3$
7.8	$8.0605 \cdot 10^2$	$6.5024 \cdot 10^3$	11.8	$2.9596 \cdot 10^2$	$8.5250 \cdot 10^3$
7.9	$7.8576 \cdot 10^2$	$6.5820 \cdot 10^3$	11.9	$2.8907 \cdot 10^2$	$8.5543 \cdot 10^3$
8.0	$7.6592 \cdot 10^2$	$6.6595 \cdot 10^3$	12.0	$2.8237 \cdot 10^2$	$8.5829 \cdot 10^3$
8.1	$7.4655 \cdot 10^2$	$6.7352 \cdot 10^3$	12.1	$2.7585 \cdot 10^2$	$8.6108 \cdot 10^3$
8.2	$7.2763 \cdot 10^2$	$6.8089 \cdot 10^3$	12.2	$2.6951 \cdot 10^2$	$8.6380 \cdot 10^3$
8.3	$7.0917 \cdot 10^2$	$6.8807 \cdot 10^3$	12.3	$2.6333 \cdot 10^2$	$8.6647 \cdot 10^3$
8.4	$6.9117 \cdot 10^2$	$6.9507 \cdot 10^3$	12.4	$2.5733 \cdot 10^2$	$8.6907 \cdot 10^3$
8.5	$6.7360 \cdot 10^2$	$7.0189 \cdot 10^3$	12.5	$2.5148 \cdot 10^2$	$8.7161 \cdot 10^3$
8.6	$6.5649 \cdot 10^2$	$7.0854 \cdot 10^3$	12.6	$2.4580 \cdot 10^2$	$8.7410 \cdot 10^3$
8.7	$6.3981 \cdot 10^2$	$7.1502 \cdot 10^3$	12.7	$2.4026 \cdot 10^2$	$8.7653 \cdot 10^3$
8.8	$6.2356 \cdot 10^2$	$7.2134 \cdot 10^3$	12.8	$2.3487 \cdot 10^2$	$8.7890 \cdot 10^3$
8.9	$6.0773 \cdot 10^2$	$7.2750 \cdot 10^3$	12.9	$2.2962 \cdot 10^2$	$8.8123 \cdot 10^3$
9.0	$5.9232 \cdot 10^2$	$7.3350 \cdot 10^3$	13.0	$2.2452 \cdot 10^2$	$8.8350 \cdot 10^3$
9.1	$5.7731 \cdot 10^2$	$7.3934 \cdot 10^3$	13.1	$2.1954 \cdot 10^2$	$8.8572 \cdot 10^3$
9.2	$5.6271 \cdot 10^2$	$7.4504 \cdot 10^3$	13.2	$2.1470 \cdot 10^2$	$8.8789 \cdot 10^3$
9.3	$5.4849 \cdot 10^2$	$7.5060 \cdot 10^3$	13.3	$2.0999 \cdot 10^2$	$8.9001 \cdot 10^3$
9.4	$5.3466 \cdot 10^2$	$7.5601 \cdot 10^3$	13.4	$2.0539 \cdot 10^2$	$8.9209 \cdot 10^3$
9.5	$5.2121 \cdot 10^2$	$7.6129 \cdot 10^3$	13.5	$2.0092 \cdot 10^2$	$8.9412 \cdot 10^3$
9.6	$5.0812 \cdot 10^2$	$7.6644 \cdot 10^3$	13.6	$1.9657 \cdot 10^2$	$8.9611 \cdot 10^3$
9.7	$4.9539 \cdot 10^2$	$7.7146 \cdot 10^3$	13.7	$1.9232 \cdot 10^2$	$8.9805 \cdot 10^3$
9.8	$4.8300 \cdot 10^2$	$7.7635 \cdot 10^3$	13.8	$1.8819 \cdot 10^2$	$8.9995 \cdot 10^3$
9.9	$4.7096 \cdot 10^2$	$7.8112 \cdot 10^3$	13.9	$1.8416 \cdot 10^2$	$9.0181 \cdot 10^3$
10.0	$4.5925 \cdot 10^2$	$7.8577 \cdot 10^3$	14.0	$1.8024 \cdot 10^2$	$9.0364 \cdot 10^3$
10.1	$4.4786 \cdot 10^2$	$7.9030 \cdot 10^3$	14.1	$1.7641 \cdot 10^2$	$9.0542 \cdot 10^3$
10.2	$4.3678 \cdot 10^2$	$7.9472 \cdot 10^3$	14.2	$1.7268 \cdot 10^2$	$9.0716 \cdot 10^3$
10.3	$4.2602 \cdot 10^2$	$7.9904 \cdot 10^3$	14.3	$1.6905 \cdot 10^2$	$9.0887 \cdot 10^3$
10.4	$4.1555 \cdot 10^2$	$8.0325 \cdot 10^3$	14.4	$1.6551 \cdot 10^2$	$9.1054 \cdot 10^3$
10.5	$4.0536 \cdot 10^2$	$8.0735 \cdot 10^3$	14.5	$1.6206 \cdot 10^2$	$9.1218 \cdot 10^3$
10.6	$3.9546 \cdot 10^2$	$8.1135 \cdot 10^3$	14.6	$1.5869 \cdot 10^2$	$9.1379 \cdot 10^3$
10.7	$3.8584 \cdot 10^2$	$8.1526 \cdot 10^3$	14.7	$1.5541 \cdot 10^2$	$9.1536 \cdot 10^3$
10.8	$3.7648 \cdot 10^2$	$8.1907 \cdot 10^3$	14.8	$1.5221 \cdot 10^2$	$9.1689 \cdot 10^3$
10.9	$3.6738 \cdot 10^2$	$8.2279 \cdot 10^3$	14.9	$1.4909 \cdot 10^2$	$9.1840 \cdot 10^3$
11.0	$3.5853 \cdot 10^2$	$8.2642 \cdot 10^3$	15.0	$1.4605 \cdot 10^2$	$9.1988 \cdot 10^3$
11.1	$3.4992 \cdot 10^2$	$8.2996 \cdot 10^3$	15.1	$1.4308 \cdot 10^2$	$9.2132 \cdot 10^3$
11.2	$3.4156 \cdot 10^2$	$8.3342 \cdot 10^3$	15.2	$1.4018 \cdot 10^2$	$9.2274 \cdot 10^3$
11.3	$3.3342 \cdot 10^2$	$8.3679 \cdot 10^3$	15.3	$1.3736 \cdot 10^2$	$9.2412 \cdot 10^3$
11.4	$3.2551 \cdot 10^2$	$8.4009 \cdot 10^3$	15.4	$1.3460 \cdot 10^2$	$9.2548 \cdot 10^3$

Table 2.2. (*continued*)

temperature 700 K

λ [µm]	$m_{\lambda,bb}$ [W/µm m²]	$M_{T,o-\lambda,bb}$ [W/m²]	λ [µm]	$m_{\lambda,bb}$ [W/µm m²]	$M_{T,o-\lambda,bb}$ [W/m²]
0.4	$1.7638\cdot10^{-12}$	$1.4564\cdot10^{-14}$			
0.5	$1.6795\cdot10^{-8}$	$2.1993\cdot10^{-10}$	4.0	$2.1563\cdot10^{3}$	$3.1025\cdot10^{3}$
0.6	$6.3794\cdot10^{-6}$	$1.2211\cdot10^{-7}$	4.1	$2.1620\cdot10^{3}$	$3.3185\cdot10^{3}$
0.7	$3.9392\cdot10^{-4}$	$1.0418\cdot10^{-5}$	4.2	$2.1614\cdot10^{3}$	$3.5347\cdot10^{3}$
0.8	$7.9337\cdot10^{-3}$	$2.7821\cdot10^{-4}$	4.3	$2.1551\cdot10^{3}$	$3.7506\cdot10^{3}$
0.9	$7.6473\cdot10^{-2}$	$3.4457\cdot10^{-3}$	4.4	$2.1436\cdot10^{3}$	$3.9655\cdot10^{3}$
1.0	$4.4316\cdot10^{-1}$	$2.5029\cdot10^{-2}$	4.5	$2.1276\cdot10^{3}$	$4.1791\cdot10^{3}$
1.1	$1.7828\cdot10^{0}$	$1.2370\cdot10^{-1}$	4.6	$2.1074\cdot10^{3}$	$4.3909\cdot10^{3}$
1.2	$5.4752\cdot10^{0}$	$4.5908\cdot10^{-1}$	4.7	$2.0838\cdot10^{3}$	$4.6005\cdot10^{3}$
1.3	$1.3703\cdot10^{1}$	$1.3692\cdot10^{0}$	4.8	$2.0570\cdot10^{3}$	$4.8075\cdot10^{3}$
1.4	$2.9265\cdot10^{1}$	$3.4438\cdot10^{0}$	4.9	$2.0275\cdot10^{3}$	$5.0118\cdot10^{3}$
1.5	$5.5157\cdot10^{1}$	$7.5668\cdot10^{0}$	5.0	$1.9958\cdot10^{3}$	$5.2130\cdot10^{3}$
1.6	$9.4059\cdot10^{1}$	$1.4910\cdot10^{1}$	5.1	$1.9621\cdot10^{3}$	$5.4109\cdot10^{3}$
1.7	$1.4789\cdot10^{2}$	$2.6877\cdot10^{1}$	5.2	$1.9269\cdot10^{3}$	$5.6053\cdot10^{3}$
1.8	$2.1754\cdot10^{2}$	$4.5016\cdot10^{1}$	5.3	$1.8903\cdot10^{3}$	$5.7962\cdot10^{3}$
1.9	$3.0279\cdot10^{2}$	$7.0906\cdot10^{1}$	5.4	$1.8528\cdot10^{3}$	$5.9834\cdot10^{3}$
2.0	$4.0242\cdot10^{2}$	$1.0605\cdot10^{2}$	5.5	$1.8144\cdot10^{3}$	$6.1667\cdot10^{3}$
2.1	$5.1437\cdot10^{2}$	$1.5180\cdot10^{2}$	5.6	$1.7755\cdot10^{3}$	$6.3462\cdot10^{3}$
2.2	$6.3605\cdot10^{2}$	$2.0925\cdot10^{2}$	5.7	$1.7363\cdot10^{3}$	$6.5218\cdot10^{3}$
2.3	$7.6454\cdot10^{2}$	$2.7924\cdot10^{2}$	5.8	$1.6968\cdot10^{3}$	$6.6935\cdot10^{3}$
2.4	$8.9685\cdot10^{2}$	$3.6229\cdot10^{2}$	5.9	$1.6573\cdot10^{3}$	$6.8612\cdot10^{3}$
2.5	$1.0301\cdot10^{3}$	$4.5864\cdot10^{2}$	6.0	$1.6178\cdot10^{3}$	$7.0249\cdot10^{3}$
2.6	$1.1617\cdot10^{3}$	$5.6825\cdot10^{2}$	6.1	$1.5786\cdot10^{3}$	$7.1847\cdot10^{3}$
2.7	$1.2893\cdot10^{3}$	$6.9084\cdot10^{2}$	6.2	$1.5397\cdot10^{3}$	$7.3406\cdot10^{3}$
2.8	$1.4110\cdot10^{3}$	$8.2591\cdot10^{2}$	6.3	$1.5012\cdot10^{3}$	$7.4927\cdot10^{3}$
2.9	$1.5253\cdot10^{3}$	$9.7279\cdot10^{2}$	6.4	$1.4631\cdot10^{3}$	$7.6409\cdot10^{3}$
3.0	$1.6309\cdot10^{3}$	$1.1307\cdot10^{3}$	6.5	$1.4256\cdot10^{3}$	$7.7853\cdot10^{3}$
3.1	$1.7271\cdot10^{3}$	$1.2987\cdot10^{3}$	6.6	$1.3887\cdot10^{3}$	$7.9260\cdot10^{3}$
3.2	$1.8134\cdot10^{3}$	$1.4758\cdot10^{3}$	6.7	$1.3523\cdot10^{3}$	$8.0631\cdot10^{3}$
3.3	$1.8896\cdot10^{3}$	$1.6610\cdot10^{3}$	6.8	$1.3167\cdot10^{3}$	$8.1965\cdot10^{3}$
3.4	$1.9556\cdot10^{3}$	$1.8533\cdot10^{3}$	6.9	$1.2818\cdot10^{3}$	$8.3264\cdot10^{3}$
3.5	$2.0116\cdot10^{3}$	$2.0518\cdot10^{3}$	7.0	$1.2476\cdot10^{3}$	$8.4529\cdot10^{3}$
3.6	$2.0579\cdot10^{3}$	$2.2553\cdot10^{3}$	7.1	$1.2141\cdot10^{3}$	$8.5760\cdot10^{3}$
3.7	$2.0950\cdot10^{3}$	$2.4630\cdot10^{3}$	7.2	$1.1814\cdot10^{3}$	$8.6957\cdot10^{3}$
3.8	$2.1234\cdot10^{3}$	$2.6740\cdot10^{3}$	7.3	$1.1494\cdot10^{3}$	$8.8122\cdot10^{3}$
3.9	$2.1436\cdot10^{3}$	$2.8874\cdot10^{3}$	7.4	$1.1182\cdot10^{3}$	$8.9256\cdot10^{3}$

Table 2.2. (*continued*)

λ [μm]	$m_{\lambda,bb}$ [W/μm m²]	$M_{T,o-\lambda,bb}$ [W/m²]	λ [μm]	$m_{\lambda,bb}$ [W/μm m²]	$M_{T,o-\lambda,bb}$ [W/m²]
7.5	$1.0878 \cdot 10^3$	$9.0359 \cdot 10^3$	11.5	$3.7407 \cdot 10^2$	$1.1684 \cdot 10^4$
7.6	$1.0582 \cdot 10^3$	$9.1432 \cdot 10^3$	11.6	$3.6492 \cdot 10^2$	$1.1721 \cdot 10^4$
7.7	$1.0293 \cdot 10^3$	$9.2476 \cdot 10^3$	11.7	$3.5604 \cdot 10^2$	$1.1757 \cdot 10^4$
7.8	$1.0012 \cdot 10^3$	$9.3491 \cdot 10^3$	11.8	$3.4741 \cdot 10^2$	$1.1792 \cdot 10^4$
7.9	$9.7380 \cdot 10^2$	$9.4478 \cdot 10^3$	11.9	$3.3903 \cdot 10^2$	$1.1827 \cdot 10^4$
8.0	$9.4717 \cdot 10^2$	$9.5438 \cdot 10^3$	12.0	$3.3089 \cdot 10^2$	$1.1860 \cdot 10^4$
8.1	$9.2128 \cdot 10^2$	$9.6373 \cdot 10^3$	12.1	$3.2298 \cdot 10^2$	$1.1893 \cdot 10^4$
8.2	$8.9611 \cdot 10^2$	$9.7281 \cdot 10^3$	12.2	$3.1529 \cdot 10^2$	$1.1925 \cdot 10^4$
8.3	$8.7165 \cdot 10^2$	$9.8165 \cdot 10^3$	12.3	$3.0782 \cdot 10^2$	$1.1956 \cdot 10^4$
8.4	$8.4787 \cdot 10^2$	$9.9025 \cdot 10^3$	12.4	$3.0056 \cdot 10^2$	$1.1986 \cdot 10^4$
8.5	$8.2478 \cdot 10^2$	$9.9861 \cdot 10^3$	12.5	$2.9351 \cdot 10^2$	$1.2016 \cdot 10^4$
8.6	$8.0235 \cdot 10^2$	$1.0067 \cdot 10^4$	12.6	$2.8665 \cdot 10^2$	$1.2045 \cdot 10^4$
8.7	$7.8058 \cdot 10^2$	$1.0147 \cdot 10^4$	12.7	$2.7999 \cdot 10^2$	$1.2073 \cdot 10^4$
8.8	$7.5943 \cdot 10^2$	$1.0224 \cdot 10^4$	12.8	$2.7350 \cdot 10^2$	$1.2101 \cdot 10^4$
8.9	$7.3890 \cdot 10^2$	$1.0298 \cdot 10^4$	12.9	$2.6720 \cdot 10^2$	$1.2128 \cdot 10^4$
9.0	$7.1898 \cdot 10^2$	$1.0371 \cdot 10^4$	13.0	$2.6107 \cdot 10^2$	$1.2154 \cdot 10^4$
9.1	$6.9964 \cdot 10^2$	$1.0442 \cdot 10^4$	13.1	$2.5511 \cdot 10^2$	$1.2180 \cdot 10^4$
9.2	$6.8087 \cdot 10^2$	$1.0511 \cdot 10^4$	13.2	$2.4932 \cdot 10^2$	$1.2205 \cdot 10^4$
9.3	$6.6266 \cdot 10^2$	$1.0578 \cdot 10^4$	13.3	$2.4368 \cdot 10^2$	$1.2230 \cdot 10^4$
9.4	$6.4499 \cdot 10^2$	$1.0644 \cdot 10^4$	13.4	$2.3819 \cdot 10^2$	$1.2254 \cdot 10^4$
9.5	$6.2784 \cdot 10^2$	$1.0707 \cdot 10^4$	13.5	$2.3285 \cdot 10^2$	$1.2278 \cdot 10^4$
9.6	$6.1121 \cdot 10^2$	$1.0769 \cdot 10^4$	13.6	$2.2766 \cdot 10^2$	$1.2301 \cdot 10^4$
9.7	$5.9507 \cdot 10^2$	$1.0830 \cdot 10^4$	13.7	$2.2260 \cdot 10^2$	$1.2323 \cdot 10^4$
9.8	$5.7940 \cdot 10^2$	$1.0888 \cdot 10^4$	13.8	$2.1768 \cdot 10^2$	$1.2345 \cdot 10^4$
9.9	$5.6421 \cdot 10^2$	$1.0946 \cdot 10^4$	13.9	$2.1290 \cdot 10^2$	$1.2367 \cdot 10^4$
10.0	$5.4947 \cdot 10^2$	$1.1001 \cdot 10^4$	14.0	$2.0823 \cdot 10^2$	$1.2388 \cdot 10^4$
10.1	$5.3516 \cdot 10^2$	$1.1056 \cdot 10^4$	14.1	$2.0369 \cdot 10^2$	$1.2408 \cdot 10^4$
10.2	$5.2128 \cdot 10^2$	$1.1108 \cdot 10^4$	14.2	$1.9928 \cdot 10^2$	$1.2429 \cdot 10^4$
10.3	$5.0782 \cdot 10^2$	$1.1160 \cdot 10^4$	14.3	$1.9497 \cdot 10^2$	$1.2448 \cdot 10^4$
10.4	$4.9475 \cdot 10^2$	$1.1210 \cdot 10^4$	14.4	$1.9078 \cdot 10^2$	$1.2468 \cdot 10^4$
10.5	$4.8207 \cdot 10^2$	$1.1259 \cdot 10^4$	14.5	$1.8670 \cdot 10^2$	$1.2486 \cdot 10^4$
10.6	$4.6976 \cdot 10^2$	$1.1306 \cdot 10^4$	14.6	$1.8272 \cdot 10^2$	$1.2505 \cdot 10^4$
10.7	$4.5782 \cdot 10^2$	$1.1353 \cdot 10^4$	14.7	$1.7885 \cdot 10^2$	$1.2523 \cdot 10^4$
10.8	$4.4623 \cdot 10^2$	$1.1398 \cdot 10^4$	14.8	$1.7507 \cdot 10^2$	$1.2541 \cdot 10^4$
10.9	$4.3498 \cdot 10^2$	$1.1442 \cdot 10^4$	14.9	$1.7139 \cdot 10^2$	$1.2558 \cdot 10^4$
11.0	$4.2407 \cdot 10^2$	$1.1485 \cdot 10^4$	15.0	$1.6781 \cdot 10^2$	$1.2575 \cdot 10^4$
11.1	$4.1347 \cdot 10^2$	$1.1527 \cdot 10^4$	15.1	$1.6432 \cdot 10^2$	$1.2592 \cdot 10^4$
11.2	$4.0318 \cdot 10^2$	$1.1568 \cdot 10^4$	15.2	$1.6091 \cdot 10^2$	$1.2608 \cdot 10^4$
11.3	$3.9319 \cdot 10^2$	$1.1607 \cdot 10^4$	15.3	$1.5759 \cdot 10^2$	$1.2624 \cdot 10^4$
11.4	$3.8349 \cdot 10^2$	$1.1646 \cdot 10^4$	15.4	$1.5435 \cdot 10^2$	$1.2639 \cdot 10^4$

Table 2.2. (*continued*)

temperature 750 K

λ [µm]	$m_{\lambda,bb}$ [W/µm m^2]	$M_{T,o-\lambda,bb}$ [W/m^2]	λ [µm]	$m_{\lambda,bb}$ [W/µm m^2]	$M_{T,o-\lambda,bb}$ [W/m^2]
0.4	$5.4226\cdot10^{-11}$	$4.8176\cdot10^{-13}$			
0.5	$2.6024\cdot10^{-7}$	$3.6708\cdot10^{-9}$	4.0	$3.0446\cdot10^{3}$	$4.9019\cdot10^{3}$
0.6	$6.2606\cdot10^{-5}$	$1.2922\cdot10^{-6}$	4.1	$3.0281\cdot10^{3}$	$5.2056\cdot10^{3}$
0.7	$2.7897\cdot10^{-3}$	$7.9646\cdot10^{-5}$	4.2	$3.0040\cdot10^{3}$	$5.5073\cdot10^{3}$
0.8	$4.3990\cdot10^{-2}$	$1.6671\cdot10^{-3}$	4.3	$2.9734\cdot10^{3}$	$5.8062\cdot10^{3}$
0.9	$3.5054\cdot10^{-1}$	$1.7088\cdot10^{-2}$	4.4	$2.9370\cdot10^{3}$	$6.1018\cdot10^{3}$
1.0	$1.7445\cdot10^{0}$	$1.0671\cdot10^{-1}$	4.5	$2.8957\cdot10^{3}$	$6.3934\cdot10^{3}$
1.1	$6.1958\cdot10^{0}$	$4.6617\cdot10^{-1}$	4.6	$2.8502\cdot10^{3}$	$6.6808\cdot10^{3}$
1.2	$1.7152\cdot10^{1}$	$1.5612\cdot10^{0}$	4.7	$2.8012\cdot10^{3}$	$6.9633\cdot10^{3}$
1.3	$3.9316\cdot10^{1}$	$4.2696\cdot10^{0}$	4.8	$2.7493\cdot10^{3}$	$7.2409\cdot10^{3}$
1.4	$7.7879\cdot10^{1}$	$9.9716\cdot10^{0}$	4.9	$2.6950\cdot10^{3}$	$7.5131\cdot10^{3}$
1.5	$1.3751\cdot10^{2}$	$2.0549\cdot10^{1}$	5.0	$2.6389\cdot10^{3}$	$7.7798\cdot10^{3}$
1.6	$2.2148\cdot10^{2}$	$3.8287\cdot10^{1}$	5.1	$2.5813\cdot10^{3}$	$8.0408\cdot10^{3}$
1.7	$3.3113\cdot10^{2}$	$6.5704\cdot10^{1}$	5.2	$2.5227\cdot10^{3}$	$8.2960\cdot10^{3}$
1.8	$4.6575\cdot10^{2}$	$1.0535\cdot10^{2}$	5.3	$2.4634\cdot10^{3}$	$8.5454\cdot10^{3}$
1.9	$6.2283\cdot10^{2}$	$1.5960\cdot10^{2}$	5.4	$2.4037\cdot10^{3}$	$8.7887\cdot10^{3}$
2.0	$7.9845\cdot10^{2}$	$2.3053\cdot10^{2}$	5.5	$2.3440\cdot10^{3}$	$9.0261\cdot10^{3}$
2.1	$9.8783\cdot10^{2}$	$3.1975\cdot10^{2}$	5.6	$2.2843\cdot10^{3}$	$9.2575\cdot10^{3}$
2.2	$1.1858\cdot10^{3}$	$4.2838\cdot10^{2}$	5.7	$2.2250\cdot10^{3}$	$9.4830\cdot10^{3}$
2.3	$1.3874\cdot10^{3}$	$5.5703\cdot10^{2}$	5.8	$2.1662\cdot10^{3}$	$9.7025\cdot10^{3}$
2.4	$1.5876\cdot10^{3}$	$7.0581\cdot10^{2}$	5.9	$2.1080\cdot10^{3}$	$9.9162\cdot10^{3}$
2.5	$1.7824\cdot10^{3}$	$8.7437\cdot10^{2}$	6.0	$2.0506\cdot10^{3}$	$1.0124\cdot10^{4}$
2.6	$1.9683\cdot10^{3}$	$1.0620\cdot10^{3}$	6.1	$1.9941\cdot10^{3}$	$1.0326\cdot10^{4}$
2.7	$2.1424\cdot10^{3}$	$1.2676\cdot10^{3}$	6.2	$1.9386\cdot10^{3}$	$1.0523\cdot10^{4}$
2.8	$2.3027\cdot10^{3}$	$1.4900\cdot10^{3}$	6.3	$1.8842\cdot10^{3}$	$1.0714\cdot10^{4}$
2.9	$2.4478\cdot10^{3}$	$1.7277\cdot10^{3}$	6.4	$1.8308\cdot10^{3}$	$1.0900\cdot10^{4}$
3.0	$2.5767\cdot10^{3}$	$1.9790\cdot10^{3}$	6.5	$1.7786\cdot10^{3}$	$1.1080\cdot10^{4}$
3.1	$2.6891\cdot10^{3}$	$2.2425\cdot10^{3}$	6.6	$1.7276\cdot10^{3}$	$1.1256\cdot10^{4}$
3.2	$2.7851\cdot10^{3}$	$2.5163\cdot10^{3}$	6.7	$1.6778\cdot10^{3}$	$1.1426\cdot10^{4}$
3.3	$2.8651\cdot10^{3}$	$2.7989\cdot10^{3}$	6.8	$1.6293\cdot10^{3}$	$1.1591\cdot10^{4}$
3.4	$2.9296\cdot10^{3}$	$3.0888\cdot10^{3}$	6.9	$1.5820\cdot10^{3}$	$1.1752\cdot10^{4}$
3.5	$2.9795\cdot10^{3}$	$3.3844\cdot10^{3}$	7.0	$1.5359\cdot10^{3}$	$1.1908\cdot10^{4}$
3.6	$3.0158\cdot10^{3}$	$3.6842\cdot10^{3}$	7.1	$1.4911\cdot10^{3}$	$1.2059\cdot10^{4}$
3.7	$3.0393\cdot10^{3}$	$3.9871\cdot10^{3}$	7.2	$1.4475\cdot10^{3}$	$1.2206\cdot10^{4}$
3.8	$3.0513\cdot10^{3}$	$4.2917\cdot10^{3}$	7.3	$1.4052\cdot10^{3}$	$1.2348\cdot10^{4}$
3.9	$3.0527\cdot10^{3}$	$4.5970\cdot10^{3}$	7.4	$1.3641\cdot10^{3}$	$1.2487\cdot10^{4}$

Table 2.2. (*continued*)

λ [μm]	$m_{\lambda,bb}$ [W/μm m²]	$M_{T,o-\lambda,bb}$ [W/m²]	λ [μm]	$m_{\lambda,bb}$ [W/μm m²]	$M_{T,o-\lambda,bb}$ [W/m²]
7.5	$1.3242\cdot10^{3}$	$1.2621\cdot10^{4}$	11.5	$4.3241\cdot10^{2}$	$1.5764\cdot10^{4}$
7.6	$1.2854\cdot10^{3}$	$1.2752\cdot10^{4}$	11.6	$4.2150\cdot10^{2}$	$1.5807\cdot10^{4}$
7.7	$1.2479\cdot10^{3}$	$1.2878\cdot10^{4}$	11.7	$4.1093\cdot10^{2}$	$1.5849\cdot10^{4}$
7.8	$1.2114\cdot10^{3}$	$1.3001\cdot10^{4}$	11.8	$4.0067\cdot10^{2}$	$1.5889\cdot10^{4}$
7.9	$1.1761\cdot10^{3}$	$1.3121\cdot10^{4}$	11.9	$3.9071\cdot10^{2}$	$1.5929\cdot10^{4}$
8.0	$1.1418\cdot10^{3}$	$1.3237\cdot10^{4}$	12.0	$3.8106\cdot10^{2}$	$1.5967\cdot10^{4}$
8.1	$1.1086\cdot10^{3}$	$1.3349\cdot10^{4}$	12.1	$3.7168\cdot10^{2}$	$1.6005\cdot10^{4}$
8.2	$1.0765\cdot10^{3}$	$1.3458\cdot10^{4}$	12.2	$3.6258\cdot10^{2}$	$1.6042\cdot10^{4}$
8.3	$1.0453\cdot10^{3}$	$1.3565\cdot10^{4}$	12.3	$3.5375\cdot10^{2}$	$1.6077\cdot10^{4}$
8.4	$1.0152\cdot10^{3}$	$1.3668\cdot10^{4}$	12.4	$3.4518\cdot10^{2}$	$1.6112\cdot10^{4}$
8.5	$9.8593\cdot10^{2}$	$1.3768\cdot10^{4}$	12.5	$3.3686\cdot10^{2}$	$1.6146\cdot10^{4}$
8.6	$9.5763\cdot10^{2}$	$1.3865\cdot10^{4}$	12.6	$3.2877\cdot10^{2}$	$1.6180\cdot10^{4}$
8.7	$9.3023\cdot10^{2}$	$1.3959\cdot10^{4}$	12.7	$3.2092\cdot10^{2}$	$1.6212\cdot10^{4}$
8.8	$9.0370\cdot10^{2}$	$1.4051\cdot10^{4}$	12.8	$3.1330\cdot10^{2}$	$1.6244\cdot10^{4}$
8.9	$8.7801\cdot10^{2}$	$1.4140\cdot10^{4}$	12.9	$3.0589\cdot10^{2}$	$1.6275\cdot10^{4}$
9.0	$8.5314\cdot10^{2}$	$1.4226\cdot10^{4}$	13.0	$2.9870\cdot10^{2}$	$1.6305\cdot10^{4}$
9.1	$8.2906\cdot10^{2}$	$1.4311\cdot10^{4}$	13.1	$2.9170\cdot10^{2}$	$1.6335\cdot10^{4}$
9.2	$8.0574\cdot10^{2}$	$1.4392\cdot10^{4}$	13.2	$2.8491\cdot10^{2}$	$1.6363\cdot10^{4}$
9.3	$7.8318\cdot10^{2}$	$1.4472\cdot10^{4}$	13.3	$2.7831\cdot10^{2}$	$1.6392\cdot10^{4}$
9.4	$7.6133\cdot10^{2}$	$1.4549\cdot10^{4}$	13.4	$2.7189\cdot10^{2}$	$1.6419\cdot10^{4}$
9.5	$7.4017\cdot10^{2}$	$1.4624\cdot10^{4}$	13.5	$2.6565\cdot10^{2}$	$1.6446\cdot10^{4}$
9.6	$7.1969\cdot10^{2}$	$1.4697\cdot10^{4}$	13.6	$2.5958\cdot10^{2}$	$1.6472\cdot10^{4}$
9.7	$6.9986\cdot10^{2}$	$1.4768\cdot10^{4}$	13.7	$2.5368\cdot10^{2}$	$1.6498\cdot10^{4}$
9.8	$6.8065\cdot10^{2}$	$1.4837\cdot10^{4}$	13.8	$2.4794\cdot10^{2}$	$1.6523\cdot10^{4}$
9.9	$6.6206\cdot10^{2}$	$1.4904\cdot10^{4}$	13.9	$2.4236\cdot10^{2}$	$1.6548\cdot10^{4}$
10.0	$6.4405\cdot10^{2}$	$1.4969\cdot10^{4}$	14.0	$2.3694\cdot10^{2}$	$1.6571\cdot10^{4}$
10.1	$6.2661\cdot10^{2}$	$1.5033\cdot10^{4}$	14.1	$2.3166\cdot10^{2}$	$1.6595\cdot10^{4}$
10.2	$6.0972\cdot10^{2}$	$1.5095\cdot10^{4}$	14.2	$2.2652\cdot10^{2}$	$1.6618\cdot10^{4}$
10.3	$5.9336\cdot10^{2}$	$1.5155\cdot10^{4}$	14.3	$2.2152\cdot10^{2}$	$1.6640\cdot10^{4}$
10.4	$5.7751\cdot10^{2}$	$1.5213\cdot10^{4}$	14.4	$2.1665\cdot10^{2}$	$1.6662\cdot10^{4}$
10.5	$5.6216\cdot10^{2}$	$1.5270\cdot10^{4}$	14.5	$2.1192\cdot10^{2}$	$1.6684\cdot10^{4}$
10.6	$5.4728\cdot10^{2}$	$1.5326\cdot10^{4}$	14.6	$2.0731\cdot10^{2}$	$1.6705\cdot10^{4}$
10.7	$5.3287\cdot10^{2}$	$1.5380\cdot10^{4}$	14.7	$2.0282\cdot10^{2}$	$1.6725\cdot10^{4}$
10.8	$5.1890\cdot10^{2}$	$1.5432\cdot10^{4}$	14.8	$1.9845\cdot10^{2}$	$1.6745\cdot10^{4}$
10.9	$5.0536\cdot10^{2}$	$1.5484\cdot10^{4}$	14.9	$1.9420\cdot10^{2}$	$1.6765\cdot10^{4}$
11.0	$4.9224\cdot10^{2}$	$1.5533\cdot10^{4}$	15.0	$1.9005\cdot10^{2}$	$1.6784\cdot10^{4}$
11.1	$4.7952\cdot10^{2}$	$1.5582\cdot10^{4}$	15.1	$1.8602\cdot10^{2}$	$1.6803\cdot10^{4}$
11.2	$4.6719\cdot10^{2}$	$1.5629\cdot10^{4}$	15.2	$1.8208\cdot10^{2}$	$1.6821\cdot10^{4}$
11.3	$4.5524\cdot10^{2}$	$1.5676\cdot10^{4}$	15.3	$1.7825\cdot10^{2}$	$1.6839\cdot10^{4}$
11.4	$4.4365\cdot10^{2}$	$1.5720\cdot10^{4}$	15.4	$1.7452\cdot10^{2}$	$1.6857\cdot10^{4}$

Table 2.2. (*continued*)

temperature 800 K

λ [μm]	$m_{\lambda,bb}$ [W/μm m²]	$M_{T,o-\lambda,bb}$ [W/m²]	λ [μm]	$m_{\lambda,bb}$ [W/μm m²]	$M_{T,o-\lambda,bb}$ [W/m²]
0.4	$1.0864 \cdot 10^{-9}$	$1.0339 \cdot 10^{-11}$			
0.5	$2.8629 \cdot 10^{-6}$	$4.3305 \cdot 10^{-8}$	4.0	$4.1208 \cdot 10^{3}$	$7.3878 \cdot 10^{3}$
0.6	$4.6182 \cdot 10^{-4}$	$1.0233 \cdot 10^{-5}$	4.1	$4.0696 \cdot 10^{3}$	$7.7974 \cdot 10^{3}$
0.7	$1.5468 \cdot 10^{-2}$	$4.7462 \cdot 10^{-4}$	4.2	$4.0104 \cdot 10^{3}$	$8.2015 \cdot 10^{3}$
0.8	$1.9690 \cdot 10^{-1}$	$8.0287 \cdot 10^{-3}$	4.3	$3.9444 \cdot 10^{3}$	$8.5992 \cdot 10^{3}$
0.9	$1.3283 \cdot 10^{0}$	$6.9750 \cdot 10^{-2}$	4.4	$3.8727 \cdot 10^{3}$	$8.9901 \cdot 10^{3}$
1.0	$5.7860 \cdot 10^{0}$	$3.8168 \cdot 10^{-1}$	4.5	$3.7963 \cdot 10^{3}$	$9.3736 \cdot 10^{3}$
1.1	$1.8428 \cdot 10^{1}$	$1.4968 \cdot 10^{0}$	4.6	$3.7163 \cdot 10^{3}$	$9.7493 \cdot 10^{3}$
1.2	$4.6585 \cdot 10^{1}$	$4.5829 \cdot 10^{0}$	4.7	$3.6334 \cdot 10^{3}$	$1.0117 \cdot 10^{4}$
1.3	$9.8884 \cdot 10^{1}$	$1.1619 \cdot 10^{1}$	4.8	$3.5483 \cdot 10^{3}$	$1.0476 \cdot 10^{4}$
1.4	$1.8338 \cdot 10^{2}$	$2.5435 \cdot 10^{1}$	4.9	$3.4617 \cdot 10^{3}$	$1.0826 \cdot 10^{4}$
1.5	$3.0583 \cdot 10^{2}$	$4.9564 \cdot 10^{1}$	5.0	$3.3741 \cdot 10^{3}$	$1.1168 \cdot 10^{4}$
1.6	$4.6859 \cdot 10^{2}$	$8.7949 \cdot 10^{1}$	5.1	$3.2862 \cdot 10^{3}$	$1.1501 \cdot 10^{4}$
1.7	$6.7036 \cdot 10^{2}$	$1.4459 \cdot 10^{2}$	5.2	$3.1982 \cdot 10^{3}$	$1.1825 \cdot 10^{4}$
1.8	$9.0668 \cdot 10^{2}$	$2.2317 \cdot 10^{2}$	5.3	$3.1105 \cdot 10^{3}$	$1.2141 \cdot 10^{4}$
1.9	$1.1707 \cdot 10^{3}$	$3.2684 \cdot 10^{2}$	5.4	$3.0235 \cdot 10^{3}$	$1.2448 \cdot 10^{4}$
2.0	$1.4542 \cdot 10^{3}$	$4.5796 \cdot 10^{2}$	5.5	$2.9375 \cdot 10^{3}$	$1.2746 \cdot 10^{4}$
2.1	$1.7486 \cdot 10^{3}$	$6.1805 \cdot 10^{2}$	5.6	$2.8526 \cdot 10^{3}$	$1.3035 \cdot 10^{4}$
2.2	$2.0453 \cdot 10^{3}$	$8.0775 \cdot 10^{2}$	5.7	$2.7691 \cdot 10^{3}$	$1.3316 \cdot 10^{4}$
2.3	$2.3370 \cdot 10^{3}$	$1.0269 \cdot 10^{3}$	5.8	$2.6870 \cdot 10^{3}$	$1.3589 \cdot 10^{4}$
2.4	$2.6170 \cdot 10^{3}$	$1.2748 \cdot 10^{3}$	5.9	$2.6067 \cdot 10^{3}$	$1.3854 \cdot 10^{4}$
2.5	$2.8802 \cdot 10^{3}$	$1.5498 \cdot 10^{3}$	6.0	$2.5280 \cdot 10^{3}$	$1.4110 \cdot 10^{4}$
2.6	$3.1227 \cdot 10^{3}$	$1.8501 \cdot 10^{3}$	6.1	$2.4512 \cdot 10^{3}$	$1.4359 \cdot 10^{4}$
2.7	$3.3417 \cdot 10^{3}$	$2.1735 \cdot 10^{3}$	6.2	$2.3763 \cdot 10^{3}$	$1.4601 \cdot 10^{4}$
2.8	$3.5356 \cdot 10^{3}$	$2.5176 \cdot 10^{3}$	6.3	$2.3033 \cdot 10^{3}$	$1.4835 \cdot 10^{4}$
2.9	$3.7036 \cdot 10^{3}$	$2.8798 \cdot 10^{3}$	6.4	$2.2322 \cdot 10^{3}$	$1.5061 \cdot 10^{4}$
3.0	$3.8458 \cdot 10^{3}$	$3.2574 \cdot 10^{3}$	6.5	$2.1631 \cdot 10^{3}$	$1.5281 \cdot 10^{4}$
3.1	$3.9628 \cdot 10^{3}$	$3.6481 \cdot 10^{3}$	6.6	$2.0959 \cdot 10^{3}$	$1.5494 \cdot 10^{4}$
3.2	$4.0557 \cdot 10^{3}$	$4.0492 \cdot 10^{3}$	6.7	$2.0307 \cdot 10^{3}$	$1.5700 \cdot 10^{4}$
3.3	$4.1257 \cdot 10^{3}$	$4.4585 \cdot 10^{3}$	6.8	$1.9674 \cdot 10^{3}$	$1.5900 \cdot 10^{4}$
3.4	$4.1746 \cdot 10^{3}$	$4.8736 \cdot 10^{3}$	6.9	$1.9061 \cdot 10^{3}$	$1.6094 \cdot 10^{4}$
3.5	$4.2041 \cdot 10^{3}$	$5.2927 \cdot 10^{3}$	7.0	$1.8467 \cdot 10^{3}$	$1.6281 \cdot 10^{4}$
3.6	$4.2158 \cdot 10^{3}$	$5.7138 \cdot 10^{3}$	7.1	$1.7891 \cdot 10^{3}$	$1.6463 \cdot 10^{4}$
3.7	$4.2116 \cdot 10^{3}$	$6.1353 \cdot 10^{3}$	7.2	$1.7333 \cdot 10^{3}$	$1.6639 \cdot 10^{4}$
3.8	$4.1933 \cdot 10^{3}$	$6.5557 \cdot 10^{3}$	7.3	$1.6794 \cdot 10^{3}$	$1.6810 \cdot 10^{4}$
3.9	$4.1625 \cdot 10^{3}$	$6.9736 \cdot 10^{3}$	7.4	$1.6272 \cdot 10^{3}$	$1.6975 \cdot 10^{4}$

λ [μm]	$m_{\lambda,bb}$ [W/μm m²]	$M_{T,o-\lambda,bb}$ [W/m²]	λ [μm]	$m_{\lambda,bb}$ [W/μm m²]	$M_{T,o-\lambda,bb}$ [W/m²]
7.5	$1.5767 \cdot 10^3$	$1.7135 \cdot 10^4$	11.5	$4.9250 \cdot 10^2$	$2.0798 \cdot 10^4$
7.6	$1.5278 \cdot 10^3$	$1.7291 \cdot 10^4$	11.6	$4.7976 \cdot 10^2$	$2.0846 \cdot 10^4$
7.7	$1.4806 \cdot 10^3$	$1.7441 \cdot 10^4$	11.7	$4.6742 \cdot 10^2$	$2.0894 \cdot 10^4$
7.8	$1.4350 \cdot 10^3$	$1.7587 \cdot 10^4$	11.8	$4.5545 \cdot 10^2$	$2.0940 \cdot 10^4$
7.9	$1.3909 \cdot 10^3$	$1.7728 \cdot 10^4$	11.9	$4.4386 \cdot 10^2$	$2.0985 \cdot 10^4$
8.0	$1.3482 \cdot 10^3$	$1.7865 \cdot 10^4$	12.0	$4.3262 \cdot 10^2$	$2.1028 \cdot 10^4$
8.1	$1.3071 \cdot 10^3$	$1.7998 \cdot 10^4$	12.1	$4.2172 \cdot 10^2$	$2.1071 \cdot 10^4$
8.2	$1.2673 \cdot 10^3$	$1.8126 \cdot 10^4$	12.2	$4.1115 \cdot 10^2$	$2.1113 \cdot 10^4$
8.3	$1.2288 \cdot 10^3$	$1.8251 \cdot 10^4$	12.3	$4.0090 \cdot 10^2$	$2.1153 \cdot 10^4$
8.4	$1.1916 \cdot 10^3$	$1.8372 \cdot 10^4$	12.4	$3.9096 \cdot 10^2$	$2.1193 \cdot 10^4$
8.5	$1.1558 \cdot 10^3$	$1.8490 \cdot 10^4$	12.5	$3.8131 \cdot 10^2$	$2.1232 \cdot 10^4$
8.6	$1.1211 \cdot 10^3$	$1.8603 \cdot 10^4$	12.6	$3.7196 \cdot 10^2$	$2.1269 \cdot 10^4$
8.7	$1.0876 \cdot 10^3$	$1.8714 \cdot 10^4$	12.7	$3.6288 \cdot 10^2$	$2.1306 \cdot 10^4$
8.8	$1.0552 \cdot 10^3$	$1.8821 \cdot 10^4$	12.8	$3.5407 \cdot 10^2$	$2.1342 \cdot 10^4$
8.9	$1.0240 \cdot 10^3$	$1.8925 \cdot 10^4$	12.9	$3.4551 \cdot 10^2$	$2.1377 \cdot 10^4$
9.0	$9.9377 \cdot 10^2$	$1.9026 \cdot 10^4$	13.0	$3.3721 \cdot 10^2$	$2.1411 \cdot 10^4$
9.1	$9.6460 \cdot 10^2$	$1.9124 \cdot 10^4$	13.1	$3.2915 \cdot 10^2$	$2.1444 \cdot 10^4$
9.2	$9.3640 \cdot 10^2$	$1.9219 \cdot 10^4$	13.2	$3.2132 \cdot 10^2$	$2.1477 \cdot 10^4$
9.3	$9.0916 \cdot 10^2$	$1.9311 \cdot 10^4$	13.3	$3.1372 \cdot 10^2$	$2.1509 \cdot 10^4$
9.4	$8.8283 \cdot 10^2$	$1.9401 \cdot 10^4$	13.4	$3.0634 \cdot 10^2$	$2.1540 \cdot 10^4$
9.5	$8.5739 \cdot 10^2$	$1.9488 \cdot 10^4$	13.5	$2.9917 \cdot 10^2$	$2.1570 \cdot 10^4$
9.6	$8.3280 \cdot 10^2$	$1.9572 \cdot 10^4$	13.6	$2.9220 \cdot 10^2$	$2.1599 \cdot 10^4$
9.7	$8.0903 \cdot 10^2$	$1.9654 \cdot 10^4$	13.7	$2.8543 \cdot 10^2$	$2.1628 \cdot 10^4$
9.8	$7.8606 \cdot 10^2$	$1.9734 \cdot 10^4$	13.8	$2.7884 \cdot 10^2$	$2.1656 \cdot 10^4$
9.9	$7.6385 \cdot 10^2$	$1.9811 \cdot 10^4$	13.9	$2.7245 \cdot 10^2$	$2.1684 \cdot 10^4$
10.0	$7.4237 \cdot 10^2$	$1.9887 \cdot 10^4$	14.0	$2.6623 \cdot 10^2$	$2.1711 \cdot 10^4$
10.1	$7.2160 \cdot 10^2$	$1.9960 \cdot 10^4$	14.1	$2.6019 \cdot 10^2$	$2.1737 \cdot 10^4$
10.2	$7.0152 \cdot 10^2$	$2.0031 \cdot 10^4$	14.2	$2.5431 \cdot 10^2$	$2.1763 \cdot 10^4$
10.3	$6.8209 \cdot 10^2$	$2.0100 \cdot 10^4$	14.3	$2.4859 \cdot 10^2$	$2.1788 \cdot 10^4$
10.4	$6.6330 \cdot 10^2$	$2.0168 \cdot 10^4$	14.4	$2.4303 \cdot 10^2$	$2.1813 \cdot 10^4$
10.5	$6.4512 \cdot 10^2$	$2.0233 \cdot 10^4$	14.5	$2.3762 \cdot 10^2$	$2.1837 \cdot 10^4$
10.6	$6.2753 \cdot 10^2$	$2.0297 \cdot 10^4$	14.6	$2.3236 \cdot 10^2$	$2.1860 \cdot 10^4$
10.7	$6.1051 \cdot 10^2$	$2.0358 \cdot 10^4$	14.7	$2.2724 \cdot 10^2$	$2.1883 \cdot 10^4$
10.8	$5.9404 \cdot 10^2$	$2.0419 \cdot 10^4$	14.8	$2.2226 \cdot 10^2$	$2.1906 \cdot 10^4$
10.9	$5.7809 \cdot 10^2$	$2.0477 \cdot 10^4$	14.9	$2.1741 \cdot 10^2$	$2.1928 \cdot 10^4$
11.0	$5.6266 \cdot 10^2$	$2.0534 \cdot 10^4$	15.0	$2.1269 \cdot 10^2$	$2.1949 \cdot 10^4$
11.1	$5.4771 \cdot 10^2$	$2.0590 \cdot 10^4$	15.1	$2.0810 \cdot 10^2$	$2.1970 \cdot 10^4$
11.2	$5.3324 \cdot 10^2$	$2.0644 \cdot 10^4$	15.2	$2.0363 \cdot 10^2$	$2.1991 \cdot 10^4$
11.3	$5.1923 \cdot 10^2$	$2.0696 \cdot 10^4$	15.3	$1.9927 \cdot 10^2$	$2.2011 \cdot 10^4$
11.4	$5.0565 \cdot 10^2$	$2.0748 \cdot 10^4$	15.4	$1.9503 \cdot 10^2$	$2.2031 \cdot 10^4$

Table 2.2. (*continued*)

temperature 850 K

λ [μm]	$m_{\lambda,bb}$ [W/μm m²]	$M_{T,o-\lambda,bb}$ [W/m²]	λ [μm]	$m_{\lambda,bb}$ [W/μm m²]	$M_{T,o-\lambda,bb}$ [W/m²]
0.4	$1.5298 \cdot 10^{-8}$	$1.5535 \cdot 10^{-10}$			
0.5	$2.3753 \cdot 10^{-5}$	$3.8379 \cdot 10^{-7}$	4.0	$5.3868 \cdot 10^{3}$	$1.0707 \cdot 10^{4}$
0.6	$2.6930 \cdot 10^{-3}$	$6.3812 \cdot 10^{-5}$	4.1	$5.2873 \cdot 10^{3}$	$1.1240 \cdot 10^{4}$
0.7	$7.0113 \cdot 10^{-2}$	$2.3031 \cdot 10^{-3}$	4.2	$5.1800 \cdot 10^{3}$	$1.1764 \cdot 10^{4}$
0.8	$7.3887 \cdot 10^{-1}$	$3.2289 \cdot 10^{-2}$	4.3	$5.0665 \cdot 10^{3}$	$1.2276 \cdot 10^{4}$
0.9	$4.3034 \cdot 10^{0}$	$2.4245 \cdot 10^{-1}$	4.4	$4.9483 \cdot 10^{3}$	$1.2777 \cdot 10^{4}$
1.0	$1.6666 \cdot 10^{1}$	$1.1809 \cdot 10^{0}$	4.5	$4.8264 \cdot 10^{3}$	$1.3266 \cdot 10^{4}$
1.1	$4.8212 \cdot 10^{1}$	$4.2113 \cdot 10^{0}$	4.6	$4.7021 \cdot 10^{3}$	$1.3742 \cdot 10^{4}$
1.2	$1.1249 \cdot 10^{2}$	$1.1914 \cdot 10^{1}$	4.7	$4.5763 \cdot 10^{3}$	$1.4206 \cdot 10^{4}$
1.3	$2.2313 \cdot 10^{2}$	$2.8259 \cdot 10^{1}$	4.8	$4.4497 \cdot 10^{3}$	$1.4657 \cdot 10^{4}$
1.4	$3.9043 \cdot 10^{2}$	$5.8435 \cdot 10^{1}$	4.9	$4.3231 \cdot 10^{3}$	$1.5096 \cdot 10^{4}$
1.5	$6.1914 \cdot 10^{2}$	$1.0840 \cdot 10^{2}$	5.0	$4.1971 \cdot 10^{3}$	$1.5522 \cdot 10^{4}$
1.6	$9.0773 \cdot 10^{2}$	$1.8426 \cdot 10^{2}$	5.1	$4.0721 \cdot 10^{3}$	$1.5936 \cdot 10^{4}$
1.7	$1.2491 \cdot 10^{3}$	$2.9171 \cdot 10^{2}$	5.2	$3.9487 \cdot 10^{3}$	$1.6337 \cdot 10^{4}$
1.8	$1.6320 \cdot 10^{3}$	$4.3546 \cdot 10^{2}$	5.3	$3.8271 \cdot 10^{3}$	$1.6725 \cdot 10^{4}$
1.9	$2.0431 \cdot 10^{3}$	$6.1904 \cdot 10^{2}$	5.4	$3.7076 \cdot 10^{3}$	$1.7102 \cdot 10^{4}$
2.0	$2.4683 \cdot 10^{3}$	$8.4455 \cdot 10^{2}$	5.5	$3.5906 \cdot 10^{3}$	$1.7467 \cdot 10^{4}$
2.1	$2.8942 \cdot 10^{3}$	$1.1127 \cdot 10^{3}$	5.6	$3.4761 \cdot 10^{3}$	$1.7820 \cdot 10^{4}$
2.2	$3.3089 \cdot 10^{3}$	$1.4230 \cdot 10^{3}$	5.7	$3.3644 \cdot 10^{3}$	$1.8162 \cdot 10^{4}$
2.3	$3.7027 \cdot 10^{3}$	$1.7738 \cdot 10^{3}$	5.8	$3.2554 \cdot 10^{3}$	$1.8493 \cdot 10^{4}$
2.4	$4.0680 \cdot 10^{3}$	$2.1626 \cdot 10^{3}$	5.9	$3.1494 \cdot 10^{3}$	$1.8813 \cdot 10^{4}$
2.5	$4.3992 \cdot 10^{3}$	$2.5862 \cdot 10^{3}$	6.0	$3.0464 \cdot 10^{3}$	$1.9123 \cdot 10^{4}$
2.6	$4.6931 \cdot 10^{3}$	$3.0412 \cdot 10^{3}$	6.1	$2.9463 \cdot 10^{3}$	$1.9423 \cdot 10^{4}$
2.7	$4.9477 \cdot 10^{3}$	$3.5235 \cdot 10^{3}$	6.2	$2.8492 \cdot 10^{3}$	$1.9713 \cdot 10^{4}$
2.8	$5.1627 \cdot 10^{3}$	$4.0294 \cdot 10^{3}$	6.3	$2.7552 \cdot 10^{3}$	$1.9993 \cdot 10^{4}$
2.9	$5.3388 \cdot 10^{3}$	$4.5548 \cdot 10^{3}$	6.4	$2.6641 \cdot 10^{3}$	$2.0264 \cdot 10^{4}$
3.0	$5.4777 \cdot 10^{3}$	$5.0959 \cdot 10^{3}$	6.5	$2.5759 \cdot 10^{3}$	$2.0526 \cdot 10^{4}$
3.1	$5.5815 \cdot 10^{3}$	$5.6491 \cdot 10^{3}$	6.6	$2.4907 \cdot 10^{3}$	$2.0779 \cdot 10^{4}$
3.2	$5.6528 \cdot 10^{3}$	$6.2111 \cdot 10^{3}$	6.7	$2.4082 \cdot 10^{3}$	$2.1024 \cdot 10^{4}$
3.3	$5.6943 \cdot 10^{3}$	$6.7787 \cdot 10^{3}$	6.8	$2.3286 \cdot 10^{3}$	$2.1261 \cdot 10^{4}$
3.4	$5.7089 \cdot 10^{3}$	$7.3491 \cdot 10^{3}$	6.9	$2.2517 \cdot 10^{3}$	$2.1490 \cdot 10^{4}$
3.5	$5.6996 \cdot 10^{3}$	$7.9197 \cdot 10^{3}$	7.0	$2.1774 \cdot 10^{3}$	$2.1711 \cdot 10^{4}$
3.6	$5.6691 \cdot 10^{3}$	$8.4883 \cdot 10^{3}$	7.1	$2.1058 \cdot 10^{3}$	$2.1925 \cdot 10^{4}$
3.7	$5.6202 \cdot 10^{3}$	$9.0529 \cdot 10^{3}$	7.2	$2.0366 \cdot 10^{3}$	$2.2132 \cdot 10^{4}$
3.8	$5.5553 \cdot 10^{3}$	$9.6118 \cdot 10^{3}$	7.3	$1.9699 \cdot 10^{3}$	$2.2333 \cdot 10^{4}$
3.9	$5.4767 \cdot 10^{3}$	$1.0163 \cdot 10^{4}$	7.4	$1.9055 \cdot 10^{3}$	$2.2526 \cdot 10^{4}$

Table 2.2. (*continued*)

λ [μm]	$m_{\lambda,bb}$ [W/μm m²]	$M_{T,o-\lambda,bb}$ [W/m²]	λ [μm]	$m_{\lambda,bb}$ [W/μm m²]	$M_{T,o-\lambda,bb}$ [W/m²]
7.5	$1.8435 \cdot 10^3$	$2.2714 \cdot 10^4$	11.5	$5.5409 \cdot 10^2$	$2.6917 \cdot 10^4$
7.6	$1.7836 \cdot 10^3$	$2.2895 \cdot 10^4$	11.6	$5.3945 \cdot 10^2$	$2.6971 \cdot 10^4$
7.7	$1.7259 \cdot 10^3$	$2.3071 \cdot 10^4$	11.7	$5.2527 \cdot 10^2$	$2.7025 \cdot 10^4$
7.8	$1.6703 \cdot 10^3$	$2.3240 \cdot 10^4$	11.8	$5.1153 \cdot 10^2$	$2.7076 \cdot 10^4$
7.9	$1.6167 \cdot 10^3$	$2.3405 \cdot 10^4$	11.9	$4.9824 \cdot 10^2$	$2.7127 \cdot 10^4$
8.0	$1.5650 \cdot 10^3$	$2.3564 \cdot 10^4$	12.0	$4.8536 \cdot 10^2$	$2.7176 \cdot 10^4$
8.1	$1.5152 \cdot 10^3$	$2.3718 \cdot 10^4$	12.1	$4.7288 \cdot 10^2$	$2.7224 \cdot 10^4$
8.2	$1.4671 \cdot 10^3$	$2.3867 \cdot 10^4$	12.2	$4.6079 \cdot 10^2$	$2.7271 \cdot 10^4$
8.3	$1.4208 \cdot 10^3$	$2.4011 \cdot 10^4$	12.3	$4.4907 \cdot 10^2$	$2.7316 \cdot 10^4$
8.4	$1.3762 \cdot 10^3$	$2.4151 \cdot 10^4$	12.4	$4.3772 \cdot 10^2$	$2.7361 \cdot 10^4$
8.5	$1.3332 \cdot 10^3$	$2.4287 \cdot 10^4$	12.5	$4.2671 \cdot 10^2$	$2.7404 \cdot 10^4$
8.6	$1.2917 \cdot 10^3$	$2.4418 \cdot 10^4$	12.6	$4.1604 \cdot 10^2$	$2.7446 \cdot 10^4$
8.7	$1.2517 \cdot 10^3$	$2.4545 \cdot 10^4$	12.7	$4.0569 \cdot 10^2$	$2.7487 \cdot 10^4$
8.8	$1.2131 \cdot 10^3$	$2.4668 \cdot 10^4$	12.8	$3.9566 \cdot 10^2$	$2.7527 \cdot 10^4$
8.9	$1.1759 \cdot 10^3$	$2.4788 \cdot 10^4$	12.9	$3.8592 \cdot 10^2$	$2.7566 \cdot 10^4$
9.0	$1.1400 \cdot 10^3$	$2.4903 \cdot 10^4$	13.0	$3.7648 \cdot 10^2$	$2.7604 \cdot 10^4$
9.1	$1.1054 \cdot 10^3$	$2.5016 \cdot 10^4$	13.1	$3.6732 \cdot 10^2$	$2.7641 \cdot 10^4$
9.2	$1.0721 \cdot 10^3$	$2.5124 \cdot 10^4$	13.2	$3.5843 \cdot 10^2$	$2.7678 \cdot 10^4$
9.3	$1.0399 \cdot 10^3$	$2.5230 \cdot 10^4$	13.3	$3.4980 \cdot 10^2$	$2.7713 \cdot 10^4$
9.4	$1.0088 \cdot 10^3$	$2.5332 \cdot 10^4$	13.4	$3.4142 \cdot 10^2$	$2.7748 \cdot 10^4$
9.5	$9.7884 \cdot 10^2$	$2.5432 \cdot 10^4$	13.5	$3.3329 \cdot 10^2$	$2.7781 \cdot 10^4$
9.6	$9.4991 \cdot 10^2$	$2.5528 \cdot 10^4$	13.6	$3.2540 \cdot 10^2$	$2.7814 \cdot 10^4$
9.7	$9.2199 \cdot 10^2$	$2.5622 \cdot 10^4$	13.7	$3.1773 \cdot 10^2$	$2.7846 \cdot 10^4$
9.8	$8.9504 \cdot 10^2$	$2.5713 \cdot 10^4$	13.8	$3.1028 \cdot 10^2$	$2.7878 \cdot 10^4$
9.9	$8.6902 \cdot 10^2$	$2.5801 \cdot 10^4$	13.9	$3.0305 \cdot 10^2$	$2.7908 \cdot 10^4$
10.0	$8.4389 \cdot 10^2$	$2.5887 \cdot 10^4$	14.0	$2.9602 \cdot 10^2$	$2.7938 \cdot 10^4$
10.1	$8.1963 \cdot 10^2$	$2.5970 \cdot 10^4$	14.1	$2.8919 \cdot 10^2$	$2.7968 \cdot 10^4$
10.2	$7.9619 \cdot 10^2$	$2.6050 \cdot 10^4$	14.2	$2.8255 \cdot 10^2$	$2.7996 \cdot 10^4$
10.3	$7.7356 \cdot 10^2$	$2.6129 \cdot 10^4$	14.3	$2.7610 \cdot 10^2$	$2.8024 \cdot 10^4$
10.4	$7.5168 \cdot 10^2$	$2.6205 \cdot 10^4$	14.4	$2.6983 \cdot 10^2$	$2.8052 \cdot 10^4$
10.5	$7.3055 \cdot 10^2$	$2.6279 \cdot 10^4$	14.5	$2.6373 \cdot 10^2$	$2.8078 \cdot 10^4$
10.6	$7.1012 \cdot 10^2$	$2.6351 \cdot 10^4$	14.6	$2.5781 \cdot 10^2$	$2.8104 \cdot 10^4$
10.7	$6.9038 \cdot 10^2$	$2.6421 \cdot 10^4$	14.7	$2.5204 \cdot 10^2$	$2.8130 \cdot 10^4$
10.8	$6.7129 \cdot 10^2$	$2.6489 \cdot 10^4$	14.8	$2.4643 \cdot 10^2$	$2.8155 \cdot 10^4$
10.9	$6.5283 \cdot 10^2$	$2.6556 \cdot 10^4$	14.9	$2.4098 \cdot 10^2$	$2.8179 \cdot 10^4$
11.0	$6.3498 \cdot 10^2$	$2.6620 \cdot 10^4$	15.0	$2.3567 \cdot 10^2$	$2.8203 \cdot 10^4$
11.1	$6.1772 \cdot 10^2$	$2.6683 \cdot 10^4$	15.1	$2.3051 \cdot 10^2$	$2.8226 \cdot 10^4$
11.2	$6.0102 \cdot 10^2$	$2.6744 \cdot 10^4$	15.2	$2.2548 \cdot 10^2$	$2.8249 \cdot 10^4$
11.3	$5.8486 \cdot 10^2$	$2.6803 \cdot 10^4$	15.3	$2.2059 \cdot 10^2$	$2.8271 \cdot 10^4$
11.4	$5.6923 \cdot 10^2$	$2.6861 \cdot 10^4$	15.4	$2.1583 \cdot 10^2$	$2.8293 \cdot 10^4$

Table 2.2. (*continued*)

temperature 900 K

λ [µm]	$m_{\lambda,bb}$ [W/µm m²]	$M_{T,o-\lambda,bb}$ [W/m²]	λ [µm]	$m_{\lambda,bb}$ [W/µm m²]	$M_{T,o-\lambda,bb}$ [W/m²]
0.4	$1.6056 \cdot 10^{-7}$	$1.7338 \cdot 10^{-9}$			
0.5	$1.5578 \cdot 10^{-4}$	$2.6795 \cdot 10^{-6}$	4.0	$6.8411 \cdot 10^{3}$	$1.5015 \cdot 10^{4}$
0.6	$1.2909 \cdot 10^{-2}$	$3.2599 \cdot 10^{-4}$	4.1	$6.6785 \cdot 10^{3}$	$1.5691 \cdot 10^{4}$
0.7	$2.6868 \cdot 10^{-1}$	$9.4158 \cdot 10^{-3}$	4.2	$6.5094 \cdot 10^{3}$	$1.6350 \cdot 10^{4}$
0.8	$2.3937 \cdot 10^{0}$	$1.1173 \cdot 10^{-1}$	4.3	$6.3359 \cdot 10^{3}$	$1.6992 \cdot 10^{4}$
0.9	$1.2235 \cdot 10^{1}$	$7.3705 \cdot 10^{-1}$	4.4	$6.1594 \cdot 10^{3}$	$1.7617 \cdot 10^{4}$
1.0	$4.2681 \cdot 10^{1}$	$3.2374 \cdot 10^{0}$	4.5	$5.9813 \cdot 10^{3}$	$1.8224 \cdot 10^{4}$
1.1	$1.1335 \cdot 10^{2}$	$1.0611 \cdot 10^{1}$	4.6	$5.8028 \cdot 10^{3}$	$1.8813 \cdot 10^{4}$
1.2	$2.4630 \cdot 10^{2}$	$2.7988 \cdot 10^{1}$	4.7	$5.6249 \cdot 10^{3}$	$1.9385 \cdot 10^{4}$
1.3	$4.5995 \cdot 10^{2}$	$6.2570 \cdot 10^{1}$	4.8	$5.4485 \cdot 10^{3}$	$1.9939 \cdot 10^{4}$
1.4	$7.6430 \cdot 10^{2}$	$1.2301 \cdot 10^{2}$	4.9	$5.2742 \cdot 10^{3}$	$2.0475 \cdot 10^{4}$
1.5	$1.1590 \cdot 10^{3}$	$2.1845 \cdot 10^{2}$	5.0	$5.1026 \cdot 10^{3}$	$2.0993 \cdot 10^{4}$
1.6	$1.6339 \cdot 10^{3}$	$3.5748 \cdot 10^{2}$	5.1	$4.9342 \cdot 10^{3}$	$2.1495 \cdot 10^{4}$
1.7	$2.1719 \cdot 10^{3}$	$5.4733 \cdot 10^{2}$	5.2	$4.7693 \cdot 10^{3}$	$2.1980 \cdot 10^{4}$
1.8	$2.7519 \cdot 10^{3}$	$7.9326 \cdot 10^{2}$	5.3	$4.6084 \cdot 10^{3}$	$2.2449 \cdot 10^{4}$
1.9	$3.3518 \cdot 10^{3}$	$1.0984 \cdot 10^{3}$	5.4	$4.4515 \cdot 10^{3}$	$2.2902 \cdot 10^{4}$
2.0	$3.9505 \cdot 10^{3}$	$1.4636 \cdot 10^{3}$	5.5	$4.2988 \cdot 10^{3}$	$2.3340 \cdot 10^{4}$
2.1	$4.5298 \cdot 10^{3}$	$1.8878 \cdot 10^{3}$	5.6	$4.1505 \cdot 10^{3}$	$2.3762 \cdot 10^{4}$
2.2	$5.0749 \cdot 10^{3}$	$2.3684 \cdot 10^{3}$	5.7	$4.0067 \cdot 10^{3}$	$2.4170 \cdot 10^{4}$
2.3	$5.5748 \cdot 10^{3}$	$2.9013 \cdot 10^{3}$	5.8	$3.8673 \cdot 10^{3}$	$2.4564 \cdot 10^{4}$
2.4	$6.0218 \cdot 10^{3}$	$3.4816 \cdot 10^{3}$	5.9	$3.7324 \cdot 10^{3}$	$2.4944 \cdot 10^{4}$
2.5	$6.4116 \cdot 10^{3}$	$4.1037 \cdot 10^{3}$	6.0	$3.6020 \cdot 10^{3}$	$2.5310 \cdot 10^{4}$
2.6	$6.7424 \cdot 10^{3}$	$4.7619 \cdot 10^{3}$	6.1	$3.4759 \cdot 10^{3}$	$2.5664 \cdot 10^{4}$
2.7	$7.0146 \cdot 10^{3}$	$5.4502 \cdot 10^{3}$	6.2	$3.3542 \cdot 10^{3}$	$2.6006 \cdot 10^{4}$
2.8	$7.2301 \cdot 10^{3}$	$6.1629 \cdot 10^{3}$	6.3	$3.2368 \cdot 10^{3}$	$2.6335 \cdot 10^{4}$
2.9	$7.3920 \cdot 10^{3}$	$6.8944 \cdot 10^{3}$	6.4	$3.1235 \cdot 10^{3}$	$2.6653 \cdot 10^{4}$
3.0	$7.5042 \cdot 10^{3}$	$7.6396 \cdot 10^{3}$	6.5	$3.0144 \cdot 10^{3}$	$2.6960 \cdot 10^{4}$
3.1	$7.5710 \cdot 10^{3}$	$8.3938 \cdot 10^{3}$	6.6	$2.9092 \cdot 10^{3}$	$2.7256 \cdot 10^{4}$
3.2	$7.5970 \cdot 10^{3}$	$9.1525 \cdot 10^{3}$	6.7	$2.8079 \cdot 10^{3}$	$2.7542 \cdot 10^{4}$
3.3	$7.5867 \cdot 10^{3}$	$9.9119 \cdot 10^{3}$	6.8	$2.7103 \cdot 10^{3}$	$2.7818 \cdot 10^{4}$
3.4	$7.5446 \cdot 10^{3}$	$1.0669 \cdot 10^{4}$	6.9	$2.6164 \cdot 10^{3}$	$2.8084 \cdot 10^{4}$
3.5	$7.4749 \cdot 10^{3}$	$1.1420 \cdot 10^{4}$	7.0	$2.5260 \cdot 10^{3}$	$2.8341 \cdot 10^{4}$
3.6	$7.3816 \cdot 10^{3}$	$1.2163 \cdot 10^{4}$	7.1	$2.4390 \cdot 10^{3}$	$2.8589 \cdot 10^{4}$
3.7	$7.2684 \cdot 10^{3}$	$1.2896 \cdot 10^{4}$	7.2	$2.3554 \cdot 10^{3}$	$2.8829 \cdot 10^{4}$
3.8	$7.1387 \cdot 10^{3}$	$1.3616 \cdot 10^{4}$	7.3	$2.2748 \cdot 10^{3}$	$2.9060 \cdot 10^{4}$
3.9	$6.9953 \cdot 10^{3}$	$1.4323 \cdot 10^{4}$	7.4	$2.1974 \cdot 10^{3}$	$2.9284 \cdot 10^{4}$

Table 2.2. (*continued*)

λ [μm]	$m_{\lambda,bb}$ [W/μm m²]	$M_{T,o-\lambda,bb}$ [W/m²]	λ [μm]	$m_{\lambda,bb}$ [W/μm m²]	$M_{T,o-\lambda,bb}$ [W/m²]
7.5	$2.1229 \cdot 10^3$	$2.9500 \cdot 10^4$	11.5	$6.1697 \cdot 10^2$	$3.4262 \cdot 10^4$
7.6	$2.0512 \cdot 10^3$	$2.9709 \cdot 10^4$	11.6	$6.0035 \cdot 10^2$	$3.4323 \cdot 10^4$
7.7	$1.9823 \cdot 10^3$	$2.9910 \cdot 10^4$	11.7	$5.8428 \cdot 10^2$	$3.4382 \cdot 10^4$
7.8	$1.9160 \cdot 10^3$	$3.0105 \cdot 10^4$	11.8	$5.6872 \cdot 10^2$	$3.4440 \cdot 10^4$
7.9	$1.8522 \cdot 10^3$	$3.0294 \cdot 10^4$	11.9	$5.5367 \cdot 10^2$	$3.4496 \cdot 10^4$
8.0	$1.7908 \cdot 10^3$	$3.0476 \cdot 10^4$	12.0	$5.3911 \cdot 10^2$	$3.4551 \cdot 10^4$
8.1	$1.7318 \cdot 10^3$	$3.0652 \cdot 10^4$	12.1	$5.2500 \cdot 10^2$	$3.4604 \cdot 10^4$
8.2	$1.6750 \cdot 10^3$	$3.0822 \cdot 10^4$	12.2	$5.1135 \cdot 10^2$	$3.4656 \cdot 10^4$
8.3	$1.6204 \cdot 10^3$	$3.0987 \cdot 10^4$	12.3	$4.9813 \cdot 10^2$	$3.4706 \cdot 10^4$
8.4	$1.5678 \cdot 10^3$	$3.1146 \cdot 10^4$	12.4	$4.8532 \cdot 10^2$	$3.4755 \cdot 10^4$
8.5	$1.5172 \cdot 10^3$	$3.1301 \cdot 10^4$	12.5	$4.7291 \cdot 10^2$	$3.4803 \cdot 10^4$
8.6	$1.4685 \cdot 10^3$	$3.1450 \cdot 10^4$	12.6	$4.6089 \cdot 10^2$	$3.4850 \cdot 10^4$
8.7	$1.4216 \cdot 10^3$	$3.1594 \cdot 10^4$	12.7	$4.4924 \cdot 10^2$	$3.4896 \cdot 10^4$
8.8	$1.3765 \cdot 10^3$	$3.1734 \cdot 10^4$	12.8	$4.3795 \cdot 10^2$	$3.4940 \cdot 10^4$
8.9	$1.3330 \cdot 10^3$	$3.1870 \cdot 10^4$	12.9	$4.2701 \cdot 10^2$	$3.4983 \cdot 10^4$
9.0	$1.2912 \cdot 10^3$	$3.2001 \cdot 10^4$	13.0	$4.1639 \cdot 10^2$	$3.5025 \cdot 10^4$
9.1	$1.2509 \cdot 10^3$	$3.2128 \cdot 10^4$	13.1	$4.0610 \cdot 10^2$	$3.5066 \cdot 10^4$
9.2	$1.2121 \cdot 10^3$	$3.2251 \cdot 10^4$	13.2	$3.9612 \cdot 10^2$	$3.5107 \cdot 10^4$
9.3	$1.1747 \cdot 10^3$	$3.2370 \cdot 10^4$	13.3	$3.8644 \cdot 10^2$	$3.5146 \cdot 10^4$
9.4	$1.1387 \cdot 10^3$	$3.2486 \cdot 10^4$	13.4	$3.7705 \cdot 10^2$	$3.5184 \cdot 10^4$
9.5	$1.1039 \cdot 10^3$	$3.2598 \cdot 10^4$	13.5	$3.6794 \cdot 10^2$	$3.5221 \cdot 10^4$
9.6	$1.0703 \cdot 10^3$	$3.2707 \cdot 10^4$	13.6	$3.5909 \cdot 10^2$	$3.5257 \cdot 10^4$
9.7	$1.0382 \cdot 10^3$	$3.2812 \cdot 10^4$	13.7	$3.5051 \cdot 10^2$	$3.5293 \cdot 10^4$
9.8	$1.0071 \cdot 10^3$	$3.2915 \cdot 10^4$	13.8	$3.4218 \cdot 10^2$	$3.5328 \cdot 10^4$
9.9	$9.7712 \cdot 10^2$	$3.3014 \cdot 10^4$	13.9	$3.3409 \cdot 10^2$	$3.5361 \cdot 10^4$
10.0	$9.4819 \cdot 10^2$	$3.3110 \cdot 10^4$	14.0	$3.2623 \cdot 10^2$	$3.5394 \cdot 10^4$
10.1	$9.2028 \cdot 10^2$	$3.3203 \cdot 10^4$	14.1	$3.1860 \cdot 10^2$	$3.5427 \cdot 10^4$
10.2	$8.9336 \cdot 10^2$	$3.3294 \cdot 10^4$	14.2	$3.1119 \cdot 10^2$	$3.5458 \cdot 10^4$
10.3	$8.6738 \cdot 10^2$	$3.3382 \cdot 10^4$	14.3	$3.0399 \cdot 10^2$	$3.5489 \cdot 10^4$
10.4	$8.4230 \cdot 10^2$	$3.3468 \cdot 10^4$	14.4	$2.9699 \cdot 10^2$	$3.5519 \cdot 10^4$
10.5	$8.1809 \cdot 10^2$	$3.3551 \cdot 10^4$	14.5	$2.9019 \cdot 10^2$	$3.5548 \cdot 10^4$
10.6	$7.9472 \cdot 10^2$	$3.3631 \cdot 10^4$	14.6	$2.8358 \cdot 10^2$	$3.5577 \cdot 10^4$
10.7	$7.7215 \cdot 10^2$	$3.3710 \cdot 10^4$	14.7	$2.7716 \cdot 10^2$	$3.5605 \cdot 10^4$
10.8	$7.5035 \cdot 10^2$	$3.3786 \cdot 10^4$	14.8	$2.7091 \cdot 10^2$	$3.5632 \cdot 10^4$
10.9	$7.2929 \cdot 10^2$	$3.3860 \cdot 10^4$	14.9	$2.6484 \cdot 10^2$	$3.5659 \cdot 10^4$
11.0	$7.0895 \cdot 10^2$	$3.3932 \cdot 10^4$	15.0	$2.5893 \cdot 10^2$	$3.5685 \cdot 10^4$
11.1	$6.8928 \cdot 10^2$	$3.4001 \cdot 10^4$	15.1	$2.5319 \cdot 10^2$	$3.5711 \cdot 10^4$
11.2	$6.7028 \cdot 10^2$	$3.4069 \cdot 10^4$	15.2	$2.4760 \cdot 10^2$	$3.5736 \cdot 10^4$
11.3	$6.5191 \cdot 10^2$	$3.4136 \cdot 10^4$	15.3	$2.4216 \cdot 10^2$	$3.5760 \cdot 10^4$
11.4	$6.3414 \cdot 10^2$	$3.4200 \cdot 10^4$	15.4	$2.3687 \cdot 10^2$	$3.5784 \cdot 10^4$

Table 2.2. (*continued*)

temperature 950 K

λ [μm]	$m_{\lambda,bb}$ [W/μm m²]	$M_{T,o-\lambda,bb}$ [W/m²]	λ [μm]	$m_{\lambda,bb}$ [W/μm m²]	$M_{T,o-\lambda,bb}$ [W/m²]
0.4	$1.3158\cdot10^{-6}$	$1.5061\cdot10^{-8}$			
0.5	$8.3821\cdot10^{-4}$	$1.5300\cdot10^{-5}$	4.0	$8.4799\cdot10^{3}$	$2.0477\cdot10^{4}$
0.6	$5.2472\cdot10^{-2}$	$1.4077\cdot10^{-3}$	4.1	$8.2386\cdot10^{3}$	$2.1313\cdot10^{4}$
0.7	$8.9382\cdot10^{-1}$	$3.3316\cdot10^{-2}$	4.2	$7.9936\cdot10^{3}$	$2.2124\cdot10^{4}$
0.8	$6.8523\cdot10^{0}$	$3.4055\cdot10^{-1}$	4.3	$7.7469\cdot10^{3}$	$2.2911\cdot10^{4}$
0.9	$3.1161\cdot10^{1}$	$2.0011\cdot10^{0}$	4.4	$7.5002\cdot10^{3}$	$2.3674\cdot10^{4}$
1.0	$9.9002\cdot10^{1}$	$8.0140\cdot10^{0}$	4.5	$7.2550\cdot10^{3}$	$2.4411\cdot10^{4}$
1.1	$2.4357\cdot10^{2}$	$2.4361\cdot10^{1}$	4.6	$7.0124\cdot10^{3}$	$2.5125\cdot10^{4}$
1.2	$4.9657\cdot10^{2}$	$6.0355\cdot10^{1}$	4.7	$6.7734\cdot10^{3}$	$2.5814\cdot10^{4}$
1.3	$8.7861\cdot10^{2}$	$1.2799\cdot10^{2}$	4.8	$6.5388\cdot10^{3}$	$2.6480\cdot10^{4}$
1.4	$1.3940\cdot10^{3}$	$2.4054\cdot10^{2}$	4.9	$6.3093\cdot10^{3}$	$2.7122\cdot10^{4}$
1.5	$2.0309\cdot10^{3}$	$4.1086\cdot10^{2}$	5.0	$6.0852\cdot10^{3}$	$2.7742\cdot10^{4}$
1.6	$2.7645\cdot10^{3}$	$6.4995\cdot10^{2}$	5.1	$5.8671\cdot10^{3}$	$2.8339\cdot10^{4}$
1.7	$3.5630\cdot10^{3}$	$9.6592\cdot10^{2}$	5.2	$5.6551\cdot10^{3}$	$2.8915\cdot10^{4}$
1.8	$4.3922\cdot10^{3}$	$1.3636\cdot10^{3}$	5.3	$5.4495\cdot10^{3}$	$2.9470\cdot10^{4}$
1.9	$5.2198\cdot10^{3}$	$1.8443\cdot10^{3}$	5.4	$5.2504\cdot10^{3}$	$3.0005\cdot10^{4}$
2.0	$6.0178\cdot10^{3}$	$2.4065\cdot10^{3}$	5.5	$5.0578\cdot10^{3}$	$3.0521\cdot10^{4}$
2.1	$6.7638\cdot10^{3}$	$3.0461\cdot10^{3}$	5.6	$4.8717\cdot10^{3}$	$3.1017\cdot10^{4}$
2.2	$7.4416\cdot10^{3}$	$3.7570\cdot10^{3}$	5.7	$4.6921\cdot10^{3}$	$3.1495\cdot10^{4}$
2.3	$8.0406\cdot10^{3}$	$4.5318\cdot10^{3}$	5.8	$4.5189\cdot10^{3}$	$3.1956\cdot10^{4}$
2.4	$8.5549\cdot10^{3}$	$5.3623\cdot10^{3}$	5.9	$4.3521\cdot10^{3}$	$3.2399\cdot10^{4}$
2.5	$8.9831\cdot10^{3}$	$6.2399\cdot10^{3}$	6.0	$4.1914\cdot10^{3}$	$3.2826\cdot10^{4}$
2.6	$9.3264\cdot10^{3}$	$7.1560\cdot10^{3}$	6.1	$4.0368\cdot10^{3}$	$3.3238\cdot10^{4}$
2.7	$9.5889\cdot10^{3}$	$8.1024\cdot10^{3}$	6.2	$3.8882\cdot10^{3}$	$3.3634\cdot10^{4}$
2.8	$9.7759\cdot10^{3}$	$9.0713\cdot10^{3}$	6.3	$3.7452\cdot10^{3}$	$3.4016\cdot10^{4}$
2.9	$9.8937\cdot10^{3}$	$1.0055\cdot10^{4}$	6.4	$3.6078\cdot10^{3}$	$3.4383\cdot10^{4}$
3.0	$9.9494\cdot10^{3}$	$1.1048\cdot10^{4}$	6.5	$3.4759\cdot10^{3}$	$3.4737\cdot10^{4}$
3.1	$9.9498\cdot10^{3}$	$1.2043\cdot10^{4}$	6.6	$3.3491\cdot10^{3}$	$3.5079\cdot10^{4}$
3.2	$9.9020\cdot10^{3}$	$1.3036\cdot10^{4}$	6.7	$3.2274\cdot10^{3}$	$3.5407\cdot10^{4}$
3.3	$9.8126\cdot10^{3}$	$1.4022\cdot10^{4}$	6.8	$3.1105\cdot10^{3}$	$3.5724\cdot10^{4}$
3.4	$9.6879\cdot10^{3}$	$1.4998\cdot10^{4}$	6.9	$2.9983\cdot10^{3}$	$3.6030\cdot10^{4}$
3.5	$9.5334\cdot10^{3}$	$1.5959\cdot10^{4}$	7.0	$2.8905\cdot10^{3}$	$3.6324\cdot10^{4}$
3.6	$9.3545\cdot10^{3}$	$1.6903\cdot10^{4}$	7.1	$2.7871\cdot10^{3}$	$3.6608\cdot10^{4}$
3.7	$9.1558\cdot10^{3}$	$1.7829\cdot10^{4}$	7.2	$2.6879\cdot10^{3}$	$3.6881\cdot10^{4}$
3.8	$8.9415\cdot10^{3}$	$1.8734\cdot10^{4}$	7.3	$2.5927\cdot10^{3}$	$3.7145\cdot10^{4}$
3.9	$8.7151\cdot10^{3}$	$1.9617\cdot10^{4}$	7.4	$2.5012\cdot10^{3}$	$3.7400\cdot10^{4}$

Table 2.2. (*continued*)

λ [μm]	$m_{\lambda,bb}$ [W/μm m²]	$M_{T,o-\lambda,bb}$ [W/m²]	λ [μm]	$m_{\lambda,bb}$ [W/μm m²]	$M_{T,o-\lambda,bb}$ [W/m²]
7.5	$2.4135\cdot10^3$	$3.7646\cdot10^4$	11.5	$6.8094\cdot10^2$	$4.2984\cdot10^4$
7.6	$2.3292\cdot10^3$	$3.7883\cdot10^4$	11.6	$6.6230\cdot10^2$	$4.3051\cdot10^4$
7.7	$2.2484\cdot10^3$	$3.8112\cdot10^4$	11.7	$6.4429\cdot10^2$	$4.3116\cdot10^4$
7.8	$2.1707\cdot10^3$	$3.8333\cdot10^4$	11.8	$6.2687\cdot10^2$	$4.3180\cdot10^4$
7.9	$2.0962\cdot10^3$	$3.8546\cdot10^4$	11.9	$6.1002\cdot10^2$	$4.3242\cdot10^4$
8.0	$2.0246\cdot10^3$	$3.8752\cdot10^4$	12.0	$5.9372\cdot10^2$	$4\ 3302\cdot10^4$
8.1	$1.9559\cdot10^3$	$3.8951\cdot10^4$	12.1	$5.7796\cdot10^2$	$4.3361\cdot10^4$
8.2	$1.8899\cdot10^3$	$3.9143\cdot10^4$	12.2	$5.6270\cdot10^2$	$4.3418\cdot10^4$
8.3	$1.8265\cdot10^3$	$3.9329\cdot10^4$	12.3	$5.4793\cdot10^2$	$4.3473\cdot10^4$
8.4	$1.7655\cdot10^3$	$3.9509\cdot10^4$	12.4	$5.3364\cdot10^2$	$4.3527\cdot10^4$
8.5	$1.7070\cdot10^3$	$3.9682\cdot10^4$	12.5	$5.1980\cdot10^2$	$4.3580\cdot10^4$
8.6	$1.6507\cdot10^3$	$3.9850\cdot10^4$	12.6	$5.0640\cdot10^2$	$4.3631\cdot10^4$
8.7	$1.5967\cdot10^3$	$4.0013\cdot10^4$	12.7	$4.9342\cdot10^2$	$4.3681\cdot10^4$
8.8	$1.5447\cdot10^3$	$4.0170\cdot10^4$	12.8	$4.8084\cdot10^2$	$4.3730\cdot10^4$
8.9	$1.4947\cdot10^3$	$4.0322\cdot10^4$	12.9	$4.6866\cdot10^2$	$4.3777\cdot10^4$
9.0	$1.4466\cdot10^3$	$4.0469\cdot10^4$	13.0	$4.5686\cdot10^2$	$4.3824\cdot10^4$
9.1	$1.4004\cdot10^3$	$4.0611\cdot10^4$	13.1	$4.4541\cdot10^2$	$4.3869\cdot10^4$
9.2	$1.3559\cdot10^3$	$4.0749\cdot10^4$	13.2	$4.3432\cdot10^2$	$4.3913\cdot10^4$
9.3	$1.3131\cdot10^3$	$4.0882\cdot10^4$	13.3	$4.2357\cdot10^2$	$4.3956\cdot10^4$
9.4	$1.2719\cdot10^3$	$4.1011\cdot10^4$	13.4	$4.1314\cdot10^2$	$4.3997\cdot10^4$
9.5	$1.2322\cdot10^3$	$4.1137\cdot10^4$	13.5	$4.0303\cdot10^2$	$4.4038\cdot10^4$
9.6	$1.1940\cdot10^3$	$4.1258\cdot10^4$	13.6	$3.9322\cdot10^2$	$4.4078\cdot10^4$
9.7	$1.1573\cdot10^3$	$4.1375\cdot10^4$	13.7	$3.8370\cdot10^2$	$4.4117\cdot10^4$
9.8	$1.1219\cdot10^3$	$4.1489\cdot10^4$	13.8	$3.7446\cdot10^2$	$4.4155\cdot10^4$
9.9	$1.0878\cdot10^3$	$4.1600\cdot10^4$	13.9	$3.6550\cdot10^2$	$4.4192\cdot10^4$
10.0	$1.0549\cdot10^3$	$4.1707\cdot10^4$	14.0	$3.5680\cdot10^2$	$4.4228\cdot10^4$
10.1	$1.0232\cdot10^3$	$4.1811\cdot10^4$	14.1	$3.4835\cdot10^2$	$4.4263\cdot10^4$
10.2	$9.9268\cdot10^2$	$4.1912\cdot10^4$	14.2	$3.4015\cdot10^2$	$4.4298\cdot10^4$
10.3	$9.6324\cdot10^2$	$4.2009\cdot10^4$	14.3	$3.3219\cdot10^2$	$4.4331\cdot10^4$
10.4	$9.3485\cdot10^2$	$4.2104\cdot10^4$	14.4	$3.2445\cdot10^2$	$4.4364\cdot10^4$
10.5	$9.0747\cdot10^2$	$4.2196\cdot10^4$	14.5	$3.1694\cdot10^2$	$4.4396\cdot10^4$
10.6	$8.8106\cdot10^2$	$4.2286\cdot10^4$	14.6	$3.0964\cdot10^2$	$4.4427\cdot10^4$
10.7	$8.5558\cdot10^2$	$4.2373\cdot10^4$	14.7	$3.0254\cdot10^2$	$4.4458\cdot10^4$
10.8	$8.3099\cdot10^2$	$4.2457\cdot10^4$	14.8	$2.9565\cdot10^2$	$4.4488\cdot10^4$
10.9	$8.0725\cdot10^2$	$4.2539\cdot10^4$	14.9	$2.8895\cdot10^2$	$4.4517\cdot10^4$
11.0	$7.8433\cdot10^2$	$4.2618\cdot10^4$	15.0	$2.8243\cdot10^2$	$4.4546\cdot10^4$
11.1	$7.6219\cdot10^2$	$4.2696\cdot10^4$	15.1	$2.7610\cdot10^2$	$4.4574\cdot10^4$
11.2	$7.4082\cdot10^2$	$4.2771\cdot10^4$	15.2	$2.6994\cdot10^2$	$4.4601\cdot10^4$
11.3	$7.2017\cdot10^2$	$4.2844\cdot10^4$	15.3	$2.6395\cdot10^2$	$4.4628\cdot10^4$
11.4	$7.0022\cdot10^2$	$4.2915\cdot10^4$	15.4	$2.5812\cdot10^2$	$4.4654\cdot10^4$

Table 2.2. (*continued*)

temperature 1000 K

λ [μm]	$m_{\lambda,bb}$ [W/μm m²]	$M_{T,o-\lambda,bb}$ [W/m²]	λ [μm]	$m_{\lambda,bb}$ [W/μm m²]	$M_{T,o-\lambda,bb}$ [W/m²]
0.4	$8.7369 \cdot 10^{-6}$	$1.0572 \cdot 10^{-7}$			
0.5	$3.8115 \cdot 10^{-3}$	$7.3629 \cdot 10^{-5}$	4.0	$1.0297 \cdot 10^{4}$	$2.7266 \cdot 10^{4}$
0.6	$1.8537 \cdot 10^{-1}$	$5.2689 \cdot 10^{-3}$	4.1	$9.9614 \cdot 10^{3}$	$2.8279 \cdot 10^{4}$
0.7	$2.6368 \cdot 10^{0}$	$1.0424 \cdot 10^{-1}$	4.2	$9.6260 \cdot 10^{3}$	$2.9258 \cdot 10^{4}$
0.8	$1.7657 \cdot 10^{1}$	$9.3183 \cdot 10^{-1}$	4.3	$9.2932 \cdot 10^{3}$	$3.0204 \cdot 10^{4}$
0.9	$7.2280 \cdot 10^{1}$	$4.9343 \cdot 10^{0}$	4.4	$8.9645 \cdot 10^{3}$	$3.1117 \cdot 10^{4}$
1.0	$2.1111 \cdot 10^{2}$	$1.8187 \cdot 10^{1}$	4.5	$8.6413 \cdot 10^{3}$	$3.1997 \cdot 10^{4}$
1.1	$4.8485 \cdot 10^{2}$	$5.1666 \cdot 10^{1}$	4.6	$8.3248 \cdot 10^{3}$	$3.2846 \cdot 10^{4}$
1.2	$9.3334 \cdot 10^{2}$	$1.2101 \cdot 10^{2}$	4.7	$8.0158 \cdot 10^{3}$	$3.3663 \cdot 10^{4}$
1.3	$1.5732 \cdot 10^{3}$	$2.4474 \cdot 10^{2}$	4.8	$7.7150 \cdot 10^{3}$	$3.4449 \cdot 10^{4}$
1.4	$2.3944 \cdot 10^{3}$	$4.4171 \cdot 10^{2}$	4.9	$7.4228 \cdot 10^{3}$	$3.5206 \cdot 10^{4}$
1.5	$3.3647 \cdot 10^{3}$	$7.2860 \cdot 10^{2}$	5.0	$7.1397 \cdot 10^{3}$	$3.5934 \cdot 10^{4}$
1.6	$4.4379 \cdot 10^{3}$	$1.1181 \cdot 10^{3}$	5.1	$6.8657 \cdot 10^{3}$	$3.6634 \cdot 10^{4}$
1.7	$5.5629 \cdot 10^{3}$	$1.6179 \cdot 10^{3}$	5.2	$6.6011 \cdot 10^{3}$	$3.7307 \cdot 10^{4}$
1.8	$6.6902 \cdot 10^{3}$	$2.2307 \cdot 10^{3}$	5.3	$6.3459 \cdot 10^{3}$	$3.7955 \cdot 10^{4}$
1.9	$7.7771 \cdot 10^{3}$	$2.9546 \cdot 10^{3}$	5.4	$6.1000 \cdot 10^{3}$	$3.8577 \cdot 10^{4}$
2.0	$8.7897 \cdot 10^{3}$	$3.7836 \cdot 10^{3}$	5.5	$5.8632 \cdot 10^{3}$	$3.9175 \cdot 10^{4}$
2.1	$9.7037 \cdot 10^{3}$	$4.7092 \cdot 10^{3}$	5.6	$5.6356 \cdot 10^{3}$	$3.9750 \cdot 10^{4}$
2.2	$1.0504 \cdot 10^{4}$	$5.7205 \cdot 10^{3}$	5.7	$5.4168 \cdot 10^{3}$	$4.0302 \cdot 10^{4}$
2.3	$1.1182 \cdot 10^{4}$	$6.8058 \cdot 10^{3}$	5.8	$5.2067 \cdot 10^{3}$	$4.0833 \cdot 10^{4}$
2.4	$1.1737 \cdot 10^{4}$	$7.9528 \cdot 10^{3}$	5.9	$5.0051 \cdot 10^{3}$	$4.1344 \cdot 10^{4}$
2.5	$1.2171 \cdot 10^{4}$	$9.1491 \cdot 10^{3}$	6.0	$4.8117 \cdot 10^{3}$	$4.1835 \cdot 10^{4}$
2.6	$1.2492 \cdot 10^{4}$	$1.0383 \cdot 10^{4}$	6.1	$4.6261 \cdot 10^{3}$	$4.2306 \cdot 10^{4}$
2.7	$1.2708 \cdot 10^{4}$	$1.1644 \cdot 10^{4}$	6.2	$4.4483 \cdot 10^{3}$	$4.2760 \cdot 10^{4}$
2.8	$1.2830 \cdot 10^{4}$	$1.2922 \cdot 10^{4}$	6.3	$4.2778 \cdot 10^{3}$	$4.3196 \cdot 10^{4}$
2.9	$1.2867 \cdot 10^{4}$	$1.4207 \cdot 10^{4}$	6.4	$4.1145 \cdot 10^{3}$	$4.3616 \cdot 10^{4}$
3.0	$1.2830 \cdot 10^{4}$	$1.5493 \cdot 10^{4}$	6.5	$3.9580 \cdot 10^{3}$	$4.4019 \cdot 10^{4}$
3.1	$1.2730 \cdot 10^{4}$	$1.6771 \cdot 10^{4}$	6.6	$3.8082 \cdot 10^{3}$	$4.4408 \cdot 10^{4}$
3.2	$1.2576 \cdot 10^{4}$	$1.8037 \cdot 10^{4}$	6.7	$3.6646 \cdot 10^{3}$	$4.4781 \cdot 10^{4}$
3.3	$1.2376 \cdot 10^{4}$	$1.9285 \cdot 10^{4}$	6.8	$3.5271 \cdot 10^{3}$	$4.5141 \cdot 10^{4}$
3.4	$1.2140 \cdot 10^{4}$	$2.0511 \cdot 10^{4}$	6.9	$3.3954 \cdot 10^{3}$	$4.5487 \cdot 10^{4}$
3.5	$1.1875 \cdot 10^{4}$	$2.1712 \cdot 10^{4}$	7.0	$3.2693 \cdot 10^{3}$	$4.5820 \cdot 10^{4}$
3.6	$1.1586 \cdot 10^{4}$	$2.2885 \cdot 10^{4}$	7.1	$3.1485 \cdot 10^{3}$	$4.6141 \cdot 10^{4}$
3.7	$1.1279 \cdot 10^{4}$	$2.4028 \cdot 10^{4}$	7.2	$3.0328 \cdot 10^{3}$	$4.6450 \cdot 10^{4}$
3.8	$1.0959 \cdot 10^{4}$	$2.5140 \cdot 10^{4}$	7.3	$2.9219 \cdot 10^{3}$	$4.6748 \cdot 10^{4}$
3.9	$1.0631 \cdot 10^{4}$	$2.6220 \cdot 10^{4}$	7.4	$2.8157 \cdot 10^{3}$	$4.7034 \cdot 10^{4}$

Table 2.2. (*continued*)

λ [µm]	$m_{\lambda,bb}$ [W/µm m²]	$M_{T,o-\lambda,bb}$ [W/m²]	λ [µm]	$m_{\lambda,bb}$ [W/µm m²]	$M_{T,o-\lambda,bb}$ [W/m²]
7.5	$2.7140 \cdot 10^3$	$4.7311 \cdot 10^4$	11.5	$7.4587 \cdot 10^2$	$5.3239 \cdot 10^4$
7.6	$2.6166 \cdot 10^3$	$4.7577 \cdot 10^4$	11.6	$7.2517 \cdot 10^2$	$5.3313 \cdot 10^4$
7.7	$2.5231 \cdot 10^3$	$4.7834 \cdot 10^4$	11.7	$7.0517 \cdot 10^2$	$5.3384 \cdot 10^4$
7.8	$2.4336 \cdot 10^3$	$4.8082 \cdot 10^4$	11.8	$6.8584 \cdot 10^2$	$5.3454 \cdot 10^4$
7.9	$2.3478 \cdot 10^3$	$4.8321 \cdot 10^4$	11.9	$6.6715 \cdot 10^2$	$5.3521 \cdot 10^4$
8.0	$2.2655 \cdot 10^3$	$4.8552 \cdot 10^4$	12.0	$6.4909 \cdot 10^2$	$5.3587 \cdot 10^4$
8.1	$2.1866 \cdot 10^3$	$4.8774 \cdot 10^4$	12.1	$6.3162 \cdot 10^2$	$5.3651 \cdot 10^4$
8.2	$2.1110 \cdot 10^3$	$4.8989 \cdot 10^4$	12.2	$6.1473 \cdot 10^2$	$5.3714 \cdot 10^4$
8.3	$2.0384 \cdot 10^3$	$4.9197 \cdot 10^4$	12.3	$5.9839 \cdot 10^2$	$5.3774 \cdot 10^4$
8.4	$1.9688 \cdot 10^3$	$4.9397 \cdot 10^4$	12.4	$5.8258 \cdot 10^2$	$5.3833 \cdot 10^4$
8.5	$1.9019 \cdot 10^3$	$4.9591 \cdot 10^4$	12.5	$5.6728 \cdot 10^2$	$5.3891 \cdot 10^4$
8.6	$1.8378 \cdot 10^3$	$4.9777 \cdot 10^4$	12.6	$5.5247 \cdot 10^2$	$5.3947 \cdot 10^4$
8.7	$1.7762 \cdot 10^3$	$4.9958 \cdot 10^4$	12.7	$5.3814 \cdot 10^2$	$5.4001 \cdot 10^4$
8.8	$1.7171 \cdot 10^3$	$5.0133 \cdot 10^4$	12.8	$5.2426 \cdot 10^2$	$5.4054 \cdot 10^4$
8.9	$1.6603 \cdot 10^3$	$5.0302 \cdot 10^4$	12.9	$5.1081 \cdot 10^2$	$5.4106 \cdot 10^4$
9.0	$1.6058 \cdot 10^3$	$5.0465 \cdot 10^4$	13.0	$4.9779 \cdot 10^2$	$5.4157 \cdot 10^4$
9.1	$1.5534 \cdot 10^3$	$5.0623 \cdot 10^4$	13.1	$4.8518 \cdot 10^2$	$5.4206 \cdot 10^4$
9.2	$1.5030 \cdot 10^3$	$5.0776 \cdot 10^4$	13.2	$4.7296 \cdot 10^2$	$5.4254 \cdot 10^4$
9.3	$1.4546 \cdot 10^3$	$5.0924 \cdot 10^4$	13.3	$4.6111 \cdot 10^2$	$5.4300 \cdot 10^4$
9.4	$1.4080 \cdot 10^3$	$5.1067 \cdot 10^4$	13.4	$4.4963 \cdot 10^2$	$5.4346 \cdot 10^4$
9.5	$1.3633 \cdot 10^3$	$5.1205 \cdot 10^4$	13.5	$4.3850 \cdot 10^2$	$5.4390 \cdot 10^4$
9.6	$1.3202 \cdot 10^3$	$5.1339 \cdot 10^4$	13.6	$4.2771 \cdot 10^2$	$5.4434 \cdot 10^4$
9.7	$1.2788 \cdot 10^3$	$5.1469 \cdot 10^4$	13.7	$4.1724 \cdot 10^2$	$5.4476 \cdot 10^4$
9.8	$1.2390 \cdot 10^3$	$5.1595 \cdot 10^4$	13.8	$4.0709 \cdot 10^2$	$5.4517 \cdot 10^4$
9.9	$1.2006 \cdot 10^3$	$5.1717 \cdot 10^4$	13.9	$3.9724 \cdot 10^2$	$5.4557 \cdot 10^4$
10.0	$1.1637 \cdot 10^3$	$5.1835 \cdot 10^4$	14.0	$3.8768 \cdot 10^2$	$5.4596 \cdot 10^4$
10.1	$1.1281 \cdot 10^3$	$5.1950 \cdot 10^4$	14.1	$3.7841 \cdot 10^2$	$5.4635 \cdot 10^4$
10.2	$1.0939 \cdot 10^3$	$5.2061 \cdot 10^4$	14.2	$3.6940 \cdot 10^2$	$5.4672 \cdot 10^4$
10.3	$1.0609 \cdot 10^3$	$5.2169 \cdot 10^4$	14.3	$3.6066 \cdot 10^2$	$5.4709 \cdot 10^4$
10.4	$1.0291 \cdot 10^3$	$5.2273 \cdot 10^4$	14.4	$3.5218 \cdot 10^2$	$5.4744 \cdot 10^4$
10.5	$9.9845 \cdot 10^2$	$5.2375 \cdot 10^4$	14.5	$3.4394 \cdot 10^2$	$5.4779 \cdot 10^4$
10.6	$9.6892 \cdot 10^2$	$5.2473 \cdot 10^4$	14.6	$3.3594 \cdot 10^2$	$5.4813 \cdot 10^4$
10.7	$9.4044 \cdot 10^2$	$5.2568 \cdot 10^4$	14.7	$3.2816 \cdot 10^2$	$5.4846 \cdot 10^4$
10.8	$9.1298 \cdot 10^2$	$5.2661 \cdot 10^4$	14.8	$3.2061 \cdot 10^2$	$5.4879 \cdot 10^4$
10.9	$8.8649 \cdot 10^2$	$5.2751 \cdot 10^4$	14.9	$3.1327 \cdot 10^2$	$5.4910 \cdot 10^4$
11.0	$8.6093 \cdot 10^2$	$5.2838 \cdot 10^4$	15.0	$3.0614 \cdot 10^2$	$5.4941 \cdot 10^4$
11.1	$8.3627 \cdot 10^2$	$5.2923 \cdot 10^4$	15.1	$2.9921 \cdot 10^2$	$5.4972 \cdot 10^4$
11.2	$8.1247 \cdot 10^2$	$5.3006 \cdot 10^4$	15.2	$2.9247 \cdot 10^2$	$5.5001 \cdot 10^4$
11.3	$7.8949 \cdot 10^2$	$5.3086 \cdot 10^4$	15.3	$2.8592 \cdot 10^2$	$5.5030 \cdot 10^4$
11.4	$7.6730 \cdot 10^2$	$5.3164 \cdot 10^4$	15.4	$2.7955 \cdot 10^2$	$5.5058 \cdot 10^4$

Table 2.2. (*continued*)

temperature 1100 K

λ [μm]	$m_{\lambda,bb}$ [W/μm m²]	$M_{T,o-\lambda,bb}$ [W/m²]	λ [μm]	$m_{\lambda,bb}$ [W/μm m²]	$M_{T,o-\lambda,bb}$ [W/m²]
0.4	$2.2987 \cdot 10^{-4}$	$3.0862 \cdot 10^{-6}$			
0.5	$5.2144 \cdot 10^{-2}$	$1.1200 \cdot 10^{-3}$	4.0	$1.4437 \cdot 10^{4}$	$4.5559 \cdot 10^{4}$
0.6	$1.6398 \cdot 10^{0}$	$5.1941 \cdot 10^{-2}$	4.1	$1.3865 \cdot 10^{4}$	$4.6974 \cdot 10^{4}$
0.7	$1.7083 \cdot 10^{1}$	$7.5431 \cdot 10^{-1}$	4.2	$1.3307 \cdot 10^{4}$	$4.8332 \cdot 10^{4}$
0.8	$9.0571 \cdot 10^{1}$	$5.3504 \cdot 10^{0}$	4.3	$1.2763 \cdot 10^{4}$	$4.9635 \cdot 10^{4}$
0.9	$3.0916 \cdot 10^{2}$	$2.3679 \cdot 10^{1}$	4.4	$1.2235 \cdot 10^{4}$	$5.0885 \cdot 10^{4}$
1.0	$7.8085 \cdot 10^{2}$	$7.5645 \cdot 10^{1}$	4.5	$1.1725 \cdot 10^{4}$	$5.2083 \cdot 10^{4}$
1.1	$1.5923 \cdot 10^{3}$	$1.9124 \cdot 10^{2}$	4.6	$1.1232 \cdot 10^{4}$	$5.3231 \cdot 10^{4}$
1.2	$2.7760 \cdot 10^{3}$	$4.0658 \cdot 10^{2}$	4.7	$1.0757 \cdot 10^{4}$	$5.4330 \cdot 10^{4}$
1.3	$4.3028 \cdot 10^{3}$	$7.5792 \cdot 10^{2}$	4.8	$1.0301 \cdot 10^{4}$	$5.5383 \cdot 10^{4}$
1.4	$6.0949 \cdot 10^{3}$	$1.2760 \cdot 10^{3}$	4.9	$9.8630 \cdot 10^{3}$	$5.6391 \cdot 10^{4}$
1.5	$8.0479 \cdot 10^{3}$	$1.9823 \cdot 10^{3}$	5.0	$9.4428 \cdot 10^{3}$	$5.7356 \cdot 10^{4}$
1.6	$1.0053 \cdot 10^{4}$	$2.8873 \cdot 10^{3}$	5.1	$9.0403 \cdot 10^{3}$	$5.8280 \cdot 10^{4}$
1.7	$1.2010 \cdot 10^{4}$	$3.9912 \cdot 10^{3}$	5.2	$8.6551 \cdot 10^{3}$	$5.9165 \cdot 10^{4}$
1.8	$1.3841 \cdot 10^{4}$	$5.2851 \cdot 10^{3}$	5.3	$8.2867 \cdot 10^{3}$	$6.0011 \cdot 10^{4}$
1.9	$1.5489 \cdot 10^{4}$	$6.7533 \cdot 10^{3}$	5.4	$7.9346 \cdot 10^{3}$	$6.0822 \cdot 10^{4}$
2.0	$1.6916 \cdot 10^{4}$	$8.3755 \cdot 10^{3}$	5.5	$7.5983 \cdot 10^{3}$	$6.1599 \cdot 10^{4}$
2.1	$1.8107 \cdot 10^{4}$	$1.0129 \cdot 10^{4}$	5.6	$7.2771 \cdot 10^{3}$	$6.2343 \cdot 10^{4}$
2.2	$1.9057 \cdot 10^{4}$	$1.1989 \cdot 10^{4}$	5.7	$6.9706 \cdot 10^{3}$	$6.3055 \cdot 10^{4}$
2.3	$1.9776 \cdot 10^{4}$	$1.3932 \cdot 10^{4}$	5.8	$6.6782 \cdot 10^{3}$	$6.3737 \cdot 10^{4}$
2.4	$2.0278 \cdot 10^{4}$	$1.5937 \cdot 10^{4}$	5.9	$6.3992 \cdot 10^{3}$	$6.4391 \cdot 10^{4}$
2.5	$2.0583 \cdot 10^{4}$	$1.7981 \cdot 10^{4}$	6.0	$6.1331 \cdot 10^{3}$	$6.5017 \cdot 10^{4}$
2.6	$2.0713 \cdot 10^{4}$	$2.0047 \cdot 10^{4}$	6.1	$5.8793 \cdot 10^{3}$	$6.5618 \cdot 10^{4}$
2.7	$2.0692 \cdot 10^{4}$	$2.2119 \cdot 10^{4}$	6.2	$5.6373 \cdot 10^{3}$	$6.6194 \cdot 10^{4}$
2.8	$2.0541 \cdot 10^{4}$	$2.4181 \cdot 10^{4}$	6.3	$5.4065 \cdot 10^{3}$	$6.6746 \cdot 10^{4}$
2.9	$2.0281 \cdot 10^{4}$	$2.6223 \cdot 10^{4}$	6.4	$5.1864 \cdot 10^{3}$	$6.7275 \cdot 10^{4}$
3.0	$1.9932 \cdot 10^{4}$	$2.8235 \cdot 10^{4}$	6.5	$4.9765 \cdot 10^{3}$	$6.7783 \cdot 10^{4}$
3.1	$1.9511 \cdot 10^{4}$	$3.0207 \cdot 10^{4}$	6.6	$4.7763 \cdot 10^{3}$	$6.8271 \cdot 10^{4}$
3.2	$1.9034 \cdot 10^{4}$	$3.2135 \cdot 10^{4}$	6.7	$4.5853 \cdot 10^{3}$	$6.8739 \cdot 10^{4}$
3.3	$1.8513 \cdot 10^{4}$	$3.4013 \cdot 10^{4}$	6.8	$4.4032 \cdot 10^{3}$	$6.9188 \cdot 10^{4}$
3.4	$1.7961 \cdot 10^{4}$	$3.5837 \cdot 10^{4}$	6.9	$4.2294 \cdot 10^{3}$	$6.9619 \cdot 10^{4}$
3.5	$1.7387 \cdot 10^{4}$	$3.7604 \cdot 10^{4}$	7.0	$4.0636 \cdot 10^{3}$	$7.0034 \cdot 10^{4}$
3.6	$1.6799 \cdot 10^{4}$	$3.9313 \cdot 10^{4}$	7.1	$3.9053 \cdot 10^{3}$	$7.0432 \cdot 10^{4}$
3.7	$1.6205 \cdot 10^{4}$	$4.0964 \cdot 10^{4}$	7.2	$3.7542 \cdot 10^{3}$	$7.0815 \cdot 10^{4}$
3.8	$1.5611 \cdot 10^{4}$	$4.2554 \cdot 10^{4}$	7.3	$3.6100 \cdot 10^{3}$	$7.1183 \cdot 10^{4}$
3.9	$1.5020 \cdot 10^{4}$	$4.4086 \cdot 10^{4}$	7.4	$3.4722 \cdot 10^{3}$	$7.1537 \cdot 10^{4}$

λ [μm]	$m_{\lambda,bb}$ [W/μm m²]	$M_{T,o-\lambda,bb}$ [W/m²]	λ [μm]	$m_{\lambda,bb}$ [W/μm m²]	$M_{T,o-\lambda,bb}$ [W/m²]
7.5	$3.3407 \cdot 10^3$	$7.1878 \cdot 10^4$	11.5	$8.7813 \cdot 10^2$	$7.9022 \cdot 10^4$
7.6	$3.2150 \cdot 10^3$	$7.2206 \cdot 10^4$	11.6	$8.5318 \cdot 10^2$	$7.9109 \cdot 10^4$
7.7	$3.0949 \cdot 10^3$	$7.2521 \cdot 10^4$	11.7	$8.2909 \cdot 10^2$	$7.9193 \cdot 10^4$
7.8	$2.9801 \cdot 10^3$	$7.2825 \cdot 10^4$	11.8	$8.0584 \cdot 10^2$	$9.9274 \cdot 10^4$
7.9	$2.8703 \cdot 10^3$	$7.3117 \cdot 10^4$	11.9	$7.8339 \cdot 10^2$	$7.9354 \cdot 10^4$
8.0	$2.7654 \cdot 10^3$	$7.3399 \cdot 10^4$	12.0	$7.6170 \cdot 10^2$	$7.9431 \cdot 10^4$
8.1	$2.6650 \cdot 10^3$	$7.3670 \cdot 10^4$	12.1	$7.4075 \cdot 10^2$	$7.9506 \cdot 10^4$
8.2	$2.5690 \cdot 10^3$	$7.3932 \cdot 10^4$	12.2	$7.2050 \cdot 10^2$	$7.9579 \cdot 10^4$
8.3	$2.4770 \cdot 10^3$	$7.4184 \cdot 10^4$	12.3	$7.0094 \cdot 10^2$	$7.9650 \cdot 10^4$
8.4	$2.3890 \cdot 10^3$	$7.4427 \cdot 10^4$	12.4	$6.8202 \cdot 10^2$	$7.9719 \cdot 10^4$
8.5	$2.3048 \cdot 10^3$	$7.4662 \cdot 10^4$	12.5	$6.6373 \cdot 10^2$	$7.9786 \cdot 10^4$
8.6	$2.2241 \cdot 10^3$	$7.4888 \cdot 10^4$	12.6	$6.4604 \cdot 10^2$	$7.9852 \cdot 10^4$
8.7	$2.1468 \cdot 10^3$	$7.5107 \cdot 10^4$	12.7	$6.2893 \cdot 10^2$	$7.9916 \cdot 10^4$
8.8	$2.0727 \cdot 10^3$	$7.5318 \cdot 10^4$	12.8	$6.1238 \cdot 10^2$	$7.9978 \cdot 10^4$
8.9	$2.0017 \cdot 10^3$	$7.5522 \cdot 10^4$	12.9	$5.9636 \cdot 10^2$	$8.0038 \cdot 10^4$
9.0	$1.9336 \cdot 10^3$	$7.5718 \cdot 10^4$	13.0	$5.8085 \cdot 10^2$	$8.0097 \cdot 10^4$
9.1	$1.8683 \cdot 10^3$	$7.5908 \cdot 10^4$	13.1	$5.6584 \cdot 10^2$	$8.0154 \cdot 10^4$
9.2	$1.8056 \cdot 10^3$	$7.6092 \cdot 10^4$	13.2	$5.5131 \cdot 10^2$	$8.0210 \cdot 10^4$
9.3	$1.7455 \cdot 10^3$	$7.6269 \cdot 10^4$	13.3	$5.3724 \cdot 10^2$	$8.0264 \cdot 10^4$
9.4	$1.6878 \cdot 10^3$	$7.6441 \cdot 10^4$	13.4	$5.2360 \cdot 10^2$	$8.0317 \cdot 10^4$
9.5	$1.6325 \cdot 10^3$	$7.6607 \cdot 10^4$	13.5	$5.1040 \cdot 10^2$	$8.0369 \cdot 10^4$
9.6	$1.5793 \cdot 10^3$	$7.6768 \cdot 10^4$	13.6	$4.9760 \cdot 10^2$	$8.0419 \cdot 10^4$
9.7	$1.5282 \cdot 10^3$	$7.6923 \cdot 10^4$	13.7	$4.8520 \cdot 10^2$	$8.0468 \cdot 10^4$
9.8	$1.4791 \cdot 10^3$	$7.7073 \cdot 10^4$	13.8	$4.7317 \cdot 10^2$	$8.0516 \cdot 10^4$
9.9	$1.4319 \cdot 10^3$	$7.7219 \cdot 10^4$	13.9	$4.6152 \cdot 10^2$	$8.0563 \cdot 10^4$
10.0	$1.3865 \cdot 10^3$	$7.7360 \cdot 10^4$	14.0	$4.5021 \cdot 10^2$	$8.0609 \cdot 10^4$
10.1	$1.3429 \cdot 10^3$	$7.7496 \cdot 10^4$	14.1	$4.3925 \cdot 10^2$	$8.0653 \cdot 10^4$
10.2	$1.3010 \cdot 10^3$	$7.7628 \cdot 10^4$	14.2	$4.2862 \cdot 10^2$	$8.0696 \cdot 10^4$
10.3	$1.2606 \cdot 10^3$	$7.7756 \cdot 10^4$	14.3	$4.1830 \cdot 10^2$	$8.0739 \cdot 10^4$
10.4	$1.2218 \cdot 10^3$	$7.7880 \cdot 10^4$	14.4	$4.0829 \cdot 10^2$	$8.0780 \cdot 10^4$
10.5	$1.1844 \cdot 10^3$	$7.8000 \cdot 10^4$	14.5	$3.9858 \cdot 10^2$	$8.0820 \cdot 10^4$
10.6	$1.1484 \cdot 10^3$	$7.8117 \cdot 10^4$	14.6	$3.8915 \cdot 10^2$	$8.0860 \cdot 10^4$
10.7	$1.1138 \cdot 10^3$	$7.8230 \cdot 10^4$	14.7	$3.7999 \cdot 10^2$	$8.0898 \cdot 10^4$
10.8	$1.0804 \cdot 10^3$	$7.8340 \cdot 10^4$	14.8	$3.7110 \cdot 10^2$	$8.0936 \cdot 10^4$
10.9	$1.0482 \cdot 10^3$	$7.8446 \cdot 10^4$	14.9	$3.6246 \cdot 10^2$	$8.0972 \cdot 10^4$
11.0	$1.0172 \cdot 10^3$	$7.8549 \cdot 10^4$	15.0	$3.5408 \cdot 10^2$	$8.1008 \cdot 10^4$
11.1	$9.8736 \cdot 10^3$	$7.8650 \cdot 10^4$	15.1	$3.4593 \cdot 10^2$	$8.1043 \cdot 10^4$
11.2	$9.5855 \cdot 10^3$	$7.8747 \cdot 10^4$	15.2	$3.3802 \cdot 10^2$	$8.1077 \cdot 10^4$
11.3	$9.3077 \cdot 10^3$	$7.8841 \cdot 10^4$	15.3	$3.3032 \cdot 10^2$	$8.1111 \cdot 10^4$
11.4	$9.0398 \cdot 10^3$	$7.8933 \cdot 10^4$	15.4	$3.2285 \cdot 10^2$	$8.1143 \cdot 10^4$

Table 2.2. (*continued*)

temperature 1200 K

λ [μm]	$m_{\lambda,bb}$ [W/μm m²]	$M_{T,o-\lambda,bb}$ [W/m²]	λ [μm]	$m_{\lambda,bb}$ [W/μm m²]	$M_{T,o-\lambda,bb}$ [W/m²]
0.4	$3.5070 \cdot 10^{-3}$	$5.1805 \cdot 10^{-5}$			
0.5	$4.6127 \cdot 10^{-1}$	$1.0926 \cdot 10^{-2}$	4.0	$1.9197 \cdot 10^{4}$	$7.1433 \cdot 10^{4}$
0.6	$1.0087 \cdot 10^{1}$	$3.5311 \cdot 10^{-1}$	4.1	$1.8328 \cdot 10^{4}$	$7.3309 \cdot 10^{4}$
0.7	$8.1063 \cdot 10^{1}$	$3.9648 \cdot 10^{0}$	4.2	$1.7490 \cdot 10^{4}$	$7.5100 \cdot 10^{4}$
0.8	$3.5376 \cdot 10^{2}$	$2.3201 \cdot 10^{1}$	4.3	$1.6686 \cdot 10^{4}$	$7.6808 \cdot 10^{4}$
0.9	$1.0379 \cdot 10^{3}$	$8.8456 \cdot 10^{1}$	4.4	$1.5916 \cdot 10^{4}$	$7.8438 \cdot 10^{4}$
1.0	$2.3224 \cdot 10^{3}$	$2.5092 \cdot 10^{2}$	4.5	$1.5179 \cdot 10^{4}$	$7.9993 \cdot 10^{4}$
1.1	$4.2891 \cdot 10^{3}$	$5.7584 \cdot 10^{2}$	4.6	$1.4475 \cdot 10^{4}$	$8.1475 \cdot 10^{4}$
1.2	$6.8851 \cdot 10^{3}$	$1.1298 \cdot 10^{3}$	4.7	$1.3803 \cdot 10^{4}$	$8.2888 \cdot 10^{4}$
1.3	$9.9521 \cdot 10^{3}$	$1.9686 \cdot 10^{3}$	4.8	$1.3163 \cdot 10^{4}$	$8.4236 \cdot 10^{4}$
1.4	$1.3278 \cdot 10^{4}$	$3.1289 \cdot 10^{3}$	4.9	$1.2553 \cdot 10^{4}$	$8.5522 \cdot 10^{4}$
1.5	$1.6647 \cdot 10^{4}$	$4.6256 \cdot 10^{3}$	5.0	$1.1973 \cdot 10^{4}$	$8.6748 \cdot 10^{4}$
1.6	$1.9873 \cdot 10^{4}$	$6.4535 \cdot 10^{3}$	5.1	$1.1421 \cdot 10^{4}$	$8.7917 \cdot 10^{4}$
1.7	$2.2813 \cdot 10^{4}$	$8.5906 \cdot 10^{3}$	5.2	$1.0897 \cdot 10^{4}$	$8.9033 \cdot 10^{4}$
1.8	$2.5376 \cdot 10^{4}$	$1.1003 \cdot 10^{3}$	5.3	$1.0399 \cdot 10^{4}$	$9.0098 \cdot 10^{4}$
1.9	$2.7511 \cdot 10^{4}$	$1.3651 \cdot 10^{4}$	5.4	$9.9254 \cdot 10^{3}$	$9.1114 \cdot 10^{4}$
2.0	$2.9204 \cdot 10^{4}$	$1.6491 \cdot 10^{4}$	5.5	$9.4759 \cdot 10^{3}$	$9.2083 \cdot 10^{4}$
2.1	$3.0468 \cdot 10^{4}$	$1.9478 \cdot 10^{4}$	5.6	$9.0491 \cdot 10^{3}$	$9.3009 \cdot 10^{4}$
2.2	$3.1330 \cdot 10^{4}$	$2.2571 \cdot 10^{4}$	5.7	$8.6438 \cdot 10^{3}$	$9.3894 \cdot 10^{4}$
2.3	$3.1830 \cdot 10^{4}$	$2.5732 \cdot 10^{4}$	5.8	$8.2589 \cdot 10^{3}$	$9.4739 \cdot 10^{4}$
2.4	$3.2014 \cdot 10^{4}$	$2.8926 \cdot 10^{4}$	5.9	$7.8935 \cdot 10^{3}$	$9.5546 \cdot 10^{4}$
2.5	$3.1925 \cdot 10^{4}$	$3.2125 \cdot 10^{4}$	6.0	$7.5465 \cdot 10^{3}$	$9.6318 \cdot 10^{4}$
2.6	$3.1609 \cdot 10^{4}$	$3.5304 \cdot 10^{4}$	6.1	$7.2169 \cdot 10^{3}$	$9.7056 \cdot 10^{4}$
2.7	$3.1106 \cdot 10^{4}$	$3.8441 \cdot 10^{4}$	6.2	$6.9039 \cdot 10^{3}$	$9.7762 \cdot 10^{4}$
2.8	$3.0454 \cdot 10^{4}$	$4.1520 \cdot 10^{4}$	6.3	$6.6066 \cdot 10^{3}$	$9.8437 \cdot 10^{4}$
2.9	$2.9685 \cdot 10^{4}$	$4.4528 \cdot 10^{4}$	6.4	$6.3241 \cdot 10^{3}$	$9.9084 \cdot 10^{4}$
3.0	$2.8828 \cdot 10^{4}$	$4.7454 \cdot 10^{4}$	6.5	$6.0556 \cdot 10^{3}$	$9.9702 \cdot 10^{4}$
3.1	$2.7908 \cdot 10^{4}$	$5.0291 \cdot 10^{4}$	6.6	$5.8004 \cdot 10^{3}$	$1.0030 \cdot 10^{5}$
3.2	$2.6945 \cdot 10^{4}$	$5.3034 \cdot 10^{4}$	6.7	$5.5578 \cdot 10^{3}$	$1.0086 \cdot 10^{5}$
3.3	$2.5955 \cdot 10^{4}$	$5.5679 \cdot 10^{4}$	6.8	$5.3271 \cdot 10^{3}$	$1.0141 \cdot 10^{5}$
3.4	$2.4954 \cdot 10^{4}$	$5.8225 \cdot 10^{4}$	6.9	$5.1077 \cdot 10^{3}$	$1.0193 \cdot 10^{5}$
3.5	$2.3953 \cdot 10^{4}$	$6.0670 \cdot 10^{4}$	7.0	$4.8989 \cdot 10^{3}$	$1.0243 \cdot 10^{5}$
3.6	$2.2960 \cdot 10^{4}$	$6.3016 \cdot 10^{4}$	7.1	$4.7002 \cdot 10^{3}$	$1.0291 \cdot 10^{5}$
3.7	$2.1983 \cdot 10^{4}$	$6.5263 \cdot 10^{4}$	7.2	$4.5110 \cdot 10^{3}$	$1.0337 \cdot 10^{5}$
3.8	$2.1028 \cdot 10^{4}$	$6.7413 \cdot 10^{4}$	7.3	$4.3308 \cdot 10^{3}$	$1.0381 \cdot 10^{5}$
3.9	$2.0098 \cdot 10^{4}$	$6.9469 \cdot 10^{4}$	7.4	$4.1592 \cdot 10^{3}$	$1.0424 \cdot 10^{5}$

Table 2.2. (*continued*)

λ	$m_{\lambda,bb}$	$M_{T,o-\lambda,bb}$	λ	$m_{\lambda,bb}$	$M_{T,o-\lambda,bb}$
|μm|	|W/μm m²|	|W/m²|	|μm|	|W/μm m²|	|W/m²|
7.5	$3.9957 \cdot 10^3$	$1.0464 \cdot 10^5$	11.5	$1.0130 \cdot 10^3$	$1.1304 \cdot 10^5$
7.6	$3.8398 \cdot 10^3$	$1.0503 \cdot 10^5$	11.6	$9.8364 \cdot 10^2$	$1.1314 \cdot 10^5$
7.7	$3.6912 \cdot 10^3$	$1.0541 \cdot 10^5$	11.7	$9.5536 \cdot 10^2$	$1.1324 \cdot 10^5$
7.8	$3.5495 \cdot 10^3$	$1.0577 \cdot 10^5$	11.8	$9.2808 \cdot 10^2$	$1.1334 \cdot 10^5$
7.9	$3.4142 \cdot 10^3$	$1.0612 \cdot 10^5$	11.9	$9.0175 \cdot 10^2$	$1.1343 \cdot 10^5$
8.0	$3.2852 \cdot 10^3$	$1.0646 \cdot 10^5$	12.0	$8.7634 \cdot 10^2$	$1.1352 \cdot 10^5$
8.1	$3.1620 \cdot 10^3$	$1.0678 \cdot 10^5$	12.1	$8.5180 \cdot 10^2$	$1.1360 \cdot 10^5$
8.2	$3.0443 \cdot 10^3$	$1.0709 \cdot 10^5$	12.2	$8.2812 \cdot 10^2$	$1.1369 \cdot 10^5$
8.3	$2.9319 \cdot 10^3$	$1.0739 \cdot 10^5$	12.3	$8.0524 \cdot 10^2$	$1.1377 \cdot 10^5$
8.4	$2.8246 \cdot 10^3$	$1.0768 \cdot 10^5$	12.4	$7.8313 \cdot 10^2$	$1.1385 \cdot 10^5$
8.5	$2.7219 \cdot 10^3$	$1.0795 \cdot 10^5$	12.5	$7.6178 \cdot 10^2$	$1.1392 \cdot 10^5$
8.6	$2.6237 \cdot 10^3$	$1.0822 \cdot 10^5$	12.6	$7.4114 \cdot 10^2$	$1.1400 \cdot 10^5$
8.7	$2.5299 \cdot 10^3$	$1.0848 \cdot 10^5$	12.7	$7.2118 \cdot 10^2$	$1.1407 \cdot 10^5$
8.8	$2.4400 \cdot 10^3$	$1.0873 \cdot 10^5$	12.8	$7\ 0189 \cdot 10^2$	$1.1414 \cdot 10^5$
8.9	$2.3541 \cdot 10^3$	$1.0897 \cdot 10^5$	12.9	$6.8323 \cdot 10^2$	$1.1421 \cdot 10^5$
9.0	$2.2718 \cdot 10^3$	$1.0920 \cdot 10^5$	13.0	$6.6519 \cdot 10^2$	$1.1428 \cdot 10^5$
9.1	$2.1930 \cdot 10^3$	$1.0942 \cdot 10^5$	13.1	$6.4773 \cdot 10^2$	$1.1435 \cdot 10^5$
9.2	$2.1175 \cdot 10^3$	$1.0964 \cdot 10^5$	13.2	$6.3083 \cdot 10^2$	$1.1441 \cdot 10^5$
9.3	$2.0451 \cdot 10^3$	$1.0984 \cdot 10^5$	13.3	$6.1448 \cdot 10^2$	$1.1447 \cdot 10^5$
9.4	$1.9758 \cdot 10^3$	$1.1004 \cdot 10^5$	13.4	$5.9865 \cdot 10^2$	$1.1453 \cdot 10^5$
9.5	$1.9093 \cdot 10^3$	$1.1024 \cdot 10^5$	13.5	$5.8332 \cdot 10^2$	$1.1459 \cdot 10^5$
9.6	$1.8455 \cdot 10^3$	$1.1043 \cdot 10^5$	13.6	$5.6847 \cdot 10^2$	$1.1465 \cdot 10^5$
9.7	$1.7843 \cdot 10^3$	$1.1061 \cdot 10^5$	13.7	$5.5409 \cdot 10^2$	$1.1470 \cdot 10^5$
9.8	$1.7256 \cdot 10^3$	$1.1078 \cdot 10^5$	13.8	$5.4016 \cdot 10^2$	$1.1476 \cdot 10^5$
9.9	$1.6693 \cdot 10^3$	$1.1095 \cdot 10^5$	13.9	$5.2666 \cdot 10^2$	$1.1481 \cdot 10^5$
10.0	$1.6151 \cdot 10^3$	$1.1112 \cdot 10^5$	14.0	$5.1357 \cdot 10^2$	$1.1486 \cdot 10^5$
10.1	$1.5632 \cdot 10^3$	$1.1128 \cdot 10^5$	14.1	$5.0089 \cdot 10^2$	$1.1492 \cdot 10^5$
10.2	$1.5132 \cdot 10^3$	$1.1143 \cdot 10^5$	14.2	$4.8859 \cdot 10^2$	$1.1497 \cdot 10^5$
10.3	$1.4652 \cdot 10^3$	$1.1158 \cdot 10^5$	14.3	$4.7667 \cdot 10^2$	$1.1501 \cdot 10^5$
10.4	$1.4191 \cdot 10^3$	$1.1172 \cdot 10^5$	14.4	$4.6510 \cdot 10^2$	$1.1506 \cdot 10^5$
10.5	$1.3747 \cdot 10^3$	$1.1186 \cdot 10^5$	14.5	$4.5389 \cdot 10^2$	$1.1511 \cdot 10^5$
10.6	$1.3321 \cdot 10^3$	$1.1200 \cdot 10^5$	14.6	$4.4300 \cdot 10^2$	$1.1515 \cdot 10^5$
10.7	$1.2910 \cdot 10^3$	$1.1213 \cdot 10^5$	14.7	$4.3244 \cdot 10^2$	$1.1519 \cdot 10^5$
10.8	$1.2515 \cdot 10^3$	$1.1226 \cdot 10^5$	14.8	$4.2218 \cdot 10^2$	$1.1524 \cdot 10^5$
10.9	$1.2135 \cdot 10^3$	$1.1238 \cdot 10^5$	14.9	$4.1223 \cdot 10^2$	$1.1528 \cdot 10^5$
11.0	$1.1769 \cdot 10^3$	$1.1250 \cdot 10^5$	15.0	$4.0257 \cdot 10^2$	$1.1531 \cdot 10^5$
11.1	$1.1416 \cdot 10^3$	$1.1261 \cdot 10^5$	15.1	$3.9319 \cdot 10^2$	$1.1536 \cdot 10^5$
11.2	$1.1076 \cdot 10^3$	$1.1273 \cdot 10^5$	15.2	$3.8407 \cdot 10^2$	$1.1540 \cdot 10^5$
11.3	$1.0749 \cdot 10^3$	$1.1284 \cdot 10^5$	15.3	$3.7522 \cdot 10^2$	$1.1544 \cdot 10^5$
11.4	$1.0434 \cdot 10^3$	$1.1294 \cdot 10^5$	15.4	$3.6662 \cdot 10^2$	$1.1547 \cdot 10^5$

Table 2.2. (*continued*)

temperature 1300 K

λ [μm]	$m_{\lambda,bb}$ [W/μm m²]	$M_{T,o-\lambda,bb}$ [W/m²]	λ [μm]	$m_{\lambda,bb}$ [W/μm m²]	$M_{T,o-\lambda,bb}$ [W/m²]
0.4	$3.5181 \cdot 10^{-2}$	$5.6787 \cdot 10^{-4}$			
0.5	$2.9178 \cdot 10^{0}$	$7.5684 \cdot 10^{-2}$	4.0	$2.4510 \cdot 10^{4}$	$1.0655 \cdot 10^{5}$
0.6	$4.6919 \cdot 10^{1}$	$1.8027 \cdot 10^{0}$	4.1	$2.3284 \cdot 10^{4}$	$1.0894 \cdot 10^{5}$
0.7	$3.0272 \cdot 10^{2}$	$6.1287 \cdot 10^{1}$	4.2	$2.2117 \cdot 10^{4}$	$1.1121 \cdot 10^{5}$
0.8	$1.1205 \cdot 10^{3}$	$8.1020 \cdot 10^{1}$	4.3	$2.1008 \cdot 10^{4}$	$1.1337 \cdot 10^{5}$
0.9	$2.8921 \cdot 10^{3}$	$2.7238 \cdot 10^{2}$	4.4	$1.9954 \cdot 10^{4}$	$1.1542 \cdot 10^{5}$
1.0	$5.8412 \cdot 10^{3}$	$6.9901 \cdot 10^{2}$	4.5	$1.8954 \cdot 10^{4}$	$1.1736 \cdot 10^{5}$
1.1	$9.9200 \cdot 10^{3}$	$1.4785 \cdot 10^{3}$	4.6	$1.8007 \cdot 10^{4}$	$1.1921 \cdot 10^{5}$
1.2	$1.4850 \cdot 10^{4}$	$2.7115 \cdot 10^{3}$	4.7	$1.7109 \cdot 10^{4}$	$1.2096 \cdot 10^{5}$
1.3	$2.0234 \cdot 10^{4}$	$4.4636 \cdot 10^{3}$	4.8	$1.6260 \cdot 10^{4}$	$1.2263 \cdot 10^{5}$
1.4	$2.5664 \cdot 10^{4}$	$6.7597 \cdot 10^{3}$	4.9	$1.5456 \cdot 10^{4}$	$1.2422 \cdot 10^{5}$
1.5	$3.0797 \cdot 10^{4}$	$9.5864 \cdot 10^{3}$	5.0	$1.4696 \cdot 10^{4}$	$1.2572 \cdot 10^{5}$
1.6	$3.5383 \cdot 10^{4}$	$1.2901 \cdot 10^{4}$	5.1	$1.3977 \cdot 10^{4}$	$1.2716 \cdot 10^{5}$
1.7	$3.9272 \cdot 10^{4}$	$1.6640 \cdot 10^{4}$	5.2	$1.3297 \cdot 10^{4}$	$1.2852 \cdot 10^{5}$
1.8	$4.2396 \cdot 10^{4}$	$2.0729 \cdot 10^{4}$	5.3	$1.2655 \cdot 10^{4}$	$1.2982 \cdot 10^{5}$
1.9	$4.4752 \cdot 10^{4}$	$2.5093 \cdot 10^{4}$	5.4	$1.2047 \cdot 10^{4}$	$1.3105 \cdot 10^{5}$
2.0	$4.6383 \cdot 10^{4}$	$2.9656 \cdot 10^{4}$	5.5	$1.1473 \cdot 10^{4}$	$1.3223 \cdot 10^{5}$
2.1	$4.7356 \cdot 10^{4}$	$3.4348 \cdot 10^{4}$	5.6	$1.0930 \cdot 10^{4}$	$1.3335 \cdot 10^{5}$
2.2	$4.7753 \cdot 10^{4}$	$3.9107 \cdot 10^{4}$	5.7	$1.0416 \cdot 10^{4}$	$1.3441 \cdot 10^{5}$
2.3	$4.7662 \cdot 10^{4}$	$4.3882 \cdot 10^{4}$	5.8	$9.9303 \cdot 10^{3}$	$1.3543 \cdot 10^{5}$
2.4	$4.7165 \cdot 10^{4}$	$4.8626 \cdot 10^{4}$	5.9	$9.4708 \cdot 10^{3}$	$1.3640 \cdot 10^{5}$
2.5	$4.6341 \cdot 10^{4}$	$5.3304 \cdot 10^{4}$	6.0	$9.0359 \cdot 10^{3}$	$1.3733 \cdot 10^{5}$
2.6	$4.5261 \cdot 10^{4}$	$5.7886 \cdot 10^{4}$	6.1	$8.6242 \cdot 10^{3}$	$1.3821 \cdot 10^{5}$
2.7	$4.3986 \cdot 10^{4}$	$6.2350 \cdot 10^{4}$	6.2	$8.2345 \cdot 10^{3}$	$1.3905 \cdot 10^{5}$
2.8	$4.2567 \cdot 10^{4}$	$6.6678 \cdot 10^{4}$	6.3	$7.8654 \cdot 10^{3}$	$1.3986 \cdot 10^{5}$
2.9	$4.1050 \cdot 10^{4}$	$7.0860 \cdot 10^{4}$	6.4	$7.5158 \cdot 10^{3}$	$1.4063 \cdot 10^{5}$
3.0	$3.9471 \cdot 10^{4}$	$7.4886 \cdot 10^{4}$	6.5	$7.1845 \cdot 10^{3}$	$1.4136 \cdot 10^{5}$
3.1	$3.7859 \cdot 10^{4}$	$7.8753 \cdot 10^{4}$	6.6	$6.8704 \cdot 10^{3}$	$1.4206 \cdot 10^{5}$
3.2	$3.6338 \cdot 10^{4}$	$8.2457 \cdot 10^{4}$	6.7	$6.5726 \cdot 10^{3}$	$1.4274 \cdot 10^{5}$
3.3	$3.4628 \cdot 10^{4}$	$8.6001 \cdot 10^{4}$	6.8	$6.2901 \cdot 10^{3}$	$1.4338 \cdot 10^{5}$
3.4	$3.3043 \cdot 10^{4}$	$8.9384 \cdot 10^{4}$	6.9	$6.0220 \cdot 10^{3}$	$1.4399 \cdot 10^{5}$
3.5	$3.1494 \cdot 10^{4}$	$9.2610 \cdot 10^{4}$	7.0	$5.7675 \cdot 10^{3}$	$1.4458 \cdot 10^{5}$
3.6	$2.9989 \cdot 10^{4}$	$9.5684 \cdot 10^{4}$	7.1	$5.5259 \cdot 10^{3}$	$1.4515 \cdot 10^{5}$
3.7	$2.8535 \cdot 10^{4}$	$9.8610 \cdot 10^{4}$	7.2	$5.2963 \cdot 10^{3}$	$1.4569 \cdot 10^{5}$
3.8	$2.7136 \cdot 10^{4}$	$1.0139 \cdot 10^{5}$	7.3	$5.0781 \cdot 10^{3}$	$1.4621 \cdot 10^{5}$
3.9	$2.5793 \cdot 10^{4}$	$1.0404 \cdot 10^{5}$	7.4	$4.8707 \cdot 10^{3}$	$1.4670 \cdot 10^{5}$

Table 2.2. (*continued*)

λ [μm]	$m_{\lambda,bb}$ [W/μm m²]	$M_{T,o-\lambda,bb}$ [W/m²]	λ [μm]	$m_{\lambda,bb}$ [W/μm m²]	$M_{T,o-\lambda,bb}$ [W/m²]
7.5	$4.6735\cdot10^3$	$1.4718\cdot10^5$	11.5	$1.1498\cdot10^3$	$1.5687\cdot10^5$
7.6	$4.4858\cdot10^3$	$1.4764\cdot10^5$	11.6	$1.1160\cdot10^3$	$1.5699\cdot10^5$
7.7	$4.3072\cdot10^3$	$1.4808\cdot10^5$	11.7	$1.0835\cdot10^3$	$1.5710\cdot10^5$
7.8	$4.1372\cdot10^3$	$1.4850\cdot10^5$	11.8	$1.0521\cdot10^3$	$1.5720\cdot10^5$
7.9	$3.9752\cdot10^3$	$1.4891\cdot10^5$	11.9	$1.0218\cdot10^3$	$1.5731\cdot10^5$
8.0	$3.8209\cdot10^3$	$1.4930\cdot10^5$	12.0	$9.9255\cdot10^2$	$1.5741\cdot10^5$
8.1	$3.6739\cdot10^3$	$1.4967\cdot10^5$	12.1	$9.6437\cdot10^2$	$1.5750\cdot10^5$
8.2	$3.5336\cdot10^3$	$1.5003\cdot10^5$	12.2	$9.3717\cdot10^2$	$1.5760\cdot10^5$
8.3	$3.3999\cdot10^3$	$1.5038\cdot10^5$	12.3	$9.1091\cdot10^2$	$1.5769\cdot10^5$
8.4	$3.2722\cdot10^3$	$1.5071\cdot10^5$	12.4	$8.8556\cdot10^2$	$1.5778\cdot10^5$
8.5	$3.1504\cdot10^3$	$1.5103\cdot10^5$	12.5	$8.6108\cdot10^2$	$1.5787\cdot10^5$
8.6	$3.0341\cdot10^3$	$1.5134\cdot10^5$	12.6	$8.3743\cdot10^2$	$1.5795\cdot10^5$
8.7	$2.9230\cdot10^3$	$1.5164\cdot10^5$	12.7	$8.1458\cdot10^2$	$1.5804\cdot10^5$
8.8	$2.8168\cdot10^3$	$1.5193\cdot10^5$	12.8	$7.9250\cdot10^2$	$1.5812\cdot10^5$
8.9	$2.7153\cdot10^3$	$1.5220\cdot10^5$	12.9	$7.7115\cdot10^2$	$1.5819\cdot10^5$
9.0	$2.6182\cdot10^3$	$1.5247\cdot10^5$	13.0	$7.5052\cdot10^2$	$1.5827\cdot10^5$
9.1	$2.5254\cdot10^3$	$1.5273\cdot10^5$	13.1	$7.3056\cdot10^2$	$1.5834\cdot10^5$
9.2	$2.4366\cdot10^3$	$1.5297\cdot10^5$	13.2	$7.1126\cdot10^2$	$1.5842\cdot10^5$
9.3	$2.3516\cdot10^3$	$1.5321\cdot10^5$	13.3	$6.9259\cdot10^2$	$1.5849\cdot10^5$
9.4	$2.2702\cdot10^3$	$1.5344\cdot10^5$	13.4	$6.7452\cdot10^2$	$1.5855\cdot10^5$
9.5	$2.1922\cdot10^3$	$1.5367\cdot10^5$	13.5	$6.5703\cdot10^2$	$1.5862\cdot10^5$
9.6	$2.1175\cdot10^3$	$1.5388\cdot10^5$	13.6	$6.4011\cdot10^2$	$1.5869\cdot10^5$
9.7	$2.0459\cdot10^3$	$1.5409\cdot10^5$	13.7	$6.2372\cdot10^2$	$1.5875\cdot10^5$
9.8	$1.9773\cdot10^3$	$1.5429\cdot10^5$	13.8	$6.0785\cdot10^2$	$1.5881\cdot10^5$
9.9	$1.9114\cdot10^3$	$1.5449\cdot10^5$	13.9	$5.9247\cdot10^2$	$1.5887\cdot10^5$
10.0	$1.8483\cdot10^3$	$1.5467\cdot10^5$	14.0	$5.7758\cdot10^2$	$1.5893\cdot10^5$
10.1	$1.7877\cdot10^3$	$1.5486\cdot10^5$	14.1	$5.6315\cdot10^2$	$1.5899\cdot10^5$
10.2	$1.7295\cdot10^3$	$1.5503\cdot10^5$	14.2	$5.4916\cdot10^2$	$1.5904\cdot10^5$
10.3	$1.6737\cdot10^3$	$1.5520\cdot10^5$	14.3	$5.3561\cdot10^2$	$1.5910\cdot10^5$
10.4	$1.6200\cdot10^3$	$1.5537\cdot10^5$	14.4	$5.2246\cdot10^2$	$1.5915\cdot10^5$
10.5	$1.5685\cdot10^3$	$1.5553\cdot10^5$	14.5	$5.0972\cdot10^2$	$1.5920\cdot10^5$
10.6	$1.5189\cdot10^3$	$1.5568\cdot10^5$	14.6	$4.9736\cdot10^2$	$1.5925\cdot10^5$
10.7	$1.4713\cdot10^3$	$1.5583\cdot10^5$	14.7	$4.8537\cdot10^2$	$1.5930\cdot10^5$
10.8	$1.4255\cdot10^3$	$1.5598\cdot10^5$	14.8	$4.7373\cdot10^2$	$1.5935\cdot10^5$
10.9	$1.3815\cdot10^3$	$1.5612\cdot10^5$	14.9	$4.6245\cdot10^2$	$1.5939\cdot10^5$
11.0	$1.3391\cdot10^3$	$1.5625\cdot10^5$	15.0	$4.5149\cdot10^2$	$1.5944\cdot10^5$
11.1	$1.2983\cdot10^3$	$1.5638\cdot10^5$	15.1	$4.4085\cdot10^2$	$1.5949\cdot10^5$
11.2	$1.2591\cdot10^3$	$1.5651\cdot10^5$	15.2	$4.3053\cdot10^2$	$1.5953\cdot10^5$
11.3	$1.2213\cdot10^3$	$1.5663\cdot10^5$	15.3	$4.2050\cdot10^2$	$1.5957\cdot10^5$
11.4	$1.1849\cdot10^3$	$1.5676\cdot10^5$	15.4	$4.1076\cdot10^2$	$1.5961\cdot10^5$

Table 2.2. (*continued*)

temperature 1400 K

λ [μm]	$m_{\lambda,bb}$ [W/μm m²]	$M_{T,o-\lambda,bb}$ [W/m²]	λ [μm]	$m_{\lambda,bb}$ [W/μm m²]	$M_{T,o-\lambda,bb}$ [W/m²]
0.4	$2.5388\cdot10^{-1}$	$4.4514\cdot10^{-3}$			
0.5	$1.4181\cdot10^{1}$	$4.0046\cdot10^{-1}$	4.0	$3.0309\cdot10^{4}$	$1.5270\cdot10^{5}$
0.6	$1.7521\cdot10^{2}$	$7.3453\cdot10^{0}$	4.1	$2.8676\cdot10^{4}$	$1.5565\cdot10^{5}$
0.7	$9.3649\cdot10^{2}$	$5.5103\cdot10^{1}$	4.2	$2.7132\cdot10^{4}$	$1.5844\cdot10^{5}$
0.8	$3.0099\cdot10^{3}$	$2.3856\cdot10^{2}$	4.3	$2.5675\cdot10^{4}$	$1.6108\cdot10^{5}$
0.9	$6.9614\cdot10^{3}$	$7.2027\cdot10^{2}$	4.4	$2.4302\cdot10^{4}$	$1.6358\cdot10^{5}$
1.0	$1.2878\cdot10^{4}$	$1.6969\cdot10^{3}$	4.5	$2.3007\cdot10^{4}$	$1.6594\cdot10^{5}$
1.1	$2.0354\cdot10^{4}$	$3.3481\cdot10^{3}$	4.6	$2.1788\cdot10^{4}$	$1.6818\cdot10^{5}$
1.2	$2.8699\cdot10^{4}$	$5.7967\cdot10^{3}$	4.7	$2.0640\cdot10^{4}$	$1.7030\cdot10^{5}$
1.3	$3.7175\cdot10^{4}$	$9.0921\cdot10^{3}$	4.8	$1.9559\cdot10^{4}$	$1.7231\cdot10^{5}$
1.4	$4.5152\cdot10^{4}$	$1.3215\cdot10^{4}$	4.9	$1.8541\cdot10^{4}$	$1.7422\cdot10^{5}$
1.5	$5.2189\cdot10^{4}$	$1.8091\cdot10^{4}$	5.0	$1.7583\cdot10^{4}$	$1.7602\cdot10^{5}$
1.6	$5.8030\cdot10^{4}$	$2.3612\cdot10^{4}$	5.1	$1.6681\cdot10^{4}$	$1.7773\cdot10^{5}$
1.7	$6.2578\cdot10^{4}$	$2.9654\cdot10^{4}$	5.2	$1.5832\cdot10^{4}$	$1.7936\cdot10^{5}$
1.8	$6.5853\cdot10^{4}$	$3.6085\cdot10^{4}$	5.3	$1.5032\cdot10^{4}$	$1.8090\cdot10^{5}$
1.9	$6.7948\cdot10^{4}$	$4.2785\cdot10^{4}$	5.4	$1.4279\cdot10^{4}$	$1.8237\cdot10^{5}$
2.0	$6.9001\cdot10^{4}$	$4.9640\cdot10^{4}$	5.5	$1.3570\cdot10^{4}$	$1.8376\cdot10^{5}$
2.1	$6.9166\cdot10^{4}$	$5.6555\cdot10^{4}$	5.6	$1.2902\cdot10^{4}$	$1.8508\cdot10^{5}$
2.2	$6.8596\cdot10^{4}$	$6.3449\cdot10^{4}$	5.7	$1.2272\cdot10^{4}$	$1.8634\cdot10^{5}$
2.3	$6.7438\cdot10^{4}$	$7.0255\cdot10^{4}$	5.8	$1.1677\cdot10^{4}$	$1.8754\cdot10^{5}$
2.4	$6.5823\cdot10^{4}$	$7.6921\cdot10^{4}$	5.9	$1.1117\cdot10^{4}$	$1.8868\cdot10^{5}$
2.5	$6.3865\cdot10^{4}$	$8.3408\cdot10^{4}$	6.0	$1.0588\cdot10^{4}$	$1.8976\cdot10^{5}$
2.6	$6.1660\cdot10^{4}$	$8.9686\cdot10^{4}$	6.1	$1.0089\cdot10^{4}$	$1.9080\cdot10^{5}$
2.7	$5.9289\cdot10^{4}$	$9.5734\cdot10^{4}$	6.2	$9.6180\cdot10^{3}$	$1.9178\cdot10^{5}$
2.8	$5.6817\cdot10^{4}$	$1.0154\cdot10^{5}$	6.3	$9.1729\cdot10^{3}$	$1.9272\cdot10^{5}$
2.9	$5.4297\cdot10^{4}$	$1.0710\cdot10^{5}$	6.4	$8.7521\cdot10^{3}$	$1.9362\cdot10^{5}$
3.0	$5.1771\cdot10^{4}$	$1.1240\cdot10^{5}$	6.5	$8.3543\cdot10^{3}$	$1.9447\cdot10^{5}$
3.1	$4.9270\cdot10^{4}$	$1.1745\cdot10^{5}$	6.6	$7.9781\cdot10^{3}$	$1.9529\cdot10^{5}$
3.2	$4.6820\cdot10^{4}$	$1.2225\cdot10^{5}$	6.7	$7.6221\cdot10^{3}$	$1.9607\cdot10^{5}$
3.3	$4.4437\cdot10^{4}$	$1.2682\cdot10^{5}$	6.8	$7.2851\cdot10^{3}$	$1.9681\cdot10^{5}$
3.4	$4.2135\cdot10^{4}$	$1.3114\cdot10^{5}$	6.9	$6.9659\cdot10^{3}$	$1.9753\cdot10^{5}$
3.5	$3.9922\cdot10^{4}$	$1.3525\cdot10^{5}$	7.0	$6.6635\cdot10^{3}$	$1.9821\cdot10^{5}$
3.6	$3.7803\cdot10^{4}$	$1.3913\cdot10^{5}$	7.1	$6.3768\cdot10^{3}$	$1.9886\cdot10^{5}$
3.7	$3.5783\cdot10^{4}$	$1.4281\cdot10^{5}$	7.2	$6.1050\cdot10^{3}$	$1.9948\cdot10^{5}$
3.8	$3.3861\cdot10^{4}$	$1.4629\cdot10^{5}$	7.3	$5.8471\cdot10^{3}$	$2.0008\cdot10^{5}$
3.9	$3.2037\cdot10^{4}$	$1.4959\cdot10^{5}$	7.4	$5.6023\cdot10^{3}$	$2.0065\cdot10^{5}$

λ [μm]	$m_{\lambda,bb}$ [W/μm m²]	$M_{T,o-\lambda,bb}$ [W/m²]	λ [μm]	$m_{\lambda,bb}$ [W/μm m²]	$M_{T,o-\lambda,bb}$ [W/m²]
7.5	$5.3699 \cdot 10^3$	$2.0120 \cdot 10^5$	11.5	$1.2883 \cdot 10^3$	$2.1221 \cdot 10^5$
7.6	$5.1491 \cdot 10^3$	$2.0173 \cdot 10^5$	11.6	$1.2500 \cdot 10^3$	$2.1233 \cdot 10^5$
7.7	$4.9393 \cdot 10^3$	$2.0223 \cdot 10^5$	11.7	$1.2130 \cdot 10^3$	$2.1246 \cdot 10^5$
7.8	$4.7399 \cdot 10^3$	$2.0272 \cdot 10^5$	11.8	$1.1774 \cdot 10^3$	$2.1258 \cdot 10^5$
7.9	$4.5502 \cdot 10^3$	$2.0318 \cdot 10^5$	11.9	$1.1431 \cdot 10^3$	$2.1269 \cdot 10^5$
8.0	$4.3697 \cdot 10^3$	$2.0363 \cdot 10^5$	12.0	$1.1100 \cdot 10^3$	$2.1280 \cdot 10^5$
8.1	$4.1978 \cdot 10^3$	$2.0405 \cdot 10^5$	12.1	$1.0781 \cdot 10^3$	$2.1291 \cdot 10^5$
8.2	$4.0342 \cdot 10^3$	$2.0447 \cdot 10^5$	12.2	$1.0474 \cdot 10^3$	$2.1302 \cdot 10^5$
8.3	$3.8783 \cdot 10^3$	$2.0486 \cdot 10^5$	12.3	$1.0177 \cdot 10^3$	$2.1312 \cdot 10^5$
8.4	$3.7298 \cdot 10^3$	$2.0524 \cdot 10^5$	12.4	$9.8903 \cdot 10^2$	$2.1322 \cdot 10^5$
8.5	$3.5881 \cdot 10^3$	$2.0561 \cdot 10^5$	12.5	$9.6137 \cdot 10^2$	$2.1332 \cdot 10^5$
8.6	$3.4530 \cdot 10^3$	$2.0596 \cdot 10^5$	12.6	$9.3467 \cdot 10^2$	$2.1342 \cdot 10^5$
8.7	$3.3241 \cdot 10^3$	$2.0630 \cdot 10^5$	12.7	$9.0888 \cdot 10^2$	$2.1351 \cdot 10^5$
8.8	$3.2011 \cdot 10^3$	$2.0662 \cdot 10^5$	12.8	$8.8397 \cdot 10^2$	$2.1360 \cdot 10^5$
8.9	$3.0836 \cdot 10^3$	$2.0694 \cdot 10^5$	12.9	$8.5990 \cdot 10^2$	$2.1369 \cdot 10^5$
9.0	$2.9713 \cdot 10^3$	$2.0724 \cdot 10^5$	13.0	$8.3664 \cdot 10^2$	$2.1377 \cdot 10^5$
9.1	$2.8641 \cdot 10^3$	$2.0753 \cdot 10^5$	13.1	$8.1415 \cdot 10^2$	$2.1385 \cdot 10^5$
9.2	$2.7616 \cdot 10^3$	$2.0781 \cdot 10^5$	13.2	$7.9241 \cdot 10^2$	$2.1393 \cdot 10^5$
9.3	$2.6635 \cdot 10^3$	$2.0808 \cdot 10^5$	13.3	$7.7139 \cdot 10^2$	$2.1401 \cdot 10^5$
9.4	$2.5697 \cdot 10^3$	$2.0835 \cdot 10^5$	13.4	$7.5106 \cdot 10^2$	$2.1409 \cdot 10^5$
9.5	$2.4800 \cdot 10^3$	$2.0860 \cdot 10^5$	13.5	$7.3139 \cdot 10^2$	$2.1416 \cdot 10^5$
9.6	$2.3941 \cdot 10^3$	$2.0884 \cdot 10^5$	13.6	$7.1235 \cdot 10^2$	$2.1423 \cdot 10^5$
9.7	$2.3118 \cdot 10^3$	$2.0908 \cdot 10^5$	13.7	$6.9393 \cdot 10^2$	$2.1430 \cdot 10^5$
9.8	$2.2330 \cdot 10^3$	$2.0930 \cdot 10^5$	13.8	$6.7609 \cdot 10^2$	$2.1437 \cdot 10^5$
9.9	$2.1574 \cdot 10^3$	$2.0952 \cdot 10^5$	13.9	$6.5882 \cdot 10^2$	$2.1444 \cdot 10^5$
10.0	$2.0850 \cdot 10^3$	$2.0974 \cdot 10^5$	14.0	$6.4210 \cdot 10^2$	$2.1450 \cdot 10^5$
10.1	$2.0156 \cdot 10^3$	$2.0994 \cdot 10^5$	14.1	$6.2590 \cdot 10^2$	$2.1457 \cdot 10^5$
10.2	$1.9490 \cdot 10^3$	$2.1014 \cdot 10^5$	14.2	$6.1021 \cdot 10^2$	$2.1463 \cdot 10^5$
10.3	$1.8851 \cdot 10^3$	$2.1033 \cdot 10^5$	14.3	$5.9500 \cdot 10^2$	$2.1469 \cdot 10^5$
10.4	$1.8238 \cdot 10^3$	$2.1052 \cdot 10^5$	14.4	$5.8026 \cdot 10^2$	$2.1475 \cdot 10^5$
10.5	$1.7649 \cdot 10^3$	$2.1070 \cdot 10^5$	14.5	$5.6597 \cdot 10^2$	$2.1480 \cdot 10^5$
10.6	$1.7084 \cdot 10^3$	$2.1087 \cdot 10^5$	14.6	$5.5212 \cdot 10^2$	$2.1486 \cdot 10^5$
10.7	$1.6541 \cdot 10^3$	$2.1104 \cdot 10^5$	14.7	$5.3868 \cdot 10^2$	$2.1492 \cdot 10^5$
10.8	$1.6019 \cdot 10^3$	$2.1120 \cdot 10^5$	14.8	$5.2566 \cdot 10^2$	$2.1497 \cdot 10^5$
10.9	$1.5517 \cdot 10^3$	$2.1136 \cdot 10^5$	14.9	$5.1302 \cdot 10^2$	$2.1502 \cdot 10^5$
11.0	$1.5035 \cdot 10^3$	$2.1151 \cdot 10^5$	15.0	$5.0075 \cdot 10^2$	$2.1507 \cdot 10^5$
11.1	$1.4571 \cdot 10^3$	$2.1166 \cdot 10^5$	15.1	$4.8885 \cdot 10^2$	$2.1512 \cdot 10^5$
11.2	$1.4124 \cdot 10^3$	$2.1180 \cdot 10^5$	15.2	$4.7730 \cdot 10^2$	$2.1517 \cdot 10^5$
11.3	$1.3695 \cdot 10^3$	$2.1194 \cdot 10^5$	15.3	$4.6609 \cdot 10^2$	$2.1522 \cdot 10^5$
11.4	$1.3281 \cdot 10^3$	$2.1208 \cdot 10^5$	15.4	$4.5520 \cdot 10^2$	$2.1526 \cdot 10^5$

Table 2.2. (*continued*)

temperature 1500 K

λ [μm]	$m_{\lambda,bb}$ [W/μm m²]	$M_{T,o-\lambda,bb}$ [W/m²]	λ [μm]	$m_{\lambda,bb}$ [W/μm m²]	$M_{T.o-\lambda,bb}$ [W/m²]
0.4	$1.4077\cdot10^{0}$	$2.6674\cdot10^{-2}$			
0.5	$5.5823\cdot10^{1}$	$1.7074\cdot10^{0}$	4.0	$3.6538\cdot10^{4}$	$2.1179\cdot10^{5}$
0.6	$5.4888\cdot10^{2}$	$2.4980\cdot10^{1}$	4.1	$3.4447\cdot10^{4}$	$2.1534\cdot10^{5}$
0.7	$2.4922\cdot10^{3}$	$1.5955\cdot10^{2}$	4.2	$3.2485\cdot10^{4}$	$2.1868\cdot10^{5}$
0.8	$7.0875\cdot10^{3}$	$6.1261\cdot10^{2}$	4.3	$3.0644\cdot10^{4}$	$2.2184\cdot10^{5}$
0.9	$1.4904\cdot10^{4}$	$1.6856\cdot10^{3}$	4.4	$2.8918\cdot10^{4}$	$2.2481\cdot10^{5}$
1.0	$2.5551\cdot10^{4}$	$3.6886\cdot10^{3}$	4.5	$2.7300\cdot10^{4}$	$2.2762\cdot10^{5}$
1.1	$3.7947\cdot10^{4}$	$6.8543\cdot10^{3}$	4.6	$2.5784\cdot10^{4}$	$2.3028\cdot10^{5}$
1.2	$5.0804\cdot10^{4}$	$1.1293\cdot10^{4}$	4.7	$2.4362\cdot10^{4}$	$2.3278\cdot10^{5}$
1.3	$6.2986\cdot10^{4}$	$1.6992\cdot10^{4}$	4.8	$2.3030\cdot10^{4}$	$2.3515\cdot10^{5}$
1.4	$7.3688\cdot10^{4}$	$2.3840\cdot10^{4}$	4.9	$2.1781\cdot10^{4}$	$2.3739\cdot10^{5}$
1.5	$8.2454\cdot10^{4}$	$3.1665\cdot10^{4}$	5.0	$2.0610\cdot10^{4}$	$2.3951\cdot10^{5}$
1.6	$8.9124\cdot10^{4}$	$4.0261\cdot10^{4}$	5.1	$1.9511\cdot10^{4}$	$2.4152\cdot10^{5}$
1.7	$9.3748\cdot10^{4}$	$4.9421\cdot10^{4}$	5.2	$1.8480\cdot10^{4}$	$2.4342\cdot10^{5}$
1.8	$9.6505\cdot10^{4}$	$5.8948\cdot10^{4}$	5.3	$1.7513\cdot10^{4}$	$2.4521\cdot10^{5}$
1.9	$9.7641\cdot10^{4}$	$6.8668\cdot10^{4}$	5.4	$1.6605\cdot10^{4}$	$2.4692\cdot10^{5}$
2.0	$9.7428\cdot10^{4}$	$7.8431\cdot10^{4}$	5.5	$1.5752\cdot10^{4}$	$2.4854\cdot10^{5}$
2.1	$9.6128\cdot10^{4}$	$8.8117\cdot10^{4}$	5.6	$1.4950\cdot10^{4}$	$2.5007\cdot10^{5}$
2.2	$9.3984\cdot10^{4}$	$9.7629\cdot10^{4}$	5.7	$1.4197\cdot10^{4}$	$2.5153\cdot10^{5}$
2.3	$9.1207\cdot10^{4}$	$1.0689\cdot10^{5}$	5.8	$1.3488\cdot10^{4}$	$2.5291\cdot10^{5}$
2.4	$8.7978\cdot10^{4}$	$1.1585\cdot10^{5}$	5.9	$1.2821\cdot10^{4}$	$2.5423\cdot10^{5}$
2.5	$8.4444\cdot10^{4}$	$1.2448\cdot10^{5}$	6.0	$1.2194\cdot10^{4}$	$2.5548\cdot10^{5}$
2.6	$8.0726\cdot10^{4}$	$1.3274\cdot10^{5}$	6.1	$1.1603\cdot10^{4}$	$2.5667\cdot10^{5}$
2.7	$7.6920\cdot10^{4}$	$1.4062\cdot10^{5}$	6.2	$1.1046\cdot10^{4}$	$2.5780\cdot10^{5}$
2.8	$7.3098\cdot10^{4}$	$1.4812\cdot10^{5}$	6.3	$1.0521\cdot10^{4}$	$2.5888\cdot10^{5}$
2.9	$6.9317\cdot10^{4}$	$1.5524\cdot10^{5}$	6.4	$1.0026\cdot10^{4}$	$2.5990\cdot10^{5}$
3.0	$6.5620\cdot10^{4}$	$1.6199\cdot10^{5}$	6.5	$9.5583\cdot10^{3}$	$2.6088\cdot10^{5}$
3.1	$6.2035\cdot10^{4}$	$1.6837\cdot10^{5}$	6.6	$9.1171\cdot10^{3}$	$2.6182\cdot10^{5}$
3.2	$5.8585\cdot10^{4}$	$1.7440\cdot10^{5}$	6.7	$8.7004\cdot10^{3}$	$2.6271\cdot10^{5}$
3.3	$5.5283\cdot10^{4}$	$1.8009\cdot10^{5}$	6.8	$8.3066\cdot10^{3}$	$2.6356\cdot10^{5}$
3.4	$5.2137\cdot10^{4}$	$1.8546\cdot10^{5}$	6.9	$7.9342\cdot10^{3}$	$2.6437\cdot10^{5}$
3.5	$4.9149\cdot10^{4}$	$1.9052\cdot10^{5}$	7.0	$7.5820\cdot10^{3}$	$2.6515\cdot10^{5}$
3.6	$4.6321\cdot10^{4}$	$1.9529\cdot10^{5}$	7.1	$7.2486\cdot10^{3}$	$2.6589\cdot10^{5}$
3.7	$4.3651\cdot10^{4}$	$1.9979\cdot10^{5}$	7.2	$6.9329\cdot10^{3}$	$2.6660\cdot10^{5}$
3.8	$4.1134\cdot10^{4}$	$2.0403\cdot10^{5}$	7.3	$6.6339\cdot10^{3}$	$2.6727\cdot10^{5}$
3.9	$3.8765\cdot10^{4}$	$2.0802\cdot10^{5}$	7.4	$6.3504\cdot10^{3}$	$2.6792\cdot10^{5}$

λ [μm]	$m_{\lambda,bb}$ [W/μm m²]	$M_{T,o-\lambda,bb}$ [W/m²]	λ [μm]	$m_{\lambda,bb}$ [W/μm m²]	$M_{T,o-\lambda,bb}$ [W/m²]
7.5	$6.0817 \cdot 10^3$	$2.6854 \cdot 10^5$	11.5	$1.4281 \cdot 10^3$	$2.8089 \cdot 10^5$
7.6	$5.8267 \cdot 10^3$	$2.6914 \cdot 10^5$	11.6	$1.3851 \cdot 10^3$	$2.8103 \cdot 10^5$
7.7	$5.5847 \cdot 10^3$	$2.6971 \cdot 10^5$	11.7	$1.3438 \cdot 10^3$	$2.8116 \cdot 10^5$
7.8	$5.3549 \cdot 10^3$	$2.7026 \cdot 10^5$	11.8	$1.3039 \cdot 10^3$	$2.8130 \cdot 10^5$
7.9	$5.1365 \cdot 10^3$	$2.7078 \cdot 10^5$	11.9	$1.2655 \cdot 10^3$	$2.8142 \cdot 10^5$
8.0	$4.9290 \cdot 10^3$	$2.7128 \cdot 10^5$	12.0	$1.2285 \cdot 10^3$	$2.8155 \cdot 10^5$
8.1	$4.7317 \cdot 10^3$	$2.7177 \cdot 10^5$	12.1	$1.1929 \cdot 10^3$	$2.8167 \cdot 10^5$
8.2	$4.5440 \cdot 10^3$	$2.7223 \cdot 10^5$	12.2	$1.1585 \cdot 10^3$	$2.8179 \cdot 10^5$
8.3	$4.3654 \cdot 10^3$	$2.7268 \cdot 10^5$	12.3	$1.1253 \cdot 10^3$	$2.8190 \cdot 10^5$
8.4	$4.1953 \cdot 10^3$	$2.7310 \cdot 10^5$	12.4	$1.0933 \cdot 10^3$	$2.8201 \cdot 10^5$
8.5	$4.0334 \cdot 10^3$	$2.7352 \cdot 10^5$	12.5	$1.0625 \cdot 10^3$	$2.8212 \cdot 10^5$
8.6	$3.8790 \cdot 10^3$	$2.7391 \cdot 10^5$	12.6	$1.0327 \cdot 10^3$	$2.8222 \cdot 10^5$
8.7	$3.7319 \cdot 10^3$	$2.7429 \cdot 10^5$	12.7	$1.0039 \cdot 10^3$	$2.8233 \cdot 10^5$
8.8	$3.5915 \cdot 10^3$	$2.7466 \cdot 10^5$	12.8	$9.7614 \cdot 10^2$	$2.8243 \cdot 10^5$
8.9	$3.4576 \cdot 10^3$	$2.7501 \cdot 10^5$	12.9	$9.4931 \cdot 10^2$	$2.8252 \cdot 10^5$
9.0	$3.3299 \cdot 10^3$	$2.7535 \cdot 10^5$	13.0	$9.2339 \cdot 10^2$	$2.8261 \cdot 10^5$
9.1	$3.2079 \cdot 10^3$	$2.7568 \cdot 10^5$	13.1	$8.9835 \cdot 10^2$	$2.8271 \cdot 10^5$
9.2	$3.0913 \cdot 10^3$	$2.7599 \cdot 10^5$	13.2	$8.7415 \cdot 10^2$	$2.8279 \cdot 10^5$
9.3	$2.9800 \cdot 10^3$	$2.7629 \cdot 10^5$	13.3	$8.5075 \cdot 10^2$	$2.8288 \cdot 10^5$
9.4	$2.8735 \cdot 10^3$	$2.7659 \cdot 10^5$	13.4	$8.2813 \cdot 10^2$	$2.8296 \cdot 10^5$
9.5	$2.7717 \cdot 10^3$	$2.7687 \cdot 10^5$	13.5	$8.0625 \cdot 10^2$	$2.8205 \cdot 10^5$
9.6	$2.6744 \cdot 10^3$	$2.7714 \cdot 10^5$	13.6	$7.8508 \cdot 10^2$	$2.8313 \cdot 10^5$
9.7	$2.5812 \cdot 10^3$	$2.7740 \cdot 10^5$	13.7	$7.6461 \cdot 10^2$	$2.8320 \cdot 10^5$
9.8	$2.4920 \cdot 10^3$	$2.7766 \cdot 10^5$	13.8	$7.4479 \cdot 10^2$	$2.8328 \cdot 10^5$
9.9	$2.4066 \cdot 10^3$	$2.7790 \cdot 10^5$	13.9	$7.2560 \cdot 10^2$	$2.8335 \cdot 10^5$
10.0	$2.3248 \cdot 10^3$	$2.7814 \cdot 10^5$	14.0	$7.0703 \cdot 10^2$	$2.8342 \cdot 10^5$
10.1	$2.2464 \cdot 10^3$	$2.7837 \cdot 10^5$	14.1	$6.8905 \cdot 10^2$	$2.8349 \cdot 10^5$
10.2	$2.1712 \cdot 10^3$	$2.7859 \cdot 10^5$	14.2	$6.7163 \cdot 10^2$	$2.8356 \cdot 10^5$
10.3	$2.0991 \cdot 10^3$	$2.7880 \cdot 10^5$	14.3	$6.5475 \cdot 10^2$	$2.8363 \cdot 10^5$
10.4	$2.0300 \cdot 10^3$	$2.7901 \cdot 10^5$	14.4	$6.3840 \cdot 10^2$	$2.8369 \cdot 10^5$
10.5	$1.9637 \cdot 10^3$	$2.7921 \cdot 10^5$	14.5	$6.2256 \cdot 10^2$	$2.8376 \cdot 10^5$
10.6	$1.9000 \cdot 10^3$	$2.7940 \cdot 10^5$	14.6	$6.0720 \cdot 10^2$	$2.8382 \cdot 10^5$
10.7	$1.8389 \cdot 10^3$	$2.7959 \cdot 10^5$	14.7	$5.9231 \cdot 10^2$	$2.8388 \cdot 10^5$
10.8	$1.7801 \cdot 10^3$	$2.7977 \cdot 10^5$	14.8	$5.7787 \cdot 10^2$	$2.8394 \cdot 10^5$
10.9	$1.7237 \cdot 10^3$	$2.7994 \cdot 10^5$	14.9	$5.6387 \cdot 10^2$	$2.8399 \cdot 10^5$
11.0	$1.6695 \cdot 10^3$	$2.8011 \cdot 10^5$	15.0	$5.5029 \cdot 10^2$	$2.8405 \cdot 10^5$
11.1	$1.6174 \cdot 10^3$	$2.8028 \cdot 10^5$	15.1	$5.3711 \cdot 10^2$	$2.8410 \cdot 10^5$
11.2	$1.5673 \cdot 10^3$	$2.8044 \cdot 10^5$	15.2	$5.2433 \cdot 10^2$	$2.8416 \cdot 10^5$
11.3	$1.5191 \cdot 10^3$	$2.8059 \cdot 10^5$	15.3	$5.1192 \cdot 10^2$	$2.8421 \cdot 10^5$
11.4	$1.4727 \cdot 10^3$	$2.8074 \cdot 10^5$	15.4	$4.9987 \cdot 10^2$	$2.8426 \cdot 10^5$

Table 2.2. (*continued*)

temperature 1600 K

λ [μm]	$m_{\lambda,bb}$ [W/μm m²]	$M_{T,o-\lambda,bb}$ [W/m²]	λ [μm]	$m_{\lambda,bb}$ [W/μm m²]	$M_{T,o-\lambda,bb}$ [W/m²]
0.4	$6.3008 \cdot 10^{0}$	$1.2846 \cdot 10^{-1}$			
0.5	$1.8515 \cdot 10^{2}$	$6.1068 \cdot 10^{0}$	4.0	$4.3144 \cdot 10^{4}$	$2.8584 \cdot 10^{5}$
0.6	$1.4907 \cdot 10^{3}$	$7.3327 \cdot 10^{1}$	4.1	$4.0552 \cdot 10^{4}$	$2.9002 \cdot 10^{5}$
0.7	$5.8684 \cdot 10^{3}$	$4.0697 \cdot 10^{2}$	4.2	$3.8133 \cdot 10^{4}$	$2.9396 \cdot 10^{5}$
0.8	$1.4995 \cdot 10^{4}$	$1.4072 \cdot 10^{3}$	4.3	$3.5875 \cdot 10^{4}$	$2.9766 \cdot 10^{5}$
0.9	$2.9014 \cdot 10^{4}$	$3.5709 \cdot 10^{3}$	4.4	$3.3768 \cdot 10^{4}$	$3.0114 \cdot 10^{5}$
1.0	$4.4535 \cdot 10^{4}$	$7.3275 \cdot 10^{3}$	4.5	$3.1801 \cdot 10^{4}$	$3.0441 \cdot 10^{5}$
1.1	$6.5452 \cdot 10^{4}$	$1.2924 \cdot 10^{4}$	4.6	$2.9965 \cdot 10^{4}$	$3.0750 \cdot 10^{5}$
1.2	$8.3745 \cdot 10^{4}$	$2.0396 \cdot 10^{4}$	4.7	$2.8251 \cdot 10^{4}$	$3.1041 \cdot 10^{5}$
1.3	$9.9926 \cdot 10^{4}$	$2.9602 \cdot 10^{4}$	4.8	$2.6650 \cdot 10^{4}$	$3.1316 \cdot 10^{5}$
1.4	$1.1314 \cdot 10^{5}$	$4.0282 \cdot 10^{4}$	4.9	$2.5154 \cdot 10^{4}$	$3.1574 \cdot 10^{5}$
1.5	$1.2307 \cdot 10^{5}$	$5.2120 \cdot 10^{4}$	5.0	$2.3756 \cdot 10^{4}$	$3.1819 \cdot 10^{5}$
1.6	$1.2978 \cdot 10^{5}$	$6.4788 \cdot 10^{4}$	5.1	$2.2748 \cdot 10^{4}$	$3.2050 \cdot 10^{5}$
1.7	$1.3359 \cdot 10^{5}$	$7.7979 \cdot 10^{4}$	5.2	$2.1226 \cdot 10^{4}$	$3.2268 \cdot 10^{5}$
1.8	$1.3491 \cdot 10^{5}$	$9.1422 \cdot 10^{4}$	5.3	$2.0081 \cdot 10^{4}$	$3.2475 \cdot 10^{5}$
1.9	$1.3419 \cdot 10^{5}$	$1.0489 \cdot 10^{5}$	5.4	$1.9009 \cdot 10^{4}$	$3.2670 \cdot 10^{5}$
2.0	$1.3186 \cdot 10^{5}$	$1.1821 \cdot 10^{5}$	5.5	$1.8005 \cdot 10^{4}$	$3.2855 \cdot 10^{5}$
2.1	$1.2833 \cdot 10^{5}$	$3.1322 \cdot 10^{5}$	5.6	$1.7064 \cdot 10^{4}$	$3.3030 \cdot 10^{5}$
2.2	$1.2393 \cdot 10^{5}$	$1.4384 \cdot 10^{5}$	5.7	$1.6181 \cdot 10^{4}$	$3.3196 \cdot 10^{5}$
2.3	$1.1892 \cdot 10^{5}$	$1.5599 \cdot 10^{5}$	5.8	$1.5352 \cdot 10^{4}$	$3.3354 \cdot 10^{5}$
2.4	$1.1354 \cdot 10^{5}$	$1.6761 \cdot 10^{5}$	5.9	$1.4574 \cdot 10^{4}$	$3.3504 \cdot 10^{5}$
2.5	$1.0797 \cdot 10^{5}$	$1.7869 \cdot 10^{5}$	6.0	$1.3844 \cdot 10^{4}$	$3.3646 \cdot 10^{5}$
2.6	$1.0234 \cdot 10^{5}$	$1.8921 \cdot 10^{5}$	6.1	$1.3157 \cdot 10^{4}$	$3.3781 \cdot 10^{5}$
2.7	$9.6752 \cdot 10^{4}$	$1.9916 \cdot 10^{5}$	6.2	$1.2511 \cdot 10^{4}$	$3.3909 \cdot 10^{5}$
2.8	$9.1283 \cdot 10^{4}$	$2.0856 \cdot 10^{5}$	6.3	$1.1903 \cdot 10^{4}$	$3.4031 \cdot 10^{5}$
2.9	$8.5985 \cdot 10^{4}$	$2.1742 \cdot 10^{5}$	6.4	$1.1330 \cdot 10^{4}$	$3.4147 \cdot 10^{5}$
3.0	$8.0897 \cdot 10^{4}$	$2.2576 \cdot 10^{5}$	6.5	$1.0791 \cdot 10^{4}$	$3.4258 \cdot 10^{5}$
3.1	$7.6941 \cdot 10^{4}$	$2.3361 \cdot 10^{5}$	6.6	$1.0282 \cdot 10^{4}$	$3.4363 \cdot 10^{5}$
3.2	$7.1429 \cdot 10^{4}$	$2.4098 \cdot 10^{5}$	6.7	$9.8028 \cdot 10^{3}$	$3.4463 \cdot 10^{5}$
3.3	$6.7069 \cdot 10^{4}$	$2.4790 \cdot 10^{5}$	6.8	$9.3503 \cdot 10^{3}$	$3.4559 \cdot 10^{5}$
3.4	$6.2958 \cdot 10^{4}$	$2.5440 \cdot 10^{5}$	6.9	$8.9230 \cdot 10^{3}$	$3.4651 \cdot 10^{5}$
3.5	$5.9093 \cdot 10^{4}$	$2.6050 \cdot 10^{5}$	7.0	$8.5193 \cdot 10^{3}$	$3.4738 \cdot 10^{5}$
3.6	$5.5467 \cdot 10^{4}$	$2.6623 \cdot 10^{5}$	7.1	$8.1378 \cdot 10^{3}$	$3.4821 \cdot 10^{5}$
3.7	$5.2070 \cdot 10^{4}$	$2.7160 \cdot 10^{5}$	7.2	$7.7770 \cdot 10^{3}$	$3.4901 \cdot 10^{5}$
3.8	$4.8891 \cdot 10^{4}$	$2.7665 \cdot 10^{5}$	7.3	$7.4356 \cdot 10^{3}$	$3.4977 \cdot 10^{5}$
3.9	$4.5920 \cdot 10^{4}$	$2.8139 \cdot 10^{5}$	7.4	$7.1123 \cdot 10^{3}$	$3.5049 \cdot 10^{5}$

Table 2.2. (*continued*)

λ [µm]	$m_{\lambda,bb}$ [W/µm m²]	$M_{T,o-\lambda,bb}$ [W/m²]	λ [µm]	$m_{\lambda,bb}$ [W/µm m²]	$M_{T,o-\lambda,bb}$ [W/m²]
7.5	$7.5414 \cdot 10^3$	$4.5125 \cdot 10^5$	11.5	$1.7108 \cdot 10^3$	$4.6631 \cdot 10^5$
7.6	$7.2154 \cdot 10^3$	$4.5198 \cdot 10^5$	11.6	$1.6584 \cdot 10^3$	$4.6648 \cdot 10^5$
7.7	$6.9065 \cdot 10^3$	$4.5269 \cdot 10^5$	11.7	$1.6080 \cdot 10^3$	$4.6664 \cdot 10^5$
7.8	$6.6138 \cdot 10^3$	$4.5337 \cdot 10^5$	11.8	$1.5596 \cdot 10^3$	$4.6680 \cdot 10^5$
7.9	$6.3362 \cdot 10^3$	$4.5401 \cdot 10^5$	11.9	$1.5129 \cdot 10^3$	$4.6695 \cdot 10^5$
8.0	$6.0728 \cdot 10^3$	$4.5463 \cdot 10^5$	12.0	$1.4679 \cdot 10^3$	$4.6710 \cdot 10^5$
8.1	$5.8228 \cdot 10^3$	$4.5523 \cdot 10^5$	12.1	$1.4246 \cdot 10^3$	$4.6725 \cdot 10^5$
8.2	$5.5854 \cdot 10^3$	$4.5580 \cdot 10^5$	12.2	$1.3829 \cdot 10^3$	$4.6739 \cdot 10^5$
8.3	$5.3599 \cdot 10^3$	$4.5635 \cdot 10^5$	12.3	$1.3427 \cdot 10^3$	$4.6752 \cdot 10^5$
8.4	$5.1455 \cdot 10^3$	$4.5687 \cdot 10^5$	12.4	$1.3039 \cdot 10^3$	$4.6766 \cdot 10^5$
8.5	$4.9415 \cdot 10^3$	$4.5738 \cdot 10^5$	12.5	$1.2665 \cdot 10^3$	$4.6779 \cdot 10^5$
8.6	$4.7475 \cdot 10^3$	$4.5786 \cdot 10^5$	12.6	$1.2305 \cdot 10^3$	$4.6791 \cdot 10^5$
8.7	$4.5628 \cdot 10^3$	$4.5833 \cdot 10^5$	12.7	$1.1957 \cdot 10^3$	$4.6803 \cdot 10^5$
8.8	$4.3869 \cdot 10^3$	$4.5877 \cdot 10^5$	12.8	$1.1621 \cdot 10^3$	$4.6815 \cdot 10^5$
8.9	$4.2194 \cdot 10^3$	$4.5920 \cdot 10^5$	12.9	$1.1297 \cdot 10^3$	$4.6826 \cdot 10^5$
9.0	$4.0596 \cdot 10^3$	$4.5962 \cdot 10^5$	13.0	$1.0984 \cdot 10^3$	$4.6838 \cdot 10^5$
9.1	$3.9073 \cdot 10^3$	$4.6002 \cdot 10^5$	13.1	$1.0682 \cdot 10^3$	$4.6848 \cdot 10^5$
9.2	$3.7620 \cdot 10^3$	$4.6040 \cdot 10^5$	13.2	$1.0390 \cdot 10^3$	$4.6859 \cdot 10^5$
9.3	$3.6234 \cdot 10^3$	$4.6077 \cdot 10^5$	13.3	$1.0108 \cdot 10^3$	$4.6869 \cdot 10^5$
9.4	$3.4910 \cdot 10^3$	$4.6112 \cdot 10^5$	13.4	$9.8351 \cdot 10^2$	$4.6879 \cdot 10^5$
9.5	$3.3646 \cdot 10^3$	$4.6147 \cdot 10^5$	13.5	$9.5717 \cdot 10^2$	$4.6889 \cdot 10^5$
9.6	$3.2437 \cdot 10^3$	$4.6180 \cdot 10^5$	13.6	$9.3169 \cdot 10^2$	$4.6898 \cdot 10^5$
9.7	$3.1283 \cdot 10^3$	$4.6211 \cdot 10^5$	13.7	$9.0705 \cdot 10^2$	$4.6907 \cdot 10^5$
9.8	$3.0178 \cdot 10^3$	$4.6242 \cdot 10^5$	13.8	$8.8322 \cdot 10^2$	$4.6916 \cdot 10^5$
9.9	$2.9122 \cdot 10^3$	$4.6272 \cdot 10^5$	13.9	$8.6016 \cdot 10^2$	$4.6925 \cdot 10^5$
10.0	$2.8111 \cdot 10^3$	$4.6300 \cdot 10^5$	14.0	$8.3785 \cdot 10^2$	$4.6934 \cdot 10^5$
10.1	$2.7143 \cdot 10^3$	$4.6328 \cdot 10^5$	14.1	$8.1626 \cdot 10^2$	$4.6942 \cdot 10^5$
10.2	$2.6217 \cdot 10^3$	$4.6355 \cdot 10^5$	14.2	$7.9535 \cdot 10^2$	$4.6950 \cdot 10^5$
10.3	$2.5329 \cdot 10^3$	$4.6381 \cdot 10^5$	14.3	$7.7511 \cdot 10^2$	$4.6958 \cdot 10^5$
10.4	$2.4478 \cdot 10^3$	$4.6405 \cdot 10^5$	14.4	$7.5550 \cdot 10^2$	$4.6965 \cdot 10^5$
10.5	$2.3663 \cdot 10^3$	$4.6429 \cdot 10^5$	14.5	$7.3651 \cdot 10^2$	$4.6973 \cdot 10^5$
10.6	$2.2881 \cdot 10^3$	$4.6453 \cdot 10^5$	14.6	$7.1811 \cdot 10^2$	$4.6980 \cdot 10^5$
10.7	$2.2130 \cdot 10^3$	$4.6475 \cdot 10^5$	14.7	$7.0028 \cdot 10^2$	$4.6987 \cdot 10^5$
10.8	$2.1410 \cdot 10^3$	$4.6497 \cdot 10^5$	14.8	$6.8300 \cdot 10^2$	$4.6994 \cdot 10^5$
10.9	$2.0719 \cdot 10^3$	$4.6518 \cdot 10^5$	14.9	$6.6625 \cdot 10^2$	$4.7001 \cdot 10^5$
11.0	$2.0056 \cdot 10^3$	$4.6538 \cdot 10^5$	15.0	$6.5000 \cdot 10^2$	$4.7007 \cdot 10^5$
11.1	$1.9418 \cdot 10^3$	$4.6558 \cdot 10^5$	15.1	$6.3425 \cdot 10^2$	$4.7014 \cdot 10^5$
11.2	$1.8806 \cdot 10^3$	$4.6577 \cdot 10^5$	15.2	$6.1897 \cdot 10^2$	$4.7020 \cdot 10^5$
11.3	$1.8218 \cdot 10^3$	$4.6596 \cdot 10^5$	15.3	$6.0414 \cdot 10^2$	$4.7026 \cdot 10^5$
11.4	$1.7652 \cdot 10^3$	$4.6614 \cdot 10^5$	15.4	$5.8976 \cdot 10^2$	$4.7032 \cdot 10^5$

Table 2.2. (*continued*)

temperature 1700 K

λ [μm]	$m_{\lambda,bb}$ [W/μm m²]	$M_{T,o-\lambda,bb}$ [W/m²]	λ [μm]	$m_{\lambda,bb}$ [W/μm m²]	$M_{T,o-\lambda,bb}$ [W/m²]
0.4	$2.3644 \cdot 10^{1}$	$5.1662 \cdot 10^{-1}$			
0.5	$5.3331 \cdot 10^{2}$	$1.8895 \cdot 10^{1}$	4.0	$5.0080 \cdot 10^{4}$	$3.7702 \cdot 10^{5}$
0.6	$3.5998 \cdot 10^{3}$	$1.9063 \cdot 10^{2}$	4.1	$4.6948 \cdot 10^{4}$	$3.8187 \cdot 10^{5}$
0.7	$1.2494 \cdot 10^{4}$	$9.3497 \cdot 10^{2}$	4.2	$4.4038 \cdot 10^{4}$	$3.8642 \cdot 10^{5}$
0.8	$2.9048 \cdot 10^{4}$	$2.9483 \cdot 10^{3}$	4.3	$4.1333 \cdot 10^{4}$	$3.9068 \cdot 10^{5}$
0.9	$5.2225 \cdot 10^{4}$	$6.9676 \cdot 10^{3}$	4.4	$3.8819 \cdot 10^{4}$	$3.9469 \cdot 10^{5}$
1.0	$7.8986 \cdot 10^{4}$	$1.3513 \cdot 10^{4}$	4.5	$3.6481 \cdot 10^{4}$	$3.9845 \cdot 10^{5}$
1.1	$1.0589 \cdot 10^{5}$	$2.2768 \cdot 10^{4}$	4.6	$3.4306 \cdot 10^{4}$	$4.0199 \cdot 10^{5}$
1.2	$1.3018 \cdot 10^{5}$	$3.4602 \cdot 10^{4}$	4.7	$3.2282 \cdot 10^{4}$	$4.0532 \cdot 10^{5}$
1.3	$1.5018 \cdot 10^{5}$	$4.8659 \cdot 10^{4}$	4.8	$3.0397 \cdot 10^{4}$	$4.0845 \cdot 10^{5}$
1.4	$1.6521 \cdot 10^{5}$	$6.4471 \cdot 10^{4}$	4.9	$2.8641 \cdot 10^{4}$	$4.1140 \cdot 10^{5}$
1.5	$1.7529 \cdot 10^{5}$	$8.1535 \cdot 10^{4}$	5.0	$2.7005 \cdot 10^{4}$	$4.1419 \cdot 10^{5}$
1.6	$1.8089 \cdot 10^{5}$	$9.9378 \cdot 10^{4}$	5.1	$2.5478 \cdot 10^{4}$	$4.1681 \cdot 10^{5}$
1.7	$1.8269 \cdot 10^{5}$	$1.1759 \cdot 10^{5}$	5.2	$2.4054 \cdot 10^{4}$	$4.1928 \cdot 10^{5}$
1.8	$1.8141 \cdot 10^{5}$	$1.3581 \cdot 10^{5}$	5.3	$2.2724 \cdot 10^{4}$	$4.2162 \cdot 10^{5}$
1.9	$1.7777 \cdot 10^{5}$	$1.5379 \cdot 10^{5}$	5.4	$2.1481 \cdot 10^{4}$	$4.2383 \cdot 10^{5}$
2.0	$1.7238 \cdot 10^{5}$	$1.7131 \cdot 10^{5}$	5.5	$2.0319 \cdot 10^{4}$	$4.2592 \cdot 10^{5}$
2.1	$1.6576 \cdot 10^{5}$	$1.8822 \cdot 10^{5}$	5.6	$1.9233 \cdot 10^{4}$	$4.2790 \cdot 10^{5}$
2.2	$1.5834 \cdot 10^{5}$	$2.0443 \cdot 10^{5}$	5.7	$1.8215 \cdot 10^{4}$	$4.2977 \cdot 10^{5}$
2.3	$1.5047 \cdot 10^{5}$	$2.1988 \cdot 10^{5}$	5.8	$1.7262 \cdot 10^{4}$	$4.3154 \cdot 10^{5}$
2.4	$1.4239 \cdot 10^{5}$	$2.3452 \cdot 10^{5}$	5.9	$1.6369 \cdot 10^{4}$	$4.3322 \cdot 10^{5}$
2.5	$1.3431 \cdot 10^{5}$	$2.4835 \cdot 10^{5}$	6.0	$1.5531 \cdot 10^{4}$	$4.3482 \cdot 10^{5}$
2.6	$1.2636 \cdot 10^{5}$	$2.6139 \cdot 10^{5}$	6.1	$1.4745 \cdot 10^{4}$	$4.3633 \cdot 10^{5}$
2.7	$1.1864 \cdot 10^{5}$	$2.7363 \cdot 10^{5}$	6.2	$1.4007 \cdot 10^{4}$	$4.3777 \cdot 10^{5}$
2.8	$1.1124 \cdot 10^{5}$	$2.8512 \cdot 10^{5}$	6.3	$1.3313 \cdot 10^{4}$	$4.3913 \cdot 10^{5}$
2.9	$1.0417 \cdot 10^{5}$	$2.9589 \cdot 10^{5}$	6.4	$1.2661 \cdot 10^{4}$	$4.4043 \cdot 10^{5}$
3.0	$9.7483 \cdot 10^{4}$	$3.0597 \cdot 10^{5}$	6.5	$1.2047 \cdot 10^{4}$	$4.4167 \cdot 10^{5}$
3.1	$9.1176 \cdot 10^{4}$	$3.1540 \cdot 10^{5}$	6.6	$1.1470 \cdot 10^{4}$	$4.4284 \cdot 10^{5}$
3.2	$8.5250 \cdot 10^{4}$	$3.2422 \cdot 10^{5}$	6.7	$1.0926 \cdot 10^{4}$	$4.4396 \cdot 10^{5}$
3.3	$7.9701 \cdot 10^{4}$	$3.3246 \cdot 10^{5}$	6.8	$1.0413 \cdot 10^{4}$	$4.4503 \cdot 10^{5}$
3.4	$7.4515 \cdot 10^{4}$	$3.4017 \cdot 10^{5}$	6.9	$9.9290 \cdot 10^{3}$	$4.4605 \cdot 10^{5}$
3.5	$6.9677 \cdot 10^{4}$	$3.4738 \cdot 10^{5}$	7.0	$9.4726 \cdot 10^{3}$	$4.4702 \cdot 10^{5}$
3.6	$6.5171 \cdot 10^{4}$	$3.5412 \cdot 10^{5}$	7.1	$9.0417 \cdot 10^{3}$	$4.4794 \cdot 10^{5}$
3.7	$6.0977 \cdot 10^{4}$	$3.6042 \cdot 10^{5}$	7.2	$8.6346 \cdot 10^{3}$	$4.4883 \cdot 10^{5}$
3.8	$5.7076 \cdot 10^{4}$	$3.6632 \cdot 10^{5}$	7.3	$8.2498 \cdot 10^{3}$	$4.4967 \cdot 10^{5}$
3.9	$5.3450 \cdot 10^{4}$	$3.7185 \cdot 10^{5}$	7.4	$7.8858 \cdot 10^{3}$	$4.5048 \cdot 10^{5}$

Table 2.2. (*continued*)

λ [µm]	$m_{\lambda,bb}$ [W/µm m²]	$M_{T,o-\lambda,bb}$ [W/m²]	λ [µm]	$m_{\lambda,bb}$ [W/µm m²]	$M_{T,o-\lambda,bb}$ [W/m²]
7.5	$6.8062 \cdot 10^3$	$3.5119 \cdot 10^5$	11.5	$1.5690 \cdot 10^3$	$3.6488 \cdot 10^5$
7.6	$6.5161 \cdot 10^3$	$3.5185 \cdot 10^5$	11.6	$1.5214 \cdot 10^3$	$3.6504 \cdot 10^5$
7.7	$6.2410 \cdot 10^3$	$3.5249 \cdot 10^5$	11.7	$1.4755 \cdot 10^3$	$3.6519 \cdot 10^5$
7.8	$5.9800 \cdot 10^3$	$3.5310 \cdot 10^5$	11.8	$1.4314 \cdot 10^3$	$3.6533 \cdot 10^5$
7.9	$5.7324 \cdot 10^3$	$3.5369 \cdot 10^5$	11.9	$1.3888 \cdot 10^3$	$3.6547 \cdot 10^5$
8.0	$5.4972 \cdot 10^3$	$3.5425 \cdot 10^5$	12.0	$1.3479 \cdot 10^3$	$3.6561 \cdot 10^5$
8.1	$5.2738 \cdot 10^3$	$3.5479 \cdot 10^5$	12.1	$1.3084 \cdot 10^3$	$3.6574 \cdot 10^5$
8.2	$5.0615 \cdot 10^3$	$3.5531 \cdot 10^5$	12.2	$1.2704 \cdot 10^3$	$3.6587 \cdot 10^5$
8.3	$4.8596 \cdot 10^3$	$3.5580 \cdot 10^5$	12.3	$1.2337 \cdot 10^3$	$3.6600 \cdot 10^5$
8.4	$4.6676 \cdot 10^3$	$3.5628 \cdot 10^5$	12.4	$1.1983 \cdot 10^3$	$3.6612 \cdot 10^5$
8.5	$4.4848 \cdot 10^3$	$3.5673 \cdot 10^5$	12.5	$1.1642 \cdot 10^3$	$3.6624 \cdot 10^5$
8.6	$4.3108 \cdot 10^3$	$3.5717 \cdot 10^5$	12.6	$1.1313 \cdot 10^3$	$3.6635 \cdot 10^5$
8.7	$4.1451 \cdot 10^3$	$3.5760 \cdot 10^5$	12.7	$1.0995 \cdot 10^3$	$3.6646 \cdot 10^5$
8.8	$3.9871 \cdot 10^3$	$3.5800 \cdot 10^5$	12.8	$1.0689 \cdot 10^3$	$3.6657 \cdot 10^5$
8.9	$3.8365 \cdot 10^3$	$3.5839 \cdot 10^5$	12.9	$1.0393 \cdot 10^3$	$3.6668 \cdot 10^5$
9.0	$3.6929 \cdot 10^3$	$3.5877 \cdot 10^5$	13.0	$1.0107 \cdot 10^3$	$3.6678 \cdot 10^5$
9.1	$3.5558 \cdot 10^3$	$3.5913 \cdot 10^5$	13.1	$9.8306 \cdot 10^2$	$3.6688 \cdot 10^5$
9.2	$3.4250 \cdot 10^3$	$3.5948 \cdot 10^5$	13.2	$9.5637 \cdot 10^2$	$3.6698 \cdot 10^5$
9.3	$3.3001 \cdot 10^3$	$3.5982 \cdot 10^5$	13.3	$9.3057 \cdot 10^2$	$3.6707 \cdot 10^5$
9.4	$3.1808 \cdot 10^3$	$3.6014 \cdot 10^5$	13.4	$9.0564 \cdot 10^2$	$3.6716 \cdot 10^5$
9.5	$3.0668 \cdot 10^3$	$3.6045 \cdot 10^5$	13.5	$8.8153 \cdot 10^2$	$3.6725 \cdot 10^5$
9.6	$2.9578 \cdot 10^3$	$3.6076 \cdot 10^5$	13.6	$8.5822 \cdot 10^2$	$3.6734 \cdot 10^5$
9.7	$2.8535 \cdot 10^3$	$3.6105 \cdot 10^5$	13.7	$8.3567 \cdot 10^2$	$3.6742 \cdot 10^5$
9.8	$2.7538 \cdot 10^3$	$3.6133 \cdot 10^5$	13.8	$8.1385 \cdot 10^2$	$3.6751 \cdot 10^5$
9.9	$2.6583 \cdot 10^3$	$3.6160 \cdot 10^5$	13.9	$7.9274 \cdot 10^2$	$3.6759 \cdot 10^5$
10.0	$2.5669 \cdot 10^3$	$3.6186 \cdot 10^5$	14.0	$7.7230 \cdot 10^2$	$3.6766 \cdot 10^5$
10.1	$2.4794 \cdot 10^3$	$3.6211 \cdot 10^5$	14.1	$7.5252 \cdot 10^2$	$3.6774 \cdot 10^5$
10.2	$2.3955 \cdot 10^3$	$3.6235 \cdot 10^5$	14.2	$7.3336 \cdot 10^2$	$3.6782 \cdot 10^5$
10.3	$2.3152 \cdot 10^3$	$3.6259 \cdot 10^5$	14.3	$7.1481 \cdot 10^2$	$3.6789 \cdot 10^5$
10.4	$2.2381 \cdot 10^3$	$3.6282 \cdot 10^5$	14.4	$6.9684 \cdot 10^2$	$3.6796 \cdot 10^5$
10.5	$2.1642 \cdot 10^3$	$3.6304 \cdot 10^5$	14.5	$6.7942 \cdot 10^2$	$3.6803 \cdot 10^5$
10.6	$2.0933 \cdot 10^3$	$3.6325 \cdot 10^5$	14.6	$6.6255 \cdot 10^2$	$3.6809 \cdot 10^5$
10.7	$2.0253 \cdot 10^3$	$3.6346 \cdot 10^5$	14.7	$6.4619 \cdot 10^2$	$3.6816 \cdot 10^5$
10.8	$1.9599 \cdot 10^3$	$3.6366 \cdot 10^5$	14.8	$6.3034 \cdot 10^2$	$3.6822 \cdot 10^5$
10.9	$1.8972 \cdot 10^3$	$3.6385 \cdot 10^5$	14.9	$6.1496 \cdot 10^2$	$3.6829 \cdot 10^5$
11.0	$1.8370 \cdot 10^3$	$3.6404 \cdot 10^5$	15.0	$6.0005 \cdot 10^2$	$3.6835 \cdot 10^5$
11.1	$1.7791 \cdot 10^3$	$3.6422 \cdot 10^5$	15.1	$5.8559 \cdot 10^2$	$3.6841 \cdot 10^5$
11.2	$1.7234 \cdot 10^3$	$3.6439 \cdot 10^5$	15.2	$5.7156 \cdot 10^2$	$3.6846 \cdot 10^5$
11.3	$1.6699 \cdot 10^3$	$3.6456 \cdot 10^5$	15.3	$5.5795 \cdot 10^2$	$3.6852 \cdot 10^5$
11.4	$1.6185 \cdot 10^3$	$3.6472 \cdot 10^5$	15.4	$5.4473 \cdot 10^2$	$3.6857 \cdot 10^5$

Table 2.2. (*continued*)

temperature 1800 K

λ [µm]	$m_{\lambda,bb}$ [W/µm m²]	$M_{T,o-\lambda,bb}$ [W/m²]	λ [µm]	$m_{\lambda,bb}$ [W/µm m²]	$M_{T,o-\lambda,bb}$ [W/m²]
0.4	$7.6598\cdot10^{1}$	$1.7876\cdot10^{0}$			
0.5	$1.3658\cdot10^{3}$	$5.1798\cdot10^{1}$	4.0	$5.7306\cdot10^{4}$	$4.8761\cdot10^{5}$
0.6	$7.8816\cdot10^{3}$	$4.4781\cdot10^{2}$	4.1	$5.3600\cdot10^{4}$	$4.9316\cdot10^{5}$
0.7	$2.4458\cdot10^{4}$	$1.9682\cdot10^{3}$	4.2	$5.0169\cdot10^{4}$	$4.9834\cdot10^{5}$
0.8	$5.2284\cdot10^{4}$	$5.7198\cdot10^{3}$	4.3	$4.6991\cdot10^{4}$	$5.0320\cdot10^{5}$
0.9	$8.8063\cdot10^{4}$	$1.2692\cdot10^{4}$	4.4	$4.4047\cdot10^{4}$	$5.0775\cdot10^{5}$
1.0	$1.2642\cdot10^{5}$	$2.3417\cdot10^{4}$	4.5	$4.1318\cdot10^{4}$	$5.1201\cdot10^{5}$
1.1	$1.6240\cdot10^{5}$	$3.7895\cdot10^{4}$	4.6	$3.8786\cdot10^{4}$	$5.1602\cdot10^{5}$
1.2	$1.9270\cdot10^{5}$	$5.5706\cdot10^{4}$	4.7	$3.6437\cdot10^{4}$	$5.1978\cdot10^{5}$
1.3	$2.1576\cdot10^{5}$	$7.6191\cdot10^{4}$	4.8	$3.4255\cdot10^{4}$	$5.2331\cdot10^{5}$
1.4	$2.3136\cdot10^{5}$	$9.8608\cdot10^{4}$	4.9	$3.2227\cdot10^{4}$	$5.2663\cdot10^{5}$
1.5	$2.4013\cdot10^{5}$	$1.2224\cdot10^{5}$	5.0	$3.0342\cdot10^{4}$	$5.2976\cdot10^{5}$
1.6	$2.4310\cdot10^{5}$	$1.4644\cdot10^{5}$	5.1	$2.8587\cdot10^{4}$	$5.3271\cdot10^{5}$
1.7	$2.4143\cdot10^{5}$	$1.7070\cdot10^{5}$	5.2	$2.6954\cdot10^{4}$	$5.3548\cdot10^{5}$
1.8	$2.3621\cdot10^{5}$	$1.9461\cdot10^{5}$	5.3	$2.5431\cdot10^{4}$	$5.3810\cdot10^{5}$
1.9	$2.2844\cdot10^{5}$	$2.1786\cdot10^{5}$	5.4	$2.4011\cdot10^{4}$	$5.4057\cdot10^{5}$
2.0	$2.1892\cdot10^{5}$	$2.4024\cdot10^{5}$	5.5	$2.2686\cdot10^{4}$	$5.4291\cdot10^{5}$
2.1	$2.0830\cdot10^{5}$	$2.6160\cdot10^{5}$	5.6	$2.1449\cdot10^{4}$	$5.4511\cdot10^{5}$
2.2	$1.9710\cdot10^{5}$	$2.8188\cdot10^{5}$	5.7	$2.0293\cdot10^{4}$	$5.4720\cdot10^{5}$
2.3	$1.8569\cdot10^{5}$	$3.0102\cdot10^{5}$	5.8	$1.9211\cdot10^{4}$	$5.4917\cdot10^{5}$
2.4	$1.7435\cdot10^{5}$	$3.1902\cdot10^{5}$	5.9	$1.8199\cdot10^{4}$	$5.5104\cdot10^{5}$
2.5	$1.6328\cdot10^{5}$	$3.3590\cdot10^{5}$	6.0	$1.7251\cdot10^{4}$	$5.5282\cdot10^{5}$
2.6	$1.5262\cdot10^{5}$	$3.5169\cdot10^{5}$	6.1	$1.6363\cdot10^{4}$	$5.5450\cdot10^{5}$
2.7	$1.4245\cdot10^{5}$	$3.6644\cdot10^{5}$	6.2	$1.5530\cdot10^{4}$	$5.5609\cdot10^{5}$
2.8	$1.3282\cdot10^{5}$	$3.8019\cdot10^{5}$	6.3	$1.4748\cdot10^{4}$	$5.5760\cdot10^{5}$
2.9	$1.2375\cdot10^{5}$	$3.9302\cdot10^{5}$	6.4	$1.4014\cdot10^{4}$	$5.5904\cdot10^{5}$
3.0	$1.1526\cdot10^{5}$	$4.0496\cdot10^{5}$	6.5	$1.3325\cdot10^{4}$	$5.6041\cdot10^{5}$
3.1	$1.0733\cdot10^{5}$	$4.1609\cdot10^{5}$	6.6	$1.2676\cdot10^{4}$	$5.6171\cdot10^{5}$
3.2	$9.9953\cdot10^{4}$	$4.2645\cdot10^{5}$	6.7	$1.2066\cdot10^{4}$	$5.6294\cdot10^{5}$
3.3	$9.3094\cdot10^{4}$	$4.3610\cdot10^{5}$	6.8	$1.1491\cdot10^{4}$	$5.6412\cdot10^{5}$
3.4	$8.6730\cdot10^{4}$	$4.4508\cdot10^{5}$	6.9	$1.0950\cdot10^{4}$	$5.6524\cdot10^{5}$
3.5	$8.0833\cdot10^{4}$	$4.5346\cdot10^{5}$	7.0	$1.0439\cdot10^{4}$	$5.6631\cdot10^{5}$
3.6	$7.5371\cdot10^{4}$	$4.6126\cdot10^{5}$	7.1	$9.9579\cdot10^{3}$	$5.6733\cdot10^{5}$
3.7	$7.0316\cdot10^{4}$	$4.6855\cdot10^{5}$	7.2	$9.5036\cdot10^{3}$	$5.6830\cdot10^{5}$
3.8	$6.5638\cdot10^{4}$	$4.7534\cdot10^{5}$	7.3	$9.0745\cdot10^{3}$	$5.6923\cdot10^{5}$
3.9	$6.1311\cdot10^{4}$	$4.8168\cdot10^{5}$	7.4	$8.6691\cdot10^{3}$	$5.7012\cdot10^{5}$

Table 2.2. (*continued*)

λ [μm]	$m_{\lambda,bb}$ [W/μm m²]	$M_{T,o-\lambda,bb}$ [W/m²]	λ [μm]	$m_{\lambda,bb}$ [W/μm m²]	$M_{T,o-\lambda,bb}$ [W/m²]
7.5	$8.2858\cdot10^3$	$5.7097\cdot10^5$	11.5	$1.8533\cdot10^3$	$5.8741\cdot10^5$
7.6	$7.9231\cdot10^3$	$5.7178\cdot10^5$	11.6	$1.7962\cdot10^3$	$5.8759\cdot10^5$
7.7	$7.5798\cdot10^3$	$5.7255\cdot10^5$	11.7	$1.7412\cdot10^3$	$5.8777\cdot10^5$
7.8	$7.2548\cdot10^3$	$5.7329\cdot10^5$	11.8	$1.6884\cdot10^3$	$5.8794\cdot10^5$
7.9	$6.9467\cdot10^3$	$5.7400\cdot10^5$	11.9	$1.6375\cdot10^3$	$5.8811\cdot10^5$
8.0	$6.6547\cdot10^3$	$5.7468\cdot10^5$	12.0	$1.5886\cdot10^3$	$5.8827\cdot10^5$
8.1	$6.3777\cdot10^3$	$5.7534\cdot10^5$	12.1	$1.5414\cdot10^3$	$5.8843\cdot10^5$
8.2	$6.1148\cdot10^3$	$5.7596\cdot10^5$	12.2	$1.4960\cdot10^3$	$5.8858\cdot10^5$
8.3	$5.8651\cdot10^3$	$5.7656\cdot10^5$	12.3	$1.4522\cdot10^3$	$5.8873\cdot10^5$
8.4	$5.6280\cdot10^3$	$5.7713\cdot10^5$	12.4	$1.4100\cdot10^3$	$5.8887\cdot10^5$
8.5	$5.4026\cdot10^3$	$5.7768\cdot10^5$	12.5	$1.3693\cdot10^3$	$5.8901\cdot10^5$
8.6	$5.1883\cdot10^3$	$5.7821\cdot10^5$	12.6	$1.3301\cdot10^3$	$5.8914\cdot10^5$
8.7	$4.9844\cdot10^3$	$5.7872\cdot10^5$	12.7	$1.2922\cdot10^3$	$5.8927\cdot10^5$
8.8	$4.7904\cdot10^3$	$5.7921\cdot10^5$	12.8	$1.2557\cdot10^3$	$5.8940\cdot10^5$
8.9	$4.6056\cdot10^3$	$5.7968\cdot10^5$	12.9	$1.2205\cdot10^3$	$5.8953\cdot10^5$
9.0	$4.4296\cdot10^3$	$5.8013\cdot10^5$	13.0	$1.1865\cdot10^3$	$5.8965\cdot10^5$
9.1	$4.2618\cdot10^3$	$5.8057\cdot10^5$	13.1	$1.1536\cdot10^3$	$5.8976\cdot10^5$
9.2	$4.1018\cdot10^3$	$5.8099\cdot10^5$	13.2	$1.1219\cdot10^3$	$5.8988\cdot10^5$
9.3	$3.9493\cdot10^3$	$5.8139\cdot10^5$	13.3	$1.0913\cdot10^3$	$5.8999\cdot10^5$
9.4	$3.8037\cdot10^3$	$5.8178\cdot10^5$	13.4	$1.0617\cdot10^3$	$5.9009\cdot10^5$
9.5	$3.6646\cdot10^3$	$5.8215\cdot10^5$	13.5	$1.0331\cdot10^3$	$5.9020\cdot10^5$
9.6	$3.5319\cdot10^3$	$5.8251\cdot10^5$	13.6	$1.0054\cdot10^3$	$5.9030\cdot10^5$
9.7	$3.4050\cdot10^3$	$5.8285\cdot10^5$	13.7	$9.7870\cdot10^2$	$5.9040\cdot10^5$
9.8	$3.2838\cdot10^3$	$5.8319\cdot10^5$	13.8	$9.5285\cdot10^2$	$5.9050\cdot10^5$
9.9	$3.1679\cdot10^3$	$5.8351\cdot10^5$	13.9	$9.2784\cdot10^2$	$5.9059\cdot10^5$
10.0	$3.0570\cdot10^3$	$5.8382\cdot10^5$	14.0	$9.0364\cdot10^2$	$5.9068\cdot10^5$
10.1	$2.9509\cdot10^3$	$5.8412\cdot10^5$	14.1	$8.8022\cdot10^2$	$5.9077\cdot10^5$
10.2	$2.8493\cdot10^3$	$5.8441\cdot10^5$	14.2	$8.5756\cdot10^2$	$5.9086\cdot10^5$
10.3	$2.7521\cdot10^3$	$5.8469\cdot10^5$	14.3	$8.3562\cdot10^2$	$5.9094\cdot10^5$
10.4	$2.6589\cdot10^3$	$5.8496\cdot10^5$	14.4	$8.1437\cdot10^2$	$5.9103\cdot10^5$
10.5	$2.5696\cdot10^3$	$5.8523\cdot10^5$	14.5	$7.9379\cdot10^2$	$5.9111\cdot10^5$
10.6	$2.4840\cdot10^3$	$5.8548\cdot10^5$	14.6	$7.7386\cdot10^2$	$5.9118\cdot10^5$
10.7	$2.4020\cdot10^3$	$5.8572\cdot10^5$	14.7	$7.5455\cdot10^2$	$5.9126\cdot10^5$
10.8	$2.3232\cdot10^3$	$5.8596\cdot10^5$	14.8	$7.3583\cdot10^2$	$5.9133\cdot10^5$
10.9	$2.2477\cdot10^3$	$5.8619\cdot10^5$	14.9	$7.1769\cdot10^2$	$5.9141\cdot10^5$
11.0	$2.1752\cdot10^3$	$5.8641\cdot10^5$	15.0	$7.0011\cdot10^2$	$5.9148\cdot10^5$
11.1	$2.1055\cdot10^3$	$5.8662\cdot10^5$	15.1	$6.8305\cdot10^2$	$5.9155\cdot10^5$
11.2	$2.0387\cdot10^3$	$5.8683\cdot10^5$	15.2	$6.6652\cdot10^2$	$5.9161\cdot10^5$
11.3	$1.9744\cdot10^3$	$5.8703\cdot10^5$	15.3	$6.5047\cdot10^2$	$5.9168\cdot10^5$
11.4	$1.9126\cdot10^3$	$5.8722\cdot10^5$	15.4	$6.3491\cdot10^2$	$5.9174\cdot10^5$

Table 2.2. (*continued*)

temperature 1900 K

λ [μm]	$m_{\lambda,bb}$ [W/μm m²]	$M_{T,o-\lambda,bb}$ [W/m²]	λ [μm]	$m_{\lambda,bb}$ [W/μm m²]	$M_{T,o-\lambda,bb}$ [W/m²]
0.4	$2.1927 \cdot 10^{2}$	$5.4488 \cdot 10^{0}$			
0.5	$3.1681 \cdot 10^{3}$	$1.2822 \cdot 10^{2}$	4.0	$6.4788 \cdot 10^{4}$	$6.2003 \cdot 10^{5}$
0.6	$1.5890 \cdot 10^{4}$	$9.6569 \cdot 10^{2}$	4.1	$6.0476 \cdot 10^{4}$	$6.2629 \cdot 10^{5}$
0.7	$4.4610 \cdot 10^{4}$	$3.8487 \cdot 10^{3}$	4.2	$5.6497 \cdot 10^{4}$	$6.3214 \cdot 10^{5}$
0.8	$8.8464 \cdot 10^{4}$	$1.0399 \cdot 10^{4}$	4.3	$5.2824 \cdot 10^{4}$	$6.3760 \cdot 10^{5}$
0.9	$1.4055 \cdot 10^{5}$	$2.1817 \cdot 10^{4}$	4.4	$4.9429 \cdot 10^{4}$	$6.4271 \cdot 10^{5}$
1.0	$1.9257 \cdot 10^{5}$	$3.8505 \cdot 10^{4}$	4.5	$4.6291 \cdot 10^{4}$	$6.4750 \cdot 10^{5}$
1.1	$2.3813 \cdot 10^{5}$	$6.0112 \cdot 10^{4}$	4.6	$4.3388 \cdot 10^{4}$	$6.5198 \cdot 10^{5}$
1.2	$2.7376 \cdot 10^{5}$	$8.5797 \cdot 10^{4}$	4.7	$4.0700 \cdot 10^{4}$	$6.5618 \cdot 10^{5}$
1.3	$2.9845 \cdot 10^{5}$	$1.1450 \cdot 10^{5}$	4.8	$3.8209 \cdot 10^{4}$	$6.6013 \cdot 10^{5}$
1.4	$3.1283 \cdot 10^{5}$	$1.4514 \cdot 10^{5}$	4.9	$3.5900 \cdot 10^{4}$	$6.6383 \cdot 10^{5}$
1.5	$3.1838 \cdot 10^{5}$	$1.7677 \cdot 10^{5}$	5.0	$3.3756 \cdot 10^{4}$	$6.6731 \cdot 10^{5}$
1.6	$3.1686 \cdot 10^{5}$	$2.0858 \cdot 10^{5}$	5.1	$3.1766 \cdot 10^{4}$	$6.7059 \cdot 10^{5}$
1.7	$3.1001 \cdot 10^{5}$	$2.3996 \cdot 10^{5}$	5.2	$2.9915 \cdot 10^{4}$	$6.7367 \cdot 10^{5}$
1.8	$2.9935 \cdot 10^{5}$	$2.7046 \cdot 10^{5}$	5.3	$2.8194 \cdot 10^{4}$	$6.7657 \cdot 10^{5}$
1.9	$2.8613 \cdot 10^{5}$	$2.9975 \cdot 10^{5}$	5.4	$2.6592 \cdot 10^{4}$	$6.7931 \cdot 10^{5}$
2.0	$2.7136 \cdot 10^{5}$	$3.2763 \cdot 10^{5}$	5.5	$2.5098 \cdot 10^{4}$	$6.8190 \cdot 10^{5}$
2.1	$2.5579 \cdot 10^{5}$	$3.5399 \cdot 10^{5}$	5.6	$2.3706 \cdot 10^{4}$	$6.8434 \cdot 10^{5}$
2.2	$2.4001 \cdot 10^{5}$	$3.7878 \cdot 10^{5}$	5.7	$2.2407 \cdot 10^{4}$	$6.8664 \cdot 10^{5}$
2.3	$2.2440 \cdot 10^{5}$	$4.0200 \cdot 10^{5}$	5.8	$2.1194 \cdot 10^{4}$	$6.8882 \cdot 10^{5}$
2.4	$2.0924 \cdot 10^{5}$	$4.2368 \cdot 10^{5}$	5.9	$2.0060 \cdot 10^{4}$	$6.9088 \cdot 10^{5}$
2.5	$1.9473 \cdot 10^{5}$	$4.4387 \cdot 10^{5}$	6.0	$1.8999 \cdot 10^{4}$	$6.9283 \cdot 10^{5}$
2.6	$1.8096 \cdot 10^{5}$	$4.6265 \cdot 10^{5}$	6.1	$1.8006 \cdot 10^{4}$	$6.9468 \cdot 10^{5}$
2.7	$1.6801 \cdot 10^{5}$	$4.8009 \cdot 10^{5}$	6.2	$1.7076 \cdot 10^{4}$	$6.9644 \cdot 10^{5}$
2.8	$1.5589 \cdot 10^{5}$	$4.9628 \cdot 10^{5}$	6.3	$1.6205 \cdot 10^{4}$	$6.9810 \cdot 10^{5}$
2.9	$1.4461 \cdot 10^{5}$	$5.1129 \cdot 10^{5}$	6.4	$1.5387 \cdot 10^{4}$	$6.9968 \cdot 10^{5}$
3.0	$1.3413 \cdot 10^{5}$	$5.2522 \cdot 10^{5}$	6.5	$1.4619 \cdot 10^{4}$	$7.0118 \cdot 10^{5}$
3.1	$1.2442 \cdot 10^{5}$	$5.3814 \cdot 10^{5}$	6.6	$1.3898 \cdot 10^{4}$	$7.0261 \cdot 10^{5}$
3.2	$1.1545 \cdot 10^{5}$	$5.5013 \cdot 10^{5}$	6.7	$1.3221 \cdot 10^{4}$	$7.0396 \cdot 10^{5}$
3.3	$1.0717 \cdot 10^{5}$	$5.6126 \cdot 10^{5}$	6.8	$1.2583 \cdot 10^{4}$	$7.0525 \cdot 10^{5}$
3.4	$9.9535 \cdot 10^{4}$	$5.7159 \cdot 10^{5}$	6.9	$1.1983 \cdot 10^{4}$	$7.0648 \cdot 10^{5}$
3.5	$9.2497 \cdot 10^{4}$	$5.8118 \cdot 10^{5}$	7.0	$1.1418 \cdot 10^{4}$	$7.0765 \cdot 10^{5}$
3.6	$8.6013 \cdot 10^{4}$	$5.9011 \cdot 10^{5}$	7.1	$1.0885 \cdot 10^{4}$	$7.0876 \cdot 10^{5}$
3.7	$8.0039 \cdot 10^{4}$	$5.9840 \cdot 10^{5}$	7.2	$1.0382 \cdot 10^{4}$	$7.0983 \cdot 10^{5}$
3.8	$7.4536 \cdot 10^{4}$	$6.0613 \cdot 10^{5}$	7.3	$9.9084 \cdot 10^{3}$	$7.1084 \cdot 10^{5}$
3.9	$6.9464 \cdot 10^{4}$	$6.1332 \cdot 10^{5}$	7.4	$9.4607 \cdot 10^{3}$	$7.1181 \cdot 10^{5}$

λ [µm]	$m_{\lambda,bb}$ [W/µm m²]	$M_{T,o-\lambda,bb}$ [W/m²]	λ [µm]	$m_{\lambda,bb}$ [W/µm m²]	$M_{T,o-\lambda,bb}$ [W/m²]
7.5	$9.0378 \cdot 10^3$	$7.1273 \cdot 10^5$	11.5	$1.9964 \cdot 10^3$	$7.3057 \cdot 10^5$
7.6	$8.6380 \cdot 10^3$	$7.1362 \cdot 10^5$	11.6	$1.9345 \cdot 10^3$	$7.3077 \cdot 10^5$
7.7	$8.2598 \cdot 10^3$	$7.1446 \cdot 10^5$	11.7	$1.8750 \cdot 10^3$	$7.3096 \cdot 10^5$
7.8	$7.9019 \cdot 10^3$	$7.1527 \cdot 10^5$	11.8	$1.8178 \cdot 10^3$	$7.3114 \cdot 10^5$
7.9	$7.5630 \cdot 10^3$	$7.1604 \cdot 10^5$	11.9	$1.7627 \cdot 10^3$	$7.3132 \cdot 10^5$
8.0	$7.2419 \cdot 10^3$	$7.1678 \cdot 10^5$	12.0	$1.7097 \cdot 10^3$	$7.3149 \cdot 10^5$
8.1	$6.9374 \cdot 10^3$	$7.1749 \cdot 10^5$	12.1	$1.6586 \cdot 10^3$	$7.3166 \cdot 10^5$
8.2	$6.6487 \cdot 10^3$	$7.1817 \cdot 10^5$	12.2	$1.6094 \cdot 10^3$	$7.3183 \cdot 10^5$
8.3	$6.3747 \cdot 10^3$	$7.1882 \cdot 10^5$	12.3	$1.5621 \cdot 10^3$	$7.3198 \cdot 10^5$
8.4	$6.1146 \cdot 10^3$	$7.1945 \cdot 10^5$	12.4	$1.5164 \cdot 10^3$	$7.3214 \cdot 10^5$
8.5	$5.8675 \cdot 10^3$	$7.2004 \cdot 10^5$	12.5	$1.4724 \cdot 10^3$	$7.3229 \cdot 10^5$
8.6	$5.6326 \cdot 10^3$	$7.2062 \cdot 10^5$	12.6	$1.4300 \cdot 10^3$	$7.3243 \cdot 10^5$
8.7	$5.4093 \cdot 10^3$	$7.2117 \cdot 10^5$	12.7	$1.3891 \cdot 10^3$	$7.3257 \cdot 10^5$
8.8	$5.1969 \cdot 10^3$	$7.2170 \cdot 10^5$	12.8	$1.3497 \cdot 10^3$	$7.3271 \cdot 10^5$
8.9	$4.9947 \cdot 10^3$	$7.2221 \cdot 10^5$	12.9	$1.3116 \cdot 10^3$	$7.3284 \cdot 10^5$
9.0	$4.8022 \cdot 10^3$	$7.2270 \cdot 10^5$	13.0	$1.2749 \cdot 10^3$	$7.3297 \cdot 10^5$
9.1	$4.6188 \cdot 10^3$	$7.2317 \cdot 10^5$	13.1	$1.2394 \cdot 10^3$	$7.3310 \cdot 10^5$
9.2	$4.4440 \cdot 10^3$	$7.2362 \cdot 10^5$	13.2	$1.2052 \cdot 10^3$	$7.3322 \cdot 10^5$
9.3	$4.2773 \cdot 10^3$	$7.2406 \cdot 10^5$	13.3	$1.1721 \cdot 10^3$	$7.3334 \cdot 10^5$
9.4	$4.1184 \cdot 10^3$	$7.2448 \cdot 10^5$	13.4	$1.1401 \cdot 10^3$	$7.3345 \cdot 10^5$
9.5	$3.9667 \cdot 10^3$	$7.2488 \cdot 10^5$	13.5	$1.1093 \cdot 10^3$	$7.3357 \cdot 10^5$
9.6	$3.8219 \cdot 10^3$	$7.2527 \cdot 10^5$	13.6	$1.0794 \cdot 10^3$	$7.3368 \cdot 10^5$
9.7	$3.6836 \cdot 10^3$	$7.2565 \cdot 10^5$	13.7	$1.0506 \cdot 10^3$	$7.3378 \cdot 10^5$
9.8	$3.5514 \cdot 10^3$	$7.2601 \cdot 10^5$	13.8	$1.0227 \cdot 10^3$	$7.3389 \cdot 10^5$
9.9	$3.4251 \cdot 10^3$	$7.2636 \cdot 10^5$	13.9	$9.9571 \cdot 10^2$	$7.3399 \cdot 10^5$
10.0	$3.3044 \cdot 10^3$	$7.2670 \cdot 10^5$	14.0	$9.6962 \cdot 10^2$	$7.3409 \cdot 10^5$
10.1	$3.1888 \cdot 10^3$	$7.2702 \cdot 10^5$	14.1	$9.4438 \cdot 10^2$	$7.3418 \cdot 10^5$
10.2	$3.0783 \cdot 10^3$	$7.2733 \cdot 10^5$	14.2	$9.1995 \cdot 10^2$	$7.3427 \cdot 10^5$
10.3	$2.9725 \cdot 10^3$	$7.2764 \cdot 10^5$	14.3	$8.9630 \cdot 10^2$	$7.3436 \cdot 10^5$
10.4	$2.8711 \cdot 10^3$	$7.2793 \cdot 10^5$	14.4	$8.7341 \cdot 10^2$	$7.3445 \cdot 10^5$
10.5	$2.7741 \cdot 10^3$	$7.2821 \cdot 10^5$	14.5	$8.5124 \cdot 10^2$	$7.3454 \cdot 10^5$
10.6	$2.6810 \cdot 10^3$	$7.2848 \cdot 10^5$	14.6	$8.2976 \cdot 10^2$	$7.3462 \cdot 10^5$
10.7	$2.5919 \cdot 10^3$	$7.2875 \cdot 10^5$	14.7	$8.0896 \cdot 10^2$	$7.3471 \cdot 10^5$
10.8	$2.5063 \cdot 10^3$	$7.2900 \cdot 10^5$	14.8	$7.8881 \cdot 10^2$	$7.3479 \cdot 10^5$
10.9	$2.4243 \cdot 10^3$	$7.2925 \cdot 10^5$	14.9	$7.6927 \cdot 10^2$	$7.3486 \cdot 10^5$
11.0	$2.3456 \cdot 10^3$	$7.2949 \cdot 10^5$	15.0	$7.5034 \cdot 10^2$	$7.3494 \cdot 10^5$
11.1	$2.2700 \cdot 10^3$	$7.2972 \cdot 10^5$	15.1	$7.3199 \cdot 10^2$	$7.3501 \cdot 10^5$
11.2	$2.1975 \cdot 10^3$	$7.2994 \cdot 10^5$	15.2	$7.1419 \cdot 10^2$	$7.3509 \cdot 10^5$
11.3	$2.1278 \cdot 10^3$	$7.3016 \cdot 10^5$	15.3	$6.9693 \cdot 10^2$	$7.3516 \cdot 10^5$
11.4	$2.0608 \cdot 10^3$	$7.3037 \cdot 10^5$	15.4	$6.8018 \cdot 10^2$	$7.3522 \cdot 10^5$

Table 2.2. (*continued*)

temperature 2000 K

λ [μm]	$m_{\lambda,bb}$ [W/μm m²]	$M_{T,o-\lambda,bb}$ [W/m²]	λ [μm]	$m_{\lambda,bb}$ [W/μm m²]	$M_{T,o-\lambda,bb}$ [W/m²]
0.4	$5.6504 \cdot 10^{2}$	$1.4909 \cdot 10^{1}$			
0.5	$6.7556 \cdot 10^{3}$	$2.9100 \cdot 10^{2}$	4.0	$7.2497 \cdot 10^{4}$	$7.7683 \cdot 10^{5}$
0.6	$2.9867 \cdot 10^{4}$	$1.9361 \cdot 10^{3}$	4.1	$6.7551 \cdot 10^{4}$	$7.8383 \cdot 10^{5}$
0.7	$7.6621 \cdot 10^{4}$	$7.0675 \cdot 10^{3}$	4.2	$6.3000 \cdot 10^{4}$	$7.9035 \cdot 10^{5}$
0.8	$1.4201 \cdot 10^{5}$	$1.7889 \cdot 10^{4}$	4.3	$5.8809 \cdot 10^{4}$	$7.9644 \cdot 10^{5}$
0.9	$2.1409 \cdot 10^{5}$	$3.5692 \cdot 10^{4}$	4.4	$5.4947 \cdot 10^{4}$	$8.0213 \cdot 10^{5}$
1.0	$2.8127 \cdot 10^{5}$	$6.0539 \cdot 10^{4}$	4.5	$5.1384 \cdot 10^{4}$	$8.0744 \cdot 10^{5}$
1.1	$3.3612 \cdot 10^{5}$	$9.1530 \cdot 10^{4}$	4.6	$4.8096 \cdot 10^{4}$	$8.1241 \cdot 10^{5}$
1.2	$3.7557 \cdot 10^{5}$	$1.2725 \cdot 10^{5}$	4.7	$4.5057 \cdot 10^{4}$	$8.1707 \cdot 10^{5}$
1.3	$3.9975 \cdot 10^{5}$	$1.6613 \cdot 10^{5}$	4.8	$4.2248 \cdot 10^{4}$	$8.2143 \cdot 10^{5}$
1.4	$4.1055 \cdot 10^{5}$	$2.0675 \cdot 10^{5}$	4.9	$3.9647 \cdot 10^{4}$	$8.2552 \cdot 10^{5}$
1.5	$4.1056 \cdot 10^{5}$	$2.4788 \cdot 10^{5}$	5.0	$3.7238 \cdot 10^{4}$	$8.2937 \cdot 10^{5}$
1.6	$4.0242 \cdot 10^{5}$	$2.8859 \cdot 10^{5}$	5.1	$3.5004 \cdot 10^{4}$	$8.3298 \cdot 10^{5}$
1.7	$3.8849 \cdot 10^{5}$	$3.2817 \cdot 10^{5}$	5.2	$3.2931 \cdot 10^{4}$	$8.3637 \cdot 10^{5}$
1.8	$3.7074 \cdot 10^{5}$	$3.6616 \cdot 10^{5}$	5.3	$3.1005 \cdot 10^{4}$	$8.3957 \cdot 10^{5}$
1.9	$3.5069 \cdot 10^{5}$	$4.0224 \cdot 10^{5}$	5.4	$2.9215 \cdot 10^{4}$	$8.4258 \cdot 10^{5}$
2.0	$3.2951 \cdot 10^{5}$	$4.3626 \cdot 10^{5}$	5.5	$2.7550 \cdot 10^{4}$	$8.4542 \cdot 10^{5}$
2.1	$3.0803 \cdot 10^{5}$	$4.6814 \cdot 10^{5}$	5.6	$2.5999 \cdot 10^{4}$	$8.4809 \cdot 10^{5}$
2.2	$2.8686 \cdot 10^{5}$	$4.9788 \cdot 10^{5}$	5.7	$2.4554 \cdot 10^{4}$	$8.5062 \cdot 10^{5}$
2.3	$2.6639 \cdot 10^{5}$	$5.2553 \cdot 10^{5}$	5.8	$2.3205 \cdot 10^{4}$	$8.5301 \cdot 10^{5}$
2.4	$2.4688 \cdot 10^{5}$	$5.5119 \cdot 10^{5}$	5.9	$2.1947 \cdot 10^{4}$	$8.5526 \cdot 10^{5}$
2.5	$2.2847 \cdot 10^{5}$	$5.7494 \cdot 10^{5}$	6.0	$2.0771 \cdot 10^{4}$	$8.5740 \cdot 10^{5}$
2.6	$2.1124 \cdot 10^{5}$	$5.9692 \cdot 10^{5}$	6.1	$1.9671 \cdot 10^{4}$	$8.5942 \cdot 10^{5}$
2.7	$1.9520 \cdot 10^{5}$	$6.1723 \cdot 10^{5}$	6.2	$1.8643 \cdot 10^{4}$	$8.6133 \cdot 10^{5}$
2.8	$1.8034 \cdot 10^{5}$	$6.3600 \cdot 10^{5}$	6.3	$1.7679 \cdot 10^{4}$	$8.6315 \cdot 10^{5}$
2.9	$1.6662 \cdot 10^{5}$	$6.5333 \cdot 10^{5}$	6.4	$1.6776 \cdot 10^{4}$	$8.6487 \cdot 10^{5}$
3.0	$1.5397 \cdot 10^{5}$	$6.6935 \cdot 10^{5}$	6.5	$1.5929 \cdot 10^{4}$	$8.6651 \cdot 10^{5}$
3.1	$1.4234 \cdot 10^{5}$	$6.8416 \cdot 10^{5}$	6.6	$1.5135 \cdot 10^{4}$	$8.6806 \cdot 10^{5}$
3.2	$1.3166 \cdot 10^{5}$	$6.9786 \cdot 10^{5}$	6.7	$1.4388 \cdot 10^{4}$	$8.6954 \cdot 10^{5}$
3.3	$1.2186 \cdot 10^{5}$	$7.1052 \cdot 10^{5}$	6.8	$1.3687 \cdot 10^{4}$	$8.7094 \cdot 10^{5}$
3.4	$1.1287 \cdot 10^{5}$	$7.2225 \cdot 10^{5}$	6.9	$1.3027 \cdot 10^{4}$	$8.7227 \cdot 10^{5}$
3.5	$1.0462 \cdot 10^{5}$	$7.3312 \cdot 10^{5}$	7.0	$1.2406 \cdot 10^{4}$	$8.7355 \cdot 10^{5}$
3.6	$9.7048 \cdot 10^{4}$	$7.4320 \cdot 10^{5}$	7.1	$1.1821 \cdot 10^{4}$	$8.7476 \cdot 10^{5}$
3.7	$9.0104 \cdot 10^{4}$	$7.5255 \cdot 10^{5}$	7.2	$1.1270 \cdot 10^{4}$	$8.7591 \cdot 10^{5}$
3.8	$8.3729 \cdot 10^{4}$	$7.6124 \cdot 10^{5}$	7.3	$1.0750 \cdot 10^{4}$	$8.7701 \cdot 10^{5}$
3.9	$7.7876 \cdot 10^{4}$	$7.6932 \cdot 10^{5}$	7.4	$1.0260 \cdot 10^{4}$	$8.7806 \cdot 10^{5}$

λ [μm]	$m_{\lambda,bb}$ [W/μm m²]	$M_{T,o-\lambda,bb}$ [W/m²]	λ [μm]	$m_{\lambda,bb}$ [W/μm m²]	$M_{T,o-\lambda,bb}$ [W/m²]
7.5	$9.7965\cdot10^3$	$8.7906\cdot10^5$	11.5	$2.1401\cdot10^3$	$8.9830\cdot10^5$
7.6	$9.3591\cdot10^3$	$8.8002\cdot10^5$	11.6	$2.0734\cdot10^3$	$8.9851\cdot10^5$
7.7	$8.9455\cdot10^3$	$8.8094\cdot10^5$	11.7	$2.0093\cdot10^3$	$8.9871\cdot10^5$
7.8	$8.5544\cdot10^3$	$8.8181\cdot10^5$	11.8	$1.9476\cdot10^3$	$8.9891\cdot10^5$
7.9	$8.1842\cdot10^3$	$8.8265\cdot10^5$	11.9	$1.8883\cdot10^3$	$8.9910\cdot10^5$
8.0	$7.8336\cdot10^3$	$8.8345\cdot10^5$	12.0	$1.8312\cdot10^3$	$8.9929\cdot10^5$
8.1	$7.5015\cdot10^3$	$8.8422\cdot10^5$	12.1	$1.7763\cdot10^3$	$8.9947\cdot10^5$
8.2	$7.1866\cdot10^3$	$8.8495\cdot10^5$	12.2	$1.7233\cdot10^3$	$8.9965\cdot10^5$
8.3	$6.8880\cdot10^3$	$8.8565\cdot10^5$	12.3	$1.6724\cdot10^3$	$8.9982\cdot10^5$
8.4	$6.6046\cdot10^3$	$8.8633\cdot10^5$	12.4	$1.6233\cdot10^3$	$8.9998\cdot10^5$
8.5	$6.3355\cdot10^3$	$8.8697\cdot10^5$	12.5	$1.5759\cdot10^3$	$9.0014\cdot10^5$
8.6	$6.0799\cdot10^3$	$8.8759\cdot10^5$	12.6	$1.5303\cdot10^3$	$9.0030\cdot10^5$
8.7	$5.8370\cdot10^3$	$8.8819\cdot10^5$	12.7	$1.4863\cdot10^3$	$9.0045\cdot10^5$
8.8	$5.6060\cdot10^3$	$8.8876\cdot10^5$	12.8	$1.4439\cdot10^3$	$9.0059\cdot10^5$
8.9	$5.3862\cdot10^3$	$8.8931\cdot10^5$	12.9	$1.4030\cdot10^3$	$9.0073\cdot10^5$
9.0	$5.1771\cdot10^3$	$8.8984\cdot10^5$	13.0	$1.3635\cdot10^3$	$9.0087\cdot10^5$
9.1	$4.9779\cdot10^3$	$8.9035\cdot10^5$	13.1	$1.3254\cdot10^3$	$9.0101\cdot10^5$
9.2	$4.7882\cdot10^3$	$8.9084\cdot10^5$	13.2	$1.2886\cdot10^3$	$9.0114\cdot10^5$
9.3	$4.6073\cdot10^3$	$8.9131\cdot10^5$	13.3	$1.2531\cdot10^3$	$9.0127\cdot10^5$
9.4	$4.4349\cdot10^3$	$8.9176\cdot10^5$	13.4	$1.2188\cdot10^3$	$9.0139\cdot10^5$
9.5	$4.2704\cdot10^3$	$8.9219\cdot10^5$	13.5	$1.1856\cdot10^3$	$9.0151\cdot10^5$
9.6	$4.1134\cdot10^3$	$8.9261\cdot10^5$	13.6	$1.1536\cdot10^3$	$9.0163\cdot10^5$
9.7	$3.9636\cdot10^3$	$8.9302\cdot10^5$	13.7	$1.1226\cdot10^3$	$9.0174\cdot10^5$
9.8	$3.8204\cdot10^3$	$8.9340\cdot10^5$	13.8	$1.0927\cdot10^3$	$9.0185\cdot10^5$
9.9	$3.6837\cdot10^3$	$8.9378\cdot10^5$	13.9	$1.0638\cdot10^3$	$9.0196\cdot10^5$
10.0	$3.5529\cdot10^3$	$8.9414\cdot10^5$	14.0	$1.0358\cdot10^3$	$9.0206\cdot10^5$
10.1	$3.4279\cdot10^3$	$8.9449\cdot10^5$	14.1	$1.0087\cdot10^3$	$9.0217\cdot10^5$
10.2	$3.3083\cdot10^3$	$8.9483\cdot10^5$	14.2	$9.8249\cdot10^2$	$9.0226\cdot10^5$
10.3	$3.1939\cdot10^3$	$8.9515\cdot10^5$	14.3	$9.5713\cdot10^2$	$9.0236\cdot10^5$
10.4	$3.0843\cdot10^3$	$8.9547\cdot10^5$	14.4	$9.3258\cdot10^2$	$9.0246\cdot10^5$
10.5	$2.9794\cdot10^3$	$8.9577\cdot10^5$	14.5	$9.0882\cdot10^2$	$9.0255\cdot10^5$
10.6	$2.8789\cdot10^3$	$8.9606\cdot10^5$	14.6	$8.8580\cdot10^2$	$9.0264\cdot10^5$
10.7	$2.7826\cdot10^3$	$8.9635\cdot10^5$	14.7	$8.6350\cdot10^2$	$9.0273\cdot10^5$
10.8	$2.6902\cdot10^3$	$8.9662\cdot10^5$	14.8	$8.4190\cdot10^2$	$9.0281\cdot10^5$
10.9	$2.6017\cdot10^3$	$8.9688\cdot10^5$	14.9	$8.2097\cdot10^2$	$9.0289\cdot10^5$
11.0	$2.5167\cdot10^3$	$8.9714\cdot10^5$	15.0	$8.0069\cdot10^2$	$9.0297\cdot10^5$
11.1	$2.4352\cdot10^3$	$8.9739\cdot10^5$	15.1	$7.8103\cdot10^2$	$9.0305\cdot10^5$
11.2	$2.3569\cdot10^3$	$8.9763\cdot10^5$	15.2	$7.6196\cdot10^2$	$9.0313\cdot10^5$
11.3	$2.2817\cdot10^3$	$8.9786\cdot10^5$	15.3	$7.4348\cdot10^2$	$9.0321\cdot10^5$
11.4	$2.2095\cdot10^3$	$8.9808\cdot10^5$	15.4	$7.2555\cdot10^2$	$9.0328\cdot10^5$

Table 2.2. (*continued*)

temperature 2100 K

λ [µm]	$m_{\lambda,bb}$ [W/µm m²]	$M_{T,o-\lambda,bb}$ [W/m²]	λ [µm]	$m_{\lambda,bb}$ [W/µm m²]	$M_{T,o-\lambda,bb}$ [W/m²]
0.4	$1.3305\cdot10^{3}$	$3.7186\cdot10^{1}$			
0.5	$1.3403\cdot10^{4}$	$6.1292\cdot10^{2}$	4.0	$8.0406\cdot10^{4}$	$9.6068\cdot10^{5}$
0.6	$5.2863\cdot10^{4}$	$3.6464\cdot10^{3}$	4.1	$7.4801\cdot10^{4}$	$9.6844\cdot10^{5}$
0.7	$1.2499\cdot10^{5}$	$1.2296\cdot10^{4}$	4.2	$6.9656\cdot10^{4}$	$9.7566\cdot10^{5}$
0.8	$2.1794\cdot10^{5}$	$2.9346\cdot10^{4}$	4.3	$6.4930\cdot10^{4}$	$9.8238\cdot10^{5}$
0.9	$3.1330\cdot10^{5}$	$5.5959\cdot10^{4}$	4.4	$6.0584\cdot10^{4}$	$9.8865\cdot10^{5}$
1.0	$3.9631\cdot10^{5}$	$9.1586\cdot10^{4}$	4.5	$5.6583\cdot10^{4}$	$9.9451\cdot10^{5}$
1.1	$4.5917\cdot10^{5}$	$1.3454\cdot10^{5}$	4.6	$5.2897\cdot10^{4}$	$9.9998\cdot10^{5}$
1.2	$5.0007\cdot10^{5}$	$1.8268\cdot10^{5}$	4.7	$4.9498\cdot10^{4}$	$1.0051\cdot10^{6}$
1.3	$5.2090\cdot10^{5}$	$2.3388\cdot10^{5}$	4.8	$4.6360\cdot10^{4}$	$1.0099\cdot10^{6}$
1.4	$5.2523\cdot10^{5}$	$2.8631\cdot10^{5}$	4.9	$4.3460\cdot10^{4}$	$1.0144\cdot10^{6}$
1.5	$5.1700\cdot10^{5}$	$3.3851\cdot10^{5}$	5.0	$4.0778\cdot10^{4}$	$1.0186\cdot10^{6}$
1.6	$4.9984\cdot10^{5}$	$3.8941\cdot10^{5}$	5.1	$3.8295\cdot10^{4}$	$1.0225\cdot10^{6}$
1.7	$4.7679\cdot10^{5}$	$4.3828\cdot10^{5}$	5.2	$3.5993\cdot10^{4}$	$1.0263\cdot10^{6}$
1.8	$4.5022\cdot10^{5}$	$4.8466\cdot10^{5}$	5.3	$3.3859\cdot10^{4}$	$1.0297\cdot10^{6}$
1.9	$4.2191\cdot10^{5}$	$5.2827\cdot10^{5}$	5.4	$3.1877\cdot10^{4}$	$1.0330\cdot10^{6}$
2.0	$3.9314\cdot10^{5}$	$5.6902\cdot10^{5}$	5.5	$3.0036\cdot10^{4}$	$1.0361\cdot10^{6}$
2.1	$3.6478\cdot10^{5}$	$6.0691\cdot10^{5}$	5.6	$2.8323\cdot10^{4}$	$1.0390\cdot10^{6}$
2.2	$3.3744\cdot10^{5}$	$6.4201\cdot10^{5}$	5.7	$2.6728\cdot10^{4}$	$1.0418\cdot10^{6}$
2.3	$3.1147\cdot10^{5}$	$6.7444\cdot10^{5}$	5.8	$2.5242\cdot10^{4}$	$1.0444\cdot10^{6}$
2.4	$2.8707\cdot10^{5}$	$7.0436\cdot10^{5}$	5.9	$2.3857\cdot10^{4}$	$1.0468\cdot10^{6}$
2.5	$2.6434\cdot10^{5}$	$7.3191\cdot10^{5}$	6.0	$2.2563\cdot10^{4}$	$1.0492\cdot10^{6}$
2.6	$2.4328\cdot10^{5}$	$7.5728\cdot10^{5}$	6.1	$2.1355\cdot10^{4}$	$1.0514\cdot10^{6}$
2.7	$2.2387\cdot10^{5}$	$7.8062\cdot10^{5}$	6.2	$2.0226\cdot10^{4}$	$1.0534\cdot10^{6}$
2.8	$2.0603\cdot10^{5}$	$8.0211\cdot10^{5}$	6.3	$1.9169\cdot10^{4}$	$1.0554\cdot10^{6}$
2.9	$1.8968\cdot10^{5}$	$8.2188\cdot10^{5}$	6.4	$1.8180\cdot10^{4}$	$1.0573\cdot10^{6}$
3.0	$1.7471\cdot10^{5}$	$8.4009\cdot10^{5}$	6.5	$1.7253\cdot10^{4}$	$1.0590\cdot10^{6}$
3.1	$1.6103\cdot10^{5}$	$8.5686\cdot10^{5}$	6.6	$1.6383\cdot10^{4}$	$1.0607\cdot10^{6}$
3.2	$1.4852\cdot10^{5}$	$8.7233\cdot10^{5}$	6.7	$1.5567\cdot10^{4}$	$1.0623\cdot10^{6}$
3.3	$1.3710\cdot10^{5}$	$8.8661\cdot10^{5}$	6.8	$1.4800\cdot10^{4}$	$1.0638\cdot10^{6}$
3.4	$1.2667\cdot10^{5}$	$8.9979\cdot10^{5}$	6.9	$1.4080\cdot10^{4}$	$1.0653\cdot10^{6}$
3.5	$1.1714\cdot10^{6}$	$9.1197\cdot10^{6}$	7.0	$1.3403\cdot10^{4}$	$1.0667\cdot10^{6}$
3.6	$1.0843\cdot10^{5}$	$9.2324\cdot10^{5}$	7.1	$1.2765\cdot10^{4}$	$1.0680\cdot10^{6}$
3.7	$1.0047\cdot10^{5}$	$9.3368\cdot10^{5}$	7.2	$1.2164\cdot10^{4}$	$1.0692\cdot10^{6}$
3.8	$9.3187\cdot10^{4}$	$9.4336\cdot10^{5}$	7.3	$1.1598\cdot10^{4}$	$1.0704\cdot10^{6}$
3.9	$8.6517\cdot10^{4}$	$9.5234\cdot10^{5}$	7.4	$1.1065\cdot10^{4}$	$1.0715\cdot10^{6}$

λ [µm]	$m_{\lambda,bb}$ [W/µm m²]	$M_{T,o-\lambda,bb}$ [W/m²]	λ [µm]	$m_{\lambda,bb}$ [W/µm m²]	$M_{T,o-\lambda,bb}$ [W/m²]
7.5	$1.0561 \cdot 10^4$	$1.0726 \cdot 10^6$	11.5	$2.2843 \cdot 10^3$	$1.0932 \cdot 10^6$
7.6	$1.0085 \cdot 10^4$	$1.0736 \cdot 10^6$	11.6	$2.2128 \cdot 10^3$	$1.0935 \cdot 10^6$
7.7	$9.6362 \cdot 10^3$	$1.0746 \cdot 10^6$	11.7	$2.1440 \cdot 10^3$	$1.0937 \cdot 10^6$
7.8	$9.2114 \cdot 10^3$	$1.0756 \cdot 10^6$	11.8	$2.0779 \cdot 10^3$	$1.0939 \cdot 10^6$
7.9	$8.8096 \cdot 10^3$	$1.0765 \cdot 10^6$	11.9	$2.0143 \cdot 10^3$	$1.0941 \cdot 10^6$
8.0	$8.4293 \cdot 10^3$	$1.0773 \cdot 10^6$	12.0	$1.9531 \cdot 10^3$	$1.0943 \cdot 10^6$
8.1	$8.0692 \cdot 10^3$	$1.0782 \cdot 10^6$	12.1	$1.8942 \cdot 10^3$	$1.0945 \cdot 10^6$
8.2	$7.7280 \cdot 10^3$	$1.0789 \cdot 10^6$	12.2	$1.8375 \cdot 10^3$	$1.0947 \cdot 10^6$
8.3	$7.4045 \cdot 10^3$	$1.0797 \cdot 10^6$	12.3	$1.7830 \cdot 10^3$	$1.0948 \cdot 10^6$
8.4	$7.0976 \cdot 10^3$	$1.0804 \cdot 10^6$	12.4	$1.7304 \cdot 10^3$	$1.0950 \cdot 10^6$
8.5	$6.8064 \cdot 10^3$	$1.0811 \cdot 10^6$	12.5	$1.6797 \cdot 10^3$	$1.0952 \cdot 10^6$
8.6	$6.5298 \cdot 10^3$	$1.0818 \cdot 10^6$	12.6	$1.6309 \cdot 10^3$	$1.0954 \cdot 10^6$
8.7	$6.2671 \cdot 10^3$	$1.0824 \cdot 10^6$	12.7	$1.5838 \cdot 10^3$	$1.0955 \cdot 10^6$
8.8	$6.0174 \cdot 10^3$	$1.0830 \cdot 10^6$	12.8	$1.5384 \cdot 10^3$	$1.0957 \cdot 10^6$
8.9	$5.7799 \cdot 10^3$	$1.0836 \cdot 10^6$	12.9	$1.4946 \cdot 10^3$	$1.0958 \cdot 10^6$
9.0	$5.5540 \cdot 10^3$	$1.0842 \cdot 10^6$	13.0	$1.4524 \cdot 10^3$	$1.0960 \cdot 10^6$
9.1	$5.3389 \cdot 10^3$	$1.0847 \cdot 10^6$	13.1	$1.4116 \cdot 10^3$	$1.0961 \cdot 10^6$
9.2	$5.1341 \cdot 10^3$	$1.0853 \cdot 10^6$	13.2	$1.3723 \cdot 10^3$	$1.0963 \cdot 10^6$
9.3	$4.9389 \cdot 10^3$	$1.0858 \cdot 10^6$	13.3	$1.3343 \cdot 10^3$	$1.0964 \cdot 10^6$
9.4	$4.7529 \cdot 10^3$	$1.0862 \cdot 10^6$	13.4	$1.2976 \cdot 10^3$	$1.0965 \cdot 10^6$
9.5	$4.5756 \cdot 10^3$	$1.0867 \cdot 10^6$	13.5	$1.2622 \cdot 10^3$	$1.0966 \cdot 10^6$
9.6	$4.4063 \cdot 10^3$	$1.0872 \cdot 10^6$	13.6	$1.2280 \cdot 10^3$	$1.0968 \cdot 10^6$
9.7	$4.2448 \cdot 10^3$	$1.0876 \cdot 10^6$	13.7	$1.1949 \cdot 10^3$	$1.0969 \cdot 10^6$
9.8	$4.0906 \cdot 10^3$	$1.0880 \cdot 10^6$	13.8	$1.1629 \cdot 10^3$	$1.0970 \cdot 10^6$
9.9	$3.9433 \cdot 10^3$	$1.0884 \cdot 10^6$	13.9	$1.1320 \cdot 10^3$	$1.0971 \cdot 10^6$
10.0	$3.8026 \cdot 10^3$	$1.0888 \cdot 10^6$	14.0	$1.1021 \cdot 10^3$	$1.0972 \cdot 10^6$
10.1	$3.6680 \cdot 10^3$	$1.0892 \cdot 10^6$	14.1	$1.0732 \cdot 10^3$	$1.0973 \cdot 10^6$
10.2	$3.5393 \cdot 10^3$	$1.0895 \cdot 10^6$	14.2	$1.0452 \cdot 10^3$	$1.0974 \cdot 10^6$
10.3	$3.4162 \cdot 10^3$	$1.0899 \cdot 10^6$	14.3	$1.0181 \cdot 10^3$	$1.0975 \cdot 10^6$
10.4	$3.2984 \cdot 10^3$	$1.0902 \cdot 10^6$	14.4	$9.9189 \cdot 10^2$	$1.0976 \cdot 10^6$
10.5	$3.1856 \cdot 10^3$	$1.0905 \cdot 10^6$	14.5	$9.6651 \cdot 10^2$	$1.0977 \cdot 10^6$
10.6	$3.0775 \cdot 10^3$	$1.0908 \cdot 10^6$	14.6	$9.4195 \cdot 10^2$	$1.0978 \cdot 10^6$
10.7	$2.9740 \cdot 10^3$	$1.0911 \cdot 10^6$	14.7	$9.1815 \cdot 10^2$	$1.0979 \cdot 10^6$
10.8	$2.8748 \cdot 10^3$	$1.0914 \cdot 10^6$	14.8	$8.9511 \cdot 10^2$	$1.0980 \cdot 10^6$
10.9	$2.7797 \cdot 10^3$	$1.0917 \cdot 10^6$	14.9	$8.7277 \cdot 10^2$	$1.0981 \cdot 10^6$
11.0	$2.6884 \cdot 10^3$	$1.0920 \cdot 10^6$	15.0	$8.5114 \cdot 10^2$	$1.0982 \cdot 10^6$
11.1	$2.6009 \cdot 10^3$	$1.0923 \cdot 10^6$	15.1	$8.3016 \cdot 10^2$	$1.0983 \cdot 10^6$
11.2	$2.5169 \cdot 10^3$	$1.0925 \cdot 10^6$	15.2	$8.0983 \cdot 10^2$	$1.0984 \cdot 10^6$
11.3	$2.4362 \cdot 10^3$	$1.0928 \cdot 10^6$	15.3	$7.9011 \cdot 10^2$	$1.0984 \cdot 10^6$
11.4	$2.3587 \cdot 10^3$	$1.0930 \cdot 10^6$	15.4	$7.7099 \cdot 10^2$	$1.0985 \cdot 10^6$

Table 2.2. (*continued*)

temperature 2200 K

λ [µm]	$m_{\lambda,bb}$ [W/µm m²]	$M_{T,o-\lambda,bb}$ [W/m²]	λ [µm]	$m_{\lambda,bb}$ [W/µm m²]	$M_{T,o-\lambda,bb}$ [W/m²]
0.4	$2.8983 \cdot 10^{3}$	$8.5605 \cdot 10^{1}$			
0.5	$2.4987 \cdot 10^{4}$	$1.2103 \cdot 10^{3}$	4.0	$8.8492 \cdot 10^{4}$	$1.1744 \cdot 10^{6}$
0.6	$8.8833 \cdot 10^{4}$	$6.5053 \cdot 10^{3}$	4.1	$8.2206 \cdot 10^{4}$	$1.1829 \cdot 10^{6}$
0.7	$1.9504 \cdot 10^{5}$	$2.0416 \cdot 10^{4}$	4.2	$7.6449 \cdot 10^{4}$	$1.1908 \cdot 10^{6}$
0.8	$3.2169 \cdot 10^{5}$	$4.6197 \cdot 10^{4}$	4.3	$7.1171 \cdot 10^{4}$	$1.1982 \cdot 10^{6}$
0.9	$4.4293 \cdot 10^{5}$	$8.4563 \cdot 10^{4}$	4.4	$6.6326 \cdot 10^{4}$	$1.2051 \cdot 10^{6}$
1.0	$5.4132 \cdot 10^{5}$	$1.3401 \cdot 10^{5}$	4.5	$6.1875 \cdot 10^{4}$	$1.2115 \cdot 10^{6}$
1.1	$6.0983 \cdot 10^{5}$	$1.9182 \cdot 10^{5}$	4.6	$5.7781 \cdot 10^{4}$	$1.2175 \cdot 10^{6}$
1.2	$6.4888 \cdot 10^{5}$	$2.5499 \cdot 10^{5}$	4.7	$5.4011 \cdot 10^{4}$	$1.2231 \cdot 10^{6}$
1.3	$6.6283 \cdot 10^{5}$	$3.2076 \cdot 10^{5}$	4.8	$5.0537 \cdot 10^{4}$	$1.2283 \cdot 10^{6}$
1.4	$6.5731 \cdot 10^{5}$	$3.8690 \cdot 10^{5}$	4.9	$4.7331 \cdot 10^{4}$	$1.2332 \cdot 10^{6}$
1.5	$6.3784 \cdot 10^{5}$	$4.5176 \cdot 10^{5}$	5.0	$4.4369 \cdot 10^{4}$	$1.2378 \cdot 10^{6}$
1.6	$6.0908 \cdot 10^{5}$	$5.1416 \cdot 10^{5}$	5.1	$4.1631 \cdot 10^{4}$	$1.2421 \cdot 10^{6}$
1.7	$5.7474 \cdot 10^{5}$	$5.7339 \cdot 10^{5}$	5.2	$3.9097 \cdot 10^{4}$	$1.2461 \cdot 10^{6}$
1.8	$5.3757 \cdot 10^{5}$	$6.2902 \cdot 10^{5}$	5.3	$3.6750 \cdot 10^{4}$	$1.2499 \cdot 10^{6}$
1.9	$4.9954 \cdot 10^{5}$	$6.8087 \cdot 10^{5}$	5.4	$3.4573 \cdot 10^{4}$	$1.2535 \cdot 10^{6}$
2.0	$4.6199 \cdot 10^{5}$	$7.2894 \cdot 10^{5}$	5.5	$3.2551 \cdot 10^{4}$	$1.2568 \cdot 10^{6}$
2.1	$4.2581 \cdot 10^{5}$	$7.7332 \cdot 10^{5}$	5.6	$3.0674 \cdot 10^{4}$	$1.2600 \cdot 10^{6}$
2.2	$3.9153 \cdot 10^{5}$	$8.1416 \cdot 10^{5}$	5.7	$2.8927 \cdot 10^{4}$	$1.2629 \cdot 10^{6}$
2.3	$3.5942 \cdot 10^{5}$	$8.5169 \cdot 10^{5}$	5.8	$2.7302 \cdot 10^{4}$	$1.2658 \cdot 10^{6}$
2.4	$3.2963 \cdot 10^{5}$	$8.8613 \cdot 10^{5}$	5.9	$2.5787 \cdot 10^{4}$	$1.2684 \cdot 10^{6}$
2.5	$3.0217 \cdot 10^{5}$	$9.1770 \cdot 10^{5}$	6.0	$2.4374 \cdot 10^{4}$	$1.2709 \cdot 10^{6}$
2.6	$2.7696 \cdot 10^{5}$	$9.4663 \cdot 10^{5}$	6.1	$2.3056 \cdot 10^{4}$	$1.2733 \cdot 10^{6}$
2.7	$2.5391 \cdot 10^{5}$	$9.7316 \cdot 10^{5}$	6.2	$2.1825 \cdot 10^{4}$	$1.2755 \cdot 10^{6}$
2.8	$2.3287 \cdot 10^{5}$	$9.9748 \cdot 10^{5}$	6.3	$2.0673 \cdot 10^{4}$	$1.2777 \cdot 10^{6}$
2.9	$2.1370 \cdot 10^{5}$	$1.0198 \cdot 10^{6}$	6.4	$1.9596 \cdot 19^{4}$	$1.2797 \cdot 10^{6}$
3.0	$1.9626 \cdot 10^{5}$	$1.0403 \cdot 10^{6}$	6.5	$1.8587 \cdot 10^{4}$	$1.2816 \cdot 10^{6}$
3.1	$1.8039 \cdot 10^{5}$	$1.0591 \cdot 10^{6}$	6.6	$1.7642 \cdot 10^{4}$	$1.2834 \cdot 10^{6}$
3.2	$1.6596 \cdot 10^{5}$	$1.0764 \cdot 10^{6}$	6.7	$1.6755 \cdot 10^{4}$	$1.2851 \cdot 10^{6}$
3.3	$1.5284 \cdot 10^{5}$	$1.0923 \cdot 10^{6}$	6.8	$1.5923 \cdot 10^{4}$	$1.2867 \cdot 10^{6}$
3.4	$1.4090 \cdot 10^{5}$	$1.1070 \cdot 10^{6}$	6.9	$1.5142 \cdot 10^{4}$	$1.2883 \cdot 10^{6}$
3.5	$1.3003 \cdot 10^{5}$	$1.1205 \cdot 10^{6}$	7.0	$1.4407 \cdot 10^{4}$	$1.2898 \cdot 10^{6}$
3.6	$1.2013 \cdot 10^{5}$	$1.1330 \cdot 10^{6}$	7.1	$1.3716 \cdot 10^{4}$	$1.2912 \cdot 10^{6}$
3.7	$1.1111 \cdot 10^{5}$	$1.1446 \cdot 10^{6}$	7.2	$1.3065 \cdot 10^{4}$	$1.2925 \cdot 10^{6}$
3.8	$1.0288 \cdot 10^{5}$	$1.1553 \cdot 10^{6}$	7.3	$1.2453 \cdot 10^{4}$	$1.2938 \cdot 10^{6}$
3.9	$9.5363 \cdot 10^{4}$	$1.1652 \cdot 10^{6}$	7.4	$1.1875 \cdot 10^{4}$	$1.2950 \cdot 10^{6}$

Table 2.2. (*continued*)

λ [µm]	$m_{\lambda,bb}$ [W/µm m²]	$M_{T,o-\lambda,bb}$ [W/m²]	λ [µm]	$m_{\lambda,bb}$ [W/µm m²]	$M_{T,o-\lambda,bb}$ [W/m²]
7.5	$1.1331 \cdot 10^4$	$1.2962 \cdot 10^6$	11.5	$2.4289 \cdot 10^3$	$1.3182 \cdot 10^6$
7.6	$1.0817 \cdot 10^4$	$1.2973 \cdot 10^6$	11.6	$2.3525 \cdot 10^3$	$1.3184 \cdot 10^6$
7.7	$1.0331 \cdot 10^4$	$1.2983 \cdot 10^6$	11.7	$2.2791 \cdot 10^3$	$1.3187 \cdot 10^6$
7.8	$9.8724 \cdot 10^3$	$1.2993 \cdot 10^6$	11.8	$2.2085 \cdot 10^3$	$1.3189 \cdot 10^6$
7.9	$9.4388 \cdot 10^3$	$1.3003 \cdot 10^6$	11.9	$2.1406 \cdot 10^3$	$1.3191 \cdot 10^6$
8.0	$9.0285 \cdot 10^3$	$1.3012 \cdot 10^6$	12.0	$2.0753 \cdot 10^3$	$1.3193 \cdot 10^6$
8.1	$8.6401 \cdot 10^3$	$1.3021 \cdot 10^6$	12.1	$2.0125 \cdot 10^3$	$1.3195 \cdot 10^6$
8.2	$8.2723 \cdot 10^3$	$1.3029 \cdot 10^6$	12.2	$1.9520 \cdot 10^3$	$1.3197 \cdot 10^6$
8.3	$7.9237 \cdot 10^3$	$1.3038 \cdot 10^6$	12.3	$1.8938 \cdot 10^3$	$1.3199 \cdot 10^6$
8.4	$7.5932 \cdot 10^3$	$1.3045 \cdot 10^6$	12.4	$1.8377 \cdot 10^3$	$1.3201 \cdot 10^6$
8.5	$7.2796 \cdot 10^3$	$1.3053 \cdot 10^6$	12.5	$1.7837 \cdot 10^3$	$1.3203 \cdot 10^6$
8.6	$6.9820 \cdot 10^3$	$1.3060 \cdot 10^6$	12.6	$1.7317 \cdot 10^3$	$1.3205 \cdot 10^6$
8.7	$6.6993 \cdot 10^3$	$1.3067 \cdot 10^6$	12.7	$1.6815 \cdot 10^3$	$1.3206 \cdot 10^6$
8.8	$6.4307 \cdot 10^3$	$1.3073 \cdot 10^6$	12.8	$1.6331 \cdot 10^3$	$1.3208 \cdot 10^6$
8.9	$6.1754 \cdot 10^3$	$1.3080 \cdot 10^6$	12.9	$1.5865 \cdot 10^3$	$1.3210 \cdot 10^6$
9.0	$5.9326 \cdot 10^3$	$1.3086 \cdot 10^6$	13.0	$1.5415 \cdot 10^3$	$1.3211 \cdot 10^6$
9.1	$5.7015 \cdot 10^3$	$1.3091 \cdot 10^6$	13.1	$1.4981 \cdot 10^3$	$1.3213 \cdot 10^6$
9.2	$5.4815 \cdot 10^3$	$1.3097 \cdot 10^6$	13.2	$1.4561 \cdot 10^3$	$1.3214 \cdot 10^6$
9.3	$5.2720 \cdot 10^3$	$1.3102 \cdot 10^6$	13.3	$1.4157 \cdot 10^3$	$1.3215 \cdot 10^6$
9.4	$5.0723 \cdot 10^3$	$1.3108 \cdot 10^6$	13.4	$1.3766 \cdot 10^3$	$1.3217 \cdot 10^6$
9.5	$4.8820 \cdot 10^3$	$1.3112 \cdot 10^6$	13.5	$1.3389 \cdot 10^3$	$1.3218 \cdot 10^6$
9.6	$4.7004 \cdot 10^3$	$1.3117 \cdot 10^6$	13.6	$1.3025 \cdot 10^3$	$1.3220 \cdot 10^6$
9.7	$4.5272 \cdot 10^3$	$1.3122 \cdot 10^6$	13.7	$1.2673 \cdot 10^3$	$1.3221 \cdot 10^6$
9.8	$4.3619 \cdot 10^3$	$1.3126 \cdot 10^6$	13.8	$1.2332 \cdot 10^3$	$1.3222 \cdot 10^6$
9.9	$4.2040 \cdot 10^3$	$1.3131 \cdot 10^6$	13.9	$1.2003 \cdot 10^3$	$1.3223 \cdot 10^6$
10.0	$4.0532 \cdot 10^3$	$1.3135 \cdot 10^6$	14.0	$1.1685 \cdot 10^3$	$1.3224 \cdot 10^6$
10.1	$3.9090 \cdot 10^3$	$1.3139 \cdot 10^6$	14.1	$1.1377 \cdot 10^3$	$1.3226 \cdot 10^6$
10.2	$3.7712 \cdot 10^3$	$1.3143 \cdot 10^6$	14.2	$1.1080 \cdot 10^3$	$1.3227 \cdot 10^6$
10.3	$3.6393 \cdot 10^3$	$1.3146 \cdot 10^6$	14.3	$1.0792 \cdot 10^3$	$1.3228 \cdot 10^6$
10.4	$3.5132 \cdot 10^3$	$1.3150 \cdot 10^6$	14.4	$1.0513 \cdot 10^3$	$1.3229 \cdot 10^6$
10.5	$3.3925 \cdot 10^3$	$1.3153 \cdot 10^6$	14.5	$1.0243 \cdot 10^3$	$1.3230 \cdot 10^6$
10.6	$3.2768 \cdot 10^3$	$1.3157 \cdot 10^6$	14.6	$9.9819 \cdot 10^2$	$1.3231 \cdot 10^6$
10.7	$3.1661 \cdot 10^3$	$1.3160 \cdot 10^6$	14.7	$9.7290 \cdot 10^2$	$1.3232 \cdot 10^6$
10.8	$3.0600 \cdot 10^3$	$1.3163 \cdot 10^6$	14.8	$9.4840 \cdot 10^2$	$1.3233 \cdot 10^6$
10.9	$2.9582 \cdot 10^3$	$1.3166 \cdot 10^6$	14.9	$9.2466 \cdot 10^2$	$1.3234 \cdot 10^6$
11.0	$2.8607 \cdot 10^3$	$1.3169 \cdot 10^6$	15.0	$9.0166 \cdot 10^2$	$1.3235 \cdot 10^6$
11.1	$2.7671 \cdot 10^3$	$1.3172 \cdot 10^6$	15.1	$8.7938 \cdot 10^2$	$1.3236 \cdot 10^6$
11.2	$2.6773 \cdot 10^3$	$1.3174 \cdot 10^6$	15.2	$8.5777 \cdot 10^2$	$1.3236 \cdot 10^6$
11.3	$2.5911 \cdot 10^3$	$1.3177 \cdot 10^6$	15.3	$8.3682 \cdot 10^2$	$1.3237 \cdot 10^6$
11.4	$2.5084 \cdot 10^3$	$1.3180 \cdot 10^6$	15.4	$8.1651 \cdot 10^2$	$1.3238 \cdot 10^6$

Table 2.2. (*continued*)

temperature 2300 K

λ [μm]	$m_{\lambda,bb}$ [W/μm m²]	$M_{T,o-\lambda,bb}$ [W/m²]	λ [μm]	$m_{\lambda,bb}$ [W/μm m²]	$M_{T,o-\lambda,bb}$ [W/m²]
0.4	$5.9001 \cdot 10^3$	$1.8380 \cdot 10^2$			
0.5	$4.4126 \cdot 10^4$	$2.2593 \cdot 10^3$	4.0	$9.6738 \cdot 10^4$	$1.4209 \cdot 10^6$
0.6	$1.4269 \cdot 10^5$	$1.1071 \cdot 10^4$	4.1	$8.9751 \cdot 10^4$	$1.4302 \cdot 10^6$
0.7	$2.9279 \cdot 10^5$	$3.2546 \cdot 10^4$	4.2	$8.3363 \cdot 10^4$	$1.4389 \cdot 10^6$
0.8	$4.5904 \cdot 10^5$	$7.0162 \cdot 10^4$	4.3	$7.7518 \cdot 10^4$	$1.4469 \cdot 10^6$
0.9	$6.0765 \cdot 10^5$	$1.2375 \cdot 10^5$	4.4	$7.2162 \cdot 10^4$	$1.4544 \cdot 10^6$
1.0	$7.1970 \cdot 10^5$	$1.9046 \cdot 10^5$	4.5	$6.7249 \cdot 10^4$	$1.4614 \cdot 10^6$
1.1	$7.9033 \cdot 10^5$	$2.6630 \cdot 10^5$	4.6	$6.2737 \cdot 10^4$	$1.4679 \cdot 10^6$
1.2	$8.2333 \cdot 10^5$	$3.4726 \cdot 10^5$	4.7	$5.8589 \cdot 10^4$	$1.4739 \cdot 10^6$
1.3	$8.2621 \cdot 10^5$	$4.2996 \cdot 10^5$	4.8	$5.4771 \cdot 10^4$	$1.4796 \cdot 10^6$
1.4	$8.0706 \cdot 10^5$	$5.1177 \cdot 10^5$	4.9	$5.1252 \cdot 10^4$	$1.4849 \cdot 10^6$
1.5	$7.7306 \cdot 10^5$	$5.9087 \cdot 10^5$	5.0	$4.8007 \cdot 10^4$	$1.4898 \cdot 10^6$
1.6	$7.2996 \cdot 10^5$	$6.6608 \cdot 10^5$	5.1	$4.5009 \cdot 10^4$	$1.4945 \cdot 10^6$
1.7	$6.8209 \cdot 10^5$	$7.3671 \cdot 10^5$	5.2	$4.2238 \cdot 10^4$	$1.4989 \cdot 10^6$
1.8	$6.3250 \cdot 10^5$	$8.0244 \cdot 10^5$	5.3	$3.9673 \cdot 10^4$	$1.5029 \cdot 10^6$
1.9	$5.8329 \cdot 10^5$	$8.6322 \cdot 10^5$	5.4	$3.7297 \cdot 10^4$	$1.5068 \cdot 10^6$
2.0	$5.3581 \cdot 10^5$	$9.1916 \cdot 10^5$	5.5	$3.5094 \cdot 10^4$	$1.5104 \cdot 10^6$
2.1	$4.9086 \cdot 10^5$	$9.7047 \cdot 10^5$	5.6	$3.3049 \cdot 10^4$	$1.5138 \cdot 10^6$
2.2	$4.4888 \cdot 10^5$	$1.0174 \cdot 10^6$	5.7	$3.1148 \cdot 10^4$	$1.5170 \cdot 10^6$
2.3	$4.1006 \cdot 10^5$	$1.0603 \cdot 10^6$	5.8	$2.9380 \cdot 10^4$	$1.5200 \cdot 10^6$
2.4	$3.7440 \cdot 10^5$	$1.0995 \cdot 10^6$	5.9	$2.7735 \cdot 10^4$	$1.5229 \cdot 10^6$
2.5	$3.4181 \cdot 10^5$	$1.1353 \cdot 10^6$	6.0	$2.6201 \cdot 10^4$	$1.5256 \cdot 10^6$
2.6	$3.1214 \cdot 10^5$	$1.1680 \cdot 10^6$	6.1	$2.4771 \cdot 10^4$	$1.5281 \cdot 10^6$
2.7	$2.8519 \cdot 10^5$	$1.1978 \cdot 10^6$	6.2	$2.3437 \cdot 10^4$	$1.5306 \cdot 10^6$
2.8	$2.6074 \cdot 10^5$	$1.2251 \cdot 10^6$	6.3	$2.2190 \cdot 10^4$	$1.5328 \cdot 10^6$
2.9	$2.3860 \cdot 10^5$	$1.2501 \cdot 10^6$	6.4	$2.1023 \cdot 10^4$	$1.5350 \cdot 10^6$
3.0	$2.1854 \cdot 10^5$	$1.2729 \cdot 10^6$	6.5	$1.9932 \cdot 10^4$	$1.5370 \cdot 10^6$
3.1	$2.0038 \cdot 10^5$	$1.2938 \cdot 10^6$	6.6	$1.8910 \cdot 10^4$	$1.5390 \cdot 10^6$
3.2	$1.8393 \cdot 10^5$	$1.3130 \cdot 10^6$	6.7	$1.7952 \cdot 10^4$	$1.5408 \cdot 10^6$
3.3	$1.6903 \cdot 10^5$	$1.3307 \cdot 10^6$	6.8	$1.7054 \cdot 10^4$	$1.5426 \cdot 10^6$
3.4	$1.5551 \cdot 10^5$	$1.3469 \cdot 10^6$	6.9	$1.6210 \cdot 10^4$	$1.5442 \cdot 10^6$
3.5	$1.4325 \cdot 10^5$	$1.3618 \cdot 10^6$	7.0	$1.5418 \cdot 10^4$	$1.5458 \cdot 10^6$
3.6	$1.3212 \cdot 10^5$	$1.3756 \cdot 10^6$	7.1	$1.4673 \cdot 10^4$	$1.5473 \cdot 10^6$
3.7	$1.8199 \cdot 10^5$	$1.3883 \cdot 10^6$	7.2	$1.3972 \cdot 10^4$	$1.5487 \cdot 10^6$
3.8	$1.1278 \cdot 10^5$	$1.4000 \cdot 10^6$	7.3	$1.3312 \cdot 10^4$	$1.5501 \cdot 10^6$
3.9	$1.0439 \cdot 10^5$	$1.4109 \cdot 10^6$	7.4	$1.2690 \cdot 10^4$	$1.5514 \cdot 10^6$

Table 2.2. (*continued*)

λ [µm]	$m_{\lambda,bb}$ [W/µm m^2]	$M_{T,o-\lambda,bb}$ [W/m^2]	λ [µm]	$m_{\lambda,bb}$ [W/µm m^2]	$M_{T,o-\lambda,bb}$ [W/m^2]
7.5	$1.2104 \cdot 10^4$	$1.5526 \cdot 10^6$	11.5	$2.5738 \cdot 10^3$	$1.5761 \cdot 10^6$
7.6	$1.1552 \cdot 10^4$	$1.5538 \cdot 10^6$	11.6	$2.4925 \cdot 10^3$	$1.5764 \cdot 10^6$
7.7	$1.1030 \cdot 10^4$	$1.5550 \cdot 10^6$	11.7	$2.4145 \cdot 10^3$	$1.5766 \cdot 10^6$
7.8	$1.0537 \cdot 10^4$	$1.5560 \cdot 10^6$	11.8	$2.3394 \cdot 10^3$	$1.5768 \cdot 10^6$
7.9	$1.0071 \cdot 10^4$	$1.5571 \cdot 10^6$	11.9	$2.2672 \cdot 10^3$	$1.5771 \cdot 10^6$
8.0	$9.6307 \cdot 10^3$	$1.5581 \cdot 10^6$	12.0	$2.1978 \cdot 10^3$	$1.5773 \cdot 10^6$
8.1	$9.2139 \cdot 10^3$	$1.5590 \cdot 10^6$	12.1	$2.1311 \cdot 10^3$	$1.5775 \cdot 10^6$
8.2	$8.8192 \cdot 10^3$	$1.5599 \cdot 10^6$	12.2	$2.0668 \cdot 10^3$	$1.5777 \cdot 10^6$
8.3	$8.4454 \cdot 10^3$	$1.5608 \cdot 10^6$	12.3	$2.0049 \cdot 10^3$	$1.5779 \cdot 10^6$
8.4	$8.0910 \cdot 10^3$	$1.5616 \cdot 10^6$	12.4	$1.9453 \cdot 10^3$	$1.5781 \cdot 10^6$
8.5	$7.7550 \cdot 10^3$	$1.5624 \cdot 10^6$	12.5	$1.8880 \cdot 10^3$	$1.5783 \cdot 10^6$
8.6	$7.4361 \cdot 10^3$	$1.5631 \cdot 10^6$	12.6	$1.8327 \cdot 10^3$	$1.5785 \cdot 10^6$
8.7	$7.1334 \cdot 10^3$	$1.5639 \cdot 10^6$	12.7	$1.7794 \cdot 10^3$	$1.5787 \cdot 10^6$
8.8	$6.8458 \cdot 10^3$	$1.5646 \cdot 10^6$	12.8	$1.7280 \cdot 10^3$	$1.5789 \cdot 10^6$
8.9	$6.5725 \cdot 10^3$	$1.5652 \cdot 10^6$	12.9	$1.6785 \cdot 10^3$	$1.5790 \cdot 10^6$
9.0	$6.3127 \cdot 10^3$	$1.5659 \cdot 10^6$	13.0	$1.6307 \cdot 10^3$	$1.5792 \cdot 10^6$
9.1	$6.0655 \cdot 10^3$	$1.5665 \cdot 10^6$	13.1	$1.5846 \cdot 10^3$	$1.5794 \cdot 10^6$
9.2	$5.8303 \cdot 10^3$	$1.5671 \cdot 10^6$	13.2	$1.5402 \cdot 10^3$	$1.5795 \cdot 10^6$
9.3	$5.6063 \cdot 10^3$	$1.5677 \cdot 10^6$	13.3	$1.4972 \cdot 10^3$	$1.5797 \cdot 10^6$
9.4	$5.3929 \cdot 10^3$	$1.5682 \cdot 10^6$	13.4	$1.4558 \cdot 10^3$	$1.5798 \cdot 10^6$
9.5	$5.1895 \cdot 10^3$	$1.5687 \cdot 10^6$	13.5	$1.4158 \cdot 10^3$	$1.5799 \cdot 10^6$
9.6	$4.9956 \cdot 10^3$	$1.5692 \cdot 10^6$	13.6	$1.3771 \cdot 10^3$	$1.5801 \cdot 10^6$
9.7	$4.8106 \cdot 10^3$	$1.5697 \cdot 10^6$	13.7	$1.3398 \cdot 10^3$	$1.5802 \cdot 10^6$
9.8	$4.6341 \cdot 10^3$	$1.5702 \cdot 10^6$	13.8	$1.3037 \cdot 10^3$	$1.5804 \cdot 10^6$
9.9	$4.4655 \cdot 10^3$	$1.5707 \cdot 10^6$	13.9	$1.2688 \cdot 10^3$	$1.5805 \cdot 10^6$
10.0	$4.3045 \cdot 10^3$	$1.5711 \cdot 10^6$	14.0	$1.2350 \cdot 10^3$	$1.5806 \cdot 10^6$
10.1	$4.1507 \cdot 10^3$	$1.5715 \cdot 10^6$	14.1	$1.2024 \cdot 10^3$	$1.5807 \cdot 10^6$
10.2	$4.0037 \cdot 10^3$	$1.5719 \cdot 10^6$	14.2	$1.1709 \cdot 10^3$	$1.5808 \cdot 10^6$
10.3	$3.8631 \cdot 10^3$	$1.5723 \cdot 10^6$	14.3	$1.1403 \cdot 10^3$	$1.5810 \cdot 10^6$
10.4	$3.7286 \cdot 10^3$	$1.5727 \cdot 10^6$	14.4	$1.1108 \cdot 10^3$	$1.5811 \cdot 10^6$
10.5	$3.5999 \cdot 10^3$	$1.5731 \cdot 10^6$	14.5	$1.0822 \cdot 10^3$	$1.5812 \cdot 10^6$
10.6	$3.4767 \cdot 10^3$	$1.5734 \cdot 10^6$	14.6	$1.0545 \cdot 10^3$	$1.5813 \cdot 10^6$
10.7	$3.3587 \cdot 10^3$	$1.5738 \cdot 10^6$	14.7	$1.0277 \cdot 10^3$	$1.5814 \cdot 10^6$
10.8	$3.2456 \cdot 10^3$	$1.5741 \cdot 10^6$	14.8	$1.0018 \cdot 10^3$	$1.5815 \cdot 10^6$
10.9	$3.1373 \cdot 10^3$	$1.5744 \cdot 10^6$	14.9	$9.7663 \cdot 10^2$	$1.5816 \cdot 10^6$
11.0	$3.0334 \cdot 10^3$	$1.5747 \cdot 10^6$	15.0	$9.5227 \cdot 10^2$	$1.5817 \cdot 10^6$
11.1	$2.9338 \cdot 10^3$	$1.5750 \cdot 10^6$	15.1	$9.2866 \cdot 10^2$	$1.5818 \cdot 10^6$
11.2	$2.8382 \cdot 10^3$	$1.5753 \cdot 10^6$	15.2	$9.0578 \cdot 10^2$	$1.5819 \cdot 10^6$
11.3	$2.7464 \cdot 10^3$	$1.5756 \cdot 10^6$	15.3	$8.8360 \cdot 10^2$	$1.5820 \cdot 10^6$
11.4	$2.6584 \cdot 10^3$	$1.5759 \cdot 10^6$	15.4	$8.6209 \cdot 10^2$	$1.5820 \cdot 10^6$

Table 2.2. (*continued*)

temperature 2400 K

λ [μm]	$m_{\lambda,bb}$ [W/μm m²]	$M_{T,o-\lambda,bb}$ [W/m²]	λ [μm]	$m_{\lambda,bb}$ [W/μm m²]	$M_{T,o-\lambda,bb}$ [W/m²]
0.4	$1.1320 \cdot 10^4$	$3.7121 \cdot 10^2$			
0.5	$7.4318 \cdot 10^4$	$4.0148 \cdot 10^3$	4.0	$1.0513 \cdot 10^5$	$1.7033 \cdot 10^6$
0.6	$2.2033 \cdot 10^5$	$1.8077 \cdot 10^4$	4.1	$9.7418 \cdot 10^4$	$1.7134 \cdot 10^6$
0.7	$4.2490 \cdot 10^5$	$5.0063 \cdot 10^4$	4.2	$9.0386 \cdot 10^4$	$1.7228 \cdot 10^6$
0.8	$6.3594 \cdot 10^5$	$1.0326 \cdot 10^5$	4.3	$8.3960 \cdot 10^4$	$1.7315 \cdot 10^6$
0.9	$8.1203 \cdot 10^5$	$1.7606 \cdot 10^5$	4.4	$7.8081 \cdot 10^4$	$1.7396 \cdot 10^6$
1.0	$9.3454 \cdot 10^5$	$2.6386 \cdot 10^5$	4.5	$7.2697 \cdot 10^4$	$1.7472 \cdot 10^6$
1.1	$1.0026 \cdot 10^6$	$3.6114 \cdot 10^5$	4.6	$6.7759 \cdot 10^4$	$1.7542 \cdot 10^6$
1.2	$1.0244 \cdot 10^6$	$4.6283 \cdot 10^5$	4.7	$6.3225 \cdot 10^4$	$1.7607 \cdot 10^6$
1.3	$1.0115 \cdot 10^6$	$5.6486 \cdot 10^5$	4.8	$5.9056 \cdot 10^4$	$1.7668 \cdot 10^6$
1.4	$9.7453 \cdot 10^5$	$6.6432 \cdot 10^5$	4.9	$5.5220 \cdot 10^4$	$1.7725 \cdot 10^6$
1.5	$9.2251 \cdot 10^5$	$7.5927 \cdot 10^5$	5.0	$5.1684 \cdot 10^4$	$1.7779 \cdot 10^6$
1.6	$8.6223 \cdot 10^5$	$8.4855 \cdot 10^5$	5.1	$4.8423 \cdot 10^4$	$1.7829 \cdot 10^6$
1.7	$7.9853 \cdot 10^5$	$9.3160 \cdot 10^5$	5.2	$4.5411 \cdot 10^4$	$1.7876 \cdot 10^6$
1.8	$7.3471 \cdot 10^5$	$1.0083 \cdot 10^6$	5.3	$4.2626 \cdot 10^4$	$1.7920 \cdot 10^6$
1.9	$6.7288 \cdot 10^5$	$1.0786 \cdot 10^6$	5.4	$4.0048 \cdot 10^4$	$1.7961 \cdot 10^6$
2.0	$6.1431 \cdot 10^5$	$1.1429 \cdot 10^6$	5.5	$3.7660 \cdot 10^4$	$1.8000 \cdot 10^6$
2.1	$5.5969 \cdot 10^5$	$1.2016 \cdot 10^6$	5.6	$3.5445 \cdot 10^4$	$1.8036 \cdot 10^6$
2.2	$5.0930 \cdot 10^5$	$1.2550 \cdot 10^6$	5.7	$3.3388 \cdot 10^4$	$1.8071 \cdot 10^6$
2.3	$4.6318 \cdot 10^5$	$1.3036 \cdot 10^6$	5.8	$3.1476 \cdot 10^4$	$1.8103 \cdot 10^6$
2.4	$4.2120 \cdot 10^5$	$1.3478 \cdot 10^6$	5.9	$2.9698 \cdot 10^4$	$1.8134 \cdot 10^6$
2.5	$3.8313 \cdot 10^5$	$1.3880 \cdot 10^6$	6.0	$2.8043 \cdot 10^4$	$1.8163 \cdot 10^6$
2.6	$3.4870 \cdot 10^5$	$1.4245 \cdot 10^6$	6.1	$2.6500 \cdot 10^4$	$1.8190 \cdot 10^6$
2.7	$3.1761 \cdot 10^5$	$1.4578 \cdot 10^6$	6.2	$2.5060 \cdot 10^4$	$1.8216 \cdot 10^6$
2.8	$2.8957 \cdot 10^5$	$1.4882 \cdot 10^6$	6.3	$2.3716 \cdot 10^4$	$1.8240 \cdot 10^6$
2.9	$2.6428 \cdot 10^5$	$1.5158 \cdot 10^6$	6.4	$2.2461 \cdot 10^4$	$1.8263 \cdot 10^6$
3.0	$2.4149 \cdot 10^5$	$1.5411 \cdot 10^6$	6.5	$2.1286 \cdot 10^4$	$1.8285 \cdot 10^6$
3.1	$2.2093 \cdot 10^5$	$1.5642 \cdot 10^6$	6.6	$2.0187 \cdot 10^4$	$1.8306 \cdot 10^6$
3.2	$2.0237 \cdot 10^5$	$1.5853 \cdot 10^6$	6.7	$1.9157 \cdot 10^4$	$1.8325 \cdot 10^6$
3.3	$1.8561 \cdot 10^5$	$1.6047 \cdot 10^6$	6.8	$1.8191 \cdot 10^4$	$1.8344 \cdot 10^6$
3.4	$1.7047 \cdot 10^5$	$1.6225 \cdot 10^6$	6.9	$1.7285 \cdot 10^4$	$1.8362 \cdot 10^6$
3.5	$1.5676 \cdot 10^5$	$1.6389 \cdot 10^6$	7.0	$1.6434 \cdot 10^4$	$1.8379 \cdot 10^6$
3.6	$1.4435 \cdot 10^5$	$1.6539 \cdot 10^6$	7.1	$1.5635 \cdot 10^4$	$1.8395 \cdot 10^6$
3.7	$1.3309 \cdot 10^5$	$1.6678 \cdot 10^6$	7.2	$1.4883 \cdot 10^4$	$1.8410 \cdot 10^6$
3.8	$1.2287 \cdot 10^5$	$1.6806 \cdot 10^6$	7.3	$1.4176 \cdot 10^4$	$1.8424 \cdot 10^6$
3.9	$1.1358 \cdot 10^5$	$1.6924 \cdot 10^6$	7.4	$1.3510 \cdot 10^4$	$1.8438 \cdot 10^6$

Table 2.2. (*continued*)

λ [µm]	$m_{\lambda,bb}$ [W/µm m²]	$M_{T,o-\lambda,bb}$ [W/m²]	λ [µm]	$m_{\lambda,bb}$ [W/µm m²]	$M_{T,o-\lambda,bb}$ [W/m²]
7.5	$1.2882 \cdot 10^4$	$1.8452 \cdot 10^6$	11.5	$2.7190 \cdot 10^3$	$1.8700 \cdot 10^6$
7.6	$1.2290 \cdot 10^4$	$1.8464 \cdot 10^6$	11.6	$2.6329 \cdot 10^3$	$1.8703 \cdot 10^6$
7.7	$1.1732 \cdot 10^4$	$1.8476 \cdot 10^6$	11.7	$2.5501 \cdot 10^3$	$1.8706 \cdot 10^6$
7.8	$1.1205 \cdot 10^4$	$1.8488 \cdot 10^6$	11.8	$2.4706 \cdot 10^3$	$1.8708 \cdot 10^6$
7.9	$1.0706 \cdot 10^4$	$1.8499 \cdot 10^6$	11.9	$2.3941 \cdot 10^3$	$1.8711 \cdot 10^6$
8.0	$1.0236 \cdot 10^4$	$1.8509 \cdot 10^6$	12.0	$2.3206 \cdot 10^3$	$1.8713 \cdot 10^6$
8.1	$9.7901 \cdot 10^3$	$1.8519 \cdot 10^6$	12.1	$2.2498 \cdot 10^3$	$1.8715 \cdot 10^6$
8.2	$9.3685 \cdot 10^3$	$1.8529 \cdot 10^6$	12.2	$2.1817 \cdot 10^3$	$1.8717 \cdot 10^6$
8.3	$8.9692 \cdot 10^3$	$1.8538 \cdot 10^6$	12.3	$2.1162 \cdot 10^3$	$1.8720 \cdot 10^6$
8.4	$8.5909 \cdot 10^3$	$1.8546 \cdot 10^6$	12.4	$2.0531 \cdot 10^3$	$1.8722 \cdot 10^6$
8.5	$8.2322 \cdot 10^3$	$1.8555 \cdot 10^6$	12.5	$1.9924 \cdot 10^3$	$1.8724 \cdot 10^6$
8.6	$7.8920 \cdot 10^3$	$1.8563 \cdot 10^6$	12.6	$1.9338 \cdot 10^3$	$1.8726 \cdot 10^6$
8.7	$7.5691 \cdot 10^3$	$1.8571 \cdot 10^6$	12.7	$1.8774 \cdot 10^3$	$1.8728 \cdot 10^6$
8.8	$7.2624 \cdot 10^3$	$1.8578 \cdot 10^6$	12.8	$1.8231 \cdot 10^3$	$1.8729 \cdot 10^6$
8.9	$6.9711 \cdot 10^3$	$1.8585 \cdot 10^6$	12.9	$1.7707 \cdot 10^3$	$1.8731 \cdot 10^6$
9.0	$6.6941 \cdot 10^3$	$1.8592 \cdot 10^6$	13.0	$1.7201 \cdot 10^3$	$1.8733 \cdot 10^6$
9.1	$6.4308 \cdot 10^3$	$1.8599 \cdot 10^6$	13.1	$1.6713 \cdot 10^3$	$1.8735 \cdot 10^6$
9.2	$6.1802 \cdot 10^3$	$1.8605 \cdot 10^6$	13.2	$1.6243 \cdot 10^3$	$1.8736 \cdot 10^6$
9.3	$5.9416 \cdot 10^3$	$1.8611 \cdot 10^6$	13.3	$1.5789 \cdot 10^3$	$1.8738 \cdot 10^6$
9.4	$5.7144 \cdot 10^3$	$1.8617 \cdot 10^6$	13.4	$1.5351 \cdot 10^3$	$1.8739 \cdot 10^6$
9.5	$5.4980 \cdot 10^3$	$1.8622 \cdot 10^6$	13.5	$1.4927 \cdot 10^3$	$1.8741 \cdot 10^6$
9.6	$5.2916 \cdot 10^3$	$1.8628 \cdot 10^6$	13.6	$1.4519 \cdot 10^3$	$1.8742 \cdot 10^6$
9.7	$5.0948 \cdot 10^3$	$1.8633 \cdot 10^6$	13.7	$1.4124 \cdot 10^3$	$1.8744 \cdot 10^6$
9.8	$4.9070 \cdot 10^3$	$1.8638 \cdot 10^6$	13.8	$1.3742 \cdot 10^3$	$1.8745 \cdot 10^6$
9.9	$4.7278 \cdot 10^3$	$1.8643 \cdot 10^6$	13.9	$1.3373 \cdot 10^3$	$1.8747 \cdot 10^6$
10.0	$4.5566 \cdot 10^3$	$1.8647 \cdot 10^6$	14.0	$1.3017 \cdot 10^3$	$1.8748 \cdot 10^6$
10.1	$4.3932 \cdot 10^3$	$1.8652 \cdot 10^6$	14.1	$1.2672 \cdot 10^3$	$1.8749 \cdot 10^6$
10.2	$4.2369 \cdot 10^3$	$1.8656 \cdot 10^6$	14.2	$1.2338 \cdot 10^3$	$1.8750 \cdot 10^6$
10.3	$4.0875 \cdot 10^3$	$1.8660 \cdot 10^6$	14.3	$1.2016 \cdot 10^3$	$1.8752 \cdot 10^6$
10.4	$3.9446 \cdot 10^3$	$1.8664 \cdot 10^6$	14.4	$1.1704 \cdot 10^3$	$1.8753 \cdot 10^6$
10.5	$3.8079 \cdot 10^3$	$1.8668 \cdot 10^6$	14.5	$1.1402 \cdot 10^3$	$1.8754 \cdot 10^6$
10.6	$3.6771 \cdot 10^3$	$1.8672 \cdot 10^6$	14.6	$1.1109 \cdot 10^3$	$1.8755 \cdot 10^6$
10.7	$3.5518 \cdot 10^3$	$1.8676 \cdot 10^6$	14.7	$1.0826 \cdot 10^3$	$1.8756 \cdot 10^6$
10.8	$3.4318 \cdot 10^3$	$1.8679 \cdot 10^6$	14.8	$1.0552 \cdot 10^3$	$1.8757 \cdot 10^6$
10.9	$3.3167 \cdot 10^3$	$1.8682 \cdot 10^6$	14.9	$1.0287 \cdot 10^3$	$1.8758 \cdot 10^6$
11.0	$3.2065 \cdot 10^3$	$1.8686 \cdot 10^6$	15.0	$1.0029 \cdot 10^3$	$1.8759 \cdot 10^6$
11.1	$3.1008 \cdot 10^3$	$1.8689 \cdot 10^6$	15.1	$9.7800 \cdot 10^2$	$1.8760 \cdot 10^6$
11.2	$2.9994 \cdot 10^3$	$1.8692 \cdot 10^6$	15.2	$9.5385 \cdot 10^2$	$1.8761 \cdot 10^6$
11.3	$2.9021 \cdot 10^3$	$1.8695 \cdot 10^6$	15.3	$9.3043 \cdot 10^2$	$1.8762 \cdot 10^6$
11.4	$2.8087 \cdot 10^3$	$1.8698 \cdot 10^6$	15.4	$9.0773 \cdot 10^2$	$1.8763 \cdot 10^6$

Table 2.2. (*continued*)

temperature 2500 K

λ [μm]	$m_{\lambda,bb}$ [W/μm m²]	$M_{T,o-\lambda,bb}$ [W/m²]	λ [μm]	$m_{\lambda,bb}$ [W/μm m²]	$M_{T,o-\lambda,bb}$ [W/m²]
0.4	$2.0617\cdot10^4$	$7.1044\cdot10^2$			
0.5	$1.2006\cdot10^5$	$6.8311\cdot10^3$	4.0	$1.1364\cdot10^5$	$2.0248\cdot10^6$
0.6	$3.2858\cdot10^5$	$2.8461\cdot10^4$	4.1	$1.0520\cdot10^5$	$2.0357\cdot10^6$
0.7	$5.9855\cdot10^5$	$7.4618\cdot10^4$	4.2	$9.7505\cdot10^4$	$2.0459\cdot10^6$
0.7	$8.5837\cdot10^5$	$1.4780\cdot10^5$	4.3	$9.0487\cdot10^4$	$2.0553\cdot10^6$
0.8	$1.0604\cdot10^6$	$2.4433\cdot10^5$	4.4	$8.4076\cdot10^4$	$2.0640\cdot10^6$
1.0	$1.1886\cdot10^6$	$3.5739\cdot10^5$	4.5	$7.8211\cdot10^4$	$2.0721\cdot10^6$
1.1	$1.2481\cdot10^6$	$4.7975\cdot10^5$	4.6	$7.2839\cdot10^4$	$2.0796\cdot10^6$
1.2	$1.2529\cdot10^6$	$6.0518\cdot10^5$	4.7	$6.7912\cdot10^4$	$2.0867\cdot10^6$
1.3	$1.2189\cdot10^6$	$7.2903\cdot10^5$	4.8	$6.3388\cdot10^4$	$2.0932\cdot10^6$
1.4	$1.1596\cdot10^6$	$8.4812\cdot10^5$	4.9	$5.9228\cdot10^4$	$2.0994\cdot10^6$
1.5	$1.0860\cdot10^6$	$9.6048\cdot10^5$	5.0	$5.5398\cdot10^4$	$2.1051\cdot10^6$
1.6	$1.0056\cdot10^6$	$1.0651\cdot10^6$	5.1	$5.1869\cdot10^4$	$2.1105\cdot10^6$
1.7	$9.2374\cdot10^5$	$1.1615\cdot10^6$	5.2	$4.8613\cdot10^4$	$2.1155\cdot10^6$
1.8	$8.4387\cdot10^5$	$1.2499\cdot10^6$	5.3	$4.5604\cdot10^4$	$2.1202\cdot10^6$
1.9	$7.6800\cdot10^5$	$1.3305\cdot10^6$	5.4	$4.2822\cdot10^4$	$2.1246\cdot10^6$
2.0	$6.9723\cdot10^5$	$1.4037\cdot10^6$	5.5	$4.0247\cdot10^4$	$2.1288\cdot10^6$
2.1	$6.3206\cdot10^5$	$1.4701\cdot10^6$	5.6	$3.7859\cdot10^4$	$2.1327\cdot10^6$
2.2	$5.7258\cdot10^5$	$1.5303\cdot10^6$	5.7	$3.5645\cdot10^4$	$2.1363\cdot10^6$
2.3	$5.1862\cdot10^5$	$1.5848\cdot10^6$	5.8	$3.3588\cdot10^4$	$2.1398\cdot10^6$
2.4	$4.6989\cdot10^5$	$1.6342\cdot10^6$	5.9	$3.1676\cdot10^4$	$2.1431\cdot10^6$
2.5	$4.2599\cdot10^5$	$1.6789\cdot10^6$	6.0	$2.9897\cdot10^4$	$2.1461\cdot10^6$
2.6	$3.8653\cdot10^5$	$1.7195\cdot10^6$	6.1	$2.8239\cdot10^4$	$2.1490\cdot10^6$
2.7	$3.5108\cdot10^5$	$1.7564\cdot10^6$	6.2	$2.6694\cdot10^4$	$2.1518\cdot10^6$
2.8	$3.1926\cdot10^5$	$1.7898\cdot10^6$	6.3	$2.5253\cdot10^4$	$2.1544\cdot10^6$
2.9	$2.9070\cdot10^5$	$1.8203\cdot10^6$	6.4	$2.3906\cdot10^4$	$2.1568\cdot10^6$
3.0	$2.6504\cdot10^5$	$1.8481\cdot10^6$	6.5	$2.2648\cdot10^4$	$2.1592\cdot10^6$
3.1	$2.4198\cdot10^5$	$1.8734\cdot10^6$	6.6	$2.1470\cdot10^4$	$2.1614\cdot10^6$
3.2	$2.2124\cdot10^5$	$1.8966\cdot10^6$	6.7	$2.0368\cdot10^4$	$2.1635\cdot10^6$
3.3	$2.0257\cdot10^5$	$1.9177\cdot10^6$	6.8	$1.9335\cdot10^4$	$2.1654\cdot10^6$
3.4	$1.8573\cdot10^5$	$1.9371\cdot10^6$	6.9	$1.8366\cdot10^4$	$2.1673\cdot10^6$
3.5	$1.7054\cdot10^5$	$1.9549\cdot10^6$	7.0	$1.7456\cdot10^4$	$2.1691\cdot10^6$
3.6	$1.5681\cdot10^5$	$1.9713\cdot10^6$	7.1	$1.6602\cdot10^4$	$2.1708\cdot10^6$
3.7	$1.4439\cdot10^5$	$1.9863\cdot10^6$	7.2	$1.5799\cdot10^4$	$2.1724\cdot10^6$
3.8	$1.3313\cdot10^5$	$2.0002\cdot10^6$	7.3	$1.5044\cdot10^4$	$2.1740\cdot10^6$
3.9	$1.2292\cdot10^5$	$2.0130\cdot10^6$	7.4	$1.4333\cdot10^4$	$2.1754\cdot10^6$

λ [µm]	$m_{\lambda,bb}$ [W/µm m²]	$M_{T,o-\lambda,bb}$ [W/m²]	λ [µm]	$m_{\lambda,bb}$ [W/µm m²]	$M_{T,o-\lambda,bb}$ [W/m²]
7.5	$1.3663 \cdot 10^4$	$2.1768 \cdot 10^6$	11.5	$2.8645 \cdot 10^3$	$2.2032 \cdot 10^6$
7.6	$1.3032 \cdot 10^4$	$2.1782 \cdot 10^6$	11.6	$2.7735 \cdot 10^3$	$2.2034 \cdot 10^6$
7.7	$1.2437 \cdot 10^4$	$2.1795 \cdot 10^6$	11.7	$2.6860 \cdot 10^3$	$2.2037 \cdot 10^6$
7.8	$1.1875 \cdot 10^4$	$2.1807 \cdot 10^6$	11.8	$2.6020 \cdot 10^3$	$2.2040 \cdot 10^6$
7.9	$1.1344 \cdot 10^4$	$2.1818 \cdot 10^6$	11.9	$2.5212 \cdot 10^3$	$2.2042 \cdot 10^6$
8.0	$1.0843 \cdot 10^4$	$2.1829 \cdot 10^6$	12.0	$2.4435 \cdot 10^3$	$2.2045 \cdot 10^6$
8.1	$1.0368 \cdot 10^4$	$2.1840 \cdot 10^6$	12.1	$2.3688 \cdot 10^3$	$2.2047 \cdot 10^6$
8.2	$9.9197 \cdot 10^3$	$2.1850 \cdot 10^6$	12.2	$2.2969 \cdot 10^3$	$2.2050 \cdot 10^6$
8.3	$9.4949 \cdot 10^3$	$2.1860 \cdot 10^6$	12.3	$2.2277 \cdot 10^3$	$2.2052 \cdot 10^6$
8.4	$9.0925 \cdot 10^3$	$2.1869 \cdot 10^6$	12.4	$2.1611 \cdot 10^3$	$2.2054 \cdot 10^6$
8.5	$8.7111 \cdot 10^3$	$2.1878 \cdot 10^6$	12.5	$2.0970 \cdot 10^3$	$2.2056 \cdot 10^6$
8.6	$8.3494 \cdot 10^3$	$2.1886 \cdot 10^6$	12.6	$2.0352 \cdot 10^3$	$2.2058 \cdot 10^6$
8.7	$8.0062 \cdot 10^3$	$2.1895 \cdot 10^6$	12.7	$1.9756 \cdot 10^3$	$2.2060 \cdot 10^6$
8.8	$7.6803 \cdot 10^3$	$2.1903 \cdot 10^6$	12.8	$1.9183 \cdot 10^3$	$2.2062 \cdot 10^6$
8.9	$7.3709 \cdot 10^3$	$2.1910 \cdot 10^6$	12.9	$1.8630 \cdot 10^3$	$2.2064 \cdot 10^6$
9.0	$7.0768 \cdot 10^3$	$2.1917 \cdot 10^6$	13.0	$1.8096 \cdot 10^3$	$2.2066 \cdot 10^6$
9.1	$6.7971 \cdot 10^3$	$2.1924 \cdot 10^6$	13.1	$1.7582 \cdot 10^3$	$2.2068 \cdot 10^6$
9.2	$6.5311 \cdot 10^3$	$2.1931 \cdot 10^6$	13.2	$1.7086 \cdot 10^3$	$2.2069 \cdot 10^6$
9.3	$6.2780 \cdot 10^3$	$2.1937 \cdot 10^6$	13.3	$1.6607 \cdot 10^3$	$2.2071 \cdot 10^6$
9.4	$6.0369 \cdot 10^3$	$2.1943 \cdot 10^6$	13.4	$1.6144 \cdot 10^3$	$2.2073 \cdot 10^6$
9.5	$5.8073 \cdot 10^3$	$2.1949 \cdot 10^6$	13.5	$1.5698 \cdot 10^3$	$2.2074 \cdot 10^6$
9.6	$5.5884 \cdot 10^3$	$2.1955 \cdot 10^6$	13.6	$1.5267 \cdot 10^3$	$2.2076 \cdot 10^6$
9.7	$5.3798 \cdot 10^3$	$2.1960 \cdot 10^6$	13.7	$1.4851 \cdot 10^3$	$2.2077 \cdot 10^6$
9.8	$5.1807 \cdot 10^3$	$2.1966 \cdot 10^6$	13.8	$1.4448 \cdot 10^3$	$2.2079 \cdot 10^6$
9.9	$4.9907 \cdot 10^3$	$2.1971 \cdot 10^6$	13.9	$1.4060 \cdot 10^3$	$2.2080 \cdot 10^6$
10.0	$4.8094 \cdot 10^3$	$2.1976 \cdot 10^6$	14.0	$1.3684 \cdot 10^3$	$2.2082 \cdot 10^6$
10.1	$4.6362 \cdot 10^3$	$2.1980 \cdot 10^6$	14.1	$1.3320 \cdot 10^3$	$2.2083 \cdot 10^6$
10.2	$4.4706 \cdot 10^3$	$2.1985 \cdot 10^6$	14.2	$1.2969 \cdot 10^3$	$2.2084 \cdot 10^6$
10.3	$4.3124 \cdot 10^3$	$2.1989 \cdot 10^6$	14.3	$1.2629 \cdot 10^3$	$2.2086 \cdot 10^6$
10.4	$4.1611 \cdot 10^3$	$2.1994 \cdot 10^6$	14.4	$1.2300 \cdot 10^3$	$2.2087 \cdot 10^6$
10.5	$4.0164 \cdot 10^3$	$2.1998 \cdot 10^6$	14.5	$1.1982 \cdot 10^3$	$2.2088 \cdot 10^6$
10.6	$3.8779 \cdot 10^3$	$2.2002 \cdot 10^6$	14.6	$1.1674 \cdot 10^3$	$2.2089 \cdot 10^6$
10.7	$3.7453 \cdot 10^3$	$2.2005 \cdot 10^6$	14.7	$1.1376 \cdot 10^3$	$2.2090 \cdot 10^6$
10.8	$3.6183 \cdot 10^3$	$2.2009 \cdot 10^6$	14.8	$1.1087 \cdot 10^3$	$2.2091 \cdot 10^6$
10.9	$3.4966 \cdot 10^3$	$2.2013 \cdot 10^6$	14.9	$1.0807 \cdot 10^3$	$2.2092 \cdot 10^6$
11.0	$3.3800 \cdot 10^3$	$2.2016 \cdot 10^6$	15.0	$1.0537 \cdot 10^3$	$2.2094 \cdot 10^6$
11.1	$3.2682 \cdot 10^3$	$2.2019 \cdot 10^6$	15.1	$1.0274 \cdot 10^3$	$2.2095 \cdot 10^6$
11.2	$3.1609 \cdot 10^3$	$2.2023 \cdot 10^6$	15.2	$1.0020 \cdot 10^3$	$2.2096 \cdot 10^6$
11.3	$3.0581 \cdot 10^3$	$2.2026 \cdot 10^6$	15.3	$9.7731 \cdot 10^2$	$2.2097 \cdot 10^6$
11.4	$2.9593 \cdot 10^3$	$2.2029 \cdot 10^6$	15.4	$9.5341 \cdot 10^2$	$2.2098 \cdot 10^6$

Table 2.2. (*continued*)

temperature 2600 K

λ [μm]	$m_{\lambda,bb}$ [W/μm m²]	$M_{T,o-\lambda,bb}$ [W/m²]	λ [μm]	$m_{\lambda,bb}$ [W/μm m²]	$M_{T,o-\lambda,bb}$ [W/m²]
0.4	$3.5855\cdot10^4$	$1.2963\cdot10^3$			
0.5	$1.8692\cdot10^5$	$1.1184\cdot10^4$	4.0	$1.2227\cdot10^5$	$2.3888\cdot10^6$
0.6	$4.7520\cdot10^5$	$4.3384\cdot10^4$	4.1	$1.1308\cdot10^5$	$2.4005\cdot10^6$
0.7	$8.2125\cdot10^5$	$1.0816\cdot10^5$	4.2	$1.0471\cdot10^5$	$2.4114\cdot10^6$
0.8	$1.1323\cdot10^6$	$2.0641\cdot10^5$	4.3	$9.7090\cdot10^4$	$2.4215\cdot10^6$
0.9	$1.3567\cdot10^6$	$3.3167\cdot10^5$	4.4	$9.0137\cdot10^4$	$2.4308\cdot10^6$
1.0	$1.4843\cdot10^6$	$4.7449\cdot10^5$	4.5	$8.3783\cdot10^4$	$2.4395\cdot10^6$
1.1	$1.5281\cdot10^6$	$6.2572\cdot10^5$	4.6	$7.7970\cdot10^4$	$2.4476\cdot10^6$
1.2	$1.5093\cdot10^6$	$7.7802\cdot10^5$	4.7	$7.2645\cdot10^4$	$2.4551\cdot10^6$
1.3	$1.4484\cdot10^6$	$9.2618\cdot10^5$	4.8	$6.7760\cdot10^4$	$2.4622\cdot10^6$
1.4	$1.3622\cdot10^6$	$1.0669\cdot10^6$	4.9	$6.3272\cdot10^4$	$2.4687\cdot10^6$
1.5	$1.2631\cdot10^6$	$1.1982\cdot10^6$	5.0	$5.9145\cdot10^4$	$2.4748\cdot10^6$
1.6	$1.1596\cdot10^6$	$1.3193\cdot10^6$	5.1	$5.5344\cdot10^4$	$2.4805\cdot10^6$
1.7	$1.0574\cdot10^6$	$1.4301\cdot10^6$	5.2	$5.1840\cdot10^4$	$2.4859\cdot10^6$
1.8	$9.5965\cdot10^5$	$1.5309\cdot10^6$	5.3	$4.8606\cdot10^4$	$2.4909\cdot10^6$
1.9	$8.6834\cdot10^5$	$1.6223\cdot10^6$	5.4	$4.5617\cdot10^4$	$2.4956\cdot10^6$
2.0	$7.8430\cdot10^5$	$1.7049\cdot10^6$	5.5	$4.2852\cdot10^4$	$2.5000\cdot10^6$
2.1	$7.0775\cdot10^5$	$1.7794\cdot10^6$	5.6	$4.0291\cdot10^4$	$2.5042\cdot10^6$
2.2	$6.3852\cdot10^5$	$1.8466\cdot10^6$	5.7	$3.7917\cdot10^4$	$2.5081\cdot10^6$
2.3	$5.7621\cdot10^5$	$1.9073\cdot10^6$	5.8	$3.5713\cdot10^4$	$2.5118\cdot10^6$
2.4	$5.2031\cdot10^5$	$1.9621\cdot10^6$	5.9	$3.3666\cdot10^4$	$2.5153\cdot10^6$
2.5	$4.7027\cdot10^5$	$2.0116\cdot10^6$	6.0	$3.1762\cdot10^4$	$2.5185\cdot10^6$
2.6	$4.2552\cdot10^5$	$2.0563\cdot10^6$	6.1	$2.9989\cdot10^4$	$2.5216\cdot10^6$
2.7	$3.8551\cdot10^5$	$2.0968\cdot10^6$	6.2	$2.8338\cdot10^4$	$2.5245\cdot10^6$
2.8	$3.4975\cdot10^5$	$2.1336\cdot10^6$	6.3	$2.6798\cdot10^4$	$2.5273\cdot10^6$
2.9	$3.1777\cdot10^5$	$2.1669\cdot10^6$	6.4	$2.5360\cdot10^4$	$2.5299\cdot10^6$
3.0	$2.8915\cdot10^5$	$2.1972\cdot10^6$	6.5	$2.4016\cdot10^4$	$2.5324\cdot10^6$
3.1	$2.6350\cdot10^5$	$2.2248\cdot10^6$	6.6	$2.2760\cdot10^4$	$2.5347\cdot10^6$
3.2	$2.4051\cdot10^5$	$2.2500\cdot10^6$	6.7	$2.1585\cdot10^4$	$2.5369\cdot10^6$
3.3	$2.1985\cdot10^5$	$2.2730\cdot10^6$	6.8	$2.0483\cdot10^4$	$2.5390\cdot10^6$
3.4	$2.0128\cdot10^5$	$2.2941\cdot10^6$	6.9	$1.9451\cdot10^4$	$2.5410\cdot10^6$
3.5	$1.8456\cdot10^5$	$2.3133\cdot10^6$	7.0	$1.8483\cdot10^4$	$2.5429\cdot10^6$
3.6	$1.6948\cdot10^5$	$2.3310\cdot10^6$	7.1	$1.7573\cdot10^4$	$2.5447\cdot10^6$
3.7	$1.5586\cdot10^5$	$2.3473\cdot10^6$	7.2	$1.6719\cdot10^4$	$2.5464\cdot10^6$
3.8	$1.4355\cdot10^5$	$2.3622\cdot10^6$	7.3	$1.5915\cdot10^4$	$2.5481\cdot10^6$
3.9	$1.3239\cdot10^5$	$2.3760\cdot10^6$	7.4	$1.5160\cdot10^4$	$2.5496\cdot10^6$

λ [µm]	$m_{\lambda,bb}$ [W/µm m²]	$M_{T,o-\lambda,bb}$ [W/m²]	λ [µm]	$m_{\lambda,bb}$ [W/µm m²]	$M_{T,o-\lambda,bb}$ [W/m²]
7.5	$1.4448 \cdot 10^4$	$2.5511 \cdot 10^6$	11.5	$3.0103 \cdot 10^3$	$2.5788 \cdot 10^6$
7.6	$1.3777 \cdot 10^4$	$2.5525 \cdot 10^6$	11.6	$2.9143 \cdot 10^3$	$2.5791 \cdot 10^6$
7.7	$1.3144 \cdot 10^4$	$2.5538 \cdot 10^6$	11.7	$2.8222 \cdot 10^3$	$2.5794 \cdot 10^6$
7.8	$1.2548 \cdot 10^4$	$2.5551 \cdot 10^6$	11.8	$2.7336 \cdot 10^3$	$2.5797 \cdot 10^6$
7.9	$1.1984 \cdot 10^4$	$2.5563 \cdot 10^6$	11.9	$2.6485 \cdot 10^3$	$2.5800 \cdot 10^6$
8.0	$1.1452 \cdot 10^4$	$2.5575 \cdot 10^6$	12.0	$2.5666 \cdot 10^3$	$2.5802 \cdot 10^6$
8.1	$1.0949 \cdot 10^4$	$2.5586 \cdot 10^6$	12.1	$2.4879 \cdot 10^3$	$2.5805 \cdot 10^6$
8.2	$1.0473 \cdot 10^4$	$2.5597 \cdot 10^6$	12.2	$2.4122 \cdot 10^3$	$2.5807 \cdot 10^6$
8.3	$1.0022 \cdot 10^4$	$2.5607 \cdot 10^6$	12.3	$2.3394 \cdot 10^3$	$2.5810 \cdot 10^6$
8.4	$9.5956 \cdot 10^3$	$2.5617 \cdot 10^6$	12.4	$2.2692 \cdot 10^3$	$2.5812 \cdot 10^6$
8.5	$9.1914 \cdot 10^3$	$2.5627 \cdot 10^6$	12.5	$2.2017 \cdot 10^3$	$2.5814 \cdot 10^6$
8.6	$8.8081 \cdot 10^3$	$2.5636 \cdot 10^6$	12.6	$2.1367 \cdot 10^3$	$2.5816 \cdot 10^6$
8.7	$8.4445 \cdot 10^3$	$2.5644 \cdot 10^6$	12.7	$2.0740 \cdot 10^3$	$2.5818 \cdot 10^6$
8.8	$8.0995 \cdot 10^3$	$2.5652 \cdot 10^6$	12.8	$2.0136 \cdot 10^3$	$2.5820 \cdot 10^6$
8.9	$7.7718 \cdot 10^3$	$2.5660 \cdot 10^6$	12.9	$1.9554 \cdot 10^3$	$2.5822 \cdot 10^6$
9.0	$7.4604 \cdot 10^3$	$2.5668 \cdot 10^6$	13.0	$1.8993 \cdot 10^3$	$2.5824 \cdot 10^6$
9.1	$7.1645 \cdot 10^3$	$2.5675 \cdot 10^6$	13.1	$1.8451 \cdot 10^3$	$2.5826 \cdot 10^6$
9.2	$6.8830 \cdot 10^3$	$2.5682 \cdot 10^6$	13.2	$1.7929 \cdot 10^3$	$2.5828 \cdot 10^6$
9.3	$6.6152 \cdot 10^3$	$2.5689 \cdot 10^6$	13.3	$1.7426 \cdot 10^3$	$2.5830 \cdot 10^6$
9.4	$6.3602 \cdot 10^3$	$2.5695 \cdot 10^6$	13.4	$1.6939 \cdot 10^3$	$2.5831 \cdot 10^6$
9.5	$6.1173 \cdot 10^3$	$2.5702 \cdot 10^6$	13.5	$1.6470 \cdot 10^3$	$2.5833 \cdot 10^6$
9.6	$5.8860 \cdot 10^3$	$2.5708 \cdot 10^6$	13.6	$1.6016 \cdot 10^3$	$2.5835 \cdot 10^6$
9.7	$5.6654 \cdot 10^3$	$2.5713 \cdot 10^6$	13.7	$1.5579 \cdot 10^3$	$2.5836 \cdot 10^6$
9.8	$5.4550 \cdot 10^3$	$2.5719 \cdot 10^6$	13.8	$1.5156 \cdot 10^3$	$2.5838 \cdot 10^6$
9.9	$5.2543 \cdot 10^3$	$2.5724 \cdot 10^6$	13.9	$1.4747 \cdot 10^3$	$2.5839 \cdot 10^6$
10.0	$5.0626 \cdot 10^3$	$2.5730 \cdot 10^6$	14.0	$1.4352 \cdot 10^3$	$2.5841 \cdot 10^6$
10.1	$4.8797 \cdot 10^3$	$2.5735 \cdot 10^6$	14.1	$1.3970 \cdot 10^3$	$2.5842 \cdot 10^6$
10.2	$4.7049 \cdot 10^3$	$2.5739 \cdot 10^6$	14.2	$1.3600 \cdot 10^3$	$2.5844 \cdot 10^6$
10.3	$4.5378 \cdot 10^3$	$2.5744 \cdot 10^6$	14.3	$1.3243 \cdot 10^3$	$2.5845 \cdot 10^6$
10.4	$4.3781 \cdot 10^3$	$2.5748 \cdot 10^6$	14.4	$1.2897 \cdot 10^3$	$2.5846 \cdot 10^6$
10.5	$4.2253 \cdot 10^3$	$2.5753 \cdot 10^6$	14.5	$1.2563 \cdot 10^3$	$2.5847 \cdot 10^6$
10.6	$4.0791 \cdot 10^3$	$2.5757 \cdot 10^6$	14.6	$1.2239 \cdot 10^3$	$2.5849 \cdot 10^6$
10.7	$3.9391 \cdot 10^3$	$2.5761 \cdot 10^6$	14.7	$1.1926 \cdot 10^3$	$2.5850 \cdot 10^6$
10.8	$3.8051 \cdot 10^3$	$2.5765 \cdot 10^6$	14.8	$1.1623 \cdot 10^3$	$2.5851 \cdot 10^6$
10.9	$3.6768 \cdot 10^3$	$2.5768 \cdot 10^6$	14.9	$1.1329 \cdot 10^3$	$2.5852 \cdot 10^6$
11.0	$3.5538 \cdot 10^3$	$2.5772 \cdot 10^6$	15.0	$1.1044 \cdot 10^3$	$2.5853 \cdot 10^6$
11.1	$3.4358 \cdot 10^3$	$2.5776 \cdot 10^6$	15.1	$1.0769 \cdot 10^3$	$2.5854 \cdot 10^6$
11.2	$3.3228 \cdot 10^3$	$2.5779 \cdot 10^6$	15.2	$1.0501 \cdot 10^3$	$2.5855 \cdot 10^6$
11.3	$3.2143 \cdot 10^3$	$2.5782 \cdot 10^6$	15.3	$1.0242 \cdot 10^3$	$2.5857 \cdot 10^6$
11.4	$3.1102 \cdot 10^3$	$2.5785 \cdot 10^6$	15.4	$9.9913 \cdot 10^2$	$2.5858 \cdot 10^6$

Table 2.2. (*continued*)

temperature 2700 K

λ [µm]	$m_{\lambda,bb}$ [W/µm m²]	$M_{T,o-\lambda,bb}$ [W/m²]	λ [µm]	$m_{\lambda,bb}$ [W/µm m²]	$M_{T,o-\lambda,bb}$ [W/m²]
0.4	$5.9851 \cdot 10^4$	$2.2670 \cdot 10^3$			
0.5	$2.8163 \cdot 10^5$	$1.7695 \cdot 10^4$	4.0	$1.3100 \cdot 10^5$	$2.7986 \cdot 10^6$
0.6	$6.6872 \cdot 10^5$	$6.4255 \cdot 10^4$	4.1	$1.2104 \cdot 10^5$	$2.8112 \cdot 10^6$
0.7	$1.1007 \cdot 10^6$	$1.5291 \cdot 10^5$	4.2	$1.1200 \cdot 10^5$	$2.8229 \cdot 10^6$
0.8	$1.4633 \cdot 10^6$	$2.8201 \cdot 10^5$	4.3	$1.0376 \cdot 10^5$	$2.8336 \cdot 10^6$
0.9	$1.7046 \cdot 10^6$	$4.4147 \cdot 10^5$	4.4	$9.6258 \cdot 10^4$	$2.8436 \cdot 10^6$
1.0	$1.8235 \cdot 10^6$	$6.1882 \cdot 10^5$	4.5	$8.9409 \cdot 10^4$	$2.8529 \cdot 10^6$
1.1	$1.8436 \cdot 10^6$	$8.0288 \cdot 10^5$	4.6	$8.3149 \cdot 10^4$	$2.8615 \cdot 10^6$
1.2	$1.7937 \cdot 10^6$	$9.8521 \cdot 10^5$	4.7	$7.7419 \cdot 10^4$	$2.8696 \cdot 10^6$
1.3	$1.6999 \cdot 10^6$	$1.1602 \cdot 10^6$	4.8	$7.2168 \cdot 10^4$	$2.8770 \cdot 10^6$
1.4	$1.5818 \cdot 10^6$	$1.3244 \cdot 10^6$	4.9	$6.7349 \cdot 10^4$	$2.8840 \cdot 10^6$
1.5	$1.4534 \cdot 10^6$	$1.4762 \cdot 10^6$	5.0	$6.2920 \cdot 10^4$	$2.8905 \cdot 10^6$
1.6	$1.3240 \cdot 10^6$	$1.6150 \cdot 10^6$	5.1	$5.8845 \cdot 10^4$	$2.8966 \cdot 10^6$
1.7	$1.1990 \cdot 10^6$	$1.7411 \cdot 10^6$	5.2	$5.5091 \cdot 10^4$	$2.9023 \cdot 10^6$
1.8	$1.0817 \cdot 10^6$	$1.8551 \cdot 10^6$	5.3	$5.1628 \cdot 10^4$	$2.9076 \cdot 10^6$
1.9	$9.7362 \cdot 10^5$	$1.9578 \cdot 10^6$	5.4	$4.8430 \cdot 10^4$	$2.9126 \cdot 10^6$
2.0	$8.7527 \cdot 10^5$	$2.0501 \cdot 10^6$	5.5	$4.5474 \cdot 10^4$	$2.9173 \cdot 10^6$
2.1	$7.8654 \cdot 10^5$	$2.1331 \cdot 10^6$	5.6	$4.2738 \cdot 10^4$	$2.9217 \cdot 10^6$
2.2	$7.0693 \cdot 10^5$	$2.2077 \cdot 10^6$	5.7	$4.0202 \cdot 10^4$	$2.9259 \cdot 10^6$
2.3	$6.3579 \cdot 10^5$	$2.2748 \cdot 10^6$	5.8	$3.7850 \cdot 10^4$	$2.9298 \cdot 10^6$
2.4	$5.7235 \cdot 10^5$	$2.3352 \cdot 10^6$	5.9	$3.5667 \cdot 10^4$	$2.9335 \cdot 10^6$
2.5	$5.1586 \cdot 10^5$	$2.3895 \cdot 10^6$	6.0	$3.3637 \cdot 10^4$	$2.9369 \cdot 10^6$
2.6	$4.6558 \cdot 10^5$	$2.4385 \cdot 10^6$	6.1	$3.1748 \cdot 10^4$	$2.9402 \cdot 10^6$
2.7	$4.2082 \cdot 10^5$	$2.4828 \cdot 10^6$	6.2	$2.9990 \cdot 10^4$	$2.9433 \cdot 10^6$
2.8	$3.8097 \cdot 10^5$	$2.5229 \cdot 10^6$	6.3	$2.8350 \cdot 10^4$	$2.9462 \cdot 10^6$
2.9	$3.4545 \cdot 10^5$	$2.5591 \cdot 10^6$	6.4	$2.6820 \cdot 10^4$	$2.9489 \cdot 10^6$
3.0	$3.1376 \cdot 10^5$	$2.5921 \cdot 10^6$	6.5	$2.5392 \cdot 10^4$	$2.9515 \cdot 10^6$
3.1	$2.8545 \cdot 10^5$	$2.6220 \cdot 10^6$	6.6	$2.4056 \cdot 10^4$	$2.9540 \cdot 10^6$
3.2	$2.6012 \cdot 10^5$	$2.6493 \cdot 10^6$	6.7	$2.2807 \cdot 10^4$	$2.9564 \cdot 10^6$
3.3	$2.3744 \cdot 10^5$	$2.6741 \cdot 10^6$	6.8	$2.1637 \cdot 10^4$	$2.9586 \cdot 10^6$
3.4	$2.1709 \cdot 10^5$	$2.6968 \cdot 10^6$	6.9	$2.0541 \cdot 10^4$	$2.9607 \cdot 10^6$
3.5	$1.9880 \cdot 10^5$	$2.7176 \cdot 10^6$	7.0	$1.9513 \cdot 10^4$	$2.9627 \cdot 10^6$
3.6	$1.8233 \cdot 10^5$	$2.7366 \cdot 10^6$	7.1	$1.8548 \cdot 10^4$	$2.9646 \cdot 10^6$
3.7	$1.6750 \cdot 10^5$	$2.7541 \cdot 10^6$	7.2	$1.7642 \cdot 10^4$	$2.9664 \cdot 10^6$
3.8	$1.5410 \cdot 10^5$	$2.7702 \cdot 10^6$	7.3	$1.6790 \cdot 10^4$	$2.9681 \cdot 10^6$
2.9	$1.4198 \cdot 10^5$	$2.7850 \cdot 10^6$	7.4	$1.5989 \cdot 10^4$	$2.9698 \cdot 10^6$

Table 2.2. (*continued*)

λ [µm]	$m_{\lambda,bb}$ [W/µm m²]	$M_{T,o-\lambda,bb}$ [W/m²]	λ [µm]	$m_{\lambda,bb}$ [W/µm m²]	$M_{T,o-\lambda,bb}$ [W/m²]
7.5	$1.5235 \cdot 10^4$	$2.9713 \cdot 10^6$	11.5	$3.1562 \cdot 10^3$	$3.0005 \cdot 10^6$
7.6	$1.4524 \cdot 10^4$	$2.9728 \cdot 10^6$	11.6	$3.0554 \cdot 10^3$	$3.0008 \cdot 10^6$
7.7	$1.3854 \cdot 10^4$	$2.9742 \cdot 10^6$	11.7	$2.9585 \cdot 10^3$	$3.0011 \cdot 10^6$
7.8	$1.3223 \cdot 10^4$	$2.9756 \cdot 10^6$	11.8	$2.8654 \cdot 10^3$	$3.0014 \cdot 10^6$
7.9	$1.2626 \cdot 10^4$	$2.9769 \cdot 10^6$	11.9	$2.7759 \cdot 10^3$	$3.0017 \cdot 10^6$
8.0	$1.2063 \cdot 10^4$	$2.9781 \cdot 10^6$	12.0	$2.6900 \cdot 10^3$	$3.0020 \cdot 10^6$
8.1	$1.1531 \cdot 10^4$	$2.9793 \cdot 10^6$	12.1	$2.6073 \cdot 10^3$	$3.0022 \cdot 10^6$
8.2	$1.1027 \cdot 10^4$	$2.9804 \cdot 10^6$	12.2	$2.5277 \cdot 10^3$	$3.0025 \cdot 10^6$
8.3	$1.0551 \cdot 10^4$	$2.9815 \cdot 10^6$	12.3	$2.4512 \cdot 10^3$	$3.0027 \cdot 10^6$
8.4	$1.0100 \cdot 10^4$	$2.9825 \cdot 10^6$	12.4	$2.3775 \cdot 10^3$	$3.0030 \cdot 10^6$
8.5	$9.6730 \cdot 10^3$	$2.9835 \cdot 10^6$	12.5	$2.3066 \cdot 10^3$	$3.0032 \cdot 10^6$
8.6	$9.2681 \cdot 10^3$	$2.9845 \cdot 10^6$	12.6	$2.2383 \cdot 10^3$	$3.0034 \cdot 10^6$
8.7	$8.8840 \cdot 10^3$	$2.9854 \cdot 10^6$	12.7	$2.1725 \cdot 10^3$	$3.0037 \cdot 10^6$
8.8	$8.5196 \cdot 10^3$	$2.9862 \cdot 10^6$	12.8	$2.1091 \cdot 10^3$	$3.0039 \cdot 10^6$
8.9	$8.1736 \cdot 10^3$	$2.9871 \cdot 10^6$	12.9	$2.0479 \cdot 10^3$	$3.0041 \cdot 10^6$
9.0	$7.8450 \cdot 10^3$	$2.9879 \cdot 10^6$	13.0	$1.9890 \cdot 10^3$	$3.0043 \cdot 10^6$
9.1	$7.5327 \cdot 10^3$	$2.9886 \cdot 10^6$	13.1	$1.9322 \cdot 10^3$	$3.0045 \cdot 10^6$
9.2	$7.2357 \cdot 10^3$	$2.9894 \cdot 10^6$	13.2	$1.8774 \cdot 10^3$	$3.0047 \cdot 10^6$
9.3	$6.9531 \cdot 10^3$	$2.9901 \cdot 10^6$	13.3	$1.8245 \cdot 10^3$	$3.0049 \cdot 10^6$
9.4	$6.6842 \cdot 10^3$	$2.9908 \cdot 10^6$	13.4	$1.7735 \cdot 10^3$	$3.0050 \cdot 10^6$
9.5	$6.4281 \cdot 10^3$	$2.9914 \cdot 10^6$	13.5	$1.7242 \cdot 10^3$	$3.0052 \cdot 10^6$
9.6	$6.1841 \cdot 10^3$	$2.9920 \cdot 10^6$	13.6	$1.6767 \cdot 10^3$	$3.0054 \cdot 10^6$
9.7	$5.9516 \cdot 10^3$	$2.9927 \cdot 10^6$	13.7	$1.6307 \cdot 10^3$	$3.0055 \cdot 10^6$
9.8	$5.7298 \cdot 10^3$	$2.9932 \cdot 10^6$	13.8	$1.5863 \cdot 10^3$	$3.0057 \cdot 10^6$
9.9	$5.5183 \cdot 10^3$	$2.9938 \cdot 10^6$	13.9	$1.5435 \cdot 10^3$	$3.0059 \cdot 10^6$
10.0	$5.3164 \cdot 10^3$	$2.9943 \cdot 10^6$	14.0	$1.5020 \cdot 10^3$	$3.0060 \cdot 10^6$
10.1	$5.1237 \cdot 10^3$	$2.9949 \cdot 10^6$	14.1	$1.4620 \cdot 10^3$	$3.0062 \cdot 10^6$
10.2	$4.9395 \cdot 10^3$	$2.9954 \cdot 10^6$	14.2	$1.4232 \cdot 10^3$	$3.0063 \cdot 10^6$
10.3	$4.7636 \cdot 10^3$	$2.9958 \cdot 10^6$	14.3	$1.3858 \cdot 10^3$	$3.0064 \cdot 10^6$
10.4	$4.5954 \cdot 10^3$	$2.9963 \cdot 10^6$	14.4	$1.3495 \cdot 10^3$	$3.0066 \cdot 10^6$
10.5	$4.4345 \cdot 10^3$	$2.9968 \cdot 10^6$	14.5	$1.3145 \cdot 10^3$	$3.0067 \cdot 10^6$
10.6	$4.2806 \cdot 10^3$	$2.9972 \cdot 10^6$	14.6	$1.2805 \cdot 10^3$	$3.0068 \cdot 10^6$
10.7	$4.1334 \cdot 10^3$	$2.9976 \cdot 10^6$	14.7	$1.2477 \cdot 10^3$	$3.0070 \cdot 10^6$
10.8	$3.9923 \cdot 10^3$	$2.9980 \cdot 10^6$	14.8	$1.2159 \cdot 10^3$	$3.0071 \cdot 10^6$
10.9	$3.8573 \cdot 10^3$	$2.9984 \cdot 10^6$	14.9	$1.1851 \cdot 10^3$	$3.0072 \cdot 10^6$
11.0	$3.7278 \cdot 10^3$	$2.9988 \cdot 10^6$	15.0	$1.1552 \cdot 10^3$	$3.0073 \cdot 10^6$
11.1	$3.6038 \cdot 10^3$	$2.9992 \cdot 10^6$	15.1	$1.1263 \cdot 10^3$	$3.0074 \cdot 10^6$
11.2	$3.4848 \cdot 10^3$	$2.9995 \cdot 10^6$	15.2	$1.0983 \cdot 10^3$	$3.0075 \cdot 10^6$
11.3	$3.3708 \cdot 10^3$	$2.9999 \cdot 10^6$	15.3	$1.0712 \cdot 10^3$	$3.0077 \cdot 10^6$
11.4	$3.2613 \cdot 10^3$	$3.0002 \cdot 10^6$	15.4	$1.0449 \cdot 10^3$	$3.0078 \cdot 10^6$

Table 2.2. (*continued*)

temperature 2800 K

λ [μm]	$m_{\lambda,bb}$ [W/μm m²]	$M_{T,o-\lambda,bb}$ [W/m²]	λ [μm]	$m_{\lambda,bb}$ [W/μm m²]	$M_{T.o-\lambda,bb}$ [W/m²]
0.4	$9.6318 \cdot 10^4$	$3.8170 \cdot 10^3$			
0.5	$4.1208 \cdot 10^5$	$9.7150 \cdot 10^4$	4.0	$1.3983 \cdot 10^5$	$3.2580 \cdot 10^6$
0.6	$9.1839 \cdot 10^5$	$9.2747 \cdot 10^4$	4.1	$1.2909 \cdot 10^5$	$3.2715 \cdot 10^6$
0.7	$1.4449 \cdot 10^6$	$2.1144 \cdot 10^5$	4.2	$1.1935 \cdot 10^5$	$3.2839 \cdot 10^6$
0.8	$1.8570 \cdot 10^6$	$3.7780 \cdot 10^5$	4.3	$1.1050 \cdot 10^5$	$3.2954 \cdot 10^6$
0.9	$2.1073 \cdot 10^6$	$5.7737 \cdot 10^5$	4.4	$1.0243 \cdot 10^5$	$3.3060 \cdot 10^6$
1.0	$2.2080 \cdot 10^6$	$7.9425 \cdot 10^5$	4.5	$9.5082 \cdot 10^4$	$3.3159 \cdot 10^6$
1.1	$2.1951 \cdot 10^6$	$1.0152 \cdot 10^6$	4.6	$8.8370 \cdot 10^4$	$3.3250 \cdot 10^6$
1.2	$2.1063 \cdot 10^6$	$1.2307 \cdot 10^6$	4.7	$8.2231 \cdot 10^4$	$3.3336 \cdot 10^6$
1.3	$1.9731 \cdot 10^6$	$1.4350 \cdot 10^6$	4.8	$7.6610 \cdot 10^4$	$3.3415 \cdot 10^6$
1.4	$1.8181 \cdot 10^6$	$1.6246 \cdot 10^6$	4.9	$7.1455 \cdot 10^4$	$3.3489 \cdot 10^6$
1.5	$1.6567 \cdot 10^6$	$1.7984 \cdot 10^6$	5.0	$6.6721 \cdot 10^4$	$3.3558 \cdot 10^6$
1.6	$1.4982 \cdot 10^6$	$1.9561 \cdot 10^6$	5.1	$6.2369 \cdot 10^4$	$3.3623 \cdot 10^6$
1.7	$1.3483 \cdot 10^6$	$2.0983 \cdot 10^6$	5.2	$5.8362 \cdot 10^4$	$3.3683 \cdot 10^6$
1.8	$1.2097 \cdot 10^6$	$2.2261 \cdot 10^6$	5.3	$5.4669 \cdot 10^4$	$3.3739 \cdot 10^6$
1.9	$1.0836 \cdot 10^6$	$2.3407 \cdot 10^6$	5.4	$5.1260 \cdot 10^4$	$3.3792 \cdot 10^6$
2.0	$9.6990 \cdot 10^5$	$2.4432 \cdot 10^6$	5.5	$4.8111 \cdot 10^4$	$3.3842 \cdot 10^6$
2.1	$8.6822 \cdot 10^5$	$2.5351 \cdot 10^6$	5.6	$4.5198 \cdot 10^4$	$3.3889 \cdot 10^6$
2.2	$7.7765 \cdot 10^5$	$2.6173 \cdot 10^6$	5.7	$4.2500 \cdot 10^4$	$3.3932 \cdot 10^6$
2.3	$6.9721 \cdot 10^5$	$2.6909 \cdot 10^6$	5.8	$3.9999 \cdot 10^4$	$3.3974 \cdot 10^6$
2.4	$6.2587 \cdot 10^5$	$2.7570 \cdot 10^6$	5.9	$3.7678 \cdot 10^4$	$3.4012 \cdot 10^6$
2.5	$5.6265 \cdot 10^5$	$2.8164 \cdot 10^6$	6.0	$3.5521 \cdot 10^4$	$3.4049 \cdot 10^6$
2.6	$5.0662 \cdot 10^5$	$2.8698 \cdot 10^6$	6.1	$3.3516 \cdot 10^4$	$3.4084 \cdot 10^6$
2.7	$4.5694 \cdot 10^5$	$2.9179 \cdot 10^6$	6.2	$3.1649 \cdot 10^4$	$3.4116 \cdot 10^6$
2.8	$4.1285 \cdot 10^5$	$2.9613 \cdot 10^6$	6.3	$2.9909 \cdot 10^4$	$3.4147 \cdot 10^6$
2.9	$3.7368 \cdot 10^5$	$3.0006 \cdot 10^6$	6.4	$2.8287 \cdot 10^4$	$3.4176 \cdot 10^6$
3.0	$3.3883 \cdot 10^5$	$3.0362 \cdot 10^6$	6.5	$2.6772 \cdot 10^4$	$3.4203 \cdot 10^6$
3.1	$3.0778 \cdot 10^5$	$3.0685 \cdot 10^6$	6.6	$2.5357 \cdot 10^4$	$3.4230 \cdot 10^6$
3.2	$2.8007 \cdot 10^5$	$3.0979 \cdot 10^6$	6.7	$2.4034 \cdot 10^4$	$3.4254 \cdot 10^6$
3.3	$2.5530 \cdot 10^5$	$3.1246 \cdot 10^6$	6.8	$2.2795 \cdot 10^4$	$3.4278 \cdot 10^6$
3.4	$2.3312 \cdot 10^5$	$3.1490 \cdot 10^6$	6.9	$2.1635 \cdot 10^4$	$3.4300 \cdot 10^6$
3.5	$2.1323 \cdot 10^5$	$3.1713 \cdot 10^6$	7.0	$2.0547 \cdot 10^4$	$3.4321 \cdot 10^6$
3.6	$1.9536 \cdot 10^5$	$3.1917 \cdot 10^6$	7.1	$1.9527 \cdot 10^4$	$3.4341 \cdot 10^6$
3.7	$1.7927 \cdot 10^5$	$3.2105 \cdot 10^6$	7.2	$1.8568 \cdot 10^4$	$3.4360 \cdot 10^6$
3.8	$1.6477 \cdot 10^5$	$3.2276 \cdot 10^6$	7.3	$1.7668 \cdot 10^4$	$3.4378 \cdot 10^6$
3.9	$1.5168 \cdot 10^5$	$3.2435 \cdot 10^6$	7.4	$1.6821 \cdot 10^4$	$3.4395 \cdot 10^6$

Table 2.2. (*continued*)

λ [μm]	$m_{\lambda,bb}$ [W/μm m²]	$M_{T,o-\lambda,bb}$ [W/m²]	λ [μm]	$m_{\lambda,bb}$ [W/μm m²]	$M_{T,o-\lambda,bb}$ [W/m²]
7.5	$1.6024 \cdot 10^4$	$3.4412 \cdot 10^6$	11.5	$3.3024 \cdot 10^3$	$3.4718 \cdot 10^6$
7.6	$1.5274 \cdot 10^4$	$3.4427 \cdot 10^6$	11.6	$3.1966 \cdot 10^3$	$3.4721 \cdot 10^6$
7.7	$1.4566 \cdot 10^4$	$3.4442 \cdot 10^6$	11.7	$3.0950 \cdot 10^3$	$3.4724 \cdot 10^6$
7.8	$1.3899 \cdot 10^4$	$3.4456 \cdot 10^6$	11.8	$3.9974 \cdot 10^3$	$3.4727 \cdot 10^6$
7.9	$1.3270 \cdot 10^4$	$3.4470 \cdot 10^6$	11.9	$2.9036 \cdot 10^3$	$3.4730 \cdot 10^6$
8.0	$1.2676 \cdot 10^4$	$3.4483 \cdot 10^6$	12.0	$2.8134 \cdot 10^3$	$3.4733 \cdot 10^6$
8.1	$1.2114 \cdot 10^4$	$3.4495 \cdot 10^6$	12.1	$2.7267 \cdot 10^3$	$3.4736 \cdot 10^6$
8.2	$1.1584 \cdot 10^4$	$3.4507 \cdot 10^6$	12.2	$2.6433 \cdot 10^3$	$3.4739 \cdot 10^6$
8.3	$1.1081 \cdot 10^4$	$3.4519 \cdot 10^6$	12.3	$2.5631 \cdot 10^3$	$3.4741 \cdot 10^6$
8.4	$1.0606 \cdot 10^4$	$3.4529 \cdot 10^6$	12.4	$2.4859 \cdot 10^3$	$3.4744 \cdot 10^6$
8.5	$1.0156 \cdot 10^4$	$3.4540 \cdot 10^6$	12.5	$2.4115 \cdot 10^3$	$3.4746 \cdot 10^6$
8.6	$9.7291 \cdot 10^3$	$3.4550 \cdot 10^6$	12.6	$2.3400 \cdot 10^3$	$3.4749 \cdot 10^6$
8.7	$9.3246 \cdot 10^3$	$3.4559 \cdot 10^6$	12.7	$2.2710 \cdot 10^3$	$3.4751 \cdot 10^6$
8.8	$8.9407 \cdot 10^3$	$3.4568 \cdot 10^6$	12.8	$2.2046 \cdot 10^3$	$3.4753 \cdot 10^6$
8.9	$8.5764 \cdot 10^3$	$3.4577 \cdot 10^6$	12.9	$2.1406 \cdot 10^3$	$3.4755 \cdot 10^6$
9.0	$8.2304 \cdot 10^3$	$3.4586 \cdot 10^6$	13.0	$2.0789 \cdot 10^3$	$3.4757 \cdot 10^6$
9.1	$7.9016 \cdot 10^3$	$3.4594 \cdot 10^6$	13.1	$2.0194 \cdot 10^3$	$3.4759 \cdot 10^6$
9.2	$7.5891 \cdot 10^3$	$3.4601 \cdot 10^6$	13.2	$1.9620 \cdot 10^3$	$3.4761 \cdot 10^6$
9.3	$7.2918 \cdot 10^3$	$3.4609 \cdot 10^6$	13.3	$1.9066 \cdot 10^3$	$3.4763 \cdot 10^6$
9.4	$7.0088 \cdot 10^3$	$3.4616 \cdot 10^6$	13.4	$1.8531 \cdot 10^3$	$3.4765 \cdot 10^6$
9.5	$6.7395 \cdot 10^3$	$3.4623 \cdot 10^6$	13.5	$1.8016 \cdot 10^3$	$3.4767 \cdot 10^6$
9.6	$6.4829 \cdot 10^3$	$3.4629 \cdot 10^6$	13.6	$1.7518 \cdot 10^3$	$3.4769 \cdot 10^6$
9.7	$6.2383 \cdot 10^3$	$3.4636 \cdot 10^6$	13.7	$1.7037 \cdot 10^3$	$3.4771 \cdot 10^6$
9.8	$6.0052 \cdot 10^3$	$3.4642 \cdot 10^6$	13.8	$1.6572 \cdot 10^3$	$3.4772 \cdot 10^6$
9.9	$5.7828 \cdot 10^3$	$3.4648 \cdot 10^6$	13.9	$1.6123 \cdot 10^3$	$3.4774 \cdot 10^6$
10.0	$5.5707 \cdot 10^3$	$3.4653 \cdot 10^6$	14.0	$1.5689 \cdot 10^3$	$3.4775 \cdot 10^6$
10.1	$5.3681 \cdot 10^3$	$3.4659 \cdot 10^6$	14.1	$1.5270 \cdot 10^3$	$3.4777 \cdot 10^6$
10.2	$5.1746 \cdot 10^3$	$3.4664 \cdot 10^6$	14.2	$1.4865 \cdot 10^3$	$3.4779 \cdot 10^6$
10.3	$4.9898 \cdot 10^3$	$3.4669 \cdot 10^6$	14.3	$1.4473 \cdot 10^3$	$3.4780 \cdot 10^6$
10.4	$4.8131 \cdot 10^3$	$3.4674 \cdot 10^6$	14.4	$1.4093 \cdot 10^3$	$3.4781 \cdot 10^6$
10.5	$4.6441 \cdot 10^3$	$3.4679 \cdot 10^6$	14.5	$1.3727 \cdot 10^3$	$3.4783 \cdot 10^6$
10.6	$4.4825 \cdot 10^3$	$3.4683 \cdot 10^6$	14.6	$1.3371 \cdot 10^3$	$3.4784 \cdot 10^6$
10.7	$4.3279 \cdot 10^3$	$3.4688 \cdot 10^6$	14.7	$1.3028 \cdot 10^3$	$3.4785 \cdot 10^6$
10.8	$4.1798 \cdot 10^3$	$3.4692 \cdot 10^6$	14.8	$1.2695 \cdot 10^3$	$3.4787 \cdot 10^6$
10.9	$4.0380 \cdot 10^3$	$3.4696 \cdot 10^6$	14.9	$1.2373 \cdot 10^3$	$3.4788 \cdot 10^6$
11.0	$3.9022 \cdot 10^3$	$3.4700 \cdot 10^6$	15.0	$1.2061 \cdot 10^3$	$3.4789 \cdot 10^6$
11.1	$3.7720 \cdot 10^3$	$3.4704 \cdot 10^6$	15.1	$1.1759 \cdot 10^3$	$3.4790 \cdot 10^6$
11.2	$3.6472 \cdot 10^3$	$3.4708 \cdot 10^6$	15.2	$1.1466 \cdot 10^3$	$3.4792 \cdot 10^6$
11.3	$3.5275 \cdot 10^3$	$3.4711 \cdot 10^6$	15.3	$1.1182 \cdot 10^3$	$3.4793 \cdot 10^6$
11.4	$3.4126 \cdot 10^3$	$3.4715 \cdot 10^6$	15.4	$1.0907 \cdot 10^3$	$3.4794 \cdot 10^6$

Table 2.2. (*continued*)

temperature 2900 K

λ [μm]	$m_{\lambda,bb}$ [W/μm m²]	$M_{T,o-\lambda,bb}$ [W/m²]	λ [μm]	$m_{\lambda,bb}$ [W/μm m²]	$M_{T,o-\lambda,bb}$ [W/m²]
0.4	$1.5000 \cdot 10^5$	$6.2112 \cdot 10^3$			
0.5	$5.8735 \cdot 10^5$	$4.0529 \cdot 10^4$	4.0	$1.4874 \cdot 10^5$	$3.7707 \cdot 10^6$
0.6	$1.2340 \cdot 10^6$	$1.3081 \cdot 10^5$	4.1	$1.3722 \cdot 10^5$	$3.7850 \cdot 10^6$
0.7	$1.8614 \cdot 10^6$	$2.8657 \cdot 10^5$	4.2	$1.2677 \cdot 10^5$	$3.7982 \cdot 10^6$
0.8	$2.3183 \cdot 10^6$	$4.9727 \cdot 10^5$	4.3	$1.1729 \cdot 10^5$	$3.8104 \cdot 10^6$
0.9	$2.5677 \cdot 10^6$	$7.4324 \cdot 10^5$	4.4	$1.0866 \cdot 10^5$	$3.8217 \cdot 10^6$
1.0	$2.6391 \cdot 10^6$	$1.0049 \cdot 10^6$	4.5	$1.0080 \cdot 10^5$	$3.8321 \cdot 10^6$
1.1	$2.5830 \cdot 10^6$	$1.2668 \cdot 10^6$	4.6	$9.3628 \cdot 10^4$	$3.8419 \cdot 10^6$
1.2	$2.4469 \cdot 10^6$	$1.5188 \cdot 10^6$	4.7	$8.7077 \cdot 10^4$	$3.8509 \cdot 10^6$
1.3	$2.2677 \cdot 10^6$	$1.7548 \cdot 10^6$	4.8	$8.1081 \cdot 10^4$	$3.8593 \cdot 10^6$
1.4	$2.0707 \cdot 10^6$	$1.9718 \cdot 10^6$	4.9	$7.5587 \cdot 10^4$	$3.8671 \cdot 10^6$
1.5	$1.8723 \cdot 10^6$	$2.1689 \cdot 10^6$	5.0	$7.0546 \cdot 10^4$	$3.8744 \cdot 10^6$
1.6	$1.6820 \cdot 10^6$	$2.3465 \cdot 10^6$	5.1	$6.5914 \cdot 10^4$	$3.8812 \cdot 10^6$
1.7	$1.5049 \cdot 10^6$	$2.5057 \cdot 10^6$	5.2	$6.1652 \cdot 10^4$	$3.8876 \cdot 10^6$
1.8	$1.3433 \cdot 10^6$	$2.6480 \cdot 10^6$	5.3	$5.7727 \cdot 10^4$	$3.8936 \cdot 10^6$
1.9	$1.1979 \cdot 10^6$	$2.7749 \cdot 10^6$	5.4	$5.4106 \cdot 10^4$	$3.8992 \cdot 10^6$
2.0	$1.0680 \cdot 10^6$	$2.8881 \cdot 10^6$	5.5	$5.0762 \cdot 10^4$	$3.9044 \cdot 10^6$
2.1	$9.5260 \cdot 10^5$	$2.9890 \cdot 10^6$	5.6	$4.7670 \cdot 10^4$	$3.9093 \cdot 10^6$
2.2	$8.5052 \cdot 10^5$	$3.0790 \cdot 10^6$	5.7	$4.4809 \cdot 10^4$	$3.9139 \cdot 10^6$
2.3	$7.6035 \cdot 10^5$	$3.1595 \cdot 10^6$	5.8	$4.2157 \cdot 10^4$	$3.9183 \cdot 10^6$
2.4	$6.8078 \cdot 10^5$	$3.2314 \cdot 10^6$	5.9	$3.9698 \cdot 10^4$	$3.9224 \cdot 10^6$
2.5	$6.1056 \cdot 10^5$	$3.2959 \cdot 10^6$	6.0	$3.7414 \cdot 10^4$	$3.9262 \cdot 10^6$
2.6	$5.4858 \cdot 10^5$	$3.3538 \cdot 10^6$	6.1	$3.5290 \cdot 10^4$	$3.9299 \cdot 10^6$
2.7	$4.9380 \cdot 10^5$	$3.4059 \cdot 10^6$	6.2	$3.3315 \cdot 10^4$	$3.9333 \cdot 10^6$
2.8	$4.4535 \cdot 10^5$	$3.4528 \cdot 10^6$	6.3	$3.1475 \cdot 10^4$	$3.9365 \cdot 10^6$
2.9	$4.0241 \cdot 10^5$	$3.4951 \cdot 10^6$	6.4	$2.9759 \cdot 10^4$	$3.9396 \cdot 10^6$
3.0	$3.6432 \cdot 10^5$	$3.5334 \cdot 10^6$	6.5	$2.8158 \cdot 10^4$	$3.9425 \cdot 10^6$
3.1	$3.3046 \cdot 10^5$	$3.5682 \cdot 10^6$	6.6	$2.6663 \cdot 10^4$	$3.9452 \cdot 10^6$
3.2	$3.0031 \cdot 10^5$	$3.5997 \cdot 10^6$	6.7	$2.5265 \cdot 10^4$	$3.9478 \cdot 10^6$
3.3	$2.7341 \cdot 10^5$	$3.6283 \cdot 10^6$	6.7	$2.3957 \cdot 10^4$	$3.9503 \cdot 10^6$
3.4	$2.4937 \cdot 10^5$	$3.6544 \cdot 10^6$	6.8	$2.2732 \cdot 10^4$	$3.9526 \cdot 10^6$
3.5	$2.2784 \cdot 10^5$	$3.6783 \cdot 10^6$	7.0	$2.1585 \cdot 10^4$	$3.9548 \cdot 10^6$
3.6	$2.0854 \cdot 10^5$	$3.7001 \cdot 10^6$	7.1	$2.0508 \cdot 10^4$	$3.9569 \cdot 10^6$
3.7	$1.9118 \cdot 10^5$	$3.7200 \cdot 10^6$	7.2	$1.9497 \cdot 10^4$	$3.9589 \cdot 10^6$
3.8	$1.7556 \cdot 10^5$	$3.7384 \cdot 10^6$	7.3	$1.8548 \cdot 10^4$	$3.9608 \cdot 10^6$
3.9	$1.6147 \cdot 10^5$	$3.7552 \cdot 10^6$	7.4	$1.7655 \cdot 10^4$	$3.9626 \cdot 10^6$

Table 2.2. (*continued*)

λ [μm]	$m_{\lambda,bb}$ [W/μm m²]	$M_{T,o-\lambda,bb}$ [W/m²]	λ [μm]	$m_{\lambda,bb}$ [W/μm m²]	$M_{T,o-\lambda,bb}$ [W/m²]
7.5	$1.6816 \cdot 10^4$	$3.9644 \cdot 10^6$	11.5	$3.4487 \cdot 10^3$	$3.9964 \cdot 10^6$
7.6	$1.6025 \cdot 10^4$	$3.9660 \cdot 10^6$	11.6	$3.3380 \cdot 10^3$	$3.9968 \cdot 10^6$
7.7	$1.5280 \cdot 10^4$	$3.9676 \cdot 10^6$	11.7	$3.2316 \cdot 10^3$	$3.9971 \cdot 10^6$
7.8	$1.4578 \cdot 10^4$	$3.9691 \cdot 10^6$	11.8	$3.1295 \cdot 10^3$	$3.9974 \cdot 10^6$
7.9	$1.3916 \cdot 10^4$	$3.9705 \cdot 10^6$	11.9	$3.0313 \cdot 10^3$	$3.9977 \cdot 10^6$
8.0	$1.3290 \cdot 10^4$	$3.9719 \cdot 10^6$	12.0	$2.9370 \cdot 10^3$	$3.9980 \cdot 10^6$
8.1	$1.2699 \cdot 10^4$	$3.9732 \cdot 10^6$	12.1	$2.8463 \cdot 10^3$	$3.9983 \cdot 10^6$
8.2	$1.2141 \cdot 10^4$	$3.9744 \cdot 10^6$	12.2	$2.7591 \cdot 10^3$	$3.9986 \cdot 10^6$
8.3	$1.1613 \cdot 10^4$	$3.9756 \cdot 10^6$	12.3	$2.6751 \cdot 10^3$	$3.9989 \cdot 10^6$
8.4	$1.1113 \cdot 10^4$	$3.9767 \cdot 10^6$	12.4	$2.5944 \cdot 10^3$	$3.9991 \cdot 10^6$
8.5	$1.0640 \cdot 10^4$	$3.9778 \cdot 10^6$	12.5	$2.5166 \cdot 10^3$	$3.9994 \cdot 10^6$
8.6	$1.0191 \cdot 10^4$	$3.9788 \cdot 10^6$	12.6	$2.4418 \cdot 10^3$	$3.9996 \cdot 10^6$
8.7	$9.7660 \cdot 10^3$	$3.9798 \cdot 10^6$	12.7	$2.3697 \cdot 10^3$	$3.9999 \cdot 10^6$
8.8	$9.3627 \cdot 10^3$	$3.9808 \cdot 10^6$	12.8	$2.3002 \cdot 10^3$	$4.0001 \cdot 10^6$
8.9	$8.9800 \cdot 10^3$	$3.9817 \cdot 10^6$	12.9	$2.2333 \cdot 10^3$	$4.0003 \cdot 10^6$
9.0	$8.6166 \cdot 10^3$	$3.9826 \cdot 10^6$	13.0	$2.1688 \cdot 10^3$	$4.0006 \cdot 10^6$
9.1	$8.2713 \cdot 10^3$	$3.9834 \cdot 10^6$	13.1	$2.1066 \cdot 10^3$	$4.0008 \cdot 10^6$
9.2	$7.9431 \cdot 10^3$	$3.9842 \cdot 10^6$	13.2	$2.0466 \cdot 10^3$	$4.0010 \cdot 10^6$
9.3	$7.6310 \cdot 10^3$	$3.9850 \cdot 10^6$	13.3	$1.9887 \cdot 10^3$	$4.0012 \cdot 10^6$
9.4	$7.3340 \cdot 10^3$	$3.9858 \cdot 10^6$	13.4	$1.9329 \cdot 10^3$	$4.0014 \cdot 10^6$
9.5	$7.0513 \cdot 10^3$	$3.9865 \cdot 10^6$	13.5	$1.8790 \cdot 10^3$	$4.0016 \cdot 10^6$
9.6	$6.7821 \cdot 10^3$	$3.9872 \cdot 10^6$	13.6	$1.8269 \cdot 10^3$	$4.0018 \cdot 10^6$
9.7	$6.5256 \cdot 10^3$	$3.9878 \cdot 10^6$	13.7	$1.7767 \cdot 10^3$	$4.0019 \cdot 10^6$
9.8	$6.2810 \cdot 10^3$	$3.9885 \cdot 10^6$	13.8	$1.7281 \cdot 10^3$	$4.0021 \cdot 10^6$
9.9	$6.0478 \cdot 10^3$	$3.9891 \cdot 10^6$	13.9	$1.6812 \cdot 10^3$	$4.0023 \cdot 10^6$
10.0	$5.8253 \cdot 10^3$	$3.9897 \cdot 10^6$	14.0	$1.6359 \cdot 10^3$	$4.0024 \cdot 10^6$
10.1	$5.6129 \cdot 10^3$	$3.9903 \cdot 10^6$	14.1	$1.5921 \cdot 10^3$	$4.0026 \cdot 10^6$
10.2	$5.4100 \cdot 10^3$	$3.9908 \cdot 10^6$	14.2	$1.5498 \cdot 10^3$	$4.0028 \cdot 10^6$
10.3	$5.2163 \cdot 10^3$	$3.9913 \cdot 10^6$	14.3	$1.5088 \cdot 10^3$	$4.0029 \cdot 10^6$
10.4	$5.0311 \cdot 10^3$	$3.9919 \cdot 10^6$	14.4	$1.4692 \cdot 10^3$	$4.0031 \cdot 10^6$
10.5	$4.8540 \cdot 10^3$	$3.9924 \cdot 10^6$	14.5	$1.4309 \cdot 10^3$	$4.0032 \cdot 10^6$
10.6	$4.6847 \cdot 10^3$	$3.9928 \cdot 10^6$	14.6	$1.3938 \cdot 10^3$	$4.0033 \cdot 10^6$
10.7	$4.5226 \cdot 10^3$	$3.9933 \cdot 10^6$	14.7	$1.3579 \cdot 10^3$	$4.0035 \cdot 10^6$
10.8	$4.3675 \cdot 10^3$	$3.9937 \cdot 10^6$	14.8	$1.3232 \cdot 10^3$	$4.0036 \cdot 10^6$
10.9	$4.2190 \cdot 10^3$	$3.9942 \cdot 10^6$	14.9	$1.2896 \cdot 10^3$	$4.0038 \cdot 10^6$
11.0	$4.0767 \cdot 10^3$	$3.9946 \cdot 10^6$	15.0	$1.2570 \cdot 10^3$	$4.0039 \cdot 10^6$
11.1	$3.9404 \cdot 10^3$	$3.9950 \cdot 10^6$	15.1	$1.2254 \cdot 10^3$	$4.0040 \cdot 10^6$
11.2	$3.8097 \cdot 10^3$	$3.9954 \cdot 10^6$	15.2	$1.1949 \cdot 10^3$	$4.0041 \cdot 10^6$
11.3	$3.6843 \cdot 10^3$	$3.9957 \cdot 10^6$	15.3	$1.1652 \cdot 10^3$	$4.0042 \cdot 10^6$
11.4	$3.5641 \cdot 10^3$	$3.9961 \cdot 10^6$	15.4	$1.1365 \cdot 10^3$	$4.0044 \cdot 10^6$

Table 2.2. (*continued*)

temperature 3000 K

λ [μm]	$m_{\lambda,bb}$ [W/μm m²]	$M_{T,o-\lambda,bb}$ [W/m²]	λ [μm]	$m_{\lambda,bb}$ [W/μm m²]	$M_{T,o-\lambda,bb}$ [W/m²]
0.4	$2.2680\cdot10^{5}$	$9.8017\cdot10^{3}$			
0.5	$8.1762\cdot10^{5}$	$5.9017\cdot10^{4}$	4.0	$1.5773\cdot10^{5}$	$4.3406\cdot10^{6}$
0.6	$1.6257\cdot10^{6}$	$1.8069\cdot10^{5}$	4.1	$1.4541\cdot10^{5}$	$4.3557\cdot10^{6}$
0.7	$2.3580\cdot10^{6}$	$3.8145\cdot10^{5}$	4.2	$1.3425\cdot10^{5}$	$4.3697\cdot10^{6}$
0.8	$2.8520\cdot10^{6}$	$6.4418\cdot10^{5}$	4.3	$1.2413\cdot10^{5}$	$4.3826\cdot10^{6}$
0.9	$3.0882\cdot10^{6}$	$9.4318\cdot10^{5}$	4.4	$1.1493\cdot10^{5}$	$4.3945\cdot10^{6}$
1.0	$3.1177\cdot10^{6}$	$1.2549\cdot10^{6}$	4.5	$1.0656\cdot10^{5}$	$4.4056\cdot10^{6}$
1.1	$3.0075\cdot10^{6}$	$1.5621\cdot10^{6}$	4.6	$9.8922\cdot10^{4}$	$4.4159\cdot10^{6}$
1.2	$2.8153\cdot10^{6}$	$1.8537\cdot10^{6}$	4.7	$9.1952\cdot10^{4}$	$4.4254\cdot10^{6}$
1.3	$2.5832\cdot10^{6}$	$2.1238\cdot10^{6}$	4.8	$8.5580\cdot10^{4}$	$4.4343\cdot10^{6}$
1.4	$2.3391\cdot10^{6}$	$2.3699\cdot10^{6}$	4.9	$7.9744\cdot10^{4}$	$4.4425\cdot10^{6}$
1.5	$2.0998\cdot10^{6}$	$2.5918\cdot10^{6}$	5.0	$7.4392\cdot10^{4}$	$4.4502\cdot10^{6}$
1.6	$1.8747\cdot10^{6}$	$2.7904\cdot10^{6}$	5.1	$6.9478\cdot10^{4}$	$4.4574\cdot10^{6}$
1.7	$1.6684\cdot10^{6}$	$2.9674\cdot10^{6}$	5.2	$6.4960\cdot10^{4}$	$4.4642\cdot10^{6}$
1.8	$1.4823\cdot10^{6}$	$3.1247\cdot10^{6}$	5.3	$6.0799\cdot10^{4}$	$4.4704\cdot10^{6}$
1.9	$1.3163\cdot10^{6}$	$3.2645\cdot10^{6}$	5.4	$5.6964\cdot10^{4}$	$4.4763\cdot10^{6}$
2.0	$1.1692\cdot10^{6}$	$3.3886\cdot10^{6}$	5.5	$5.3425\cdot10^{4}$	$4.4818\cdot10^{6}$
2.1	$1.0395\cdot10^{6}$	$3.4989\cdot10^{6}$	5.6	$5.0154\cdot10^{4}$	$4.4870\cdot10^{6}$
2.2	$9.2538\cdot10^{5}$	$3.5970\cdot10^{6}$	5.7	$4.7128\cdot10^{4}$	$4.4919\cdot10^{6}$
2.3	$8.2508\cdot10^{5}$	$3.6844\cdot10^{6}$	5.8	$4.4325\cdot10^{4}$	$4.4965\cdot10^{6}$
2.4	$7.3696\cdot10^{5}$	$3.7624\cdot10^{6}$	5.9	$4.1726\cdot10^{4}$	$4.5008\cdot10^{6}$
2.5	$6.5951\cdot10^{5}$	$3.8322\cdot10^{6}$	6.0	$3.9313\cdot10^{4}$	$4.5048\cdot10^{6}$
2.6	$5.9137\cdot10^{5}$	$3.8947\cdot10^{6}$	6.1	$3.7072\cdot10^{4}$	$4.5086\cdot10^{6}$
2.7	$5.3135\cdot10^{5}$	$3.9507\cdot10^{6}$	6.2	$3.4987\cdot10^{4}$	$4.5122\cdot10^{6}$
2.8	$4.7840\cdot10^{5}$	$4.0012\cdot10^{6}$	6.3	$3.3045\cdot10^{4}$	$4.5156\cdot10^{6}$
2.9	$4.3162\cdot10^{5}$	$4.0466\cdot10^{6}$	6.4	$3.1236\cdot10^{4}$	$4.5188\cdot10^{6}$
3.0	$3.9020\cdot10^{5}$	$4.0877\cdot10^{6}$	6.5	$2.9549\cdot10^{4}$	$4.5219\cdot10^{6}$
3.1	$3.5346\cdot10^{5}$	$4.1248\cdot10^{6}$	6.6	$2.7973\cdot10^{4}$	$4.5247\cdot10^{6}$
3.2	$3.2082\cdot10^{5}$	$4.1585\cdot10^{6}$	6.7	$2.6500\cdot10^{4}$	$4.5275\cdot10^{6}$
3.3	$2.9175\cdot10^{5}$	$4.1891\cdot10^{6}$	6.8	$2.5123\cdot10^{4}$	$4.5300\cdot10^{6}$
3.4	$2.6581\cdot10^{5}$	$4.2169\cdot10^{6}$	6.9	$2.3833\cdot10^{4}$	$4.5325\cdot10^{6}$
3.5	$2.4262\cdot10^{5}$	$4.2423\cdot10^{6}$	7.0	$2.2625\cdot10^{4}$	$4.5348\cdot10^{6}$
3.6	$2.2185\cdot10^{5}$	$4.2655\cdot10^{6}$	7.1	$2.1492\cdot10^{4}$	$4.5370\cdot10^{6}$
3.7	$2.0321\cdot10^{5}$	$4.2868\cdot10^{6}$	7.2	$2.0429\cdot10^{4}$	$4.5391\cdot10^{6}$
3.8	$1.8645\cdot10^{5}$	$4.3062\cdot10^{6}$	7.3	$1.9430\cdot10^{4}$	$4.5411\cdot10^{6}$
3.9	$1.7135\cdot10^{5}$	$4.3241\cdot10^{6}$	7.4	$1.8492\cdot10^{4}$	$4.5430\cdot10^{6}$

Table 2. 2. (*continued*)

λ [μm]	$m_{\lambda,bb}$ [W/μm m²]	$M_{T,o-\lambda,bb}$ [W/m²]	λ [μm]	$m_{\lambda,bb}$ [W/μm m²]	$M_{T,o-\lambda,bb}$ [W/m²]
7.5	$1.7609 \cdot 10^4$	$4.5448 \cdot 10^6$	11.5	$3.5952 \cdot 10^3$	$4.5783 \cdot 10^6$
7.6	$1.6778 \cdot 10^4$	$4.5465 \cdot 10^6$	11.6	$3.4795 \cdot 10^3$	$4.5787 \cdot 10^6$
7.7	$1.5996 \cdot 10^4$	$4.5482 \cdot 10^6$	11.7	$3.3684 \cdot 10^3$	$4.5790 \cdot 10^6$
7.8	$1.5258 \cdot 10^4$	$4.5497 \cdot 10^6$	11.8	$3.2617 \cdot 10^3$	$4.5794 \cdot 10^6$
7.9	$1.4562 \cdot 10^4$	$4.5512 \cdot 10^6$	11.9	$3.1592 \cdot 10^3$	$4.5797 \cdot 10^6$
8.0	$1.3906 \cdot 10^4$	$4.5526 \cdot 10^6$	12.0	$3.0607 \cdot 10^3$	$4.5800 \cdot 10^6$
8.1	$1.3286 \cdot 10^4$	$4.5540 \cdot 10^6$	12.1	$2.9660 \cdot 10^3$	$4.5803 \cdot 10^6$
8.2	$1.2699 \cdot 10^4$	$4.5553 \cdot 10^6$	12.2	$2.8749 \cdot 10^3$	$4.5806 \cdot 10^6$
8.3	$1.2145 \cdot 10^4$	$4.5565 \cdot 10^6$	12.3	$2.7873 \cdot 10^3$	$4.5809 \cdot 10^6$
8.4	$1.1621 \cdot 10^4$	$4.5577 \cdot 10^6$	12.4	$2.7030 \cdot 10^3$	$4.5811 \cdot 10^6$
8.5	$1.1124 \cdot 10^4$	$4.5589 \cdot 10^6$	12.5	$2.6218 \cdot 10^3$	$4.5814 \cdot 10^6$
8.6	$1.0654 \cdot 10^4$	$4.5599 \cdot 10^6$	12.6	$2.5437 \cdot 10^3$	$4.5817 \cdot 10^6$
8.7	$1.0208 \cdot 10^4$	$4.5610 \cdot 10^6$	12.7	$2.4684 \cdot 10^3$	$4.5819 \cdot 10^6$
8.8	$9.7855 \cdot 10^3$	$4.5620 \cdot 10^6$	12.8	$2.3960 \cdot 10^3$	$4.5822 \cdot 10^6$
8.9	$9.3843 \cdot 10^3$	$4.5629 \cdot 10^6$	12.9	$2.3261 \cdot 10^3$	$4.5824 \cdot 10^6$
9.0	$9.0034 \cdot 10^3$	$4.5639 \cdot 10^6$	13.0	$2.2588 \cdot 10^3$	$4.5826 \cdot 10^6$
9.1	$8.6416 \cdot 10^3$	$4.5647 \cdot 10^6$	13.1	$2.1939 \cdot 10^3$	$4.5828 \cdot 10^6$
9.2	$8.2978 \cdot 10^3$	$4.5656 \cdot 10^6$	13.2	$2.1313 \cdot 10^3$	$4.5831 \cdot 10^6$
9.3	$7.9708 \cdot 10^3$	$4.5664 \cdot 10^6$	13.3	$2.0709 \cdot 10^3$	$4.5833 \cdot 10^6$
9.4	$7.6598 \cdot 10^3$	$4.5672 \cdot 10^6$	13.4	$2.0126 \cdot 10^3$	$4.5835 \cdot 10^6$
9.5	$7.3637 \cdot 10^3$	$4.5679 \cdot 10^6$	13.5	$1.9564 \cdot 10^3$	$4.5837 \cdot 10^6$
9.6	$7.0818 \cdot 10^3$	$4.5687 \cdot 10^6$	13.6	$1.9021 \cdot 10^3$	$4.5839 \cdot 10^6$
9.7	$6.8132 \cdot 10^3$	$4.5694 \cdot 10^6$	13.7	$1.8497 \cdot 10^3$	$4.5841 \cdot 10^6$
9.8	$6.5572 \cdot 10^3$	$4.5700 \cdot 10^6$	13.8	$1.7991 \cdot 10^3$	$4.5842 \cdot 10^6$
9.9	$6.3131 \cdot 10^3$	$4.5707 \cdot 10^6$	13.9	$1.7502 \cdot 10^3$	$4.5844 \cdot 10^6$
10.0	$6.0802 \cdot 10^3$	$4.5713 \cdot 10^6$	14.0	$1.7029 \cdot 10^3$	$4.5846 \cdot 10^6$
10.1	$5.8580 \cdot 10^3$	$4.5719 \cdot 10^6$	14.1	$1.6573 \cdot 10^3$	$4.5847 \cdot 10^6$
10.2	$5.6458 \cdot 10^3$	$4.5725 \cdot 10^6$	14.2	$1.6131 \cdot 10^3$	$4.5849 \cdot 10^6$
10.3	$5.4431 \cdot 10^3$	$4.5730 \cdot 10^6$	14.3	$1.5704 \cdot 10^3$	$4.5851 \cdot 10^6$
10.4	$5.2494 \cdot 10^3$	$4.5735 \cdot 10^6$	14.4	$1.5291 \cdot 10^3$	$4.5852 \cdot 10^6$
10.5	$5.0642 \cdot 10^3$	$4.5741 \cdot 10^6$	14.5	$1.4892 \cdot 10^3$	$4.5854 \cdot 10^6$
10.6	$4.8871 \cdot 10^3$	$4.5746 \cdot 10^6$	14.6	$1.4505 \cdot 10^3$	$4.5855 \cdot 10^6$
10.7	$4.7176 \cdot 10^3$	$4.5750 \cdot 10^6$	14.7	$1.4131 \cdot 10^3$	$4.5857 \cdot 10^6$
10.8	$4.5555 \cdot 10^3$	$4.5755 \cdot 10^6$	14.8	$1.3769 \cdot 10^3$	$4.5858 \cdot 10^6$
10.9	$4.4002 \cdot 10^3$	$4.5759 \cdot 10^6$	14.9	$1.3419 \cdot 10^3$	$4.5859 \cdot 10^6$
11.0	$4.2515 \cdot 10^3$	$4.5764 \cdot 10^6$	15.0	$1.3079 \cdot 10^3$	$4.5861 \cdot 10^6$
11.1	$4.1089 \cdot 10^3$	$4.5768 \cdot 10^6$	15.1	$1.2750 \cdot 10^3$	$4.5862 \cdot 10^6$
11.2	$3.9724 \cdot 10^3$	$4.5772 \cdot 10^6$	15.2	$1.2432 \cdot 10^3$	$4.5863 \cdot 10^6$
11.3	$3.8414 \cdot 10^3$	$4.5776 \cdot 10^6$	15.3	$1.2123 \cdot 10^3$	$4.5864 \cdot 10^6$
11.4	$3.7158 \cdot 10^3$	$4.5780 \cdot 10^6$	15.4	$1.1824 \cdot 10^3$	$4.5866 \cdot 10^6$

Table 2.2. (*continued*)

temperature 3100 K

λ [μm]	$m_{\lambda,bb}$ [W/μm m^2]	$M_{T,o-\lambda,bb}$ [W/m^2]	λ [μm]	$m_{\lambda,bb}$ [W/μm m^2]	$M_{T,o-\lambda,bb}$ [W/m^2]
0.4	$3.3390 \cdot 10^5$	$1.5044 \cdot 10^4$			
0.5	$1.1141 \cdot 10^6$	$8.4033 \cdot 10^4$	4.0	$1.6679 \cdot 10^5$	$4.9716 \cdot 10^6$
0.6	$2.1041 \cdot 10^6$	$2.4492 \cdot 10^5$	4.1	$1.5366 \cdot 10^5$	$4.9876 \cdot 10^6$
0.7	$2.9420 \cdot 10^6$	$4.9952 \cdot 10^5$	4.2	$1.4178 \cdot 10^5$	$5.0024 \cdot 10^6$
0.8	$3.4623 \cdot 10^6$	$8.2254 \cdot 10^5$	4.3	$1.3102 \cdot 10^5$	$5.0160 \cdot 10^6$
0.9	$3.6707 \cdot 10^6$	$1.1815 \cdot 10^6$	4.4	$1.2124 \cdot 10^5$	$5.0286 \cdot 10^6$
1.0	$3.6444 \cdot 10^6$	$1.5489 \cdot 10^6$	4.5	$1.1235 \cdot 10^5$	$5.0403 \cdot 10^6$
1.1	$3.4684 \cdot 10^6$	$1.9054 \cdot 10^6$	4.6	$1.0425 \cdot 10^5$	$5.0511 \cdot 10^6$
1.2	$3.2109 \cdot 10^6$	$2.2398 \cdot 10^6$	4.7	$9.6856 \cdot 10^4$	$5.0612 \cdot 10^6$
1.3	$2.9192 \cdot 10^6$	$2.5465 \cdot 10^6$	4.8	$9.0103 \cdot 10^4$	$5.0705 \cdot 10^6$
1.4	$2.6227 \cdot 10^6$	$2.8235 \cdot 10^6$	4.9	$8.3922 \cdot 10^4$	$5.0792 \cdot 10^6$
1.5	$2.3388 \cdot 10^6$	$3.0714 \cdot 10^6$	5.0	$7.8258 \cdot 10^4$	$5.0873 \cdot 10^6$
1.6	$2.0761 \cdot 10^6$	$3.2920 \cdot 10^6$	5.1	$7.3060 \cdot 10^4$	$5.0949 \cdot 10^6$
1.7	$1.8384 \cdot 10^6$	$3.4875 \cdot 10^6$	5.2	$6.8282 \cdot 10^4$	$5.1019 \cdot 10^6$
1.8	$1.6262 \cdot 10^6$	$3.6605 \cdot 10^6$	5.3	$6.3886 \cdot 10^4$	$5.1085 \cdot 10^6$
1.9	$1.4386 \cdot 10^6$	$3.8136 \cdot 10^6$	5.4	$5.9836 \cdot 10^4$	$5.1147 \cdot 10^6$
2.0	$1.2735 \cdot 10^6$	$3.9490 \cdot 10^6$	5.5	$5.6099 \cdot 10^4$	$5.1205 \cdot 10^6$
2.1	$1.1288 \cdot 10^6$	$4.0689 \cdot 10^6$	5.6	$5.2647 \cdot 10^4$	$5.1260 \cdot 10^6$
2.2	$1.0021 \cdot 10^6$	$4.1753 \cdot 10^6$	5.7	$4.9455 \cdot 10^4$	$5.1311 \cdot 10^6$
2.3	$8.9129 \cdot 10^5$	$4.2699 \cdot 10^6$	5.8	$4.6500 \cdot 10^4$	$5.1359 \cdot 10^6$
2.4	$7.9432 \cdot 10^5$	$4.3541 \cdot 10^6$	5.9	$4.3761 \cdot 10^4$	$5.1404 \cdot 10^6$
2.5	$7.0940 \cdot 10^5$	$4.4292 \cdot 10^6$	6.0	$4.1220 \cdot 10^4$	$5.1446 \cdot 10^6$
2.6	$6.3493 \cdot 10^5$	$4.4963 \cdot 10^6$	6.1	$3.8859 \cdot 10^4$	$5.1486 \cdot 10^6$
2.7	$5.6953 \cdot 10^5$	$4.5564 \cdot 10^6$	6.2	$3.6664 \cdot 10^4$	$5.1524 \cdot 10^6$
2.8	$5.1198 \cdot 10^5$	$4.6105 \cdot 10^6$	6.3	$3.4621 \cdot 10^4$	$5.1560 \cdot 10^6$
2.9	$4.6125 \cdot 10^5$	$4.6591 \cdot 10^6$	6.4	$3.2718 \cdot 10^4$	$5.1593 \cdot 10^6$
3.0	$4.1643 \cdot 10^5$	$4.7029 \cdot 10^6$	6.5	$3.0943 \cdot 10^4$	$5.1625 \cdot 10^6$
3.1	$3.7677 \cdot 10^5$	$4.7425 \cdot 10^6$	6.6	$2.9287 \cdot 10^4$	$5.1655 \cdot 10^6$
3.2	$3.4158 \cdot 10^5$	$4.7784 \cdot 10^6$	6.7	$2.7739 \cdot 10^4$	$5.1684 \cdot 10^6$
3.3	$3.1029 \cdot 10^5$	$4.8110 \cdot 10^6$	6.8	$2.6291 \cdot 10^4$	$5.1711 \cdot 10^6$
3.4	$2.8243 \cdot 10^5$	$4.8406 \cdot 10^6$	6.9	$2.4937 \cdot 10^4$	$5.1736 \cdot 10^6$
3.5	$2.5755 \cdot 10^5$	$4.8675 \cdot 10^6$	7.0	$2.3668 \cdot 10^4$	$5.1760 \cdot 10^6$
3.6	$2.3530 \cdot 10^5$	$4.8922 \cdot 10^6$	7.1	$2.2479 \cdot 10^4$	$5.1784 \cdot 10^6$
3.7	$2.1535 \cdot 10^5$	$4.9147 \cdot 10^6$	7.2	$2.1363 \cdot 10^4$	$5.1805 \cdot 10^6$
3.8	$1.9744 \cdot 10^5$	$4.9353 \cdot 10^6$	7.3	$2.0315 \cdot 10^4$	$5.1826 \cdot 10^6$
3.9	$1.8132 \cdot 10^5$	$4.9542 \cdot 10^6$	7.4	$1.9330 \cdot 10^4$	$5.1846 \cdot 10^6$

Table 2.2. (*continued*)

λ [µm]	$m_{\lambda,bb}$ [W/µm m²]	$M_{T,o-\lambda,bb}$ [W/m²]	λ [µm]	$m_{\lambda,bb}$ [W/µm m²]	$M_{T,o-\lambda,bb}$ [W/m²]
7.5	$1.8405 \cdot 10^4$	$5.1865 \cdot 10^6$	11.5	$3.7419 \cdot 10^3$	$5.2215 \cdot 10^6$
7.6	$1.7533 \cdot 10^4$	$5.1883 \cdot 10^6$	11.6	$3.6212 \cdot 10^3$	$5.2218 \cdot 10^6$
7.7	$1.6713 \cdot 10^4$	$5.1900 \cdot 10^6$	11.7	$3.5054 \cdot 10^3$	$5.2222 \cdot 10^6$
7.8	$1.5940 \cdot 10^4$	$5.1916 \cdot 10^6$	11.8	$3.3941 \cdot 10^3$	$5.2225 \cdot 10^6$
7.9	$1.5211 \cdot 10^4$	$5.1932 \cdot 10^6$	11.9	$3.2872 \cdot 10^3$	$5.2229 \cdot 10^6$
8.0	$1.4523 \cdot 10^4$	$5.1947 \cdot 10^6$	12.0	$3.1845 \cdot 10^3$	$5.2232 \cdot 10^6$
8.1	$1.3873 \cdot 10^4$	$5.1961 \cdot 10^6$	12.1	$3.0858 \cdot 10^3$	$5.2235 \cdot 10^6$
8.2	$1.3259 \cdot 10^4$	$5.1975 \cdot 10^6$	12.2	$2.9908 \cdot 10^3$	$5.2238 \cdot 10^6$
8.3	$1.2679 \cdot 10^4$	$5.1987 \cdot 10^6$	12.3	$2.8995 \cdot 10^3$	$5.2241 \cdot 10^6$
8.4	$1.2130 \cdot 10^4$	$5.2000 \cdot 10^6$	12.4	$2.8117 \cdot 10^3$	$5.2244 \cdot 10^6$
8.5	$1.1610 \cdot 10^4$	$5.2012 \cdot 10^6$	12.5	$2.7271 \cdot 10^3$	$5.2247 \cdot 10^6$
8.6	$1.1118 \cdot 10^4$	$5.2023 \cdot 10^6$	12.6	$2.6457 \cdot 10^3$	$5.2249 \cdot 10^6$
8.7	$1.0651 \cdot 10^4$	$5.2034 \cdot 10^6$	12.7	$2.5673 \cdot 10^3$	$5.2252 \cdot 10^6$
8.8	$1.0209 \cdot 10^4$	$5.2044 \cdot 10^6$	12.8	$2.4918 \cdot 10^3$	$5.2254 \cdot 10^6$
8.9	$9.7892 \cdot 10^3$	$5.2054 \cdot 10^6$	12.9	$2.4190 \cdot 10^3$	$5.2257 \cdot 10^6$
9.0	$9.3908 \cdot 10^3$	$5.2064 \cdot 10^6$	13.0	$2.3489 \cdot 10^3$	$5.2259 \cdot 10^6$
9.1	$9.0125 \cdot 10^3$	$5.2073 \cdot 10^6$	13.1	$2.2812 \cdot 10^3$	$5.2262 \cdot 10^6$
9.2	$8.6529 \cdot 10^3$	$5.2082 \cdot 10^6$	13.2	$2.2160 \cdot 10^3$	$5.2264 \cdot 10^6$
9.3	$8.3111 \cdot 10^3$	$5.2090 \cdot 10^6$	13.3	$2.1531 \cdot 10^3$	$5.2266 \cdot 10^6$
9.4	$7.9860 \cdot 10^3$	$5.2099 \cdot 10^6$	13.4	$2.0925 \cdot 10^3$	$5.2268 \cdot 10^6$
9.5	$7.6765 \cdot 10^3$	$5.2106 \cdot 10^6$	13.5	$2.0339 \cdot 10^3$	$5.2270 \cdot 10^6$
9.6	$7.3819 \cdot 10^3$	$5.2114 \cdot 10^6$	13.6	$1.9774 \cdot 10^3$	$5.2272 \cdot 10^6$
9.7	$7.1013 \cdot 10^3$	$5.2121 \cdot 10^6$	13.7	$1.9228 \cdot 10^3$	$5.2274 \cdot 10^6$
9.8	$6.8338 \cdot 10^3$	$5.2128 \cdot 10^6$	13.8	$1.8701 \cdot 10^3$	$5.2276 \cdot 10^6$
9.9	$6.5788 \cdot 10^3$	$5.2135 \cdot 10^6$	13.9	$1.8192 \cdot 10^3$	$5.2278 \cdot 10^6$
10.0	$6.3355 \cdot 10^3$	$5.2141 \cdot 10^6$	14.0	$1.7700 \cdot 10^3$	$5.2280 \cdot 10^6$
10.1	$6.1034 \cdot 10^3$	$5.2148 \cdot 10^6$	14.1	$1.7224 \cdot 10^3$	$5.2281 \cdot 10^6$
10.2	$5.8818 \cdot 10^3$	$5.2154 \cdot 10^6$	14.2	$1.6765 \cdot 10^3$	$5.2283 \cdot 10^6$
10.3	$5.6701 \cdot 10^3$	$5.2159 \cdot 10^6$	14.3	$1.6320 \cdot 10^3$	$5.2285 \cdot 10^6$
10.4	$5.4679 \cdot 10^3$	$5.2165 \cdot 10^6$	14.4	$1.5891 \cdot 10^3$	$5.2286 \cdot 10^6$
10.5	$5.2746 \cdot 10^3$	$5.2170 \cdot 10^6$	14.5	$1.5475 \cdot 10^3$	$5.2288 \cdot 10^6$
10.6	$5.0897 \cdot 10^3$	$5.2175 \cdot 10^6$	14.6	$1.5073 \cdot 10^3$	$5.2289 \cdot 10^6$
10.7	$4.9129 \cdot 10^3$	$5.2180 \cdot 10^6$	14.7	$1.4683 \cdot 10^3$	$5.2291 \cdot 10^6$
10.8	$4.7436 \cdot 10^3$	$5.2185 \cdot 10^6$	14.8	$1.4307 \cdot 10^3$	$5.2292 \cdot 10^6$
10.9	$4.5816 \cdot 10^3$	$5.2190 \cdot 10^6$	14.9	$1.3942 \cdot 10^3$	$5.2294 \cdot 10^6$
11.0	$4.4264 \cdot 10^3$	$5.2194 \cdot 10^6$	15.0	$1.3589 \cdot 10^3$	$5.2295 \cdot 10^6$
11.1	$4.2777 \cdot 10^3$	$5.2199 \cdot 10^6$	15.1	$1.3246 \cdot 10^3$	$5.2296 \cdot 10^6$
11.2	$4.1352 \cdot 10^3$	$5.2203 \cdot 10^6$	15.2	$1.2915 \cdot 10^3$	$5.2298 \cdot 10^6$
11.3	$3.9986 \cdot 10^3$	$5.2207 \cdot 10^6$	15.3	$1.2594 \cdot 10^3$	$5.2299 \cdot 10^6$
11.4	$3.8676 \cdot 10^3$	$5.2211 \cdot 10^6$	15.4	$1.2283 \cdot 10^3$	$5.2300 \cdot 10^6$

Table 2.2. (*continued*)

temperature 3200 K

λ [μm]	$m_{\lambda,bb}$ [W/μm m²]	$M_{T,o-\lambda,bb}$ [W/m²]	λ [μm]	$m_{\lambda,bb}$ [W/μm m²]	$M_{T,o-\lambda,bb}$ [W/m²]
0.4	$4.7984\cdot10^{5}$	$2.2515\cdot10^{4}$			
0.5	$1.4891\cdot10^{6}$	$1.1724\cdot10^{5}$	4.0	$1.7591\cdot10^{5}$	$5.6680\cdot10^{6}$
0.6	$2.6798\cdot10^{6}$	$3.2634\cdot10^{5}$	4.1	$1.6197\cdot10^{5}$	$5.6849\cdot10^{6}$
0.7	$3.6204\cdot10^{6}$	$6.4451\cdot10^{5}$	4.2	$1.4936\cdot10^{5}$	$5.7005\cdot10^{6}$
0.8	$4.1530\cdot10^{6}$	$1.0366\cdot10^{6}$	4.3	$1.3795\cdot10^{5}$	$5.7148\cdot10^{6}$
0.9	$4.3170\cdot10^{6}$	$1.4628\cdot10^{6}$	4.4	$1.2759\cdot10^{5}$	$5.7281\cdot10^{6}$
1.0	$4.2197\cdot10^{6}$	$1.8913\cdot10^{6}$	4.5	$1.1817\cdot10^{5}$	$5.7404\cdot10^{6}$
1.1	$3.9656\cdot10^{6}$	$2.3015\cdot10^{6}$	4.6	$1.0960\cdot10^{5}$	$5.7517\cdot10^{6}$
1.2	$3.6334\cdot10^{6}$	$2.6818\cdot10^{6}$	4.7	$1.0179\cdot10^{5}$	$5.7623\cdot10^{6}$
1.3	$3.2749\cdot10^{6}$	$3.0273\cdot10^{6}$	4.8	$9.4648\cdot10^{4}$	$5.7721\cdot10^{6}$
1.4	$2.9210\cdot10^{6}$	$3.3370\cdot10^{6}$	4.9	$8.8121\cdot10^{4}$	$5.7813\cdot10^{6}$
1.5	$2.5887\cdot10^{6}$	$3.6122\cdot10^{6}$	5.0	$8.2141\cdot10^{4}$	$5.7898\cdot10^{6}$
1.6	$2.2857\cdot10^{6}$	$3.8557\cdot10^{6}$	5.1	$7.6657\cdot10^{4}$	$5.7977\cdot10^{6}$
1.7	$2.0147\cdot10^{6}$	$4.0705\cdot10^{6}$	5.2	$7.1619\cdot10^{4}$	$5.8051\cdot10^{6}$
1.8	$1.7749\cdot10^{6}$	$4.2597\cdot10^{6}$	5.3	$6.6986\cdot10^{4}$	$5.8120\cdot10^{6}$
1.9	$1.5645\cdot10^{6}$	$4.4264\cdot10^{6}$	5.4	$6.2718\cdot10^{4}$	$5.8185\cdot10^{6}$
2.0	$1.3806\cdot10^{6}$	$4.5735\cdot10^{6}$	5.5	$5.8783\cdot10^{4}$	$5.8246\cdot10^{6}$
2.1	$1.2202\cdot10^{6}$	$4.7033\cdot10^{6}$	5.6	$5.5150\cdot10^{4}$	$5.8303\cdot10^{6}$
2.2	$1.0806\cdot10^{6}$	$4.8182\cdot10^{6}$	5.7	$5.1791\cdot10^{4}$	$5.8356\cdot10^{6}$
2.3	$9.5887\cdot10^{5}$	$4.9200\cdot10^{6}$	5.8	$4.8683\cdot10^{4}$	$5.8406\cdot10^{6}$
2.4	$8.5279\cdot10^{5}$	$5.0105\cdot10^{6}$	5.9	$4.5803\cdot10^{4}$	$5.8454\cdot10^{6}$
2.5	$7.6018\cdot10^{5}$	$5.0910\cdot10^{6}$	6.0	$4.3132\cdot10^{4}$	$5.8498\cdot10^{6}$
2.6	$6.7922\cdot10^{5}$	$5.1629\cdot10^{6}$	6.1	$4.0652\cdot10^{4}$	$5.8540\cdot10^{6}$
2.7	$6.0829\cdot10^{5}$	$5.2272\cdot10^{6}$	6.2	$3.8347\cdot10^{4}$	$5.8579\cdot10^{6}$
2.8	$5.4604\cdot10^{5}$	$5.2849\cdot10^{6}$	6.3	$3.6202\cdot10^{4}$	$5.8617\cdot10^{6}$
2.9	$4.9127\cdot10^{5}$	$5.3367\cdot10^{6}$	6.4	$3.4204\cdot10^{4}$	$5.8652\cdot10^{6}$
3.0	$4.4300\cdot10^{5}$	$5.3833\cdot10^{6}$	6.5	$3.2342\cdot10^{4}$	$5.8685\cdot10^{6}$
3.1	$4.0034\cdot10^{5}$	$5.4255\cdot10^{6}$	6.6	$3.0604\cdot10^{4}$	$5.8717\cdot10^{6}$
3.2	$3.6256\cdot10^{5}$	$5.4636\cdot10^{6}$	6.7	$2.8980\cdot10^{4}$	$5.8746\cdot10^{6}$
3.3	$3.2903\cdot10^{5}$	$5.4981\cdot10^{6}$	6.8	$2.7463\cdot10^{4}$	$5.8775\cdot10^{6}$
3.4	$2.9921\cdot10^{5}$	$5.5295\cdot10^{6}$	6.9	$2.6043\cdot10^{4}$	$5.8801\cdot10^{6}$
3.5	$2.7262\cdot10^{5}$	$5.5581\cdot10^{6}$	7.0	$2.4714\cdot10^{4}$	$5.8827\cdot10^{6}$
3.6	$2.4886\cdot10^{5}$	$5.5841\cdot10^{6}$	7.1	$2.3468\cdot10^{4}$	$5.8851\cdot10^{6}$
3.7	$2.2759\cdot10^{5}$	$5.6079\cdot10^{6}$	7.2	$2.2299\cdot10^{4}$	$5.8874\cdot10^{6}$
3.8	$2.0851\cdot10^{5}$	$5.6297\cdot10^{6}$	7.3	$2.1202\cdot10^{4}$	$5.8895\cdot10^{6}$
3.9	$1.9136\cdot10^{5}$	$5.6497\cdot10^{6}$	7.4	$2.0171\cdot10^{4}$	$5.8916\cdot10^{6}$

Table 2.2. (*continued*)

λ [μm]	$m_{\lambda,bb}$ [W/μm m²]	$M_{T,o-\lambda,bb}$ [W/m²]	λ [μm]	$m_{\lambda,bb}$ [W/μm m²]	$M_{T,o-\lambda,bb}$ [W/m²]
7.5	$1.9202 \cdot 10^4$	$5.8936 \cdot 10^6$	11.5	$3.8886 \cdot 10^3$	$5.9300 \cdot 10^6$
7.6	$1.8290 \cdot 10^4$	$5.8955 \cdot 10^6$	11.6	$3.7630 \cdot 10^3$	$5.9304 \cdot 10^6$
7.7	$1.7431 \cdot 10^4$	$5.8972 \cdot 10^6$	11.7	$3.6424 \cdot 10^3$	$5.9308 \cdot 10^6$
7.8	$1.6623 \cdot 10^4$	$5.8989 \cdot 10^6$	11.8	$3.5266 \cdot 10^3$	$5.9311 \cdot 10^6$
7.9	$1.5860 \cdot 10^4$	$5.9006 \cdot 10^6$	11.9	$3.4153 \cdot 10^3$	$5.9315 \cdot 10^6$
8.0	$1.5141 \cdot 10^4$	$5.9021 \cdot 10^6$	12.0	$3.3084 \cdot 10^3$	$5.9318 \cdot 10^6$
8.1	$1.4461 \cdot 10^4$	$5.9036 \cdot 10^6$	12.1	$3.2057 \cdot 10^3$	$5.9321 \cdot 10^6$
8.2	$1.3820 \cdot 10^4$	$5.9050 \cdot 10^6$	12.2	$3.1069 \cdot 10^3$	$5.9324 \cdot 10^6$
8.3	$1.3213 \cdot 10^4$	$5.9064 \cdot 10^6$	12.3	$3.0119 \cdot 10^3$	$5.9327 \cdot 10^6$
8.4	$1.2639 \cdot 10^4$	$5.9076 \cdot 10^6$	12.4	$2.9204 \cdot 10^3$	$5.9330 \cdot 10^6$
8.5	$1.2096 \cdot 10^4$	$5.9089 \cdot 10^6$	12.5	$2.8324 \cdot 10^3$	$5.9333 \cdot 10^6$
8.6	$1.1582 \cdot 10^4$	$5.9101 \cdot 10^6$	12.6	$2.7477 \cdot 10^3$	$5.9336 \cdot 10^6$
8.7	$1.1095 \cdot 10^4$	$5.9112 \cdot 10^6$	12.7	$2.6662 \cdot 10^3$	$5.9339 \cdot 10^6$
8.8	$1.0633 \cdot 10^4$	$5.9123 \cdot 10^6$	12.8	$2.5876 \cdot 10^3$	$5.9341 \cdot 10^6$
8.9	$1.0195 \cdot 10^4$	$5.9133 \cdot 10^6$	12.9	$2.5119 \cdot 10^3$	$5.9344 \cdot 10^6$
9.0	$9.7788 \cdot 10^3$	$5.9143 \cdot 10^6$	13.0	$2.4390 \cdot 10^3$	$5.9346 \cdot 10^6$
9.1	$9.3839 \cdot 10^3$	$5.9153 \cdot 10^6$	13.1	$2.3687 \cdot 10^3$	$5.9349 \cdot 10^6$
9.2	$9.0086 \cdot 10^3$	$5.9162 \cdot 10^6$	13.2	$2.3009 \cdot 10^3$	$5.9351 \cdot 10^6$
9.3	$8.6519 \cdot 10^3$	$5.9171 \cdot 10^6$	13.3	$2.2354 \cdot 10^3$	$5.9353 \cdot 10^6$
9.4	$8.3126 \cdot 10^3$	$5.9179 \cdot 10^6$	13.4	$2.1724 \cdot 10^3$	$5.9356 \cdot 10^6$
9.5	$7.9898 \cdot 10^3$	$5.9187 \cdot 10^6$	13.5	$2.1115 \cdot 10^3$	$5.9358 \cdot 10^6$
9.6	$7.6824 \cdot 10^3$	$5.9195 \cdot 10^6$	13.6	$2.0527 \cdot 10^3$	$5.9360 \cdot 10^6$
9.7	$7.3897 \cdot 10^3$	$5.9203 \cdot 10^6$	13.7	$1.9960 \cdot 10^3$	$5.9362 \cdot 10^6$
9.8	$7.1107 \cdot 10^3$	$5.9210 \cdot 10^6$	13.8	$1.9412 \cdot 10^3$	$5.9364 \cdot 10^6$
9.9	$6.8448 \cdot 10^3$	$5.9217 \cdot 10^6$	13.9	$1.8882 \cdot 10^3$	$5.9366 \cdot 10^6$
10.0	$6.5911 \cdot 10^3$	$5.9224 \cdot 10^6$	14.0	$1.8371 \cdot 10^3$	$5.9368 \cdot 10^6$
10.1	$6.3491 \cdot 10^3$	$5.9230 \cdot 10^6$	14.1	$1.7877 \cdot 10^3$	$5.9369 \cdot 10^6$
10.2	$6.1181 \cdot 10^3$	$5.9236 \cdot 10^6$	14.2	$1.7399 \cdot 10^3$	$5.9371 \cdot 10^6$
10.3	$5.8975 \cdot 10^3$	$5.9242 \cdot 10^6$	14.3	$1.6937 \cdot 10^3$	$5.9373 \cdot 10^6$
10.4	$5.6867 \cdot 10^3$	$5.9248 \cdot 10^6$	14.4	$1.6490 \cdot 10^3$	$5.9374 \cdot 10^6$
10.5	$5.4852 \cdot 10^3$	$5.9254 \cdot 10^6$	14.5	$1.6058 \cdot 10^3$	$5.9376 \cdot 10^6$
10.6	$5.2926 \cdot 10^3$	$5.9259 \cdot 10^6$	14.6	$1.5640 \cdot 10^3$	$5.9378 \cdot 10^6$
10.7	$5.1083 \cdot 10^3$	$5.9264 \cdot 10^6$	14.7	$1.5236 \cdot 10^3$	$5.9379 \cdot 10^6$
10.8	$4.9319 \cdot 10^3$	$5.9269 \cdot 10^6$	14.8	$1.4844 \cdot 10^3$	$5.9381 \cdot 10^6$
10.9	$4.7631 \cdot 10^3$	$5.9274 \cdot 10^6$	14.9	$1.4465 \cdot 10^3$	$5.9382 \cdot 10^6$
11.0	$4.6015 \cdot 10^3$	$5.9279 \cdot 10^6$	15.0	$1.4098 \cdot 10^3$	$5.9384 \cdot 10^6$
11.1	$4.4466 \cdot 10^3$	$5.9283 \cdot 10^6$	15.1	$1.3743 \cdot 10^3$	$5.9385 \cdot 10^6$
11.2	$4.2982 \cdot 10^3$	$5.9288 \cdot 10^6$	15.2	$1.3399 \cdot 10^3$	$5.9386 \cdot 10^6$
11.3	$4.1560 \cdot 10^3$	$5.9292 \cdot 10^6$	15.3	$1.3065 \cdot 10^3$	$5.9388 \cdot 10^6$
11.4	$4.0195 \cdot 10^3$	$5.9296 \cdot 10^6$	15.4	$1.2742 \cdot 10^3$	$5.9389 \cdot 10^6$

Table 2.2. (*continued*)

temperature 3300 K

λ [μm]	$m_{\lambda,bb}$ [W/μm m^2]	$M_{T,o-\lambda,bb}$ [W/m^2]	λ [μm]	$m_{\lambda,bb}$ [W/μm m^2]	$M_{T,o-\lambda,bb}$ [W/m^2]
0.4	$6.7457\cdot10^5$	$3.2933\cdot10^4$			
0.5	$1.9557\cdot10^6$	$1.6057\cdot10^5$	4.0	$1.8509\cdot10^5$	$6.4341\cdot10^6$
0.6	$3.3635\cdot10^6$	$4.2810\cdot10^5$	4.1	$1.7033\cdot10^5$	$6.4518\cdot10^6$
0.7	$4.3999\cdot10^6$	$8.2043\cdot10^5$	4.2	$1.5699\cdot10^5$	$6.4682\cdot10^6$
0.8	$4.9275\cdot10^6$	$1.2909\cdot10^5$	4.3	$1.4492\cdot10^5$	$6.4833\cdot10^6$
0.9	$5.0282\cdot10^6$	$1.7916\cdot10^5$	4.4	$1.3397\cdot10^5$	$6.4972\cdot10^6$
1.0	$4.8436\cdot10^6$	$2.2870\cdot10^6$	4.5	$1.2403\cdot10^5$	$6.5101\cdot10^6$
1.1	$4.4986\cdot10^6$	$2.7550\cdot10^6$	4.6	$1.1498\cdot10^5$	$6.5220\cdot10^6$
1.2	$4.0822\cdot10^6$	$3.1844\cdot10^6$	4.7	$1.0674\cdot10^5$	$6.5331\cdot10^6$
1.3	$3.6499\cdot10^6$	$3.5710\cdot10^6$	4.8	$9.9215\cdot10^4$	$6.5434\cdot10^6$
1.4	$3.2335\cdot10^6$	$3.9149\cdot10^6$	4.9	$9.2337\cdot10^4$	$6.5530\cdot10^6$
1.5	$2.8491\cdot10^6$	$4.2187\cdot10^6$	5.0	$8.6041\cdot10^4$	$6.5619\cdot10^6$
1.6	$2.5032\cdot10^6$	$4.4860\cdot10^6$	5.1	$8.0269\cdot10^4$	$6.5702\cdot10^6$
1.7	$2.1968\cdot10^6$	$4.7207\cdot10^6$	5.2	$7.4969\cdot10^4$	$6.5780\cdot10^6$
1.8	$1.9281\cdot10^6$	$4.9266\cdot10^6$	5.3	$7.0096\cdot10^4$	$6.5852\cdot10^6$
1.9	$1.6939\cdot10^6$	$5.1074\cdot10^6$	5.4	$6.5611\cdot10^4$	$6.5920\cdot10^6$
2.0	$1.4903\cdot10^6$	$5.2664\cdot10^6$	5.5	$6.1477\cdot10^4$	$6.5983\cdot10^6$
2.1	$1.3138\cdot10^6$	$5.4064\cdot10^6$	5.6	$5.7661\cdot10^4$	$6.6043\cdot10^6$
2.2	$1.1606\cdot10^6$	$5.5299\cdot10^6$	5.7	$5.4135\cdot10^4$	$6.6099\cdot10^6$
2.3	$1.0277\cdot10^6$	$5.6392\cdot10^6$	5.8	$5.0873\cdot10^4$	$6.6151\cdot10^6$
2.4	$9.1227\cdot10^5$	$5.7361\cdot10^6$	5.9	$4.7852\cdot10^4$	$6.6201\cdot10^6$
2.5	$8.1178\cdot10^5$	$5.8222\cdot10^6$	6.0	$4.5050\cdot10^4$	$6.6247\cdot10^6$
2.6	$7.2416\cdot10^5$	$5.8989\cdot10^6$	6.1	$4.2450\cdot10^4$	$6.6291\cdot10^6$
2.7	$6.4759\cdot10^5$	$5.9674\cdot10^6$	6.2	$4.0034\cdot10^4$	$6.6332\cdot10^6$
2.8	$5.8053\cdot10^5$	$6.0287\cdot10^6$	6.3	$3.7787\cdot10^4$	$6.6371\cdot10^6$
2.9	$5.2166\cdot10^5$	$6.0837\cdot10^6$	6.4	$3.5694\cdot10^4$	$6.6408\cdot10^6$
3.0	$4.6986\cdot10^5$	$6.1333\cdot10^6$	6.5	$3.3744\cdot10^4$	$6.6442\cdot10^6$
3.1	$4.2416\cdot10^5$	$6.1779\cdot10^6$	6.6	$3.1924\cdot10^4$	$6.6475\cdot10^6$
3.2	$3.8376\cdot10^5$	$6.2183\cdot10^6$	6.7	$3.0225\cdot10^4$	$6.6506\cdot10^6$
3.3	$3.4795\cdot10^5$	$6.2548\cdot10^6$	6.8	$2.8637\cdot10^4$	$6.6536\cdot10^6$
3.4	$3.1614\cdot10^5$	$6.2880\cdot10^6$	6.9	$2.7152\cdot10^4$	$6.6563\cdot10^6$
3.5	$2.8781\cdot10^5$	$6.3182\cdot10^6$	7.0	$2.5761\cdot10^4$	$6.6590\cdot10^6$
3.6	$2.6253\cdot10^5$	$6.3456\cdot10^6$	7.1	$2.4458\cdot10^4$	$6.6615\cdot10^6$
3.7	$2.3993\cdot10^5$	$6.3707\cdot10^6$	7.2	$2.3237\cdot10^4$	$6.6639\cdot10^6$
3.8	$2.1967\cdot10^5$	$6.3937\cdot10^6$	7.3	$2.2090\cdot10^4$	$6.6661\cdot10^6$
3.9	$2.0147\cdot10^5$	$6.4147\cdot10^6$	7.4	$2.1013\cdot10^4$	$6.6683\cdot10^6$

Table 2.2. (*continued*)

λ [µm]	$m_{\lambda,bb}$ [W/µm m²]	$M_{T,o-\lambda,bb}$ [W/m²]	λ [µm]	$m_{\lambda,bb}$ [W/µm m²]	$M_{T,o-\lambda,bb}$ [W/m²]
7.5	$2.0000 \cdot 10^4$	$6.6704 \cdot 10^6$	11.5	$4.0355 \cdot 10^3$	$6.7082 \cdot 10^6$
7.6	$1.9048 \cdot 10^4$	$6.6723 \cdot 10^6$	11.6	$3.9049 \cdot 10^3$	$6.7086 \cdot 10^6$
7.7	$1.8151 \cdot 10^4$	$6.6742 \cdot 10^6$	11.7	$3.7795 \cdot 10^3$	$6.7090 \cdot 10^6$
7.8	$1.7307 \cdot 10^4$	$6.6759 \cdot 10^6$	11.8	$3.6592 \cdot 10^3$	$6.7094 \cdot 10^6$
7.9	$1.6511 \cdot 10^4$	$6.6776 \cdot 10^6$	11.9	$3.5435 \cdot 10^3$	$6.7097 \cdot 10^6$
8.0	$1.5760 \cdot 10^4$	$6.6792 \cdot 10^6$	12.0	$3.4324 \cdot 10^3$	$6.7101 \cdot 10^6$
8.1	$1.5051 \cdot 10^4$	$6.6808 \cdot 10^6$	12.1	$3.3257 \cdot 10^3$	$6.7104 \cdot 10^6$
8.2	$1.4381 \cdot 10^4$	$6.6822 \cdot 10^6$	12.2	$3.2230 \cdot 10^3$	$6.7108 \cdot 10^6$
8.3	$1.3748 \cdot 10^4$	$6.6837 \cdot 10^6$	12.3	$3.1243 \cdot 10^3$	$6.7111 \cdot 10^6$
8.4	$1.3150 \cdot 10^4$	$6.6850 \cdot 10^6$	12.4	$3.0293 \cdot 10^3$	$6.7114 \cdot 10^6$
8.5	$1.2583 \cdot 10^4$	$6.6863 \cdot 10^6$	12.5	$2.9379 \cdot 10^3$	$6.7117 \cdot 10^6$
8.6	$1.2047 \cdot 10^4$	$6.6875 \cdot 10^6$	12.6	$2.8499 \cdot 10^3$	$6.7120 \cdot 10^6$
8.7	$1.1539 \cdot 10^4$	$6.6887 \cdot 10^6$	12.7	$2.7651 \cdot 10^3$	$6.7122 \cdot 10^6$
8.8	$1.1058 \cdot 10^4$	$6.6898 \cdot 10^6$	12.8	$2.6835 \cdot 10^3$	$6.7125 \cdot 10^6$
8.9	$1.0601 \cdot 10^4$	$6.6909 \cdot 10^6$	12.9	$2.6049 \cdot 10^3$	$6.7128 \cdot 10^6$
9.0	$1.0167 \cdot 10^4$	$6.6919 \cdot 10^6$	13.0	$2.5292 \cdot 10^3$	$6.7130 \cdot 10^6$
9.1	$9.7557 \cdot 10^3$	$6.6929 \cdot 10^6$	13.1	$2.4561 \cdot 10^3$	$6.7133 \cdot 10^6$
9.2	$9.3647 \cdot 10^3$	$6.6939 \cdot 10^6$	13.2	$2.3857 \cdot 10^3$	$6.7135 \cdot 10^6$
9.3	$8.9931 \cdot 10^3$	$6.6948 \cdot 10^6$	13.3	$2.3178 \cdot 10^3$	$6.7138 \cdot 10^6$
9.4	$8.6396 \cdot 10^3$	$6.6957 \cdot 10^6$	13.4	$2.2523 \cdot 10^3$	$6.7140 \cdot 10^6$
9.5	$8.3033 \cdot 10^3$	$6.6965 \cdot 10^6$	13.5	$2.1891 \cdot 10^3$	$6.7142 \cdot 10^6$
9.6	$7.9833 \cdot 10^3$	$6.6974 \cdot 10^6$	13.6	$2.1281 \cdot 10^3$	$6.7144 \cdot 10^6$
9.7	$7.6784 \cdot 10^3$	$6.6981 \cdot 10^6$	13.7	$2.0691 \cdot 10^3$	$6.7146 \cdot 10^6$
9.8	$7.3879 \cdot 10^3$	$6.6989 \cdot 10^6$	13.8	$2.0123 \cdot 10^3$	$6.7148 \cdot 10^6$
9.9	$7.1111 \cdot 10^3$	$6.6996 \cdot 10^6$	13.9	$1.9573 \cdot 10^3$	$6.7150 \cdot 10^6$
10.0	$6.8470 \cdot 10^3$	$6.7003 \cdot 10^6$	14.0	$1.9042 \cdot 10^3$	$6.7152 \cdot 10^6$
10.1	$6.5951 \cdot 10^3$	$6.7010 \cdot 10^6$	14.1	$1.8529 \cdot 10^3$	$6.7154 \cdot 10^6$
10.2	$6.3546 \cdot 10^3$	$6.7016 \cdot 10^6$	14.2	$1.8033 \cdot 10^3$	$6.7156 \cdot 10^6$
10.3	$6.1250 \cdot 10^3$	$6.7022 \cdot 10^6$	14.3	$1.7554 \cdot 10^3$	$6.7158 \cdot 10^6$
10.4	$5.9057 \cdot 10^3$	$6.7029 \cdot 10^6$	14.4	$1.7091 \cdot 10^3$	$6.7160 \cdot 10^6$
10.5	$5.6960 \cdot 10^3$	$6.7034 \cdot 10^6$	14.5	$1.6642 \cdot 10^3$	$6.7161 \cdot 10^6$
10.6	$5.4956 \cdot 10^3$	$6.7040 \cdot 10^6$	14.6	$1.6208 \cdot 10^3$	$6.7163 \cdot 10^6$
10.7	$5.3039 \cdot 10^3$	$6.7045 \cdot 10^6$	14.7	$1.5789 \cdot 10^3$	$6.7164 \cdot 10^6$
10.8	$5.1205 \cdot 10^3$	$6.7050 \cdot 10^6$	14.8	$1.5382 \cdot 10^3$	$6.7166 \cdot 10^6$
10.9	$4.9449 \cdot 10^3$	$6.7056 \cdot 10^6$	14.9	$1.4989 \cdot 10^3$	$6.7167 \cdot 10^6$
11.0	$4.7768 \cdot 10^3$	$6.7060 \cdot 10^6$	15.0	$1.4608 \cdot 10^3$	$6.7169 \cdot 10^6$
11.1	$4.6157 \cdot 10^3$	$6.7065 \cdot 10^6$	15.1	$1.4240 \cdot 10^3$	$6.7170 \cdot 10^6$
11.2	$4.4614 \cdot 10^3$	$6.7070 \cdot 10^6$	15.2	$1.3882 \cdot 10^3$	$6.7172 \cdot 10^6$
11.3	$4.3134 \cdot 10^3$	$6.7074 \cdot 10^6$	15.3	$1.3536 \cdot 10^3$	$6.7173 \cdot 10^6$
11.4	$4.1716 \cdot 10^3$	$6.7078 \cdot 10^6$	15.4	$1.3201 \cdot 10^3$	$6.7175 \cdot 10^6$

Table 2.2. (*continued*)

temperature 3400 K

λ [μm]	$m_{\lambda,bb}$ [W/μm m²]	$M_{T,o-\lambda,bb}$ [W/m²]	λ [μm]	$m_{\lambda,bb}$ [W/μm m²]	$M_{T,o-\lambda,bb}$ [W/m²]
0.4	$9.2952 \cdot 10^5$	$4.7172 \cdot 10^4$			
0.5	$2.5275 \cdot 10^6$	$2.1621 \cdot 10^5$	4.0	$1.9433 \cdot 10^5$	$7.2742 \cdot 10^6$
0.6	$4.1656 \cdot 10^6$	$5.5363 \cdot 10^5$	4.1	$1.7873 \cdot 10^5$	$7.2928 \cdot 10^6$
0.7	$5.2866 \cdot 10^6$	$1.0315 \cdot 10^6$	4.2	$1.6465 \cdot 10^5$	$7.3100 \cdot 10^6$
0.8	$5.7885 \cdot 10^6$	$1.5901 \cdot 10^6$	4.3	$1.5192 \cdot 10^5$	$7.3258 \cdot 10^6$
0.9	$5.8052 \cdot 10^6$	$2.1730 \cdot 10^6$	4.4	$1.4038 \cdot 10^5$	$7.3404 \cdot 10^6$
1.0	$5.5160 \cdot 10^6$	$2.7409 \cdot 10^6$	4.5	$1.2991 \cdot 10^5$	$7.3539 \cdot 10^6$
1.1	$5.0670 \cdot 10^6$	$3.2709 \cdot 10^6$	4.6	$1.2039 \cdot 10^5$	$7.3664 \cdot 10^6$
1.2	$4.5565 \cdot 10^6$	$3.7523 \cdot 10^6$	4.7	$1.1171 \cdot 10^5$	$7.3780 \cdot 10^6$
1.3	$4.0434 \cdot 10^6$	$4.1822 \cdot 10^6$	4.8	$1.0380 \cdot 10^5$	$7.3888 \cdot 10^6$
1.4	$3.5595 \cdot 10^6$	$4.5620 \cdot 10^6$	4.9	$9.6570 \cdot 10^4$	$7.3988 \cdot 10^6$
1.5	$3.1195 \cdot 10^6$	$4.8956 \cdot 10^6$	5.0	$8.9955 \cdot 10^4$	$7.4081 \cdot 10^6$
1.6	$2.7280 \cdot 10^6$	$5.1875 \cdot 10^6$	5.1	$8.3894 \cdot 10^4$	$7.4168 \cdot 10^6$
1.7	$2.3845 \cdot 10^6$	$5.4427 \cdot 10^6$	5.2	$7.8330 \cdot 10^4$	$7.4249 \cdot 10^6$
1.8	$2.0855 \cdot 10^6$	$5.6659 \cdot 10^6$	5.3	$7.3218 \cdot 10^4$	$7.4325 \cdot 10^6$
1.9	$1.8264 \cdot 10^6$	$5.8612 \cdot 10^6$	5.4	$6.8513 \cdot 10^4$	$7.4396 \cdot 10^6$
2.0	$1.6025 \cdot 10^6$	$6.0323 \cdot 10^6$	5.5	$6.4179 \cdot 10^4$	$7.4462 \cdot 10^6$
2.1	$1.4092 \cdot 10^6$	$6.1827 \cdot 10^6$	5.6	$6.0179 \cdot 10^4$	$7.4524 \cdot 10^6$
2.2	$1.2422 \cdot 10^6$	$6.3151 \cdot 10^6$	5.7	$5.6485 \cdot 10^4$	$7.4582 \cdot 10^6$
2.3	$1.0978 \cdot 10^6$	$6.4319 \cdot 10^6$	5.8	$5.3069 \cdot 10^4$	$7.4637 \cdot 10^6$
2.4	$9.7270 \cdot 10^5$	$6.5353 \cdot 10^6$	5.9	$4.9906 \cdot 10^4$	$7.4689 \cdot 10^6$
2.5	$8.6414 \cdot 10^5$	$6.6270 \cdot 10^6$	6.0	$4.6974 \cdot 10^4$	$7.4737 \cdot 10^6$
2.6	$7.6972 \cdot 10^5$	$6.7086 \cdot 10^6$	6.1	$4.4253 \cdot 10^4$	$7.4783 \cdot 10^6$
2.7	$6.8739 \cdot 10^5$	$6.7813 \cdot 10^6$	6.2	$4.1725 \cdot 10^4$	$7.4825 \cdot 10^6$
2.8	$6.1544 \cdot 10^5$	$6.8464 \cdot 10^6$	6.3	$3.9375 \cdot 10^4$	$7.4866 \cdot 10^6$
2.9	$5.5239 \cdot 10^5$	$6.9047 \cdot 10^6$	6.4	$3.7187 \cdot 10^4$	$7.4904 \cdot 10^6$
3.0	$4.9700 \cdot 10^5$	$6.9571 \cdot 10^6$	6.5	$3.5149 \cdot 10^4$	$7.4940 \cdot 10^6$
3.1	$4.4822 \cdot 10^5$	$7.0043 \cdot 10^6$	6.6	$3.3247 \cdot 10^4$	$7.4975 \cdot 10^6$
3.2	$4.0515 \cdot 10^5$	$7.0469 \cdot 10^6$	6.7	$3.1472 \cdot 10^4$	$7.5007 \cdot 10^6$
3.3	$3.6703 \cdot 10^5$	$7.0855 \cdot 10^6$	6.8	$2.9814 \cdot 10^4$	$7.5038 \cdot 10^6$
3.4	$3.3321 \cdot 10^5$	$7.1205 \cdot 10^6$	6.9	$2.8263 \cdot 10^4$	$7.5067 \cdot 10^6$
3.5	$3.0312\ 10^5$	$7.1523\ 10^6$	7.0	$2.6811\ 10^4$	$7.5094\ 10^6$
3.6	$2.7631\ 10^5$	$7.1812\ 10^6$	7.1	$2.5451\ 10^4$	$7.5120\ 10^6$
3.7	$2.5235\ 10^5$	$7.2076\ 10^6$	7.2	$2.4176\ 10^4$	$7.5145\ 10^6$
3.8	$2.3089\ 10^5$	$7.2318\ 10^6$	7.3	$2.2980\ 10^4$	$7.5169\ 10^6$
3.9	$2.1164\ 10^5$	$7.2539\ 10^6$	7.4	$2.1856\ 10^4$	$7.5191\ 10^6$

Table 2.2. (*continued*)

λ [µm]	$m_{\lambda,bb}$ [W/µm m²]	$M_{T,o-\lambda,bb}$ [W/m²]	λ [µm]	$m_{\lambda,bb}$ [W/µm m²]	$M_{T,o-\lambda,bb}$ [W/m²]
7.5	$2.0800 \cdot 10^4$	$7.5212 \cdot 10^6$	11.5	$4.1825 \cdot 10^3$	$7.5606 \cdot 10^6$
7.6	$1.9807 \cdot 10^4$	$7.5233 \cdot 10^6$	11.6	$4.0469 \cdot 10^3$	$7.5610 \cdot 10^6$
7.7	$1.8872 \cdot 10^4$	$7.5252 \cdot 10^6$	11.7	$3.9168 \cdot 10^3$	$7.5614 \cdot 10^6$
7.8	$1.7992 \cdot 10^4$	$7.5270 \cdot 10^6$	11.8	$3.7918 \cdot 10^3$	$7.5618 \cdot 10^6$
7.9	$1.7162 \cdot 10^4$	$7.5288 \cdot 10^6$	11.9	$3.6718 \cdot 10^3$	$7.5621 \cdot 10^6$
8.0	$1.6379 \cdot 10^4$	$7.5305 \cdot 10^6$	12.0	$3.5565 \cdot 10^3$	$7.5625 \cdot 10^6$
8.1	$1.5641 \cdot 10^4$	$7.5321 \cdot 10^6$	12.1	$3.4457 \cdot 10^3$	$7.5628 \cdot 10^6$
8.2	$1.4943 \cdot 10^4$	$7.5336 \cdot 10^6$	12.2	$3.3392 \cdot 10^3$	$7.5632 \cdot 10^6$
8.3	$1.4284 \cdot 10^4$	$7.5351 \cdot 10^6$	12.3	$3.2367 \cdot 10^3$	$7.5635 \cdot 10^6$
8.4	$1.3661 \cdot 10^4$	$7.5365 \cdot 10^6$	12.4	$3.1382 \cdot 10^3$	$7.5638 \cdot 10^6$
8.5	$1.3071 \cdot 10^4$	$7.5378 \cdot 10^6$	12.5	$3.0433 \cdot 10^3$	$7.5641 \cdot 10^6$
8.6	$1.2513 \cdot 10^4$	$7.5391 \cdot 10^6$	12.6	$2.9521 \cdot 10^3$	$7.5644 \cdot 10^6$
8.7	$1.1984 \cdot 10^4$	$7.5403 \cdot 10^6$	12.7	$2.8642 \cdot 10^3$	$7.5647 \cdot 10^6$
8.8	$1.1483 \cdot 10^4$	$7.5415 \cdot 10^6$	12.8	$2.7795 \cdot 10^3$	$7.5650 \cdot 10^6$
8.9	$1.1007 \cdot 10^4$	$7.5426 \cdot 10^6$	12.9	$2.6980 \cdot 10^3$	$7.5653 \cdot 10^6$
9.0	$1.0556 \cdot 10^4$	$7.5437 \cdot 10^6$	13.0	$2.6194 \cdot 10^3$	$7.5655 \cdot 10^6$
9.1	$1.0128 \cdot 10^4$	$7.5447 \cdot 10^6$	13.1	$2.5437 \cdot 10^3$	$7.5658 \cdot 10^6$
9.2	$9.7212 \cdot 10^3$	$7.5457 \cdot 10^6$	13.2	$2.4706 \cdot 10^3$	$7.5661 \cdot 10^6$
9.3	$9.3346 \cdot 10^3$	$7.5466 \cdot 10^6$	13.3	$2.4002 \cdot 10^3$	$7.5663 \cdot 10^6$
9.4	$8.9670 \cdot 10^3$	$7.5476 \cdot 10^6$	13.4	$2.3323 \cdot 10^3$	$7.5665 \cdot 10^6$
9.5	$8.6173 \cdot 10^3$	$7.5484 \cdot 10^6$	13.5	$2.2667 \cdot 10^3$	$7.5668 \cdot 10^6$
9.6	$8.2844 \cdot 10^3$	$7.5493 \cdot 10^6$	13.6	$2.2034 \cdot 10^3$	$7.5670 \cdot 10^6$
9.7	$7.9674 \cdot 10^3$	$7.5501 \cdot 10^6$	13.7	$2.1424 \cdot 10^3$	$7.5672 \cdot 10^6$
9.8	$7.6654 \cdot 10^3$	$7.5509 \cdot 10^6$	13.8	$2.0834 \cdot 10^3$	$7.5674 \cdot 10^6$
9.9	$7.3776 \cdot 10^3$	$7.5516 \cdot 10^6$	13.9	$2.0264 \cdot 10^3$	$7.5676 \cdot 10^6$
10.0	$7.1031 \cdot 10^3$	$7.5524 \cdot 10^6$	14.0	$1.9714 \cdot 10^3$	$7.5678 \cdot 10^6$
10.1	$6.8413 \cdot 10^3$	$7.5530 \cdot 10^6$	14.1	$1.9182 \cdot 10^3$	$7.5680 \cdot 10^6$
10.2	$6.5914 \cdot 10^3$	$7.5537 \cdot 10^6$	14.2	$1.8668 \cdot 10^3$	$7.5682 \cdot 10^6$
10.3	$6.3528 \cdot 10^3$	$7.5544 \cdot 10^6$	14.3	$1.8171 \cdot 10^3$	$7.5684 \cdot 10^6$
10.4	$6.1248 \cdot 10^3$	$7.5550 \cdot 10^6$	14.4	$1.7691 \cdot 10^3$	$7.5686 \cdot 10^6$
10.5	$5.9070 \cdot 10^3$	$7.5556 \cdot 10^6$	14.5	$1.7226 \cdot 10^3$	$7.5687 \cdot 10^6$
10.6	$5.6988 \cdot 10^3$	$7.5562 \cdot 10^6$	14.6	$1.6777 \cdot 10^3$	$7.5689 \cdot 10^6$
10.7	$5.4996 \cdot 10^3$	$7.5567 \cdot 10^6$	14.7	$1.6342 \cdot 10^3$	$7.5691 \cdot 10^6$
10.8	$5.3091 \cdot 10^3$	$7.5573 \cdot 10^6$	14.8	$1.5921 \cdot 10^3$	$7.5692 \cdot 10^6$
10.9	$5.1268 \cdot 10^3$	$7.5578 \cdot 10^6$	14.9	$1.5513 \cdot 10^3$	$7.5694 \cdot 10^6$
11.0	$4.9521 \cdot 10^3$	$7.5583 \cdot 10^6$	15.0	$1.5119 \cdot 10^3$	$7.5695 \cdot 10^6$
11.1	$4.7849 \cdot 10^3$	$7.5588 \cdot 10^6$	15.1	$1.4736 \cdot 10^3$	$7.5697 \cdot 10^6$
11.2	$4.6246 \cdot 10^3$	$7.5592 \cdot 10^6$	15.2	$1.4366 \cdot 10^3$	$7.5698 \cdot 10^6$
11.3	$4.4710 \cdot 10^3$	$7.5597 \cdot 10^6$	15.3	$1.4008 \cdot 10^3$	$7.5700 \cdot 10^6$
11.4	$4.3237 \cdot 10^3$	$7.5601 \cdot 10^6$	15.4	$1.3660 \cdot 10^3$	$7.5701 \cdot 10^6$

Table 2. 2. (*continued*)

temperature 3500 K

λ [μm]	$m_{\lambda,bb}$ [W/μm m²]	$M_{T,o-\lambda,bb}$ [W/m²]	λ [μm]	$m_{\lambda,bb}$ [W/μm m²]	$M_{T,o-\lambda,bb}$ [W/m²]
0.4	$1.2576\cdot10^{6}$	$6.6285\cdot10^{4}$			
0.5	$3.2191\cdot10^{6}$	$2.8665\cdot10^{5}$	4.0	$2.0362\cdot10^{5}$	$8.1929\cdot10^{6}$
0.6	$5.0966\cdot10^{6}$	$7.0668\cdot10^{5}$	4.1	$1.8718\cdot10^{5}$	$8.2124\cdot10^{6}$
0.7	$6.2861\cdot10^{6}$	$1.2824\cdot10^{6}$	4.2	$1.7236\cdot10^{5}$	$8.2304\cdot10^{6}$
0.8	$6.7384\cdot10^{6}$	$1.9391\cdot10^{6}$	4.3	$1.5896\cdot10^{5}$	$8.2469\cdot10^{6}$
0.9	$6.6486\cdot10^{6}$	$2.6120\cdot10^{6}$	4.4	$1.4682\cdot10^{5}$	$8.2622\cdot10^{6}$
1.0	$6.2367\cdot10^{6}$	$3.2581\cdot10^{6}$	4.5	$1.3582\cdot10^{5}$	$8.2763\cdot10^{6}$
1.1	$5.6701\cdot10^{6}$	$3.8542\cdot10^{6}$	4.6	$1.2581\cdot10^{5}$	$8.2894\cdot10^{6}$
1.2	$5.0557\cdot10^{6}$	$4.3906\cdot10^{6}$	4.7	$1.1671\cdot10^{5}$	$8.3015\cdot10^{6}$
1.3	$4.4549\cdot10^{6}$	$4.8659\cdot10^{6}$	4.8	$1.0840\cdot10^{5}$	$8.3128\cdot10^{6}$
1.4	$3.8986\cdot10^{6}$	$5.2831\cdot10^{6}$	4.9	$1.0082\cdot10^{5}$	$8.3232\cdot10^{6}$
1.5	$3.3994\cdot10^{6}$	$5.6475\cdot10^{6}$	5.0	$9.3883\cdot10^{4}$	$8.3330\cdot10^{6}$
1.6	$2.9599\cdot10^{6}$	$5.9650\cdot10^{6}$	5.1	$8.7530\cdot10^{4}$	$8.3420\cdot10^{6}$
1.7	$2.5774\cdot10^{6}$	$6.2414\cdot10^{6}$	5.2	$8.1703\cdot10^{4}$	$8.3505\cdot10^{6}$
1.8	$2.2468\cdot10^{6}$	$6.4822\cdot10^{6}$	5.3	$7.6349\cdot10^{4}$	$8.3584\cdot10^{6}$
1.9	$1.9620\cdot10^{6}$	$6.6922\cdot10^{6}$	5.4	$7.1424\cdot10^{4}$	$8.3658\cdot10^{6}$
2.0	$1.7171\cdot10^{6}$	$6.8759\cdot10^{6}$	5.5	$6.6888\cdot10^{4}$	$8.3727\cdot10^{6}$
2.1	$1.5065\cdot10^{6}$	$7.0368\cdot10^{6}$	5.6	$6.2705\cdot10^{4}$	$8.3792\cdot10^{6}$
2.2	$1.3252\cdot10^{6}$	$7.1782\cdot10^{6}$	5.7	$5.8842\cdot10^{4}$	$8.3852\cdot10^{6}$
2.3	$1.1690\cdot10^{6}$	$7.3027\cdot10^{6}$	5.8	$5.5271\cdot10^{4}$	$8.3909\cdot10^{6}$
2.4	$1.0340\cdot10^{6}$	$7.4127\cdot10^{6}$	5.9	$5.1965\cdot10^{4}$	$8.3963\cdot10^{6}$
2.5	$9.1721\cdot10^{5}$	$7.5101\cdot10^{6}$	6.0	$4.8901\cdot10^{4}$	$8.4013\cdot10^{6}$
2.6	$8.1585\cdot10^{5}$	$7.5966\cdot10^{6}$	6.1	$4.6060\cdot10^{4}$	$8.4061\cdot10^{6}$
2.7	$7.2766\cdot10^{5}$	$7.6737\cdot10^{6}$	6.2	$4.3421\cdot10^{4}$	$8.4106\cdot10^{6}$
2.8	$6.5073\cdot10^{5}$	$7.7425\cdot10^{6}$	6.3	$4.0967\cdot10^{4}$	$8.4148\cdot10^{6}$
2.9	$5.8343\cdot10^{5}$	$7.8041\cdot10^{6}$	6.4	$3.8684\cdot10^{4}$	$8.4188\cdot10^{6}$
3.0	$5.2440\cdot10^{5}$	$7.8595\cdot10^{6}$	6.5	$3.6557\cdot10^{4}$	$8.4225\cdot10^{6}$
3.1	$4.7249\cdot10^{5}$	$7.9093\cdot10^{6}$	6.6	$3.4573\cdot10^{4}$	$8.4261\cdot10^{6}$
3.2	$4.2672\cdot10^{5}$	$7.9542\cdot10^{6}$	6.7	$3.2722\cdot10^{4}$	$8.4294\cdot10^{6}$
3.3	$3.8626\cdot10^{5}$	$7.9948\cdot10^{6}$	6.8	$3.0993\cdot10^{4}$	$8.4326\cdot10^{6}$
3.4	$3.5040\cdot10^{5}$	$8.0316\cdot10^{6}$	6.9	$2.9376\cdot10^{4}$	$8.4356\cdot10^{6}$
3.5	$3.1854\cdot10^{5}$	$8.0650\cdot10^{6}$	7.0	$2.7863\cdot10^{4}$	$8.4385\cdot10^{6}$
3.6	$2.9017\cdot10^{5}$	$8.0954\cdot10^{6}$	7.1	$2.6446\cdot10^{4}$	$8.4412\cdot10^{6}$
3.7	$2.6484\cdot10^{5}$	$8.1231\cdot10^{6}$	7.2	$2.5117\cdot10^{4}$	$8.4438\cdot10^{6}$
3.8	$2.4218\cdot10^{5}$	$8.1485\cdot10^{6}$	7.3	$2.3871\cdot10^{4}$	$8.4462\cdot10^{6}$
3.9	$2.2187\cdot10^{5}$	$8.1716\cdot10^{6}$	7.4	$2.2701\cdot10^{4}$	$8.4486\cdot10^{6}$

Table 2.2. (*continued*)

λ [μm]	$m_{\lambda,bb}$ [W/μm m²]	$M_{T,o-\lambda,bb}$ [W/m²]	λ [μm]	$m_{\lambda,bb}$ [W/μm m²]	$M_{T,o-\lambda,bb}$ [W/m²]
7.5	$2.1601 \cdot 10^4$	$8.4508 \cdot 10^6$	11.5	$4.3295 \cdot 10^3$	$8.4916 \cdot 10^6$
7.6	$2.0567 \cdot 10^4$	$8.4529 \cdot 10^6$	11.6	$4.1890 \cdot 10^3$	$8.4920 \cdot 10^6$
7.7	$1.9594 \cdot 10^4$	$8.4549 \cdot 10^6$	11.7	$4.0541 \cdot 10^3$	$8.4924 \cdot 10^6$
7.8	$1.8678 \cdot 10^4$	$8.4568 \cdot 10^6$	11.8	$3.9246 \cdot 10^3$	$8.4928 \cdot 10^6$
7.9	$1.7814 \cdot 10^4$	$8.4586 \cdot 10^6$	11.9	$3.8002 \cdot 10^3$	$8.4932 \cdot 10^6$
8.0	$1.7000 \cdot 10^4$	$8.4604 \cdot 10^6$	12.0	$3.6807 \cdot 10^3$	$8.4936 \cdot 10^6$
8.1	$1.6232 \cdot 10^4$	$8.4620 \cdot 10^6$	12.1	$3.5659 \cdot 10^3$	$8.4939 \cdot 10^6$
8.2	$1.5506 \cdot 10^4$	$8.4636 \cdot 10^6$	12.2	$3.4555 \cdot 10^3$	$8.4943 \cdot 10^6$
8.3	$1.4821 \cdot 10^4$	$8.4651 \cdot 10^6$	12.3	$3.3493 \cdot 10^3$	$8.4946 \cdot 10^6$
8.4	$1.4173 \cdot 10^4$	$8.4666 \cdot 10^6$	12.4	$3.2472 \cdot 10^3$	$8.4949 \cdot 10^6$
8.5	$1.3560 \cdot 10^4$	$8.4680 \cdot 10^6$	12.5	$3.1489 \cdot 10^3$	$8.4953 \cdot 10^6$
8.6	$1.2979 \cdot 10^4$	$8.4693 \cdot 10^6$	12.6	$3.0543 \cdot 10^3$	$8.4956 \cdot 10^6$
8.7	$1.2429 \cdot 10^4$	$8.4706 \cdot 10^6$	12.7	$2.9632 \cdot 10^3$	$8.4959 \cdot 10^6$
8.8	$1.1908 \cdot 10^4$	$8.4718 \cdot 10^6$	12.8	$2.8756 \cdot 10^3$	$8.4962 \cdot 10^6$
8.9	$1.1414 \cdot 10^4$	$8.4729 \cdot 10^6$	12.9	$2.7911 \cdot 10^3$	$8.4965 \cdot 10^6$
9.0	$1.0946 \cdot 10^4$	$8.4741 \cdot 10^6$	13.0	$2.7097 \cdot 10^3$	$8.4967 \cdot 10^6$
9.1	$1.0501 \cdot 10^4$	$8.4751 \cdot 10^6$	13.1	$2.6312 \cdot 10^3$	$8.4970 \cdot 10^6$
9.2	$1.0078 \cdot 10^4$	$8.4762 \cdot 10^6$	13.2	$2.5556 \cdot 10^3$	$8.4973 \cdot 10^6$
9.3	$9.6765 \cdot 10^3$	$8.4771 \cdot 10^6$	13.3	$2.4826 \cdot 10^3$	$8.4975 \cdot 10^6$
9.4	$9.2947 \cdot 10^3$	$8.4781 \cdot 10^6$	13.4	$2.4123 \cdot 10^3$	$8.4977 \cdot 10^6$
9.5	$8.9315 \cdot 10^3$	$8.4790 \cdot 10^6$	13.5	$2.3444 \cdot 10^3$	$8.4980 \cdot 10^6$
9.6	$8.5858 \cdot 10^3$	$8.4799 \cdot 10^6$	13.6	$2.2789 \cdot 10^3$	$8.4982 \cdot 10^6$
9.7	$8.2567 \cdot 10^3$	$8.4807 \cdot 10^6$	13.7	$2.2156 \cdot 10^3$	$8.4984 \cdot 10^6$
9.8	$7.9432 \cdot 10^3$	$8.4815 \cdot 10^6$	13.8	$2.1546 \cdot 10^3$	$8.4987 \cdot 10^6$
9.9	$7.6444 \cdot 10^3$	$8.4823 \cdot 10^6$	13.9	$2.0956 \cdot 10^3$	$8.4989 \cdot 10^6$
10.0	$7.3595 \cdot 10^3$	$8.4831 \cdot 10^6$	14.0	$2.0386 \cdot 10^3$	$8.4991 \cdot 10^6$
10.1	$7.0877 \cdot 10^3$	$8.4838 \cdot 10^6$	14.1	$1.9835 \cdot 10^3$	$8.4993 \cdot 10^6$
10.2	$6.8283 \cdot 10^3$	$8.4845 \cdot 10^6$	14.2	$1.9303 \cdot 10^3$	$8.4995 \cdot 10^6$
10.3	$6.5807 \cdot 10^3$	$8.4851 \cdot 10^6$	14.3	$1.8789 \cdot 10^3$	$8.4997 \cdot 10^6$
10.4	$6.3442 \cdot 10^3$	$8.4858 \cdot 10^6$	14.4	$1.8292 \cdot 10^3$	$8.4998 \cdot 10^6$
10.5	$6.1182 \cdot 10^3$	$8.4864 \cdot 10^6$	14.5	$1.7811 \cdot 10^3$	$8.5000 \cdot 10^6$
10.6	$5.9022 \cdot 10^3$	$8.4870 \cdot 10^6$	14.6	$1.7345 \cdot 10^3$	$8.5002 \cdot 10^6$
10.7	$5.6956 \cdot 10^3$	$8.4876 \cdot 10^6$	14.7	$1.6895 \cdot 10^3$	$8.5004 \cdot 10^6$
10.8	$5.4979 \cdot 10^3$	$8.4882 \cdot 10^6$	14.8	$1.6459 \cdot 10^3$	$8.5005 \cdot 10^6$
10.9	$5.3088 \cdot 10^3$	$8.4887 \cdot 10^6$	14.9	$1.6037 \cdot 10^3$	$8.5007 \cdot 10^6$
11.0	$5.1277 \cdot 10^3$	$8.4892 \cdot 10^6$	15.0	$1.5629 \cdot 10^3$	$8.5009 \cdot 10^6$
11.1	$4.9542 \cdot 10^3$	$8.4897 \cdot 10^6$	15.1	$1.5234 \cdot 10^3$	$8.5010 \cdot 10^6$
11.2	$4.7880 \cdot 10^3$	$8.4902 \cdot 10^6$	15.2	$1.4851 \cdot 10^3$	$8.5012 \cdot 10^6$
11.3	$4.6287 \cdot 10^3$	$8.4907 \cdot 10^6$	15.3	$1.4479 \cdot 10^3$	$8.5013 \cdot 10^6$
11.4	$4.4760 \cdot 10^3$	$8.4911 \cdot 10^6$	15.4	$1.4120 \cdot 10^3$	$8.5015 \cdot 10^6$

Table 2. 2. (*continued*)

temperature 3600 K

λ [μm]	$m_{\lambda,bb}$ [W/μm m²]	$M_{T,o-\lambda,bb}$ [W/m²]	λ [μm]	$m_{\lambda,bb}$ [W/μm m²]	$M_{T,o-\lambda,bb}$ [W/m²]
0.4	$1.6731 \cdot 10^6$	$9.1517 \cdot 10^4$			
0.5	$4.0453 \cdot 10^6$	$3.7468 \cdot 10^5$	4.0	$2.1295 \cdot 10^5$	$9.1950 \cdot 10^6$
0.6	$6.1664 \cdot 10^6$	$8.9129 \cdot 10^5$	4.1	$1.9567 \cdot 10^5$	$9.2154 \cdot 10^6$
0.7	$7.4037 \cdot 10^6$	$1.5777 \cdot 10^6$	4.2	$1.8010 \cdot 10^5$	$9.2342 \cdot 10^6$
0.8	$7.7793 \cdot 10^6$	$2.3431 \cdot 10^6$	4.3	$1.6603 \cdot 10^5$	$9.2514 \cdot 10^6$
0.9	$7.5588 \cdot 10^6$	$3.1137 \cdot 10^6$	4.4	$1.5329 \cdot 10^5$	$9.2674 \cdot 10^6$
1.0	$7.0053 \cdot 10^6$	$3.8438 \cdot 10^6$	4.5	$1.4175 \cdot 10^5$	$9.2821 \cdot 10^6$
1.1	$6.3071 \cdot 10^6$	$4.5101 \cdot 10^6$	4.6	$1.3126 \cdot 10^5$	$9.2958 \cdot 10^6$
1.2	$5.5792 \cdot 10^6$	$5.1043 \cdot 10^6$	4.7	$1.2172 \cdot 10^5$	$9.3084 \cdot 10^6$
1.3	$4.8838 \cdot 10^6$	$5.6270 \cdot 10^6$	4.8	$1.1302 \cdot 10^5$	$9.3202 \cdot 10^6$
1.4	$4.2501 \cdot 10^6$	$6.0831 \cdot 10^6$	4.9	$1.0508 \cdot 10^5$	$9.3311 \cdot 10^6$
1.5	$3.6884 \cdot 10^6$	$6.4794 \cdot 10^6$	5.0	$9.7824 \cdot 10^4$	$9.3412 \cdot 10^6$
1.6	$3.1985 \cdot 10^6$	$6.8232 \cdot 10^6$	5.1	$9.1179 \cdot 10^4$	$9.3506 \cdot 10^6$
1.7	$2.7754 \cdot 10^6$	$7.1214 \cdot 10^6$	5.2	$8.5085 \cdot 10^4$	$9.3594 \cdot 10^6$
1.8	$2.4119 \cdot 10^6$	$7.3803 \cdot 10^6$	5.3	$7.9489 \cdot 10^4$	$9.3677 \cdot 10^6$
1.9	$2.1004 \cdot 10^6$	$7.6055 \cdot 10^6$	5.4	$7.4343 \cdot 10^4$	$9.3754 \cdot 10^6$
2.0	$1.8338 \cdot 10^6$	$7.8018 \cdot 10^6$	5.5	$6.9605 \cdot 10^4$	$9.3826 \cdot 10^6$
2.1	$1.6054 \cdot 10^6$	$7.9735 \cdot 10^6$	5.6	$6.5237 \cdot 10^4$	$9.3893 \cdot 10^6$
2.2	$1.4095 \cdot 10^6$	$8.1240 \cdot 10^6$	5.7	$6.1205 \cdot 10^4$	$9.3956 \cdot 10^6$
2.3	$1.2412 \cdot 10^6$	$8.2563 \cdot 10^6$	5.8	$5.7477 \cdot 10^4$	$9.4015 \cdot 10^6$
2.4	$1.0962 \cdot 10^6$	$8.3730 \cdot 10^6$	5.9	$5.4029 \cdot 10^4$	$9.4071 \cdot 10^6$
2.5	$9.7094 \cdot 10^5$	$8.4762 \cdot 10^6$	6.0	$5.0834 \cdot 10^4$	$9.4124 \cdot 10^6$
2.6	$8.6251 \cdot 10^5$	$8.5677 \cdot 10^6$	6.1	$4.7870 \cdot 10^4$	$9.4173 \cdot 10^6$
2.7	$7.6836 \cdot 10^5$	$8.6492 \cdot 10^6$	6.2	$4.5119 \cdot 10^4$	$9.4219 \cdot 10^6$
2.8	$6.8636 \cdot 10^5$	$8.7218 \cdot 10^6$	6.3	$4.2562 \cdot 10^4$	$9.4263 \cdot 10^6$
2.9	$6.1475 \cdot 10^5$	$8.7868 \cdot 10^6$	6.4	$4.0183 \cdot 10^4$	$9.4305 \cdot 10^6$
3.0	$5.5204 \cdot 10^5$	$8.8451 \cdot 10^6$	6.5	$3.7967 \cdot 10^4$	$9.4344 \cdot 10^6$
3.1	$4.9696 \cdot 10^5$	$8.8974 \cdot 10^6$	6.6	$3.5902 \cdot 10^4$	$9.4381 \cdot 10^6$
3.2	$4.4846 \cdot 10^5$	$8.9447 \cdot 10^6$	6.7	$3.3974 \cdot 10^4$	$9.4415 \cdot 10^6$
3.3	$4.0563 \cdot 10^5$	$8.9873 \cdot 10^6$	6.8	$3.2174 \cdot 10^4$	$9.4449 \cdot 10^6$
3.4	$3.6771 \cdot 10^5$	$9.0260 \cdot 10^6$	6.9	$3.0491 \cdot 10^4$	$9.4480 \cdot 10^6$
3.5	$3.3406 \cdot 10^5$	$9.0610 \cdot 10^6$	7.0	$2.8916 \cdot 10^4$	$9.4510 \cdot 10^6$
3.6	$3.0411 \cdot 10^5$	$9.0929 \cdot 10^6$	7.1	$2.7442 \cdot 10^4$	$9.4538 \cdot 10^6$
3.7	$2.7741 \cdot 10^5$	$9.1219 \cdot 10^6$	7.2	$2.6060 \cdot 10^4$	$9.4564 \cdot 10^6$
3.8	$2.5354 \cdot 10^5$	$9.1485 \cdot 10^6$	7.3	$2.4764 \cdot 10^4$	$9.4590 \cdot 10^6$
3.9	$2.3215 \cdot 10^5$	$9.1727 \cdot 10^6$	7.4	$2.3547 \cdot 10^4$	$9.4614 \cdot 10^6$

Table 2.2. (*continued*)

λ [µm]	$m_{\lambda,bb}$ [W/µm m^2]	$M_{T,o-\lambda,bb}$ [W/m^2]	λ [µm]	$m_{\lambda,bb}$ [W/µm m^2]	$M_{T,o-\lambda,bb}$ [W/m^2]
7.5	$2.2403\cdot10^4$	$9.4637\cdot10^6$	11.5	$4.4767\cdot10^3$	$9.5059\cdot10^6$
7.6	$2.1328\cdot10^4$	$9.4637\cdot10^6$	11.6	$4.3312\cdot10^3$	$9.5064\cdot10^6$
7.7	$2.0317\cdot10^4$	$9.4680\cdot10^6$	11.7	$4.1915\cdot10^3$	$9.5068\cdot10^6$
7.8	$1.9365\cdot10^4$	$9.4699\cdot10^6$	11.8	$4.0574\cdot10^3$	$9.5072\cdot10^6$
7.9	$1.8468\cdot10^4$	$9.4718\cdot10^6$	11.9	$3.9287\cdot10^3$	$9.5076\cdot10^6$
8.0	$1.7622\cdot10^4$	$9.4736\cdot10^6$	12.0	$3.8049\cdot10^3$	$9.5080\cdot10^6$
8.1	$1.6824\cdot10^4$	$9.4754\cdot10^6$	12.1	$3.6861\cdot10^3$	$9.5084\cdot10^6$
8.2	$1.6070\cdot10^4$	$9.4770\cdot10^6$	12.2	$3.5718\cdot10^3$	$9.5087\cdot10^6$
8.3	$1.5358\cdot10^4$	$9.4786\cdot10^6$	12.3	$3.4619\cdot10^3$	$9.5091\cdot10^6$
8.4	$1.4685\cdot10^4$	$9.4801\cdot10^6$	12.4	$3.3562\cdot10^3$	$9.5094\cdot10^6$
8.5	$1.4048\cdot10^4$	$9.4815\cdot10^6$	12.5	$3.2545\cdot10^3$	$9.5098\cdot10^6$
8.6	$1.3446\cdot10^4$	$9.4829\cdot10^6$	12.6	$3.1566\cdot10^3$	$9.5101\cdot10^6$
8.7	$1.2875\cdot10^4$	$9.4842\cdot10^6$	12.7	$3.0624\cdot10^3$	$9.5104\cdot10^6$
8.8	$1.2335\cdot10^4$	$9.4855\cdot10^6$	12.8	$2.9716\cdot10^3$	$9.5107\cdot10^6$
8.9	$1.1822\cdot10^4$	$9.4867\cdot10^6$	12.9	$2.8842\cdot10^3$	$9.5110\cdot10^6$
9.0	$1.1335\cdot10^4$	$9.4878\cdot10^6$	13.0	$2.8000\cdot10^3$	$9.5113\cdot10^6$
9.1	$1.0874\cdot10^4$	$9.4889\cdot10^6$	13.1	$2.7188\cdot10^3$	$9.5115\cdot10^6$
9.2	$1.0435\cdot10^4$	$9.4900\cdot10^6$	13.2	$2.6406\cdot10^3$	$9.5118\cdot10^6$
9.3	$1.0019\cdot10^4$	$9.4910\cdot10^6$	13.3	$2.5651\cdot10^3$	$9.5121\cdot10^6$
9.4	$9.6227\cdot10^3$	$9.4920\cdot10^6$	13.4	$2.4923\cdot10^3$	$9.5123\cdot10^6$
9.5	$9.2460\cdot10^3$	$9.4929\cdot10^6$	13.5	$2.4221\cdot10^3$	$9.5126\cdot10^6$
9.6	$8.8875\cdot10^3$	$9.4939\cdot10^6$	13.6	$2.3543\cdot10^3$	$9.5128\cdot10^6$
9.7	$8.5462\cdot10^3$	$9.4947\cdot10^6$	13.7	$2.2889\cdot10^3$	$9.5130\cdot10^6$
9.8	$8.2212\cdot10^3$	$9.4956\cdot10^6$	13.8	$2.2258\cdot10^3$	$9.5133\cdot10^6$
9.9	$7.9114\cdot10^3$	$9.4964\cdot10^6$	13.9	$2.1648\cdot10^3$	$9.5135\cdot10^6$
10.0	$7.6160\cdot10^3$	$9.4971\cdot10^6$	14.0	$2.1058\cdot10^3$	$9.5137\cdot10^6$
10.1	$7.3343\cdot10^3$	$9.4979\cdot10^6$	14.1	$2.0489\cdot10^3$	$9.5139\cdot10^6$
10.2	$7.0654\cdot10^3$	$9.4986\cdot10^6$	14.2	$1.9939\cdot10^3$	$9.5141\cdot10^6$
10.3	$6.8088\cdot10^3$	$9.4993\cdot10^6$	14.3	$1.9407\cdot10^3$	$9.5143\cdot10^6$
10.4	$6.5637\cdot10^3$	$9.5000\cdot10^6$	14.4	$1.8892\cdot10^3$	$9.5145\cdot10^6$
10.5	$6.3295\cdot10^3$	$9.5006\cdot10^6$	14.5	$1.8395\cdot10^3$	$9.5147\cdot10^6$
10.6	$6.1057\cdot10^3$	$9.5012\cdot10^6$	14.6	$1.7914\cdot10^3$	$9.5149\cdot10^6$
10.7	$5.8916\cdot10^3$	$9.5018\cdot10^6$	14.7	$1.7448\cdot10^3$	$9.5150\cdot10^6$
10.8	$5.6869\cdot10^3$	$9.5024\cdot10^6$	14.8	$1.6998\cdot10^3$	$9.5152\cdot10^6$
10.9	$5.4909\cdot10^3$	$9.5030\cdot10^6$	14.9	$1.6562\cdot10^3$	$9.5154\cdot10^6$
11.0	$5.3033\cdot10^3$	$9.5035\cdot10^6$	15.0	$1.6140\cdot10^3$	$9.5155\cdot10^6$
11.1	$5.1236\cdot10^3$	$9.5040\cdot10^6$	15.1	$1.5731\cdot10^3$	$9.5157\cdot10^6$
11.2	$4.9515\cdot10^3$	$9.5045\cdot10^6$	15.2	$1.5335\cdot10^3$	$9.5159\cdot10^6$
11.3	$4.7865\cdot10^3$	$9.5050\cdot10^6$	15.3	$1.4951\cdot10^3$	$9.5160\cdot10^6$
11.4	$4.6284\cdot10^3$	$9.5055\cdot10^6$	15.4	$1.4580\cdot10^3$	$9.5161\cdot10^6$

Table 2. 2. (*continued*)

temperature 3700 K

λ [μm]	$m_{\lambda,bb}$ [W/μm m²]	$M_{T,o-\lambda,bb}$ [W/m²]	λ [μm]	$m_{\lambda,bb}$ [W/μm m²]	$M_{T,o-\lambda,bb}$ [W/m²]
0.4	$2.1918\cdot10^{6}$	$1.2432\cdot10^{5}$			
0.5	$5.0213\cdot10^{6}$	$4.8336\cdot10^{5}$	4.0	$2.2232\cdot10^{5}$	$1.0285\cdot10^{7}$
0.6	$7.3845\cdot10^{6}$	$1.1118\cdot10^{6}$	4.1	$2.0420\cdot10^{5}$	$1.0306\cdot10^{7}$
0.7	$8.6438\cdot10^{6}$	$1.9226\cdot10^{6}$	4.2	$1.8787\cdot10^{5}$	$1.0326\cdot10^{7}$
0.8	$8.9127\cdot10^{6}$	$2.8073\cdot10^{6}$	4.3	$1.7312\cdot10^{5}$	$1.0344\cdot10^{7}$
0.9	$8.5356\cdot10^{6}$	$3.6837\cdot10^{6}$	4.4	$1.5978\cdot10^{5}$	$1.0361\cdot10^{7}$
1.0	$7.8211\cdot10^{6}$	$4.5033\cdot10^{6}$	4.5	$1.4770\cdot10^{5}$	$1.0376\cdot10^{7}$
1.1	$6.9775\cdot10^{6}$	$5.2437\cdot10^{6}$	4.6	$1.3672\cdot10^{5}$	$1.0390\cdot10^{7}$
1.2	$6.1261\cdot10^{6}$	$5.8986\cdot10^{6}$	4.7	$1.2774\cdot10^{5}$	$1.0403\cdot10^{7}$
1.3	$5.3293\cdot10^{6}$	$6.4708\cdot10^{6}$	4.8	$1.1765\cdot10^{5}$	$1.0416\cdot10^{7}$
1.4	$4.6136\cdot10^{6}$	$6.9672\cdot10^{6}$	4.9	$1.0936\cdot10^{5}$	$1.0427\cdot10^{7}$
1.5	$3.9861\cdot10^{6}$	$7.3964\cdot10^{6}$	5.0	$1.0178\cdot10^{5}$	$1.0438\cdot10^{6}$
1.6	$3.4434\cdot10^{6}$	$7.7672\cdot10^{6}$	5.1	$9.4837\cdot10^{4}$	$1.0447\cdot10^{7}$
1.7	$2.9780\cdot10^{6}$	$8.0877\cdot10^{6}$	5.2	$8.8476\cdot10^{4}$	$1.0457\cdot10^{7}$
1.8	$2.5805\cdot10^{6}$	$8.3651\cdot10^{6}$	5.3	$8.2637\cdot10^{4}$	$1.0465\cdot10^{7}$
1.9	$2.2415\cdot10^{6}$	$8.6057\cdot10^{6}$	5.4	$7.7270\cdot10^{4}$	$1.0473\cdot10^{7}$
2.0	$1.9525\cdot10^{6}$	$8.8150\cdot10^{6}$	5.5	$7.2329\cdot10^{4}$	$1.0481\cdot10^{7}$
2.1	$1.7059\cdot10^{6}$	$8.9976\cdot10^{6}$	5.6	$6.7775\cdot10^{4}$	$1.0488\cdot10^{7}$
2.2	$1.4950\cdot10^{6}$	$9.1574\cdot10^{6}$	5.7	$6.3572\cdot10^{4}$	$1.0494\cdot10^{7}$
2.3	$1.3143\cdot10^{6}$	$9.2976\cdot10^{6}$	5.8	$5.9689\cdot10^{4}$	$1.0500\cdot10^{7}$
2.4	$1.1591\cdot10^{6}$	$9.4211\cdot10^{6}$	5.9	$5.6097\cdot10^{4}$	$1.0506\cdot10^{7}$
2.5	$1.0253\cdot10^{6}$	$9.5302\cdot10^{6}$	6.0	$5.2770\cdot10^{4}$	$1.0512\cdot10^{7}$
2.6	$9.0967\cdot10^{5}$	$9.6268\cdot10^{6}$	6.1	$4.9685\cdot10^{4}$	$1.0517\cdot10^{7}$
2.7	$8.0945\cdot10^{5}$	$9.7126\cdot10^{6}$	6.2	$4.6821\cdot10^{4}$	$1.0522\cdot10^{7}$
2.8	$7.2233\cdot10^{5}$	$9.7891\cdot10^{6}$	6.3	$4.4160\cdot10^{4}$	$1.0526\cdot10^{7}$
2.9	$6.4635\cdot10^{5}$	$9.8575\cdot10^{6}$	6.4	$4.1685\cdot10^{4}$	$1.0530\cdot10^{7}$
3.0	$5.7990\cdot10^{5}$	$9.9187\cdot10^{6}$	6.5	$3.9380\cdot10^{4}$	$1.0534\cdot10^{7}$
3.1	$5.2162\cdot10^{5}$	$9.9737\cdot10^{6}$	6.6	$3.7232\cdot10^{4}$	$1.0538\cdot10^{7}$
3.2	$4.7035\cdot10^{5}$	$1.0023\cdot10^{7}$	6.7	$3.5228\cdot10^{4}$	$1.0542\cdot10^{7}$
3.3	$4.2513\cdot10^{5}$	$1.0068\cdot10^{7}$	6.8	$3.3357\cdot10^{4}$	$1.0545\cdot10^{7}$
3.4	$3.8513\cdot10^{5}$	$1.0108\cdot10^{7}$	6.9	$3.1608\cdot10^{4}$	$1.0549\cdot10^{7}$
3.5	$3.4967\cdot10^{5}$	$1.0145\cdot10^{7}$	7.0	$2.9971\cdot10^{4}$	$1.0552\cdot10^{7}$
3.6	$3.1814\cdot10^{5}$	$1.0179\cdot10^{7}$	7.1	$2.8439\cdot10^{4}$	$1.0555\cdot10^{7}$
3.7	$2.9005\cdot10^{5}$	$1.0209\cdot10^{7}$	7.2	$2.7004\cdot10^{4}$	$1.0557\cdot10^{7}$
3.8	$2.6495\cdot10^{5}$	$1.0237\cdot10^{7}$	7.3	$2.5657\cdot10^{4}$	$1.0560\cdot10^{7}$
3.9	$2.4248\cdot10^{5}$	$1.0262\cdot10^{7}$	7.4	$2.4394\cdot10^{4}$	$1.0562\cdot10^{7}$

Table 2. 2. (*continued*)

λ [μm]	$m_{\lambda,bb}$ [W/μm m²]	$M_{T,o-\lambda,bb}$ [W/m²]	λ [μm]	$m_{\lambda,bb}$ [W/μm m²]	$M_{T,o-\lambda,bb}$ [W/m²]
7.5	$2.3207\cdot10^{4}$	$1.0565\cdot10^{7}$	11.5	$4.6240\cdot10^{3}$	$1.0609\cdot10^{7}$
7.6	$2.2091\cdot10^{4}$	$1.0567\cdot10^{7}$	11.6	$4.4734\cdot10^{3}$	$1.0609\cdot10^{7}$
7.7	$2.1041\cdot10^{4}$	$1.0569\cdot10^{7}$	11.7	$4.3290\cdot10^{3}$	$1.0609\cdot10^{7}$
7.8	$2.0053\cdot10^{4}$	$1.0571\cdot10^{7}$	11.8	$4.1903\cdot10^{3}$	$1.0610\cdot10^{7}$
7.9	$1.9122\cdot10^{4}$	$1.0573\cdot10^{7}$	11.9	$4.0572\cdot10^{3}$	$1.0610\cdot10^{7}$
8.0	$1.8244\cdot10^{4}$	$1.0575\cdot10^{7}$	12.0	$3.9292\cdot10^{3}$	$1.0611\cdot10^{7}$
8.1	$1.7416\cdot10^{4}$	$1.0577\cdot10^{7}$	12.1	$3.8063\cdot10^{3}$	$1.0611\cdot10^{7}$
8.2	$1.6634\cdot10^{4}$	$1.0579\cdot10^{7}$	12.2	$3.6882\cdot10^{3}$	$1.0611\cdot10^{7}$
8.3	$1.5896\cdot10^{4}$	$1.0580\cdot10^{7}$	12.3	$3.5745\cdot10^{3}$	$1.0612\cdot10^{7}$
8.4	$1.5198\cdot10^{4}$	$1.0582\cdot10^{7}$	12.4	$3.4653\cdot10^{3}$	$1.0612\cdot10^{7}$
8.5	$1.4538\cdot10^{4}$	$1.0583\cdot10^{7}$	12.5	$3.3601\cdot10^{3}$	$1.0612\cdot10^{7}$
8.6	$1.3913\cdot10^{4}$	$1.0585\cdot10^{7}$	12.6	$3.2590\cdot10^{3}$	$1.0613\cdot10^{7}$
8.7	$1.3322\cdot10^{4}$	$1.0596\cdot10^{7}$	12.7	$3.1616\cdot10^{3}$	$1.0613\cdot10^{7}$
8.8	$1.2761\cdot10^{4}$	$1.0587\cdot10^{7}$	12.8	$3.0678\cdot10^{3}$	$1.0613\cdot10^{7}$
8.9	$1.2230\cdot10^{4}$	$1.0589\cdot10^{7}$	12.9	$2.9774\cdot10^{3}$	$1.0614\cdot10^{7}$
9.0	$1.1726\cdot10^{4}$	$1.0590\cdot10^{7}$	13.0	$2.8904\cdot10^{3}$	$1.0614\cdot10^{7}$
9.1	$1.1247\cdot10^{4}$	$1.0591\cdot10^{7}$	13.1	$2.8065\cdot10^{3}$	$1.0614\cdot10^{7}$
9.2	$1.0793\cdot10^{4}$	$1.0592\cdot10^{7}$	13.2	$2.7256\cdot10^{3}$	$1.0615\cdot10^{7}$
9.3	$1.0361\cdot10^{4}$	$1.0593\cdot10^{7}$	13.3	$2.6476\cdot10^{3}$	$1.0615\cdot10^{7}$
9.4	$9.9510\cdot10^{3}$	$1.0594\cdot10^{7}$	13.4	$2.5724\cdot10^{3}$	$1.0615\cdot10^{7}$
9.5	$9.5608\cdot10^{3}$	$1.0595\cdot10^{7}$	13.5	$2.4999\cdot10^{3}$	$1.0615\cdot10^{7}$
9.6	$9.1895\cdot10^{3}$	$1.0596\cdot10^{7}$	13.6	$2.4298\cdot10^{3}$	$1.0616\cdot10^{7}$
9.7	$8.8360\cdot10^{3}$	$1.0597\cdot10^{7}$	13.7	$2.3622\cdot10^{3}$	$1.0616\cdot10^{7}$
9.8	$8.4994\cdot10^{3}$	$1.0598\cdot10^{7}$	13.8	$2.2970\cdot10^{3}$	$1.0616\cdot10^{7}$
9.9	$8.1786\cdot10^{3}$	$1.0599\cdot10^{7}$	13.9	$2.2340\cdot10^{3}$	$1.0616\cdot10^{7}$
10.0	$7.8727\cdot10^{3}$	$1.0599\cdot10^{7}$	14.0	$2.1731\cdot10^{3}$	$1.0617\cdot10^{7}$
10.1	$7.5811\cdot10^{3}$	$1.0600\cdot10^{7}$	14.1	$2.1143\cdot10^{3}$	$1.0617\cdot10^{7}$
10.2	$7.3028\cdot10^{3}$	$1.0601\cdot10^{7}$	14.2	$2.0574\cdot10^{3}$	$1.0617\cdot10^{7}$
10.3	$7.0371\cdot10^{3}$	$1.0602\cdot10^{7}$	14.3	$2.0025\cdot10^{3}$	$1.0617\cdot10^{7}$
10.4	$6.7834\cdot10^{3}$	$1.0602\cdot10^{7}$	14.4	$1.9494\cdot10^{3}$	$1.0617\cdot10^{7}$
10.5	$6.5410\cdot10^{3}$	$1.0603\cdot10^{7}$	14.5	$1.8980\cdot10^{3}$	$1.0618\cdot10^{7}$
10.6	$6.3093\cdot10^{3}$	$1.0604\cdot10^{7}$	14.6	$1.8483\cdot10^{3}$	$1.0618\cdot10^{7}$
10.7	$6.0878\cdot10^{3}$	$1.0604\cdot10^{7}$	14.7	$1.8002\cdot10^{3}$	$1.0618\cdot10^{7}$
10.8	$5.8759\cdot10^{3}$	$1.0605\cdot10^{7}$	14.8	$1.7537\cdot10^{3}$	$1.0618\cdot10^{7}$
10.9	$5.6731\cdot10^{3}$	$1.0605\cdot10^{7}$	14.9	$1.7086\cdot10^{3}$	$1.0618\cdot10^{7}$
11.0	$5.4790\cdot10^{3}$	$1.0606\cdot10^{7}$	15.0	$1.6650\cdot10^{3}$	$1.0618\cdot10^{7}$
11.1	$5.2932\cdot10^{3}$	$1.0607\cdot10^{7}$	15.1	$1.6228\cdot10^{3}$	$1.0619\cdot10^{7}$
11.2	$5.1151\cdot10^{3}$	$1.0607\cdot10^{7}$	15.2	$1.5819\cdot10^{3}$	$1.0619\cdot10^{7}$
11.3	$4.9445\cdot10^{3}$	$1.0608\cdot10^{7}$	15.3	$1.5423\cdot10^{3}$	$1.0619\cdot10^{7}$
11.4	$4.7809\cdot10^{3}$	$1.0608\cdot10^{7}$	15.4	$1.5039\cdot10^{3}$	$1.0619\cdot10^{7}$

Table 2.2. (*continued*)

temperature 3800 K

λ [μm]	$m_{\lambda,bb}$ [W/μm m²]	$M_{T,o-\lambda,bb}$ [W/m²]	λ [μm]	$m_{\lambda,bb}$ [W/μm m²]	$M_{T,o-\lambda,bb}$ [W/m²]
0.4	$2.8309\cdot10^{6}$	$1.6639\cdot10^{5}$			
0.5	$6.1622\cdot10^{6}$	$6.1608\cdot10^{5}$	4.0	$2.3174\cdot10^{5}$	$1.1469\cdot10^{7}$
0.6	$8.7603\cdot10^{6}$	$1.3728\cdot10^{6}$	4.1	$2.1276\cdot10^{5}$	$1.1491\cdot10^{7}$
0.7	$1.0011\cdot10^{7}$	$2.3223\cdot10^{6}$	4.2	$1.9567\cdot10^{5}$	$1.1511\cdot10^{7}$
0.8	$1.0140\cdot10^{7}$	$3.3373\cdot10^{6}$	4.3	$1.8024\cdot10^{5}$	$1.1530\cdot10^{7}$
0.9	$9.5790\cdot10^{6}$	$4.3273\cdot10^{6}$	4.4	$1.6630\cdot10^{5}$	$1.1547\cdot10^{7}$
1.0	$8.6834\cdot10^{6}$	$5.2421\cdot10^{6}$	4.5	$1.5367\cdot10^{5}$	$1.1563\cdot10^{7}$
1.1	$7.6802\cdot10^{6}$	$6.0605\cdot10^{6}$	4.6	$1.4221\cdot10^{5}$	$1.1578\cdot10^{7}$
1.2	$6.6957\cdot10^{6}$	$6.7788\cdot10^{6}$	4.7	$1.3179\cdot10^{5}$	$1.1592\cdot10^{7}$
1.3	$5.7908\cdot10^{6}$	$7.4024\cdot10^{6}$	4.8	$1.2230\cdot10^{5}$	$1.1604\cdot10^{7}$
1.4	$4.9886\cdot10^{6}$	$7.9404\cdot10^{6}$	4.9	$1.1365\cdot10^{5}$	$1.1616\cdot10^{7}$
1.5	$4.2920\cdot10^{6}$	$8.4036\cdot10^{6}$	5.0	$1.0574\cdot10^{5}$	$1.1627\cdot10^{7}$
1.6	$3.6944\cdot10^{6}$	$8.8021\cdot10^{6}$	5.1	$9.8505\cdot10^{4}$	$1.1637\cdot10^{7}$
1.7	$3.1851\cdot10^{6}$	$9.1454\cdot10^{6}$	5.2	$9.1876\cdot10^{4}$	$1.1647\cdot10^{7}$
1.8	$2.7524\cdot10^{6}$	$9.4417\cdot10^{6}$	5.3	$8.5793\cdot10^{4}$	$1.1656\cdot10^{7}$
1.9	$2.3851\cdot10^{6}$	$9.6981\cdot10^{6}$	5.4	$8.0203\cdot10^{4}$	$1.1664\cdot10^{7}$
2.0	$2.0732\cdot10^{6}$	$9.9206\cdot10^{6}$	5.5	$7.5058\cdot10^{4}$	$1.1672\cdot10^{7}$
2.1	$1.8079\cdot10^{6}$	$1.0114\cdot10^{7}$	5.6	$7.0318\cdot10^{4}$	$1.1679\cdot10^{7}$
2.2	$1.5817\cdot10^{6}$	$1.0283\cdot10^{7}$	5.7	$6.5945\cdot10^{4}$	$1.1686\cdot10^{7}$
2.3	$1.3884\cdot10^{6}$	$1.0432\cdot10^{7}$	5.8	$6.1905\cdot10^{4}$	$1.1692\cdot10^{7}$
2.4	$1.2227\cdot10^{6}$	$1.0562\cdot10^{7}$	5.9	$5.8169\cdot10^{4}$	$1.1698\cdot10^{7}$
2.5	$1.0802\cdot10^{6}$	$1.0677\cdot10^{7}$	6.0	$5.4709\cdot10^{4}$	$1.1704\cdot10^{7}$
2.6	$9.5729\cdot10^{5}$	$1.0779\cdot10^{7}$	6.1	$5.1502\cdot10^{4}$	$1.1709\cdot10^{7}$
2.7	$8.5092\cdot10^{5}$	$1.0869\cdot10^{7}$	6.2	$4.8526\cdot10^{4}$	$1.1714\cdot10^{7}$
2.8	$7.5860\cdot10^{5}$	$1.0949\cdot10^{7}$	6.3	$4.5761\cdot10^{4}$	$1.1719\cdot10^{7}$
2.9	$6.7820\cdot10^{5}$	$1.1021\cdot10^{7}$	6.4	$4.3190\cdot10^{4}$	$1.1723\cdot10^{7}$
3.0	$6.0797\cdot10^{5}$	$1.1085\cdot10^{7}$	6.5	$4.0796\cdot10^{4}$	$1.1728\cdot10^{7}$
3.1	$5.4644\cdot10^{5}$	$1.1143\cdot10^{7}$	6.6	$3.8565\cdot10^{4}$	$1.1732\cdot10^{7}$
3.2	$4.9238\cdot10^{5}$	$1.1195\cdot10^{7}$	6.7	$3.6484\cdot10^{4}$	$1.1735\cdot10^{7}$
3.3	$4.4474\cdot10^{5}$	$1.1242\cdot10^{7}$	6.8	$3.4541\cdot10^{4}$	$1.1739\cdot10^{7}$
3.4	$4.0265\cdot10^{5}$	$1.1284\cdot10^{7}$	6.9	$3.2726\cdot10^{4}$	$1.1742\cdot10^{7}$
3.5	$3.6536\cdot10^{5}$	$1.1322\cdot10^{7}$	7.0	$3.1028\cdot10^{4}$	$1.1745\cdot10^{7}$
3.6	$3.3224\cdot10^{5}$	$1.1357\cdot10^{7}$	7.1	$2.9438\cdot10^{4}$	$1.1748\cdot10^{7}$
3.7	$3.0274\cdot10^{5}$	$1.1389\cdot10^{7}$	7.2	$2.7949\cdot10^{4}$	$1.1751\cdot10^{7}$
3.8	$2.7641\cdot10^{5}$	$1.1418\cdot10^{7}$	7.3	$2.6552\cdot10^{4}$	$1.1754\cdot10^{7}$
3.9	$2.5286\cdot10^{5}$	$1.1444\cdot10^{7}$	7.4	$2.5242\cdot10^{4}$	$1.1757\cdot10^{7}$

Table 2.2. (*continued*)

λ [µm]	$m_{\lambda,bb}$ [W/µm m²]	$M_{T,o-\lambda,bb}$ [W/m²]	λ [µm]	$m_{\lambda,bb}$ [W/µm m²]	q $M_{T,o-\lambda,bb}$ [W/m²]
7.5	$2.4011\cdot10^4$	$1.1759\cdot10^7$	11.5	$4.7713\cdot10^3$	$1.1804\cdot10^7$
7.6	$2.2854\cdot10^4$	$1.1761\cdot10^7$	11.6	$4.6158\cdot10^3$	$1.1805\cdot10^7$
7.7	$2.1766\cdot10^4$	$1.1764\cdot10^7$	11.7	$4.4665\cdot10^3$	$1.1805\cdot10^7$
7.8	$2.0741\cdot10^4$	$1.1766\cdot10^7$	11.8	$4.3233\cdot10^3$	$1.1806\cdot10^7$
7.9	$1.9777\cdot10^4$	$1.1768\cdot10^7$	11.9	$4.1857\cdot10^3$	$1.1806\cdot10^7$
8.0	$1.8867\cdot10^4$	$1.1770\cdot10^7$	12.0	$4.0536\cdot10^3$	$1.1806\cdot10^7$
8.1	$1.8009\cdot10^4$	$1.1772\cdot10^7$	12.1	$3.9266\cdot10^3$	$1.1807\cdot10^7$
8.2	$1.7199\cdot10^4$	$1.1773\cdot10^7$	12.2	$3.8046\cdot10^3$	$1.1807\cdot10^7$
8.3	$1.6434\cdot10^4$	$1.1775\cdot10^7$	12.3	$3.6872\cdot10^3$	$1.1808\cdot10^7$
8.4	$1.5711\cdot10^4$	$1.1777\cdot10^7$	12.4	$3.5744\cdot10^3$	$1.1808\cdot10^7$
8.5	$1.5028\cdot10^4$	$1.1778\cdot10^7$	12.5	$3.4658\cdot10^3$	$1.1808\cdot10^7$
8.6	$1.4381\cdot10^4$	$1.1780\cdot10^7$	12.6	$3.3613\cdot10^3$	$1.1809\cdot10^7$
8.7	$1.3768\cdot10^4$	$1.1781\cdot10^7$	12.7	$3.2608\cdot10^3$	$1.1809\cdot10^7$
8.8	$1.3188\cdot10^4$	$1.1782\cdot10^7$	12.8	$3.1639\cdot10^3$	$1.1809\cdot10^7$
8.9	$1.2638\cdot10^4$	$1.1784\cdot10^7$	12.9	$3.0706\cdot10^3$	$1.1810\cdot10^7$
9.0	$1.2116\cdot10^4$	$1.1785\cdot10^7$	13.0	$2.9808\cdot10^3$	$1.1810\cdot10^7$
9.1	$1.1621\cdot10^4$	$1.1786\cdot10^7$	13.1	$2.8942\cdot10^3$	$1.1810\cdot10^7$
9.2	$1.1151\cdot10^4$	$1.1787\cdot10^7$	13.2	$2.8107\cdot10^3$	$1.1810\cdot10^7$
9.3	$1.0704\cdot10^4$	$1.1788\cdot10^7$	13.3	$2.7302\cdot10^3$	$1.1811\cdot10^7$
9.4	$1.0279\cdot10^4$	$1.1789\cdot10^7$	13.4	$2.6525\cdot10^3$	$1.1811\cdot10^7$
9.5	$9.8757\cdot10^3$	$1.1790\cdot10^7$	13.5	$2.5776\cdot10^3$	$1.1811\cdot10^7$
9.6	$9.4916\cdot10^3$	$1.1791\cdot10^7$	13.6	$2.5054\cdot10^3$	$1.1812\cdot10^7$
9.7	$9.1260\cdot10^3$	$1.1792\cdot10^7$	13.7	$2.4356\cdot10^3$	$1.1812\cdot10^7$
9.8	$8.7777\cdot10^3$	$1.1793\cdot10^7$	13.8	$2.3682\cdot10^3$	$1.1812\cdot10^7$
9.9	$8.4459\cdot10^3$	$1.1794\cdot10^7$	13.9	$2.3032\cdot10^3$	$1.1812\cdot10^7$
10.0	$8.1296\cdot10^3$	$1.1795\cdot10^7$	14.0	$2.2404\cdot10^3$	$1.1812\cdot10^7$
10.1	$7.8280\cdot10^3$	$1.1796\cdot10^7$	14.1	$2.1797\cdot10^3$	$1.1813\cdot10^7$
10.2	$7.5402\cdot10^3$	$1.1796\cdot10^7$	14.2	$2.1210\cdot10^3$	$1.1813\cdot10^7$
10.3	$7.2655\cdot10^3$	$1.1797\cdot10^7$	14.3	$2.0643\cdot10^3$	$1.1813\cdot10^7$
10.4	$7.0032\cdot10^3$	$1.1798\cdot10^7$	14.4	$2.0095\cdot10^3$	$1.1813\cdot10^7$
10.5	$6.7526\cdot10^3$	$1.1799\cdot10^7$	14.5	$1.9565\cdot10^3$	$1.1814\cdot10^7$
10.6	$6.5131\cdot10^3$	$1.1799\cdot10^7$	14.6	$1.9052\cdot10^3$	$1.1814\cdot10^7$
10.7	$6.2841\cdot10^3$	$1.1800\cdot10^7$	14.7	$1.8556\cdot10^3$	$1.1814\cdot10^7$
10.8	$6.0651\cdot10^3$	$1.1800\cdot10^7$	14.8	$1.8076\cdot10^3$	$1.1814\cdot10^7$
10.9	$5.8555\cdot10^3$	$1.1801\cdot10^7$	14.9	$1.7611\cdot10^3$	$1.1814\cdot10^7$
11.0	$5.6549\cdot10^3$	$1.1802\cdot10^7$	15.0	$1.7161\cdot10^3$	$1.1814\cdot10^7$
11.1	$5.4628\cdot10^3$	$1.1802\cdot10^7$	15.1	$1.6726\cdot10^3$	$1.1815\cdot10^7$
11.2	$5.2788\cdot10^3$	$1.1803\cdot10^7$	15.2	$1.6304\cdot10^3$	$1.1815\cdot10^7$
11.3	$5.1024\cdot10^3$	$1.1803\cdot10^7$	15.3	$1.5895\cdot10^3$	$1.1815\cdot10^7$
11.4	$4.9334\cdot10^3$	$1.1804\cdot10^7$	15.4	$1.5499\cdot10^3$	$1.1815\cdot10^7$

Table 2.2. (*continued*)

temperature 3900 K

λ [μm]	$m_{\lambda,bb}$ [W/μm m²]	$M_{T,o-\lambda,bb}$ [W/m²]	λ [μm]	$m_{\lambda,bb}$ [W/μm m²]	$M_{T,o-\lambda,bb}$ [W/m²]
0.4	$3.6086 \cdot 10^6$	$2.1963 \cdot 10^5$			
0.5	$7.4836 \cdot 10^6$	$7.7650 \cdot 10^5$	4.0	$2.4119 \cdot 10^5$	$1.2750 \cdot 10^7$
0.6	$1.0302 \cdot 10^7$	$1.6791 \cdot 10^6$	4.1	$2.2135 \cdot 10^5$	$1.2773 \cdot 10^7$
0.7	$1.1507 \cdot 10^7$	$2.7822 \cdot 10^6$	4.2	$2.0349 \cdot 10^5$	$1.2794 \cdot 10^7$
0.8	$1.1461 \cdot 10^7$	$3.9388 \cdot 10^6$	4.3	$1.8739 \cdot 10^5$	$1.2814 \cdot 10^7$
0.9	$1.0689 \cdot 10^7$	$5.0504 \cdot 10^6$	4.4	$1.7284 \cdot 10^5$	$1.2832 \cdot 10^7$
1.0	$9.5913 \cdot 10^6$	$6.0658 \cdot 10^6$	4.5	$1.5966 \cdot 10^5$	$1.2849 \cdot 10^7$
1.1	$8.4145 \cdot 10^6$	$6.9661 \cdot 10^6$	4.6	$1.4771 \cdot 10^5$	$1.2864 \cdot 10^7$
1.2	$7.2873 \cdot 10^6$	$7.7504 \cdot 10^6$	4.7	$1.3685 \cdot 10^5$	$1.2878 \cdot 10^7$
1.3	$6.2678 \cdot 10^6$	$8.4272 \cdot 10^6$	4.8	$1.2696 \cdot 10^5$	$1.2891 \cdot 10^7$
1.4	$5.3745 \cdot 10^6$	$9.0082 \cdot 10^6$	4.9	$1.1794 \cdot 10^5$	$1.2904 \cdot 10^7$
1.5	$4.6058 \cdot 10^6$	$9.5062 \cdot 10^6$	5.0	$1.0971 \cdot 10^5$	$1.2915 \cdot 10^7$
1.6	$3.9511 \cdot 10^6$	$9.9332 \cdot 10^6$	5.1	$1.0218 \cdot 10^5$	$1.2926 \cdot 10^7$
1.7	$3.3964 \cdot 10^6$	$1.0300 \cdot 10^7$	5.2	$9.5284 \cdot 10^4$	$1.2935 \cdot 10^7$
1.8	$2.9275 \cdot 10^6$	$1.0615 \cdot 10^7$	5.3	$8.8956 \cdot 10^4$	$1.2945 \cdot 10^7$
1.9	$2.5311 \cdot 10^6$	$1.0888 \cdot 10^7$	5.4	$8.3142 \cdot 10^4$	$1.2953 \cdot 10^7$
2.0	$2.1957 \cdot 10^6$	$1.1124 \cdot 10^7$	5.5	$7.7794 \cdot 10^4$	$1.2961 \cdot 10^7$
2.1	$1.9113 \cdot 10^6$	$1.1329 \cdot 10^7$	5.6	$7.2867 \cdot 10^4$	$1.2969 \cdot 10^7$
2.2	$1.6695 \cdot 10^6$	$1.1507 \cdot 10^7$	5.7	$6.8323 \cdot 10^4$	$1.2976 \cdot 10^7$
2.3	$1.4633 \cdot 10^6$	$1.1664 \cdot 10^7$	5.8	$6.4126 \cdot 10^4$	$1.2982 \cdot 10^7$
2.4	$1.2870 \cdot 10^6$	$1.1801 \cdot 10^7$	5.9	$6.0245 \cdot 10^4$	$1.2989 \cdot 10^7$
2.5	$1.1357 \cdot 10^6$	$1.1922 \cdot 10^7$	6.0	$5.6653 \cdot 10^4$	$1.2995 \cdot 10^7$
2.6	$1.0053 \cdot 10^6$	$1.2029 \cdot 10^7$	6.1	$5.3323 \cdot 10^4$	$1.3000 \cdot 10^7$
2.7	$8.9274 \cdot 10^5$	$1.2123 \cdot 10^7$	6.2	$5.0234 \cdot 10^4$	$1.3005 \cdot 10^7$
2.8	$7.9515 \cdot 10^5$	$1.2208 \cdot 10^7$	6.3	$4.7364 \cdot 10^4$	$1.3010 \cdot 10^7$
2.9	$7.1027 \cdot 10^5$	$1.2283 \cdot 10^7$	6.4	$4.4696 \cdot 10^4$	$1.3015 \cdot 10^7$
3.0	$6.3623 \cdot 10^5$	$1.2350 \cdot 10^7$	6.5	$4.2213 \cdot 10^4$	$1.3019 \cdot 10^7$
3.1	$5.7143 \cdot 10^5$	$1.2410 \cdot 10^7$	6.6	$3.9899 \cdot 10^4$	$1.3023 \cdot 10^7$
3.2	$5.1455 \cdot 10^5$	$1.2465 \cdot 10^7$	6.7	$3.7742 \cdot 10^4$	$1.3027 \cdot 10^7$
3.3	$4.6447 \cdot 10^5$	$1.2514 \cdot 10^7$	6.8	$3.5728 \cdot 10^4$	$1.3031 \cdot 10^7$
3.4	$4.2027 \cdot 10^5$	$1.2558 \cdot 10^7$	6.9	$3.3846 \cdot 10^4$	$1.3034 \cdot 10^7$
3.5	$3.8114 \cdot 10^5$	$1.2598 \cdot 10^7$	7.0	$3.2086 \cdot 10^4$	$1.3037 \cdot 10^7$
3.6	$3.4640 \cdot 10^5$	$1.2634 \cdot 10^7$	7.1	$3.0438 \cdot 10^4$	$1.3041 \cdot 10^7$
3.7	$3.1550 \cdot 10^5$	$1.2667 \cdot 10^7$	7.2	$2.8895 \cdot 10^4$	$1.3044 \cdot 10^7$
3.8	$2.8793 \cdot 10^5$	$1.2697 \cdot 10^7$	7.3	$2.7448 \cdot 10^4$	$1.3046 \cdot 10^7$
3.9	$2.6328 \cdot 10^5$	$1.2725 \cdot 10^7$	7.4	$2.6091 \cdot 10^4$	$1.3049 \cdot 10^7$

Table 2.2. (*continued*)

λ [μm]	$m_{\lambda,bb}$ [W/μm m^2]	$M_{T,o-\lambda,bb}$ [W/m^2]	λ [μm]	$m_{\lambda,bb}$ [W/μm m^2]	$M_{T,o-\lambda,bb}$ [W/m^2]
7.5	$2.4816 \cdot 10^4$	$1.3052 \cdot 10^7$	11.5	$4.9187 \cdot 10^3$	$1.3098 \cdot 10^7$
7.6	$2.3618 \cdot 10^4$	$1.3054 \cdot 10^7$	11.6	$4.7582 \cdot 10^3$	$1.3099 \cdot 10^7$
7.7	$2.2491 \cdot 10^4$	$1.3056 \cdot 10^7$	11.7	$4.6042 \cdot 10^3$	$1.3099 \cdot 10^7$
7.8	$2.1431 \cdot 10^4$	$1.3059 \cdot 10^7$	11.8	$4.4563 \cdot 10^3$	$1.3100 \cdot 10^7$
7.9	$2.0432 \cdot 10^4$	$1.3061 \cdot 10^7$	11.9	$4.3144 \cdot 10^3$	$1.3100 \cdot 10^7$
8.0	$1.9490 \cdot 10^4$	$1.5063 \cdot 10^7$	12.0	$4.1780 \cdot 10^3$	$1.3100 \cdot 10^7$
8.1	$1.8603 \cdot 10^4$	$1.3065 \cdot 10^7$	12.1	$4.0470 \cdot 10^3$	$1.3101 \cdot 10^7$
8.2	$1.7765 \cdot 10^4$	$1.3066 \cdot 10^7$	12.2	$3.9211 \cdot 10^3$	$1.3101 \cdot 10^7$
8.3	$1.6973 \cdot 10^4$	$1.3068 \cdot 10^7$	12.3	$3.8000 \cdot 10^3$	$1.3102 \cdot 10^7$
8.4	$1.6225 \cdot 10^4$	$1.3070 \cdot 10^7$	12.4	$3.6836 \cdot 10^3$	$1.3102 \cdot 10^7$
8.5	$1.5518 \cdot 10^4$	$1.3071 \cdot 10^7$	12.5	$3.5716 \cdot 10^3$	$1.3102 \cdot 10^7$
8.6	$1.4849 \cdot 10^4$	$1.3073 \cdot 10^7$	12.6	$3.4638 \cdot 10^3$	$1.3103 \cdot 10^7$
8.7	$1.4215 \cdot 10^4$	$1.3074 \cdot 10^7$	12.7	$3.3600 \cdot 10^3$	$1.3103 \cdot 10^7$
8.8	$1.3615 \cdot 10^4$	$1.3076 \cdot 10^7$	12.8	$3.2601 \cdot 10^3$	$1.3103 \cdot 10^7$
8.9	$1.3046 \cdot 10^4$	$1.3077 \cdot 10^7$	12.9	$3.1639 \cdot 10^3$	$1.3104 \cdot 10^7$
9.0	$1.2507 \cdot 10^4$	$1.3078 \cdot 10^7$	13.0	$3.0712 \cdot 10^3$	$1.3104 \cdot 10^7$
9.1	$1.1995 \cdot 10^4$	$1.3080 \cdot 10^7$	13.1	$2.9819 \cdot 10^3$	$1.3104 \cdot 10^7$
9.2	$1.1509 \cdot 10^4$	$1.3081 \cdot 10^7$	13.2	$2.8958 \cdot 10^3$	$1.3105 \cdot 10^7$
9.3	$1.1047 \cdot 10^4$	$1.3082 \cdot 10^7$	13.3	$2.8127 \cdot 10^3$	$1.3105 \cdot 10^7$
9.4	$1.0608 \cdot 10^4$	$1.3083 \cdot 10^7$	13.4	$2.7327 \cdot 10^3$	$1.3105 \cdot 10^7$
9.5	$1.0191 \cdot 10^4$	$1.3084 \cdot 10^7$	13.5	$2.6554 \cdot 10^3$	$1.3106 \cdot 10^7$
9.6	$9.7940 \cdot 10^3$	$1.3085 \cdot 10^7$	13.6	$2.5809 \cdot 10^3$	$1.3106 \cdot 10^7$
9.7	$9.4161 \cdot 10^3$	$1.3086 \cdot 10^7$	13.7	$2.5090 \cdot 10^3$	$1.3106 \cdot 10^7$
9.8	$9.0563 \cdot 10^3$	$1.3087 \cdot 10^7$	13.8	$2.4395 \cdot 10^3$	$1.3106 \cdot 10^7$
9.9	$8.7135 \cdot 10^3$	$1.3088 \cdot 10^7$	13.9	$2.3724 \cdot 10^3$	$1.3107 \cdot 10^7$
10.0	$8.3867 \cdot 10^3$	$1.3089 \cdot 10^7$	14.0	$2.3077 \cdot 10^3$	$1.3107 \cdot 10^7$
10.1	$8.0751 \cdot 10^3$	$1.3089 \cdot 10^7$	14.1	$2.2451 \cdot 10^3$	$1.3107 \cdot 10^7$
10.2	$7.7778 \cdot 10^3$	$1.3090 \cdot 10^7$	14.2	$2.1846 \cdot 10^3$	$1.3107 \cdot 10^7$
10.3	$7.4941 \cdot 10^3$	$1.3091 \cdot 10^7$	14.3	$2.1261 \cdot 10^3$	$1.3107 \cdot 10^7$
10.4	$7.2231 \cdot 10^3$	$1.3092 \cdot 10^7$	14.4	$2.0696 \cdot 10^3$	$1.3108 \cdot 10^7$
10.5	$6.9643 \cdot 10^3$	$1.3092 \cdot 10^7$	14.5	$2.0150 \cdot 10^3$	$1.3108 \cdot 10^7$
10.6	$6.7170 \cdot 10^3$	$1.3093 \cdot 10^7$	14.6	$1.9621 \cdot 10^3$	$1.3108 \cdot 10^7$
10.7	$6.4805 \cdot 10^3$	$1.3094 \cdot 10^7$	14.7	$1.9110 \cdot 10^3$	$1.3108 \cdot 10^7$
10.8	$6.2543 \cdot 10^3$	$1.3094 \cdot 10^7$	14.8	$1.8615 \cdot 10^3$	$1.3108 \cdot 10^7$
10.9	$6.0379 \cdot 10^3$	$1.3095 \cdot 10^7$	14.9	$1.8136 \cdot 10^3$	$1.3109 \cdot 10^7$
11.0	$5.8308 \cdot 10^3$	$1.3096 \cdot 10^7$	15.0	$1.7672 \cdot 10^3$	$1.3109 \cdot 10^7$
11.1	$5.6325 \cdot 10^3$	$1.3096 \cdot 10^7$	15.1	$1.7223 \cdot 10^3$	$1.3109 \cdot 10^7$
11.2	$5.4425 \cdot 10^3$	$1.3097 \cdot 10^7$	15.2	$1.6789 \cdot 10^3$	$1.3109 \cdot 10^7$
11.3	$5.2605 \cdot 10^3$	$1.3097 \cdot 10^7$	15.3	$1.6368 \cdot 10^3$	$1.3109 \cdot 10^7$
11.4	$5.0860 \cdot 10^3$	$1.3098 \cdot 10^7$	15.4	$1.5960 \cdot 10^3$	$1.3109 \cdot 10^7$

Table 2.2. (*continued*)

temperature 4000 K

λ [μm]	$m_{\lambda,bb}$ [W/μm m²]	$M_{T,o-\lambda,bb}$ [W/m²]	λ [μm]	$m_{\lambda,bb}$ [W/μm m²]	$M_{T,o-\lambda,bb}$ [W/m²]
0.4	$4.5445 \cdot 10^6$	$2.8623 \cdot 10^5$			
0.5	$9.0007 \cdot 10^6$	$9.6862 \cdot 10^5$	4.0	$2.5068 \cdot 10^5$	$1.4135 \cdot 10^7$
0.6	$1.2018 \cdot 10^7$	$2.0359 \cdot 10^6$	4.1	$2.2997 \cdot 10^5$	$1.4159 \cdot 10^7$
0.7	$1.3138 \cdot 10^7$	$3.3080 \cdot 10^6$	4.2	$2.1135 \cdot 10^5$	$1.4181 \cdot 10^7$
0.8	$1.2877 \cdot 10^7$	$4.6174 \cdot 10^6$	4.3	$1.9456 \cdot 10^5$	$1.4202 \cdot 10^7$
0.9	$1.1864 \cdot 10^7$	$5.8586 \cdot 10^6$	4.4	$1.7939 \cdot 10^5$	$1.4220 \cdot 10^7$
1.0	$1.0544 \cdot 10^7$	$6.9802 \cdot 10^6$	4.5	$1.6567 \cdot 10^5$	$1.4237 \cdot 10^7$
1.1	$9.1795 \cdot 10^6$	$7.9660 \cdot 10^6$	4.6	$1.5322 \cdot 10^5$	$1.4253 \cdot 10^7$
1.2	$7.9001 \cdot 10^6$	$8.8190 \cdot 10^6$	4.7	$1.4192 \cdot 10^5$	$1.4268 \cdot 10^7$
1.3	$6.7595 \cdot 10^6$	$9.5507 \cdot 10^6$	4.8	$1.3163 \cdot 10^5$	$1.4282 \cdot 10^7$
1.4	$5.7708 \cdot 10^6$	$1.0176 \cdot 10^7$	4.9	$1.2225 \cdot 10^5$	$1.4295 \cdot 10^7$
1.5	$4.9271 \cdot 10^6$	$1.0710 \cdot 10^7$	5.0	$1.1369 \cdot 10^5$	$1.4306 \cdot 10^7$
1.6	$4.2132 \cdot 10^6$	$1.1166 \cdot 10^7$	5.1	$1.0587 \cdot 10^5$	$1.4317 \cdot 10^7$
1.7	$3.6117 \cdot 10^6$	$1.1556 \cdot 10^7$	5.2	$9.8699 \cdot 10^4$	$1.4327 \cdot 10^7$
1.8	$3.1055 \cdot 10^6$	$1.1891 \cdot 10^7$	5.3	$9.2125 \cdot 10^4$	$1.4337 \cdot 10^7$
1.9	$2.6793 \cdot 10^6$	$1.2180 \cdot 10^7$	5.4	$8.6087 \cdot 10^4$	$1.4346 \cdot 10^7$
2.0	$2.3199 \cdot 10^6$	$1.2429 \cdot 10^7$	5.5	$8.0534 \cdot 10^4$	$1.4354 \cdot 10^7$
2.1	$2.0160 \cdot 10^6$	$1.2646 \cdot 10^7$	5.6	$7.5420 \cdot 10^4$	$1.4362 \cdot 10^7$
2.2	$1.7583 \cdot 10^6$	$1.2834 \cdot 10^7$	5.7	$8.0704 \cdot 10^4$	$1.4369 \cdot 10^7$
2.3	$1.5391 \cdot 10^6$	$1.2999 \cdot 10^7$	5.8	$6.6350 \cdot 10^4$	$1.4376 \cdot 10^7$
2.4	$1.3519 \cdot 10^6$	$1.3143 \cdot 10^7$	5.9	$6.2324 \cdot 10^4$	$1.4383 \cdot 10^7$
2.5	$1.1916 \cdot 10^6$	$1.3270 \cdot 10^7$	6.0	$5.8599 \cdot 10^4$	$1.4389 \cdot 10^7$
2.6	$1.0538 \cdot 10^6$	$1.3382 \cdot 10^7$	6.1	$5.5147 \cdot 10^4$	$1.4394 \cdot 10^7$
2.7	$9.3489 \cdot 10^5$	$1.3481 \cdot 10^7$	6.2	$5.1944 \cdot 10^4$	$1.4400 \cdot 10^7$
2.8	$8.3197 \cdot 10^5$	$1.3569 \cdot 10^7$	6.3	$4.8970 \cdot 10^4$	$1.4405 \cdot 10^7$
2.9	$7.4257 \cdot 10^5$	$1.3648 \cdot 10^7$	6.4	$4.6205 \cdot 10^4$	$1.4410 \cdot 10^7$
3.0	$6.6466 \cdot 10^5$	$1.3718 \cdot 10^7$	6.5	$4.3633 \cdot 10^4$	$1.4414 \cdot 10^7$
3.1	$5.9656 \cdot 10^5$	$1.3781 \cdot 10^7$	6.6	$4.1236 \cdot 10^4$	$1.4418 \cdot 10^7$
3.2	$5.3684 \cdot 10^5$	$1.3838 \cdot 10^7$	6.7	$3.9001 \cdot 10^4$	$1.4422 \cdot 10^7$
3.3	$4.8431 \cdot 10^5$	$1.3889 \cdot 10^7$	6.8	$3.6915 \cdot 10^4$	$1.4426 \cdot 10^7$
3.4	$4.3797 \cdot 10^5$	$1.3935 \cdot 10^7$	6.9	$3.4967 \cdot 10^4$	$1.4430 \cdot 10^7$
3.5	$3.9698 \cdot 10^5$	$1.3977 \cdot 10^7$	7.0	$3.3145 \cdot 10^4$	$1.4433 \cdot 10^7$
3.6	$3.6063 \cdot 10^5$	$1.4015 \cdot 10^7$	7.1	$3.1440 \cdot 10^4$	$1.4436 \cdot 10^7$
3.7	$3.2831 \cdot 10^5$	$1.4049 \cdot 10^7$	7.2	$2.9843 \cdot 10^4$	$1.4439 \cdot 10^7$
3.8	$2.9949 \cdot 10^5$	$1.4080 \cdot 10^7$	7.3	$2.8346 \cdot 10^4$	$1.4442 \cdot 10^7$
3.9	$2.7374 \cdot 10^5$	$1.4109 \cdot 10^7$	7.4	$2.6941 \cdot 10^4$	$1.4445 \cdot 10^7$

Table 2.2. (*continued*)

λ [μm]	$m_{\lambda,bb}$ [W/μm m²]	$M_{T,o-\lambda,bb}$ [W/m²]	λ [μm]	$m_{\lambda,bb}$ [W/μm m²]	$M_{T,o-\lambda,bb}$ [W/m²]
7.5	$2.5622 \cdot 10^4$	$1.4448 \cdot 10^7$	11.5	$5.0662 \cdot 10^3$	$1.4496 \cdot 10^7$
7.6	$2.4383 \cdot 10^4$	$1.4450 \cdot 10^7$	11.6	$4.9006 \cdot 10^3$	$1.4496 \cdot 10^7$
7.7	$2.3217 \cdot 10^4$	$1.4453 \cdot 10^7$	11.7	$4.7418 \cdot 10^3$	$1.4497 \cdot 10^7$
7.8	$2.2121 \cdot 10^4$	$1.4455 \cdot 10^7$	11.8	$4.5894 \cdot 10^3$	$1.4497 \cdot 10^7$
7.9	$2.1088 \cdot 10^4$	$1.4457 \cdot 10^7$	11.9	$4.4430 \cdot 10^3$	$1.4498 \cdot 10^7$
8.0	$2.0115 \cdot 10^4$	$1.4459 \cdot 10^7$	12.0	$4.3025 \cdot 10^3$	$1.4498 \cdot 10^7$
8.1	$1.9197 \cdot 10^4$	$1.4461 \cdot 10^7$	12.1	$4.1674 \cdot 10^3$	$1.4498 \cdot 10^7$
8.2	$1.8330 \cdot 10^4$	$1.4463 \cdot 10^7$	12.2	$4.0376 \cdot 10^3$	$1.4499 \cdot 10^7$
8.3	$1.7512 \cdot 10^4$	$1.4465 \cdot 10^7$	12.3	$3.9128 \cdot 10^3$	$1.4499 \cdot 10^7$
8.4	$1.6739 \cdot 10^4$	$1.4466 \cdot 10^7$	12.4	$3.7928 \cdot 10^3$	$1.4500 \cdot 10^7$
8.5	$1.6009 \cdot 10^4$	$1.4468 \cdot 10^7$	12.5	$3.6773 \cdot 10^3$	$1.4500 \cdot 10^7$
8.6	$1.5317 \cdot 10^4$	$1.4470 \cdot 10^7$	12.6	$3.5662 \cdot 10^3$	$1.4500 \cdot 10^7$
8.7	$1.4663 \cdot 10^4$	$1.4471 \cdot 10^7$	12.7	$3.4593 \cdot 10^3$	$1.4501 \cdot 10^7$
8.8	$1.4043 \cdot 10^4$	$1.4472 \cdot 10^7$	12.8	$3.3564 \cdot 10^3$	$1.4501 \cdot 10^7$
8.9	$1.3455 \cdot 10^4$	$1.4474 \cdot 10^7$	12.9	$3.2572 \cdot 10^3$	$1.4501 \cdot 10^7$
9.0	$1.2898 \cdot 10^4$	$1.4475 \cdot 10^7$	13.0	$3.1617 \cdot 10^3$	$1.4502 \cdot 10^7$
9.1	$1.2369 \cdot 10^4$	$1.4476 \cdot 10^7$	13.1	$3.0696 \cdot 10^3$	$1.4502 \cdot 10^7$
9.2	$1.1867 \cdot 10^4$	$1.4478 \cdot 10^7$	13.2	$2.9809 \cdot 10^3$	$1.4503 \cdot 10^7$
9.3	$1.1390 \cdot 10^4$	$1.4479 \cdot 10^7$	13.3	$2.8954 \cdot 10^3$	$1.4503 \cdot 10^7$
9.4	$1.0937 \cdot 10^4$	$1.4480 \cdot 10^7$	13.4	$2.8128 \cdot 10^3$	$1.4503 \cdot 10^7$
9.5	$1.0506 \cdot 10^4$	$1.4481 \cdot 10^7$	13.5	$2.7333 \cdot 10^3$	$1.4503 \cdot 10^7$
9.6	$1.0097 \cdot 10^4$	$1.4482 \cdot 10^7$	13.6	$2.6565 \cdot 10^3$	$1.4503 \cdot 10^7$
9.7	$9.7065 \cdot 10^3$	$1.4483 \cdot 10^7$	13.7	$2.5824 \cdot 10^3$	$1.4504 \cdot 10^7$
9.8	$9.3350 \cdot 10^3$	$1.4484 \cdot 10^7$	13.8	$2.5108 \cdot 10^3$	$1.4504 \cdot 10^7$
9.9	$8.9812 \cdot 10^3$	$1.4485 \cdot 10^7$	13.9	$2.4417 \cdot 10^3$	$1.4504 \cdot 10^7$
10.0	$8.6439 \cdot 10^3$	$1.4486 \cdot 10^7$	14.0	$2.3750 \cdot 10^3$	$1.4504 \cdot 10^7$
10.1	$8.3223 \cdot 10^3$	$1.4487 \cdot 10^7$	14.1	$2.3105 \cdot 10^3$	$1.4505 \cdot 10^7$
10.2	$8.0155 \cdot 10^3$	$1.4487 \cdot 10^7$	14.2	$2.2482 \cdot 10^3$	$1.4505 \cdot 10^7$
10.3	$7.7227 \cdot 10^3$	$1.4488 \cdot 10^7$	14.3	$2.1880 \cdot 10^3$	$1.4505 \cdot 10^7$
10.4	$7.4432 \cdot 10^3$	$1.4489 \cdot 10^7$	14.4	$2.1298 \cdot 10^3$	$1.4505 \cdot 10^7$
10.5	$7.1761 \cdot 10^3$	$1.4490 \cdot 10^7$	14.5	$2.0735 \cdot 10^3$	$1.4506 \cdot 10^7$
10.6	$6.9210 \cdot 10^3$	$1.4490 \cdot 10^7$	14.6	$2.0191 \cdot 10^3$	$1.4606 \cdot 10^7$
10.7	$6.6770 \cdot 10^3$	$1.4491 \cdot 10^7$	14.7	$1.9664 \cdot 10^3$	$1.4506 \cdot 10^7$
10.8	$6.4437 \cdot 10^3$	$1.4492 \cdot 10^7$	14.8	$1.9154 \cdot 10^3$	$1.4506 \cdot 10^7$
10.9	$6.2205 \cdot 10^3$	$1.4492 \cdot 10^7$	14.9	$1.8661 \cdot 10^3$	$1.4506 \cdot 10^7$
11.0	$6.0068 \cdot 10^3$	$1.4493 \cdot 10^7$	15.0	$1.8184 \cdot 10^3$	$1.4507 \cdot 10^7$
11.1	$5.8023 \cdot 10^3$	$1.4494 \cdot 10^7$	15.1	$1.7721 \cdot 10^3$	$1.4507 \cdot 10^7$
11.2	$5.6064 \cdot 10^3$	$1.4494 \cdot 10^7$	15.2	$1.7274 \cdot 10^3$	$1.4507 \cdot 10^7$
11.3	$5.4186 \cdot 10^3$	$1.4495 \cdot 10^7$	15.3	$1.6840 \cdot 10^3$	$1.4507 \cdot 10^7$
11.4	$5.2387 \cdot 10^3$	$1.4495 \cdot 10^7$	15.4	$1.6420 \cdot 10^3$	$1.4507 \cdot 10^7$

Table 2.2. (*continued*)

temperature 4100 K

λ [μm]	$m_{\lambda,bb}$ [W/μm m^2]	$M_{T,o-\lambda,bb}$ [W/m^2]	λ [μm]	$m_{\lambda,bb}$ [W/μm m^2]	$M_{T,o-\lambda,bb}$ [W/m^2]
0.4	$5.6591 \cdot 10^6$	$3.6863\ 10^5$			
0.5	$1.0729 \cdot 10^7$	$1.1967 \cdot 10^6$	4.0	$2.6019 \cdot 10^5$	$1.5629 \cdot 10^7$
0.6	$1.3916 \cdot 10^7$	$2.4487 \cdot 10^6$	4.1	$2.3862 \cdot 10^5$	$1.5654 \cdot 10^7$
0.7	$1.4904 \cdot 10^7$	$3.9056 \cdot 10^6$	4.2	$2.1922 \cdot 10^5$	$1.5677 \cdot 19^7$
0.8	$1.4389 \cdot 10^7$	$5.3794 \cdot 10^6$	4.3	$2.0175 \cdot 10^5$	$1.5698 \cdot 10^7$
0.9	$1.3103 \cdot 10^7$	$6.7579 \cdot 10^6$	4.4	$1.8597 \cdot 10^5$	$1.5717 \cdot 10^7$
1.0	$1.1541 \cdot 10^7$	$7.9910 \cdot 10^6$	4.5	$1.7169 \cdot 10^5$	$1.5735 \cdot 10^7$
1.1	$9.9744 \cdot 10^6$	$9.0661 \cdot 10^6$	4.6	$1.5875 \cdot 10^5$	$1.5752 \cdot 10^7$
1.2	$8.5333 \cdot 10^6$	$9.9902 \cdot 10^6$	4.7	$1.4700 \cdot 10^5$	$1.5767 \cdot 10^7$
1.3	$7.2655 \cdot 10^6$	$1.0779 \cdot 10^7$	4.8	$1.3631 \cdot 10^5$	$1.5781 \cdot 10^7$
1.4	$6.1771 \cdot 10^6$	$1.1449 \cdot 10^7$	4.9	$1.2657 \cdot 10^5$	$1.5794 \cdot 10^7$
1.5	$5.2555 \cdot 10^6$	$1.2020 \cdot 10^7$	5.0	$1.1768 \cdot 10^5$	$1.5806 \cdot 10^7$
1.6	$4.4805 \cdot 10^6$	$1.2505 \cdot 10^7$	5.1	$1.0956 \cdot 10^5$	$1.5818 \cdot 10^7$
1.7	$3.8308 \cdot 10^6$	$1.2920 \cdot 10^7$	5.2	$1.0212 \cdot 10^5$	$1.5828 \cdot 10^7$
1.8	$3.2864 \cdot 10^6$	$1.3275 \cdot 10^7$	5.3	$9.5300 \cdot 10^4$	$1.5838 \cdot 10^7$
1.9	$2.8297 \cdot 10^6$	$1.3580 \cdot 10^7$	5.4	$8.9038 \cdot 10^4$	$1.5847 \cdot 10^7$
2.0	$2.4457 \cdot 10^6$	$1.3843 \cdot 10^7$	5.5	$8.3279 \cdot 10^4$	$1.5856 \cdot 10^7$
2.1	$2.1219 \cdot 10^6$	$1.4071 \cdot 10^7$	5.6	$7.7978 \cdot 10^4$	$1.5864 \cdot 10^7$
2.2	$1.8480 \cdot 10^6$	$1.4269 \cdot 10^7$	5.7	$7.3090 \cdot 10^4$	$1.5872 \cdot 10^7$
2.3	$1.6155 \cdot 10^6$	$1.4442 \cdot 10^7$	5.8	$6.8577 \cdot 10^4$	$1.5879 \cdot 10^7$
2.4	$1.4174 \cdot 10^6$	$1.4594 \cdot 10^7$	5.9	$6.4407 \cdot 10^4$	$1.5885 \cdot 10^7$
2.5	$1.2480 \cdot 10^6$	$1.4727 \cdot 10^7$	6.0	$6.0548 \cdot 10^4$	$1.5892 \cdot 10^7$
2.6	$1.1026 \cdot 10^6$	$1.4844 \cdot 10^7$	6.1	$5.6973 \cdot 10^4$	$1.5898 \cdot 10^7$
2.7	$9.7734 \cdot 10^5$	$1.4948 \cdot 10^7$	6.2	$5.3657 \cdot 10^4$	$1.5903 \cdot 10^7$
2.8	$8.6903 \cdot 10^5$	$1.5040 \cdot 10^7$	6.3	$5.0578 \cdot 10^4$	$1.5908 \cdot 10^7$
2.9	$7.7507 \cdot 10^5$	$1.5122 \cdot 10^7$	6.4	$4.7716 \cdot 10^4$	$1.5913 \cdot 10^7$
3.0	$6.9327 \cdot 10^5$	$1.5195 \cdot 10^7$	6.5	$4.5054 \cdot 10^4$	$1.5918 \cdot 10^7$
3.1	$6.2183 \cdot 10^5$	$1.5261 \cdot 10^7$	6.6	$4.2574 \cdot 10^4$	$1.5922 \cdot 10^7$
3.2	$5.5924 \cdot 10^5$	$1.5320 \cdot 10^7$	6.7	$4.0262 \cdot 10^4$	$1.5926 \cdot 10^7$
3.3	$5.0424 \cdot 10^5$	$1.5373 \cdot 10^7$	6.8	$3.8104 \cdot 10^4$	$1.5930 \cdot 10^7$
3.4	$4.5575 \cdot 10^5$	$1.5421 \cdot 10^7$	6.9	$3.6089 \cdot 10^4$	$1.5934 \cdot 10^7$
3.5	$4.1290 \cdot 10^5$	$1.5464 \cdot 10^7$	7.0	$3.4205 \cdot 10^4$	$1.5938 \cdot 10^7$
3.6	$3.7492 \cdot 10^5$	$1.5504 \cdot 10^7$	7.1	$3.2442 \cdot 10^4$	$1.5941 \cdot 10^7$
3.7	$3.4116 \cdot 10^5$	$1.5540 \cdot 10^7$	7.2	$3.0791 \cdot 10^4$	$1.5944 \cdot 10^7$
3.8	$3.1109 \cdot 10^5$	$1.5572 \cdot 10^7$	7.3	$2.9243 \cdot 10^4$	$1.5947 \cdot 10^7$
3.9	$2.8423 \cdot 10^5$	$1.5602 \cdot 10^7$	7.4	$2.7792 \cdot 10^4$	$1.5950 \cdot 10^7$

Table 2. 2. (*continued*)

λ [μm]	$m_{\lambda,bb}$ [W/μm m²]	$M_{T,o-\lambda,bb}$ [W/m²]	λ [μm]	$m_{\lambda,bb}$ [W/μm m²]	$M_{T,o-\lambda,bb}$ [W/m²]
7.5	$2.6429 \cdot 10^4$	$1.5953 \cdot 10^7$	11.5	$5.2137 \cdot 10^3$	$1.6002 \cdot 10^7$
7.6	$2.5148 \cdot 10^4$	$1.5955 \cdot 10^7$	11.6	$5.0432 \cdot 10^3$	$1.6003 \cdot 10^7$
7.7	$2.3944 \cdot 10^4$	$1.5958 \cdot 10^7$	11.7	$4.8796 \cdot 10^3$	$1.6003 \cdot 10^7$
7.8	$2.2811 \cdot 10^4$	$1.5960 \cdot 10^7$	11.8	$4.7225 \cdot 10^3$	$1.6004 \cdot 10^7$
7.9	$2.1744 \cdot 10^4$	$1.5962 \cdot 10^7$	11.9	$4.5718 \cdot 10^3$	$1.6004 \cdot 10^7$
8.0	$2.0739 \cdot 10^4$	$1.5964 \cdot 10^7$	12.0	$4.4270 \cdot 10^3$	$1.6005 \cdot 10^7$
8.1	$1.9791 \cdot 10^4$	$1.5966 \cdot 10^7$	12.1	$4.2879 \cdot 10^3$	$1.6005 \cdot 10^7$
8.2	$1.8897 \cdot 10^4$	$1.5968 \cdot 10^7$	12.2	$4.1542 \cdot 10^3$	$1.6005 \cdot 10^7$
8.3	$1.8052 \cdot 10^4$	$1.5970 \cdot 10^7$	12.3	$4.1542 \cdot 10^3$	$1.6006 \cdot 10^7$
8.4	$1.7254 \cdot 10^4$	$1.5972 \cdot 10^7$	12.4	$3.9021 \cdot 10^3$	$1.6006 \cdot 10^7$
8.5	$1.6500 \cdot 10^4$	$1.5974 \cdot 10^7$	12.5	$3.7831 \cdot 10^3$	$1.6007 \cdot 10^7$
8.6	$1.5786 \cdot 10^4$	$1.5975 \cdot 10^7$	12.6	$3.6688 \cdot 10^3$	$1.6007 \cdot 10^7$
8.7	$1.5110 \cdot 10^4$	$1.5977 \cdot 10^7$	12.7	$3.5586 \cdot 10^3$	$1.6007 \cdot 10^7$
8.8	$1.4470 \cdot 10^4$	$1.5978 \cdot 10^7$	12.8	$3.4526 \cdot 10^3$	$1.6008 \cdot 10^7$
8.9	$1.3864 \cdot 10^4$	$1.5980 \cdot 10^7$	12.9	$3.3505 \cdot 10^3$	$1.6008 \cdot 10^7$
9.0	$1.3289 \cdot 10^4$	$1.5981 \cdot 10^7$	13.0	$3.2522 \cdot 10^3$	$1.6008 \cdot 10^7$
9.1	$1.2743 \cdot 10^4$	$1.5982 \cdot 10^7$	13.1	$3.1574 \cdot 10^3$	$1.6009 \cdot 10^7$
9.2	$1.2226 \cdot 10^4$	$1.5983 \cdot 10^7$	13.2	$3.0660 \cdot 10^3$	$1.6009 \cdot 10^7$
9.3	$1.1734 \cdot 10^4$	$1.5985 \cdot 10^7$	13.3	$2.9780 \cdot 10^3$	$1.6009 \cdot 10^7$
9.4	$1.1266 \cdot 10^4$	$1.5986 \cdot 10^7$	13.4	$2.8930 \cdot 10^3$	$1.6010 \cdot 10^7$
9.5	$1.0822 \cdot 10^4$	$1.5987 \cdot 10^7$	13.5	$2.8111 \cdot 10^3$	$1.6010 \cdot 10^7$
9.6	$1.0399 \cdot 10^4$	$1.5988 \cdot 10^7$	13.6	$2.7321 \cdot 10^3$	$1.6010 \cdot 10^7$
9.7	$9.9970 \cdot 10^3$	$1.5989 \cdot 10^7$	13.7	$2.6558 \cdot 10^3$	$1.6010 \cdot 10^7$
9.8	$9.6139 \cdot 10^3$	$1.5990 \cdot 10^7$	13.8	$2.5821 \cdot 10^3$	$1.6011 \cdot 10^7$
9.9	$9.2490 \cdot 10^3$	$1.5991 \cdot 10^7$	13.9	$2.5110 \cdot 10^3$	$1.6011 \cdot 10^7$
10.0	$8.9013 \cdot 10^3$	$1.5992 \cdot 10^7$	14.0	$2.4423 \cdot 10^3$	$1.6011 \cdot 10^7$
10.1	$8.5697 \cdot 10^3$	$1.5993 \cdot 10^7$	14.1	$2.3760 \cdot 10^3$	$1.6011 \cdot 10^7$
10.2	$8.2534 \cdot 10^3$	$1.5994 \cdot 10^7$	14.2	$2.3119 \cdot 10^3$	$1.6012 \cdot 10^7$
10.3	$7.9515 \cdot 10^3$	$1.5994 \cdot 10^7$	14.3	$2.2499 \cdot 10^3$	$1.6012 \cdot 10^7$
10.4	$7.6633 \cdot 10^3$	$1.5995 \cdot 10^7$	14.4	$2.1900 \cdot 10^3$	$1.6012 \cdot 10^7$
10.5	$7.3881 \cdot 10^3$	$1.5996 \cdot 10^7$	14.5	$2.1320 \cdot 10^3$	$1.6012 \cdot 10^7$
10.6	$7.1250 \cdot 10^3$	$1.5997 \cdot 10^7$	14.6	$2.0760 \cdot 10^3$	$1.6012 \cdot 10^7$
10.7	$6.8736 \cdot 10^3$	$1.5997 \cdot 10^7$	14.7	$2.0218 \cdot 10^3$	$1.6013 \cdot 10^7$
10.8	$6.6331 \cdot 10^3$	$1.5998 \cdot 10^7$	14.8	$1.9694 \cdot 10^3$	$1.6013 \cdot 10^7$
10.9	$6.4031 \cdot 10^3$	$1.5999 \cdot 10^7$	14.9	$1.9186 \cdot 10^3$	$1.6013 \cdot 10^7$
11.0	$6.1829 \cdot 10^3$	$1.5999 \cdot 10^7$	15.0	$1.8695 \cdot 10^3$	$1.6013 \cdot 10^7$
11.1	$5.9722 \cdot 10^3$	$1.6000 \cdot 10^7$	15.1	$1.8219 \cdot 10^3$	$1.6013 \cdot 10^7$
11.2	$5.7703 \cdot 10^3$	$1.6000 \cdot 10^7$	15.2	$1.7759 \cdot 10^3$	$1.6014 \cdot 10^7$
11.3	$5.5768 \cdot 10^3$	$1.6001 \cdot 10^7$	15.3	$1.7312 \cdot 10^3$	$1.6014 \cdot 10^7$
11.4	$5.3914 \cdot 10^3$	$1.6002 \cdot 10^7$	15.4	$1.6880 \cdot 10^3$	$1.6014 \cdot 10^7$

Table 2.2. (*continued*)

temperature 4200 K

λ [μm]	$m_{\lambda,bb}$ [W/μm m²]	$M_{T,o-\lambda,bb}$ [W/m²]	λ [μm]	$m_{\lambda,bb}$ [W/μm m²]	$M_{T,o-\lambda,bb}$ [W/m²]
0.4	$6.9740 \cdot 10^{6}$	$4.6953 \cdot 10^{5}$			
0.5	$1.2682 \cdot 10^{7}$	$1.4654 \cdot 10^{6}$	4.0	$2.6974 \cdot 10^{5}$	$1.7237 \cdot 10^{7}$
0.6	$1.6002 \cdot 10^{7}$	$2.9229 \cdot 10^{6}$	4.1	$2.4730 \cdot 10^{5}$	$1.7263 \cdot 10^{7}$
0.7	$1.6807 \cdot 10^{7}$	$4.5810 \cdot 10^{6}$	4.2	$2.2712 \cdot 10^{5}$	$1.7287 \cdot 10^{7}$
0.8	$1.5995 \cdot 10^{7}$	$6.2306 \cdot 10^{6}$	4.3	$2.0895 \cdot 10^{5}$	$1.7309 \cdot 10^{7}$
0.9	$1.4407 \cdot 10^{7}$	$7.7545 \cdot 10^{6}$	4.4	$1.9256 \cdot 10^{5}$	$1.7329 \cdot 10^{7}$
1.0	$1.2580 \cdot 10^{7}$	$9.1044 \cdot 10^{6}$	4.5	$1.7773 \cdot 10^{5}$	$1.7347 \cdot 10^{7}$
1.1	$1.0798 \cdot 10^{7}$	$1.0272 \cdot 10^{7}$	4.6	$1.6429 \cdot 10^{5}$	$1.7364 \cdot 10^{7}$
1.2	$9.1863 \cdot 10^{6}$	$1.1270 \cdot 10^{7}$	4.7	$1.5209 \cdot 10^{5}$	$1.7380 \cdot 10^{7}$
1.3	$7.7850 \cdot 10^{6}$	$1.2117 \cdot 10^{7}$	4.8	$1.4100 \cdot 10^{5}$	$1.7395 \cdot 10^{7}$
1.4	$6.5930 \cdot 10^{6}$	$1.2834 \cdot 10^{7}$	4.9	$1.3090 \cdot 10^{5}$	$1.7408 \cdot 10^{7}$
1.5	$5.5907 \cdot 10^{6}$	$1.3441 \cdot 10^{7}$	5.0	$1.2168 \cdot 10^{5}$	$1.7421 \cdot 10^{7}$
1.6	$4.7527 \cdot 10^{6}$	$1.3957 \cdot 10^{7}$	5.1	$1.1326 \cdot 10^{5}$	$1.7433 \cdot 10^{7}$
1.7	$4.0535 \cdot 10^{6}$	$1.4397 \cdot 10^{7}$	5.2	$1.0555 \cdot 10^{5}$	$1.7444 \cdot 10^{7}$
1.8	$3.4699 \cdot 10^{6}$	$1.4772 \cdot 10^{7}$	5.3	$9.8481 \cdot 10^{4}$	$1.7454 \cdot 10^{7}$
1.9	$2.9820 \cdot 10^{6}$	$1.5094 \cdot 10^{7}$	5.4	$9.1994 \cdot 10^{4}$	$1.7463 \cdot 10^{7}$
2.0	$2.5730 \cdot 10^{6}$	$1.5371 \cdot 10^{7}$	5.5	$8.6029 \cdot 10^{4}$	$1.7472 \cdot 10^{7}$
2.1	$2.2290 \cdot 10^{6}$	$1.5611 \cdot 10^{7}$	5.6	$8.0539 \cdot 10^{4}$	$1.7481 \cdot 10^{7}$
2.2	$1.9387 \cdot 10^{6}$	$1.5818 \cdot 10^{7}$	5.7	$7.5479 \cdot 10^{4}$	$1.7488 \cdot 10^{7}$
2.3	$1.6927 \cdot 10^{6}$	$1.6000 \cdot 10^{7}$	5.8	$7.0808 \cdot 10^{4}$	$1.7496 \cdot 10^{7}$
2.4	$1.4835 \cdot 10^{6}$	$1.6158 \cdot 10^{7}$	5.9	$6.6493 \cdot 10^{4}$	$1.7502 \cdot 10^{7}$
2.5	$1.3049 \cdot 10^{6}$	$1.6297 \cdot 10^{7}$	6.0	$6.2500 \cdot 10^{4}$	$1.7509 \cdot 10^{7}$
2.6	$1.1518 \cdot 10^{6}$	$1.6420 \cdot 10^{7}$	6.1	$5.8801 \cdot 10^{4}$	$1.7515 \cdot 10^{7}$
2.7	$1.0201 \cdot 10^{6}$	$1.6528 \cdot 10^{7}$	6.2	$5.5372 \cdot 10^{4}$	$1.7521 \cdot 10^{7}$
2.8	$9.0632 \cdot 10^{5}$	$1.6625 \cdot 10^{7}$	6.3	$5.2188 \cdot 10^{4}$	$1.7526 \cdot 10^{7}$
2.9	$8.0775 \cdot 10^{5}$	$1.6710 \cdot 10^{7}$	6.4	$4.9229 \cdot 10^{4}$	$1.7531 \cdot 10^{7}$
3.0	$7.2203 \cdot 10^{5}$	$1.6787 \cdot 10^{7}$	6.5	$4.6477 \cdot 10^{4}$	$1.7536 \cdot 10^{7}$
3.1	$6.4723 \cdot 10^{5}$	$1.6855 \cdot 10^{7}$	6.6	$4.3913 \cdot 10^{4}$	$1.7540 \cdot 10^{7}$
3.2	$5.8175 \cdot 10^{5}$	$1.6916 \cdot 10^{7}$	6.7	$4.1524 \cdot 10^{4}$	$1.7545 \cdot 10^{7}$
3.3	$5.2426 \cdot 10^{5}$	$1.6972 \cdot 10^{7}$	6.8	$3.9295 \cdot 10^{4}$	$1.7549 \cdot 10^{7}$
3.4	$4.7361 \cdot 10^{5}$	$1.7021 \cdot 10^{7}$	6.9	$3.7213 \cdot 10^{4}$	$1.7552 \cdot 10^{7}$
3.5	$4.2888 \cdot 10^{5}$	$1.7067 \cdot 10^{7}$	7.0	$3.5266 \cdot 10^{4}$	$1.7556 \cdot 10^{7}$
3.6	$3.8926 \cdot 10^{5}$	$1.7107 \cdot 10^{7}$	7.1	$3.3445 \cdot 10^{4}$	$1.7559 \cdot 10^{7}$
3.7	$3.5407 \cdot 10^{5}$	$1.7145 \cdot 10^{7}$	7.2	$3.1740 \cdot 10^{4}$	$1.7563 \cdot 10^{7}$
3.8	$3.2273 \cdot 10^{5}$	$1.7178 \cdot 10^{7}$	7.3	$3.0142 \cdot 10^{4}$	$1.7566 \cdot 10^{7}$
3.9	$2.9476 \cdot 10^{5}$	$1.7209 \cdot 10^{7}$	7.4	$2.8643 \cdot 10^{4}$	$1.7569 \cdot 10^{7}$

Table 2.2. (*continued*)

λ [µm]	$m_{\lambda,bb}$ [W/µm m^2]	$M_{T,o-\lambda,bb}$ [W/m^2]
7.5	$2.7236\cdot10^4$	$1.7572\cdot10^7$
7.6	$2.5914\cdot10^4$	$1.7574\cdot10^7$
7.7	$2.4672\cdot10^4$	$1.7577\cdot10^7$
7.8	$2.3502\cdot10^4$	$1.7579\cdot10^7$
7.9	$2.2402\cdot10^4$	$1.7581\cdot10^7$
8.0	$2.1364\cdot10^4$	$1.7584\cdot10^7$
8.1	$2.0386\cdot10^4$	$1.7586\cdot10^7$
8.2	$1.9463\cdot10^4$	$1.7588\cdot10^7$
8.3	$1.8592\cdot10^4$	$1.7590\cdot10^7$
8.4	$1.7769\cdot10^4$	$1.7591\cdot10^7$
8.5	$1.6991\cdot10^4$	$1.7593\cdot10^7$
8.6	$1.6255\cdot10^4$	$1.7595\cdot10^7$
8.7	$1.5558\cdot10^4$	$1.7596\cdot10^7$
8.8	$1.4899\cdot10^4$	$1.7598\cdot10^7$
8.9	$1.4273\cdot10^4$	$1.7599\cdot10^7$
9.0	$1.3681\cdot10^4$	$1.7601\cdot10^7$
9.1	$1.3118\cdot10^4$	$1.7602\cdot10^7$
9.2	$1.2584\cdot10^4$	$1.7603\cdot10^7$
9.3	$1.2077\cdot10^4$	$1.7604\cdot10^7$
9.4	$1.1596\cdot10^4$	$1.7606\cdot10^7$
9.5	$1.1138\cdot10^4$	$1.7607\cdot10^7$
9.6	$1.0702\cdot10^4$	$1.7608\cdot10^7$
9.7	$1.0288\cdot10^4$	$1.7609\cdot10^7$
9.8	$9.8930\cdot10^3$	$1.7610\cdot10^7$
9.9	$9.5170\cdot10^3$	$1.7611\cdot10^7$
10.0	$9.1588\cdot10^3$	$1.7612\cdot10^7$
10.1	$8.8172\cdot10^3$	$1.7613\cdot10^7$
10.2	$8.4914\cdot10^3$	$1.7614\cdot10^7$
10.3	$8.1804\cdot10^3$	$1.7614\cdot10^7$
10.4	$7.8836\cdot10^3$	$1.7615\cdot10^7$
10.5	$7.6001\cdot10^3$	$1.7616\cdot10^7$
10.6	$7.3292\cdot10^3$	$1.7617\cdot10^7$
10.7	$7.0703\cdot10^3$	$1.7617\cdot10^7$
10.8	$6.8227\cdot10^3$	$1.7618\cdot10^7$
10.9	$6.5858\cdot10^3$	$1.7619\cdot10^7$
11.0	$6.3591\cdot10^3$	$1.7619\cdot10^7$
11.1	$6.1421\cdot10^3$	$1.7620\cdot10^7$
11.2	$5.9342\cdot10^3$	$1.7621\cdot10^7$
11.3	$5.7351\cdot10^3$	$1.7621\cdot10^7$
11.4	$5.5442\cdot10^3$	$1.7622\cdot10^7$
11.5	$5.3612\cdot10^3$	$1.7622\cdot10^7$
11.6	$5.1857\cdot10^3$	$1.7623\cdot10^7$
11.7	$5.0173\cdot10^3$	$1.7623\cdot10^7$
11.8	$4.8557\cdot10^3$	$1.7624\cdot10^7$
11.9	$4.7006\cdot10^3$	$1.7624\cdot10^7$
12.0	$4.5515\cdot10^3$	$1.7625\cdot10^7$
12.1	$4.4084\cdot10^3$	$1.7625\cdot10^7$
12.2	$4.2708\cdot10^3$	$1.7626\cdot10^7$
12.3	$4.1385\cdot10^3$	$1.7626\cdot10^7$
12.4	$4.0113\cdot10^3$	$1.7626\cdot10^7$
12.5	$3.8890\cdot10^3$	$1.7627\cdot10^7$
12.6	$3.7713\cdot10^3$	$1.7627\cdot10^7$
12.7	$3.6580\cdot10^3$	$1.7628\cdot10^7$
12.8	$3.5489\cdot10^3$	$1.7628\cdot10^7$
12.9	$3.4439\cdot10^3$	$1.7628\cdot10^7$
13.0	$3.3427\cdot10^3$	$1.7629\cdot10^7$
13.1	$3.2452\cdot10^3$	$1.7629\cdot10^7$
13.2	$3.1512\cdot10^3$	$1.7629\cdot10^7$
13.3	$3.0606\cdot10^3$	$1.7629\cdot10^7$
13.4	$2.9733\cdot10^3$	$1.7630\cdot10^7$
13.5	$2.8890\cdot10^3$	$1.7630\cdot10^7$
13.6	$2.8077\cdot10^3$	$1.7630\cdot10^7$
13.7	$2.7292\cdot10^3$	$1.7631\cdot10^7$
13.8	$2.6535\cdot10^3$	$1.7631\cdot10^7$
13.9	$2.5803\cdot10^3$	$1.7631\cdot10^7$
14.0	$2.5097\cdot10^3$	$1.7631\cdot10^7$
14.1	$2.4414\cdot10^3$	$1.7632\cdot10^7$
14.2	$2.3755\cdot10^3$	$1.7632\cdot10^7$
14.3	$2.3118\cdot10^3$	$1.7632\cdot10^7$
14.4	$2.2502\cdot10^3$	$1.7632\cdot10^7$
14.5	$2.1906\cdot10^3$	$1.7632\cdot10^7$
14.6	$2.1330\cdot10^3$	$1.7633\cdot10^7$
14.7	$2.0773\cdot10^3$	$1.7633\cdot10^7$
14.8	$2.0233\cdot10^3$	$1.7633\cdot10^7$
14.9	$1.9712\cdot10^3$	$1.7633\cdot10^7$
15.0	$1.9206\cdot10^3$	$1.7633\cdot10^7$
15.1	$1.8717\cdot10^3$	$1.7634\cdot10^7$
15.2	$1.8244\cdot10^3$	$1.7634\cdot10^7$
15.3	$1.7785\cdot10^3$	$1.7634\cdot10^7$
15.4	$1.7341\cdot10^3$	$1.7634\cdot10^7$

Table 2.2. (*continued*)

temperature 4300 K

λ [μm]	$m_{\lambda,bb}$ [W/μm m²]	$M_{T,o-\lambda,bb}$ [W/m²]	λ [μm]	$m_{\lambda,bb}$ [W/μm m²]	$M_{T,o-\lambda,bb}$ [W/m²]
0.4	$8.5113\cdot10^{6}$	$5.9195\cdot10^{5}$			
0.5	$1.4875\cdot10^{7}$	$1.7795\cdot10^{6}$	4.0	$2.7931\cdot10^{5}$	$1.8966\cdot10^{7}$
0.6	$1.8283\cdot10^{7}$	$3.4646\cdot10^{6}$	4.1	$2.5599\cdot10^{5}$	$1.8992\cdot10^{7}$
0.7	$1.8850\cdot10^{7}$	$5.3405\cdot10^{6}$	4.2	$2.3504\cdot10^{5}$	$1.9017\cdot10^{7}$
0.8	$1.7696\cdot10^{7}$	$7.1776\cdot10^{6}$	4.3	$2.1618\cdot10^{5}$	$1.9039\cdot10^{7}$
0.9	$1.5774\cdot10^{7}$	$8.8545\cdot10^{6}$	4.4	$1.9916\cdot10^{5}$	$1.9060\cdot10^{7}$
1.0	$1.3662\cdot10^{7}$	$1.0326\cdot10^{7}$	4.5	$1.8378\cdot10^{5}$	$1.9079\cdot10^{7}$
1.1	$1.1650\cdot10^{7}$	$1.1590\cdot10^{7}$	4.6	$1.6984\cdot10^{5}$	$1.9097\cdot10^{7}$
1.2	$9.8581\cdot10^{6}$	$1.2664\cdot10^{7}$	4.7	$1.5720\cdot10^{5}$	$1.9113\cdot10^{7}$
1.3	$8.3176\cdot10^{6}$	$1.3570\cdot10^{7}$	4.8	$1.4570\cdot10^{5}$	$1.9128\cdot10^{7}$
1.4	$7.0180\cdot10^{6}$	$1.4335\cdot10^{7}$	4.9	$1.3524\cdot10^{5}$	$1.9142\cdot10^{7}$
1.5	$5.9324\cdot10^{6}$	$1.4981\cdot10^{7}$	5.0	$1.2569\cdot10^{5}$	$1.9155\cdot10^{7}$
1.6	$5.0295\cdot10^{6}$	$1.5528\cdot10^{7}$	5.1	$1.1696\cdot10^{5}$	$1.9168\cdot10^{7}$
1.7	$4.2795\cdot10^{6}$	$1.5992\cdot10^{7}$	5.2	$1.0898\cdot10^{5}$	$1.9179\cdot10^{7}$
1.8	$3.6559\cdot10^{6}$	$1.6388\cdot10^{7}$	5.3	$1.0167\cdot10^{5}$	$1.9189\cdot10^{7}$
1.9	$3.1362\cdot10^{6}$	$1.6727\cdot10^{7}$	5.4	$9.4954\cdot10^{4}$	$1.9199\cdot10^{7}$
2.0	$2.7017\cdot10^{6}$	$1.7018\cdot10^{7}$	5.5	$8.8783\cdot10^{4}$	$1.9208\cdot10^{7}$
2.1	$2.3372\cdot10^{6}$	$1.7269\cdot10^{7}$	5.6	$8.3105\cdot10^{4}$	$1.9217\cdot10^{7}$
2.2	$2.0301\cdot10^{6}$	$1.7487\cdot10^{7}$	5.7	$7.7871\cdot10^{4}$	$1.9225\cdot10^{7}$
2.3	$1.7705\cdot10^{6}$	$1.7677\cdot10^{7}$	5.8	$7.3042\cdot10^{4}$	$1.9232\cdot10^{7}$
2.4	$1.5501\cdot10^{6}$	$1.7843\cdot10^{7}$	5.9	$6.8581\cdot10^{4}$	$1.9240\cdot10^{7}$
2.5	$1.3622\cdot10^{6}$	$1.7988\cdot10^{7}$	6.0	$5.4454\cdot10^{4}$	$1.9246\cdot10^{7}$
2.6	$1.2013\cdot10^{6}$	$1.8116\cdot10^{7}$	6.1	$6.0632\cdot10^{4}$	$1.9252\cdot10^{7}$
2.7	$1.0631\cdot10^{6}$	$1.8229\cdot10^{7}$	6.2	$5.7089\cdot10^{4}$	$1.9258\cdot10^{7}$
2.8	$9.4384\cdot10^{5}$	$1.8329\cdot10^{7}$	6.3	$5.3800\cdot10^{4}$	$1.9264\cdot10^{7}$
2.9	$8.4062\cdot10^{5}$	$1.8418\cdot10^{7}$	6.4	$5.0744\cdot10^{4}$	$1.9269\cdot10^{7}$
3.0	$7.5094\cdot10^{5}$	$1.8498\cdot10^{7}$	6.5	$4.7901\cdot10^{4}$	$1.9274\cdot10^{7}$
3.1	$6.7275\cdot10^{5}$	$1.8569\cdot10^{7}$	6.6	$4.5254\cdot10^{4}$	$1.9279\cdot10^{7}$
3.2	$6.0437\cdot10^{5}$	$1.8633\cdot10^{7}$	6.7	$4.2788\cdot10^{4}$	$1.9283\cdot10^{7}$
3.3	$5.4436\cdot10^{5}$	$1.8690\cdot10^{7}$	6.8	$4.0486\cdot10^{4}$	$1.9287\cdot10^{7}$
3.4	$4.9154\cdot10^{5}$	$1.8742\cdot10^{7}$	6.9	$3.8337\cdot10^{4}$	$1.9291\cdot10^{7}$
3.5	$4.4492\cdot10^{5}$	$1.8789\cdot10^{7}$	7.0	$3.6329\cdot10^{4}$	$1.9295\cdot10^{7}$
3.6	$4.0365\cdot10^{5}$	$1.8831\cdot10^{7}$	7.1	$3.4450\cdot10^{4}$	$1.9298\cdot10^{7}$
3.7	$3.6702\cdot10^{5}$	$1.8869\cdot10^{7}$	7.2	$3.2690\cdot10^{4}$	$1.9302\cdot10^{7}$
3.8	$3.3441\cdot10^{5}$	$1.8904\cdot10^{7}$	7.3	$3.1042\cdot10^{4}$	$1.9305\cdot10^{7}$
3.9	$3.0533\cdot10^{5}$	$1.8936\cdot10^{7}$	7.4	$2.9496\cdot10^{4}$	$1.9308\cdot10^{7}$

Table 2. 2. (*continued*)

λ [µm]	$m_{\lambda,bb}$ [W/µm m²]	$M_{T,o-\lambda,bb}$ [W/m²]	λ [µm]	$m_{\lambda,bb}$ [W/µm m²]	$M_{T,o-\lambda,bb}$ [W/m²]
7.5	$2.8044 \cdot 10^4$	$1.9311 \cdot 10^7$	11.5	$5.5089 \cdot 10^3$	$1.9363 \cdot 10^7$
7.6	$2.6681 \cdot 10^4$	$1.9314 \cdot 10^7$	11.6	$5.3284 \cdot 10^3$	$1.9364 \cdot 10^7$
7.7	$2.5400 \cdot 10^4$	$1.9316 \cdot 10^7$	11.7	$5.1552 \cdot 10^3$	$1.9364 \cdot 10^7$
7.8	$2.4194 \cdot 10^4$	$1.9319 \cdot 10^7$	11.8	$4.9889 \cdot 10^3$	$1.9365 \cdot 10^7$
7.9	$2.3059 \cdot 10^4$	$1.9321 \cdot 10^7$	11.9	$4.8294 \cdot 10^3$	$1.9365 \cdot 10^7$
8.0	$2.1990 \cdot 10^4$	$1.9323 \cdot 10^7$	12.0	$4.6761 \cdot 10^3$	$1.9365 \cdot 10^7$
8.1	$2.0982 \cdot 10^4$	$1.9325 \cdot 10^7$	12.1	$4.5289 \cdot 10^3$	$1.9366 \cdot 10^7$
8.2	$2.0031 \cdot 10^4$	$1.9327 \cdot 10^7$	12.2	$4.3874 \cdot 10^3$	$1.9366 \cdot 10^7$
8.3	$1.9133 \cdot 10^4$	$1.9329 \cdot 10^7$	12.3	$4.2514 \cdot 10^3$	$1.9367 \cdot 10^7$
8.4	$1.8285 \cdot 10^4$	$1.9331 \cdot 10^7$	12.4	$4.1207 \cdot 10^3$	$1.9367 \cdot 10^7$
8.5	$1.7483 \cdot 10^4$	$1.9333 \cdot 10^7$	12.5	$3.9949 \cdot 10^3$	$1.9368 \cdot 10^7$
8.6	$1.6724 \cdot 10^4$	$1.9335 \cdot 10^7$	12.6	$3.8739 \cdot 10^3$	$1.9368 \cdot 10^7$
8.7	$1.6007 \cdot 10^4$	$1.9336 \cdot 10^7$	12.7	$3.7574 \cdot 10^3$	$1.9368 \cdot 10^7$
8.8	$1.5327 \cdot 10^4$	$1.9338 \cdot 10^7$	12.8	$3.6452 \cdot 10^3$	$1.9369 \cdot 10^7$
8.9	$1.4683 \cdot 10^4$	$1.9339 \cdot 10^7$	12.9	$3.5373 \cdot 10^3$	$1.9369 \cdot 10^7$
9.0	$1.4072 \cdot 10^4$	$1.9341 \cdot 10^7$	13.0	$3.4332 \cdot 10^3$	$1.9369 \cdot 10^7$
9.1	$1.3493 \cdot 10^4$	$1.9342 \cdot 10^7$	13.1	$3.3330 \cdot 10^3$	$1.9370 \cdot 10^7$
9.2	$1.2943 \cdot 10^4$	$1.9343 \cdot 10^7$	13.2	$3.2364 \cdot 10^3$	$1.9370 \cdot 10^7$
9.3	$1.2421 \cdot 10^4$	$1.9345 \cdot 10^7$	13.3	$3.1433 \cdot 10^3$	$1.9370 \cdot 10^7$
9.4	$1.1925 \cdot 10^4$	$1.9346 \cdot 10^7$	13.4	$3.0535 \cdot 10^3$	$1.9371 \cdot 10^7$
9.5	$1.1454 \cdot 10^4$	$1.9347 \cdot 10^7$	13.5	$2.9669 \cdot 10^3$	$1.9371 \cdot 10^7$
9.6	$1.1005 \cdot 10^4$	$1.9348 \cdot 10^7$	13.6	$2.8833 \cdot 10^3$	$1.9371 \cdot 10^7$
9.7	$1.0578 \cdot 10^4$	$1.9349 \cdot 10^7$	13.7	$2.8027 \cdot 10^3$	$1.9372 \cdot 10^7$
9.8	$1.0172 \cdot 10^4$	$1.9350 \cdot 10^7$	13.8	$2.7248 \cdot 10^3$	$1.9372 \cdot 10^7$
9.9	$9.7851 \cdot 10^3$	$1.9351 \cdot 10^7$	13.9	$2.6496 \cdot 10^3$	$1.9372 \cdot 10^7$
10.0	$9.4164 \cdot 10^3$	$1.9352 \cdot 10^7$	14.0	$2.5770 \cdot 10^3$	$1.9372 \cdot 10^7$
10.1	$9.0648 \cdot 10^3$	$1.9353 \cdot 10^7$	14.1	$2.5069 \cdot 10^3$	$1.9373 \cdot 10^7$
10.2	$8.7294 \cdot 10^3$	$1.9354 \cdot 10^7$	14.2	$2.4392 \cdot 10^3$	$1.9373 \cdot 10^7$
10.3	$8.4094 \cdot 10^3$	$1.9355 \cdot 10^7$	14.3	$2.3737 \cdot 10^3$	$1.9373 \cdot 10^7$
10.4	$8.1040 \cdot 10^3$	$1.9356 \cdot 10^7$	14.4	$2.3104 \cdot 10^3$	$1.9373 \cdot 10^7$
10.5	$7.8122 \cdot 10^3$	$1.9356 \cdot 10^7$	14.5	$2.2492 \cdot 10^3$	$1.9373 \cdot 10^7$
10.6	$7.5335 \cdot 10^3$	$1.9357 \cdot 10^7$	14.6	$2.1900 \cdot 10^3$	$1.9374 \cdot 10^7$
10.7	$7.2670 \cdot 10^3$	$1.9358 \cdot 10^7$	14.7	$2.1327 \cdot 10^3$	$1.9374 \cdot 10^7$
10.8	$7.0123 \cdot 10^3$	$1.9359 \cdot 10^7$	14.8	$2.0773 \cdot 10^3$	$1.9374 \cdot 10^7$
10.9	$6.7686 \cdot 10^3$	$1.9359 \cdot 10^7$	14.9	$2.0237 \cdot 10^3$	$1.9374 \cdot 10^7$
11.0	$6.5354 \cdot 10^3$	$1.9360 \cdot 10^7$	15.0	$1.9718 \cdot 10^3$	$1.9374 \cdot 10^7$
11.1	$6.3121 \cdot 10^3$	$1.9361 \cdot 10^7$	15.1	$1.9216 \cdot 10^3$	$1.9375 \cdot 10^7$
11.2	$6.0983 \cdot 10^3$	$1.9361 \cdot 10^7$	15.2	$1.8729 \cdot 10^3$	$1.9375 \cdot 10^7$
11.3	$5.8934 \cdot 10^3$	$1.9362 \cdot 10^7$	15.3	$1.8258 \cdot 10^3$	$1.9375 \cdot 10^7$
11.4	$5.6971 \cdot 10^3$	$1.9362 \cdot 10^7$	15.4	$1.7801 \cdot 10^3$	$1.9375 \cdot 10^7$

Table 2.2. (*continued*)

temperature 4400 K

λ [μm]	$m_{\lambda,bb}$ [W/μm m²]	$M_{T,o-\lambda,bb}$ [W/m²]	λ [μm]	$m_{\lambda,bb}$ [W/μm m²]	$M_{T,o-\lambda,bb}$ [W/m²]
0.4	$1.0294\cdot10^{7}$	$7.3916\cdot10^{5}$			
0.5	$1.7322\cdot10^{7}$	$2.1441\cdot10^{6}$	4.0	$2.8891\cdot10^{5}$	$2.0820\cdot10^{7}$
0.6	$2.0764\cdot10^{7}$	$4.0798\cdot10^{6}$	4.1	$2.6471\cdot10^{5}$	$2.0847\cdot10^{7}$
0.7	$2.1034\cdot10^{7}$	$6.1905\cdot10^{6}$	4.2	$2.4298\cdot10^{5}$	$2.0873\cdot10^{7}$
0.8	$1.9491\cdot10^{7}$	$8.2266\cdot10^{6}$	4.3	$2.2342\cdot10^{5}$	$2.0896\cdot10^{7}$
0.9	$1.7202\cdot10^{7}$	$1.0064\cdot10^{7}$	4.4	$2.0578\cdot10^{5}$	$2.0917\cdot10^{7}$
1.0	$1.4784\cdot10^{7}$	$1.1663\cdot10^{7}$	4.5	$1.8984\cdot10^{5}$	$2.0937\cdot10^{7}$
1.1	$1.2529\cdot10^{7}$	$1.3027\cdot10^{7}$	4.6	$1.7541\cdot10^{5}$	$2.0955\cdot10^{7}$
1.2	$1.0548\cdot10^{7}$	$1.4178\cdot10^{7}$	4.7	$1.6231\cdot10^{5}$	$2.0972\cdot10^{7}$
1.3	$8.8628\cdot10^{6}$	$1.5146\cdot10^{7}$	4.8	$1.5041\cdot10^{5}$	$2.0988\cdot10^{7}$
1.4	$7.4517\cdot10^{6}$	$1.5960\cdot10^{7}$	4.9	$1.3958\cdot10^{5}$	$2.1002\cdot10^{7}$
1.5	$6.2802\cdot10^{6}$	$1.6644\cdot10^{7}$	5.0	$1.2970\cdot10^{5}$	$2.1016\cdot10^{7}$
1.6	$5.3108\cdot10^{6}$	$1.7222\cdot10^{7}$	5.1	$1.2068\cdot10^{5}$	$2.1028\cdot10^{7}$
1.7	$4.5088\cdot10^{6}$	$1.7712\cdot10^{7}$	5.2	$1.1242\cdot10^{5}$	$2.1040\cdot10^{7}$
1.8	$3.8443\cdot10^{6}$	$1.8129\cdot10^{7}$	5.3	$1.0486\cdot10^{5}$	$2.1051\cdot10^{7}$
1.9	$3.2921\cdot10^{6}$	$1.8485\cdot10^{7}$	5.4	$9.7919\cdot10^{4}$	$2.1061\cdot10^{7}$
2.0	$2.8317\cdot10^{6}$	$1.8790\cdot10^{7}$	5.5	$9.1541\cdot10^{4}$	$2.1070\cdot10^{7}$
2.1	$2.4464\cdot10^{6}$	$1.9054\cdot10^{7}$	5.6	$8.5674\cdot10^{4}$	$2.1079\cdot10^{7}$
2.2	$2.1224\cdot10^{6}$	$1.9282\cdot10^{7}$	5.7	$8.0267\cdot10^{4}$	$2.1088\cdot10^{7}$
2.3	$1.8490\cdot10^{6}$	$1.9480\cdot10^{7}$	5.8	$7.5279\cdot10^{4}$	$2.1095\cdot10^{7}$
2.4	$1.6172\cdot10^{6}$	$1.9653\cdot10^{7}$	5.9	$7.0672\cdot10^{4}$	$2.1103\cdot10^{7}$
2.5	$1.4198\cdot10^{6}$	$1.9804\cdot10^{7}$	6.0	$6.6411\cdot10^{4}$	$2.1109\cdot10^{7}$
2.6	$1.2511\cdot10^{6}$	$1.9938\cdot10^{7}$	6.1	$6.2465\cdot10^{4}$	$2.1161\cdot10^{7}$
2.7	$1.1063\cdot10^{6}$	$2.0055\cdot10^{7}$	6.2	$5.8808\cdot10^{4}$	$2.1122\cdot10^{7}$
2.8	$9.8156\cdot10^{5}$	$2.0160\cdot10^{7}$	6.3	$5.5413\cdot10^{4}$	$2.1128\cdot10^{7}$
2.9	$8.7365\cdot10^{5}$	$2.0252\cdot10^{7}$	6.4	$5.2260\cdot10^{4}$	$2.1133\cdot10^{7}$
3.0	$7.7998\cdot10^{5}$	$2.0335\cdot10^{7}$	6.5	$4.9327\cdot10^{4}$	$2.1138\cdot10^{7}$
3.1	$6.9839\cdot10^{5}$	$2.0409\cdot10^{7}$	6.6	$4.6597\cdot10^{4}$	$2.1143\cdot10^{7}$
3.2	$6.2707\cdot10^{5}$	$2.0475\cdot10^{7}$	6.7	$4.4052\cdot10^{4}$	$2.1147\cdot10^{7}$
3.3	$5.6454\cdot10^{5}$	$2.0534\cdot10^{7}$	6.8	$4.1679\cdot10^{4}$	$2.1152\cdot10^{7}$
3.4	$5.0954\cdot10^{5}$	$2.0588\cdot10^{7}$	6.9	$3.9463\cdot10^{4}$	$2.1156\cdot10^{7}$
3.5	$4.6102\cdot10^{5}$	$2.0636\cdot10^{7}$	7.0	$3.7392\cdot10^{4}$	$2.1160\cdot10^{7}$
3.6	$4.1809\cdot10^{5}$	$2.0680\cdot10^{7}$	7.1	$3.5455\cdot10^{4}$	$2.1163\cdot10^{7}$
3.7	$3.8000\cdot10^{5}$	$2.0720\cdot10^{7}$	7.2	$3.3641\cdot10^{4}$	$2.1167\cdot10^{7}$
3.8	$3.4613\cdot10^{5}$	$2.0756\cdot10^{7}$	7.3	$3.1942\cdot10^{4}$	$2.1170\cdot10^{7}$
3.9	$3.1592\cdot10^{5}$	$2.0789\cdot10^{7}$	7.4	$3.0349\cdot10^{4}$	$2.1173\cdot10^{7}$

Table 2. 2. (*continued*)

λ [μm]	$m_{\lambda,bb}$ [W/μm m^2]	$M_{T,o-\lambda,bb}$ [W/m^2]	λ [μm]	$m_{\lambda,bb}$ [W/μm m^2]	$M_{T,o-\lambda,bb}$ [W/m^2]
7.5	$2.8853 \cdot 10^{4}$	$2.1176 \cdot 10^{7}$	11.5	$5.6565 \cdot 10^{3}$	$2.1230 \cdot 10^{7}$
7.6	$2.7449 \cdot 10^{4}$	$2.1179 \cdot 10^{7}$	11.6	$5.4710 \cdot 10^{3}$	$2.1230 \cdot 10^{7}$
7.7	$2.6128 \cdot 10^{4}$	$2.1181 \cdot 10^{7}$	11.7	$5.2930 \cdot 10^{3}$	$2.1231 \cdot 10^{7}$
7.8	$2.4886 \cdot 10^{4}$	$2.1184 \cdot 10^{7}$	11.8	$5.1222 \cdot 19^{3}$	$2.1231 \cdot 10^{7}$
7.9	$2.3717 \cdot 10^{4}$	$2.1186 \cdot 10^{7}$	11.9	$4.9582 \cdot 10^{3}$	$2.1232 \cdot 10^{7}$
8.0	$2.2616 \cdot 10^{4}$	$2.1189 \cdot 10^{7}$	12.0	$4.8008 \cdot 10^{3}$	$2.1232 \cdot 10^{7}$
8.1	$2.1578 \cdot 10^{4}$	$2.1191 \cdot 10^{7}$	12.1	$4.6495 \cdot 10^{3}$	$2.1233 \cdot 10^{7}$
8.2	$2.0598 \cdot 10^{4}$	$2.1193 \cdot 10^{7}$	12.2	$4.5041 \cdot 10^{3}$	$2.1233 \cdot 10^{7}$
8.3	$1.9673 \cdot 10^{4}$	$2.1195 \cdot 10^{7}$	12.3	$4.3644 \cdot 10^{3}$	$2.1234 \cdot 10^{7}$
8.4	$1.8800 \cdot 10^{4}$	$2.1197 \cdot 10^{7}$	12.4	$4.2300 \cdot 10^{3}$	$2.1234 \cdot 10^{7}$
8.5	$1.7975 \cdot 10^{4}$	$2.1199 \cdot 10^{7}$	12.5	$4.1008 \cdot 10^{3}$	$2.1234 \cdot 10^{7}$
8.6	$1.7194 \cdot 10^{4}$	$2.1200 \cdot 10^{7}$	12.6	$3.9764 \cdot 10^{3}$	$2.1235 \cdot 10^{7}$
8.7	$1.6455 \cdot 10^{4}$	$2.1202 \cdot 10^{7}$	12.7	$3.8568 \cdot 10^{3}$	$2.1235 \cdot 10^{7}$
8.8	$1.5756 \cdot 10^{4}$	$2.1204 \cdot 10^{7}$	12.8	$3.7416 \cdot 10^{3}$	$2.1235 \cdot 10^{7}$
8.9	$1.5093 \cdot 10^{4}$	$2.1205 \cdot 10^{7}$	12.9	$3.6307 \cdot 10^{3}$	$2.1236 \cdot 10^{7}$
9.0	$1.4464 \cdot 10^{4}$	$2.1207 \cdot 10^{7}$	13.0	$3.5238 \cdot 10^{3}$	$2.1236 \cdot 10^{7}$
9.1	$1.3868 \cdot 10^{4}$	$2.1208 \cdot 10^{7}$	13.1	$3.4209 \cdot 10^{3}$	$2.1237 \cdot 10^{7}$
9.2	$1.3302 \cdot 10^{4}$	$2.1209 \cdot 10^{7}$	13.2	$3.3216 \cdot 10^{3}$	$2.1237 \cdot 10^{7}$
9.3	$1.2765 \cdot 10^{4}$	$2.1211 \cdot 10^{7}$	13.3	$3.2260 \cdot 10^{3}$	$2.1237 \cdot 10^{7}$
9.4	$1.2255 \cdot 10^{4}$	$2.1212 \cdot 10^{7}$	13.4	$3.1337 \cdot 10^{3}$	$2.1238 \cdot 10^{7}$
9.5	$1.1770 \cdot 10^{4}$	$2.1213 \cdot 10^{7}$	13.5	$3.0448 \cdot 10^{3}$	$2.1238 \cdot 10^{7}$
9.6	$1.1308 \cdot 10^{4}$	$2.1214 \cdot 10^{7}$	13.6	$2.9590 \cdot 10^{3}$	$2.1238 \cdot 10^{7}$
9.7	$1.0869 \cdot 10^{4}$	$2.1215 \cdot 10^{7}$	13.7	$2.8761 \cdot 10^{3}$	$2.1238 \cdot 10^{7}$
9.8	$1.0451 \cdot 10^{4}$	$2.1217 \cdot 10^{7}$	13.8	$2.7962 \cdot 10^{3}$	$2.1239 \cdot 10^{7}$
9.9	$1.0053 \cdot 10^{4}$	$2.1218 \cdot 10^{7}$	13.9	$2.7190 \cdot 10^{3}$	$2.1239 \cdot 10^{7}$
10.0	$9.6741 \cdot 10^{3}$	$2.1219 \cdot 10^{7}$	14.0	$2.6444 \cdot 10^{3}$	$2.1239 \cdot 10^{7}$
10.1	$9.3125 \cdot 10^{3}$	$2.1219 \cdot 10^{7}$	14.1	$2.5724 \cdot 10^{3}$	$2.1239 \cdot 10^{7}$
10.2	$8.9676 \cdot 10^{3}$	$2.1220 \cdot 10^{7}$	14.2	$2.5028 \cdot 10^{3}$	$2.1240 \cdot 10^{7}$
10.3	$8.6385 \cdot 10^{3}$	$2.1221 \cdot 10^{7}$	14.3	$2.4356 \cdot 10^{3}$	$2.1240 \cdot 10^{7}$
10.4	$8.3244 \cdot 10^{3}$	$2.1222 \cdot 10^{7}$	14.4	$2.3706 \cdot 10^{3}$	$2.1240 \cdot 10^{7}$
10.5	$8.0244 \cdot 10^{3}$	$2.1223 \cdot 10^{7}$	14.5	$2.3077 \cdot 10^{3}$	$2.1240 \cdot 10^{7}$
10.6	$7.7378 \cdot 10^{3}$	$2.1224 \cdot 10^{7}$	14.6	$2.2470 \cdot 10^{3}$	$2.1241 \cdot 10^{7}$
10.7	$7.4639 \cdot 10^{3}$	$2.1224 \cdot 10^{7}$	14.7	$2.1882 \cdot 10^{3}$	$2.1241 \cdot 10^{7}$
10.8	$7.2019 \cdot 10^{3}$	$2.1225 \cdot 10^{7}$	14.8	$2.1313 \cdot 10^{3}$	$2.1241 \cdot 10^{7}$
10.9	$6.9514 \cdot 10^{3}$	$2.1226 \cdot 10^{7}$	14.9	$2.0762 \cdot 10^{3}$	$2.1241 \cdot 10^{7}$
11.0	$6.7116 \cdot 10^{3}$	$2.1227 \cdot 10^{7}$	15.0	$2.0230 \cdot 10^{3}$	$2.1241 \cdot 10^{7}$
11.1	$6.4821 \cdot 10^{3}$	$2.1227 \cdot 10^{7}$	15.1	$1.9714 \cdot 10^{3}$	$2.1242 \cdot 10^{7}$
11.2	$6.2624 \cdot 10^{3}$	$2.1228 \cdot 10^{7}$	15.2	$1.9214 \cdot 10^{3}$	$2.1242 \cdot 10^{7}$
11.3	$6.0518 \cdot 10^{3}$	$2.1228 \cdot 10^{7}$	15.3	$1.8730 \cdot 10^{3}$	$2.1242 \cdot 10^{7}$
11.4	$5.8500 \cdot 10^{3}$	$2.1229 \cdot 10^{7}$	15.4	$1.8262 \cdot 10^{3}$	$2.1242 \cdot 10^{7}$

Table 2.2. (*continued*)

temperature 4500 K

λ [μm]	$m_{\lambda,bb}$ [W/μm m²]	$M_{T,o-\lambda,bb}$ [W/m²]	λ [μm]	$m_{\lambda,bb}$ [W/μm m²]	$M_{T,o-\lambda,bb}$ [W/m²]
0.4	$1.2345 \cdot 10^{7}$	$9.1474 \cdot 10^{5}$			
0.5	$2.0036 \cdot 10^{7}$	$2.5649 \cdot 10^{6}$	4.0	$2.9854 \cdot 10^{5}$	$2.2805 \cdot 10^{7}$
0.6	$2.3451 \cdot 10^{7}$	$4.7748 \cdot 10^{6}$	4.1	$2.7345 \cdot 10^{5}$	$2.2834 \cdot 10^{7}$
0.7	$2.3359 \cdot 10^{7}$	$7.1375 \cdot 10^{6}$	4.2	$2.5094 \cdot 10^{5}$	$2.2860 \cdot 10^{7}$
0.8	$2.1378 \cdot 10^{7}$	$9.3843 \cdot 10^{6}$	4.3	$2.3068 \cdot 10^{5}$	$2.2884 \cdot 10^{7}$
0.9	$1.8691 \cdot 10^{7}$	$1.1390 \cdot 10^{7}$	4.4	$2.1242 \cdot 10^{5}$	$2.2906 \cdot 10^{7}$
1.0	$1.5945 \cdot 10^{7}$	$1.3121 \cdot 10^{7}$	4.5	$1.9592 \cdot 10^{5}$	$2.2927 \cdot 10^{7}$
1.1	$1.3434 \cdot 10^{7}$	$1.4587 \cdot 10^{7}$	4.6	$1.8098 \cdot 10^{5}$	$2.2946 \cdot 10^{7}$
1.2	$1.1256 \cdot 10^{7}$	$1.5819 \cdot 10^{7}$	4.7	$1.6744 \cdot 10^{5}$	$2.2963 \cdot 10^{7}$
1.3	$9.4199 \cdot 10^{6}$	$1.6850 \cdot 10^{7}$	4.8	$1.5513 \cdot 10^{5}$	$2.2979 \cdot 10^{7}$
1.4	$7.8938 \cdot 10^{6}$	$1.7713 \cdot 10^{7}$	4.9	$1.4393 \cdot 10^{5}$	$2.2994 \cdot 10^{7}$
1.5	$6.6339 \cdot 10^{6}$	$1.8438 \cdot 10^{7}$	5.0	$1.3572 \cdot 10^{5}$	$2.3008 \cdot 10^{7}$
1.6	$5.5963 \cdot 10^{6}$	$1.9047 \cdot 10^{7}$	5.1	$1.2440 \cdot 10^{5}$	$2.3021 \cdot 10^{7}$
1.7	$4.7411 \cdot 10^{6}$	$1.9563 \cdot 10^{7}$	5.2	$1.1587 \cdot 10^{5}$	$2.3033 \cdot 10^{7}$
1.8	$4.0349 \cdot 10^{6}$	$2.0001 \cdot 10^{7}$	5.3	$1.0805 \cdot 10^{5}$	$2.3044 \cdot 10^{7}$
1.9	$3.4498 \cdot 10^{6}$	$2.0374 \cdot 10^{7}$	5.4	$1.0089 \cdot 10^{5}$	$2.3054 \cdot 10^{7}$
2.0	$2.9631 \cdot 10^{6}$	$2.0694 \cdot 10^{7}$	5.5	$9.4303 \cdot 10^{4}$	$2.3064 \cdot 10^{7}$
2.1	$2.5565 \cdot 10^{6}$	$2.0969 \cdot 10^{7}$	5.6	$8.8246 \cdot 10^{4}$	$2.3073 \cdot 10^{7}$
2.2	$2.2154 \cdot 10^{6}$	$2.1207 \cdot 10^{7}$	5.7	$8.2666 \cdot 10^{4}$	$2.3082 \cdot 10^{7}$
2.3	$1.9280 \cdot 10^{6}$	$2.1414 \cdot 10^{7}$	5.8	$7.7519 \cdot 10^{4}$	$2.3090 \cdot 10^{7}$
2.4	$1.6847 \cdot 10^{6}$	$2.1594 \cdot 10^{7}$	5.9	$7.2765 \cdot 10^{4}$	$2.3097 \cdot 10^{7}$
2.5	$1.4778 \cdot 10^{6}$	$2.1752 \cdot 10^{7}$	6.0	$6.8370 \cdot 10^{4}$	$2.3104 \cdot 10^{7}$
2.6	$1.3012 \cdot 10^{6}$	$2.1891 \cdot 10^{7}$	6.1	$6.4300 \cdot 10^{4}$	$2.3111 \cdot 10^{7}$
2.7	$1.1498 \cdot 10^{6}$	$2.2013 \cdot 10^{7}$	6.2	$6.0528 \cdot 10^{4}$	$2.3117 \cdot 10^{7}$
2.8	$1.0195 \cdot 10^{6}$	$2.2121 \cdot 10^{7}$	6.3	$5.7029 \cdot 10^{4}$	$2.3123 \cdot 10^{7}$
2.9	$9.0683 \cdot 10^{5}$	$2.2218 \cdot 10^{7}$	6.4	$5.3777 \cdot 10^{4}$	$2.3129 \cdot 10^{7}$
3.0	$8.0915 \cdot 10^{5}$	$2.2303 \cdot 10^{7}$	6.5	$5.0754 \cdot 10^{4}$	$2.3134 \cdot 10^{7}$
3.1	$7.2413 \cdot 10^{5}$	$2.2380 \cdot 10^{7}$	6.6	$4.7940 \cdot 10^{4}$	$2.3139 \cdot 10^{7}$
3.2	$6.4987 \cdot 10^{5}$	$2.2449 \cdot 10^{7}$	6.7	$4.5318 \cdot 10^{4}$	$2.3143 \cdot 10^{7}$
3.3	$5.8480 \cdot 10^{5}$	$2.2510 \cdot 10^{7}$	6.8	$4.2873 \cdot 10^{4}$	$2.3148 \cdot 10^{7}$
3.4	$5.2760 \cdot 10^{5}$	$2.2566 \cdot 10^{7}$	6.9	$4.0589 \cdot 10^{4}$	$2.3152 \cdot 10^{7}$
3.5	$4.7717 \cdot 10^{5}$	$2.2616 \cdot 10^{7}$	7.0	$3.8456 \cdot 10^{4}$	$2.3156 \cdot 10^{7}$
3.6	$4.3257 \cdot 10^{5}$	$2.2661 \cdot 10^{7}$	7.1	$3.6461 \cdot 10^{4}$	$2.3160 \cdot 10^{7}$
3.7	$3.9303 \cdot 10^{5}$	$2.2703 \cdot 10^{7}$	7.2	$3.4593 \cdot 10^{4}$	$2.3163 \cdot 10^{7}$
3.8	$3.5787 \cdot 10^{5}$	$2.2740 \cdot 10^{7}$	7.3	$3.2843 \cdot 10^{4}$	$2.3167 \cdot 10^{7}$
3.9	$3.2654 \cdot 10^{5}$	$2.2774 \cdot 10^{7}$	7.4	$3.1202 \cdot 10^{4}$	$2.3170 \cdot 10^{7}$

Table 2.2. (*continued*)

λ [μm]	$m_{\lambda,bb}$ [W/μm m²]	$M_{T,o-\lambda,bb}$ [W/m²]	λ [μm]	$m_{\lambda,bb}$ [W/μm m²]	$M_{T,o-\lambda,bb}$ [W/m²]
7.5	$2.9663 \cdot 10^4$	$2.3173 \cdot 10^7$	11.5	$5.8043 \cdot 10^3$	$2.3228 \cdot 10^7$
7.6	$2.8217 \cdot 10^4$	$2.3176 \cdot 10^7$	11.6	$5.6137 \cdot 10^3$	$2.3229 \cdot 10^7$
7.7	$2.6857 \cdot 10^4$	$2.3178 \cdot 10^7$	11.7	$5.4309 \cdot 10^3$	$2.3229 \cdot 10^7$
7.8	$2.5579 \cdot 10^4$	$2.3181 \cdot 10^7$	11.8	$5.2555 \cdot 10^3$	$2.3230 \cdot 10^7$
7.9	$2.4376 \cdot 10^4$	$2.3184 \cdot 10^7$	11.9	$5.0871 \cdot 10^3$	$2.3230 \cdot 10^7$
8.0	$2.3242 \cdot 10^4$	$2.3186 \cdot 10^7$	12.0	$4.9254 \cdot 10^3$	$2.3231 \cdot 10^7$
8.1	$2.2174 \cdot 10^4$	$2.3188 \cdot 10^7$	12.1	$4.7701 \cdot 10^3$	$2.3231 \cdot 10^7$
8.2	$2.1166 \cdot 10^4$	$2.3190 \cdot 10^7$	12.2	$4.6208 \cdot 10^3$	$2.3232 \cdot 10^7$
8.3	$2.0215 \cdot 10^4$	$2.3192 \cdot 10^7$	12.3	$4.4774 \cdot 10^3$	$2.3232 \cdot 10^7$
8.4	$1.9316 \cdot 10^4$	$2.3194 \cdot 10^7$	12.4	$4.3394 \cdot 10^3$	$2.3232 \cdot 10^7$
8.5	$1.8467 \cdot 10^4$	$2.3196 \cdot 10^7$	12.5	$4.2067 \cdot 10^3$	$2.3233 \cdot 10^7$
8.6	$1.7664 \cdot 10^4$	$2.3198 \cdot 10^7$	12.6	$4.0791 \cdot 10^3$	$2.3233 \cdot 10^7$
8.7	$1.6904 \cdot 10^4$	$2.3200 \cdot 10^7$	12.7	$3.9562 \cdot 10^3$	$2.3234 \cdot 10^7$
8.8	$1.6184 \cdot 10^4$	$2.3201 \cdot 10^7$	12.8	$3.8380 \cdot 10^3$	$2.3234 \cdot 10^7$
8.9	$1.5503 \cdot 10^4$	$2.3203 \cdot 10^7$	12.9	$3.7241 \cdot 10^3$	$2.3234 \cdot 10^7$
9.0	$1.4856 \cdot 10^4$	$2.3205 \cdot 10^7$	13.0	$3.6144 \cdot 10^3$	$2.3235 \cdot 10^7$
9.1	$1.4244 \cdot 10^4$	$2.3206 \cdot 10^7$	13.1	$3.5087 \cdot 10^3$	$2.3235 \cdot 10^7$
9.2	$1.3662 \cdot 10^4$	$2.3207 \cdot 10^7$	13.2	$3.4069 \cdot 10^3$	$2.3235 \cdot 10^7$
9.3	$1.3110 \cdot 10^4$	$2.3209 \cdot 10^7$	13.3	$3.3087 \cdot 10^3$	$2.3236 \cdot 10^7$
9.4	$1.2585 \cdot 10^4$	$2.3210 \cdot 10^7$	13.4	$3.2140 \cdot 10^3$	$2.3236 \cdot 10^7$
9.5	$1.2086 \cdot 10^4$	$2.3211 \cdot 10^7$	13.5	$3.1227 \cdot 10^3$	$2.3236 \cdot 10^7$
9.6	$1.1612 \cdot 10^4$	$2.3212 \cdot 10^7$	13.6	$3.0346 \cdot 10^3$	$2.3237 \cdot 10^7$
9.7	$1.1160 \cdot 10^4$	$2.3214 \cdot 10^7$	13.7	$2.9496 \cdot 10^3$	$2.3237 \cdot 10^7$
9.8	$1.0731 \cdot 10^4$	$2.3215 \cdot 10^7$	13.8	$2.8676 \cdot 10^3$	$2.3237 \cdot 10^7$
9.9	$1.0322 \cdot 10^4$	$2.3216 \cdot 10^7$	13.9	$2.7883 \cdot 10^3$	$2.3238 \cdot 10^7$
10.0	$9.9319 \cdot 10^3$	$2.3217 \cdot 10^7$	14.0	$2.7118 \cdot 10^3$	$2.3238 \cdot 10^7$
10.1	$9.5603 \cdot 10^3$	$2.3218 \cdot 10^7$	14.1	$2.6379 \cdot 10^3$	$2.3238 \cdot 10^7$
10.2	$9.2059 \cdot 10^3$	$2.3219 \cdot 10^7$	14.2	$2.5665 \cdot 10^3$	$2.3238 \cdot 10^7$
10.3	$8.8677 \cdot 10^3$	$2.3219 \cdot 10^7$	14.3	$2.4975 \cdot 10^3$	$2.3239 \cdot 10^7$
10.4	$8.5449 \cdot 10^3$	$2.3220 \cdot 10^7$	14.4	$2.4308 \cdot 10^3$	$2.3239 \cdot 10^7$
10.5	$8.2367 \cdot 10^3$	$2.3221 \cdot 10^7$	14.5	$2.3663 \cdot 10^3$	$2.3239 \cdot 10^7$
10.6	$7.9422 \cdot 10^3$	$2.3222 \cdot 10^7$	14.6	$2.3040 \cdot 10^3$	$2.3239 \cdot 10^7$
10.7	$7.6608 \cdot 10^3$	$2.3223 \cdot 10^7$	14.7	$2.2436 \cdot 10^3$	$2.3240 \cdot 10^7$
10.8	$7.3917 \cdot 10^3$	$2.3223 \cdot 10^7$	14.8	$2.1853 \cdot 10^3$	$2.3240 \cdot 10^7$
10.9	$7.1343 \cdot 10^3$	$2.3224 \cdot 10^7$	14.9	$2.1288 \cdot 10^3$	$2.3240 \cdot 10^7$
11.0	$6.8880 \cdot 10^3$	$2.3225 \cdot 10^7$	15.0	$2.0741 \cdot 10^3$	$2.3240 \cdot 10^7$
11.1	$6.6523 \cdot 10^3$	$2.3226 \cdot 10^7$	15.1	$2.0212 \cdot 10^3$	$2.3240 \cdot 10^7$
11.2	$6.4265 \cdot 10^3$	$2.3226 \cdot 10^7$	15.2	$1.9700 \cdot 10^3$	$2.3241 \cdot 10^7$
11.3	$6.2102 \cdot 10^3$	$2.3227 \cdot 10^7$	15.3	$1.9203 \cdot 10^3$	$2.3241 \cdot 10^7$
11.4	$6.0029 \cdot 10^3$	$2.3227 \cdot 10^7$	15.4	$1.8722 \cdot 10^3$	$2.3241 \cdot 10^7$

Table 2.2. (*continued*)

temperature 4600 K

λ [μm]	$m_{\lambda,bb}$ [W/μm m²]	$M_{T,o-\lambda,bb}$ [W/m²]	λ [μm]	$m_{\lambda,bb}$ [W/μm m²]	$M_{T,o-\lambda,bb}$ [W/m²]
0.4	$1.4689 \cdot 10^7$	$1.1229 \cdot 10^6$			
0.5	$2.3030 \cdot 10^7$	$3.0476 \cdot 10^6$	4.0	$3.0818 \cdot 10^5$	$2.4929 \cdot 10^7$
0.6	$2.6347 \cdot 10^7$	$5.5565 \cdot 10^6$	4.1	$2.8222 \cdot 10^5$	$2.4959 \cdot 10^7$
0.7	$2.5826 \cdot 10^7$	$8.1887 \cdot 10^6$	4.2	$2.5891 \cdot 10^5$	$2.4986 \cdot 10^7$
0.8	$2.3359 \cdot 10^7$	$1.0658 \cdot 10^7$	4.3	$2.3796 \cdot 10^5$	$2.5011 \cdot 10^7$
0.9	$2.0240 \cdot 10^7$	$1.2839 \cdot 10^7$	4.4	$2.1907 \cdot 10^5$	$2.5033 \cdot 10^7$
1.0	$1.7146 \cdot 10^7$	$1.4707 \cdot 10^7$	4.5	$2.0201 \cdot 10^5$	$2.5055 \cdot 10^7$
1.1	$1.4364 \cdot 10^7$	$1.6279 \cdot 10^7$	4.6	$1.8657 \cdot 10^5$	$2.5074 \cdot 10^7$
1.2	$1.1981 \cdot 10^7$	$1.7593 \cdot 10^7$	4.7	$1.7257 \cdot 10^5$	$2.5092 \cdot 10^7$
1.3	$9.9885 \cdot 10^6$	$1.8688 \cdot 10^7$	4.8	$1.5986 \cdot 10^5$	$2.5108 \cdot 10^7$
1.4	$8.3437 \cdot 10^6$	$1.9602 \cdot 10^7$	4.9	$1.4829 \cdot 10^5$	$2.5124 \cdot 10^7$
1.5	$6.9932 \cdot 10^6$	$2.0367 \cdot 10^7$	5.0	$1.3775 \cdot 10^5$	$2.5138 \cdot 10^7$
1.6	$5.8857 \cdot 10^6$	$2.1009 \cdot 10^7$	5.1	$1.2812 \cdot 10^5$	$2.5151 \cdot 10^7$
1.7	$4.9764 \cdot 10^6$	$2.1551 \cdot 10^7$	5.2	$1.1932 \cdot 10^5$	$2.5164 \cdot 10^7$
1.8	$4.2277 \cdot 10^6$	$2.2010 \cdot 10^7$	5.3	$1.1125 \cdot 10^5$	$2.5175 \cdot 10^7$
1.9	$3.6090 \cdot 10^6$	$2.2400 \cdot 10^7$	5.4	$1.0386 \cdot 10^5$	$2.5186 \cdot 10^7$
2.0	$3.0956 \cdot 10^6$	$2.2735 \cdot 10^7$	5.5	$9.7068 \cdot 10^4$	$2.5196 \cdot 10^7$
2.1	$2.6676 \cdot 10^6$	$2.3022 \cdot 10^7$	5.6	$9.0821 \cdot 10^4$	$2.5205 \cdot 10^7$
2.2	$2.3092 \cdot 10^6$	$2.3271 \cdot 10^7$	5.7	$8.5067 \cdot 10^4$	$2.5214 \cdot 10^7$
2.3	$2.0076 \cdot 10^6$	$2.3486 \cdot 10^7$	5.8	$7.9761 \cdot 10^4$	$2.5222 \cdot 10^7$
2.4	$1.7527 \cdot 10^6$	$2.3674 \cdot 10^7$	5.9	$7.4861 \cdot 10^4$	$2.5230 \cdot 10^7$
2.5	$1.5362 \cdot 10^6$	$2.3838 \cdot 10^7$	6.0	$7.0331 \cdot 10^4$	$2.5237 \cdot 10^7$
2.6	$1.3516 \cdot 10^6$	$2.3982 \cdot 10^7$	6.1	$6.6137 \cdot 10^4$	$2.5244 \cdot 10^7$
2.7	$1.1935 \cdot 10^6$	$2.4109 \cdot 10^7$	6.2	$6.2251 \cdot 10^4$	$2.5251 \cdot 10^7$
2.8	$1.0575 \cdot 10^6$	$2.4221 \cdot 10^7$	6.3	$5.8645 \cdot 10^4$	$2.5257 \cdot 10^7$
2.9	$9.4017 \cdot 10^5$	$2.4321 \cdot 10^7$	6.4	$5.5296 \cdot 10^4$	$2.5262 \cdot 10^7$
3.0	$8.3844 \cdot 10^5$	$2.4410 \cdot 10^7$	6.5	$5.2183 \cdot 10^4$	$2.5268 \cdot 10^7$
3.1	$7.4997 \cdot 10^5$	$2.4489 \cdot 10^7$	6.6	$4.9285 \cdot 10^4$	$2.5273 \cdot 10^7$
3.2	$6.7275 \cdot 10^5$	$2.4560 \cdot 10^7$	6.7	$4.6585 \cdot 10^4$	$2.5278 \cdot 10^7$
3.3	$6.0512 \cdot 10^5$	$2.4624 \cdot 10^7$	6.8	$4.4067 \cdot 10^4$	$2.5282 \cdot 10^7$
3.4	$5.4572 \cdot 10^5$	$2.4682 \cdot 10^7$	6.9	$4.1717 \cdot 10^4$	$2.5286 \cdot 10^7$
3.5	$4.9337 \cdot 10^5$	$2.4733 \cdot 10^7$	7.0	$3.9521 \cdot 10^4$	$2.5290 \cdot 10^7$
3.6	$4.4710 \cdot 10^5$	$2.4780 \cdot 10^7$	7.1	$3.7467 \cdot 10^4$	$2.5294 \cdot 10^7$
3.7	$4.0609 \cdot 10^5$	$2.4823 \cdot 10^7$	7.2	$3.5545 \cdot 10^4$	$2.5298 \cdot 10^7$
3.8	$3.6965 \cdot 10^5$	$2.4862 \cdot 10^7$	7.3	$3.3745 \cdot 10^4$	$2.5301 \cdot 10^7$
3.9	$3.3718 \cdot 10^5$	$2.4897 \cdot 10^7$	7.4	$3.2057 \cdot 10^4$	$2.5305 \cdot 10^7$

Table 2.2. (*continued*)

λ [μm]	$m_{\lambda,bb}$ [W/μm m²]	$M_{T,o-\lambda,bb}$ [W/m²]	λ [μm]	$m_{\lambda,bb}$ [W/μm m²]	$M_{T,o-\lambda,bb}$ [W/m²]
7.5	$3.0472 \cdot 10^4$	$2.5308 \cdot 10^7$	11.5	$5.9520 \cdot 10^3$	$2.5364 \cdot 10^7$
7.6	$2.8985 \cdot 10^4$	$2.5311 \cdot 10^7$	11.6	$5.7565 \cdot 10^3$	$2.5365 \cdot 10^7$
7.7	$2.7587 \cdot 10^4$	$2.5314 \cdot 10^7$	11.7	$5.5689 \cdot 10^3$	$2.5365 \cdot 10^7$
7.8	$2.6272 \cdot 10^4$	$2.5316 \cdot 10^7$	11.8	$5.3889 \cdot 10^3$	$2.5366 \cdot 10^7$
7.9	$2.5035 \cdot 10^4$	$2.5319 \cdot 10^7$	11.9	$5.2161 \cdot 10^3$	$2.5367 \cdot 10^7$
8.0	$2.3869 \cdot 10^4$	$2.5321 \cdot 10^7$	12.0	$5.0501 \cdot 10^3$	$2.5367 \cdot 10^7$
8.1	$2.2770 \cdot 10^4$	$2.5324 \cdot 10^7$	12.1	$4.8907 \cdot 10^3$	$2.5368 \cdot 10^7$
8.2	$2.1734 \cdot 10^4$	$2.5326 \cdot 10^7$	12.2	$4.7376 \cdot 10^3$	$2.5368 \cdot 10^7$
8.3	$2.0756 \cdot 10^4$	$2.5328 \cdot 10^7$	12.3	$4.5903 \cdot 10^3$	$2.5368 \cdot 10^7$
8.4	$1.9832 \cdot 10^4$	$2.5330 \cdot 10^7$	12.4	$4.4488 \cdot 10^3$	$2.5369 \cdot 10^7$
8.5	$1.8959 \cdot 10^4$	$2.5332 \cdot 10^7$	12.5	$4.3127 \cdot 10^3$	$2.5369 \cdot 10^7$
8.6	$1.8134 \cdot 10^4$	$2.5334 \cdot 10^7$	12.6	$4.1817 \cdot 10^3$	$2.5370 \cdot 10^7$
8.7	$1.7353 \cdot 10^4$	$2.5335 \cdot 10^7$	12.7	$4.0557 \cdot 10^3$	$2.5370 \cdot 10^7$
8.8	$1.6613 \cdot 10^4$	$2.5337 \cdot 10^7$	12.8	$3.9344 \cdot 10^3$	$2.5371 \cdot 10^7$
8.9	$1.5913 \cdot 10^4$	$2.5339 \cdot 10^7$	12.9	$3.8175 \cdot 10^3$	$2.5371 \cdot 10^7$
9.0	$1.5249 \cdot 10^4$	$2.5340 \cdot 10^7$	13.0	$3.7050 \cdot 10^3$	$2.5371 \cdot 10^7$
9.1	$1.4619 \cdot 10^4$	$2.5342 \cdot 10^7$	13.1	$3.5966 \cdot 10^3$	$2.5372 \cdot 10^7$
9.2	$1.4021 \cdot 10^4$	$2.5343 \cdot 10^7$	13.2	$3.4921 \cdot 10^3$	$2.5372 \cdot 10^7$
9.3	$1.3454 \cdot 10^4$	$2.5345 \cdot 10^7$	13.3	$3.3914 \cdot 10^3$	$2.5372 \cdot 10^7$
9.4	$1.2915 \cdot 10^4$	$2.5346 \cdot 10^7$	13.4	$3.2943 \cdot 10^3$	$2.5373 \cdot 10^7$
9.5	$1.2402 \cdot 10^4$	$2.5347 \cdot 10^7$	13.5	$3.2006 \cdot 10^3$	$2.5373 \cdot 10^7$
9.6	$1.1915 \cdot 10^4$	$2.5348 \cdot 10^7$	13.6	$3.1103 \cdot 10^3$	$2.5373 \cdot 10^7$
9.7	$1.1452 \cdot 10^4$	$2.5349 \cdot 10^7$	13.7	$3.0231 \cdot 10^3$	$2.5374 \cdot 10^7$
9.8	$1.1010 \cdot 10^4$	$2.5351 \cdot 10^7$	13.8	$2.9390 \cdot 10^3$	$2.5374 \cdot 10^7$
9.9	$1.0590 \cdot 10^4$	$2.5352 \cdot 10^7$	13.9	$2.8577 \cdot 10^3$	$2.5374 \cdot 10^7$
10.0	$1.0190 \cdot 10^4$	$2.5353 \cdot 10^7$	14.0	$2.7792 \cdot 10^3$	$2.5374 \cdot 10^7$
10.1	$9.8082 \cdot 10^3$	$2.5354 \cdot 10^7$	14.1	$2.7035 \cdot 10^3$	$2.5375 \cdot 10^7$
10.2	$9.4442 \cdot 10^3$	$2.5355 \cdot 10^7$	14.2	$2.6302 \cdot 10^3$	$2.5375 \cdot 10^7$
10.3	$9.0970 \cdot 10^3$	$2.5356 \cdot 10^7$	14.3	$2.5595 \cdot 10^3$	$2.5375 \cdot 10^7$
10.4	$8.7655 \cdot 10^3$	$2.5356 \cdot 10^7$	14.4	$2.4911 \cdot 10^3$	$2.5375 \cdot 10^7$
10.5	$8.4490 \cdot 10^3$	$2.5357 \cdot 10^7$	14.5	$2.4249 \cdot 10^3$	$2.5376 \cdot 10^7$
10.6	$8.1467 \cdot 10^3$	$2.5358 \cdot 10^7$	14.6	$2.3610 \cdot 10^3$	$2.5376 \cdot 10^7$
10.7	$7.8578 \cdot 10^3$	$2.5359 \cdot 10^7$	14.7	$2.2991 \cdot 10^3$	$2.5376 \cdot 10^7$
10.8	$7.5815 \cdot 10^3$	$2.5360 \cdot 10^7$	14.8	$2.2393 \cdot 10^3$	$2.5376 \cdot 10^7$
10.9	$7.3173 \cdot 10^3$	$2.5360 \cdot 10^7$	14.9	$2.1814 \cdot 10^3$	$2.5377 \cdot 10^7$
11.0	$7.0644 \cdot 10^3$	$2.5361 \cdot 10^7$	15.0	$2.1253 \cdot 10^3$	$2.5377 \cdot 10^7$
11.1	$6.8224 \cdot 10^3$	$2.5362 \cdot 10^7$	15.1	$2.0711 \cdot 10^3$	$2.5377 \cdot 10^7$
11.2	$6.5907 \cdot 10^3$	$2.5362 \cdot 10^7$	15.2	$2.0185 \cdot 10^3$	$2.5377 \cdot 10^7$
11.3	$6.3687 \cdot 10^3$	$2.5363 \cdot 10^7$	15.3	$1.9676 \cdot 10^3$	$2.5377 \cdot 10^7$
11.4	$6.1560 \cdot 10^3$	$2.5364 \cdot 10^7$	15.4	$1.9183 \cdot 10^3$	$2.5378 \cdot 10^7$

Table .2. 2 (*continued*)

temperature 4700 K

λ [µm]	$m_{\lambda,bb}$ [W/µm m²]	$M_{T,o-\lambda,bb}$ [W/m²]	λ [µm]	$m_{\lambda,bb}$ [W/µm m²]	$M_{T,o-\lambda,bb}$ [W/m²]
0.4	$1.7349 \cdot 10^{7}$	$1.3669 \cdot 10^{6}$			
0.5	$2.6316 \cdot 10^{7}$	$3.5975 \cdot 10^{6}$	4.0	$3.1785 \cdot 10^{5}$	$2.7197 \cdot 10^{7}$
0.6	$2.9456 \cdot 10^{7}$	$6.4307 \cdot 10^{6}$	4.1	$2.9099 \cdot 10^{5}$	$2.7227 \cdot 10^{7}$
0.7	$2.8434 \cdot 10^{7}$	$9.3499 \cdot 10^{6}$	4.2	$2.6690 \cdot 10^{5}$	$2.7255 \cdot 10^{7}$
0.8	$2.5430 \cdot 10^{7}$	$1.2053 \cdot 10^{7}$	4.3	$2.4524 \cdot 10^{5}$	$2.7281 \cdot 10^{7}$
0.9	$2.1847 \cdot 10^{7}$	$1.4417 \cdot 10^{7}$	4.4	$2.2573 \cdot 10^{5}$	$2.7304 \cdot 10^{7}$
1.0	$1.8384 \cdot 10^{7}$	$1.6426 \cdot 10^{7}$	4.5	$2.0810 \cdot 10^{5}$	$2.7326 \cdot 10^{7}$
1.1	$1.5319 \cdot 10^{7}$	$1.8108 \cdot 10^{7}$	4.6	$1.9216 \cdot 10^{5}$	$2.7346 \cdot 10^{7}$
1.2	$1.2722 \cdot 10^{7}$	$1.9506 \cdot 10^{7}$	4.7	$1.7771 \cdot 10^{5}$	$2.7364 \cdot 10^{7}$
1.3	$1.0568 \cdot 10^{7}$	$2.0667 \cdot 10^{7}$	4.8	$1.6459 \cdot 10^{5}$	$2.7381 \cdot 10^{7}$
1.4	$8.8013 \cdot 10^{6}$	$2.1632 \cdot 10^{7}$	4.9	$1.5265 \cdot 10^{5}$	$2.7397 \cdot 10^{7}$
1.5	$7.3578 \cdot 10^{6}$	$2.2438 \cdot 10^{7}$	5.0	$1.4178 \cdot 10^{5}$	$2.7412 \cdot 10^{7}$
1.6	$6.1790 \cdot 10^{6}$	$2.3113 \cdot 10^{7}$	5.1	$1.3185 \cdot 10^{5}$	$2.7425 \cdot 10^{7}$
1.7	$5.2144 \cdot 10^{6}$	$2.3681 \cdot 10^{7}$	5.2	$1.2277 \cdot 10^{5}$	$2.7438 \cdot 10^{7}$
1.8	$4.4225 \cdot 10^{6}$	$2.4161 \cdot 10^{7}$	5.3	$1.1446 \cdot 10^{5}$	$2.7450 \cdot 10^{7}$
1.9	$3.7698 \cdot 10^{6}$	$2.4570 \cdot 10^{7}$	5.4	$1.0684 \cdot 10^{5}$	$2.7461 \cdot 10^{7}$
2.0	$3.2293 \cdot 10^{6}$	$2.4919 \cdot 10^{7}$	5.5	$9.9837 \cdot 10^{4}$	$2.7471 \cdot 10^{7}$
2.1	$2.7796 \cdot 10^{6}$	$2.5219 \cdot 10^{7}$	5.6	$9.3399 \cdot 10^{4}$	$2.7481 \cdot 10^{7}$
2.2	$2.4036 \cdot 10^{6}$	$2.5477 \cdot 10^{7}$	5.7	$8.7471 \cdot 10^{4}$	$2.7490 \cdot 10^{7}$
2.3	$2.0877 \cdot 10^{6}$	$2.5701 \cdot 10^{7}$	5.8	$8.2005 \cdot 10^{4}$	$2.7499 \cdot 10^{7}$
2.4	$1.8210 \cdot 10^{6}$	$2.5896 \cdot 10^{7}$	5.9	$7.6959 \cdot 10^{4}$	$2.7506 \cdot 10^{7}$
2.5	$1.5949 \cdot 10^{6}$	$2.6067 \cdot 10^{7}$	6.0	$7.2293 \cdot 10^{4}$	$2.7514 \cdot 10^{7}$
2.6	$1.4023 \cdot 10^{6}$	$2.6216 \cdot 10^{7}$	6.1	$6.7976 \cdot 10^{4}$	$2.7521 \cdot 10^{7}$
2.7	$1.2374 \cdot 10^{6}$	$2.6348 \cdot 10^{7}$	6.2	$6.3975 \cdot 10^{4}$	$2.7528 \cdot 10^{7}$
2.8	$1.0958 \cdot 10^{6}$	$2.6465 \cdot 10^{7}$	6.3	$6.0263 \cdot 10^{4}$	$2.7534 \cdot 10^{7}$
2.9	$9.7364 \cdot 10^{5}$	$2.6568 \cdot 10^{7}$	6.4	$5.6817 \cdot 10^{4}$	$2.7540 \cdot 10^{7}$
3.0	$8.6785 \cdot 10^{5}$	$2.6660 \cdot 10^{7}$	6.5	$5.3613 \cdot 10^{4}$	$2.7545 \cdot 10^{7}$
3.1	$7.7590 \cdot 10^{5}$	$2.6742 \cdot 10^{7}$	6.6	$5.0631 \cdot 10^{4}$	$2.7550 \cdot 10^{7}$
3.2	$6.9570 \cdot 10^{5}$	$2.6816 \cdot 10^{7}$	6.7	$4.7853 \cdot 10^{4}$	$2.7555 \cdot 10^{7}$
3.3	$6.2552 \cdot 10^{5}$	$2.6881 \cdot 10^{7}$	6.8	$4.5263 \cdot 10^{4}$	$2.7560 \cdot 10^{7}$
3.4	$5.6389 \cdot 10^{5}$	$2.6941 \cdot 10^{7}$	6.9	$4.2845 \cdot 10^{4}$	$2.7564 \cdot 10^{7}$
3.5	$5.0961 \cdot 10^{5}$	$2.6994 \cdot 10^{7}$	7.0	$4.0587 \cdot 10^{4}$	$2.7568 \cdot 10^{7}$
3.6	$4.6166 \cdot 10^{5}$	$2.7043 \cdot 10^{7}$	7.1	$3.8475 \cdot 10^{4}$	$2.7572 \cdot 10^{7}$
3.7	$4.1919 \cdot 10^{5}$	$2.7087 \cdot 10^{7}$	7.2	$3.6498 \cdot 10^{4}$	$2.7576 \cdot 10^{7}$
3.8	$3.8146 \cdot 10^{5}$	$2.7127 \cdot 10^{7}$	7.3	$3.4647 \cdot 10^{4}$	$2.7580 \cdot 10^{7}$
3.9	$3.4785 \cdot 10^{5}$	$2.7163 \cdot 10^{7}$	7.4	$3.2911 \cdot 10^{4}$	$2.7583 \cdot 10^{7}$

Table 2.2. (*continued*)

λ [μm]	$m_{\lambda,bb}$ [W/μm m²]	$M_{T,o-\lambda,bb}$ [W/m²]	λ [μm]	$m_{\lambda,bb}$ [W/μm m²]	$M_{T,o-\lambda,bb}$ [W/m²]
7.5	$3.1283 \cdot 10^4$	$2.7586 \cdot 10^7$	11.5	$6.0998 \cdot 10^3$	$2.7644 \cdot 10^7$
7.6	$2.9754 \cdot 10^4$	$2.7589 \cdot 10^7$	11.6	$5.8993 \cdot 10^3$	$2.7645 \cdot 10^7$
7.7	$2.8317 \cdot 10^4$	$2.7592 \cdot 10^7$	11.7	$5.7069 \cdot 10^3$	$2.7645 \cdot 10^7$
7.8	$2.6966 \cdot 10^4$	$2.7595 \cdot 10^7$	11.8	$5.5222 \cdot 10^3$	$2.7646 \cdot 10^7$
7.9	$2.5694 \cdot 10^4$	$2.7598 \cdot 10^7$	11.9	$5.3450 \cdot 10^3$	$2.7647 \cdot 10^7$
8.0	$2.4496 \cdot 10^4$	$2.7600 \cdot 10^7$	12.0	$5.1749 \cdot 10^3$	$2.7647 \cdot 10^7$
8.1	$2.3367 \cdot 10^4$	$2.7602 \cdot 10^7$	12.1	$5.0114 \cdot 10^3$	$2.7648 \cdot 10^7$
8.2	$2.2303 \cdot 10^4$	$2.7605 \cdot 10^7$	12.2	$4.8543 \cdot 10^3$	$2.7648 \cdot 10^7$
8.3	$2.1298 \cdot 10^4$	$2.7607 \cdot 10^7$	12.3	$4.7034 \cdot 10^3$	$2.7648 \cdot 10^7$
8.4	$2.0349 \cdot 10^4$	$2.7609 \cdot 10^7$	12.4	$4.5582 \cdot 10^3$	$2.7649 \cdot 10^7$
8.5	$1.9452 \cdot 10^4$	$2.7611 \cdot 10^7$	12.5	$4.4187 \cdot 10^3$	$2.7649 \cdot 10^7$
8.6	$1.8604 \cdot 10^4$	$2.7613 \cdot 10^7$	12.6	$4.2844 \cdot 10^3$	$2.7650 \cdot 10^7$
8.7	$1.7802 \cdot 10^4$	$2.7615 \cdot 10^7$	12.7	$4.1552 \cdot 10^3$	$2.7650 \cdot 10^7$
8.8	$1.7043 \cdot 10^4$	$2.7616 \cdot 10^7$	12.8	$4.0308 \cdot 10^3$	$2.7651 \cdot 10^7$
8.9	$1.6323 \cdot 10^4$	$2.7618 \cdot 10^7$	12.9	$3.9110 \cdot 10^3$	$2.7651 \cdot 10^7$
9.0	$1.5641 \cdot 10^4$	$2.7620 \cdot 10^7$	13.0	$3.7956 \cdot 10^3$	$2.7651 \cdot 10^7$
9.1	$1.4995 \cdot 10^4$	$2.7621 \cdot 10^7$	13.1	$3.6845 \cdot 10^3$	$2.7652 \cdot 10^7$
9.2	$1.4381 \cdot 10^4$	$2.7623 \cdot 10^7$	13.2	$3.5774 \cdot 10^3$	$2.7652 \cdot 10^7$
9.3	$1.3798 \cdot 10^4$	$2.7624 \cdot 10^7$	13.3	$3.4742 \cdot 10^3$	$2.7652 \cdot 10^7$
9.4	$1.3245 \cdot 10^4$	$2.7625 \cdot 10^7$	13.4	$3.3746 \cdot 10^3$	$2.7653 \cdot 10^7$
9.5	$1.2719 \cdot 10^4$	$2.7627 \cdot 10^7$	13.5	$3.2786 \cdot 10^3$	$2.7653 \cdot 10^7$
9.6	$1.2219 \cdot 10^4$	$2.7628 \cdot 10^7$	13.6	$3.1860 \cdot 10^3$	$2.7653 \cdot 10^7$
9.7	$1.1743 \cdot 10^4$	$2.7629 \cdot 10^7$	13.7	$3.0966 \cdot 10^3$	$2.7654 \cdot 10^7$
9.8	$1.1290 \cdot 10^4$	$2.7630 \cdot 10^7$	13.8	$3.0104 \cdot 10^3$	$2.7654 \cdot 10^7$
9.9	$1.0859 \cdot 10^4$	$2.7631 \cdot 10^7$	13.9	$2.9271 \cdot 10^3$	$2.7654 \cdot 10^7$
10.0	$1.0448 \cdot 10^4$	$2.7632 \cdot 10^7$	14.0	$2.8467 \cdot 10^3$	$2.7655 \cdot 10^7$
10.1	$1.0056 \cdot 10^4$	$2.7633 \cdot 10^7$	14.1	$2.7690 \cdot 10^3$	$2.7655 \cdot 10^7$
10.2	$9.6826 \cdot 10^3$	$2.7634 \cdot 10^7$	14.2	$2.6939 \cdot 10^3$	$2.7655 \cdot 10^7$
10.3	$9.3263 \cdot 10^3$	$2.7635 \cdot 10^7$	14.3	$2.6214 \cdot 10^3$	$2.7655 \cdot 10^7$
10.4	$8.9862 \cdot 10^3$	$2.7636 \cdot 10^7$	14.4	$2.5513 \cdot 10^3$	$2.7656 \cdot 10^7$
10.5	$8.6615 \cdot 10^3$	$2.7637 \cdot 10^7$	14.5	$2.4835 \cdot 10^3$	$2.7656 \cdot 10^7$
10.6	$8.3513 \cdot 10^3$	$2.7638 \cdot 10^7$	14.6	$2.4180 \cdot 10^3$	$2.7656 \cdot 10^7$
10.7	$8.0548 \cdot 10^3$	$2.7639 \cdot 10^7$	14.7	$2.3546 \cdot 10^3$	$2.7656 \cdot 10^7$
10.8	$7.7714 \cdot 10^3$	$2.7640 \cdot 10^7$	14.8	$2.2953 \cdot 10^3$	$2.7657 \cdot 10^7$
10.9	$7.5003 \cdot 10^3$	$2.7640 \cdot 10^7$	14.9	$2.2339 \cdot 10^3$	$2.7657 \cdot 10^7$
11.0	$7.2409 \cdot 10^3$	$2.7641 \cdot 10^7$	15.0	$2.1765 \cdot 10^3$	$2.7657 \cdot 10^7$
11.1	$6.9926 \cdot 10^3$	$2.7642 \cdot 10^7$	15.1	$2.1209 \cdot 10^3$	$2.7657 \cdot 10^7$
11.2	$6.7549 \cdot 10^3$	$2.7642 \cdot 10^7$	15.2	$2.0671 \cdot 10^3$	$2.7657 \cdot 10^7$
11.3	$6.5272 \cdot 10^3$	$2.7643 \cdot 10^7$	15.3	$2.0149 \cdot 10^3$	$2.7658 \cdot 10^7$
11.4	$6.3090 \cdot 10^3$	$2.7644 \cdot 10^7$	15.4	$1.9644 \cdot 10^3$	$2.7658 \cdot 10^7$

Table 2.2. (*continued*)

temperature 4800 K

λ [μm]	$m_{\lambda,bb}$ [W/μm m²]	$M_{T,o-\lambda,bb}$ [W/m²]	λ [μm]	$m_{\lambda,bb}$ [W/μm m²]	$M_{T,o-\lambda,bb}$ [W/m²]
0.4	$2.0350 \cdot 10^7$	$1.6521 \cdot 10^6$			
0.5	$2.9905 \cdot 10^7$	$4.2217 \cdot 10^6$	4.0	$3.2754 \cdot 10^5$	$2.9615 \cdot 10^7$
0.6	$3.2782 \cdot 10^7$	$7.4052 \cdot 10^6$	4.1	$2.9979 \cdot 10^5$	$2.9646 \cdot 10^7$
0.7	$3.1185 \cdot 10^7$	$1.0629 \cdot 10^7$	4.2	$2.7491 \cdot 10^5$	$2.9675 \cdot 10^7$
0.8	$2.7591 \cdot 10^7$	$1.3577 \cdot 10^7$	4.3	$2.5254 \cdot 10^5$	$2.9701 \cdot 10^7$
0.9	$2.3511 \cdot 10^7$	$1.6132 \cdot 10^7$	4.4	$2.3240 \cdot 10^5$	$2.9725 \cdot 10^7$
1.0	$1.9658 \cdot 10^7$	$1.8287 \cdot 10^7$	4.5	$2.1421 \cdot 10^5$	$2.9747 \cdot 10^7$
1.1	$1.6298 \cdot 10^7$	$2.0080 \cdot 10^7$	4.6	$1.9777 \cdot 10^5$	$2.9768 \cdot 10^7$
1.2	$1.3478 \cdot 10^7$	$2.1565 \cdot 10^7$	4.7	$1.8286 \cdot 10^5$	$2.9787 \cdot 10^7$
1.3	$1.1158 \cdot 10^7$	$2.2793 \cdot 10^7$	4.8	$1.6933 \cdot 10^5$	$2.9805 \cdot 10^7$
1.4	$9.2662 \cdot 10^6$	$2.3811 \cdot 10^7$	4.9	$1.5702 \cdot 10^5$	$2.9821 \cdot 10^7$
1.5	$7.7275 \cdot 10^6$	$2.4658 \cdot 10^7$	5.0	$1.4581 \cdot 10^5$	$2.9836 \cdot 10^7$
1.6	$6.4758 \cdot 10^6$	$2.5366 \cdot 10^7$	5.1	$1.3558 \cdot 10^5$	$2.9850 \cdot 10^7$
1.7	$5.4549 \cdot 10^6$	$2.5960 \cdot 10^7$	5.2	$1.2623 \cdot 10^5$	$2.9863 \cdot 10^7$
1.8	$4.6192 \cdot 10^6$	$2.6463 \cdot 10^7$	5.3	$1.1767 \cdot 10^5$	$2.9875 \cdot 10^7$
1.9	$3.9319 \cdot 10^6$	$2.6889 \cdot 10^7$	5.4	$1.0982 \cdot 10^5$	$2.9887 \cdot 10^7$
2.0	$3.3640 \cdot 10^6$	$2.7253 \cdot 10^7$	5.5	$1.0261 \cdot 10^5$	$2.9897 \cdot 10^7$
2.1	$2.8923 \cdot 10^6$	$2.7565 \cdot 10^7$	5.6	$9.5980 \cdot 10^4$	$2.9907 \cdot 10^7$
2.2	$2.4986 \cdot 10^6$	$2.7834 \cdot 10^7$	5.7	$8.9878 \cdot 10^4$	$2.9916 \cdot 10^7$
2.3	$2.1683 \cdot 10^6$	$2.8067 \cdot 10^7$	5.8	$8.4252 \cdot 10^4$	$2.9925 \cdot 10^7$
2.4	$1.8898 \cdot 10^6$	$2.8269 \cdot 10^7$	5.9	$7.9058 \cdot 10^4$	$2.9933 \cdot 10^7$
2.5	$1.6539 \cdot 10^6$	$2.8446 \cdot 10^7$	6.0	$7.4258 \cdot 10^4$	$2.9941 \cdot 10^7$
2.6	$1.4531 \cdot 10^6$	$2.8601 \cdot 10^7$	6.1	$6.9816 \cdot 10^4$	$2.9948 \cdot 10^7$
2.7	$1.2815 \cdot 10^6$	$2.8738 \cdot 10^7$	6.2	$6.5700 \cdot 10^4$	$2.9955 \cdot 10^7$
2.8	$1.1342 \cdot 10^6$	$2.8858 \cdot 10^7$	6.3	$6.1883 \cdot 10^4$	$2.9961 \cdot 10^7$
2.9	$1.0072 \cdot 10^6$	$2.8965 \cdot 10^7$	6.4	$5.8338 \cdot 10^4$	$2.9967 \cdot 10^7$
3.0	$8.9736 \cdot 10^5$	$2.9060 \cdot 10^7$	6.5	$5.5043 \cdot 10^4$	$2.9973 \cdot 10^7$
3.1	$8.0193 \cdot 10^5$	$2.9145 \cdot 10^7$	6.6	$5.1978 \cdot 10^4$	$2.9978 \cdot 10^7$
3.2	$7.1873 \cdot 10^5$	$2.9221 \cdot 10^7$	6.7	$4.9122 \cdot 10^4$	$2.9983 \cdot 10^7$
3.3	$6.4597 \cdot 10^5$	$2.9289 \cdot 10^7$	6.8	$4.6459 \cdot 10^4$	$2.9988 \cdot 10^7$
3.4	$5.8211 \cdot 10^5$	$2.9351 \cdot 10^7$	6.9	$4.3974 \cdot 10^4$	$2.9993 \cdot 10^7$
3.5	$5.2590 \cdot 10^5$	$2.9406 \cdot 10^7$	7.0	$4.1653 \cdot 10^4$	$2.9997 \cdot 10^7$
3.6	$4.7627 \cdot 10^5$	$2.9456 \cdot 10^7$	7.1	$3.9483 \cdot 10^4$	$3.0001 \cdot 10^7$
3.7	$4.3231 \cdot 10^5$	$2.9501 \cdot 10^7$	7.2	$3.7452 \cdot 10^4$	$3.0005 \cdot 10^7$
3.8	$3.9329 \cdot 10^5$	$2.9543 \cdot 10^7$	7.3	$3.5550 \cdot 10^4$	$3.0008 \cdot 10^7$
3.9	$3.5855 \cdot 10^5$	$2.9580 \cdot 10^7$	7.4	$3.3767 \cdot 10^4$	$3.0012 \cdot 10^7$

Table 2.2. (*continued*)

λ [μm]	$m_{\lambda,bb}$ [W/μm m^2]	$M_{T,o-\lambda,bb}$ [W/m^2]	λ [μm]	$m_{\lambda,bb}$ [W/μm m^2]	$M_{T,o-\lambda,bb}$ [W/m^2]
7.5	$3.2094\cdot10^4$	$3.0015\cdot10^7$	11.5	$6.2477\cdot10^3$	$3.0075\cdot10^7$
7.6	$3.0523\cdot10^4$	$3.0018\cdot10^7$	11.6	$6.0421\cdot10^3$	$3.0075\cdot10^7$
7.7	$2.9047\cdot10^4$	$3.0021\cdot10^7$	11.7	$5.8449\cdot10^3$	$3.0076\cdot10^7$
7.8	$2.7659\cdot10^4$	$3.0024\cdot10^7$	11.8	$5.6557\cdot10^3$	$3.0076\cdot10^7$
7.9	$2.6353\cdot10^4$	$3.0027\cdot10^7$	11.9	$5.4740\cdot10^3$	$3.0077\cdot10^7$
8.0	$2.5123\cdot10^4$	$3.0029\cdot10^7$	12.0	$5.2996\cdot10^3$	$3.0078\cdot10^7$
8.1	$2.3964\cdot10^4$	$3.0032\cdot10^7$	12.1	$5.1321\cdot10^3$	$3.0078\cdot10^7$
8.2	$2.2871\cdot10^4$	$3.0034\cdot10^7$	12.2	$4.9711\cdot10^3$	$3.0079\cdot10^7$
8.3	$2.1840\cdot10^4$	$3.0036\cdot10^7$	12.3	$4.8164\cdot10^3$	$3.0079\cdot10^7$
8.4	$2.0866\cdot10^4$	$3.0039\cdot10^7$	12.4	$4.6677\cdot10^3$	$3.0080\cdot10^7$
8.5	$1.9945\cdot10^4$	$3.0041\cdot10^7$	12.5	$4.5247\cdot10^3$	$3.0080\cdot10^7$
8.6	$1.9075\cdot10^4$	$3.0042\cdot10^7$	12.6	$4.3871\cdot10^3$	$3.0080\cdot10^7$
8.7	$1.8252\cdot10^4$	$3.0044\cdot10^7$	12.7	$4.2547\cdot10^3$	$3.0081\cdot10^7$
8.8	$1.7472\cdot10^4$	$3.0046\cdot10^7$	12.8	$4.1272\cdot10^3$	$3.0081\cdot10^7$
8.9	$1.6734\cdot10^4$	$3.0048\cdot10^7$	12.9	$4.0045\cdot10^3$	$3.0082\cdot10^7$
9.0	$1.6034\cdot10^4$	$3.0049\cdot10^7$	13.0	$3.8863\cdot10^3$	$3.0082\cdot10^7$
9.1	$1.5370\cdot10^4$	$3.0051\cdot10^7$	13.1	$3.7724\cdot10^3$	$3.0082\cdot10^7$
9.2	$1.4741\cdot10^4$	$3.0053\cdot10^7$	13.2	$3.6627\cdot10^3$	$3.0083\cdot10^7$
9.3	$1.4143\cdot10^4$	$3.0054\cdot10^7$	13.3	$3.5569\cdot10^3$	$3.0083\cdot10^7$
9.4	$1.3575\cdot10^4$	$3.0055\cdot10^7$	13.4	$3.4549\cdot10^3$	$3.0083\cdot10^7$
9.5	$1.3036\cdot10^4$	$3.0057\cdot10^7$	13.5	$3.3566\cdot10^3$	$3.0084\cdot10^7$
9.6	$1.2522\cdot10^4$	$3.0058\cdot10^7$	13.6	$3.2617\cdot10^3$	$3.0084\cdot10^7$
9.7	$1.2034\cdot10^4$	$3.0059\cdot10^7$	13.7	$3.1702\cdot10^3$	$3.0084\cdot10^7$
9.8	$1.1570\cdot10^4$	$3.0060\cdot10^7$	13.8	$3.0818\cdot10^3$	$3.0085\cdot10^7$
9.9	$1.1127\cdot10^4$	$3.0061\cdot10^7$	13.9	$2.9965\cdot10^3$	$3.0085\cdot10^7$
10.0	$1.0706\cdot10^4$	$3.0063\cdot10^7$	14.0	$2.9141\cdot10^3$	$3.0085\cdot10^7$
10.1	$1.0304\cdot10^4$	$3.0064\cdot10^7$	14.1	$2.8345\cdot10^3$	$3.0086\cdot10^7$
10.2	$9.9212\cdot10^3$	$3.0065\cdot10^7$	14.2	$2.7577\cdot10^3$	$3.0086\cdot10^7$
10.3	$9.5558\cdot10^3$	$3.0066\cdot10^7$	14.3	$2.6834\cdot10^3$	$3.0086\cdot10^7$
10.4	$9.2070\cdot10^3$	$3.0066\cdot10^7$	14.4	$2.6116\cdot10^3$	$3.0086\cdot10^7$
10.5	$8.8740\cdot10^3$	$3.0067\cdot10^7$	14.5	$2.5422\cdot10^3$	$3.0087\cdot10^7$
10.6	$8.5559\cdot10^3$	$3.0068\cdot10^7$	14.6	$2.4750\cdot10^3$	$3.0087\cdot10^7$
10.7	$8.2519\cdot10^3$	$3.0069\cdot10^7$	14.7	$2.4101\cdot10^3$	$3.0087\cdot10^7$
10.8	$7.9613\cdot10^3$	$3.0070\cdot10^7$	14.8	$2.3473\cdot10^3$	$3.0087\cdot10^7$
10.9	$7.6833\cdot10^3$	$3.0071\cdot10^7$	14.9	$2.2865\cdot10^3$	$3.0088\cdot10^7$
11.0	$7.4174\cdot10^3$	$3.0071\cdot10^7$	15.0	$2.2277\cdot10^3$	$3.0088\cdot10^7$
11.1	$7.1629\cdot10^3$	$3.0072\cdot10^7$	15.1	$2.1708\cdot10^3$	$3.0088\cdot10^7$
11.2	$6.9192\cdot10^3$	$3.0073\cdot10^7$	15.2	$2.1156\cdot10^3$	$3.0088\cdot10^7$
11.3	$6.6858\cdot10^3$	$3.0073\cdot10^7$	15.3	$2.0622\cdot10^3$	$3.0088\cdot10^7$
11.4	$6.4621\cdot10^3$	$3.0074\cdot10^7$	15.4	$2.0105\cdot10^3$	$3.0089\cdot10^7$

Table 2.2. (*continued*)

temperature 4900 K

λ [μm]	$m_{\lambda,bb}$ [W/μm m²]	$M_{T,o-\lambda,bb}$ [W/m²]	λ [μm]	$m_{\lambda,bb}$ [W/μm m²]	$M_{T,o-\lambda,bb}$ [W/m²]
0.4	$2.3715 \cdot 10^{7}$	$1.9831 \cdot 10^{6}$			
0.5	$3.3808 \cdot 10^{7}$	$4.9264 \cdot 10^{6}$	4.0	$3.3724 \cdot 10^{5}$	$3.2189 \cdot 10^{7}$
0.6	$3.6327 \cdot 10^{7}$	$8.4869 \cdot 10^{6}$	4.1	$3.0860 \cdot 10^{5}$	$3.2221 \cdot 10^{7}$
0.7	$3.4076 \cdot 10^{7}$	$1.2033 \cdot 10^{7}$	4.2	$2.8293 \cdot 10^{5}$	$3.2251 \cdot 10^{7}$
0.8	$2.9841 \cdot 10^{7}$	$1.5237 \cdot 10^{7}$	4.3	$2.5986 \cdot 10^{5}$	$3.2278 \cdot 10^{7}$
0.9	$2.5230 \cdot 10^{7}$	$1.7990 \cdot 10^{7}$	4.4	$2.3908 \cdot 10^{5}$	$3.2303 \cdot 10^{7}$
1.0	$2.0967 \cdot 10^{7}$	$2.0296 \cdot 10^{7}$	4.5	$2.2033 \cdot 10^{5}$	$3.2326 \cdot 10^{7}$
1.1	$1.7299 \cdot 10^{7}$	$2.2204 \cdot 10^{7}$	4.6	$2.0338 \cdot 10^{5}$	$3.2347 \cdot 10^{7}$
1.2	$1.4250 \cdot 10^{7}$	$2.3776 \cdot 10^{7}$	4.7	$1.8802 \cdot 10^{5}$	$3.2367 \cdot 10^{7}$
1.3	$1.1759 \cdot 10^{7}$	$2.5072 \cdot 10^{7}$	4.8	$1.7408 \cdot 10^{5}$	$3.2385 \cdot 10^{7}$
1.4	$9.7380 \cdot 10^{6}$	$2.6143 \cdot 10^{7}$	4.9	$1.6140 \cdot 10^{5}$	$3.2402 \cdot 10^{7}$
1.5	$8.1021 \cdot 10^{6}$	$2.7033 \cdot 10^{7}$	5.0	$1.4985 \cdot 10^{5}$	$3.2417 \cdot 10^{7}$
1.6	$6.7761 \cdot 10^{6}$	$2.7774 \cdot 10^{7}$	5.1	$1.3932 \cdot 10^{5}$	$3.2432 \cdot 10^{7}$
1.7	$5.6980 \cdot 10^{6}$	$2.8396 \cdot 10^{7}$	5.2	$1.2969 \cdot 10^{5}$	$3.2445 \cdot 10^{7}$
1.8	$4.8177 \cdot 10^{6}$	$2.8920 \cdot 10^{7}$	5.3	$1.2088 \cdot 10^{5}$	$3.2458 \cdot 10^{7}$
1.9	$4.0954 \cdot 10^{6}$	$2.9365 \cdot 10^{7}$	5.4	$1.1280 \cdot 10^{5}$	$3.2469 \cdot 10^{7}$
2.0	$3.4997 \cdot 10^{6}$	$2.9744 \cdot 10^{7}$	5.5	$1.0538 \cdot 10^{5}$	$3.2480 \cdot 10^{7}$
2.1	$3.0059 \cdot 10^{6}$	$3.0068 \cdot 10^{7}$	5.6	$9.8564 \cdot 10^{4}$	$3.2490 \cdot 10^{7}$
2.2	$2.5942 \cdot 10^{6}$	$3.0347 \cdot 10^{7}$	5.7	$9.2287 \cdot 10^{4}$	$3.2500 \cdot 10^{7}$
2.3	$2.2493 \cdot 10^{6}$	$3.0589 \cdot 10^{7}$	5.8	$8.6501 \cdot 10^{4}$	$3.2509 \cdot 10^{7}$
2.4	$1.9589 \cdot 10^{6}$	$3.0799 \cdot 10^{7}$	5.9	$8.1160 \cdot 10^{4}$	$3.2517 \cdot 10^{7}$
2.5	$1.7132 \cdot 10^{6}$	$3.0982 \cdot 10^{7}$	6.0	$7.6224 \cdot 10^{4}$	$3.2525 \cdot 10^{7}$
2.6	$1.5043 \cdot 10^{6}$	$3.1143 \cdot 10^{7}$	6.1	$7.1658 \cdot 10^{4}$	$3.2532 \cdot 10^{7}$
2.7	$1.3258 \cdot 10^{6}$	$3.1284 \cdot 10^{7}$	6.2	$6.7427 \cdot 10^{4}$	$3.2539 \cdot 10^{7}$
2.8	$1.1728 \cdot 10^{6}$	$3.1409 \cdot 10^{7}$	6.3	$6.3504 \cdot 10^{4}$	$3.2546 \cdot 10^{7}$
2.9	$1.0410 \cdot 10^{6}$	$3.1520 \cdot 10^{7}$	6.4	$5.9861 \cdot 10^{4}$	$3.2552 \cdot 10^{7}$
3.0	$9.2698 \cdot 10^{5}$	$3.1618 \cdot 10^{7}$	6.5	$5.6475 \cdot 10^{4}$	$3.2558 \cdot 10^{7}$
3.1	$8.2803 \cdot 10^{5}$	$3.1705 \cdot 10^{7}$	6.6	$5.3325 \cdot 10^{4}$	$3.2563 \cdot 10^{7}$
3.2	$7.4183 \cdot 10^{5}$	$3.1784 \cdot 10^{7}$	6.7	$5.0392 \cdot 10^{4}$	$3.2568 \cdot 10^{7}$
3.3	$6.6648 \cdot 10^{5}$	$3.1854 \cdot 10^{7}$	6.8	$4.7656 \cdot 10^{4}$	$3.2573 \cdot 10^{7}$
3.4	$6.0038 \cdot 10^{5}$	$3.1917 \cdot 10^{7}$	6.9	$4.5104 \cdot 10^{4}$	$3.2578 \cdot 10^{7}$
3.5	$5.4223 \cdot 10^{5}$	$3.1974 \cdot 10^{7}$	7.0	$4.2720 \cdot 10^{4}$	$3.2582 \cdot 10^{7}$
3.6	$4.9090 \cdot 10^{5}$	$3.2026 \cdot 10^{7}$	7.1	$4.0492 \cdot 10^{4}$	$3.2587 \cdot 10^{7}$
3.7	$4.4547 \cdot 10^{5}$	$3.2073 \cdot 10^{7}$	7.2	$3.8406 \cdot 10^{4}$	$3.2590 \cdot 10^{7}$
3.8	$4.0515 \cdot 10^{5}$	$3.2115 \cdot 10^{7}$	7.3	$3.6453 \cdot 10^{4}$	$3.2594 \cdot 10^{7}$
3.9	$3.6926 \cdot 10^{5}$	$3.2154 \cdot 19^{7}$	7.4	$3.4622 \cdot 10^{4}$	$3.2598 \cdot 10^{7}$

Table 2. 2. (*continued*)

λ [μm]	$m_{\lambda,bb}$ [W/μm m²]	$M_{T,o-\lambda,bb}$ [W/m²]	λ [μm]	$m_{\lambda,bb}$ [W/μm m²]	$M_{T,o-\lambda,bb}$ [W/m²]
7.5	$3.2905 \cdot 10^4$	$3.2601 \cdot 10^7$	11.5	$6.3956 \cdot 10^3$	$3.2662 \cdot 10^7$
7.6	$3.1293 \cdot 10^4$	$3.2604 \cdot 10^7$	11.6	$6.1849 \cdot 10^3$	$3.2663 \cdot 10^7$
7.7	$2.9778 \cdot 10^4$	$3.2607 \cdot 10^7$	11.7	$5.9829 \cdot 10^3$	$3.2663 \cdot 10^7$
7.8	$2.8354 \cdot 10^4$	$3.2610 \cdot 10^7$	11.8	$5.7891 \cdot 10^3$	$3.2664 \cdot 10^7$
7.9	$2.7013 \cdot 10^4$	$3.2613 \cdot 10^7$	11.9	$5.6030 \cdot 10^3$	$3.2664 \cdot 10^7$
8.0	$2.5751 \cdot 10^4$	$3.2616 \cdot 10^7$	12.0	$5.4244 \cdot 10^3$	$3.2665 \cdot 10^7$
8.1	$2.4562 \cdot 10^4$	$3.2618 \cdot 10^7$	12.1	$5.2528 \cdot 10^3$	$3.2666 \cdot 10^7$
8.2	$2.3440 \cdot 10^4$	$3.2620 \cdot 10^7$	12.2	$5.0880 \cdot 10^3$	$3.2666 \cdot 10^7$
8.3	$2.2382 \cdot 10^4$	$3.2623 \cdot 10^7$	12.3	$4.9295 \cdot 10^3$	$3.2667 \cdot 10^7$
8.4	$2.1383 \cdot 10^4$	$3.2625 \cdot 10^7$	12.4	$4.7772 \cdot 10^3$	$3.2667 \cdot 10^7$
8.5	$2.0438 \cdot 10^4$	$3.2627 \cdot 10^7$	12.5	$4.6307 \cdot 10^3$	$3.2667 \cdot 10^7$
8.6	$1.9546 \cdot 10^4$	$3.2629 \cdot 10^7$	12.6	$4.4898 \cdot 10^3$	$3.2668 \cdot 10^7$
8.7	$1.8701 \cdot 10^4$	$3.2631 \cdot 10^7$	12.7	$4.3542 \cdot 10^3$	$3.2668 \cdot 10^7$
8.8	$1.7902 \cdot 10^4$	$3.2633 \cdot 10^7$	12.8	$4.2237 \cdot 10^3$	$3.2669 \cdot 10^7$
8.9	$1.7144 \cdot 10^4$	$3.2634 \cdot 10^7$	12.9	$4.0980 \cdot 10^3$	$3.2669 \cdot 10^7$
9.0	$1.6427 \cdot 10^4$	$3.2636 \cdot 10^7$	13.0	$3.9769 \cdot 10^3$	$3.2670 \cdot 10^7$
9.1	$1.5746 \cdot 10^4$	$3.2638 \cdot 10^7$	13.1	$3.8603 \cdot 10^3$	$3.2670 \cdot 10^7$
9.2	$1.5101 \cdot 10^4$	$3.2639 \cdot 10^7$	13.2	$3.7480 \cdot 10^3$	$3.2670 \cdot 10^7$
9.3	$1.4488 \cdot 10^4$	$3.2641 \cdot 10^7$	13.3	$3.6397 \cdot 10^3$	$3.2671 \cdot 10^7$
9.4	$1.3906 \cdot 10^4$	$3.2642 \cdot 10^7$	13.4	$3.5353 \cdot 10^3$	$3.2671 \cdot 10^7$
9.5	$1.3352 \cdot 10^4$	$3.2644 \cdot 10^7$	13.5	$3.4346 \cdot 10^3$	$3.2671 \cdot 10^7$
9.6	$1.2826 \cdot 10^4$	$3.2645 \cdot 10^7$	13.6	$3.3374 \cdot 10^3$	$3.2672 \cdot 10^7$
9.7	$1.2326 \cdot 10^4$	$3.2646 \cdot 10^7$	13.7	$3.2437 \cdot 10^3$	$3.2672 \cdot 10^7$
9.8	$1.1850 \cdot 10^4$	$3.2647 \cdot 10^7$	13.8	$3.1532 \cdot 10^3$	$3.2672 \cdot 10^7$
9.9	$1.1396 \cdot 10^4$	$3.2648 \cdot 10^7$	13.9	$3.0659 \cdot 10^3$	$3.2673 \cdot 10^7$
10.0	$1.0964 \cdot 10^4$	$3.2650 \cdot 10^7$	14.0	$2.9815 \cdot 10^3$	$3.2673 \cdot 10^7$
10.1	$1.0552 \cdot 10^4$	$3.2651 \cdot 10^7$	14.1	$2.9001 \cdot 10^3$	$3.2673 \cdot 10^7$
10.2	$1.0160 \cdot 10^4$	$3.2652 \cdot 10^7$	14.2	$2.8214 \cdot 10^3$	$3.2673 \cdot 10^7$
10.3	$9.7852 \cdot 10^3$	$3.2653 \cdot 10^7$	14.3	$2.7454 \cdot 10^3$	$3.2674 \cdot 10^7$
10.4	$9.4278 \cdot 10^3$	$3.2654 \cdot 10^7$	14.4	$2.6719 \cdot 10^3$	$3.2674 \cdot 10^7$
10.5	$9.0865 \cdot 10^3$	$3.2654 \cdot 10^7$	14.5	$2.6008 \cdot 10^3$	$3.2674 \cdot 10^7$
10.6	$8.7605 \cdot 10^3$	$3.2655 \cdot 10^7$	14.6	$2.5321 \cdot 10^3$	$3.2675 \cdot 10^7$
10.7	$8.4490 \cdot 10^3$	$3.2656 \cdot 10^7$	14.7	$2.4656 \cdot 10^3$	$3.2675 \cdot 10^7$
10.8	$8.1512 \cdot 10^3$	$3.2657 \cdot 10^7$	14.8	$2.4013 \cdot 10^3$	$3.2675 \cdot 10^7$
10.9	$7.8664 \cdot 10^3$	$3.2658 \cdot 10^7$	14.9	$2.3391 \cdot 10^3$	$3.2675 \cdot 10^7$
11.0	$7.5940 \cdot 10^3$	$3.2659 \cdot 10^7$	15.0	$2.2789 \cdot 10^3$	$3.2675 \cdot 10^7$
11.1	$7.3332 \cdot 10^3$	$3.2659 \cdot 10^7$	15.1	$2.2206 \cdot 10^3$	$3.2676 \cdot 10^7$
11.2	$7.0835 \cdot 10^3$	$3.2660 \cdot 10^7$	15.2	$2.1642 \cdot 10^3$	$3.2676 \cdot 10^7$
11.3	$6.8443 \cdot 10^3$	$3.2661 \cdot 10^7$	15.3	$2.1095 \cdot 10^3$	$3.2676 \cdot 10^7$
11.4	$6.6152 \cdot 10^3$	$3.2661 \cdot 10^7$	15.4	$2.0566 \cdot 10^3$	$3.2676 \cdot 10^7$

Table 2. 2. (*continued*)

temperature 5000 K

λ [µm]	$m_{\lambda,bb}$ [W/µm m²]	$M_{T,o-\lambda,bb}$ [W/m²]	λ [µm]	$m_{\lambda,bb}$ [W/µm m²]	$M_{T,o-\lambda,bb}$ [W/m²]
0.4	$2.7468 \cdot 10^7$	$2.3648 \cdot 10^6$			
0.5	$3.8035 \cdot 10^7$	$5.7183 \cdot 10^6$	4.0	$3.4696 \cdot 10^5$	$3.4927 \cdot 10^7$
0.6	$4.0094 \cdot 10^7$	$9.6829 \cdot 10^6$	4.1	$3.1743 \cdot 10^5$	$3.4960 \cdot 10^7$
0.7	$3.7108 \cdot 10^7$	$1.3570 \cdot 10^7$	4.2	$2.9096 \cdot 10^5$	$3.4991 \cdot 10^7$
0.8	$3.2178 \cdot 10^7$	$1.7041 \cdot 10^7$	4.3	$2.6718 \cdot 10^5$	$3.5019 \cdot 10^7$
0.9	$2.7004 \cdot 10^7$	$1.9998 \cdot 10^7$	4.4	$2.4577 \cdot 10^5$	$3.5044 \cdot 10^7$
1.0	$2.2311 \cdot 10^7$	$2.2459 \cdot 10^7$	4.5	$2.2646 \cdot 10^5$	$3.5068 \cdot 10^7$
1.1	$1.8322 \cdot 10^7$	$2.4484 \cdot 10^7$	4.6	$2.0900 \cdot 10^5$	$3.5090 \cdot 10^7$
1.2	$1.5036 \cdot 10^7$	$2.6147 \cdot 10^7$	4.7	$1.9318 \cdot 10^5$	$3.5110 \cdot 10^7$
1.3	$1.2369 \cdot 10^7$	$2.7512 \cdot 10^7$	4.8	$1.7883 \cdot 10^5$	$3.5128 \cdot 10^7$
1.4	$1.0216 \cdot 10^7$	$2.8638 \cdot 10^7$	4.9	$1.6578 \cdot 10^5$	$3.5146 \cdot 10^7$
1.5	$8.4813 \cdot 10^6$	$2.9569 \cdot 10^7$	5.0	$1.5390 \cdot 10^5$	$3.5161 \cdot 10^7$
1.6	$7.0797 \cdot 10^6$	$3.0345 \cdot 10^7$	5.1	$1.4306 \cdot 10^5$	$3.5176 \cdot 10^7$
1.7	$5.9435 \cdot 10^6$	$3.0994 \cdot 10^7$	5.2	$1.3316 \cdot 10^5$	$3.5190 \cdot 10^7$
1.8	$5.0180 \cdot 10^6$	$3.1541 \cdot 10^7$	5.3	$1.2409 \cdot 10^5$	$3.5203 \cdot 10^7$
1.9	$4.2602 \cdot 10^6$	$3.2003 \cdot 10^7$	5.4	$1.1578 \cdot 10^5$	$3.5215 \cdot 10^7$
2.0	$3.6365 \cdot 10^6$	$3.2397 \cdot 10^7$	5.5	$1.0816 \cdot 10^5$	$3.5226 \cdot 10^7$
2.1	$3.1201 \cdot 10^6$	$3.2734 \cdot 10^7$	5.6	$1.0115 \cdot 10^5$	$3.5237 \cdot 10^7$
2.2	$2.6904 \cdot 10^6$	$3.3024 \cdot 10^7$	5.7	$9.4698 \cdot 10^4$	$3.5246 \cdot 10^7$
2.3	$2.3308 \cdot 10^6$	$3.3274 \cdot 10^7$	5.8	$8.8751 \cdot 10^4$	$3.5256 \cdot 10^7$
2.4	$2.0284 \cdot 10^6$	$3.3492 \cdot 10^7$	5.9	$8.3263 \cdot 10^4$	$3.5264 \cdot 10^7$
2.5	$1.7727 \cdot 10^6$	$3.3682 \cdot 10^7$	6.0	$7.8192 \cdot 10^4$	$3.5272 \cdot 10^7$
2.6	$1.5556 \cdot 10^6$	$3.3848 \cdot 10^7$	6.1	$7.3501 \cdot 10^4$	$3.5280 \cdot 10^7$
2.7	$1.3703 \cdot 10^6$	$3.3994 \cdot 10^7$	6.2	$6.9155 \cdot 10^4$	$3.5287 \cdot 10^7$
2.8	$1.2115 \cdot 10^6$	$3.4123 \cdot 10^7$	6.3	$6.5126 \cdot 10^4$	$3.5294 \cdot 10^7$
2.9	$1.0748 \cdot 10^6$	$3.4237 \cdot 10^7$	6.4	$6.1385 \cdot 10^4$	$3.5300 \cdot 10^7$
3.0	$9.5669 \cdot 10^5$	$3.4338 \cdot 10^7$	6.5	$5.7908 \cdot 10^4$	$3.5306 \cdot 10^7$
3.1	$8.5422 \cdot 10^5$	$3.4429 \cdot 10^7$	6.6	$5.4674 \cdot 10^4$	$3.5311 \cdot 10^7$
3.2	$7.6500 \cdot 10^5$	$3.4510 \cdot 10^7$	6.7	$5.1662 \cdot 10^4$	$3.5317 \cdot 10^7$
3.3	$6.8704 \cdot 10^5$	$3.4582 \cdot 10^7$	6.8	$4.8855 \cdot 10^4$	$3.5322 \cdot 10^7$
3.4	$6.1870 \cdot 10^5$	$3.4647 \cdot 10^7$	6.9	$4.6235 \cdot 10^4$	$3.5327 \cdot 10^7$
3.5	$5.5860 \cdot 10^5$	$3.4706 \cdot 10^7$	7.0	$4.3788 \cdot 10^4$	$3.5331 \cdot 10^7$
3.6	$5.0557 \cdot 10^5$	$3.4759 \cdot 10^7$	7.1	$4.1501 \cdot 10^4$	$3.5335 \cdot 10^7$
3.7	$4.5866 \cdot 10^5$	$3.4807 \cdot 10^7$	7.2	$3.9361 \cdot 10^4$	$3.5339 \cdot 10^7$
3.8	$4.1703 \cdot 10^5$	$3.4851 \cdot 10^7$	7.3	$3.7357 \cdot 10^4$	$3.5343 \cdot 10^7$
3.9	$3.8000 \cdot 10^5$	$3.4891 \cdot 10^7$	7.4	$3.5479 \cdot 10^4$	$3.5347 \cdot 10^7$

Table 2.2. (*continued*)

λ [μm]	$m_{\lambda,bb}$ [W/μm m²]	$M_{T,o-\lambda,bb}$ [W/m²]	λ [μm]	$m_{\lambda,bb}$ [W/μm m²]	$M_{T,o-\lambda,bb}$ [W/m²]
7.5	$3.3717 \cdot 10^4$	$3.5350 \cdot 10^7$	11.5	$6.5435 \cdot 10^3$	$3.5413 \cdot 10^7$
7.6	$3.2063 \cdot 10^4$	$3.5354 \cdot 10^7$	11.6	$6.3278 \cdot 10^3$	$3.5413 \cdot 10^7$
7.7	$3.0509 \cdot 10^4$	$3.5357 \cdot 10^7$	11.7	$6.1210 \cdot 10^3$	$3.5414 \cdot 10^7$
7.8	$2.9048 \cdot 10^4$	$3.5360 \cdot 10^7$	11.8	$5.9226 \cdot 10^3$	$3.5415 \cdot 10^7$
7.9	$2.7674 \cdot 10^4$	$3.5362 \cdot 10^7$	11.9	$5.7321 \cdot 10^3$	$3.5415 \cdot 10^7$
8.0	$2.6379 \cdot 10^4$	$3.5365 \cdot 10^7$	12.0	$5.5492 \cdot 10^3$	$3.5416 \cdot 10^7$
8.1	$2.5160 \cdot 10^4$	$3.5368 \cdot 10^7$	12.1	$5.3735 \cdot 10^3$	$3.5416 \cdot 10^7$
8.2	$2.4010 \cdot 10^4$	$3.5370 \cdot 10^7$	12.2	$5.2948 \cdot 10^3$	$3.5417 \cdot 10^7$
8.3	$2.2924 \cdot 10^4$	$3.5372 \cdot 10^7$	12.3	$5.0426 \cdot 10^3$	$3.5417 \cdot 10^7$
8.4	$2.1900 \cdot 10^4$	$3.5375 \cdot 10^7$	12.4	$4.8867 \cdot 10^3$	$3.5418 \cdot 10^7$
8.5	$2.0932 \cdot 10^4$	$3.5377 \cdot 10^7$	12.5	$4.7367 \cdot 10^3$	$3.5418 \cdot 10^7$
8.6	$2.0017 \cdot 10^4$	$3.5379 \cdot 10^7$	12.6	$4.5925 \cdot 10^3$	$3.5419 \cdot 10^7$
8.7	$1.9151 \cdot 10^4$	$3.5381 \cdot 10^7$	12.7	$4.4537 \cdot 10^3$	$3.5419 \cdot 10^7$
8.8	$1.8331 \cdot 10^4$	$3.5383 \cdot 10^7$	12.8	$4.3201 \cdot 10^3$	$3.5420 \cdot 10^7$
8.9	$1.7555 \cdot 10^4$	$3.5384 \cdot 10^7$	12.9	$4.1915 \cdot 10^3$	$3.5420 \cdot 10^7$
9.0	$1.6820 \cdot 10^4$	$3.5386 \cdot 10^7$	13.0	$4.0676 \cdot 10^3$	$3.5420 \cdot 10^7$
9.1	$1.6122 \cdot 10^4$	$3.5388 \cdot 10^7$	13.1	$3.9483 \cdot 10^3$	$3.5421 \cdot 10^7$
9.2	$1.5461 \cdot 10^4$	$3.5389 \cdot 10^7$	13.2	$3.8333 \cdot 10^3$	$3.5421 \cdot 10^7$
9.3	$1.4833 \cdot 10^4$	$3.5391 \cdot 10^7$	13.3	$3.7225 \cdot 10^3$	$3.5422 \cdot 10^7$
9.4	$1.4236 \cdot 10^4$	$3.5392 \cdot 10^7$	13.4	$3.6156 \cdot 10^3$	$3.5422 \cdot 10^7$
9.5	$1.3669 \cdot 10^4$	$3.5394 \cdot 10^7$	13.5	$3.5126 \cdot 10^3$	$3.5422 \cdot 10^7$
9.6	$1.3130 \cdot 10^4$	$3.5395 \cdot 10^7$	13.6	$3.4132 \cdot 10^3$	$3.5423 \cdot 10^7$
9.7	$1.2617 \cdot 10^4$	$3.5396 \cdot 10^7$	13.7	$3.3172 \cdot 10^3$	$3.5423 \cdot 10^7$
9.8	$1.2129 \cdot 10^4$	$3.5398 \cdot 10^7$	13.8	$3.2247 \cdot 10^3$	$3.5423 \cdot 10^7$
9.9	$1.1665 \cdot 10^4$	$3.5399 \cdot 10^7$	13.9	$3.1353 \cdot 10^3$	$3.5424 \cdot 10^7$
10.0	$1.1222 \cdot 10^4$	$3.5400 \cdot 10^7$	14.0	$3.0490 \cdot 10^3$	$3.5424 \cdot 10^7$
10.1	$1.0801 \cdot 10^4$	$3.5401 \cdot 10^7$	14.1	$2.9657 \cdot 10^3$	$3.5424 \cdot 10^7$
10.2	$1.0398 \cdot 10^4$	$3.5402 \cdot 10^7$	14.2	$2.8851 \cdot 10^3$	$3.5424 \cdot 10^7$
10.3	$1.0015 \cdot 10^4$	$3.5403 \cdot 10^7$	14.3	$2.8073 \cdot 10^3$	$3.5425 \cdot 10^7$
10.4	$9.6487 \cdot 10^3$	$3.5404 \cdot 10^7$	14.4	$2.7321 \cdot 10^3$	$3.5425 \cdot 10^7$
10.5	$9.2991 \cdot 10^3$	$3.5405 \cdot 10^7$	14.5	$2.6594 \cdot 10^3$	$3.5425 \cdot 10^7$
10.6	$8.9652 \cdot 10^3$	$3.5406 \cdot 10^7$	14.6	$2.5891 \cdot 10^3$	$3.5426 \cdot 10^7$
10.7	$8.6462 \cdot 10^3$	$3.5407 \cdot 10^7$	14.7	$2.5211 \cdot 10^3$	$3.5426 \cdot 10^7$
10.8	$8.3413 \cdot 10^3$	$3.5408 \cdot 10^7$	14.8	$2.4554 \cdot 10^3$	$3.5426 \cdot 10^7$
10.9	$8.0496 \cdot 10^3$	$3.5408 \cdot 10^7$	14.9	$2.3917 \cdot 10^3$	$3.5426 \cdot 10^7$
11.0	$7.7706 \cdot 10^3$	$3.5409 \cdot 10^7$	15.0	$2.3301 \cdot 10^3$	$3.5426 \cdot 10^7$
11.1	$7.5036 \cdot 10^3$	$3.5410 \cdot 10^7$	15.1	$2.2705 \cdot 10^3$	$3.5427 \cdot 10^7$
11.2	$7.2479 \cdot 10^3$	$3.5411 \cdot 10^7$	15.2	$2.2128 \cdot 10^3$	$3.5427 \cdot 10^7$
11.3	$7.0030 \cdot 10^3$	$3.5411 \cdot 10^7$	15.3	$2.1569 \cdot 10^3$	$3.5427 \cdot 10^7$
11.4	$6.7684 \cdot 10^3$	$3.5412 \cdot 10^7$	15.4	$2.1027 \cdot 10^3$	$3.5427 \cdot 10^7$

Part III. TABLES OF RADIATIVE PROPERTIES OF MATERIALS

3.1. TABLES OF RADIATIVE PROPERTIES OF METALS AND ALLOYS

Table 3.1.1. Hemispherical total emissivity ε_T of metals and alloys

State of surface	T [K]	ε_T
Aluminium; Al		
polished surface, 98% AL [12, 76, 78]	294÷423	0.04
	473	0.05
	773	0.08
electrolytically polished surface [103]	175	0.010
	200	0.012
	225	0.014
	250	0.016
	275	0.016
	300	0.019
	350	0.020
	400	0.022
	450	0.025
	500	0.027
	550	0.029
	600	0.032
	650	0.035
	700	0.037
surface grinded [12]	303	0.05
surface polished weakly oxidized [12]	313	0.05
surface raw [12]	299	0.071
surface rough [12]	299	0.05
surface very rough [12]	313	0.07
surface oxidized [12]	313÷363	0.05÷0.2
surface very oxidized [12]	323	0.2
Antimony, Sb		
surface polished [76, 78]	313÷363	0.28
	423	0.29
	473	0.30
	523	0.33
	1373	0.40
Cast iron, carbon		
surface polished [12]	313÷363	0.21
cast surface [12]	293÷373	0.94
roughness of surface rms = 2.5÷5.0 μm [12]	293	0.15
	308	0.19
	323	0.27
	333	0.39
surface oxidized [12]	313÷363	0.63
surface rough and oxidized [12]	313	0.98
Chromium; Cr		
surface polished [12, 132]	313÷363	0.08÷0.11
	423	0.071
surface rolled [12]	373	0.08

Table 3.1.1. (*continued*)

State of surface	T[K]	ε_T
Cobalt; Co		
surface smooth		
[131]	750	0.125
	875	0.15
	1000	0.175
	1125	0.2
	1250	0.23
Copper; Cu		
surface polished, investigated in vacuum		
[1, 2, 19, 26, 44, 92, 131, 132]	50	0.022
	250	0.023
	500	0.031
	625	0.038
	750	0.047
	875	0.054
	1125	0.061
surface rough [12]	313	0.74
surface mat [12]	313÷363	0.22
surface grinded [12]	293÷323	0.07
surface oxidized [12]	313÷363	0.56÷0.7
surface strongly oxidized [12]	293÷323	0.76÷0.88
surface black oxidized [12]	313÷363	0.92÷0.91
Copper - tin alloy (red bronze)		
surface rough [12]	323	0.55
Copper - zinc alloy (brass)		
surface polished [12]	293÷363	0.05
surface rolled [12]	293÷313	0.06÷0.07
surface roughly [12]	295	0.20
surface grinded [12]	313÷363	0.21
surface mat [12]	323÷373	0.21÷0.22
surface oxidized		
[76, 78]	313	0.46
	363	0.48
	423	0.50
	473	0.53
	523	0.56
	773	0.75
Gold; Au		
surface polished		
[19]	50	0.013
	125	0.017
	250	0.023
	375	0.029
	500	0.035
	625	0.040
	750	0.045
	875	0.050
	1000	0.056
	1125	0.062
Hafnium; Hf		
surface smooth		
[131]	1125	0.28
	1375	0.29
	1625	0.305
	1875	0.32
	2000	0.325
Iridium; Ir		
surface smooth		
[131]	1000	0.11
	1500	0.14
	2000	0.175
	2500	0.21
Iron; Fe		
surface polished		
[16, 26, 55, 61, 131, 144]	300	0.07
	500	0.13
	700	0.205
	900	0.275
	1100	0.34
surface polished electrolytic iron [12]	313÷363	0.06
surface weakly oxidized electrolytic iron [12]	313÷363	0.78

Table 3.1.1. (*continued*)

State of surface	T[K]	ε_T
surface oxidized [16, 55, 61, 131, 144]	300	0.43
	500	0.50
	700	0.56
	900	0.60
	1100	0.635
surface oxidized mat, forged iron [12]	313	0.95
Iron - tungsten alloy		
phase $\alpha + \varepsilon$ [45]	1150÷1250	0.352
phase $\alpha + \xi$ [45]	1320÷1470	0.34
phase α [45]	1550÷1720	0.32
Lead; Pb		
surface polished [12]	313÷363	0.05
surface rough [12]	313	0.43
surface oxidized [12]	293÷313	0.28
Mercury; Hg		
clean [12]	313÷373	0.09÷0.12
Molybdenum; Mo		
surface polished [2, 10, 12, 19, 27, 47, 131]	313÷363	0.06
	1100	0.10
	1200	0.12
	1300	0.125
	1400	0.135
	1500	0.15
	1600	0.168
	1700	0.175
	1800	0.185
	1900	0.20
	2000	0.213
	2100	0.222
	2200	0.235
	2300	0.24
	2500	0.26
	2600	0.265
	2700	0.276

State of surface	T[K]	ε_T
Molybdenum - rhenium alloy MoRe8 surface smooth, investigated in vacuum [131]	1200	0.14
	1400	0.16
	1600	0.18
	1800	0.205
	2000	0.23
	2200	0.25
Molybdenum - rhenium alloy MoRe20 surface smooth, investigated in vacuum [131]	1200	0.135
	1400	0.155
	1600	0.175
	1800	0.20
	2000	0.22
	2200	0.235
Molybdenum - rhenium alloy MoRe47 surface smooth, investigated in vacuum [131]	1200	0.13
	1400	0.15
	1600	0.173
	1800	0.195
	2000	0.218
	2200	0.234
Molybdenum - tungsten alloy MoW20 surface polished [36, 131]	2400÷3000	0.3
Molybdenum - tungsten alloy MoW30 surface polished [36, 131]	2400÷3000	0.285
Nickel; Ni		
surface polished [26, 45, 131]	250	0.053
	375	0.055
	500	0.058
	625	0.102
	750	0.12
	875	0.13
	1000	0.149
	1125	0.16
	1250	0.173
	1375	0.187
	1500	0.195

Table 3.1.1. (*continued*)

State of surface	T[K]	ε_T
surface smooth oxidized [16, 45]	50	0.385
	100	0.392
	200	0.402
	300	0.43
	400	0.457
	500	0.48
	600	0.515
	700	0.53
	800	0.56
	900	0.60
	1000	
surface grinded [12]	373	0.05
surface lustreless [12]	293	0.11
Nickel - chromium - iron alloy; nimonic 95-BS; ЭИ607 - GOST surface polished, oxidized for 30 min at temperature 1366 K [131]	500	0.88
	700	0.896
	900	0.91
	1100	0.92
	1300	0.925
	1400	0.925
Nickel - rhenium alloy NiRe8 surface smooth [131]	1200	0.168
	1250	0.173
	1300	0.178
	1350	0.184
	1400	0.19
	1450	0.195
Nickel - rhenium alloy NiRe11 surface smooth [131]	1200	0.18
	1250	0.185
	1300	0.19
	1350	0.194
	1400	0.199
	1450	0.203
Nickel - rhenium alloy NiRe16 surface smooth [131]	1200	0.19
	1250	0.195
	1300	0.20
	1350	0.248
	1400	0.296
	1450	0.214
Nickel - rhenium alloy NiRe24 surface smooth [131]	1200	0.199
	1250	0.204
	1300	0.209
	1350	0.215
	1400	0.22
	1450	0.225
Nickel - rhenium alloy NiRe30 surface smooth [131]	1200	0.197
	1250	0.202
	1300	0.207
	1350	0.213
	1400	0.217
	1450	0.223
Niobium; Nb surface smooth, roughness rms = 2.2 μm [131]	1000	0.07
	1125	0.11
	1250	0.136
	1375	0.160
	1500	0.17
	1625	0.18
	1750	0.19
	1875	0.20
	2000	0.213
	2125	0.225
	2250	0.23
	2375	0.242
	2500	0.252
Niobium - tungsten alloy NbW10 surface smooth [131, 142]	2000÷2600	0.267

Table 3.1.1. *(continued)*

State of surface	T[K]	ε_T
Niobium - tungsten alloy NbW15 surface smooth [131, 142]	2000	0.262
	2100	0.261
	2200	0.261
	2300	0.26
	2400	0.26
	2500	0.259
	2600	0.258
surface smooth [131, 142]	2000	0.255
	2200	0.259
	2400	0.262
	2600	0.266
Palladium; Pd wire thin [131]	1125	0.124
	1250	0.14
	1375	0.16
	1500	0.173
	1625	0.18
	1750	0.183
Platinum; Pt surface smooth, roughness rms = 0.04-0.14 μm [110, 146]	300	0.04
	375	0.05
	500	0.064
	625	0.08
	750	0.096
	875	0.117
	1000	0.128
	1125	0.137
	1250	0.154
	1375	0.167
	1500	0.175
	1625	0.185
	1750	0.194
	1800	0.196
wire [23]	300	0.04
	500	0.052
	700	0.09
	900	0.12
	1100	0.148
	1300	0.17
	1500	0.18
Platinum - rhodium alloy PtRh10 wire heating in protective atmosphere or vacuum [131]	300	0.08
	500	0.095
	700	0.12
	900	0.135
	1100	0.165
	1300	0.18
	1500	0.206
	1700	0.22
Rhodium; Rh surface smooth [93, 131]	1250	0.12
	1375	0.135
	1500	0.14
	1625	0.152
	1750	0.161
	1825	0.17
	2000	0.177
	2125	0.18
Silver; Ag surface polished [12]	313÷363	0.01
surface oxidized [12]	293÷313	0.02÷0.03
surface smooth [22, 131]	50	0.013
	250	0.018
	500	0.025
	750	0.035
	1000	0.044
	1100	0.048
surface mechanically polished [103]	410	0.017
	560	0.021
	710	0.024
	860	0.026
Steel, chromium - nickel - aluminium, stainless, H17N9J - PN surface smooth, roughness rms = 0.03 μm [131]	400	0.05
	500	0.05
	600	0.05
	700	0.05

Table 3.1.1. (*continued*)

State of surface	T[K]	ε_T
	800	0.06
	900	0.08
	1000	0.10
Steel, chromium - nickel, stainless, H25N20 - PN; 310-AISI; 30310-SAE surface smooth after storage [131]	800	0.23
	1000	0.26
	1200	0.28
	1400	0.29
surface oxidized for 1 h at temperature 1255 K [131]	300	0.465
	500	0.535
	700	0.60
	900	0.65
	1100	0.685
Steel, chromium - nickel - molybdenum, stainless, H17N5M3 - PN, surface polished [131]	550	0.13
	600	0.132
	700	0.14
	800	0.16
	900	0.185
	1000	0.23
	1100	0.29
	1200	0.37
Steel, chromium - nickel - niobium, stainless, OH18N12Nb - PN; 821 Grade Nb - BS surface polished, oxidized for 30 min at temperature 1366 K [45, 131]	600	0.88
	700	0.89
	800	0.895
	900÷1300	0.90
Steel, chromium - nickel, stainless, OH18N10 - PN; 801 Grade C-BS surface sandblasted, oxidized at temperature 1255 K [131]	500	0.71
	700	0.78
	900	0.92
	1100	0.96
	1300	0.96
Steel, chromium - nickel, stainless, OH18N10T - PN; 821 Grade Ti-BS surface smooth, measurements in air [131]	400	0.09
	500	0.095
	600	0.10
	700	0.105
	800	0.11
	900	0.14
	1000	0.23
	1100	0.38
	1200	0.45
surface oxidized for 3 h at temperature 1473 K [131]	600	0.51
	700	0.57
	800	0.605
	900	0.63
	1000	0.66
	1100	0.675
	1200	0.685
Steel, chromium - nickel stainless 304-AISI surface mechanically and electrolytically polished [108]	340	0.1043
	380	0.1110
	420	0.1177
	460	0.1239
	500	0.1302
	540	0.1359
	580	0.1416
	620	0.1469
	660	0.1521
	700	0.1571
	740	0.1617
	780	0.1665
	820	0.1709
	860	0.1756
	900	0.1791
	940	0.1829
	980	0.1866
	1020	0.1900
	1060	0.1932
	1100	0.1963

Table 3.1.1. (*continued*)

State of surface	T[K]	ε_T
Steel, high chromium, stainless, H28 - PN surface polished [45, 131]	400	0.11
	500	0.14
	600	0.16
	700	0.18
surface rough, rms = 3 μm [45, 131]	400	0.44
	500	0.50
	600	0.53
	700	0.54
surface rough, rms = 4 μm [45, 131]	400	0.56
	500	0.58
	600	0.59
	700	0.60
surface rough, rms = 8 μm [45, 131]	400	0.59
	500	0.64
	600	0.67
	700	0.675
Tantalum, Ta surface smooth nonoxidized [19, 44, 104, 131, 144]	1000	0.132
	1100	0.141
	1200	0.149
	1300	0.158
	1400	0.168
	1500	0.177
	1600	0.186
	1700	0.196
	1800	0.205
	1900	0.215
	2000	0.224
	2100	0.253
	2200	0.242
	2300	0.251
	2400	0.259
	2500	0.267
	2600	0.274
	2700	0.282
	2800	0.289
	2900	0.24
	3000	0.300
	3100	0.306

State of surface	T[K]	ε_T
	3200	0.311
	3300	0.316
Tin; Sn surface polished [12]	297÷373	0.04÷0.06
coating on the sheet steel [12]	293÷313	0.08
Titanium; Ti surface with residual impurities of TiO_2 and TiC [76, 131, 144, 150]	875	0.505
	1000	0.52
	1125	0.53
	1250	0.546
	1375	0.563
	1500	0.577
	1625	0.595
	1750	0.613
Titanium - aluminium alloy surface polished [45]	100	0.11
	300	0.15
	500	0.184
	700	0.222
	900	0.266
surface weakly oxidized [45]	100	0.314
	300	0.418
	500	0.595
	700	0.63
	900	0.666
	1100	0.68
surface strongly oxidized [45]	100	0.644
	300	0.66
	500	0.681
	700	0.702
	900	0.724
	1100	0.736
Tungsten; W monocrystal, surface 100 electrolytically polished [156]	343	0.0219
	379	0.0248
	421	0.0274
	461	0.0309
	501	0.0342

Table 3.1.1. (*continued*)

State of surface	T[K]	ε_T
	560	0.0397
	616	0.0450
	669	0.0507
	721	0.0573
	772	0.0632
	822	0.0696
	901	0.0812
	989	0.0944
	1077	0.1083
	1121	0.1150
	1165	0.1230
	1210	0.1319
	1254	0.1391
surface electrolytically polished measurement in vacuum [151]	180	0.0151
	200	0.0164
	220	0.0178
	240	0.0191
	260	0.0203
	280	0.0216
	300	0.0228
	400	0.0286
	440	0.0305
	480	0.0330
	520	0.0362
	560	0.0401
	600	0.0446
	640	0.0484
	680	0.0536
	720	0.0590
	760	0.0645
	800	0.0702
	840	0.0758
	880	0.0809
	920	0.0865
	960	0.0926
	1000	0.0992
surface smooth nonoxidized [12, 131. 146]	313	0.04
	363	0.04
	400	0.04
	600	0.058
	800	0.08
	1000	0.105
	1200	0.13
	1400	0.16
	1600	0.195
	1800	0.225
	2000	0.252
	2200	0.27
	2400	0.285

State of surface	T[K]	ε_T
	2600	0.305
	2800	0.318
	3000	0.325
	3200	0.337
	3400	0.348
surface wire - smooth [65]	300	0.03
	800	0.12
	1300	0.21
	1800	0.33
	2300	0.46
	2500	0.49
Tungsten - rhenium alloy WRe27 surface grinded [131]	1600	0.213
	1800	0.23
	2000	0.25
	2200	0.274
	2400	0.29
	2600	0.312
	2800	0.335
	3000	0.35
surface electrolytically polished [76]	313÷423	0.06
	473	0.07
	523	0.08
	773	0.12
	1373	0.22
Vanadium; V purity 99.7-99.9%, surface smooth, roughness rms = 0.3 μm [131, 144]	1125	0.168
	1250	0.18
	1375	0.199
	1500	0.213
	1625	0.225
	1750	0.237
	1825	0.246
Zinc, Zn surface bright [12]	298	0.04÷0.06
surface polished [12]	313÷373	0.02÷0.05
surface oxidized [12]	293÷363	0.27÷0.21

Table 3.1.1. (*continued*)

State of surface	T[K]	ε_T
surface mat [12]	293÷366	0.28÷0.20
Zirconium; Zr surface polished [131]	1125	0.41
	1250	0.42
	1375	0.43
	1500	0.435
	1625	0.45
	1750	0.46
	1875	0.47
	2000	0.475

Table 3.1.2. Total emissivity in normal direction $\varepsilon_{T,n}$ of metals and alloys

State of surface	T[K]	$\varepsilon_{T,n}$
Aluminium; Al		
surface polished [10, 19, 22, 40, 44, 93, 104, 109, 112, 115, 131, 132]	50	0.0075
	125	0.01
	250	0.024
	375	0.030
	500	0.039
	625	0.046
	750	0.057
	800	0.06
	900	0.065
foil [132]	293	0.04
	640	0.04
surface grinded [132]	373	0.095
surface rolled [2, 19, 26, 27]	443	0.039
	773	0.050
surface rough [1, 19, 40, 44]	300	0.055
	323÷773	0.06÷0.07
surface raw [19]	300	0.07
surface oxidized at temperature 873 K [19, 47, 55]	200	0.11
	600	0.19
surface oxidized for 30 min at temperature 800 K [1, 16, 19, 44, 132]	273	0.36
	373	0.37
	473	0.378
	573	0.386
	673	0.393
	773	0.4
	873	0.414
surface oxidized at temperature of measurement, roughness γ_w = 1 [122]	570	0.122
surface oxidized at temperature of measurement, roughness γ_w = 1.3 [122]	570	0.165
surface oxidized at temperature of measurement, roughness γ_w = 1.6 [122]	570	0.200
surface oxidized at temperature of measurement, roughness γ_w = 1.9 [122]	570	0.230
surface oxidized at temperature of measurement, roughness γ_w = 2.2 [122]	570	0.255
surface oxidized at temperature of measurement, roughness γ_w = 2.5 [122]	570	0.280
surface oxidized at temperature of measurement, roughness γ_w = 2.8 [122]	570	0.305

Table 3.1.2. (*continued*)

State of surface	T[K]	$\varepsilon_{T,n}$
surface oxidized at temperature of measurement, roughness γ_w = 3.1 [122]	570	0.325
surface oxidized at temperature of measurement, roughness γ_w = 3.4 [122]	570	0.350
surface oxidized at temperature of measurement, roughness γ_w = 3.7 [122]	570	0.365
surface oxidized at temperature of measurement, roughness γ_w = 4 [122]	570	0.380
surface very rough [12]	313	0.07
surface black-oxidized [78]	313÷363	0.2
	423	0.21
	473	0.22
	523	0.23
	773	0.33
plates of roof [19]	320	0.216
coating of aluminium on copper at temperature 873 K [19, 104]	473	0.18
	873	0.19
coating of aluminium on steel at temperature 873 K [19]	473	0.52
	873	0.57
Aluminium - copper - magnesium alloy, AlCu2Mg2Ni1Si; PA29N - PN; H18 - BS (duralumin) after rough treatment [140]	623	0.30
	673	0.28
	723	0.28
	773	0.33
Aluminium - copper - magnesium alloy AlCu4MgA; PA25 - PN (duralumin) after rough treatment [140]	623	0.30
	673	0.28
	723	0.28
	773	0.31
Aluminium - copper - magnesium alloy AlCu4Mg1; PA7N - PN; 2024 - ASTM (duralumin) [45]	100	0.08
	200	0.09
	300	0.10
	400	0.11
	500	0.12
	600	0.13
	700	0.14
	800	0.15
surface polished [131]	773	0.01
	873	0.02
	973	0.025
	1073	0.03
	1173	0.035
surface nonpolished [131]	673	0.06
	773	0.06
	873	0.065
	923	0.065
surface etched in air [131]	573	0.38
	673	0.35
	773	0.34
	873	0.31
	973	0.29
surface shaved [140]	623	0.25
	673	0.27
	723	0.30
	773	0.37

Table 3.1.2. (*continued*)

State of surface	T[K]	$\varepsilon_{T,n}$
Aluminium high - silicon alloy cast polished [19]	423	0.186
Aluminium - zinc alloy; 2Cu2Pa9 - PN; 7075 - ASTM [45]	100	0.08
	200	0.088
	300	0.1
	400	0.11
	500	0.12
	600	0.14
	700	0.15
	800	0.16
Antimony, Sb powder [10, 19, 106]	373	0.82
Berylium; Be surface clean [22]	423	0.16
	643	0.21
	753	0.26
	863	0.30
surface anodized [22]	423	0.90
	643	0.88
	753	0.85
	863	0.82
hot moulded sample, bright polish with few pits, measurement in vacuum, surface roughness rms = 0.19 µm [7]	400	0.023
	450	0.027
	500	0.031
	600	0.041
	650	0.047
	700	0.053
	750	0.060
	800	0.068
	850	0.076
Bismuth; Bi surface nonoxidized [10, 19, 26, 27, 44, 47, 104, 132]	300	0.048
	400	0.061
Cast iron, carbon surface polished [19, 37, 47, 76]	700÷1293	0.144÷0.377
surface machined, roughness rms = 2.5+5 µm [146]	293	0.20÷0.30
surface cast [12]	293÷373	0.81
surface turned [19, 37, 44, 146]	295	0.435
	1100	0.60
	1263	0.70
surface oxidized, temperature 873 K [1, 19, 37, 55]	473÷873	0.64÷0.78
surface strongly oxidized [19]	400÷800	0.78÷0.82
casting strongly oxidized [1, 19, 44]	323	0.81
surface strongly oxidized [146]	293	0.94
billet strongly oxidized [19]	373÷1273	0.80÷0.95
strongly oxidized and rough surface [146]	293÷313	0.98÷0.95
surface nonoxidized in liquid state [10, 51, 131]	1373÷1873	0.28÷0.45
surface oxidized in liquid state [51]	1523÷1853	0.90÷0.95
Chromium; Cr surface polished [16, 19, 26, 44, 45, 131, 132]	300÷500	0.07
	600	0.08
	700	0.1
	800	0.13
	900	0.18
	1000	0.25
	1100	0.32
	1200	0.39
	1300	0.58
surface rolled [10, 19]	500	0.175
	600	0.212

Table 3.1.2. (*continued*)

State of surface	T[K]	$\varepsilon_{T,n}$
	700	0.250
	800	0.287
	900	0.325
	1000	0.362
Chromium - nickel - zinc alloy; CrNi25Zn20 - PN* surface grey oxidized [10, 19]	300	0.262
Copper; Cu surface polished, investigated in vacuum [1, 2, 12, 19, 26, 44, 92, 131]	293÷373	0.02÷0.03
	500	0.014
	625	0.013
	750	0.012
	875	0.015
	1000	0.018
	1125	0.021
surface grinded [12]	293÷323	0.07
surface scraped for polish [1, 19, 132]	295	0.072
surface rough cleaning [132]	310	0.15
surface mat [12, 76]	293	0.37
	313÷363	0.22
surface slightly oxidized, insignificant lustreless [19, 26]	293	0.037
surface oxidized [45]	75	0.665
	100	0.64
	200	0.55
	300	0.475
	400	0.42
	500	0.393
	600	0.39
	700	0.395
	800	0.41
	900	0.46
	925	0.48
surface oxidized [1, 19, 44]	323	0.6÷0.7
surface oxidized [19, 26]	400	0.76
surface oxidized at temperature up to 873 K [1, 19]	473	0.57÷0.88
surface oxidized at temperature above 873 K [19, 53]	875	0.55
	1273	0.6
surface grinded, oxidized for 20 min at temperature of measurement (see column 3) [140]	893	0.49
	929	0.52
	973	0.61
	1023	0.68
	1073	0.86
surface grinded, oxidized for 80 min at temperature of measurement (see column 3) [140]	893	0.645
	923	0.725
	973	0.74
	1023	0.81
	1073	0.92
surface grinded, oxidized for 160 min at temperature of measurement (see column 3) [140]	893	0.725
	923	0.77
	973	0.795
	1023	0.86
	1073	0.923
surface grinded oxidized for 240 min at temperature of measurement (see column 3) [140]	893	0.74
	923	0.795
	973	0.81
	1023	0.88
	1073	0.928

Table 3.1.2. (*continued*)

State of surface	T[K]	$\varepsilon_{T,n}$
surface grinded oxidized for 320 min at temperature of measurement (see column 3) [140]	893	0.77
	923	0.805
	973	0.82
	1023	0.90
	1073	0.93
surface grinded, oxidized for 400 min at temperature of measurement (see column 3) [140]	893	0.80
	923	0.815
	973	0.83
	1023	0.91
surface grey oxidized [19, 26, 132]	293	0.78
surface black oxidized [44]	323	0.88
liquid state [19, 44, 53, 55]	1350	0.16
	1550	0.13
electrolytic powder [19, 106]	293÷1573	0.76
surface oxidized [76, 78]	313÷423	0.56
	473	0.57
	523	0.61
	773	0.88
surface black oxidized [76, 78]	313	0.92
	363	0.91
	423	0.90
	473	0.89
	523	0.83
	773	0.77
Copper - aluminium alloy; Ba5 - PN; Alloy 606 - ASTM (aluminium bronze) surface polished [45, 131]	50	0.03
	200	0.025
	400	0.015
	500	0.01
	600	0.015
	700	0.02
	800	0.025
	900	0.03
	1000	0.035
	1100	0.04
	1200	0.045
	1300	0.05
surface polished, oxidized in air for 30 min at temperature 900 K [45, 131]	100	0.11
	200	0.09
	300	0.08
	400	0.08
	500	0.09
	600	0.10
	700	0.11
	800	0.12
	900	0.13
	1000	0.14
	1100	0.145
	1200	0.155
	1300	0.16
powder [1, 19]	293	0.6
Copper - manganese - nickel alloy; manganin wire [19, 26, 109]	390	0.048÷0.057
Copper - tin alloy (red bronze) surface polished [1, 10, 19, 44]	323	0.1
surface rough [1, 19]	323÷423	0.55
powder [1]	323÷423	0.76÷0.8
Copper - zinc alloy (brass) surface polished [1, 2, 19, 44, 76, 81, 132]	293	0.05
	473÷643	0.028÷0.031
surface grinded [132]	366	0.09
sheet rolled [19, 44]	293	0.06

Table 3.1.2. (*continued*)

State of surface	T[K]	$\varepsilon_{T,n}$
surface treated with abrasive paper [19, 44, 55, 146]	293	0.2
surface mat [1, 2, 19, 44, 132]	293÷630	0.2÷0.22
surface oxidized at temperature 873 K [1, 19, 44, 53, 55, 132]	473	0.61
	873	0.59
Copper - zinc alloy CuZn30; M70 - PN; Alloy 260-ASTM (brass without alloying components)		
surface polished [131]	500÷650	0.03
surface polished, oxidized at temperature 973 K [131]	500	0.2
	550	0.22
	600	0.24
	700	0.27
	800	0.30
surface oxidized at temperature 873 K [131]	450÷900	0.6
Gold; Au		
surface polished [19, 26, 44, 131, 132]	250	0.019
	375	0.020
	500	0.021
	625	0.022
	750	0.023
	875	0.025
	1000	0.026
	1125	0.028
	1250	0.029
surface mat [12]	293	0.47

State of surface	T[K]	$\varepsilon_{T,n}$
Hafnium; Hf		
sample vacuum are melted, surface polished, roughness rms = 0.17 μm, measurement in vacuum [7]	400	0.088
	450	0.100
	500	0.115
	550	0.131
	600	0.149
	650	0.167
	700	0.186
	750	0.206
	800	0.227
	850	0.248
Iron; Fe		
surface smooth [45, 51]	100	0.021
	300	0.078
	500	0.137
	700	0.208
	900	0.32
	1100	0.486
surface polished electrolytically [19, 132]	310	0.05
	450	0.052
	500	0.064
	533	0.07
surface polished [19, 132]	673÷1273	0.14÷0.38
surface freshly grinded [55]	293	0.242
surface treated with abrasive paper [26, 44, 132]	293	0.24
surface clean forged [132]	310÷533	0.28
freshly casted [132]	310	0.44
surface oxidized at temperature of measurement in air [16, 55, 61, 131, 144]	600	0.225
	700	0.335
	800	0.445
	900	0.55

Table 3.1.2. (*continued*)

State of surface	T[K]	$\varepsilon_{T,n}$
surface oxidized [45, 51]	100	0.27
	300	0.416
	500	0.48
	700	0.536
	900	0.58
	1100	0.63
surface red oxidized [26]	293	0.61
surface strongly corroded naturally [26]	293	0.85
surface hot rolled [26]	293	0.77
surface casted, oxidized at temperature 873 K [26, 132]	473÷873	0.64÷0.78
surface casted [55]	1200÷1400	0.87÷0.95
surface casted rough and strongly oxidized [132]	315÷533	0.95
surface oxidized for 8 h at temperature 1373 K [16, 17]	573	0.41
	773	0.48
	973	0.53
	1173	0.565
	1373	0.59
	1573	0.60
surface oxidized for 8 h at temperature 1473 K [16, 17]	573	0.485
	773	0.54
	973	0.575
	1173	0.605
	1373	0.63
	1573	0.65
surface oxidized for 8 h at temperature 1573 K [16, 17]	573	0.56
	773	0.60
	973	0.63
	1173	0.65
	1373	0.66
	1573	0.67
Lead, Pb		
surface polished [19, 104, 132]	310÷400	0.056
	500	0.074
	520÷530	0.08
rough nonoxidized [76, 132]	315	0.43
surface grey, oxidized [1, 19, 26, 55, 76, 104]	293÷313	0.28
surface oxidized at temperature 473 K [1, 19, 44, 55, 76, 104]	473	0.63
Magnesium; Mg		
powder [19, 106]	393	0.86
Mercury; Hg [1, 19, 44, 55, 76, 104]	273	0.09
	373	0.12
Molybdenum, Mo		
surface smooth [2, 10, 19, 27, 47, 131]	500	0.03
	750	0.05
	1000	0.08
	1250	0.11
	1500	0.135
	1750	0.165
	2000	0.19
	2250	0.22
	2500	0.25
	2750	0.28
fibre [2, 10, 19, 47]	1000	0.096
	2873	0.292
Nickel; Ni		
surface polished [19, 26, 44, 81, 104, 131]	269	0.04
	373	0.045
	500	0.052
	675	0.078
	750	0.09
	875	0.113
	1000	0.13
	1125	0.145
	1250	0.17

Table 3.1.2. (*continued*)

State of surface	T[K]	$\varepsilon_{T,n}$
surface polished, oxidized at temperature 873 K [140]	573	0.25
	673	0.31
	773	0.38
	873	0.44
surface polished, oxidized at temperature 1173 K [140]	573	0.5
	673	0.51
	773	0.53
	873	0.57
	973	0.65
surface mat [12]	293	0.11
nickel wire [1, 19, 44]	460	0.096
	1273	0.186
surface rolled, oxidized at temperature 873 K [1, 19, 53, 55]	473	0.37
	873	0.48
	1473	0.85
surface grinded [12]	373	0.04
nickel plated steel sheet [19, 40, 53, 55]	300	0.11÷0.40
nickel plated cast iron [1, 19, 53, 55]	323	0.5
surface bright rough [26]	373	0.046
powder [19, 106]	300	0.78
Nickel - chromium alloy, N80H20-PN (nichrome) surface rolled [10, 19, 44]	973	0.21
surface sandblasted, and oxidized at temperature of measurement [10, 19, 44]	973	0.70
surface oxidized for a sufficiently long time at temperature 873 K [19]	473	0.41
	873	0.46
surface rolled, oxidized for 60 min at temperature 1373 K [16, 55, 86]	700	0.68
	900	0.70
	1100	0.725
	1300	0.76
surface rolled, oxidized for 30 min at temperature 1373 K [16, 55, 85, 86]	700	0.65
	900	0.67
	1100	0.695
	1300	0.755
surface rolled, oxidized for 20 min at temperature 1373 K [16, 55, 86]	700	0.61
	900	0.64
	1110	0.665
	1300	0.695
surface rolled, oxidized for 10 min at temperature 1373 K [16, 55, 86]	700	0.55
	900	0.585
	1100	0.61
	1300	0.64
surface rolled, oxidized for 5 min at temperature 1373 K [16, 55, 86]	700	0.5
	900	0.535
	1100	0.565
	1300	0.60
surface rolled, oxidized for 60 min at temperature 1273 K [16, 55, 86]	700	0.55
	900	0.605
	1100	0.66

Table 3.1.2. (*continued*)

State of surface	T[K]	$\varepsilon_{T,n}$	State of surface	T[K]	$\varepsilon_{T,n}$
surface rolled, oxidized for 40 min at temperature 1273 K [16, 55, 86]	700	0.54	surface rolled, oxidized for 20 min at temperature 1173 K [16, 55, 86]	700	0.44
	900	0.585		900	0.49
	1100	0.64		1100	0.545
surface rolled, oxidized for 30 min at temperature 1273 K [16, 55, 86]	700	0.515	surface rolled, oxidized for 10 min at temperature 1173 K [16, 55, 86]	700	0.4
	900	0.56		900	0.465
	1100	0.62		1100	0.525
surface rolled, oxidized for 20 min at temperature 1273 K [16, 55, 86]	700	0.475	surface rolled, oxidized for 5 min at temperature 1173 K [16, 55, 86]	700	0.39
	900	0.53		900	0.44
	1100	0.585		1100	0.50
surface rolled, oxidized for 10 min at temperature 1273 K [16, 55, 86]	700	0.46	surface rolled, oxidized for 60 min at temperature 1073 K [16, 55, 86]	700	0.45
	900	0.505		900	0.50
	1100	0.555		1000	0.53
surface rolled, oxidized for 5 min at temperature 1273 K [16, 55, 86]	700	0.44	surface rolled, oxidized for 30 min at temperature 1073 K [16, 55, 86]	700	0.43
	900	0.475		900	0.475
	1100	0.525		1000	0.50
surface rolled, oxidized for 60 min at temperature 1173 K [16, 55, 86]	700	0.48	surface rolled, oxidized for 20 min at temperature 1073 K [16, 55, 86]	700	0.4
	900	0.55		900	0.45
	1100	0.605		1000	0.47
surface rolled, oxidized for 45 min at temperature 1173 K [16, 55, 86]	700	0.465	surface rolled, oxidized for 10 min at temperature 1073 K [16, 55, 86]	700	0.38
	900	0.525		900	0.43
	1100	0.58		1000	0.46
surface rolled, oxidized for 30 min at temperature 1173 K [16, 55, 86]	700	0.45	surface rolled, oxidized for 5 min at temperature 1973 K [16, 55, 86]	700	0.37
	900	0.50		900	0.42
	1100	0.56		1000	0.45

Table 3.1.2. (*continued*)

State of surface	T[K]	$\varepsilon_{T,n}$
surface rolled, oxidized for 60 min at temperature 973 K [16, 55, 86]	700	0.38
	900	0.445
surface rolled, oxidized for 20 min at temperature 973 K [16, 55, 86]	700	0.35
	900	0.42
surface rolled, oxidized for 10 min at temperature 973 K [16, 55, 86]	700	0.335
	900	0.50
surface rolled, oxidized for 5 min at temperature 973 K [16, 55, 86]	700	0.31
	900	0.375
Nickel - chromium - iron alloy; inconel surface polished [45]	300÷500	0.21
Nickel - chromium - iron alloy; inconel B surface polished [45]	100	0.17
	300	0.175
	500	0.18
	700	0.182
	900	0.189
	1100	0.22
	1300	0.226
	1500	0.222
Nickel - chromium - iron alloy; inconel X surface polished [45]	100	0.056
	300	0.062
	500	0.87
	700	0.12
	900	0.151
	1100	0.185
	1300	0.22
surface oxidized [45]	100	0.075
	300	0.111
	500	0.152
	700	0.19
	900	0.231
	1100	0.27
	1300	0.325
Nickel - chromium - iron alloy; nicrothal 20-BK surface rolled clean [1, 16]	573	0.305
surface rolled, oxidized for 5 s at temperature 773 K [1, 16]	773	0.315
surface rolled, oxidized for 5 s at temperature 973 K [1, 16]	973	0.33
surface rolled, oxidized for 5 s at temperature 1073 K [1, 16]	1073	0.37
surface rolled, oxidized for 5 s at temperature 1173 K [1, 16]	1173	0.44
surface rolled, oxidized for 5 s at temperature 1273 K [1, 16]	1273	0.53
surface rolled, oxidized for 5 s at temperature 1373 K [1, 16]	1375	0.625
surface rolled, oxidized for 5 s at temperature 1473 K [1, 16]	1473	0.64
surface rolled, oxidized for 1 min at temperature 1073 K [1, 16]	1073	0.46
surface rolled, oxidized for 1 min at temperature 1173 K [1, 16]	1173	0.545

Table 3.1.2. (*continued*)

State of surface	T[K]	$\varepsilon_{T,n}$
surface rolled, oxidized for 1 min at temperature 1273 K [1, 16]	1273	0.63
surface rolled, oxidized for 1 min at temperature 1373 K [1, 16]	1373	0.72
surface rolled, oxidized for 1 min at temperature 1473 K [1, 16]	1473	0.775
surface rolled, oxidized for 10 min at temperature 1073 K [1, 16]	1073	0.53
surface rolled, oxidized for 10 min at temperature 1173 K [1, 16]	1173	0.63
surface rolled, oxidized for 10 min at temperature 1273 K [1, 16]	1273	0.724
surface rolled, oxidized for 10 min at temperature 1373 K [1, 16]	1373	0.80
surface rolled, oxidized for 10 min at temperature 1473 K [1, 16]	1473	0.85
surface rolled, oxidized for 30 min at temperature 1073 K [1, 16]	1073	0.625
surface rolled oxidized for 30 min at temperature 1173 K [1, 16]	1173	0.74
surface rolled, oxidized for 30 min at temperature 1273 K [1, 16]	1273	0.812
surface rolled, oxidized for 30 min at temperature 1373 K [1, 16]	1373	0.865
surface rolled, oxidized for 30 min at temperature 1473 K [1, 16]	1473	0.895
surface rolled, oxidized for 8 h at temperature 1073 K [1, 16]	1073	0.73
surface rolled, oxidized for 8 h at temperature 1173 K [1, 16]	1173	0.805
surface rolled, oxidized for 8 h at temperature 1273 K [1, 16]	1273	0.865
surface rolled, oxidized for 8 h at temperature 1373 K [1, 16]	1373	0.914
surface rolled, oxidized for 8 h at temperature 1473 K [1, 16]	1473	0.96
surface rolled, oxidized for 8 h at temperature 1073 K [1, 16]	573	0.32
	773	0.52
	973	0.64
	1173	0.73
	1373	0.74
	1573	0.75
surface rolled, oxidized for 8 h at temperature 1173 K [1, 16]	573	0.44
	773	0.62
	973	0.74
	1173	0.085
	1373	0.84
	1573	0.845

Table 3.1.2. (*continued*)

State of surface	T[K]	$\varepsilon_{T,n}$
surface rolled, oxidized for 8 h at temperature 1273 K [1, 16]	573	0.52
	773	0.72
	973	0.82
	1173	0.875
	1373	0.88
	1573	0.88
surface rolled, oxidized for 8 h at temperature 1373 K [1, 16]	573	0.67
	773	0.80
	973	0.88
	1173	0.905
	1373	0.913
	1573	0.924
surface rolled, oxidized for 8 h at temperature 1473 K [1, 16]	573	0.806
	773	0.88
	973	0.925
	1173	0.935
	1373	0.94
	1573	0.045
surface rolled, oxidized for 8 h at temperature 1573 K [1, 16]	573	0.95
	773	0.97
	973÷1573	0.98
Nickel - chromium - iron alloy; nikrothal 30-BK surface oxidized for 1.5 h at temperature 1073 K [1, 16]	1073	0.61
surface oxidized for 3 h at temperature 1073 K [1, 16]	1073	0.72
surface oxidized for 6 h at temperature 1073 K [1, 16]	1073	0.81
surface oxidized for 9 h at temperature 1073 K [1, 16]	1073	0.87
surface oxidized for 12 h at temperature 1073 K [1, 16]	1073	0.88
surface oxidized for more than 3 h at temperature 1423 K [1, 16]	1423	0.91
Nickel - chromium - iron alloy; nikrothal 40-BK surface rolled, [1, 16]	573÷873	0.3
surface rolled, oxidized for 5 s at temperature 973 K [1, 16]	973	0.327
surface rolled, oxidized for 5 s at temperature 1173 K [1, 16]	1173	0.48
surface rolled, oxidized for 5 s at temperature 1373 K [1, 16]	1373	0.59
surface rolled, oxidized for 5 s at temperature 1573 K [1, 16]	1573	0.61
surface oxidized for 1 min at temperature 1073 K [1, 16]	1073	0.445
surface oxidized for 1 min at temperature 1173 K [1, 16]	1173	0.53
surface oxidized for 1 min at temperature 1373 K [1, 16]	1373	0.715

Table 3.1.2. (*continued*)

State of surface	T[K]	$\varepsilon_{T,n}$
surface oxidized for 1 min at temperature 1573 K [1, 16]	1573	0.83
surface oxidized for 20 min at temperature 1073 K [1, 16]	1073	0.54
surface oxidized for 20 min at temperature 1173 K [1, 16]	1173	0.63
surface oxidized for 20 min at temperature 1373 K [1, 16]	1373	0.82
surface oxidized for 20 min at temperature 1573 K [1, 16]	1573	0.91
surface oxidized for 8 h at temperature 1073 K [1, 16]	1073	0.65
surface oxidized for 8 h at temperature 1273 K [1, 16]	1273	0.79
surface oxidized for 8 h at temperature 1473 K [1, 16]	1473	0.914
surface oxidized for 8 h at temperature 1073 K [1, 16]	573	0.35
	773	0.53
	973	0.58
	1173	0.62
	1373	0.65
	1573	0.67
surface oxidized for 8 h at temperature 1173 K [1, 16]	573	0.42
	773	0.60
	973	0.665
	1173	0.69
	1373	0.815
	1573	0.83
surface oxidized for 8 h at temperature 1273 K [1, 16]	573	0.54
	773	0.72
	973	0.77
	1173	0.79
	1373	0.805
	1573	0.82
surface oxidized for 8 h at temperature 1373 K [1, 16]	573	0.72
	773	0.87
	973÷1573	0.90
surface oxidized for 8 h at temperature 1473 K [1, 16]	573	0.97
	773÷1573	0.987
Nickel - chromium - iron alloy, nikrothal 60-BK surface rolled [1, 16]	573÷673	0.24
surface rolled, oxidized for 5 s at temperature 773 K [1, 16]	773	0.25
surface rolled, oxidized for 5 s at temperature 973 K [1, 16]	973	0.37
surface rolled, oxidized for 5 s at temperature 1173 K [1, 16]	1173	0.46
surface rolled, oxidized for 5 s at temperature 1373 K [1, 16]	1373	0.475
surface rolled, oxidized for 5 s at temperature 1573 K [1, 16]	1573	0.48

Table 3.1.2. (*continued*)

State of surface	T[K]	$\varepsilon_{T,n}$	State of surface	T[K]	$\varepsilon_{T,n}$
surface rolled, oxidized for 1 min at temperature 1073 K [1, 16]	1073	0.53	surface rolled, oxidized for 8 h at temperature 1073 K [1, 16, 19]	573	0.25
surface rolled, oxidized for 1 min at temperature 1173 K [1, 16]	1173	0.62		773	0.45
				973	0.56
				1173	0.63
				1373	0.66
				1573	0.67
surface rolled, oxidized for 1 min at temperature 1373 K [1, 16]	1373	0.736	surface rolled, oxidized for 8 h at temperature 1173 K [1, 16]	573	0.31
surface rolled, oxidized for 1 min at temperature 1573 K [1, 16]	1573	0.805		773	0.50
				973	0.60
				1173	0.66
				1373	0.693
				1573	0.71
surface rolled, oxidized for 20 min at temperature 1073 K [1, 16]	1073	0.635	surface rolled, oxidized for 8 h at temperature 1273 K [1, 16, 19]	573	0.37
surface rolled, oxidized for 20 min at temperature 1173 K [1, 16]	1173	0.68		773	0.44
				973	0.65
				1173	0.70
				1373	0.73
				1573	0.735
surface rolled, oxidized for 20 min at temperature 1373 K [1, 16]	1373	0.775	surface rolled, oxidized for 8 h at temperature 1373 K [1, 16, 19]	573	0.43
surface rolled oxidized for 20 min at temperature 1573 K [1, 16]	1573	0.84		773	0.615
				973	0.71
				1173	0.665
				1373	0.67
				1573	0.675
surface oxidized for 8 h at temperature 1073 K [1, 16]	1073	0.68	surface rolled, oxidized for 8 h at temperature 1473 K [1, 16, 19]	573	0.53
surface oxidized for 8 h at temperature 1173 K [1, 16]	1173	0.735		773	0.69
				973	0.77
				1173	0.80
				1373	0.81
				1573	0.82
surface oxidized for 8 h at temperature 1373 K [1, 16]	1373	0.82	surface rolled, oxidized for 8 h at temperature 1573 K [1, 16, 19]	573	0.69
				773	0.80

Table 3.1.2. (*continued*)

State of surface	T [K]	$\varepsilon_{T,n}$
	973	0.855
	1173	0.875
	1373÷1573	0.885
Nickel - chromium - iron alloy, nikrothal 70-BK surface rolled [1, 16]	573÷823	0.275
surface rolled, oxidized for 5 s at temperature 973 K [1, 16]	973	0.335
surface rolled, oxidized for 5 s at temperature 1173 K [1, 16]	1173	0.505
surface rolled, oxidized for 5 s at temperature 1373 K [1, 16]	1373	0.62
surface rolled oxidized for 5s at temperature 1573 K [1, 16]	1573	0.63
surface rolled, oxidized for 1 min at temperature 1073 K [1, 16]	1073	0.45
surface rolled, oxidized for 1 min at temperature 1173 K [1, 16]	1173	0.54
surface rolled, oxidized for 1 min at temperature 1373 K [1, 16]	1373	0.66
surface rolled, oxidized for 1 min at temperature 1573 K [1, 16]	1573	0.75
surface rolled, oxidized for 20 min at temperature 1073 K [1, 16]	1073	0.55
surface rolled, oxidized for 20 min at temperature 1173 K [1, 16]	1173	0.62
surface rolled, oxidized for 20 min at temperature 1373 K [1, 16]	1373	0.72
surface rolled, oxidized for 20 min at temperature 1573 K [1,16]	1573	0.80
surface oxidized for 8 h at temperature 1073 K [1, 16]	573	0.275
	773	0.43
	973	0.53
	1173	0.595
	1373	0.625
	1573	0.63
surface oxidized for 8 h at temperature 1173 K [1, 16]	573	0.47
	773	0.55
	973	0.615
	1173	0.66
	1373	0.69
	1573	0.70
surface oxidized for 8 h at temperature 1273 K [1, 16]	573	0.61
	773	0.67
	973	0.70
	1173	0.73
	1373	0.74
	1573	0.745
surface oxidized for 8 h at temperature 1373 K [1, 16]	573	0.73
	773	0.78
	973	0.80
	1173	0.81
	1373	0.82
	1573	0.83
surface oxidized for 8 h at temperature 1473 K [1, 16]	573	0.84
	973	0.80
	1173	0.81
	1373	0.82
	1573	0.83

Table 3.1.2. (*continued*)

State of surface	T[K]	$\varepsilon_{T,n}$
surface oxidized for 8 h at temperature 1573 K [1, 16]	573	0.90
	773	0.92
	973	0.93
	1173	0.933
	1373	0.936
	1573	0.94
Nickel - chromium - iron alloy; nikrothal 80-BK surface rolled, [1, 16, 17]	573÷873	0.22
surface rolled, oxidized for 5 s at temperature 973 K [1, 16, 17]	973	0.23
surface rolled, oxidized for 5 s at temperature 1173 K [1, 16, 17]	1173	0.36
surface rolled, oxidized for 5 s at temperature 1373 K [1, 16, 17]	1373	0.578
surface rolled, oxidized for 5 s at temperature 1573 K [1, 16, 17]	1573	0.655
surface rolled, oxidized for 1 min at temperature 1073 K [1, 16, 17]	1073	0.415
surface rolled, oxidized for 1 min at temperature 1173 K [1, 16, 17]	1173	0.49
surface rolled, oxidized for 1 min at temperature 1373 K [1, 16, 17]	1373	0.66
surface rolled, oxidized for 1 min at temperature 1573 K [1, 16, 17]	1573	0.755
surface rolled, oxidized for 20 min at temperature 1073 K [1, 16, 17] .	1073	0.475
surface rolled, oxidized for 20 min at temperature 1173 K [1, 16, 17]	1173	0.55
surface rolled, oxidized for 20 min at temperature 1373 K [1, 16, 17]	1373	0.745
surface rolled, oxidized for 20 min at temperature 1573 K [1, 16, 17]	1573	0.88
surface rolled, oxidized for 1 h at temperature 1073 K [1, 16, 17]	1073	0.52
surface rolled, oxidized for 1 h at temperature 1173 K [1, 16, 17]	1173	0.595
surface rolled, oxidized for 1 h at temperature 1373 K [1, 16, 17]	1373	0.775
surface rolled, oxidized for 1 h at temperature 1573 K [1, 16, 17]	1573	0.92
surface oxidized for 8 h at temperature 1073 K [1, 16, 17]	1073	0.56
surface oxidized for 8 h at temperature 1173 K [1, 16, 17]	1173	0.635
surface oxidized for 8 h at temperature 1273 K [1, 16, 17]	1273	0.74
surface oxidized for 8 h at temperature 1373 K [1, 16, 17]	1373	0.835
surface oxidized for 8 h at temperature 1473 K [1, 16, 17]	1473	0.96

Table 3.1.2. (*continued*)

State of surface	T[K]	$\varepsilon_{T,n}$
surface oxidized for 8 h at temperature 1573 K [1, 16, 17]	1573	0.98
surface oxidized at temperature 1073 K for 8 h [1, 16, 17]	573	0.44
	773	0.49
	973	0.53
	1173	0.57
	1373	0.62
	1573	0.64
surface oxidized at temperature 1173 K for 8 h [1, 16, 17]	573	0.565
	773	0.63
	973	0.67
	1173	0.695
	1379	0.71
	1573	0.72
surface oxidized at temperature 1273 K for 8 h [1, 16, 17]	573	0.66
	773	0.72
	973	0.75
	1173	0.77
	1373	0.78
	1573	0.785
surface oxidized at temperature 1373 K for 8 h [1, 16, 17]	573	0.78
	773	0.82
	973	0.84
	1173	0.856
	1373	0.862
	1573	0.87
surface oxidized at temperature 1473 K for 8 h [1, 16, 17]	573	0.91
	773	0.95
	973÷1573	0.96
surface oxidized at temperature 1573 K for 8 h [1, 16, 17]	573÷773	0.96

State of surface	T[K]	$\varepsilon_{T,n}$
Nickel - chromium - iron alloy, nimonic 75-BS, ЭИ435-GOST**		
polished surface [45, 131, 140]	500	0.12
	600	0.136
	700	0.146
	800	0.16
	900	0.17
	1000	0.182
	1100	0.197
surface rolled [45, 131, 140]	500	0.173
	600	0.184
	700	0.196
	800	0.208
	900	0.219
	1000	0.23
	1100	0.244
surface shot blasted [45, 131, 140]	500	0.516
	600	0.541
	700	0.568
	800	0.60
	900	0.623
	1000	0.652
	1100	0.682
surface polished, oxidized at temperature 873 K [45, 131, 140]	573	0.17
	673	0.18
	773	0.20
	873	0.22
	973	0.26
	1073	0.32
surface shot blasted, oxidized at temperature 873 K [45, 131, 140]	573	0.57
	673	0.63
	773	0.67
	873	0.69
	973	0.72
	1073	0.76
surface polished, oxidized at temperature 1173 K [45, 131, 140]	573	0.44
	673	0.47
	773	0.49

Table 3.1.2. (*continued*)

State of surface	T[K]	$\varepsilon_{T,n}$
	873	0.51
	973	0.53
	1073	0.55
Nickel - chromium - iron alloy; nimonic; ЭИ481-GOST surface oxidized for 20 min at temperature 1073 K [140]	1073	0.63
surface oxidized for 40 min at temperature 1073 K [140]	1073	0.65
surface oxidized for 80 min at temperature 1073 K [140]	1073	0.685
surface oxidized for 120 min at temperature 1073 K [140]	1073	0.725
surface oxidized for 160 min at temperature 1073 K [140]	1073	0.755
surface oxidized for 200 min at temperature 1073 K [140]	1073	0.77
surface oxidized for 220 min at temperature 1073 K [140]	1073	0.78
surface oxidized for time over 40 min at temperature 1173 K [140]	1173	0.935
cold rolled surface, oxidized at temperature 1173 K [45, 131, 140]	573	0.42
	673	0.46
	773	0.49
	873	0.52
	973	0.54
	1073	0.57

State of surface	T[K]	$\varepsilon_{T,n}$
surface shot blasted, oxidized at temperature 1173 K [45, 131, 140]	573	0.78
	673	0.62
	773	0.66
	873	0.69
	973	0.73
	1073	0.76
surface polished, oxidized at temperature 1473 K [45, 131, 140]	573	0.72
	673	0.73
	773	0.74
	873	0.75
	973	0.80
	1073	0.88
surface rolled, oxidized at temperature 1473 K [45, 131, 140]	573	0.69
	673	0.695
	773	0.70
	873	0.71
	973	0.75
	1073	0.81
surface shot blasted, oxidized at temperature 1473 K [45, 131, 140]	573	0.91
	673	0.88
	773	0.91
	873	0.93
	973	0.99
	1073	0.995
Nickel - chromium - iron alloy; nimonic 95-BS; ЭИ607-GOST surface polished [131]	500	0.51
	700	0.51
	900	0.54
	1100	0.58
	1200	0.61
Nickel - copper - iron alloy; monel surface polished [45]	850÷900	0.3

Table 3.1.2. (*continued*)

State of surface	T[K]	$\varepsilon_{T,n}$
surface oxidized [19, 45]	500	0.582
	600	0.576
	700	0.568
	800	0.559
	900	0.553
	1000	0.542
Nickel - copper - aluminium alloy; monel K surface polished [45]	100	0.132
	300	0.147
	500	0.162
	700	0.18
	900	0.198
	1100	0.23
	1300	0.268
	1400	0.29
surface cast [45]	100	0.154
	300	0.177
	500	0.193
	700	0.213
	900	0.238
	1100	0.28
	1300	0.324
	1500	0.38
surface oxidized [45]	1100÷1300	0.755
Nickel - molybdenum - chromium alloy; hastelloy C surface smooth [45]	100	0.08
	300	0.09
	500	0.11
	700	0.14
	900	0.18
	1100	0.22
Nickel - molybdenum - iron alloy; hastelloy B surface smooth [45]	100	0.05
	300	0.082
	500	0.12
	700	0.15
	900	0.179
	1100	0.20
	1300	0.211

State of surface	T[K]	$\varepsilon_{T,n}$
Niobium; Nb sample vacuum arc melted, surface polished roughness rms = 0.19 μm, measurement in vacuum [7]	400	0.052
	450	0.056
	500	0.059
	550	0.063
	600	0.067
	650	0.071
	700	0.076
	750	0.081
	800	0.087
	850	0.098
surface smooth [131]	1000	0.06
	1250	0.12
	1500	0.148
	1750	0.17
	2000	0.185
	2250	0.21
	2500	0.225
	2600	0.23
Palladium; Pd surface polished [131]	750	0.06
	875	0.074
	1000	0.088
	1125	0.105
	1250	0.12
	1375	0.134
	1500	0.15
	1625	0.168
Platinum; Pt purity 99, 99%, surface smooth, roughness rms = 0.13 μm [19, 27, 112, 131]	500	0.04
	625	0.055
	750	0.07
	875	0.09
	1000	0.11
	1125	0.124
	1250	0.133
	1375	0.144
	1500	0.155
	1625	0.163
	1750	0.172
	1875	0.179

Table 3.1.2. (*continued*)

State of surface	T[K]	$\varepsilon_{T,n}$
wire		
[19, 44, 104]	323÷473	0.06÷0.07
	500	0.073
	773	0.1÷0.16
	1273	0.16
	1650	0.18
tape [19, 104]	1200	0.12
	1390	0.17
Platinum - rhodium alloy PtRh10		
surface smooth, heating in protective atmosphere [131]	700	0.107
	1100	0.112
	1500	0.115
surface smooth, heating in air [131]	1500	0.102
	1700	0.126
	1900	0.126
	2100	0.105
Rhenium; Re		
surface of small roughness [79, 131]	1000	0.14
	1125	0.178
	1250	0.189
	1375	0.199
	1500	0.213
	1625	0.226
	1750	0.242
	1875	0.251
	2000	0.263
	2125	0.263
	2250	0.284
	2375	0.292
	2500	0.302
	2625	0.313
	2750	0.32
	2875	0.324
Rhodium; Rh		
surface smooth [80, 131]	500	0.025
	625	0.03
	750	0.04
	875	0.05
	1000	0.062
	1125	0.07
	1250	0.076
	1375	0.082

State of surface	T[K]	$\varepsilon_{T,n}$
	1500	0.09
	1625	0.103
Silver; Ag		
surface hight-polished [19, 37, 104]	293÷500	0.0198
	900	0.0342
surface polished [19, 26, 44, 104]	293	0.025
	640	0.031
Steel, carbon		
polished sheat steel, measurement in air [19]	1023÷1323	0.52÷0.56
grinded sheat steel, measurement in air [19, 44, 55, 146]	1213	0.52
surface freshly treated with abrasive paper [19]	293	0.24
surface rolled [19, 44]	293	0.24
surface rolled after storage [19, 44]	323	0.56
surface hot-rolled [19]	293÷400	0.77÷0.60
surface oxidized [19, 55]	373	0.74
	400	0.78
	800	0.82
	473÷873	0.80
surface oxidized at temperature 873 K [19, 47, 55]	473	0.79
	873	0.79
surface very strongly oxidized [19, 44, 55]	323	0.88
	500	0.98
surface red oxidized [19]	293	0.61÷0.85
liquid state [51]		0.37÷0.4

Table 3.1.2. (*continued*)

State of surface	T[K]	$\varepsilon_{T,n}$
Steel, carbon, 10-PN; C1010-AISI surface smooth, oxidized for 2 min at temperature of measurement [15]	573	0.22
	673	0.27
	773	0.73
	873	0.97
surface smooth, oxidized for 8 h at temperature of measurement [15]	573	0.87
	673	0.93
	773	0.96
	873	0.98
Steel, carbon; 15-PN; C1015-AISI surface oxidized at temperature 1073 K [131]	600	0.86
	800	0.89
	1000	0.91
	1200	0.93
Steel, carbon, 20-PN; C1020-AISI surface oxidized at temperature above 1000 K [45, 131]	800	0.92
	1000	0.93
Steel, carbon; 50-PN; C1050-AISI surface oxidized at temperature 1073 K [131]	600	0.86
	800	0.89
	1000	0.91
	1200	0.93
Steel, cast surface polished [19, 55]	1043	0.52
	1313	0.56
Steel, chromium-aluminium, heat-resistant, H17J5-PN; baidonal surface oxidized for 5 s at temperature of measurement [16, 17]	573÷773	0.173
	973	0.18
	1173	0.23
	1273	0.295
	1373	0.36
	1473	0.375
	1573	0.38
surface oxidized for 1 min at temperature of measurement [16, 17]	1073	0.20
	1173	0.235
	1273	0.305
	1373	0.42
	1473	0.525
	1573	0.595
surface oxidized for 30 min at temperature of measurement [16, 17]	1073	0.22
	1173	0.27
	1273	0.37
	1373	0.48
	1473	0.58
	1573	0.645
surface oxidized for 8 h at temperature of measurement [16, 17]	1073	0.23
	1173	0.295
	1273	0.43
	1373	0.55
	1473	0.64
	1573	0.70
surface oxidized for 8 h at temperature 1273 K [16, 17]	573	0.34
	773	0.37
	973	0.405
	1173	0.44
	1373	0.46
	1573	0.47
surface oxidized for 8 h at temperature 1323 K [16, 17]	573	0.405
	773	0.44
	973	0.575
	1173	0.515

Table 3.1.2. (*continued*)

State of surface	T [K]	$\varepsilon_{T,n}$
	1373	0.55
	1573	0.56
surface oxidized for 8 h at temperature 1473 K [16, 17]	573	0.55
	773	0.565
	973	0.585
	1173	0.603
	1373	0.626
	1573	0.64
surface oxidized for 8 h at temperature 1573 K [16, 17]	573	0.605
	773	0.62
	973	0.645
	1173	0.66
	1373	0.675
	1573	0.70
Steel, chromium-aluminium, heat-resistant, (kanthal Al) surface oxidized for 5 s at temperature of measurement [16, 17]	573	0.23
	773	0.27
	973	0.34
	1173	0.40
	1373	0.46
	1573	0.495
	1673	0.50
surface oxidized for 10 min at temperature of measurement [16, 17]	1073	0.375
	1173	0.42
	1273	0.45
	1373	0.48
	1473	0.51
	1573	0.53
	1673	0.535
surface oxidized for 30 min at temperature of measurement [16, 17]	1073	0.39
	1173	0.43
	1273	0.47
	1373	0.51
	1473	0.55
	1573	0.58
	1673	0.60
surface oxidized for 4÷8 h at temperature of measurement [16, 17]	1073	0.40
	1173	0.44
	1273	0.48
	1373	0.54
	1473	0.59
	1573	0.64
surface oxidized for 8 h at temperature 1073 K [16, 17]	573	0.26
	773	0.34
	973	0.398
	1173	0.44
	1373	0.47
	1573	0.49
surface oxidized for 8 h at temperature 1173 K [16, 17]	573	0.29
	773	0.37
	973	0.43
	1173	0.475
	1373	0.51
	1573	0.525
surface oxidized for 8 h at temperature 1273 K [16, 17]	573	0.34
	773	0.42
	973	0.48
	1173	0.52
	1373	0.55
	1573	0.565
Steel, chromium-aluminium, heat-resistant, (kanthal DSD) surface oxidized for 4 h at temperature 1273 K [111]	1073	0.425

Table 3.1.2. (*continued*)

State of surface	T[K]	$\varepsilon_{T,n}$
Steel, chromium molybdenum, heat-resistant, H5M-PN; 51501-SAE		
surface oxidized for about 2 min at temperature of measurement [14, 15]	573	0.23
	673	0.24
	773	0.38
	873	0.86
	973	0.98
surface oxidized for about 8 h at temperature of measurement [14, 15]	573	0.30
	673	0.63
	773	0.85
	873	0.89
	973	0.95
Steel, chromium-nickel-molybdenum, stainless, OOH17N14M-PN; AM-350		
surface smooth [45]	100	0.03
	300	0.08
	500	0.08
	700	0.09
	900	0.19
	1100	0.37
	1300	0.57
Steel, chromium-nickel-molybdenum, stainless, OH17N14M2-PN; 316LC-AISI		
surface smooth oxidized [45]	100	0.11
	300	0.12
	500	0.14
	700	0.18
	900	0.246
	1100	0.362
Steel, chromium-nickel-niobium, stainless, OH18N12Nb-PN; 821 Grade Nb-BS		
surface polished [45]	100	0.17
	300	0.17
	500	0.17
	700	0.182
	900	0.21
	1100	0.34
	1300	0.68
surface clean [45, 131]	100	0.33
	300	0.35
	500	0.37
	700	0.39
	900	0.404
	1100	0.45
	1200	0.60
surface weakly oxidized [45, 131]	500	0.50
	700	0.54
	900	0.58
	1100	0.62
Steel, chromium-nickel, stainless, H18N9-PN; 302-AISI; 30302-SAE		
surface polished [10, 19, 45, 131]	300	0.17
	400	0.18
	500	0.193
	600	0.204
	700	0.23
	800	0.26
	900	0.28
surface polished, oxidized at temperature 873 K [10, 19, 45, 131]	600	0.4
	700	0.4
	800	0.42
	900	0.44
	1000	0.49
surface polished, oxidized at temperature 1069 K [10, 19, 45, 131]	300÷700	0.835
surface polished, oxidized at temperature 1173 K [10, 19, 45, 131]	600÷1100	0.85÷0.86
surface sandblasted [10, 19, 45, 131]	600	0.405
	700	0.43
	800	0.465
	900	0.49

Table 3.1.2. (*continued*)

State of surface	T[K]	$\varepsilon_{T,n}$
surface sandblasted, oxidized at temperature 873 K [10, 19, 45, 131]	600	0.6
	700	0.625
	800	0.65
	900	0.675
	1000	0.70
	1050	0.715
surface sandblasted, oxidized at temperature 1173 K [10, 19, 45, 131]	600	0.85
	700	0.86
	800	0.88
	900	0.905
	1000	0.925
	1050	0.935
surface rolled [10, 19, 45, 131]	600	0.465
	700	0.47
	800	0.48
	900	0.49
surface rolled, oxidized at temperature 873 K [10, 19, 45, 131]	600	0.75
	700	0.77
	800	0.80
	900	0.825
	1000	0.83
	1100	0.86
surface rolled, oxidized at temperature 1173 K [10, 19, 45, 131]	600	0.79
	700	0.81
	800	0.83
	900	0.85
	1000	0.87
	1100	0.90
Steel, chromium-nickel, stainless, N-155 surface polished [45]	100	0.03
	300	0.06
	500	0.09
	700	0.11
	900	0.15
	1100	0.215
surface oxidized [45]	100	0.08
	300	0.098
	500	0.134
	700	0.183
	900	0.242
	1100	0.308
	1300	0.374
Steel, chromium-nickel, stainless, PH15-7Mo [45]	100	0.03
	300	0.05
	700	0.10
	900	0.14
Steel, chromium-nickel, stainless, OH18N10-PN; 801 Grade C-BS surface oxidized after storage [131]	600	0.45
	700	0.39
	800	0.38
	900	0.41
Steel, chromium-nickel, stainless, OH18N10T-PN; 821 Grade Ti-BS surface smooth, clean [45]	100	0.04
	300	0.055
	500	0.10
	700	0.163
	900	0.24
	1100	0.34
	1300	0.49
surface oxidized at temperature 647 K [131]	300÷700	0.32
surface polished electrolytically, oxidized at temperature 1255 K [131]	700	0.66
	800	0.68
	900	0.70
	1000	0.71
	1100	0.72
	1200	0.73
	1300	0.735

Table 3.1.2. (*continued*)

State of surface	T[K]	$\varepsilon_{T,n}$
Steel, chromium-nickel, stainless, 1H18N9T-PN; 321-AISI; 30321-SAE surface polished, oxidized for 8 h at temperature of measurement [114, 115, 117, 122, 157]	773	0.15
	873	0.16
	973	0.22
	1073	0.50
	1173	0.66
	1273	0.69
surface polished, measurement in air [131, 157]	400	0.16
	500	0.164
	600	0.168
	700	0.172
	800	0.18
	900	0.213
	1000	0.30
	1100	0.52
	1150	0.64
surface polished, oxidized for 1 h at temperature 1173 K [131, 157]	350	0.48
	500	0.50
	700	0.54
	900	0.59
	1100	0.64
	1150	0.66
surface polished, oxidized for 2 h at temperature 1173 K [131, 157]	350	0.62
	500	0.64
	700	0.66
	900	0.68
	1100	0.70
	1150	0.705
surface polished, oxidized for 30 min at temperature 1373 K [131, 157]	350	0.69
	500	0.71
	700	0.72
	900	0.730
	1100	0.74
	1150	0.75
surface smooth, oxidized for 10 h at temperature 923 K [131, 157]	500	0.21
	600	0.24
	700	0.26
	800	0.28
surface nonoxidized, relative roughness $\gamma_w^* = 1.05$ [114, 115]	773	0.20
surface nonoxidized, relative roughness $\gamma_w^* = 1.1$ [114, 115]	773	0.27
surface nonoxidized, relative roughness $\gamma_w^* = 1.2$ [114, 115]	773	0.31
surface nonoxidized, relative roughness $\gamma_w^* = 1.3$ [114, 115]	773	0.35
surface nonoxidized, relative roughness $\gamma_w^* = 1.6$ [114, 115]	773	0.37
surface oxidized at temperature 973 K, relative roughness $\gamma_w^* = 1.05$ [114, 115]	773	0.31
surface oxidized at temperature 973 K, relative roughness $\gamma_w^* = 1.1$ [114, 115]	773	0.36
surface oxidized at temperature 973 K, relative roughness $\gamma_w^* = 1.2$ [114, 115]	773	0.41

Table 3.1.2. (*continued*)

State of surface	T[K]	$\varepsilon_{T,n}$
surface oxidized at temperature 973 K, relative roughness $\gamma_w^* = 1.3$ [114, 115]	773	0.45
surface oxidized at temperature 973 K, relative roughness $\gamma_w^* = 1.6$ [114, 115]	773	0.48
surface oxidized at temperature 1073 K, relative roughness $\gamma_w^* = 1.05$ [114, 115]	773	0.6
surface oxidized at temperature 1073 K, relative roughness $\gamma_w^* = 1.1$ [114, 115]	773	0.66
surface oxidized at temperature 1973 K, relative roughness $\gamma_w^* = 1.2$ [114, 115]	773	0.69
surface oxidized at temperature 1073 K, relative roughness $\gamma_w^* = 1.3$ [114, 115]	773	0.71
surface oxidized at temperature 1073 K, relative roughness $\gamma_w^* = 1.6$ [114, 115]	773	0.74
surface oxidized at temperature 1173 K, relative roughness $\gamma_w^* = 1.05$ [114, 115]	773	0.75
surface oxidized at temperature 1173 K, relative roughness $\gamma_w^* = 1.1$ [114, 115]	773	0.79
surface oxidized at temperature 1173 K, relative roughness $\gamma_w^* = 1.2$ [114, 115]	773	0.82
surface oxidized at temperature 1173 K, relative roughness $\gamma_w^* = 1.3$ [114, 115]	773	0.85
surface oxidized at temperature 1173 K, relative roughness $\gamma_w^* = 1.6$ [114, 115]	773	0.86
Steel, chromium-nickel, stainless, 303		
surface polished [45]	200÷800	0.30
surface oxidized [45]	600	0.746
	800	0.80
	1000	0.836
	1200	0.86
	1400	0.87
Steel, chromium, stainless, 1H13-PN; En56A-BS		
surface oxidized for about 2 min at temperature of measurement [14, 15]	573	0.23
	773	0.26
	973	0.36
	1073	0.44
	1173	0.53
	1273	0.64
surface oxidized for about 8 h at temperature of measurement [14, 15]	573	0.24
	773	0.29
	973	0.46
	1073	0.65
	1173	0.92
	1273	0.96

Table 3.1.2. (*continued*)

State of surface	T[K]	$\varepsilon_{T,n}$
Steel, chromium, stainless, H15-PN; 51430-SAE surface oxidized for about 2 min at temperature of measurement [14, 15]	573	0.23
	673	0.26
	773	0.60
	873	0.96
	973	0.99
surface oxidized for about 8 h at temperature of measurement [14, 15]	573	0.36
	673	0.85
	773	0.95
	873	0.98
	973	0.99
Steel, chromium-tungsten- -nickel-vanadium; WWN1-PN; H21-AISI surface grinded, oxidized for 40 min at temperature of measurement [140]	1073	0.44
	1173	0.77
	1223	0.91
surface grinded, oxidized for 80 min at temperature of measurement [140]	1073	0.47
	1173	0.91
	1223	0.955
surface grinded, oxidized for 120 min at temperature of measurement [140]	1073	0.49
	1173	0.93
	1223	0.96
surface grinded, oxidized for 160 min at temperature of measurement [140]	1073	0.51
	1223	0.96
surface grinded, oxidized for 180 min at temperature of measurement [140]	1073	0.52
	1223	0.96
Steel, high carbon, cast liquid state (C ⩾ 0.66%) [51]	1623÷1823	0.35÷0.38
Steel, high chromium, stainless, H17-PN; En60-BS surface oxidized for about 2 min at temperature of measurement [14, 15]	573	0.22
	773	0.25
	973	0.34
	1073	0.41
	1173	0.50
	1273	0.64
surface oxidized for 8 h at temperature of measurement [14, 15]	773	0.27
	873	0.33
	973	0.41
	1073	0.59
	1173	0.91
	1273	0.93
Steel, high chromium, stainless, H25T-PN; surface oxidized for about 2 min at temperature of measurement [15]	573	0.184
	623	0.195
	673	0.200
	723	0.209
	773	0.216
	823	0.221
	873	0.232
	923	0.242
	973	0.261
	1023	0.279
	1073	0.300

Table 3.1.2. (*continued*)

State of surface	T[K]	$\varepsilon_{T,n}$
	1123	0.326
	1173	0.368
	1223	0.437
	1273	0.577
surface oxidized for 8 h at temperature of measurement [15]	773	0.253
	823	0.268
	873	0.284
	923	0.316
	973	0.363
	1023	0.505
	1073	0.584
	1123	0.637
	1173	0.694
	1223	0.742
	1273	0.784
Steel, high chromium, stainless, H28-PN surface smooth, rms = 0.03 μm [45, 131]	400	0.09
	500	0.10
	600	0.12
	700	0.14
	800	0.15
	900	0.18
	1000	0.22
	1100	0.30
	1200	0.44
surface roughness rms = 0.25 μm [45, 131]	400	0.13
	500	0.16
	600	0.17
	700	0.19
	800	0.22
	900	0.26
	1000	0.30
	1100	0.37
	1150	0.42
Steel, low-carbon surface polished [45]	100	0.06
	300	0.10
	500	0.14
	700	0.176
	900	0.214

State of surface	T[K]	$\varepsilon_{T,n}$
	1100	0.256
	1300	0.298
surface oxidized for 15 min at temperature 1225 K [45, 131]	600	0.915
	800	0.968
	1000	0.994
	1100	0.986
Tantalum, Ta sample electron-beam melted, surface polished with some scratches and a few pits, roughness rms = 0.19 μm measurements in vacuum [7]	400	0.052
	450	0.055
	500	0.059
	550	0.064
	600	0.068
	650	0.073
	700	0.078
	750	0.084
	800	0.091
	850	0.099
surface smooth nonoxidized [19, 44, 104, 131, 144]	1300	0.132
	1400	0.145
	1500	0.157
	1600	0.170
	1700	0.181
	1800	0.191
	1900	0.202
	2000	0.212
	2100	0.222
	2200	0.231
	2300	0.241
	2400	0.250
	2500	0.260
surface oxidized [45]	100	0.315
	200	0.37
	400÷800	0.42
Tantalum-tungsten alloy TaW10 surface polished [131]	1600÷2800	0.349

Table 3.1.2. (*continued*)

State of surface	T[K]	$\varepsilon_{T,n}$
Tantalum-tungsten alloy TaW30 surface polished [45]	1600÷2800	0.328
Thorium; Th surface smooth, temperature 1975 K [45]	1975	0.35
Tin; Sn surface bright [19, 44, 47]	293÷323	0.04÷0.06
surface polished [12]	297÷373	0.06
tinned steel sheet [19]	300	0.056÷0.086
Titanium; Ti sample electron-beam melted, surface polished with some pits, roughness rms = 0.18 μm, measurements in vacuum [7]	400	0.151
	450	0.158
	500	0.162
	550	0.168
	600	0.174
	650	0.180
	700	0.187
	750	0.194
	800	0.202
	850	0.210
surface polished [19, 44]	473	0.15
	773	0.20
	1273	0.36
surface oxidized at temperature 813 K [19, 44]	473	0.40
	573	0.30
	1273	0.60
Tungsten; W surface smooth [39, 44, 131, 156]	473	0.05
	500	0.053
	873	0.10
	1300	0.13
	1600	0.175
	1900	0.214
	2200	0.249
	2500	0.275
	2800	0.30
	3000	0.313
fibre [19, 44, 156]	3573	0.39
using fibre [19, 156]	300	0.032
	3573	0.35
Vanadium; V sample vacuum arc-melted, surface polished roughness rms = 0.17 μm, measurement in vacuum [7]	400	0.061
	450	0.064
	500	0.068
	550	0.073
	600	0.080
	650	0.088
	700	0.098
	750	0.112
	800	0.130
	850	0.153
Zinc; Zn surface polished [19, 44]	473÷573	0.04÷0.05
surface dark, mat [12, 19, 78]	293÷363	0.25÷0.21
	553	0.21
surface grey oxidized [19, 26]	293	0.23÷0.28
surface oxidized at temperature of measurement 1273-1473 K [19]	1273÷1473	0.5÷0.6
surface oxidized at temperature of measurement 673 K [1, 12, 19, 44]	293÷363	0.23÷0.29
	673	0.11
sheet [19]	500	0.045
bright zinc plating sheet [10, 19]	300	0.043÷0.064

Table 3.1.2. (*continued*)

State of surface	T[K]	$\varepsilon_{T,n}$
grey oxidized zinc plating sheet (old) [1, 19]	293	0.28
powder [106]	1273÷1473	0.5÷0.6
powder oxidized [19]		0.82
Zirconium; Zr		
sample electron-beam melted, surface polished roughness rms = 0.19 μm, measurement in vacuum [7]	400	0.088
	450	0.100
	500	0.115
	550	0.131
	600	0.149
	650	0.167
	700	0.186
	750	0.206
	800	0.227
	850	0.248

Table 3.1.3. Spectral emissivity in normal direction $\varepsilon_{\lambda,n}$ of metals and alloys

State of surface	λ [μm]	$\varepsilon_{\lambda,n}$
Aluminium; Al surface polished [45, 131]	3	0.10
	4	0.26
	5	0.48
	6	0.50
	7	0.56
	8	0.60
	9	0.61
layer formed by vacuum sublimation, thickness 0.04 μm [129]	0.4	0.360
	0.5	0.320
	0.6	0.292
	0.7	0.260
layer formed by vacuum sublimation, thickness 0.08 μm [129]	0.4	0.493
	0.5	0.376
	0.6	0.273
	0.7	0.227
layer formed by vacuum sublimation, thickness 0.12 μm [129]	0.4	0.492
	0.5	0.400
	0.6	0.284
	0.7	0.240
layer formed by vacuum sublimation, thickness 0.12 μm [129]	0.4	0.447
	0.5	0.361
	0.6	0.264
	0.7	0.217
plated aluminium on the duralumin [45, 76]	0.25	0.48
	0.5	0.28
	0.75	0.24
	1.0	0.2
	1.25	0.14
	1.5	0.12
	1.75	0.09
	2.0	0.05
	2.25	0.005
	2.5	0.005
	2.75	0.005
oxidized plated aluminium on the duralumin [45, 76]	0.25	0.95
	0.5	0.71
	0.75	0.62
	1.0	0.48
	1.25	0.35
	1.5	0.28
	1.75	0.22
	2.0	0.16
	2.25	0.14
	2.5	0.1
	2.75	0.1
Aluminium-magnesium alloy AlMg3-PN; PA11N-PN [45]	0.25	0.4
	0.30	0.26
	0.35	0.23
	0.40	0.21
	0.45	0.20
	0.50	0.20

Table 3.1.3. (*continued*)

State of surface	λ[μm]	$\varepsilon_{\lambda,n}$
Aluminium-silver alloy AlAg2 [45]	0.2	0.1
	0.3	0.08
	0.4	0.09
	0.5	0.01
Aluminium-silver alloy - 3-12% Ag [45]	0.35	0.04
	0.4	0.06
	0.45	0.07
	0.50	0.09
	0.60	0.09
Antimony; Sb vapour coating of thickness 0.01 μm [129]	0.6	0.340
layer formed by vacuum sublimation, thickness 0.02 μm [129]	0.6	0.447
layer formed by vacuum sublimation, thickness 0.03 μm [129]	0.6	0.527
layer formed by vacuum sublimation, thickness 0.04 μm [129]	0.6	0.560
layer formed by vacuum sublimation, thickness 0.06 μm [129]	0.6	0.553
layer formed by vacuum sublimation, thickness 0.08 μm [129]	0.6	0.553
Boron; B [45]	1.0	0.72
Chromium; Cr surface polished [45, 131]	0.16	0.86
	0.18	0.84
	0.2	0.64
	0.3	0.53
	0.4	0.44
	0.5	0.42
	0.6	0.43
	0.7	0.44
	0.8	0.42
	0.9	0.42
	1.0	0.44
	1.1	0.43
	1.2	0.40
	1.3	0.35
	1.4	0.32
	1.6	0.32
	1.8	0.33
	1.9	0.30
	2.0	0.27
	2.1	0.24
	2.2	0.21
	2.3	0.19
	2.4	0.17
	2.5	0.09
	2.6	0.08
	2.7	0.09
	2.8	0.12
layer formed by vacuum sublimation, thickness 0.01 μm [129]	0.6	0.016
layer formed by vacuum sublimation, thickness 0.02 μm [129]	0.6	0.120
layer formed by vacuum sublimation, thickness 0.03 μm [129]	0.6	0.267
layer formed by vacuum sublimation, thickness 0.04 μm [129]	0.6	0.313
layer formed by vacuum sublimation, thickness 0.06 μm [129]	0.6	0.380
layer formed by vacuum sublimation, thickness 0.08 μm [129]	0.6	0.400
Cobalt; Co surface smooth, temperature 1200 K; point X for $\lambda \cong 1.45$ μm [131, 154]	0.52	0.44
	0.56	0.415

Table 3.1.3. (*continued*)

State of surface	λ[μm]	$\varepsilon_{\lambda,n}$
	0.58	0.37
	0.60	0.375
	0.62	0.425
	0.64	0.40
surface smooth, temperature 1070 K [45, 155]	1.25	0.27
	1.5	0.25
	1.75	0.23
	2.0	0.21
	2.25	0.20
	2.5	0.19
	2.75	0.18
	3.0	0.18
surface smooth, temperature 1500 K [131, 155]	0.52	0.36
	0.54	0.35
	0.56	0.34
	0.58	0.35
	0.60	0.34
	0.62	0.335
	0.64	0.33
surface smooth, temperature 1600 K [45, 155]	1.25	0.26
	1.5	0.25
	1.75	0.24
	2.0	0.23
	2.25	0.22
	2.5	0.21
	2.75	0.205
	3.0	0.2
Copper; Cu surface polished, temperature of measurement 1174 K, point X for λ ≊ 0.55 μm [41, 80, 92, 131]	2.0	0.065
	2.5	0.055
	3.0	0.045
	4.0	0.03
surface polished, temperature of measurement 973 K [41, 80, 92, 131]	1.0	0.065
	2.0	0.05
	2.5	0.04
	3.0	0.038
	4.0	0.035

State of surface	λ[μm]	$\varepsilon_{\lambda,n}$
surface polished, temperature of measurement 1242 K [41, 80, 92, 131]	1.0÷6.0	0.03
	8.0	0.035
	10.0	0.04
surface oxidized [45]	0.3÷0.7	0.86
	0.8	0.82
	0.9	0.58
	1.0	0.51
	1.2	0.46
	1.4	0.4
	1.6	0.34
	1.8	0.30
	2.0	0.3
	2.2	0.28
	2.4	0.26
	2.6	0.08
Copper-aluminium alloy BA5-PN; alloy 606-ASTM (aluminium bronze) surface polished [45]	0.22	0.8
	0.3	0.78
	0.4	0.74
	0.5	0.57
	0.6	0.34
	0.7	0.25
	0.8	0.22
	1.0	0.17
	1.2	0.16
	1.4	0.13
	1.6	0.11
	1.8	0.09
	2.0	0.07
	2.2	0.05
	2.4	0.03
	2.6	0.02
surface oxidized [45]	0.3	0.93
	0.4	0.89
	0.6	0.85
	0.7	0.82
	0.8	0.66
	0.9	0.56
	1.0	0.55
	1.2	0.54
	1.4	0.49
	1.6	0.45
	1.8	0.41

Table 3.1.3. (*continued*)

State of surface	λ[μm]	$\varepsilon_{\lambda,n}$
	2.0	0.37
	2.2	0.35
	2.4	0.32
	2.6	0.31
Copper-manganese - nickel alloy; manganin [45]	0.4	0.6
	0.45	0.5
	0.5	0.41
	0.55	0.32
	0.6	0.31
	0.65	0.29
	0.7	0.28
Copper-zinc alloy (brass) surface clean [45]	0.35	0.68
	0.4	0.68
	0.5	0.56
	0.6	0.25
	0.7	0.24
Gold; Au coating of thickness 0.01 μm [33, 63, 127, 129]	10	0.0095
	20	0.0120
	30	0.0122
	40	0.0119
	50	0.0115
	60	0.0113
	70	0.0107
	80	0.0105
	90	0.0103
	100	0.0101
	200	0.0085
	300	0.0080
	400	0.0075
	500	0.0073
	600	0.0068
	700	0.0066
	800	0.0064
	900	0.0063
	1000	0.0062
	2000	0.0054
	3000	0.0052
	4000	0.0050
coating of thickness 0.03 μm [33, 63, 127, 129]	10	0.0044
	20	0.0054
	30	0.0058
	40	0.0057
	50	0.0056
	60	0.0054
	70	0.0053
	80	0.0052
	90	0.0050
	100	0.0049
	200	0.0042
	300	0.0038
	400	0.0037
	500	0.0036
	600	0.0034
	700	0.0033
	800	0.0032
	900	0.0031
	1000	0.0030
	2000	0.0027
	3000	0.0025
	4000	0.0024
coating of thickness 0.04 μm [33, 63, 127, 129]	10	0.0052
	20	0.0037
	30	0.0039
	40	0.0038
	50	0.0037
	60	0.0035
	70	0.0033
	80	0.0032
	90	0.0031
	100	0.0030
	200	0.0027
	300	0.0026
	400	0.0024
	500	0.0023
	600	0.0021
	700	0.0020
	800	0.0019
	900	0.0018
	1000	0.0017
	2000	0.0016
	3000	0.0015
	4000	0.0014
coating of thickness 0.06 μm [33, 63, 127, 129]	10	0.0034
	20	0.0033
	30	0.0031
	40	0.0028
	50	0.0026
	60	0.0025
	70	0.0024
	80	0.0023

Table 3.1.3. (*continued*)

State of surface	$\lambda[\mu m]$	$\varepsilon_{\lambda,n}$
	90	0.0021
	100	0.0020
	200	0.0016
	300	0.0014
	400	0.0013
	500	0.0012
	600	0.00105
	700	0.00100
	800	0.00095
	900	0.00093
	1000	0.00090
	2000	0.00078
	3000	0.00071
	4000	0.00070
coating of thickness 25 μm [33, 63, 127, 129]	10	0.00360
	20	0.00340
	30	0.00300
	40	0.00280
	50	0.00270
	60	0.00260
	70	0.00240
	80	0.00220
	90	0.00210
	100	0.00200
	200	0.00150
	300	0.00120
	400	0.00097
	500	0.00085
	600	0.00076
	700	0.00070
	800	0.00065
	900	0.00060
	1000	0.00057
	2000	0.00036
	3000	0.00027
	4000	0.00020
layer formed by vacuum sublimation, thickness 0.01 μm [129]	0.4	0.131
	0.5	0.144
layer formed by vacuum sublimation, thickness 0.02 μm [129]	0.4	0.303
	0.5	0.234

State of surface	$\lambda[\mu m]$	$\varepsilon_{\lambda,n}$
layer formed by vacuum sublimation, thickness 0.03 μm [129]	0.4	0.382
	0.5	0.297
surface polished [45, 131]	1	0.60
	2	0.66
	3	0.70
	4	0.74
	5	0.77
	6	0.78
	7	0.81
	8	0.87
	9	0.88
Hafnium; Hf surface smooth [131]	0.5	0.47
	0.6	0.462
	0.7	0.44
Iridium; Ir surface smooth, temperature 300 K; point X is between $\lambda = 0.2 \div 0.8$ μm [131]	0.6	0.33
	0.8	0.255
	1.0	0.22
	1.2	0.185
	1.4	0.165
	1.6	0.14
	1.8	0.12
	2.0	0.10
surface smooth, temperature 1100 K [131]	1.0	0.22
	1.2	0.2
	1.4	0.18
	1.6	0.17
	1.8	0.15
	2.0	0.14
surface smooth, temperature 1500 K [131]	0.8	0.215
	1.0	0.24
	1.2	0.21
	1.4	0.19
	1.6	0.18
	1.8	0.17
	2.0	0.16

Table 3.1.3. (*continued*)

State of surface	λ[μm]	$\varepsilon_{\lambda,n}$
surface smooth, temperature 2000 K [131]	1.0	0.24
	1.2	0.23
	1.4	0.21
	1.6	0.195
	1.8	0.185
	2.0	0.18
surface smooth, temperature 2400 K [131]	0.6	0.36
	0.7	0.33
	0.8	0.28
	1.0	0.245
	1.2	0.235
	1.4	0.22
	1.6	0.205
	1.8	0.2
	2.0	0.19
Iron; Fe surface smooth nonoxidized, room temperature [45, 101, 127, 131, 153, 155]	0.3	0.71
	0.4	0.56
	0.6	0.52
	0.8	0.48
	1.0	0.46
	1.4	0.40
	1.8	0.29
	2.2	0.22
	2.6	0.16
	3.0	0.15
	5.0	0.075
	7.0	0.05
	9.0	0.04
	11.0	0.03
	13.0	0.025
	15.0	0.025
surface smooth nonoxidized, temperature 1120 K [101, 127, 131]	1.0	0.298
	3.0	0.185
	5.0	0.15
	7.0	0.13
	9.0	0.12
	11.0	0.115
	13.0	0.105
	15.0	0.101
liquid state [101]	1.5	0.60
	2.0	0.52
	3.0	0.37
	4.0	0.28
surface oxidized [14, 45, 153, 155]	0.3	0.93
	0.4	0.85
	0.8	0.90
	1.6	0.90
	2.0	0.82
	2.4	0.83
	2.7	0.90
Molybdenum; Mo surface smooth, temperature 1000 K, point X for λ = 0.9 μm [41, 131]	0.4	0.458
	0.5	0.438
	0.6	0.417
	0.7	0.394
	0.8	0.367
	0.9	0.333
	1.0	0.302
	1.1	0.270
	1.2	0.239
	1.3	0.213
	1.4	0.188
	1.5	0.168
	1.6	0.150
	1.7	0.136
	1.8	0.125
	1.9	0.115
	2.0	0.106
	2.2	0.092
	2.4	0.082
	2.6	0.074
	2.8	0.068
	3.0	0.063
	3.2	0.059
	3.4	0.055
	3.6	0.052
	3.8	0.049
	4.0	0.046
	4.2	0.044
	4.4	0.041
	4.6	0.039
	4.8	0.037
	5.0	0.035
surface smooth, temperature 1200 K [41, 131]	0.4	0.448
	0.5	0.429
	0.6	0.410

Table 3.1.3. (*continued*)

State of surface	$\lambda[\mu m]$	$\varepsilon_{\lambda,n}$
	0.7	0.389
	0.8	0.363
	0.9	0.333
	1.0	0.306
	1.1	0.278
	1.2	0.252
	1.3	0.229
	1.4	0.208
	1.5	0.191
	1.6	0.175
	1.7	0.162
	1.8	0.150
	1.9	0.139
	2.0	0.130
	2.2	0.115
	2.4	0.102
	2.6	0.094
	2.8	0.087
	3.0	0.081
	3.2	0.076
	3.4	0.071
	3.6	0.067
	3.8	0.064
	4.0	0.061
	4.2	0.058
	4.4	0.055
	4.6	0.053
	4.8	0.051
	5.0	0.049
surface smooth, temperature 1400 K [131]	0.4	0.440
	0.5	0.422
	0.6	0.403
	0.7	0.383
	0.8	0.361
	0.9	0.333
	1.0	0.310
	1.1	0.286
	1.2	0.263
	1.3	0.243
	1.4	0.225
	1.5	0.209
	1.6	0.195
	1.7	0.182
	1.8	0.171
	1.9	0.160
	2.0	0.151
	2.2	0.135
	2.4	0.122
	2.6	0.111
	2.8	0.103
	3.0	0.096
	3.2	0.090
	3.4	0.085
	3.6	0.081
	3.8	0.077
	4.0	0.073
	4.2	0.069
	4.4	0.066
	4.6	0.064
	4.8	0.061
	5.0	0.059
surface smooth, temperature 1600 K [131]	0.4	0.432
	0.5	0.415
	0.6	0.397
	0.7	0.378
	0.8	0.358
	0.9	0.333
	1.0	0.312
	1.1	0.291
	1.2	0.270
	1.3	0.252
	1.4	0.236
	1.5	0.220
	1.6	0.208
	1.7	0.195
	1.8	0.184
	1.9	0.174
	2.0	0.165
	2.2	0.149
	2.4	0.136
	2.6	0.125
	2.8	0.116
	3.0	0.108
	3.2	0.102
	3.4	0.097
	3.6	0.092
	3.8	0.088
	4.0	0.084
	4.2	0.079
	4.4	0.076
	4.6	0.073
	4.8	0.071
	5.0	0.068
surface smooth, temperature 1800 K [131]	0.4	0.425
	0.5	0.409
	0.6	0.392
	0.7	0.374
	0.8	0.355
	0.9	0.333

Table 3.1.3. (*continued*)

State of surface	λ [μm]	$\varepsilon_{\lambda,n}$
	1.0	0.315
	1.1	0.296
	1.2	0.277
	1.3	0.261
	1.4	0.245
	1.5	0.231
	1.6	0.219
	1.7	0.207
	1.8	0.195
	1.9	0.186
	2.0	0.178
	2.2	0.163
	2.4	0.149
	2.6	0.138
	2.8	0.129
	3.0	0.121
	3.2	0.114
	3.4	0.108
	3.6	0.102
	3.8	0.097
	4.0	0.093
	4.2	0.088
	4.4	0.085
	4.6	0.082
	4.8	0.079
	5.0	0.077
surface smooth, temperature 2000 K [131]	0.4	0.419
	0.5	0.403
	0.6	0.387
	0.7	0.370
	0.8	0.352
	0.9	0.333
	1.0	0.317
	1.1	0.300
	1.2	0.384
	1.3	0.269
	1.4	0.255
	1.5	0.242
	1.6	0.230
	1.7	0.219
	1.8	0.208
	1.9	0.199
	2.0	0.191
	2.2	0.176
	2.4	0.162
	2.6	0.151
	2.8	0.141
	3.0	0.133
	3.2	0.125
	3.4	0.119
	3.6	0.112
	3.8	0.107
	4.0	0.102
	4.2	0.097
	4.4	0.093
	4.6	0.089
	4.8	0.086
	5.0	0.084
Nickel, Ni surface polished at room temperature [45]	0.3	0.65
	0.4	0.5
	0.5	0.38
	0.6	0.33
	0.8	0.29
	1.0	0.27
	1.2	0.25
	1.4	0.214
	1.6	0.18
	1.8	0.15
	2.0	0.112
	2.2	0.098
	2.4	0.078
	2.6	0.062
	2.8	0.06
	3.2	0.042
surface rolled clean, room temperature [45]	0.3	0.73
	0.4	0.59
	0.5	0.5
	0.6	0.43
	0.8	0.358
	1.0	0.32
	1.2	0.3
	1.4	0.24
	1.6	0.22
	1.8	0.19
	2.0	0.16
	2.4	0.11
	2.6	0.08
surface polished, temperature 300 K [127]	1.0	0.27
	2.0	0.17
	4.0	0.08
	6.0	0.05
	8.0	0.03
	10.0	0.028
	20.0	0.016

Table 3.1.3. (*continued*)

State of surface	$\lambda[\mu m]$	$\varepsilon_{\lambda,n}$	State of surface	$\lambda[\mu m]$	$\varepsilon_{\lambda,n}$
surface polished, temperature 1400 K [127]			Nickel-chromium-iron alloy; inconel X surface polished [45]		
	1.0	0.28		0.3	0.72
	2.0	0.18		0.4	0.68
	4.0	0.13		0.6	0.52
	6.0	0.087		0.8	0.42
	8.0	0.075		1.0	0.36
	10.0	0.067		1.2	0.31
surface oxidized [45]				1.4	0.29
	0.3	0.85		1.6	0.27
	0.4	0.818		1.8	0.23
	0.6	0.798		2.0	0.18
	0.8	0.792		2.2	0.16
	1.0	0.76		2.4	0.17
	1.2	0.776		2.6	0.18
	1.4	0.75		2.7	0.15
	1.6	0.718	surface casted or rolled clean [45]		
	1.8	0.69		0.3	0.97
	2.0	0.68		0.4	0.84
	2.2	0.644		0.6	0.76
	2.4	0.592		0.8	0.71
	2.6	0.528		1.0	0.64
layer formed by vacuum sublimation, thickness 0.01 μm [129]	0.6	0.216		1.2	0.58
				1.4	0.56
				1.6	0.50
layer formed by vacuum sublimation, thickness 0.02 μm [129]	0.6	0.336		1.8	0.45
				2.0	0.39
				2.2	0.34
				2.4	0.32
				2.6	0.31
layer formed by vacuum sublimation, thickness 0.03 μm [129]	0.6	0.416	surface oxidized [45]		
				0.3	0.98
				0.4	0.91
				0.6	0.89
layer formed by vacuum sublimation, thickness 0.04 μm [129]	0.6	0.437		0.8	0.91
				1.0	0.90
				1.2	0.88
				1.4	0.85
layer formed by vacuum sublimation, thickness 0.06 μm [129]	0.6	0.447		1.6	0.78
				1.8	0.75
				2.0	0.703
				2.2	0.70
layer formed by vacuum sublimation, thickness 0.08 μm [129]	0.6	0.447		2.4	0.702
				2.6	0.72
				2.7	0.732
			Nickel-copper-aluminium alloy; monel K [45]		
				0.4	0.53
				0.5	0.45
				0.6	0.40

Table 3.1.3. (*continued*)

State of surface	λ[μm]	$\varepsilon_{\lambda,n}$
	0.7	0.362
	0.8	0.34
Nickel-molybdenum-chromium alloy; hastelloy C surface smooth, [45]	0.3	0.70
	0.4	0.56
	0.5	0.51
	0.75	0.44
	1.0	0.39
	1.25	0.37
	1.50	0.35
	1.75	0.30
	2.0	0.24
	2.25	0.22
	2.50	0.19
Nickel-molybdenum-iron alloy; hastelloy B surface smooth [45]	0.3	0.62
	0.4	0.54
	0.6	0.46
	0.8	0.44
	1.0	0.42
	1.2	0.40
	1.4	0.40
	1.8	0.32
	2.0	0.28
	2.2	0.26
	2.4	0.23
	2.6	0.23
surface oxidized [45]	0.3	0.89
	0.4	0.89
	0.6	0.896
	0.8	0.882
	1.0	0.87
	1.2	0.877
	1.4÷1.6	0.854
	1.8	0.86
	2.0	0.83
	2.2	0.82
	2.4	0.81
	2.6	0.78
Niobium; Nb surface smooth, temperature about 1800 K [131]	0.5	0.45
	1.0	0.31
	1.5	0.22
	2.0	0.193
	2.5	0.18
	3.0	0.166
	4.0	0.142
	5.0	0.12
Niobium-molybdenum alloy NbMo10 surface smooth, temperature of measurement 2000 K [142]	0.665	0.349
	1.0	0.273
	3.0	0.144
Niobium-molybdenum alloy NbMo20 surface smooth, temperature of measurement 2000 K [142]	0.665	0.349
	1.0	0.275
	3.0	0.144
Niobium-molybdenum alloy NbMo30 surface smooth, temperature of measurement 2000 K [142]	0.665	0.350
	1.0	0.284
	3.0	0.144
Niobium-molybdenum alloy NbMo40 surface smooth, temperature of measurement 2000 K [142]	0.655	0.353
	1.0	0.286
	3.0	0.145
Niobium-molybdenum alloy NbMo50 surface smooth, temperature of measurement 2000 K [142]	0.665	0.353
	1.0	0.289
	3.0	0.145
Niobium-molybdenum alloy NbMo60 surface smooth, temperature of measurement 2000 K [142]	0.655	0.353
	1.0	0.291
	3.0	0.142

Table 3.1.3. (*continued*)

State of surface	λ [μm]	$\varepsilon_{\lambda,n}$
Niobium-molybdenum alloy NbMo70 surface smooth, temperature of measurement 2000 K [142]	0.655	0.351
	1.0	0.293
	3.0	0.139
Niobium-molybdenum alloy NbMo80 surface smooth, temperature of measurement 2000 K [142]	0.655	0.350
	1.0	0.294
	3.0	0.135
Niobium-molybdenum alloy NbMo90 surface smooth, temperature of measurement 2000 K [142]	0.655	0.350
	1.0	0.290
	3.0	0.130
Palladium; Pd [45]	0.3	0.59
	0.5	0.41
	0.75	0.33
	1.0	0.28
	1.25	0.255
	1.5	0.22
	1.75	0.18
	2.0	0.13
	2.5	0.11
layer formed by vacuum sublimation, thickness 0.01 μm [129]	0.6	0.120
layer formed by vacuum sublimation, thickness 0.02 μm [129]	0.6	0.300
layer formed by vacuum sublimation, thickness 0.03 μm [129]	0.6	0.400
layer formed by vacuum sublimation, thickness 0.04 μm [129]	0.6	0.416
layer formed by vacuum sublimation, thickness 0.06 μm [129]	0.6	0.420
layer formed by vacuum sublimation, thickness 0.08 μm [129]	0.6	0.420
Platinum; Pt surface smooth [80, 112, 127, 132]	0.3	0.6
	0.4	0.52
	0.5	0.43
	0.6	0.38
	0.8	0.30
	1.0	0.25
	1.5	0.19
	2.0	0.17
	3.0	0.14
	4.0	0.11
	5.0	0.10
	6.0	0.08
	8.0	0.07
	10.0	0.06
	12.0	0.05
	15.0	0.04
Rhenium; Re surface nonoxidized smooth, temperature 1100 K, point X for λ = 0.9 μm [79, 131]	0.4	0.48
	0.6	0.465
	0.8	0.41
	0.9	0.375
	1.0	0.358
	1.2	0.338
	1.5	0.315
	2.0	0.29
surface nonoxidized smooth, temperature 1810 K, point X for λ = 0.9 μm [79, 131]	0.4	0.43
	0.6	0.425
	0.8	0.395
	0.9	0.375
	1.0	0.355
	1.2	0.322
	1.5	0.29
	2.0	0.255
	2.5	0.232

Table 3.1.3. (*continued*)

State of surface	λ [μm]	$\varepsilon_{\lambda,n}$
surface nonoxidized smooth, temperature 2388 K [79]	0.4	0.423
	0.6	0.425
	0.8	0.44
	0.9	0.375
	1.0	0.357
	1.2	0.335
	1.5	0.30
	2.0	0.272
	2.5	0.248
surface nonoxidized smooth, temperature 3045 K [79, 131]	0.4	0.412
	0.6	0.405
	0.8	0.38
	0.9	0.375
	1.0	0.36
	1.2	0.34
	1.5	0.315
	2.0	0.28
	2.5	0.264
	3.0	0.249
Rhodium; Rh [45]	0.3	0.5
	0.5	0.348
	0.75	0.268
	1.0	0.22
	1.25	0.203
	1.5	0.19
	2.0	0.09
	2.25	0.04
	2.5	0.027
	2.70	0.001
Silver, Ag surface polished [7, 45, 131]	6	0.82
	7	0.85
	8	0.87
	9	0.88
Silver-aluminium alloy AgAl1 phase [45]	0.25	0.71
	0.3	0.82
	0.35	0.68
	0.4	0.24
	0.45	0.12
	0.5	0.09
Silver-aluminium alloy AgAl1,7 phase [45]	0.25	0.60
	0.3	0.40
	0.35	0.39
	0.4	0.39
	0.45÷0.5	0.40
Silver-aluminium alloy AgAl3,5 phase [45]	0.25	0.65
	0.3	0.68
	0.35	0.45
	0.4	0.44
	0.45	0.12
	0.50	0.09
Silver-aluminium alloy AgAl12 phase [45]	0.25	0.6
	0.30	0.39
	0.35	0.38
	0.40	0.37
	0.45	0.36
	0.50	0.35
Silver-zinc alloy AgZn7 surface polished [45]	0.4	0.32
	0.5	0.30
	0.6	0.28
	0.7	0.26
Silver-zinc alloy AgZn37 surface at temperature 300 K [45]	0.35	0.67
	0.4	0.64
	0.45	0.63
	0.5	0.63
	0.55	0.58
	0.6	0.42
	0.65	0.31
	0.7	0.30
surface at temperature 526 K [45]	0.35	0.62
	0.4	0.57
	0.45	0.54
	0.50	0.52
	0.55	0.49
	0.6	0.45
	0.65	0.41
	0.7	0.39

Table 3.1.3. (*continued*)

State of surface	λ [μm]	$\varepsilon_{\lambda,n}$	State of surface	λ [μm]	$\varepsilon_{\lambda,n}$
Steel, chromium-nickel-aluminium, stainless, H17N9J-PN surface smooth, oxidized at temperature 900 K for 30 min [131]	1.0	0.34	surface oxidized, measurement at room temperature [45]	0.3	0.94
	1.1	0.325		0.4	0.93
	1.2	0.31		0.6	0.91
	1.3	0.298		0.8	0.888
	1.4	0.28		1.0	0.82
				1.4	0.75
				1.8	0.68
				2.2	0.635
				2.6	0.61
Steel, chromium-nickel, stainless, H18N9-PN; 302-AISI; 30302-SAE surface polished [45, 108, 157]	0.4	0.43	surface oxidized at temperature 1123 K [45]	0.66	0.75
	0.5	0.36	surface oxidized, temperature 1450 K [45]	0.66	0.58
	0.6	0.34			
	0.7	0.32			
surface polished and matted [45, 108, 157]	0.4	0.54	Steel, chromium-nickel-molybdenum, stainless, OOH17N14M-PN; AM-350 surface smooth [45]	0.3	0.7
	0.5	0.47		0.4	0.58
	0.6	0.43		0.6	0.46
	0.7	0.42		0.8	0.43
Steel, chromium-nickel-molybdenum, stainless, OH17N14M2-PN; 316LC-AISI surface smooth, room temperature [45]	0.3	0.69		1.0	0.40
	0.4	0.56		1.4	0.40
	0.6	0.49		1.8	0.30
	0.8	0.45		2.2	0.24
	1.0	0.43		2.6	0.22
	1.4	0.40	Steel, chromium-nickel, stainless, N-155-ASTM surface polished, room temperature of measurement [45]	0.3	0.63
	1.8	0.31		0.4	0.54
	2.2	0.26		0.5	0.46
	2.6	0.22		0.6	0.42
surface at temperature 1073 K [45]	0.66	0.38		0.8	0.38
				1.0	0.35
				1.2	0.33
surface at temperature 1450 K [45]	0.66	0.33		1.6	0.29
				2.0	0.23
				2.4	0.19
				2.7	0.19

Table 3.1.3. (*continued*)

State of surface	λ[μm]	$\varepsilon_{\lambda,n}$
surface clean, room temperature of measurement [45]	0.3	0.66
	0.4	0.59
	0.5	0.53
	0.6	0.50
	0.8	0.49
	1.0	0.45
	1.2	0.42
	1.6	0.39
	2.0	0.356
	2.4	0.294
	2.7	0.28
surface oxidized after storage, room temperature of measurement [45]	0.3	0.84
	0.4	0.66
	0.5	0.60
	0.6	0.57
	0.8	0.556
	1.0	0.534
	1.2	0.51
	1.6	0.48
	2.0	0.42
	2.4	0.36
	2.7	0.4
surface oxidized, room temperature of measurement [45]	0.3	0.98
	0.4	0.89
	0.5	0.87
	0.6	0.84
	0.8	0.72
	1.0	0.66
	1.2	0.66
	1.6	0.67
	2.0	0.65
	2.4	0.61
	2.7	0.60
surface polished at temperature 1073 K [45]	0.63	0.35
surface oxidized, temperature of measurement 1073 K [45]	0.63	0.71

State of surface	λ[μm]	$\varepsilon_{\lambda,n}$
surface polished, temperature of measurement 1473 K [45]	0.63	0.29
surface oxidized, temperature of measurement 1473 K [45]	0.63	0.715
Steel, chromium-nickel, stainless, PH 15-7Mo-ASTM [45]	0.3	0.65
	0.4	0.55
	0.5	0.52
	0.6	0.50
	0.8	0.46
	1.0	0.42
	1.2	0.40
	1.6	0.34
	2.0	0.27
	2.4	0.22
	2.7	0.21
Steel, chromium-nickel, stainless, PH17-7-ASTM [45]	0.3	0.65
	0.4	0.55
	0.5	0.52
	0.6	0.50
	0.8	0.46
	1.0	0.42
	1.2	0.40
	1.6	0.34
	2.0	0.27
	2.4	0.22
	2.7	0.21
Steel, chromium-nickel, stainless, OH18N10T-PN; 821 Grade Ti-BS surface smooth, roughness rms = 2 μm [45]	0.3	0.68
	0.4	0.53
	0.6	0.45
	0.8	0.42
	1.0	0.38
	1.2	0.36
	1.6	0.29

Table 3.1.3. (*continued*)

State of surface	λ [μm]	$\varepsilon_{\lambda,n}$
	2.0	0.24
	2.4	0.19
	2.8	0.22
surface oxidized [45]	0.3	0.92
	0.4	0.90
	0.6	0.884
	0.8	0.89
	1.0	0.893
	1.2	0.88
	1.6	0.878
	2.0	0.84
	2.4	0.79
	2.7	0.79
Steel, chromium-nickel, stainless, 1H18N9T-PN; 321-AISI; 30321-SAE surface polished, oxidized at temperature 873 K [119]	1.0	0.2
	3.0	0.19
	5.0	0.16
	7.0	0.11
	9.0	0.06
surface polished, oxidized at temperature 1023 K [119]	1.0	0.46
	3.0	0.35
	5.0	0.33
	7.0	0.28
	9.0	0.26
surface polished, oxidized at temperature 1073 K [119]	1.0	0.6
	3.0	0.5
	5.0	0.53
	7.0	0.50
	9.0	0.42
surface polished, oxidized at temperature 1173 K [119]	1.0	0.89
	3.0	0.78
	5.0	0.82
	7.0	0.87
	9.0	0.75
surface polished, oxidized at temperature 1273 K [119]	1.0	0.96
	3.0	0.93
	5.0	0.93
	7.0	0.97
	9.0	0.89
Steel, high chromium, stainless, H28-PN surface smooth, rms = 0.03 μm [45]	0.8	0.42
	0.9	0.41
	1.0	0.395
	1.2	0.385
	1.4	0.37
Tantalum, Ta surface smooth at temperature 523 K [131, 144]	4.0	0.105
	5.0	0.09
	7.0	0.073
	9.0	0.06
	11.0	0.054
	13.0	0.048
	15.0	0.03
surface smooth at temperature 1200 K [131, 144]	0.400	0.525
	0.425	0.523
	0.450	0.514
	0.475	0.516
	0.500	0.510
	0.525	0.503
	0.550	0.494
	0.575	0.484
	0.600	0.473
	0.625	0.461
	0.6563	0.445
	0.675	0.435
	0.700	0.421
	0.725	0.407
	0.750	0.394
	0.760	0.384
	0.780	0.376
	0.800	0.363
	0.820	0.350
	0.840	0.337
	0.860	0.326

Table 3.1.3. (*continued*)

State of surface	λ[μm]	$\varepsilon_{\lambda,n}$
	0.900	0.304
	0.940	0.286
	1.000	0.262
	1.10	0.234
	1.20	0.214
	1.30	0.197
	1.40	0.183
	1.50	0.172
	1.60	0.164
	1.70	0.159
	1.80	0.155
	1.90	0.152
	2.00	0.148
	2.20	0.142
	2.40	0.136
	2.60	0.131
	2.80	0.126
	3.00	0.123
	3.20	0.119
	3.40	0.116
	3.60	0.113
	3.80	0.110
	4.00	0.108
surface smooth, temperature 1366 K [131, 144]	4.0	0.12
	5.0	0.105
	7.0	0.09
	9.0	0.08
	11.0	0.075
	13.0	0.073
	15.0	0.069
surface smooth, temperature 1400 K [131, 144]	0.400	0.521
	0.425	0.519
	0.450	0.514
	0.475	0.508
	0.500	0.502
	0.525	0.495
	0.550	0.486
	0.575	0.476
	0.600	0.465
	0.625	0.454
	0.6563	0.439
	0.675	0.429
	0.700	0.416
	0.725	0.403
	0.750	0.390
	0.760	0.384
	0.780	0.372
	0.800	0.361
	0.820	0.350
	0.840	0.339
	0.860	0.329
	0.900	0.310
	0.940	0.293
	1.000	0.271
	1.10	0.245
	1.20	0.226
	1.30	0.210
	1.40	0.196
	1.50	0.186
	1.60	0.178
	1.70	0.172
	1.80	0.167
	1.90	0.164
	2.00	0.160
	2.20	0.154
	2.40	0.148
	2.60	0.143
	2.80	0.138
	3.00	0.134
	3.20	0.130
	3.40	0.126
	3.60	0.123
	3.80	0.121
	4.00	0.118
surface smooth, temperature 1600 K [131, 144]	0.400	0.516
	0.425	0.512
	0.450	0.507
	0.475	0.501
	0.500	0.495
	0.525	0.487
	0.550	0.479
	0.575	0.469
	0.600	0.458
	0.625	0.447
	0.6563	0.433
	0.675	0.424
	0.700	0.412
	0.725	0.400
	0.750	0.387
	0.760	0.382
	0.780	0.370
	0.800	0.361
	0.820	0.351
	0.840	0.341
	0.860	0.332
	0.900	0.316
	0.940	0.300
	1.000	0.281
	1.10	0.256

Table 3.1.3. (*continued*)

State of surface	λ[μm]	$\varepsilon_{\lambda,n}$
	1.20	0.238
	1.30	0.223
	1.40	0.210
	1.50	0.199
	1.60	0.191
	1.70	0.184
	1.80	0.180
	1.90	0.176
	2.00	0.172
	2.20	0.165
	2.40	0.160
	2.60	0.154
	2.80	0.149
	3.00	0.145
	3.20	0.141
	3.40	0.137
	3.60	0.134
	3.80	0.131
	4.00	0.128
surface smooth at temperature 1800 K [131, 144]	0.400	0.511
	0.425	0.507
	0.450	0.501
	0.475	0.494
	0.500	0.487
	0.525	0.480
	0.550	0.471
	0.575	0.462
	0.600	0.451
	0.625	0.441
	0.6563	0.428
	0.675	0.420
	0.700	0.408
	0.725	0.397
	0.750	0.385
	0.760	0.380
	0.780	0.368
	0.800	0.361
	0.820	0.352
	0.840	0.344
	0.860	0.336
	0.900	0.321
	0.940	0.307
	1.000	0.290
	1.10	0.267
	1.20	0 251
	1.30	0.236
	1.40	0.223
	1.50	0.212
	1.60	0.204
	1.70	0.197
	1.80	0.192
	1.90	0.188
	2.00	0.184
	2.20	0.177
	2.40	0.171
	2.60	0.166
	2.80	0.161
	3.00	0.156
	3.20	0.152
	3.40	0.148
	3.60	0.144
	3.80	0.141
	4.00	0.138
surface smooth at temperature 1923 K [131, 144]	4.00	0.135
	5.00	0.125
	7.00	0.11
	9.00	0.102
	11.00	0.093
	13.00	0.087
	15.00	0.083
surface smooth at temperature 2000 K [131, 144]	0.400	0.507
	0.425	0.501
	0.450	0.495
	0.475	0.487
	0.500	0.480
	0.525	0.472
	0.550	0.463
	0.575	0.455
	0.600	0.444
	0.625	0.435
	0.6563	0.423
	0.675	0.416
	0.700	0.405
	0.725	0.394
	0.750	0.383
	0.760	0.379
	0.780	0.368
	0.800	0.362
	0.820	0.354
	0.840	0.346
	0.860	0.340
	0.900	0.327
	0.940	0.314
	1.000	0.299
	1.10	0.278
	1.20	0.262
	1.30	0.250
	1.40	0.236
	1.50	0.226

Table 3.1.3. (*continued*)

State of surface	λ [μm]	$\varepsilon_{\lambda,n}$
	1.60	0.217
	1.70	0.210
	1.80	0.205
	1.90	0.200
	2.00	0.196
	2.20	0.189
	2.40	0.183
	2.60	0.177
	2.80	0.172
	3.00	0.167
	3.20	0.163
	3.40	0.159
	3.60	0.155
	3.80	0.151
	4.00	0.148
surface smooth at temperature 2200 K [131, 144]	0.400	0.502
	0.425	0.496
	0.450	0.488
	0.475	0.480
	0.500	0.472
	0.525	0.464
	0.550	0.456
	0.575	0.448
	0.600	0.438
	0.625	0.430
	0.6563	0.419
	0.675	0.412
	0.700	0.402
	0.725	0.392
	0.750	0.382
	0.760	0.378
	0.780	0.370
	0.800	0.363
	0.820	0.356
	0.840	0.350
	0.860	0.344
	0.900	0.333
	0.940	0.322
	1.000	0.308
	1.10	0.289
	1.20	0.274
	1.30	0.261
	1.40	0.250
	1.50	0.239
	1.60	0.231
	1.70	0.223
	1.80	0.217
	1.90	0.212
	2.00	0.208
	2.20	0.201
	2.40	0.194
	2.60	0.189
	2.80	0.183
	3.00	0.179
	3.20	0.174
	3.40	0.170
	3.60	0.165
	3.80	0.162
	4.00	0.158
surface smooth at temperature 2400 K [131, 144]	0.400	0.498
	0.425	0.491
	0.450	0.482
	0.475	0.473
	0.500	0.464
	0.525	0.456
	0.550	0.448
	0.575	0.440
	0.600	0.432
	0.625	0.424
	0.6563	0.414
	0.675	0.408
	0.700	0.399
	0.725	0.390
	0.750	0.381
	0.760	0.378
	0.780	0.372
	0.800	0.366
	0.820	0.360
	0.840	0.354
	0.860	0.349
	0.900	0.338
	0.940	0.329
	1.000	0.317
	1.10	0.300
	1.20	0.286
	1.30	0.274
	1.40	0.263
	1.50	0.253
	1.60	0.244
	1.70	0.236
	1.80	0.230
	1.90	0.224
	2.00	0.220
	2.20	0.212
	2.40	0.206
	2.60	0.200
	2.80	0.195
	3.00	0.190
	3.20	0.185
	3.40	0.180
	3.60	0.176
	3.80	0.172
	4.00	0.168

Table 3.1.3. (*continued*)

State of surface	$\lambda[\mu m]$	$\varepsilon_{\lambda,n}$
surface smooth at temperature 2473 K [131, 144]	4.00	0.155
	5.00	0.14
	7.00	0.125
	9.00	0.118
	11.00	0.11
	13.00	0.105
	15.00	0.10
surface oxidized [45]	0.3	0.86
	0.4	0.842
	0.6	0.84
	0.8	0.83
	1.0	0.82
	1.2	0.816
	1.6	0.79
	2.0	0.76
	2.4	0.71
	2.7	0.73
Tantalum-tungsten alloy TaW10 surface smooth, temperature of measurement 2400 K [142]	0.665	0.393
	1.0	0.294
	3.0	0.180
	4.0	0.176
Tantalum-tungsten alloy TaW30 surface smooth, temperature of measurement 2200 K [142]	0.4	0.468
	0.65	0.407
	1.0	0.331
	2.0	0.209
	4.0	0.159
surface smooth, temperature of annealing 2400 K for 20 h, temperature of measurement 2200 K [142]	0.4	0.469
	0.65	0.411
	1.0	0.339
	2.0	0.221
	4.0	0.173
surface smooth, temperature of annealing 2400 K for 70 h, temperature of measurement 2200 K [142]	0.4	0.470
	0.65	0.423
	1.0	0.361
	2.0	0.251
	4.0	0.207
Tantalum-tungsten alloy TaW50 surface smooth, temperature of measurement 2400 K [142]	0.665	0.401
	1.0	0.343
	3.0	0.176
	4.0	0.164
Tantalum-tungsten alloy TaW90 surface smooth, temperature of measurement 2400 K [142]	0.665	0.415
	1.0	0.362
	3.0	0.170
	4.0	0.158
Titanium; Ti surface formed by vacuum sublimation [45]	0.2	0.70
	0.4	0.49
	0.6	0.42
	0.8	0.41
	1.0	0.42
	2.0	0.37
	3.0	0.29
	4.0	0.18
	6.0	0.105
	8.0	0.09
	10.0	0.08
surface oxidized [45]	0.2	1.0
	0.4	1.0
	0.6	0.85
	0.8	0.62
	1.0	0.54
	2.0	0.42
	3.0	0.37

Table 3.1.3. (*continued*)

State of surface	λ [μm]	$\varepsilon_{\lambda,n}$
	4.0	0.30
	6.0	0.19
	8.0	0.14
	10.0	0.12
surface smooth, the measurement in vacuum, temperature 1200 K [43]	0.53	0.63
	0.56	0.615
	0.59	0.64
	0.62	0.59
	0.65	0.55
surface smooth, the measurement in vaccum, temperature 1300 K [43]	0.53	0.575
	0.56	0.545
	0.59	0.565
	0.62	0.55
	0.65	0.53
surface smooth, the measurement in vacuum, temperature 1400 K [43]	0.53	0.54
	0.56	0.495
	0.59	0.52
	0.62	0.52
	0.65	0.52
surface smooth, the measurement in vacuum, temperature 1500 K [43]	0.53	0.505
	0.56	0.475
	0.59	0.495
	0.62	0.50
	0.65	0.515
Titanium-aluminium-vanadium alloy surface polished [43]	0.3	0.64
	0.5	0.55
	0.75	0.49
	1.0	0.46
	1.25	0.44
	1.5	0.40
	1.75	0.364
	2.0	0.332
	2.25	0.30
	2.5	0.28
	2.75	0.276

State of surface	λ [μm]	$\varepsilon_{\lambda,n}$
surface oxidized [45]	0.3	0.882
	0.5	0.887
	1.0	0.872
	1.5	0.85
	2.0	0.818
	2.5	0.778
	2.75	0.78
Tungsten; W surface smooth nonoxidized at temperature 523 K [35, 45, 47, 73, 131, 144, 156]	5.0	0.049
	6.0	0.042
	7.0	0.04
	9.0	0.03
	12.0	0.027
	15.0	0.02
surface smooth nonoxidized at temperature 1200 K [35, 45, 47, 73, 131, 144, 156]	0.300	0.486
	0.325	0.486
	0.350	0.485
	0.375	0.484
	0.400	0.482
	0.425	0.481
	0.450	0.479
	0.475	0.477
	0.500	0.474
	0.525	0.471
	0.550	0.468
	0.575	0.464
	0.600	0.461
	0.625	0.458
	0.650	0.454
	0.6563	0.453
	0.675	0.450
	0.700	0.446
	0.725	0.442
	0.750	0.438
	0.800	0.428
	0.850	0.418
	0.900	0.408
	0.950	0.397
	1.000	0.386
	1.1	0.364
	1.2	0.342
	1.28	0.322
	1.3	0.317
	1.4	0.295
	1.5	0.277

Table 3.1.3. (*continued*)

State of surface	λ[μm]	$\varepsilon_{\lambda,n}$
	1.6	0.259
	1.7	0.241
	1.8	0.222
	1.9	0.204
	2.0	0.186
	2.1	0.172
	2.2	0.160
	2.3	0.150
	2.4	0.142
	2.5	0.135
	2.6	0.129
	2.7	0.124
	2.8	0.119
	2.9	0.115
	3.0	0.112
	3.2	0.105
	3.4	0.100
	3.6	0.095
	3.8	0.091
	4.0	0.086
	4.2	0.082
	4.4	0.080
	4.6	0.079
	4.8	0.079
	5.0	0.078
	6.0	0.074
	9.0	0.063
surface smooth nonoxidized, temperature 1366 K [35, 45, 47, 73, 131, 144, 156]	5.0	0.076
	7.0	0.064
	9.0	0.058
	11.0	0.053
	13.0	0.05
	15.0	0.05
surface smooth nonoxidized, temperature 1400 K [35, 45, 47, 73, 131, 144, 156]	0.300	0.483
	0.325	0.483
	0.350	0.482
	0.375	0.481
	0.400	0.479
	0.425	0.478
	0.450	0.476
	0.475	0.473
	0.500	0.470
	0.525	0.467
	0.550	0.464
	0.575	0.459
	0.600	0.456

State of surface	λ[μm]	$\varepsilon_{\lambda,n}$
	0.625	0.453
	0.650	0.449
	0.6563	0.448
	0.675	0.445
	0.700	0.411
	0.725	0.437
	0.750	0.433
	0.800	0.423
	0.900	0.403
	0.950	0.393
	1.000	0.382
	1.1	0.362
	1.2	0.340
	1.28	0.322
	1.3	0.317
	1.4	0.296
	1.5	0.279
	1.6	0.262
	1.7	0.245
	1.8	0.228
	1.9	0.211
	2.0	0.195
	2.1	0.182
	2.2	0.171
	2.3	0.162
	2.4	0.154
	2.5	0.147
	2.6	0.141
	2.7	0.135
	2.8	0.131
	2.9	0.127
	3.0	0.123
	3.2	0.116
	3.4	0.111
	3.6	0.106
	3.8	0.102
	4.0	0.097
	4.2	0.093
	4.4	0.091
	4.6	0.090
	4.8	0.089
	5.0	0.088
surface smooth nonoxidized, temperature 1427 K [35, 45, 47, 73, 131, 144, 156]	6.0	0.083
	7.0	0.08
	9.0	0.06
	11.0	0.06
	13.0	0.06

Table 3.1.3. (*continued*)

State of surface	$\lambda[\mu m]$	$\varepsilon_{\lambda,n}$
surface smooth nonoxidized, temperature 1600 K [35, 45, 47, 73, 131, 144, 156]	0.300	0.480
	0.325	0.480
	0.350	0.479
	0.375	0.478
	0.400	0.476
	0.425	0.475
	0.450	0.472
	0.475	0.470
	0.500	0.466
	0.525	0.463
	0.550	0.459
	0.575	0.455
	0.600	0.451
	0.625	0.448
	0.650	0.444
	0.6563	0.443
	0.675	0.440
	0.700	0.436
	0.725	0.432
	0.750	0.428
	0.800	0.418
	0.850	0.409
	0.900	0.399
	0.950	0.388
	1.000	0.378
	1.1	0.359
	1.2	0.339
	1.28	0.322
	1.3	0.318
	1.4	0.298
	1.5	0.281
	1.6	0.265
	1.7	0.249
	1.8	0.234
	1.9	0.218
	2.0	0.204
	2.1	0.192
	2.2	0.182
	2.3	0.173
	2.4	0.165
	2.5	0.158
	2.6	0.152
	2.7	0.147
	2.8	0.142
	2.9	0.138
	3.0	0.134
	3.2	0.128
	3.4	0.122
	3.6	0.118
	3.8	0.113
	4.0	0.108
	4.2	0.104
	4.4	0.102
	4.6	0.100
	4.8	0.099
	5.0	0.098
surface smooth nonoxidized, temperature 1800 K [35, 45, 47, 73, 131, 144, 156]	0.300	0.477
	0.325	0.477
	0.350	0.476
	0.375	0.475
	0.400	0.473
	0.425	0.472
	0.450	0.469
	0.475	0.466
	0.500	0.462
	0.525	0.458
	0.550	0.454
	0.575	0.450
	0.600	0.466
	0.625	0.433
	0.650	0.439
	0.6563	0.438
	0.675	0.435
	0.700	0.431
	0.725	0.427
	0.750	0.422
	0.800	0.413
	0.850	0.404
	0.900	0.394
	0.950	0.384
	1.000	0.375
	1.1	0.356
	1.2	0.337
	1.28	0.322
	1.3	0.318
	1.4	0.299
	1.5	0.283
	1.6	0.268
	1.7	0.253
	1.8	0.239
	1.9	0.226
	2.0	0.213
	2.1	0.202
	2.2	0.192
	2.3	0.184
	2.4	0.177
	2.5	0.170
	2.6	0.164
	2.7	0.158
	2.8	0.154

Table 3.1.3. (*continued*)

State of surface	λ [μm]	$\varepsilon_{\lambda,n}$
	2.9	0.150
	3.0	0.146
	3.2	0.139
	3.4	0.134
	3.6	0.129
	3.8	0.124
	4.0	0.119
	4.2	0.115
	4.4	0.112
	4.6	0.1111
	4.8	0.109
	5.0	0.108
surface smooth nonoxidized at temperature 1923 K [35, 45, 47, 73, 131, 144, 156]	5.0	0.102
	6.0	0.099
	7.0	0.09
	9.0	0.08
	11.0	0.073
	13.0	0.07
	15.0	0.065
surface smooth nonoxidized at temperature 2000 K [35, 45, 47, 73, 131, 144, 156]	0.300	0.474
	0.325	0.474
	0.350	0.473
	0.375	0.472
	0.400	0.470
	0.425	0.468
	0.450	0.466
	0.475	0.462
	0.500	0.459
	0.525	0.454
	0.550	0.450
	0.575	0.445
	0.600	0.441
	0.625	0.438
	0.650	0.434
	0.6563	0.433
	0.675	0.430
	0.700	0.426
	0.725	0.422
	0.750	0.418
	0.800	0.408
	0.850	0.399
	0.900	0.390
	0.950	0.380
	1.000	0.371
	1.1	0.353
	1.2	0.336
	1.28	0.322
	1.3	0.318
	1.4	0.301
	1.5	0.285
	1.6	0.271
	1.7	0.257
	1.8	0.245
	1.9	0.233
	2.0	0.222
	2.1	0.212
	2.2	0.203
	2.3	0.195
	2.4	0.188
	2.5	0.182
	2.6	0.176
	2.7	0.170
	2.8	0.165
	2.9	0.161
	3.0	0.157
	3.2	0.150
	3.4	0.145
	3.6	0.140
	3.8	0.135
	4.0	0.130
	4.2	0.126
	4.4	0.123
	4.6	0.121
	4.8	0.119
	5.0	0.117
surface smooth nonoxidized at temperature 2200 K [35, 45, 47, 73, 131, 144, 156]	0.300	0.471
	0.325	0.471
	0.350	0.470
	0.375	0.469
	0.4000	0.467
	0.425	0.465
	0.450	0.462
	0.475	0.459
	0.500	0.455
	0.525	0.450
	0.550	0.445
	0.575	0.441
	0.600	0.436
	0.625	0.433
	0.650	0.429
	0.6563	0.428
	0.675	0.425
	0.700	0.421
	0.725	0.417
	0.750	0.412
	0.800	0.404
	0.850	0.394

Table 3.1.3. (*continued*)

State of surface	λ[μm]	$\varepsilon_{\lambda,n}$
	0.900	0.385
	0.950	0.376
	1.000	0.367
	1.1	0.350
	1.2	0.334
	1.28	0.322
	1.3	0.319
	1.4	0.302
	1.5	0.287
	1.6	0.274
	1.7	0.262
	1.8	0.251
	1.9	0.240
	2.0	0.231
	2.1	0.222
	2.3	0.207
	2.4	0.200
	2.5	0.194
	2.6	0.187
	2.7	0.182
	2.8	0.177
	2.9	0.172
	3.0	0.169
	3.2	0.162
	3.4	0.156
	3.6	0.151
	3.8	0.146
	4.0	0.141
	4.2	0.137
	4.4	0.134
	4.6	0.131
	4.8	0.129
	5.0	0.127
surface smooth nonoxidized at temperature 2400 K [35, 45, 47, 73, 131, 144, 156]	0.300	0.468
	0.325	0.468
	0.350	0.467
	0.375	0.466
	0.400	0.464
	0.426	0.462
	0.450	0.459
	0.475	0.455
	0.500	0.450
	0.525	0.446
	0.550	0.441
	0.575	0.436
	0.600	0.431
	0.625	0.428
	0.650	0.424
	0.6563	0.423
	0.675	0.420
	0.700	0.416
	0.725	0.412
	0.750	0.407
	0.800	0.399
	0.850	0.390
	0.900	0.381
	0.950	0.372
	1.000	0.364
	1.1	0.347
	1.2	0.333
	1.28	0.322
	1.3	0.319
	1.4	0.304
	1.5	0.290
	1.6	0.277
	1.7	0.266
	1.8	0.256
	1.9	0.247
	2.0	0.239
	2.1	0.232
	2.2	0.225
	2.3	0.218
	2.4	0.212
	2.5	0.205
	2.6	0.199
	2.7	0.193
	2.8	0.189
	2.9	0.184
	3.0	0.180
	3.2	0.173
	3.4	0.167
	3.6	0.162
	3.8	0.157
	4.0	0.152
	4.2	0.148
	4.4	0.145
	4.6	0.142
	4.8	0.139
	5.0	0.136
surface smooth nonoxidized at temperature 2477 K [35, 45, 47, 73, 131, 144, 156]	5.0	0.125
	6.0	0.12
	7.0	0.112
	9.0	0.102
	11.0	0.093
	13.0	0.087
	15.0	0.085

Table 3.1.3. (*continued*)

State of surface	λ[μm]	$\varepsilon_{\lambda,n}$
surface smooth nonoxidized at temperature 2600 K [35, 45, 47, 73, 131, 144, 156]	0.300	0.465
	0.325	0.465
	0.350	0.464
	0.375	0.463
	0.400	0.461
	0.425	0.458
	0.450	0.456
	0.475	0.452
	0.500	0.447
	0.525	0.442
	0.550	0.436
	0.575	0.431
	0.600	0.426
	0.625	0.423
	0.650	0.419
	0.6563	0.418
	0.675	0.415
	0.700	0.411
	0.725	0.406
	0.750	0.402
	0.800	0.394
	0.850	0.385
	0.900	0.376
	0.950	0.368
	1.000	0.360
	1.1	0.344
	1.2	0.331
	1.28	0.322
	1.3	0.319
	1.4	0.305
	1.5	0.292
	1.6	0.280
	1.7	0.270
	1.8	0.262
	1.9	0.255
	2.0	0.248
	2.1	0.242
	2.2	0.236
	2.3	0.229
	2.4	0.223
	2.5	0.217
	2.6	0.211
	2.7	0.205
	2.8	0.200
	2.9	0.196
	3.0	0.191
	3.2	0.184
	3.4	0.178
	3.6	0.173
	3.8	0.168
	4.0	0.163
	4.2	0.159
	4.4	0.155
	4.6	0.152
	4.8	0.149
	5.0	0.146
roughness rms = 0.38 μm, temperature 2800 K [143]	1.0	0.36
	2.0	0.28
	3.0	0.26
	4.0	0.245
	5.0	0.235
roughness rms = 0.58 μm. temperature 2800 K [143]	0.3	0.45
	0.4	0.453
	0.5	0.449
	0.6	0.441
	0.8	0.40
	1.0	0.37
	2.0	0.31
	3.0	0.305
	4.0	0.305
	5.0	0.305
roughness rms = 1.14 μm, temperature 2800 K [143]	0.3	0.46
	0.4	0.475
	0.5	0.47
	0.6	0.46
	0.8	0.44
	1.0	0.41
	2.0	0.383
	3.0	0.395
	4.0	0.40
	5.0	0.42
Vanadium; V polycrystalline, surface smooth, temperature 295 K [131, 144]	2.0	0.16
	2.5	0.135
	3.0	0.115
	4.0	0.099
	6.0	0.07
	8.0	0.065

Table 3.1.3. (*continued*)

State of surface	λ[μm]	$\varepsilon_{\lambda,n}$
polycrystalline, surface smooth at temperature 523 K [131, 144]	2.0	0.25
	2.5	0.21
	3.0	0.17
	4.0	0.14
	6.0	0.102
	8.0	0.10
	10.0	0.098
Ytterbium; Yb surface smooth, room temperature [131]	0.25	0.26
	0.3	0.4
	0.5	0.62
	0.6	0.65
	0.7	0.65
	0.8	0.65
	1.0	0.67
	2.0	0.77
	3.0	0.81
	4.0	0.82
Zirconium; Zr surface smooth, temperature 1400 K, point X for λ ≅ 6.0 μm [131]	0.45	0.64
	0.5	0.6
	0.6	0.57
	0.8	0.5
	1.0	0.47
	1.5	0.42
	2.0	0.37
	3.0	0.35
	4.0	0.32
surface smooth, temperature 1600 K [131]	0.4	0.64
	0.5	0.55
	0.6	0.52
	0.8	0.47
	1.0	0.43
	2.0	0.36
	3.0	0.34
	4.0	0.31
surface smooth, temperature 2000 K [131]	0.35	0.62
	0.4	0.56
	0.5	0.52
	0.6	0.49
	0.8	0.46
	1.0	0.425
	2.0	0.34
	3.0	0.32
	4.0	0.30

Table 3.1.4. Spectral (red) emissivity in normal direction $\varepsilon_{\lambda=0.65,n}$ of metals and alloys

State of surface	T[K]	$\varepsilon_{\lambda=0.65,n}$
Berylium; Be		
solid state [53]		0.61
liquid state [53]		0.61
Cast iron, carbon [51]	1523	0.93
	1573	0.69
	1673	0.40
	1773	0.25
	1873	0.17
	1923	0.17
Cast iron, silicon 25% Si [51]	1623	0.85
	1673	0.62
	1773	0.36
	1873	0.20
Cobalt; Co surface smooth [45, 131, 153, 155]	1000÷1700	0.36
Cobalt-nickel alloy CoNi35 surface smooth [153]	1250÷1450	0.33
Copper-aluminium alloy BA5-PN; alloy 606-ASTM (aluminium bronze)		
surface polished [45, 131]	1100	0.32
	1200	0.30
surface oxidized [45, 131]	1100	0.73
	1200	0.715
	1300	0.71

State of surface	T[K]	$\varepsilon_{\lambda=0.65,n}$
Copper-manganese nickel alloy; manganin surface polished [45]	293	0.73
Copper-nickel-manganese alloy; constantan solid state [53]		0.35
Erbium; Er		
solid state [53]		0.55
liquid state [53]		0.38
Gold; Au surface smooth [53, 131]	1125÷1400	0.075
Hafnium; Hf surface smooth [131]	1500÷1875	0.445÷0.45
Iridium; Ir surface smooth [131]	1000÷2500	0.28
Iron; Fe surface smooth [45, 51]	800	0.38
	1100	0.376
	1300	0.36
	1500	0.339
	1700	0.32
	1873	0.28
	1973	0.195
	2073	0.105
Iron-cobalt alloy [68]	1000÷1600	0.4÷0.3

Table 3.1.4. (*continued*)

State of surface	T[K]	$\varepsilon_{\lambda=0.65,n}$
Iron-tungsten alloy [68]	1300÷1700	0.33÷0.31
Magnesium; Mg		
solid state [53]		0.59
liquid state [53]		0.59
Molybdenum; Mo		
surface smooth [131]	1000	0.422
	1250	0.41
	1500	0.395
	1750	0.383
	2000	0.378
	2250	0.37
	2500	0.363
	2750	0.355
	3000	0.349
surface smooth [131]	1000	0.413
	1125	0.409
	1250	0.405
	1375	0.40
	1500	0.395
	1625	0.39
	1750	0.387
	1875	0.382
	2000	0.377
	2125	0.374
	2250	0.37
	2375	0.366
	2500	0.362
	2625	0.359
	2750	0.354
	2875	0.35
Molybdenum-tungsten alloy MoW20 surface polished [36, 131]	2400÷3000	0.33
Molybdenum-tungsten alloy MoW25 surface polished [36, 131]	1200	0.46
	1400	0.445
	1600	0.435
	1800	0.42
	2000	0.41
	2200	0.40
	2400	0.40
	2600	0.41
Molibdenum-tungsten alloy MoW30 surface polished [36, 131]	2400÷3000	0.377
Molybdenum-tungsten alloy MoW37 surface polished [36, 131]	1200	0.485
	1400	0.47
	1600	0.46
	1800	0.45
	2000	0.437
	2200	0.428
	2400	0.42
	2600	0.425
Molybdenum-tungsten alloy MoW87 surface polished [36, 131]	1200	0.535
	1400	0.515
	1600	0.495
	1800	0.48
	2000	0.465
	2200	0.448
	2400	0.43
	2600	0.435
	2800	0.45
Nickel; Ni		
surface polished [45]	300	0.36
	1100	0.36
	1200	0.35
	1400	0.334
	1600	0.322
	1800	0.318
surface rolled [45]	1100	0.558
	1200	0.561
	1300	0.568
	1400	0.574
surface oxidized [45]	1000	0.882
	1100	0.868
	1200	0.854
	1300	0.84
	1400	0.83
	1500	0.82

Table 3.1.4. (*continued*)

State of surface	T[K]	$\varepsilon_{\lambda=0.65,n}$
Nickel-chromium-iron-alloy; inconel X surface polished [45]	1100	0.44
	1200	0.429
	1300	0.414
	1400	0.402
	1500	0.389
surface casted or rolled clean [45]	1100	0.775
	1200	0.715
	1300	0.64
	1400	0.575
	1500	0.50
surface oxidized [45]	1100	0.89
	1200	0.864
	1300	0.839
	1400	0.815
	1500	0.788
Nickel-chromium-iron alloy; ЭИ607-GOST; nimonic 95-BS surface polished [131]	1100	0.44
	1300	0.41
	1500	0.39
surface natural state [131]	1100	0.76
	1200	0.69
	1300	0.62
	1400	0.57
	1500	0.51
Nickel-molybdenum-chromium alloy; hastelloy C surface smooth [45]	1150	0.50
	1200	0.44
	1300	0.40
	1400	0.38
	1500	0.365
Nickel-molybdenum alloy; hastelloy B surface smooth [45]	1100	0.378
	1200	0.362
	1300	0.345
	1400	0.33
	1500	0.314
surface oxidized [45]	1100	0.69
	1200	0.682
	1300	0.676
	1400	0.67
	1500	0.662
Niobium; Nb surface smooth [131]	1000	0.368
	1500	0.362
	2000	0.355
	2500	0.35
Niobium-tungsten alloy NbW10 surface smooth [95, 131 , 142]	2000÷2600	0.25
Osmium; Os [140]	1300	0.52
	1400	0.47
	1600	0.42
	1800	0.38
	2000	0.37
	2200	0.37
	2400	0.38
	2500	0.395
Palladium; Pd surface smooth [131]	1000	0.15
	1125	0.13
	1250	0.12
	1375	0.10
	1500	0.08
	1625	0.065
	1750	0.05
Rhenium; Re [79, 95, 131]	1000	0.255
	1250	0.249
	1500	0.234
	1750	0.227
	2000	0.221
	2250	0.213
	2500	0.201
	2750	0.19
	2875	0.183

Table 3.1.4. (*continued*)

State of surface	T[K]	$\varepsilon_{\lambda=0.65,n}$
Rhodium; Rh surface smooth [80, 131]	1125	0.105
	1250	0.12
	1375	0.13
	1500	0.14
	1625	0.15
	1750	0.16
	1875	0.17
	2000	0.175
	2125	0.18
Ruthenium; Ru [140]	1200	0.45
	1400	0.37
	1600	0.33
	1800	0.32
	2000	0.31
	2200	0.32
	2400	0.325
	2500	0.33
Silver; Ag surface smooth [22, 131]	1000÷1250	0.055
Steel, chromium-nickel-molybdenum, stainless, H17N5M3-PN surface smooth roughness rms = 0.03 μm the measurement in vacuum [131]	1100	0.77
	1200	0.72
	1300	0.67
	1400	0.62
	1500	0.57
Steel, chromium-nickel-molybdenum, stainless, OH17N14M2-PN; 316LC-AISI surface smooth [45]	1100	0.386
	1200	0.372
	1300	0.358
	1400	0.343
	1500	0.33
surface oxidized [44]	1100	0.79
	1200	0.70
	1300	0.642
	1400	0.606
	1500	0.57
Steel, chromium-nickel-molybdenum, stainless, OOH17N14M-PN; AM-350 surface smooth [45]	1100	0.39
	1300	0.36
	1500	0.334
Steel, chromium-nickel, stainless, N-155-ASTM surface polished [45]	1100	0.36
	1200	0.34
	1300	0.32
	1400	0.30
	1500	0.28
surface clean [45]	1100	0.53
	1300	0.526
	1500	0.522
surface oxidized [45]	1100	0.714
	1300	0.707
	1500	0.70
Steel, chromium-nickel, stainless, PH-15-7-Mo-ASTM [45]	1100	0.39
	1200	0.382
	1300	0.376
	1400	0.368
	1500	0.36
Steel, chromium-nickel, stainless, PH 17-7-ASTM [45]	1100	0.356
	1200	0.339

Table 3.1.4. (*continued*)

State of surface	T[K]	$\varepsilon_{\lambda=0.65,n}$
	1300	0.32
	1400	0.322
	1500	0.294
Steel, chromium-nickel, stainless, OH18N10T-PN; 821 Grade Ti-BS surface smooth [131]	1100	0.39
	1200	0.37
	1300	0.35
	1400	0.33
	1500	0.31
surface oxidized [45]	1100	0.68
	1200	0.64
	1300	0.60
	1400	0.558
	1500	0.514
Tantalum; Ta surface oxidized [45]	1000	0.799
	1200	0.889
	1400	0.88
	1600	0.87
Tantalum-niobium alloy TaNb1 surface smooth [131]	1400	0.448
	1500	0.444
	1600	0.440
	1700	0.435
	1800	0.430
	1900	0.427
	2000	0.422
Tantalum-niobium alloy TaNb10 surface smooth [131]	1400	0.413
	1500	0.411
	1600	0.410
	1700	0.408
	1800	0.407
	1900	0.404
	2000	0.402
Tantalum-niobium alloy TaNb20 surface smooth [131]	1400	0.406
	1500	0.403
	1600	0.401
	1700	0.399
	1800	0.397
	1900	0.395
	2000	0.392
Tantalum-niobium alloy TaNb30 surface smooth [131]	1400	0.399
	1500	0.398
	1600	0.397
	1700	0.395
	1800	0.393
	1900	0.392
	2000	0.391
Tantalum-niobium alloy TaNb40 surface smooth [131]	1400	0.390
	1600	0.389
	1800	0.388
	2000	0.386
Tantalum-niobium alloy TaNb50 surface smooth [131]	1400	0.388
	1600	0.387
	1800	0.386
	2000	0.385
Tantalum-niobium alloy TaNb60 surface smooth [131]	1400	0.388
	1600	0.386
	1800	0.384
	2000	0.380
Tantalum-niobium alloy TaNb95 surface smooth [131]	1400	0.369
	1600	0.366
	1800	0.363
	2000	0.360
Tantalum-niobium alloy TaNb99 surface smooth [131]	1400÷2000	0.360

Table 3.1.4. (*continued*)

State of surface	T[K]	$\varepsilon_{\lambda=0.65,n}$
Tantalum-tungsten alloy TaW10 [131]	1700	0.43
	1900	0.42
	2100	0.415
	2300	0.41
	2500	0.40
	2700	0.39
	2900	0.38
	3100	0.37
Tantalum-tungsten alloy TaW30 [131]	1700	0.43
	1900	0.42
	2100	0.415
	2300	0.41
	2500	0.40
	2700	0.39
	2900	0.38
	3100	0.37
Thorium; Th [10, 45]	1813	0.35
	2063	0.40
	2313	0.62
Titanium; Ti surface smooth [131, 144, 150]	1000	0.505
	1125	0.495
	1250	0.485
	1375	0.478
	1500	0.472
	1625	0.465
	1750	0.46
	1825	0.45
Titanium-aluminium alloy surface polished [45]	300	0.52
	1100	0.52
	1300	0.486
	1500	0.45
	1700	0.418
surface clean [45]	1100	0.57
	1300	0.56
	1500	0.543
	1700	0.532
surface oxidized [45]	300	0.874
	500	0.81
	700	0.754
	900	0.71
	1100	0.68
	1300	0.662
	1500	0.642
	1700	0.63
Tungsten-rhenium alloy WRe27 surface grinded [131]	1600	0.435
	1800	0.43
	2000	0.425
	2200	0.420
	2400	0.415
	2600	0.41
	2800	0.405
	3000	0.40
Uranium, U solid state [45]		0.26
liquid state [53, 61]		0.34
Vanadium; V surface smooth polycrystalline [131, 144]	1125	0.213
	1250	0.206
	1375	0.201
	1500	0.195
	1625	0.188
	1750	0.182
	1825	0.176
Yttrium; Y solid state [53]		0.35
liquid state [53]		0.35
Zirconium; Zr surface polished [131]	1000	0.45
	1125	0.44
	1250	0.435
	1375	0.43
	1500	0.428
	1675	0.425
	1750	0.42
	1875	0.413
	2000	0.41

Table 3.1.5. Bidirectional spectral reflectivity in normal direction $r_{\lambda,n}$ of metals and alloys

State of surface	λ[µm]	$r_{\lambda,n}$
Aluminium; Al surface polished room temperature [80, 93, 109, 112, 131, 149]	0.15	0.83
	0.2	0.86
	0.25	0.91
	0.3	0.92
	0.4	0.918
	0.5	0.915
	0.6	0.905
	0.7	0.89
	0.8	0.86
	1.0	0.92
	1.2	0.95
	1.4	0.96
	1.6	0.97
	1.8	0.975
	2.0	0.978
	3.0	0.981
	5.0	0.982
	8÷20	0.984
surface polished temperature 599 K [80, 93, 109, 112, 131, 149]	2.0	0.928
	3.0	0.95
	5.0	0.962
	6.0	0.964
surface polished temperature 697 K [80, 93, 109, 112, 131, 149]	1.6	0.908
	2.0	0.92
	3.0	0.94
	4.0	0.95
	6.0	0.96
	10.0	0.96
	14.0	0.97
surface polished temperature 805 K [80, 93, 109, 112, 131, 149]	1.6	0.874
	2.0	0.90
	4.0	0.938
	6.0	0.95
	8.0	0.954
	10.0	0.942
	12.0	0.96
	14.0	0.965
casting [45, 131]	0.25	0.48
	0.30	0.66
	0.35	0.72
	0.40	0.75
	0.45	0.77
	0.50	0.79
	0.55	0.80
	0.575	0.80
powder [24]	0.6	0.68
surface hard anodized [45, 131]	0.5	0.07
	0.75	0.07
	1.0	0.08
	2.0	0.14
	2.6	0.22
	3.0	0.07
	4.0	0.24
	5.0	0.30
	6.0	0.17
	7.0	0.06
	9.0	0.05
	11.0	0.08
	15.0	0.17
	19÷23	0.20

Table 3.1.5. (*continued*)

State of surface	λ[μm]	$r_{\lambda,n}$
Aluminium-magnesium alloy MgAl30 casting [45, 131]	0.25	0.38
	0.30	0.50
	0.35	0.60
	0.40	0.67
	0.45	0.70
	0.50	0.74
	0.55	0.77
	0.575	0.78
Antimony; Sb surface polished [24, 64]	0.6	0.53
	1.0	0.55
	1.2	0.56
	1.4	0.57
	1.6	0.58
	1.8	0.59
	2.0	0.595
	2.2	0.605
	2.4	0.615
	2.6	0.63
	2.8	0.64
	3.0	0.645
	3.2	0.655
	3.4	0.67
	3.6	0.68
	3.8	0.69
	4.0	0.70
	9.0	0.72
Barium; Ba surface smooth room temperature [131]	0.3	0.28
	0.4	0.38
	0.6	0.51
	0.8	0.58
	1.0	0.58
	1.2	0.58
	1.4	0.58
	1.6	0.60
	1.8	0.61
	2.0	0.62
	3.0	0.67
	4.0	0.71
Berylium, Be hot moulded sample, bright polish with few pits roughness rms = 0.19 μm, measurement in vacuum temperature of measurement 298 K [7]	0.4	0.428
	0.5	0.433
	0.6	0.438
	0.7	0.444
	0.8	0.460
	0.9	0.478
	1.0	0.500
	1.5	0.747
	2.0	0.849
	3.0	0.912
	4.0	0.928
	5.0	0.942
	6.0	0.952
	8.0	0.964
	10.0	0.971
	15.0	0.975
	20.0	0.978
	30.0	0.987
	40.0	0.992
surface smooth room temperature [131]	0.3	0.4
	0.35	0.39
	0.4	0.38
	0.5	0.39
	0.6	0.41
	0.8	0.45
	1.0	0.51
	1.2	0.60
	1.4	0.70
	1.6	0.81
	1.8	0.88
	2.0	0.93
Cadmium; Cd [64]	1.0	0.72
	2.0	0.87
	4.0	0.96
	7.0	0.98
	10.0	0.98
	12.0	0.99

Table 3.1.5. (*continued*)

State of surface	λ [μm]	$r_{\lambda,n}$
Cobalt; Co surface smooth, room temperature [64, 131]	1.0	0.67
	2.0	0.72
	3.0	0.8
	4.0	0.88
	6.0	0.94
	8.0	0.96
	12.0	0.97
	22.0	0.98
Copper; Cu surface polished electrolytically or formed by vacuum sublimation, room temperature [41, 80, 92, 112, 131]	0.1	0.09
	0.15	0.13
	0.2	0.28
	0.25	0.39
	0.3	0.32
	0.35	0.38
	0.4	0.41
	0.5	0.57
	0.6	0.80
	0.7	0.97
	0.8	0.98
	1.0÷20	0.985
surface mechanically polished, room temperature [41 80, 92, 112, 131]	0.35	0.2
	0.40	0.22
	0.50	0.32
	0.60	0.44
	0.70	0.61
	0.80	0.77
	1.0	0.90
	1.2	0.95
	1.4	0.96
	1.6	0.97
	1.8÷20	0.98
surface polished, temperature of measurement 573 K [41, 80, 92, 112, 131]	0.45	0.6
	0.50	0.64
	0.55	0.7
	0.6	0.97
	0.65	0.98
surface polished, temperature of measurement 1242 K [41, 80, 92, 112, 131]	0.5	0.6
	0.55	0.7
	0.6	0.82
	0.65	0.88
	0.7	0.92
Copper-tin alloy (red bronze) surface polished [24, 64]	0.251	0.30
	0.385	0.53
	0.500	0.63
	0.76	0.65
	1.0	0.7
	2.0	0.8
	3.0	0.86
	4.0	0.88
	9.0	0.93
Europium; Eu surface smooth, room temperature [131]	0.3	0.48
	0.4	0.54
	0.5	0.59
	0.6	0.62
	0.7	0.63
	0.8	0.64
	1.0	0.63
	1.2	0.62
	1.4	0.62
	1.6	0.64
	1.8	0.65
	2.0	0.68
	3.0	0.71
	4.0	0.74
Gadolinium; Gd surface smooth, room temperature [131]	0.35	0.5
	0.4	0.52
	0.5	0.58
	0.6	0.6
	0.7	0.63
	0.8	0.64
	1.0	0.67
	1.2	0.71

State of surface	λ[μm]	$r_{\lambda,n}$
Gold; Au		
surface at room temperature [41, 45, 80, 93, 101, 112, 131, 149]	0.1	0.15
	0.15	0.18
	0.2	0.21
	0.3	0.365
	0.4	0.38
	0.5	0.50
	0.6	0.92
	0.7	0.97
	0.8	0.98
	1.0	0.982
	2÷4	0.985
	4÷10	0.988
surface at temperature 583 K [45, 101, 131]	0.45	0.38
	0.5	0.50
	0.55	0.60
	0.6	0.90
	0.64	0.94
surface at temperature 843 K [45, 101, 131]	0.45	0.39
	0.5	0.50
	0.55	0.77
	0.6	0.87
	0.64	0.91
surface at temperature 1193 K [45, 101, 131]	0.45	0.41
	0.5	0.5
	0.55	0.72
	0.6	0.835
	0.64	0.885
surface at temperature 1273 K [45, 101, 131]	0.45	0.35
	0.5	0.43
	0.55	0.64
	0.6	0.75
	0.7	0.85
thickness of layer 150 Å [33, 136, 149]	10.6	0.99
thickness of layer 80 Å [33, 136, 149]	10.6	0.98
thickness of layer 70 Å [33, 136, 149]	10.6	0.89
thickness of layer 60 Å [33, 136, 141]	10.6	0.82
thickness of layer 55 Å on base $r_{\lambda,n}$ = 0.19 [33, 136, 141]	10.6	0.71
thickness of layer 50 Å on base $r_{\lambda,n}$ = 0.19 [33, 136, 141]	10.6	0.51
thickness of layer 45 Å on base $r_{\lambda,n}$ = 0.19 [33, 136, 141]	10.6	0.92
thickness of layer 40 Å on base $r_{\lambda,n}$ = 0.19 [33, 136, 141]	10.6	0.23
Hafnium; Hf		
sample vacuum are melted, surface polished, roughness rms = 0.19 μm, temperature of measurement 298 K [7]	0.4	0.371
	0.5	0.419
	0.6	0.450
	0.7	0.475
	0.8	0.495
	0.9	0.507
	1.0	0.528
	1.5	0.590
	2.0	0.648
	3.0	0.760
	4.0	0.825
	5.0	0.870
	6.0	0.887
	8.0	0.908
	10.0	0.919
	15.0	0.931
	20.0	0.942
	30.0	0.947
	40.0	0.956

Table 3.1.5. (*continued*)

State of surface	λ[μm]	$r_{\lambda,n}$
Iridium, Ir [64]	1.0	0.78
	2.0	0.87
	4.0	0.94
	7.0	0.95
	10.0	0.96
	12.0	0.96
Iron; Fe [64]	0.50	0.55
	0.60	0.57
	0.70	0.59
	1.00	0.65
	2.00	0.78
	3.00	0.84
	4.00	0.89
	9.00	0.94
Lutetium; Lu surface smooth, room temperature [131]	0.3	0.3
	0.4	0.59
	0.5	0.58
	0.6	0.61
	0.7	0.62
	0.8	0.63
	1.0	0.68
	1.2	0.71
Magnesium; Mg surface smooth, room temperature [45, 64, 131]	0.25	0.35
	0.3	0.38
	0.35	0.5
	0.4	0.6
	0.5	0.71
	0.6	0.76
	1.0	0.74
	2.0	0.77
	3.0	0.80
	4.0	0.83
	9.0	0.93
Mercury; Hg [45, 64]	0.45	0.728
	0.50	0.709
	0.55	0.712
	0.60	0.699
	0.65	0.715
	0.70	0.728

State of surface	λ[μm]	$r_{\lambda,n}$
Molybdenum; Mo [64]	0.5	0.46
	0.6	0.48
	0.8	0.52
	1.0	0.58
	2.0	0.82
	4.0	0.90
	7.0	0.93
	10.0	0.94
	12.0	0.95
Nickel; Ni surface polished, room temperature [80, 112, 131, 144, 153, 154]	0.3	0.21
	0.35	0.26
	0.4	0.3
	0.5	0.37
	0.6	0.48
	0.8	0.59
	1.0	0.64
	2.0	0.79
	3.0	0.85
	4.0	0.89
	6.0	0.93
	8.0	0.96
	10.0	0.98
	20.0.	0.99
Niobium; Nb sample vacuum are melted, surface polished, roughness rms = 0.19 μm, temperature of measurement 298 K [7]	0.4	0.467
	0.5	0.486
	0.6	0.519
	0.7	0.560
	0.8	0.600
	0.9	0.650
	1.0	0.700
	1.5	0.843
	2.0	0.883
	3.0	0.912
	4.0	0.925
	5.0	0.928
	6.0	0.937
	8.0	0.945
	10.0	0.950
	15.0	0.956
	20.0	0.970
	30.0	0.980
	40.0	0.995

Table 3.1.5. (*continued*)

State of surface	λ[μm]	$r_{\lambda,n}$
surface smooth, room temperature [131]	2.0	0.77
	2.5	0.82
	3.0	0.865
	3.5	0.895
	4.0	0.925
	5.0	0.965
	6.0	0.975
	7.0	0.99
	8.0	0.985
	10.0	0.98
Palladium; Pd surface smooth, room temperature [64]	0.1	0.10
	0.2	0.30
	0.3	0.41
	0.4	0.44
	0.6	0.59
	0.8	0.67
	1.0	0.71
	2.0	0.82
	3.0	0.86
	4.0	0.90
	7.0	0.94
	10.0	0.97
	12.0	0.97
surface smooth, temperature 1370 K [131]	2.0	0.71
	3.0	0.80
	4.0	0.82
surface smooth, temperature 1628 K [131]	2.0	0.68
	3.0	0.70
	4.0	0.76
Platinum; Pt [64]	0.251	0.338
	0.288	0.383
	0.305	0.398
	0.326	0.414
	0.357	0.434
	0.385	0.454
	0.420	0.518
	0.450	0.547
	0.50	0.584
	0.55	0.611
	0.60	0.642
	0.65	0.665
	0.70	0.699
	0.80	0.703
	1.00	0.729
	2.00	0.806
	3.00	0.888
	4.00	0.915
	9.00	0.954
Rhodium; Rh surface polished [24, 64]	0.5	0.76
	0.6	0.77
	0.76	0.81
	1.0	0.84
	2.0	0.91
	3.0	0.92
	4.0	0.925
	5.0	0.93
	10.0	0.96
surface formed by vacuum sublimation [149]	0.25	0.6
	0.3	0.694
	0.4	0.76
	0.5	0.76
	0.6	0.79
	0.7	0.81
	0.8	0.83
	0.9	0.84
	1.0	0.85
	2.0	0.905
	3.0	0.93
	5.0	0.95
	7.0	0.96
	10.0	0.96
Silver; Ag surface formed by vacuum sublimation, liquid state [45, 80, 101, 109, 112, 131, 140, 149]	0.1	0.07
	0.15	0.06
	0.2	0.18
	0.25	0.22
	0.3	0.13
	0.32	0.05
	0.35	0.30
	0.4	0.87
	0.5	0.95
	0.6	0.98
	0.8	0.985
	1÷20	0.99

Table 3.1.5. (*continued*)

State of surface	λ[μm]	$r_{\lambda,n}$
surface formed by vacuum sublimation, thickness 300 Å [45, 80, 101, 109, 112, 131, 140, 149]	0.4	0.46
	0.5	0.64
	0.6	0.805
	0.7	0.82
layer formed by vacuum sublimation, thickness 0.004 μm [129, 135]	0.4	0.040
	0.5	0.062
	0.6	0.033
	0.7	0.027
	10.6	0.038
layer formed by vacuum sublimation, thickness 0.0065 μm [129, 136]	0.4	0.067
	0.5	0.129
	0.6	0.111
	0.7	0.084
	10.6	0.191
layer formed by vacuum sublimation, thickness 0.008 μm [129, 136]	0.4	0.089
	0.5	0.173
	0.6	0.200
	0.7	0.178
	10.6	0.737
layer formed by vacuum sublimation, thickness 0.0095 μm [129, 136]	0.4	0.111
	0.5	0.200
	0.6	0.278
	0.7	0.271
	10.6	0.802
layer formed by vacuum sublimation, thickness 0.01 μm [129]	0.4	0.112
	0.5	0.200
	0.6	0.276
	0.7	0.300
vapour coating of thickness 0.0125 μm [129]	0.4	0.144
	0.5	0.258
	0.6	0.367
	0.7	0.384
layer formed by vacuum sublimation, thickness 0.0140 μm [129]	0.4	0.173
	0.5	0.311
	0.6	0.429
	0.7	0.456
layer formed by vacuum sublimation, thickness 0.0170 μm [129, 136]	0.4	0.204
	0.5	0.411
	0.6	0.533
	0.7	0.578
	10.6	0.811
layer formed by vacuum sublimation, thickness 0.02 μm [129]	0.4	0.318
	0.5	0.476
	0.6	0.629
	0.7	0.663
layer formed by vacuum sublimation, thickness 0.022 μm [129]	0.4	0.356
	0.5	0.544
	0.6	0.689
	0.7	0.722
layer formed by vacuum sublimation, thickness 0.03 μm [129]	0.4	0.522
	0.5	0.684
	0.6	0.800
	0.7	0.827
layer formed by vacuum sublimation, thickness 0.045 μm [129]	0.4	0.800
	0.5	0.869
	0.6	0.916
	0.7	0.929

Table 3.1.5. (*continued*)

State of surface	λ [μm]	$r_{\lambda,n}$
layer formed by vacuum sublimation, thickness 0.055 μm [129]	0.4	0.889
	0.5	0.933
	0.6	0.960
	0.7	0.956
Steel, carbon surface polished, measurement at room temperature [131]	0.3	0.43
	0.4	0.54
	0.5	0.56
	0.6	0.59
	0.7	0.60
	0.8	0.61
	1.0	0.65
	3.0	0.81
	5.0	0.89
	7.0	0.91
	10.0	0.93
	14.0	0.95
Steel, chromium-nickel, stainless, 1H18N9T-PN; 321-AISI; 30321-SAE surface nonoxidized smooth [119]	2	0.458
	3	0.517
	4	0.583
	5	0.625
	6	0.642
	7	0.630
	8	0.655
	10	0.683
	12	0.708
	14	0.732
	15	0.767
surface oxidized at temperature 873 K [119]	2	0.320
	3	0.442
	4	0.547
	5	0.567
	6	0.575
	7	0.592
	8	0.600
	10	0.608
	12	0.663
	14	0.682
	15	0.732
surface oxidized at temperature 973 K [119]	2	0.233
	3	0.345
	4	0.425
	5	0.468
	6	0.483
	7	0.492
	8	0.513
	10	0.515
	12	0.592
	14	0.617
	15	0.683
surface oxidized at temperature 973 K [119]	2	0.047
	3	0.097
	4	0.242
	5	0.330
	6	0.392
	7	0.408
	8	0.442
	10	0.463
	12	0.525
	14	0.550
	15	0.617
surface oxidized at temperature 1073 K [119]	2	0.017
	3	0.017
	4	0.020
	5	0.030
	6	0.058
	7	0.110
	8	0.150
	10	0.215
	12	0.250
	14	0.253
	15	0.417
surface oxidized at temperature 1123 K [119]	2	0.033
	3	0.036
	4	0.035
	5	0.032
	6	0.047
	7	0.052
	8	0.050
	10	0.035

Table 3.1.5. (*continued*)

State of surface	λ[μm]	$r_{\lambda,n}$	State of surface	λ[μm]	$r_{\lambda,n}$
	12	0.118		4.0	0.93
	14	0.160		7.0	0.94
	15	0.245		12.0	0.95
Strontium; Sr			Tin; Sn		
surface smooth,			[64]	1.0	0.54
room temperature				2.0	0.61
[131]	0.25	0.33		4.0	0.72
	0.3	0.45		7.0	0.81
	0.35	0.58		10.0	0.84
	0.4	0.61		12.0	0.85
	0.5	0.68	Titanium; Ti		
	0.6	0.71	sample electron- beam melted,		
	0.8	0.75	surface polished with some pits,		
	1.0	0.76	roughness rms = 0.18 μm,		
	1.2	0.77	temperature of		
	1.4	0.775	measurement 298 K		
	1.6	0.78	[7, 151]	0.4	0.347
	1.8	0.79		0.5	0.376
	2.0	0.80		0.6	0.400
	3.0	0.88		0.7	0.452
	4.0	0.90		0.8	0.495
Tantalum; Ta				0.9	0.517
sample electron-beam melted,				1.0	0.536
surface polished with some				1.5	0.605
scratches and o few pits,				2.0	0.652
roughness rms = 0.19 μm,				3.0	0.712
temperature of				4.0	0.748
measurement 298 K				5.0	0.771
[7]	0.4	0.409		6.0	0.790
	0.5	0.404		8.0	0.819
	0.6	0.400		10.0	0.843
	0.7	0.500		15.0	0.878
	0.8	0.600		20.0	0.900
	0.9	0.676		30.0	0.918
	1.0	0.731		40.0	0.922
	1.5	0.852	surface polished		
	2.0	0.888	[76, 131, 144]	1.0	0.55
	3.0	0.900		2.0	0.61
	4.0	0.905		4.0	0.75
	5.0	0.919		6.0	0.85
	6.0	0.924		8.0	0.88
	8.0	0.943		10.0	0.89
	10.0	0.950		14.0	0.90
	15.0	0.971		20.0	0.91
	20.0	0.976		24.0	0.91
	30.0	0.981			
	40.0	0.986			
	0.5	0.38			
	0.6	0.45			
	0.8	0.64			
	1.0	0.78			
	2.0	0.90			

Table 3.1.5. (*continued*)

State of surface	λ[μm]	$r_{\lambda,n}$
surface oxidized for 303 h at temperature 813 K [76, 131, 144]	1.0	0.25
	2.0	0.31
	4.0	0.44
	6.0	0.65
	8.0	0.52
	10.0	0.60
	12.0	0.68
	14.0	0.76
	16.0	0.76
	18.0	0.73
	20.0	0.58
	22.0	0.56
	24.0	0.66
	25.0	0.68
Tungsten; W [64]	0.50	0.49
	0.60	0.51
	0.70	0.54
	1.00	0.62
	2.00	0.85
	3.00	0.90
	4.00	0.93
	9.00	0.95
Vanadium; V sample vacuum arc-melted, surface polished, roughness rms = 0.17 μm, temperature of measurement 298 K [7]	0.4	0.571
	0.5	0.576
	0.6	0.583
	0.7	0.593
	0.8	0.595
	0.9	0.605
	1.0	0.624
	1.5	0.783
	2.0	0.860
	3.0	0.895
	4.0	0.912
	5.0	0.924
	6.0	0.931
	8.0	0.943
	10.0	0.948
	15.0	0.950
	20.0	0.951
	30.0	0.952
	40.0	0.958
Zinc; Zn surface polished [24, 64]	0.6	0.41
	0.8	0.44
	1.0	0.36
	1.2	0.50
	1.4	0.65
	1.6	0.75
	1.8	0.82
	2.0	0.92
	4.0	0.97
	7.0	0.98
	10.0	0.98
	12.0	0.99
Zirconium; Zr surface polished [131]	2.0	0.72
	3.0	0.77
	4.0	0.86
	5.0	0.9
	6.0	0.92
	7.0	0.93
	9.0	0.94
	11.0	0.95
	13.0	0.956
	15.0	0.965
	16.0	0.97
sample electron-beam melted, surface polished, roughness rms = 0.19 μm, temperature of measurement 298 K [7]	0.4	0.390
	0.5	0.446
	0.6	0.481
	0.7	0.512
	0.8	0.547
	0.9	0.561
	1.0	0.583
	1.5	0.644
	2.0	0.684
	3.0	0.735
	4.0	0.776
	5.0	0.825
	6.0	0.867
	8.0	0.900
	10.0	0.915
	15.0	0.944
	20.0	0.950
	30.0	0.971
	40.0	0.974

Table 3.1.6. Spectral transmissivity in normal direction $\tau_{\lambda,n}$ of metals and alloys

State of surface and thickness of layer	λ[μm]	$\tau_{\lambda,n}$	State of surface and thickness of layer	λ[μm]	$\tau_{\lambda,n}$
Silver; Ag					
layer formed by vacuum sublimation, thickness 100 Å [33, 136, 141]	0.4	0.76	layer formed by vacuum sublimation, thickness 110 Å [129]	0.4	0.693
	0.5	0.59		0.5	0.400
	0.6	0.4		0.6	0.444
	0.7	0.37		0.7	0.633
layer formed by vacuum sublimation, thickness 200 Å [33, 136, 141]	0.4	0.63	layer formed by vacuum sublimation, thickness 150 Å [129]	0.4	0.678
	0.5	0.46		0.5	0.373
	0.6	0.34		0.6	0.364
	0.7	0.29		0.7	0.444
layer formed by vacuum sublimation, thickness 300 Å [33, 136, 141]	0.4	0.42	layer formed by vacuum sublimation, thickness 180 Å [129]	0.4	0.656
	0.5	0.24		0.5	0.333
	0.6	0.15		0.6	0.211
	0.7	0.13		0.7	0.322
layer formed by vacuum sublimation, thickness 40 Å [129]	0.4	0.900	layer formed by vacuum sublimation, thickness 200 Å [129]	0.4	0.618
	0.5	0.822		0.5	0.307
	0.6	0.912		0.6	0.173
	0.7	0.960		0.7	0.144
layer formed by vacuum sublimation, thickness 70 Å [129]	0.4	0.738	layer formed by vacuum sublimation, thickness 260 Å [129]	0.4	0.533
	0.5	0.511		0.5	0.271
	0.6	0.711		0.6	0.144
	0.7	0.842		0.7	0.122

Table 3.1.6. (*continued*)

State of surface and thickness of layer	λ [μm]	$\tau_{\lambda,n}$	
layer formed by vacuum sublimation, thickness 300 Å [129]	0.4	0.444	
	0.5	0.222	
	0.6	0.116	
	0.7	0.100	
layer formed by vacuum sublimation, thickness 380 Å [129]	0.4	0.287	
	0.5	0.122	
	0.6	0.067	
	0.7	0.056	
layer formed by vacuum sublimation, thickness 430 Å [129]	0.4	0.151	
	0.5	0.064	
	0.6	0.031	
	0.7	0.033	
layer formed by vacuum sublimation, thickness 650 Å [129]	0.4	0.018	
	0.5	0.011	
	0.6	0.009	
	0.7	0.010	

3.2. TABLES OF RADIATIVE PROPERTIES OF INORGANIC MATERIALS

Table 3.2.1. Hemispherical total emissivity ε_T of inorganic materials

State of surface	T[K]	ε_T
Asbestos		
plate [12]	293÷373	0.93÷0.96
paper [12]	313	0.93÷0.94
fabric [12]	313	0.96
Asbestos, cement [12]	313	0.96
Brick, building, red [12]	293÷313	0.93
Ceramics, foreproof 9606 [45]	100	0.87
	300	0.86
	500	0.84
	700	0.77
	900	0.625
	1100	0.66
	1300	0.63
	1500	0.59
	1600	0.58
Ceramics, fireproof 9608 [45]	100÷1000	0.85
	1100	0.82
	1200	0.78
	1300	0.725
	1400	0.665
Cerium oxide; CeO_2 density 6.878 g/cm^3 [140]	1400	0.56
	1600	0.44
	1800	0.4
Clay [19, 26]	343	0.86
Copper oxide; CuO powder [76, 78]	123	0.96
	673	0.85
	2973	0.70
Copper trioxide; Cu_2O_3 blue powder [19]	123	0.94
	323	0.87
	673	0.86
	2973	0.97
Corundum; Al_2O_3 coarse-grained [12]	313÷353	0.84÷0.86
Electrographite; Acheson graphite; carbon; C density 1.5÷1.8 g/cm^3 $\Delta\varepsilon_T$ = 0.2 [131]	1200	0.77
	1600	0.79
	2000	0.805
	2400	0.81
	2800	0.82
	3200	0.83
Germanium; Ge polished surface [45]	1000	0.56
	1100	0.55
	1200	0.53

Table 3.2.1. (*continued*)

State of surface	T[K]	ε_T
Glass		
thickness of plate 1 mm, $72\%\ SiO_2 + 13\%\ (Na_2O + K_2O) + 11\%\ CaO + 4\%\ (R_2O_3 + MgO_2)$		
[19, 26, 44, 131]	300	0.77
	500	0.72
	700	0.64
	900	0.56
	1100	0.47
	1300	0.38
	1500	0.33
thickness of plate 2 mm, $12\%\ SiO_2 + 13\%\ (Na_2O + K_2O) + 11\%\ CaO + 4\%\ (R_2O_3 + MgO_2)$		
[19, 26, 44, 131]	300	0.91
	500	0.89
	700	0.83
	900	0.73
	1100	0.61
	1300	0.50
	1500	0.43
thickness of plate 5 mm, $72\%\ SiO_2 + 13\%\ (Na_2O + K_2O) + 11\%\ CaO + 4\%\ (R_2O_3 + MgO_2)$		
[19, 26, 44, 131]	300	0.93
	500	0.92
	700	0.89
	900	0.80
	1100	0.68
	1300	0.57
	1500	0.50
thickness of plate 10 mm, $72\%\ SiO_2 + 13\%\ (Na_2O + K_2O) + 11\%\ CaO + 4\%\ (R_2O_3 + MgO_2)$		
[19, 26, 44, 131]	300	0.93
	500	0.92
	700	0.90
	900	0.86
	1100	0.76
	1300	0.63
	1500	0.56
thickness of plate 20 mm, $72\%\ SiO_2 + 13\%\ (Na_2O + K_2O) + 11\%\ CaO + 4\%\ (R_2O_3 + MgO_2)$		
[19, 26, 44, 131]	300	0.93
	500	0.93
	700	0.92
	900	0.88
	1100	0.78
	1300	0.69
	1500	0.63
thickness of plate 100 mm, $72\%\ SiO_2 + 13\%\ (Na_2O + K_2O) + 11\%\ CaO + 4\%\ (R_2O_3 + MgO_2)$		
[19, 26, 44, 131]	300	0.93
	500	0.93
	700	0.93
	900	0.90
	1100	0.86
	1300	0.81
	1500	0.83
Graphite; polycrystalline, carbon; C		
powder [19, 106]	343	0.97
α-surface parallel to surface of deposition; $\Delta\varepsilon_T = \pm 0.1$		
[131]	1200	0.89
	1500	0.71
	1800	0.76
	2100	0.80
	2400	0.84
c-surface perpendicular to surface of deposition		
[131]	1200	0.89
	1300	0.86
	1500	0.84
	1700	0.87
	1900	0.92
Gypsum; hydrated calcium sulphate; $CaSO_4 \cdot 2H_2O$		
[12]	293	0.90
Haematite; bloodstone; iron oxide; Fe_2O_3		
[16, 81]	400	0.56
	600	0.60
	800	0.66
	1000	0.72
	1200	0.79
	1400	0.84
	1600	0.90
red powder		
[76, 78]	123	0.91
	323	0.96
	673	0.70
	2973	0.59

Table 3.2.1. (*continued*)

State of surface	T[K]	ε_T
Hafnium oxide; HfO_2		
[131]	1800	0.61
	2000	0.64
	2200	0.74
	2400	0.82
	2600	0.83
Lead chromate (chrome yellow; lead chrome); $PbCrO_4$ powder		
[76, 78]	123	0.93
	323	0.95
	673	0.59
Lime mortar		
[12]	283÷361	0.93
Lime plaster		
[12]	283÷361	0.93
Magnesite; magnesium carbonate; $MgCO_3$ powder		
[19, 76, 78, 106]	123	0.91
	323	0.96
	673	0.89
	1373÷2973	0.11
Marble white		
[76, 78]	313	0.95
	523	0.94
	773	0.93
Nickel oxide; NiO smooth surface		
[131, 141]	600	0.36
	800	0.41
	1000	0.47
	1200	0.56
Niobium carbide; NbC $\Delta\varepsilon_T = 0.5$		
[131]	1400	0.42
	1800	0.43
	2200	0.44
	2600	0.45
	3000	0.47
Quartz, melted powder		
[19, 44, 131]	313	0.90
	523	0.65
	773	0.50
Sandstone		
[76, 78]	313	0.83
	523	0.90
	773	0.90
Silicon; Si polished surface		
[76]	313÷2973	0.72
Silite; SiC		
[12, 131, 133]	600	0.88
	800	0.89
	1000	0.90
	1200	0.885
	1400	0.87
	1600	0.85
	2000	0.87
	2200	0.90
	2400	0.94
Tantalum carbide; TaC $\Delta\varepsilon_T = \pm 0.25$		
[131]	1200	0.45
	1600	0.46
	2000	0.48
	2400	0.495
	2800	0.51
	3600	0.545
Tantalum oxide; Ta_2O_5 smooth surface		
[131]	1273	0.29
Titanium carbide; TiC $\Delta\varepsilon_T = 0.25$		
[10, 131]	1200÷3000	0.6
Uranium carbide; UC_2 95% U, 5% Ci ≃ 0.02% polished surface		
[131]	1600÷3000	0.417
Water; H_2O thickness of layer-more than 0.1 mm		
[12]	273÷373	0.95÷0.96
Zinc white; zinc oxide; ZnO		
[76, 78]	123	0.95
	323	0.97
	673	0.91
	1373	0.57
	2973	0.14

Table 3.2.1. (*continued*)

State of surface	T[K]	ε_T
Zirconium boride; ZrB_2 moulded sample [131]	800	0.37
	1000	0.47
	1200	0.60
	1400	0.75
	1600	0.78
	1800	0.75
	2000	0.74
	2200	0.73
	2400	0.69
	2600	0.66
	2800	0.62
Zirconium carbide; ZrC $\Delta\varepsilon_T = \pm 0.2$ [131]	1200÷3400	0.61

Table 3.2.2. Total emissivity in normal direction $\varepsilon_{T,n}$ of inorganic materials

State of surface	T[K]	$\varepsilon_{T,n}$
Aluminium chloride; $AlCl_3$		
powder [106, 131]	293	0.66
Aluminium hydroxide; aluminium trihydrate; $Al(OH)_3$		
powder [106, 131]	293	0.28
Aluminium oxide LA 603; Al_2O_3		
95% Al_2O_3 [45, 131]	100	0.72
	500	0.73
	1000	0.62
	1500	0.48
	1800	0.44
Ammonia; NH_3		
optical path 61000 Pa•m [19, 25, 100, 110]	250	0.70
	500	0.68
	750	0.55
	1000	0.43
	1250	0.28
optical path 3000 Pa•m [19, 25, 110]	250	0.68
	500	0.64
	750	0.51
	1000	0.35
	1250	0.23
optical path 21000 Pa•m [19, 25, 110]	250	0.60
	500	0.55
	750	0.45
	1000	0.30
	1250	0.18
optical path 15000 Pa•m [19, 25, 110]	250	0.54
	500	0.49
	750	0.39
	1000	0.26
	1250	0.16
optical path 9100 Pa•m [19, 25, 110]	250	0.44
	500	0.39
	750	0.29
	1000	0.19
	1250	0.095
optical path 6100 Pa•m [19, 25, 110]	250	0.39
	500	0.33
	750	0.23
	1000	0.15
	1250	0.074
optical path 3000 Pa•m [19, 25, 110]	250	0.28
	500	0.23
	750	0.17
	1000	0.09
	1250	0.04
optical path 2100 Pa•m [19, 25, 110]	250	0.22
	500	0.18
	750	0.12
	1000	0.068
	1250	0.031

Table 3.2.2. (*continued*)

State of surface	T[K]	$\varepsilon_{T,n}$
optical path 1500 Pa•m [19, 25, 110]	250	0.19
	500	0.15
	750	0.095
	1000	0.055
	1250	0.023
optical path 1200 Pa•m [19, 25, 110]	250	0.17
	500	0.13
	750	0.080
	1000	0.044
	11250	0.019
optical path 610 Pa•m [19, 25, 110]	250	0.10
	500	0.085
	750	0.053
	1000	0.026
	1250	0.0089
optical path 460 Pa•m [19, 25, 110]	250	0.09
	500	0.069
	750	0.041
	1000	0.025
	1250	0.0072
optical path 900 Pa•m [19, 25, 110]	250	0.07
	500	0.055
	750	0.032
	1000	0.014
	1250	0.0048
optical path 210 Pa•m [19, 25, 110]	250	0.055
	500	0.042
	750	0.023
	1000	0.0083
	1250	0.0032
Asbestos [45, 48]	300	0.88
	400	0.87
	500	0.84
	600	0.815
	700	0.78
	800	0.76
	900	0.73
	1000	0.70
	1500	0.57
	2000	0.48
	3000	0.43
	4000	0.38
	5000	0.36
	6000	0.36
cardboard [2, 19, 44, 48, 55, 104]	293÷373	0.96
paper [19, 22, 55, 76, 104]	313÷673	0.93÷0.95
powder [17, 106]	313÷673	0.40÷0.60
anticorrosive mass [12]	290÷350	0.59÷0.72
schist [19]	293	0.96
Barium chloride; $BaCl_2$ powder [12, 91, 131]	293	0.65
Barium sulphide; BaS powder [12, 131]	293	0.45÷0.54
Berylium oxide; BeO a sintered sample, density 2.85 g/cm^3, surface polished [128, 131]	1200	0.35
	1400	0.40
	1600	0.44
	1800	0.48
	2000	0.50
Berylium sulphate; $BeSO_4$ powder [12, 131]	293	0.85
Boron carbide; B_4C [10]	1023	0.84
	1123	0.83
	1223	0.85
	1323	0.86
	1423	0.86
	1523	0.87
	1623	0.87

Table 3.2.2. (*continued*)

State of surface	T[K]	$\varepsilon_{T,n}$
	1723	0.88
	1923	0.88
Boron nitride; BN surface grinded [10, 131]	800	0.90
	1000	0.86
	1200	0.80
	1400	0.75
	1600	0.70
	1800	0.645
	2000	0.59
	2200	0.53
	2400	0.48
Brick, burnt, red surface rough, room temperature [2, 26, 44, 76]	293	0.88÷0.93
Brick, calcium silicate [10, 19]	1273	0.85
	1500	0.66
Brick, chamotte [1, 2, 19, 44, 55, 152]	293	0.85
	1253	0.75
	1473	0.59
Brick, chrome-magnesite [94]	1173	0.87
	1273	0.84
	1373	0.81
	1473	0.78
	1573	0.75
	1673	0.70
Brick, chrome-magnesite, refractory surface smooth [45, 94]	1100	0.862
	1200	0.86
	1300	0.84
	1400	0.82
	1500	0.78
	1600	0.74
	1700	0.696
	1800	0.644
Brick, corundum [2, 10]	1273	0.46
Brick, dinas surface smooth [91, 94]	1173	0.83
	1273	0.81

State of surface	T[K]	$\varepsilon_{T,n}$
	1373	0.79
	1473	0.76
	1573	0.71
	1673	0.67
	1773	0.64
rough, surface nonglazed [1, 2, 19, 44, 55, 152]	1253	0.80
rough, surface glazed [2, 19, 44, 55, 152]	1363	0.85
Brick, dinas, refractory surface smooth [45, 94]	1173	0.79
	1273	0.78
	1373	0.76
	1473	0.72
	1573	0.69
	1673	0.66
	1773	0.64
Brick, magnesite [19, 22, 67]	653	0.39
	1273	0.39
	1673	0.38
Brick, magnesite-silicate with iron oxide [94]	1173	0.71
	1273	0.73
	1373	0.74
	1473	0.745
	1573	0.745
	1673	0.74
	1773	0.73
Brick, sillimanite; $Al(AlSiO_5)$ 33% SiO_2 and 64% Al_2O_3 [19, 67, 76]	1373	0.66
	1663	0.39
	1773	0.29
Calcite; calcspar; calcium carbonate; $CaCO_3$ powder [106, 131]	400	0.3

Table 3.2.2. (*continued*)

State of surface	T[K]	$\varepsilon_{T,n}$
Calcium nitrate; $Ca(NO_3)_2$ powder [106, 131]	293	0.58
Carbon dioxide; CO_2		
optical path $2 \cdot 10^5$ Pa·m [19]	273	0.200
	473	0.196
	673	0.200
	873	0.219
	1073	0.238
	1273	0.231
	1473	0.225
	1673	0.196
	1873	0.182
	2073	0.164
	2273	0.146
optical path 10^5 Pa·m [19]	273	0.193
	473	0.175
	673	0.179
	873	0.193
	1073	0.206
	1273	0.204
	1473	0.193
	1673	0.175
	1873	0.157
	2073	0.135
	2273	0.116
optical path $6 \cdot 10^4$ Pa·m [19]	273	0.164
	473	0.150
	673	0.153
	873	0.175
	1073	0.189
	1273	0.186
	1473	0.171
	1673	0.154
	1873	0.134
	2073	0.114
	2273	0.100
optical path $4 \cdot 10^4$ Pa·m [19]	273	0.150
	473	0.139
	673	0.145
	873	0.161
	1073	0.169
	1273	0.164
	1473	0.149
	1673	0.132
	1873	0.111
	2073	0.093
	2273	0.083
optical path $2 \cdot 10^4$ Pa·m [19]	273	0.129
	473	0.121
	673	0.125
	873	0.139
	1073	0.143
	1273	0.136
	1473	0.119
	1673	0.096
	1873	0.086
	2073	0.070
	2273	0.062
optical path 10^4 Pa·m [19]	273	0.100
	473	0.096
	673	0.100
	873	0.113
	1073	0.121
	1273	0.111
	1473	0.097
	1673	0.083
	1873	0.065
	2073	0.056
	2273	0.048
optical path $6 \cdot 10^3$ Pa·m [19]	273	0.093
	473	0.084
	673	0.090
	873	0.100
	1073	0.100
	1273	0.093
	1473	0.080
	1673	0.065
	1873	0.052
	2073	0.045
	2273	0.037

Table 3.2.2. (*continued*)

State of surface	T[K]	$\varepsilon_{T,n}$	State of surface	T[K]	$\varepsilon_{T,n}$
optical path $3 \cdot 10^3$ Pa·m [19]	273	0.078	optical path $5 \cdot 10^2$ Pa·m [19]	273	0.044
	473	0.068		473	0.035
	673	0.070		673	0.036
	873	0.080		873	0.038
	1073	0.080		1073	0.036
	1273	0.072		1273	0.032
	1473	0.062		1473	0.026
	1673	0.050		1673	0.022
	1873	0.040		1873	0.018
	2073	0.033		2073	0.014
	2273	0.027		2273	0.011
optical path $2 \cdot 10^3$ Pa·m [19]	273	0.068	optical path $3 \cdot 10^2$ Pa·m [19]	273	0.03
	473	0.058		473	0.028
	673	0.062		673	0.029
	873	0.068		873	0.03
	1073	0.067		1073	0.028
	1273	0.060		1273	0.024
	1473	0.052		1473	0.02
	1673	0.043		1673	0.016
	1873	0.035		1873	0.013
	2073	0.029		2073	0.0089
	2273	0.023		2273	0.0076
optical path 10^3 Pa·m [19]	273	0.050	optical path $2 \cdot 10^2$ Pa·m [19]	273	0.024
	473	0.046		473	0.023
	673	0.049		673	0.024
	873	0.053		873	0.024
	1073	0.052		1073	0.022
	1273	0.046		1273	0.019
	1473	0.038		1473	0.015
	1673	0.032		1673	0.012
	1873	0.026		1873	0.0093
	2073	0.020		2073	0.007
	2273	0.016		2273	0.0055
optical path $7 \cdot 10^2$ Pa·m [19]	273	0.046	optical path $1.6 \cdot 10^2$ Pa·m [19]	273	-
	473	0.040		473	-
	673	0.043		673	-
	873	0.045		873	0.02
	1073	0.045		1073	0.018
	1273	0.044		1273	0.015
	1473	0.032		1473	0.012
	1673	0.027		1673	0.0098
	1873	0.020		1873	0.0073
	2073	0.016		2073	0.0056
	2273	0.014		2273	0.0045

Table 3.2.2. (*continued*)

State of surface	T[K]	$\varepsilon_{T,n}$	State of surface	T[K]	$\varepsilon_{T,n}$
optical path $10^2 \cdot$ Pa•m [19]	873	0.015	optical path $6.1 \cdot 10^2$ Pa•m [19]	500	0.027
	1073	0.014		750	0.061
	1273	0.011		1000	0.068
	1473	0.0084		1250	0.053
	1673	0.0064	optical path $3 \cdot 10^3$ Pa•m [19]	500	0.021
	1873	0.0052		750	0.049
	2073	0.0039		1000	0.058
	2273	0.0028		1250	0.046
optical path 70 Pa•m [19]	873	0.011	optical path $1.8 \cdot 10^3$ Pa•m [19]	500	0.018
	1073	0.009		750	0.041
	1273	0.0074		1000	0.050
	1473	0.0058		1250	0.041
	1673	0.0048	optical path $1.2 \cdot 10^3$ Pa•m [19]	500	0.018
	1873	0.0036		750	0.035
optical path 50 Pa•m [19]	873	0.0076		1000	0.042
	1073	0.0062		1250	0.038
	1273	0.0051	optical path $9.1 \cdot 10^2$ Pa•m [19]	500	0.014
	1473	0.0032		750	0.030
Carbon oxide; CO optical path $6.1 \cdot 10^3$ Pa•m [19]	500	0.054		1000	0.038
	750	0.11		1250	0.034
	1000	0.12	optical path $6.1 \cdot 10^2$ Pa•m [19]	500	0.0115
	1250	0.083		750	0.026
optical path $3 \cdot 10^4$ Pa•m [19]	500	0.045		1000	0.033
	750	0.091		1250	0.029
	1000	0.10	optical path $4.6 \cdot 10^2$ Pa•m [19]	500	0.010
	1250	0.075		750	0.022
optical path $1.5 \cdot 10^4$ Pa•m [19]	500	0.037		1000	0.029
	750	0.080		1250	0.025
	1000	0.085	optical path $3 \cdot 10^2$ Pa•m [19]	500	0.008
	1250	0.066		750	0.0175
optical path $9.1 \cdot 10^2$ Pa•m [19]	500	0.032		1000	0.023
	750	0.069		1250	0.020
	1000	0.079			
	1250	0.060			

State of surface	T[K]	$\varepsilon_{T,n}$
optical path		
$1.8 \cdot 10^2$ Pa•m		
[19]	500	0.0059
	750	0.0126
	1000	0.017
	1250	0.015
optical path		
$1.2 \cdot 10^2$ Pa•m		
[19]	750	0.0025
	1000	0.012
	1250	0.011
Cassiterite;		
tin oxide; SnO_2		
powder		
[19]	293	0.4
Cement		
[2, 28]	300	0.90
Cement, fireproof		
[94]	1173	0.62
	1273	0.57
	1373	0.53
	1473	0.48
	1573	0.44
	1673	0.41
	1773	0.38
Ceramics, mullite,		
heat-insulating		
before burning		
[94]	1173	0.71
	1273	0.70
	1373	0.68
	1473	0.66
	1573	0.71
	1673	0.73
	1773	0.73
Ceramics,		
sillimanite with		
25% pottery clay		
before burning		
[94]	1173	0.56
	1273	0.57
	1373	0.575
	1473	0.61
	1573	0.63
	1673	0.64
	1773	0.645
after burning		
[94]	1173	0.51
	1273	0.52
	1373	0.52
	1473	0.52
	1573	0.61
Cerium oxide; CeO_2		
density 6.878 g/cm^3		
[45, 131]	700	0.51
	800	0.45
	900	0.41
	1000	0.37
	1100	0.30
	1200	0.73
	1400	0.91
	1600	0.94
	1800	0.95
	2000	0.94
	2200	0.93
Chile saltpetre;		
chile nitrate;		
sodium nitrate;		
$NaNO_3$		
powder		
[106, 131]	293	0.36
Chromium oxide; Cr_2O_3		
green powder		
[76, 131]	123	0.92
	323	0.95
	673	0.67
	2973	0.55
sintered sample of powder		
with grain-size distribution		
0.5-1.5 µm	1000	0.683
[131]	1100	0.674
	1200	0.668
	1300	0.664
	1400	0.659
	1500	0.653
	1600	0.647
sintered sample of powder		
with grain-size distribution		
1.5-8.0 µm		
[131]	1000	0.847
	1100	0.853
	1200	0.859
	1300	0.865
	1400	0.871
	1500	0.875
	1600	0.886

Table 3.2.2. (*continued*)

State of surface	T[K]	$\varepsilon_{T,n}$
Clay red, burned, rough [19, 44]	300÷343	0.88÷0.93
	350	0.91
Concrete smooth surface [45, 48]	300	0.93
Copper oxide; CuO powder [19, 55]	1073÷1373	0.66÷0.54
Copper sulphate; cupric sulphate; $CuSO_4$ powder [19, 106]	1373÷1573	0.92
Cork [28, 55]	300	0.79
	400	0.76
Corundum, sintered, aluminium oxide; Al_2O_3 smooth surface [88]	1173	0.44
	1273	0.44
	1373	0.40
	1473	0.39
	1573	0.385
	1673	0.375
	1773	0.365
[19, 63, 106, 131]	400	0.94
	600	0.71
	800	0.57
	1000	0.48
	1200	0.44
	1400	0.42
	1600	0.40
rough surface [19, 26, 106]	353	0.84
very rough surface, coarse-grained [26, 106]	313÷353	0.85÷0.86
powder on paper or rough tile [19]	1273	0.46
Cuprite acetate; $Cu(C_2H_3O_2)_2$ powder [19, 106]	1373÷1573	0.84
Cuprite; red copper ore; ruby copper; copper oxide; Cu_2O red powder [12, 19, 106]	1373÷1573	0.6÷0.7
Diatomaceous earth powder [106, 131]	353	0.25
Electrographite; Acheson graphite; carbon; C rough surface [45, 131]	200	0.84
	600	0.94
	1000	0.98
	1400	0.83
	1800	0.81
	2000	0.80
ground surface [19]	400÷900	0.81÷0.79
charcoal, powder [19, 106, 131]	293	0.96
Fluorite; calcium fluoride; CaF_2 natural powder [19, 106, 131]	300	0.3÷0.4
Gadolinium boride, GdB [10]	1123÷1223	0.61
	1323÷1523	0.62
	1623÷1823	0.63
	1923	0.64
Glass [12]	293÷373	0.96÷0.90
Glass, ampouled, aluminium thickness of plate from 4.76 mm to 12.7 mm [45]	100	0.886
	200	0.89
	300	0.89
	400	0.89

Table 3.2.2. (*continued*)

State of surface	T[K]	$\varepsilon_{T,n}$
	500	0.80
	600	0.88
	700	0.87
	800	0.86
	900	0.832
Glass, borosilicate-Corning 7900 thickness of plate 12.7 mm [45]	100÷1100	0.87
Glass, borosilicate-Corning 7940 thickness of plate 12.7 mm [45]	100	0.885
	300	0.88
	500	0.82
	700	0.79
	900	0.755
	1100	0.67
Glass, borosilicate-Pittsburgh 3255 thickness of plate from 6.35 mm to 12.7 mm [45]	100	0.87
	300	0.885
	500	0.905
	700	0.92
	800	0.93
Glass, borosilicate-Pyrex thickness of plate 12.7 mm [45]	100	0.865
	200	0.865
	400	0.855
	600	0.85
	800	0.837
	1000	0.818
	1200	0.79
Glass, metallized, electricity conductive; Battelle 54726 [45]	100	0.665
	300	0.66
	500	0.648
	700	0.625
	900	0.568
	950	0.55

State of surface	T[K]	$\varepsilon_{T,n}$
Glass, metallized, electricity conductive; LOF 81 E 19778 [45]	100	0.53
	300	0.472
	400	0.46
	500	0.472
	600	0.595
	700	0.78
Glass, metallized, electricity conductive; LOF PB 19195 [45]	100	0.55
	300	0.55
	500	0.545
	600	0.542
	700	0.54
	800	0.535
	850	0.53
Glass, soda, lime thickness of plate 12.7 mm [45, 139]	100÷500	0.86
	600	0.85
	700	0.83
	800	0.814
	850	0.795
Granite ground surface [19]	293	0.427
Gypsum; hydrated calcium sulphate; $CaSO_4 2H_2O$ [1, 44, 55]	300	0.8÷0.93
Haematite; bloodstone; iron oxide; Fe_2O_3 [24]	1100	0.57
	1300	0.69
	1500	0.80
liquid state [51]		0.53
Hafnium bromide; HfB [10]	850	0.85
	950	0.86
	1050	0.87
	1150	0.88
	1250	0.89

Table 3.2.2. (*continued*)

State of surface	T[K]	$\varepsilon_{T,n}$
	1350	0.90
	1450	0.92
	1550	0.92
	1650	0.94
Hafnium carbide; HfC smooth uniform surface [131]	1500	0.40
	1750	0.41
	2000	0.42
	2250	0.43
	2500	0.44
	2750	0.45
	3000	0.46
fine-grained sand [131]	1500	0.83
	1750	0.83
	2000	0.80
	2250	0.70
	2500	0.61
	2750	0.58
	3000	0.63
Hafnium nitride; HfN hot-moulded, density 12.8 g/cm^3 [131]	1600	0.60
	1800	0.65
	2000	0.75
	2200	0.86
[131]	2200	0.66
	2400	0.64
	2600	0.62
	2800	0.595
Hafnium oxide; HfO 90% HfO_2, 5% Y_2O_3 and 5% ZrO_3 [131]	1800	0.64
	2000	0.67
	2200	0.71
	2400	0.75
sintered sample, density 9.55 g/cm^3 [128, 131]	1200	0.81
	1400	0.82
	1600	0.82
	1800	0.80
	2000	0.81
	2200	0.89
	2400	0.88
Ice smooth surface [19, 26, 44, 109]	273	0.918÷0.966
rough surface hoarfrost-coated [19, 26, 109]	273	0.986
Iron sulphate; $FeSO_4$ powder [19, 106, 131]	200	0.6
Kaolin; porcelain clay; China clay [19, 106]	353	0.3
Lanthanum bromide [10]	1023	0.68
	1123	0.68
	1223	0.69
	1323	0.69
	1423	0.69
	1523	0.70
	1623	0.70
	1723	0.70
	1823	0.71
Lead acetate; $Pb(C_2H_5O_2)_2$ powder [19, 102]	473	0.7
Lead oxide, yellow; PbO powder [19, 106]	473	0.29
Lead sulphide; $PbSO_4$ powder [19, 106]	473	0.13÷0.22
Lime; calcium hydrate; calcium hydroxide; $Ca(OH)_2$ white rough mortar [19]	293÷473	0.93
plastered masonry [19]	273÷473	0.93
Lime, burnt; quicklime; calcium oxide; CaO powder [131]	1123÷1550	0.27

Table 3.2.2. (*continued*)

State of surface	T[K]	$\varepsilon_{T,n}$
Lime mortar [12]	283÷366	0.93
Lime plaster rough surface [26, 44, 96]	273	0.93
	1293	0.91÷0.93
Limestone; $CaCO_3$ [37, 76, 78]	313	0.95
	523	0.83
	773	0.75
Lithium carbonate; Li_2CO_3		
powder [106, 131]	473÷873	0.59÷0.61
Magnesite; magnesium carbonate; $MgCO_3$		
polished stone [19]	1663	0.39
tile [19, 196]	1273÷1573	0.38
Manganese carbonate; $MnCO_3$		
powder [106, 131]	393	0.44
Marble, grey polished surface [2, 19, 44, 55, 146]	300	0.93
Mica		
thick plate [19, 76]	300	0.72
agglomerate powder of mica silicone [19, 86]	323÷1273	0.81÷0.85
Minium; red lead; lead oxide; Pb_3O_4		
powder [19, 26, 44, 106]	293÷373	0.93
Molybdenum silicide; $MoSi_2$ [131]	800	0.40
	1000	0.45
	1200	0.50
	1400	0.55
	1600	0.61

State of surface	T[K]	$\varepsilon_{T,n}$
Mullite; $3Al_2O_3 \cdot 2SiO_2$ [45]	1100	0.72
	1200	0.715
	1300	0.705
	1400	0.695
	1500	0.68
	1600	0.67
	1700	0.655
	1800	0.638
	1900	0.62
Neodymium boride [10]	1123	0.56
	1223	0.56
	1323	0.56
	1423	0.57
	1523	0.58
	1623	0.58
	1723	0.58
	1823	0.58
	1923	0.59
Nickel oxide; NiO smooth surface [16, 45, 131]	200	0.38
	400	0.47
	600	0.55
	800	0.64
	1000	0.73
	1200	0.82
	1400	0.91
powder [19, 44, 55, 106]	773	0.52
	923	0.59
	1273÷1523	0.75
	1523	0.80
Periclase; magnesium oxide; MgO moulded powder [40, 41, 131]	20	0.73
	400	0.69
	600	0.57
	800	0.52
	1000	0.42
	1200	0.35
	1400	0.30
	1600	0.28
	1800	0.29
	2000	0.35
	2200	0.49

Table 3.2.2. (*continued*)

State of surface	T[K]	$\varepsilon_{T,n}$
Platinum black [78]	123	0.92
	313÷363	0.93
	423	0.94
	473	0.95
	523	0.96
	773÷2973	0.97
Porcelain white glazed [1, 19, 26, 44, 55, 81]	300	0.92
	523	0.99
white [19]		0.70÷0.75
Potassium carbonate; K_2CO_3 [106, 131]	400	0.4
Quartz, melted rough surface [19, 44]	293	0.93
smooth surface, thickness of plate 6.35 mm [131]	500	0.79
	600	0.77
$\Delta\varepsilon_{T,n}$ = 0.03	800	0.70
	1000	0.61
	1200	0.51
	1400	0.47
	1500	0.47
smooth surface, thickness of plate 12.7 mm $\Delta\varepsilon_{T,n}$ = 0.03 [131]	100÷500	0.82
	600	0.81
	800	0.79
	1000	0.69
Rubidium chloride; RbCl powder [106, 131]	273÷473	0.36
Samarium boride [10]	1123	0.71
	1273	0.7
	1323	0.7
	1423	0.69
	1523	0.69
	1623	0.68
	1723	0.68
	1823	0.67
	1923	0.67
Samarium oxide; Sm_2O_3 [140]	573	0.44
	773	0.49
	973	0.50
	1173	0.48
	1373	0.48
	1573	0.56
	1773	0.56
Sand [44, 106]	293	0.60
Silica monocrystalline; silicon dioxide; SiO_2 [24, 131]	1100	0.37
	1300	0.38
	1500	0.40
	1600	0.405
Silica polycrystalline; silicon dioxide; SiO_2 [97]	353	0.48
Silicone reinforced [45]	75	0.83
	100	0.832
	200	0.848
	300	0.862
	400	0.874
	500	0.891
	600	0.90
Silicon nitride; Si_3N_4 ground surface [132]	600	0.90
	800	0.86
	1000	0.83
	1200	0.80
	1400	0.77
Silicon; Si powder [19, 97]	353	0.30

Table 3.2.2. (*continued*)

State of surface	T[K]	$\varepsilon_{T,n}$
powder, granularity d = 10 μm (98% Si) [97]	1283÷1853	0.42÷0.35
powder, granularity d = 70÷600 μm (98% Si) [97]	1283	0.62
	1853	0.46
sand [19]		0.60
sandstone, smooth ground surface [19]		0.576
Sillimanite with clay 25% [45]	1200	0.51
	1300	0.518
	1400	0.52
	1500	0.52
	1600	0.532
	1700	0.54
	1800	0.54
	1900	0.545
Slag, blast-furnace [51]		0.55÷0.75
Slag, boiler [1, 19, 44, 109]	273÷373	0.97÷0.93
	473÷773	0.89÷0.78
	873÷1473	0.76÷0.70
	1673÷2073	0.69÷0.67
Smithsonite; zinc spar; zinc carbonate; $ZnCO_3$ [106, 131]	1273÷1473	0.24
Snow [10, 19, 44]	273	0.8
Soda ash; calcinated soda; sodium carbonate; $NaCO_3$ powder [106, 131]	293	0.4÷0.8
Sodium carbide; Na_2C_2 [10]	1123÷1923	0.91
Strontium hydroxide; strontium hydrate; $Sr(OH)_2$ powder [106, 131]	373	0.16
Sulphur dioxide; SO_2 optical path $6.1 \cdot 10^4$ [55, 110]	500	0.41
	750	0.43
	1000	0.40
	1250	0.34
	1500	0.29
	1750	0.23
optical path $3 \cdot 10^4$ Pa·m [55, 110]	500	0.35
	750	0.35
	1000	0.30
	1250	0.25
	1500	0.149
	1750	0.13
optical path $1.5 \cdot 10^4$ Pa·m [55, 110]	500	0.28
	750	0.26
	1000	0.21
	1250	0.18
	1500	0.13
	1750	0.11
optical path $9.1 \cdot 10^3$ Pa·m [55, 110]	500	0.23
	750	0.19
	1000	0.16
	1250	0.12
	1500	0.091
	1750	0.08
optical path $6.1 \cdot 10^3$ Pa·m [55, 110]	500	0.19
	750	0.16
	1000	0.13
	1250	0.09
	1500	0.074
	1750	0.06
optical path $4.6 \cdot 10^3$ Pa·m [55, 110]	500	0.16
	750	0.147
	1000	0.10

Table 3.2.2. (*continued*)

State of surface	T[K]	$\varepsilon_{T,n}$
	1250	0.08
	1500	0.06
	1750	0.048
optical path $3 \cdot 10^3$ Pa·m [55, 110]	500	0.13
	750	0.10
	1000	0.075
	1250	0.058
	1500	0.043
	1750	0.033
optical path $2.4 \cdot 10^3$ Pa·m [55, 110]	500	0.11
	750	0.088
	1000	0.067
	1250	0.050
	1500	0.037
	1750	0.028
optical path $1.8 \cdot 10^3$ Pa·m [55, 110]	500	0.09
	750	0.072
	1000	0.059
	1250	0.040
	1500	0.030
	1750	0.023
optical path $1.5 \cdot 10^3$ Pa·m [55, 110]	500	0.079
	750	0.061
	1000	0.045
	1250	0.034
	1500	0.025
	1750	0.018
optical path $1.2 \cdot 10^3$ Pa·m [55, 110]	500	0.068
	750	0.052
	1000	0.038
	1250	0.028
	1500	0.019
	1750	0.016
optical path $9.1 \cdot 10^2$ Pa·m [55, 110]	500	0.054
	750	0.044
	1000	0.03
	1250	0.021
	1500	0.015
	1750	0.012

State of surface	T[K]	$\varepsilon_{T,n}$
optical path $6.1 \cdot 10^2$ Pa·m [55, 110]	500	0.038
	750	0.030
	1000	0.020
	1250	0.015
	1500	0.012
	1750	0.008
optical path $4.6 \cdot 10^2$ Pa·m [55, 110]	500	0.029
	750	0.021
	1000	0.015
	1250	0.013
	1500	0.008
	1750	0.0062
optical path $3 \cdot 10^2$ Pa·m [55, 110]	500	0.019
	750	0.016
	1000	0.013
	1250	0.0076
	1500	0.0058
	1750	0.0045
optical path $2.4 \cdot 10^2$ Pa·m [55, 110]	500	0.016
	750	0.013
	1000	0.0085
	1250	0.006
	1500	0.0048
	1750	0.0036
optical path $1.8 \cdot 10^2$ Pa·m [55, 110]	500	0.012
	750	0.095
	1000	0.064
	1250	0.0049
	1500	0.0035
optical path $1.2 \cdot 10^2$ Pa·m [55, 110]	750	0.0066
	1000	0.0048
	1250	0.0031
optical path 6.1 Pa·m [55, 110]	750	0.0032

Table 3.2.2. (*continued*)

State of surface	T[K]	$\varepsilon_{T,n}$
Talc; $Mg_3(Si_4O_{10})(OH)_2$ fine-grained powder [106, 131]	293	0.24
Tantalum nitride; TaN hot-moulted sample, density 13.7 g/cm^3 [131]	1600÷2300	0.72÷0.73
Thallium carbonate; Tl_2CO_3 powder [106, 131]	1273	0.32
Thorium nitrate; $Th(NO_3)_4$ powder [19, 106, 131]	1273	0.56
Thorium oxide; ThO_2 sintered from powder has lower values		
monocrystalline-bigger values [45, 131]	600	0.61
	800	0.52
	1000	0.43
	1200	0.37
	1400	0.37
	1600	0.405
	1800	0.49
	2000	0.61
	2200	0.68
	2400	0.71
powder [19, 47, 106]	1273	0.15
Tile, concrete [22, 28, 131]	1273	0.63
Tile, red [24]	293	0.93
Tile, silicate [94]	1173	0.80
	1273	0.79
	1373	0.76
	1473	0.73
	1573	0.71
	1673	0.68
	1773	0.64
Titanium nitride; TiN ground surface [131]	1200	0.80
	1400	0.85
	1800	0.79
	2200	0.61
	2600	0.35
	3000	0.30
Titanium oxide; TiO_2 smooth surface [131]	1223	0.31
powder [19, 44, 106]	1223	0.20
Tungsten oxide; WO_3 powder [19, 106, 131]	1773÷2273	0.8
Water; hydrogen oxide; H_2O thickness of layer l = 0.1 mm [19, 44, 55, 67]	273÷373	0.95÷0.96
film on metallic surface [19]	293	0.98
Water vapour; hydrogen oxide; H_2O optical path 10^6 Pa•m [19, 25]	273	0.700
	473	0.650
	673	0.625
	873	0.600
	1073	0.571
	1273	0.533
	1473	0.478
	1673	0.444
	1873	0.383
	2073	0.355
	2273	0.325
	3000	0.644
	4000	0.668
optical path $4\cdot10^5$ Pa•m [19, 25]	273	0.676
	473	0.575
	673	0.550
	873	0.538
	1073	0.500
	1273	0.444
	1473	0.400
	1673	0.366

Table 3.2.2. (*continued*)

State of surface	T[K]	$\varepsilon_{T,n}$
	1873	0.325
	2073	0.294
	2273	0.269
optical path $2 \cdot 10^5$ Pa·m [19, 25]	273	0.555
	473	0.500
	673	0.470
	873	0.460
	1073	0.440
	1273	0.366
	1473	0.333
	1673	
	1873	0.294
	2073	0.238
	2273	0.213
	3000	0.281
	4000	0.230
optical path 10^5 Pa·m [19, 25]	273	0.480
	473	0.430
	673	0.400
	873	0.383
	1073	0.342
	1273	0.300
	1473	0.275
	1673	0.238
	1873	0.213
	2073	0.190
	2273	0.177
	3000	0.190
	4000	0.135
optical path $6.5 \cdot 10^4$ Pa·m [19, 25]	273	0.440
	473	0.375
	673	0.350
	873	0.333
	1073	0.300
	1273	0.263
	1473	0.231
	1673	0.197
	1873	0.180
	2073	0.163
	2273	0.147
optical path $4.5 \cdot 10^4$ Pa·m [19, 25]	273	0.400
	473	0.325
	673	0.300
	873	0.288
	1073	0.263
	1273	0.225
	1473	0.190
	1673	0.173
	1873	0.153
	2073	0.133
	2273	0.113
optical path $3 \cdot 10^4$ Pa·m [19, 25]	273	0.350
	473	0.288
	673	0.275
	873	0.250
	1073	0.213
	1273	0.187
	1473	0.167
	1673	0.137
	1873	0.120
	2073	0.100
	2273	0.090
optical path $2 \cdot 10^4$ Pa·m [19, 25]	273	0.308
	473	0.250
	673	0.231
	873	0.200
	1073	0.180
	1273	0.153
	1473	0.133
	1673	0.113
	1873	0.093
	2073	0.080
	2273	0.070
optical path $1.4 \cdot 10^4$ Pa·m [19, 25]	273	0.269
	473	0.219
	673	0.193
	873	0.179
	1073	0.146
	1273	0.123
	1473	0.100
	1673	0.087
	1873	0.074
	2073	0.063
	2273	0.054
optical path 10^4 Pa·m [19, 25]	273	0.231
	473	0.187
	673	0.177
	873	0.140

Table 3.2.2. (*continued*)

State of surface	T[K]	$\varepsilon_{T,n}$
	1073	0.117
	1273	0.095
	1473	0.083
	1673	0.070
	1873	0.059
	2073	0.050
	2273	0.042
	3000	0.042
	4000	0.025
optical path $8 \cdot 10^3$ Pa·m [19, 25]	273	0.206
	473	0.187
	673	0.153
	873	0.147
	1073	0.123
	1273	0.095
	1473	0.082
	1673	0.059
	1873	0.050
	2073	0.041
	2273	0.035
optical path $4.5 \cdot 10^3$ Pa·m [19, 25]	273	0.183
	473	0.150
	673	0.140
	873	0.100
	1073	0.082
	1273	0.070
	1473	0.058
	1673	0.048
	1873	0.039
	2073	0.034
	2273	0.021
optical path $3 \cdot 10^3$ Pa·m [19, 25]	273	0.143
	473	0.098
	673	0.082
	873	0.065
	1073	0.054
	1273	0.043
	1473	0.035
	1673	0.028
	1873	0.023
	2073	0.018
	2273	0.015
optical path $2 \cdot 10^3$ Pa·m [19, 25]	273	0.107
	473	0.076
	673	0.063
	873	0.050
	1073	0.039
	1273	0.031
	1473	0.025
	1673	0.019
	1873	0.016
	2073	0.013
	2273	0.0085
optical path $1.5 \cdot 10^3$ Pa·m [19, 25]	273	0.085
	473	0.064
	673	0.054
	873	0.041
	1073	0.033
	1273	0.025
	1473	0.019
	1673	0.016
	1873	0.012
	2073	0.0088
optical path 10^3 Pa·m [19, 25]	273	0.067
	473	0.047
	673	0.039
	873	0.030
	1073	0.024
	1273	0.019
	1473	0.015
	1673	0.010
	1873	0.008
	3000	0.0053
	4000	0.0047
optical path $8 \cdot 10^2$ Pa·m [19, 25]	273	0.058
	473	0.043
	673	0.034
	873	0.025
	1073	0.019
	1273	0.015
	1473	0.011
	1673	0.0078

Table 3.2.2. (*continued*)

State of surface	T[K]	$\varepsilon_{T,n}$	State of surface	T[K]	$\varepsilon_{T,n}$
optical path $6 \cdot 10^2$ Pa•m [19, 25]	273	0.049		1323	0.65
	473	0.037		1423	0.66
	673	0.028		1523	0.66
	873	0.020		1623	0.66
	1073	0.016		1723	0.67
	1273	0.012		1823	0.67
	1473	0.0085		1923	0.68
optical path $5 \cdot 10^2$ Pa•m [19, 25]	273	0.043	Zinc nitrate; $Zn(NO_3)_2$ powder [106, 131]	1273÷1473	0.73
	473	0.032	Zirconium boride; ZrB [10]	1123	0.86
	673	0.024		1223	0.87
	873	0.018		1323	0.88
	1073	0.013		1423	0.80
	1273	0.0095		1523	0.91
optical path $4 \cdot 10^2$ Pa•m [19, 25]	273	0.035		1623	0.92
	473	0.028		1723	0.91
	673	0.020		1823	0.93
	873	0.016		1923	0.95
	1073	0.011	Zirconium nitride; ZrN hot-moulded sample [131]	1800	0.40
	1273	0.0078		2000	0.50
optical path $3 \cdot 10^2$ Pa•m [19, 25]	273	0.029		2200	0.68
	473	0.021		2300	0.73
	673	0.016	Zirconium oxide; ZrO_2 powder [106, 131]	1273÷1473	0.16÷0.20
	873	0.011	moulded sample [24, 131]	20	0.81
	1073	0.008		400	0.75
optical path $2 \cdot 10^2$ Pa•m [19, 25]	273	0.022		600	0.65
	473	0.016		800	0.53
	673	0.011		1000	0.42
	873	0.0078		1200	0.37
optical path $1.5 \cdot 10^2$ Pa•m [19, 25]	273	0.018		1400	0.37
	473	0.013		1600	0.39
	673	0.0083		1800	0.46
optical path 10^2 Pa•m [19, 25]	273	0.013		2000	0.55
	473	0.0084		2200	0.62
Yttrium boride; YB [10]	1123	0.63		2400	0.66
	1223	0.64		2600	0.69
			Zirconium silicide; $ZrSi_2$ powder [106, 131]	1273÷1473	0.36÷0.42

Table 3.2.3. Spectral emissivity in normal direction $\varepsilon_{\lambda,n}$ of inorganic materials

State of surface	λ [μm]	$\varepsilon_{\lambda,n}$
Aluminium nitride density 2.04 g/cm^3; temperature of measurement 1223 K [131]	1.0	0.79
	1.5	0.69
	2.0	0.72
	3.0	0.74
	4.0	0.77
	5.0	0.78
	6.0	0.80
	8.0	0.93
	10.0	0.96
Beryllium oxide; BeO smooth surface [45, 128, 131]	0.3	0.81
	0.4	0.58
	0.6	0.41
	0.8	0.37
	1.0	0.36
	1.4	0.31
	1.8	0.29
	2.2	0.28
	2.5	0.10
	3.0	0.12
	4.0	0.32
	5.0	0.59
	6.0	0.73
	7.0	0.90
	8.0	0.90
	9.0	0.52
	10.0	0.39
	20.0	0.32
Boron nitride; BN polished surface, density 2.0 g/cm^3, BN-97%, temperature of nitride 873÷1353 K [131]	2.0	0.30
	3.0	0.58
	4.0	0.91
	5.0	0.96
	6.0	0.95
	7.0	0.57
	8.0	0.70
	9.0	0.87
	10.0	0.90
	20.0	0.91
Boron silicide; B_4Si content of B_4Si-95%, temperature of sintering 1423 K, density 1.32 g/cm^3 [131]	1.0	0.84
	2.0	0.86
	3.0	0.79
	4.0	0.76
	5.0	0.72
	6.0	0.70
	7.0	0.695
	8.0	0.74
	9.0	0.76
	10.0	0.75
content of B_4Si-95% temperature of sintering 1623 K, density 1.32 g/cm^3 [131]	1.0	0.985
	2.0	0.99
	3.0	0.995
	4.0	0.995
	5.0	0.995
	6.0	0.98
	7.0	0.95
	8.0	0.82
	9.0	0.86
	10.0	0.89

Table 3.2.3. (*continued*)

State of surface	λ[μm]	$\varepsilon_{\lambda,n}$
Brick, building, red temperature 373+573 K [10]	2.5	0.45
	3.0	0.53
	3.5	0.50
	4.0	0.54
	5.0	0.77
	6.0	0.87
	7.0	0.89
	8.0	0.88
	9.0	0.87
	10.0	0.86
	11.0	0.84
	13.0	0.79
Brick, calcium-silicate [10, 19]	2.5	0.68
	3.0	0.69
	4.0	0.79
	5.0	0.84
	6.0	0.87
	7.0	0.89
	8.0	0.89
	10.0	0.87
	12.0	0.84
Brick, chamotte temperature 373+573 K [10, 140]	2.5	0.32
	3.0	0.31
	3.5	0.34
	4.0	0.49
	4.5	0.70
	5.0	0.83
	6.0	0.90
	8.0	0.92
	10.0	0.89
	12.0	0.91
Brick, chamotte, porous temperature 373+573 K [10]	2.5	0.26
	3.0	0.22
	3.5	0.23
	4.0	0.32
	4.5	0.50
	5.0	0.61
	6.0	0.73
	7.0	0.74
	9.0	0.76
	11.0	0.77
	13.0	0.77

State of surface	λ[μm]	$\varepsilon_{\lambda,n}$
Brick, Dinas smooth surface, temperature 373+573 K [10, 140]	2.5	0.22
	3.0	0.43
	3.5	0.54
	4.0	0.69
	5.0	0.93
	6.0	0.96
	7.0	0.95
	8.0	0.87
	8.5	0.83
	9.0	0.84
	10.0	0.92
	11.0	0.96
	12.0	0.90
	13.0	0.836
Carbon dioxide; CO_2 temperature of gas 294 K [21, 29, 30, 55, 60, 110]	1.74	0.000
	1.82	0.136
	1.92	0.255
	2.00	0.118
	2.07	0.000
	2.47	0.000
	2.56	0.045
	2.58	0.200
	2.61	0.400
	2.73	0.800
	2.79	0.600
	2.86	0.300
	2.94	0.073
	3.01	0.000
	3.92	0.000
	4.00	0.100
	4.06	0.400
	4.14	0.800
	4.31	0.945
	4.50	0.600
	4.65	0.029
	4.70	0.183
	4.94	0.000
	5.19	0.017
	5.59	0.000
	9.02	0.000
	9.61	0.021
	10.00	0.014
	11.18	0.021
	12.94	0.000
	14.12	0.021
	15.00	0.400
	16.47	1.000
	18.47	1.000

Table 3.2.3. (*continued*)

State of surface	λ [μm]	$\varepsilon_{\lambda,n}$
	18.82	0.600
	19.18	0.021
	20.00	0.000
temperature of gas 833 K [21, 29, 30, 55, 60, 110]	1.74	0.000
	1.82	0.200
	1.93	0.345
	2.00	0.273
	2.09	0.100
	2.17	0.000
	2.47	0.000
	2.56	0.045
	2.58	0.200
	2.61	0.400
	2.70	0.800
	2.77	0.911
	2.86	0.730
	2.97	0.400
	3.06	0.100
	3.17	0.000
	3.92	0.000
	4.00	0.100
	4.06	0.400
	4.14	0.800
	4.20	0.948
	4.41	0.960
	4.64	0.930
	4.71	0.448
	4.82	0.417
	5.00	0.156
	5.07	0.029
	5.39	0.035
	5.59	0.000
	8.82	0.000
	9.22	0.870
	9.41	0.782
	9.80	0.800
	10.02	0.564
	11.11	0.682
	11.67	0.600
	12.44	0.715
	13.33	0.536
	15.29	1.000
	20.00	0.000
thickness of layer 1 mm, pressure 1290 Pa [21, 29, 30, 55, 60, 110]	1.725	0.000
	1.740	0.016
	1.763	0.024
	1.771	0.020
	1.787	0.060
	1.820	0.271
	1.823	0.265
	1.837	0.458
	1.842	0.412
	1.846	0.431
	1.852	0.388
	1.859	0.388
	1.865	0.376
	1.873	0.529
	1.892	0.267
	1.895	0.279
	1.901	0.275
	1.910	0.400
	1.915	0.368
	1.920	0.384
	1.927	0.344
	1.943	0.224
	1.959	0.120
	1.965	0.100
	1.969	0.061
	2.030	0.000
thickness of layer 1 mm, pressure 8620 Pa [21, 29, 30, 55, 60, 110]	1.725	0.009
	1.740	0.019
	1.763	0.038
	1.771	0.030
	1.781	0.120
	1.787	0.129
	1.801	0.348
	1.807	0.400
	1.817	0.563
	1.822	0.554
	1.837	0.810
	1.842	0.750
	1.847	0.758
	1.852	0.753
	1.855	0.754
	1.863	0.704
	1.873	0.868
	1.890	0.567
	1.897	0.613
	1.908	0.763
	1.915	0.708
	1.918	0.725
	1.931	0.571
	1.935	0.573
	1.941	0.500
	1.946	0.492
	1.953	0.352

Table 3.2.3. (*continued*)

State of surface	λ [μm]	$\varepsilon_{\lambda,n}$
	1.957	0.254
	1.964	0.234
	1.971	0.220
	1.978	0.100
	1.982	0.071
	1.992	0.058
	1.997	0.038
	2.005	0.029
	2.010	0.018
	2.025	0.010
	2.035	0.000
Carbon oxide; CO temperature 1200 K optical path 10^3 Pa•m [110]	4.40	0.046
	4.50	0.280
	4.55	0.331
	4.65	0.109
	4.76	0.223
	5.00	0.097
	5.20	0.011
temperature 1200 K, optical path 10^4 Pa•m [110]	4.35	0.040
	4.50	0.960
	4.55	0.983
	4.65	0.680
	4.76	0.926
	5.00	0.600
	5.26	0.034
temperature 1200 K, optical path 10^5 Pa•m [110]	4.35	0.269
	4.50	1.000
	5.00	1.000
	5.50	0.023
temperature 1200 K, optical path 10^6 Pa•m [110]	4.25	0.063
	4.35	1.000
	4.50	1.000
	5.00	1.000
	5.50	0.103
temperature 3000 K, optical path 10^3 Pa•m [110]	4.35	0.023
	4.50	0.097
	5.00	0.051
	5.50	0.028
temperature 3000 K, optical path 10^4 Pa•m [110]	4.25	0.046
	4.50	0.651
	4.73	0.337
	4.92	0.463
	5.00	0.434
	5.50	0.103
	6.00	0.009
temperature 3000 K, optical path 10^5 Pa•m [110]	4.17	0.017
	4.35	0.926
	4.50	1.000
	5.00	1.000
	5.50	0.589
	6.00	0.103
	6.25	0.017
temperature 3000 K, optical path 10^6 Pa•m [110]	4.16	0.018
	4.35	1.000
	4.50	1.000
	5.00	1.000
	5.50	1.000
	6.00	0.486
	6.25	0.109
Carbon silicide temperature of surface 1243 K [144]	2	0.75
	4	0.78
	6	0.83
	8	0.86
	10	0.84
	12	0.77
	14	0.78
Cement, Portland [40]	1.0	0.62
	2.0	0.46
	3.0	0.76
	4.0	0.60
	5.0	0.61
	6.0	0.74
	7.0	0.86
	8.0	0.81
	9.0	0.85
	10.0	0.85
	12.0	0.82
	14.0	0.81

State of surface	λ[μm]	$\varepsilon_{\lambda,n}$
Chromium oxide; Cr_2O_3 temperature of sintering 2123 K, density 3.29 g/cm^3, 99.5% Cr_2O_3 [131]	1-6	0.68
	7	0.70
	8	0.75
	9	0.78
	10	0.81
temperature of sintering 2173 K, density 3.15 g/cm^3, 99.5% Cr_2O_3, temperature of measurement 1273 K [131]	1.0	0.89
	2.0	0.90
	3.0	0.92
	4.0	0.96
	5.0	0.92
	6.0	0.91
	7.0	0.91
	8.0	0.90
	9.0	0.89
	10.0	0.91
Chromium silicide; Cr_3Si 99% Cr_3Si, sintered at temperature 1473 K, density 3.22 g/cm^3, temperature of measurement 298 K [131]	1.0	0.79
	2.0	0.82
	3.0	0.83
	4.0	0.82
	6.0	0.81
	8.0	0.83
	10.0	0.86
99% Cr_3Si, sintered at temperature 1373 K, density 3.18 g/cm^3, temperature of measurement 298 K [131]	1.0	0.74
	2.0	0.80
	3.0	0.80
	4.0	0.75
	6.0	0.70
	8.0	0.67
	10.0	0.66

State of surface	λ[μm]	$\varepsilon_{\lambda,n}$
Clay [10]	2.5÷4.0	0.83
	5.0	0.87
	6.0	0.92
	7.0	0.93
	9.0	0.90
	11.0	0.87
	13.0	0.86
Corundum sintered; aluminium oxide; Al_2O_3 99.5%$<Al_2O_3<$100%, temperature of measurement 1330 K [113, 131]	1.0	0.07
	3.0	0.10
	4.0	0.27
	5.0	0.59
	6.0	0.81
	7.0	0.92
	8.0	0.98
	9.0	0.99
	10.0	0.99
	12.0	0.83
	14.0	0.80
	15.0	0.73
99.5% Al_2O_3 100%, temperature of measurement 1713 K [113, 131]	1.0	0.15
	3.0	0.16
	4.0	0.21
	5.0	0.65
	6.0	0.84
	7.0	0.93
	8.0	0.99
	9.0	0.99
	10.0	0.99
	12.0	0.89
	14.0	0.87
	15.0	0.83
moulded and sintered powder at temperature 2123 K, density 3.35 g/cm^3, [131]	1.0	0.12
	2.0	0.12
	3.0	0.13
	4.0	0.17
	5.0	0.55
	6.0	0.84
	8.0	0.96
	10.0	0.97
	15.0	0.45

Table 3.2.3. (*continued*)

State of surface	λ[μm]	$\varepsilon_{\lambda,n}$
temperature of measurement 923 K [131]	1.0	0.41
	2.0	0.48
	3.0	0.52
	4.0	0.54
	5.0	0.59
	6.0	0.89
	8.0	0.98
	10.0	0.98
	15.0	0.72
crystal, temperature of measurement 203 K, $\varepsilon_{\lambda,n}$ is smaller at lower temperature for λ > 25 μm [63, 131]	15.0	0.14
	17.5	0.74
	20.0	0.13
	25.0	0.42
	30.0	0.64
	35.0	0.65
	40.0	0.66
Corundum sintered fr. Coralum 86% Al_2O_3, temperature of measurement 1330 K [113]	1.0	0.29
	2.0	0.28
	3.0	0.30
	4.0	0.51
	5.0	0.83
	6.0	0.90
	8.0	0.94
	10.0	0.92
	12.0	0.86
	14.0	0.82
	15.0	0.80
temperature of measurement 1720 K [113]	1.0	0.40
	2.0	0.37
	3.0	0.36
	4.0	0.56
	5.0	0.85
	6.0	0.90
	8.0	0.95
	10.0	0.92
	12.0	0.86
	14.0	0.82
	15.0	0.81
Electrographite; Acheson graphite; carbon; C rough surface, temperature 293 K [131]	1.0	0.87
	2.0	0.83
	3.0	0.78
	4.0	0.75
	5.0	0.73
temperature 300÷2300 K, $\Delta\varepsilon_{\lambda,n}$ = 0.15 [131]	1.0	0.94
	2.0	0.90
	4.0	0.85
	8.0	0.76
	10.0	0.74
	12.0	0.74
Galena; lead glance; lead sulphide; PbS [45, 152]	20.0	0.66
	40.0	0.89
	60.0	0.18
	80.0	0.19
	100.0	0.26
	120.0	0.21
	140.0	0.22
	160.0	0.25
	180.0	0.28
	200.0	0.27
Germanium; Ge polished surface [45, 152]	2.0	0.56
	2.5	0.52
	5.0	0.45
	7.5	0.41
	10.0	0.39
	12.5	0.37
	15.0	0.35
	20.0	0.33
	25.0	0.32
	37.5	0.33
	50.0	0.36
	75.0	0.41
	100.0	0.42
	125.0	0.44
	150.0	0.50
	175.0	0.54
	200.0	0.55

Table 3.2.3. (*continued*)

State of surface	$\lambda[\mu m]$	$\varepsilon_{\lambda,n}$
Glass thickness of plate 3 mm, temperature 1273 K, 72% SiO_2 13% (Na_2O+K_2O), 11% CaO, 4% (R_2O_3+MgO) [45, 131]	2.0	0.02
	3.0	0.81
	4.0	0.83
	5.0	0.94
	6.0	0.96
	7.0	0.98
	8.0	0.98
	9.0	0.80
	10.0	0.80
	11.0	0.89
	12.0	0.97
thickness of plate 6 mm, temperature 1273 K [45, 131]	2.0	0.20
	3.0	0.90
	4.0	0.91
	5.0	0.99
	6.0	0.96
	7.0	0.98
	8.0	0.98
	9.0	0.80
	10.0	0.80
	11.0	0.89
	12.0	0.97
Glass, ampouled, aluminium, Corning 1723 [45]	0.32	0.92
	0.36	0.39
	0.4	0.16
	0.6	0.03
	1.0	0.13
	1.4	0.11
	1.8	0.08
	2.2	0.15
	2.4	0.22
	2.6	0.40
Glass, Borosilicate, Corning 7740 [45]	0.3	0.35
	0.4	0.14
	0.6	0.07
	0.8	0.10
	1.0	0.43
	1.5	0.31
	1.8	0.23
	2.0	0.25
	2.2	0.31
	2.6	0.57
Glass, Borosilicate, PPG 3235 [45]	0.3	0.35
	0.4	0.05
	0.6	0.02
	1.0	0.02
	1.5	0.04
	2.0	0.05
	2.2	0.08
	2.4	0.15
	2.6	0.31
Glass, Borosilicate, Pyrex 774 [45]	2.8	0.76
	3.0	0.65
	3.2	0.58
	3.4	0.74
	3.6	0.85
	4.0	0.89
	5.0	0.91
	6.0	0.94
	7.0	0.96
	8.0	0.95
	9.0	0.72
	9.1	0.70
	9.5	0.77
	10.0	0.86
	11.0	0.92
	12.0	0.93
Glass, Borosilicate, Vicor thickness of plate 1.6 mm temperature 811 K [131]	2.5	0.15
	3.0	0.78
	3.5	0.23
	4.0	0.86
	5.0	0.97
	7.0	0.97
	9.0	0.62
	11.0	0.90
	13.0	0.91
	14.0	0.90
thickness of plate 1.6 mm temperature 500 K [131]	2.0	0.04
	3.0	0.05
	4.0	0.54

Table 3.2.3. (*continued*)

State of surface	λ[μm]	$\varepsilon_{\lambda,n}$
	5.0	0.75
	7.0	0.73
	9.0	0.20
	11.0	0.79
	12.5	0.63
	14.0	0.75
Glass, bottle, dark green thickness of plate 1.59 mm [45]	2.4	0.09
	2.6	0.14
	2.8	0.60
	3.0	0.56
	3.5	0.59
	4.0	0.63
	4.2	0.76
	4.4	0.94
	4.6	0.95
	5.0	0.96
	6.0	0.97
	8.0	0.96
	9.0	0.85
	9.5	0.78
	10.0	0.82
	11.0	0.87
	12.0	0.93
	12.5	0.93
Glass, bottle, light green thickness of plate 1.59 mm [45]	2.4	0.05
	2.6	0.06
	2.8	0.40
	3.0	0.33
	3.5	0.34
	4.0	0.44
	4.2	0.53
	4.4	0.85
	4.6	0.90
	5.0	0.92
	6.0	0.94
	8.0	0.96
	9.0	0.85
	9.5	0.78
	10.0	0.82
	11.0	0.87
	12.0	0.93
	12.5	0.93
Glass, lead, oscillograph [45]	2.5	0.02
	2.8	0.40
	3.0	0.41
	3.5	0.55
	4.0	0.70
	4.5	0.87
	5.0	0.90
	6.0	0.93
	7.0	0.94
	8.0	0.95
	9.0	0.81
	9.5	0.70
	10.0	0.74
	11.0	0.84
	12.0	0.90
Glass, lead, 21% PbO [45]	8.0	0.97
	9.0	0.80
	9.5	0.66
	10.0	0.70
	11.0	0.87
	12.0	0.90
	13.0	0.91
Glass, lead, 30% PbO [45]	8.5	0.97
	9.0	0.88
	9.5	0.75
	10.0	0.67
	11.0	0.79
	12.0	0.88
	13.0	0.90
Glass, lead, 35% PbO [45]	7.5	0.95
	8.0	0.96
	8.5	0.95
	9.0	0.90
	9.5	0.81
	10.0	0.73
	11.0	0.79
	12.0	0.81
	13.0	0.84
Glass, lead, 50% PbO [45]	7.5	0.94
	8.0	0.95
	8.5	0.95
	9.0	0.95
	10.0	0.79
	11.0	0.72
	12.0	0.74
	13.0	0.80

State of surface	$\lambda[\mu m]$	$\varepsilon_{\lambda,n}$
Glass, lead, 67% PbO [45]	7.5	0.90
	8.0	0.92
	9.0	0.93
	10.0	0.95
	11.0	0.74
	12.0	0.76
	13.0	0.82
Glass, metallized, electricity conductive; Battelle 54726 [45]	0.28	0.60
	0.32	0.38
	0.36	0.20
	0.40	0.13
	0.5	0.05
	0.6	0.05
	0.8	0.13
	1.0	0.16
	1.4	0.20
	1.8	0.33
	2.2	0.43
	2.6	0.53
Glass, metallized, electricity conductive; LOF PB 19195 [45]	0.3	0.59
	0.4	0.15
	0.5	0.06
	0.6	0.06
	0.8	0.25
	1.0	0.29
	1.4	0.32
	1.8	0.39
	2.2	0.50
	2.6	0.59
	2.7	0.67
Glass, metallized, electricity conductive; LOF 81 E 19778 [45]	0.32	0.52
	0.40	0.32
	0.50	0.22
	0.60	0.23
	0.8	0.37
	1.0	0.442
	1.2	0.443
	1.6	0.41
	2.0	0.46
	2.4	0.47
	2.7	0.59
Graphite, polycrystalline; carbon; C		
α-surface at temperature of deposition 2073 K, temperature of measurement 1000÷1400 K [131]	1.0	0.99
	2.0	0.73
	4.0	0.57
	6.0	0.48
	8.0	0.31
	10.0	0.34
	12.0	0.30
α-surface at temperature of deposition 2273 K, temperature of measurement 1000÷1400 K [131]	2.0	0.57
	4.0	0.45
	6.0	0.36
	8.0	0.23
	10.0	0.22
	11.0	0.21
C-surface, temperature of measurement 1000÷1400 K [131]	1.0	0.83
	2.0	0.76
	4.0	0.64
	6.0	0.54
	8.0	0.52
	10.0	0.50
	12.0	0.48
Haematite; blood stone, iron oxide; Fe_2O_3 temperature of measurement 1330 K [113]	1.0	0.78
	3.0	0.85
	5.0	0.90
	7.0	0.91
	9.0	0.98
	11.0	0.95
	13.0	0.95
	15.0	0.93
temperature of measurement 1723 K [113]	1.0	0.90
	3.0	0.91
	5.0	0.95
	7.0	0.94
	11.0	0.89

Table 3.2.3. (*continued*)

State of surface	λ [μm]	$\varepsilon_{\lambda,n}$
	13.0	0.93
	15.0	0.85
Hafnium nitride; HfN hot-moulded, temperature of measurement 1512 K [131]	0.4	0.97
	0.5	0.88
	0.6	0.83
	0.8	0.80
	1.0	0.74
	2.0	0.66
	3.0	0.61
	4.0	0.54
hot-moulded, temperature of measurement 1809 K [131]	0.4	0.93
	0.5	0.88
	0.6	0.87
	0.8	0.84
	1.0	0.82
	2.0	0.71
	3.0	0.65
	4.0	0.58
hot-moulded, temperature of measurement 2096 K [131]	0.4	0.90
	0.5	0.89
	0.6	0.88
	0.8	0.86
	1.0	0.84
	2.0	0.80
	3.0	0.78
	4.0	0.70
hot-moulded, temperature of measurement 2192 K [131]	0.4	0.88
	0.5	0.89
	0.6	0.89
	0.8	0.89
	1.0	0.88
	2.0	0.83
	3.0	0.81
	4.0	0.79
Hydrochloric acid; muriatic acid; HCl temperature 600 K, optical path 10^3 Pa•m [110]	3.33	0.100
	3.48	0.030
	3.61	0.067
	3.86	0.017
temperature 600 K, optical path 10^4 Pa•m [110]	3.15	0.016
	3.33	0.658
	3.50	0.283
	3.64	0.567
	4.00	0.100
temperature 600 K, optical path 10^5 Pa•m [110]	3.17	0.033
	3.30	0.983
	3.50	0.950
	4.00	0.600
	4.40	0.042
temperature 600 K, optical path 10^6 Pa•m [110]	3.17	0.200
	3.33	1.000
	4.00	1.000
	4.50	0.200
	4.80	0.025
temperature 5000 K, optical path 10^4 Pa•m [110]	3.00	0.058
	4.00	0.158
	5.00	0.117
	6.00	0.083
	6.67	0.050
temperature 5000 K, optical path 10^6 Pa•m [110]	3.00	0.533
	4.00	0.800
	5.00	0.675
	6.00	0.541
	6.67	0.417
Hydrogen fluoride; HF temperature 300 K, optical path 10^3 Pa•m [110]	2.35	0.024
	2.45	0.247
	2.50	0.200
	2.53	0.035
	2.60	0.194
	2.76	0.047
temperature 300 K, optical path 10^4 Pa•m [110]	2.31	0.029
	2.45	0.941
	2.50	0.882
	2.53	0.456

Table 3.2.3. (*continued*)

State of surface	λ [μm]	$\varepsilon_{\lambda,n}$
	2.60	0.876
	2.86	0.047
temperature 300 K, optical path 10^5 Pa•m [110]	2.30	0.029
	2.50	1.000
	2.76	0.947
	2.86	0.400
	3.00	0.011
temperature 300 K, optical path 10^6 Pa•m [110]	2.29	0.029
	2.50	1.000
	2.76	1.000
	2.86	0.976
	3.00	0.082
temperature 3000 K, optical path 10^3 Pa•m [110]	2.30	0.024
	2.50	0.035
	2.86	0.033
	3.00	0.029
temperature 3000 K, optical path 10^4 Pa•m [110]	2.30	0.147
	2.50	0.218
	2.86	0.241
	3.00	0.206
	3.33	0.147
	4.00	0.059
	5.00	0.012
temperature 3000 K, optical path 10^5 Pa•m [110]	2.28	0.331
	2.48	0.937
	2.50	0.909
	2.55	0.855
	2.86	0.926
	3.00	0.891
	4.00	0.366
	5.00	0.109
	6.67	0.029
temperature 3000 K, optical path 10^6 Pa•m [110]	2.28	0.834
	2.50	1.000
	3.00	1.000
	4.00	0.989
	5.00	0.634
	6.67	0.200

State of surface	λ [μm]	$\varepsilon_{\lambda,n}$
Indium antimonide; InSb polished surface, temperature 233 K [45, 152]	20	0.62
	40	0.70
	45	0.74
	48	0.76
	51	0.92
	54	0.10
	57	0.10
	60	0.50
	63	0.54
	66	0.58
	69	0.60
	80	0.66
	100	0.75
	120	0.78
	140	0.82
	160	0.84
	180	0.82
	200	0.77
polished surface, temperature 300 K [45, 152]	20	0.66
	40	0.74
	45	0.78
	48	0.82
	51	0.88
	54	0.17
	57	0.18
	60	0.54
	63	0.60
	66	0.62
	69	0.64
	80	0.71
	100	0.78
	109	0.81
	112	0.82
	115	0.84
	118	0.85
	120	0.86
	123	0.82
	126	0.76
	140	0.54
	160	0.50
	180	0.54
	200	0.64
polished surface, temperature 317 K [45, 152]	66	0.66
	69	0.70
	72	0.72

Table 3.2.3. (*continued*)

State of surface	λ[μm]	$\varepsilon_{\lambda,n}$
	75	0.74
	78	0.75
	80	0.76
	100	0.84
	120	0.56
	140	0.37
	160	0.32
	180	0.39
	200	0.49
polished surface, temperature 378 K [45, 152]	20	0.70
	40	0.81
	45	0.86
	48	0.88
	51	0.91
	54	0.24
	57	0.24
	60	0.62
	63	0.68
	66	0.74
	69	0.78
	72	0.83
	75	0.86
	78	0.86
	80	0.80
	84	0.64
	100	0.34
	120	0.22
	140	0.21
	160	0.23
	180	0.28
	200	0.36
polished surface, temperature 400 K [45, 152]	66	0.78
	69	0.84
	72	0.86
	75	0.74
	78	0.62
	80	0.54
	82	0.48
	100	0.28
	120	0.20
Lead selenide; PbSe [45, 152]	20	0.67
	30	0.77
	40	0.86
	50	0.87
	60	0.54
	70	0.34
	80	0.32
	100	0.39
	120	0.34
	160	0.38
	200	0.39
Limestone; $CaCO_3$ [10]	2.5	0.74
	3.0	0.78
	4.0	0.82
	5.0	0.86
	6.0	0.89
	7.0	0.91
	9.0	0.90
	11.0	0.89
	13.0	0.89
Magnesite; magnesium carbonate; $MgCO_3$ [10, 24]	2.5	0.35
	3.0	0.29
	4.0	0.31
	5.0	0.44
	6.0	0.74
	7.0	0.79
	8.0	0.80
	9.0	0.87
	10.0	0.95
	11.0	0.96
	12.0	0.94
	13.0	0.90
Magnesium cyanogen; magnesium dicyan n type at temperature 5 K [45]	3.0	0.97
	4.0	0.79
	5.0	0.45
	6.0	0.42
	7.0	0.43
	7.4	0.48
	8.0	0.39
	9.0	0.35
	10.0	0.33
	11.0	0.32
	12.0	0.31
	13.0	0.30
	14.0	0.30
	15.0	0.29

Table 3.2.3. (*continued*)

State of surface	λ[μm]	$\varepsilon_{\lambda,n}$
n type at temperature 300 K [45]	4.0	0.99
	5.0	0.98
	6.0	0.98
	7.0	0.99
	8.0÷14.0	0.998
p type at temperature 5 K [45]	3.0	0.97
	4.0	0.73
	5.0	0.58
	6.0	0.50
	7.0	0.45
	8.0	0.43
	9.0	0.41
	10.0	0.39
	11.0	0.37
	12.0	0.35
	13.0	0.32
	14.0	0.30
	15.0	0.26
p type at temperature 300 K [45]	3.0	0.97
	4.0	0.63
	5.0	0.43
	6.0	0.33
	7.0	0.29
	8.0	0.30
	9.0	0.33
	10.0	0.36
	11.0	0.39
	12.0	0.42
	13.0	0.37
	14.0	0.53
	15.0	0.59
Molybdenum silicide; $MoSi_2$ [131]	1.0	0.86
	2.0	0.90
	3.0	0.91
	4.0	0.91
	6.0	0.89
	8.0	0.81
	9.0	0.68
	10.0	0.80
	15.0	0.82
Mullite; $3Al_2O_3 \cdot 2SiO_2$ temperature 1720 K, the same values at temperature 1330 K [113]	1.0	0.31
	2.0	0.33
	3.0	0.35
	4.0	0.52
	5.0	0.89
	7.0	0.98
	9.0	0.95
	11.0	0.94
	13.0	0.94
	15.0	0.93
Nickel oxide; NiO moulded powder, sintering at temperature 1673 K, density 5.28 g/cm^3 [131]	1.0	0.77
	2.0	0.76
	4.0	0.73
	6.0	0.73
	8.0	0.78
	10.0	0.81
	15.0	0.91
Niobium carbide; NbC temperature of measurement 1793÷2490 K [131]	0.5	0.45
	1.0	0.34
	2.0	0.32
	3.0	0.30
	4.0	0.28
	5.0	0.28
Nitrogen oxide; NO temperature 3000 K, optical path 10^3 Pa•m [110]	5.00	0.013
	5.27	0.029
	5.55	0.016
	5.88	0.014
temperature 3000 K, optical path 10^4 Pa•m [110]	5.00	0.104
	5.27	0.205
	5.55	0.144
	5.88	0.120
	6.25	0.046
	6.67	0.016

Table 3.2.3. (*continued*)

State of surface	λ[μm]	$\varepsilon_{\lambda,n}$
temperature 3000 K, optical path 10^5 Pa•m [110]	5.00	0.661
	5.18	0.897
	5.27	0.896
	5.41	0.718
	5.55	0.776
	5.88	0.709
	6.25	0.360
	6.67	0.096
	7.14	0.016
temperature 3000 K, optical path 10^6 Pa•m [110]	5.00	1.000
	5.27	1.000
	5.55	1.000
	5.88	1.000
	6.25	0.979
	6.67	0.616
	7.14	0.163
	7.67	0.021
temperature 7000 K, optical path 10^4 Pa•m [110]	5.00	0.032
	5.27	0.064
	5.55	0.048
	5.88	0.048
	6.25	0.042
	6.67	0.024
	7.14	0.014
	7.67	0.016
temperature 7000 K, optical path 10^5 Pa•m [110]	5.00	0.312
	5.27	0.496
	5.55	0.410
	5.88	0.376
	6.25	0.336
	6.67	0.240
	7.14	0.141
	7.67	0.077
temperature 7000 K, optical path 10^6 Pa•m [110]	5.00	0.976
	5.27	0.990
	5.55	0.989
	5.88	0.984
	6.25	0.976
	6.67	0.936
	7.14	0.776
	7.67	0.544
Periclase, magnesium oxide, MgO [45]	0.30	0.70
	0.40	0.29
	0.50	0.19
	0.75	0.12
	1.0	0.09
	1.5	0.11
	2.0	0.11
	2.5	0.09
moulded powder, temperature of measurement 1330 K [113]	1.0	0.23
	2.0	0.16
	3.0	0.13
	4.0	0.20
	5.0	0.38
	6.0	0.57
	7.0	0.76
	8.0	0.85
	9.0	0.92
	10.0	0.95
	12.0	0.95
	14.0	0.86
	15.0	0.79
moulded powder, temperature of measurement 1800 K [113]	1.0	0.41
	2.0	0.38
	3.0	0.37
	4.0	0.43
	5.0	0.51
	6.0	0.77
	7.0	0.85
	8.0	0.89
	9.0	0.92
	10.0	0.95
	12.0	0.95
	14.0	0.91
	15.0	0.83
Silica, polycrystalline; silicon dioxide; SiO_2 temperature of sintering 1823 K, density 1.53 g/cm^3, temperature of measurement 1273 K [131]	1.0	0.04
	2.0	0.06
	3.0	0.07
	4.0	0.34

Table 3.2.3. (*continued*)

State of surface	λ[μm]	$\varepsilon_{\lambda,n}$
	5.0	0.83
	6.0	0.91
	8.0	0.96
	10.0	0.85
	15.0	0.96
Silicon; Si ground surface, temperature of measurement 1273 K [45]	0.1	0.79
	0.15	0.62
	0.2	0.51
ground surface, temperature of measurement 1683 K [45]	0.14	0.76
	0.2	0.61
Silicone reinforced [45]	0.3	0.90
	0.4	0.58
	0.5	0.41
	0.6	0.35
	0.8	0.42
	1.2	0.51
	1.6	0.51
	2.0	0.49
	2.2	0.69
	2.4	0.79
	2.6	0.79
Silicon nitride; Si_3N_4 smooth surface, density 1.82 g/cm^3 [131]	1.0	0.79
	2.0	0.82
	4.0	0.93
	6.0	0.94
	8.0	0.975
	10.0	0.87
	15.0	0.93
Silicon oxide sintered; SiO smooth surface, thickness of plate 0.1 μm, temperature of measurement 873 K [131]	2.0	0.32
	3.0	0.38
	4.0	0.33
	5.0	0.29
	6.0	0.22
	8.0	0.21
	10.0	0.37
	15.0	0.21
smooth surface, temperature of measurement 1273 K [131]	1.0	0.67
	2.0	0.39
	3.0	0.55
	4.0	0.61
	5.0	0.55
	6.0	0.33
	8.0	0.33
	10.0	0.50
	15.0	0.21
Silite temperature of measurement 873 K [66, 131, 144]	2.0	0.895
	4.0	0.95
	6.0	0.94
	8.0	0.94
	10.0	0.945
	12.0	0.61
	14.0	0.795
temperature of measurement 1243 K [66, 131, 144]	2.0	0.74
	4.0	0.80
	6.0	0.835
	8.0	0.86
	10.0	0.85
	12.0	0.78
	14.0	0.77
Silite; Carbolithe, fr. 74% SiC [16]	1.0	0.88
	2.0	0.93
	3.0	0.96
	4.0	0.96
	6.0	0.98
	8.0	0.99
	10.0	0.97
	12.0	0.91
	14.0	0.96
	15.0	0.97
Silite; globar, am. [133]	2.0	0.73
	4.0	0.78
	6.0	0.84
	9.0	0.86
	11.0	0.85

Table 3.2.3. (*continued*)

State of surface	λ [μm]	$\varepsilon_{\lambda,n}$
	13.0	0.78
	15.0	0.79
Soil [40]	1.0	0.76
	2.0	0.72
	2.5	0.76
	3.0	0.94
	3.5	0.91
	4.0	0.80
	5.0	0.86
	6.0	0.96
	8.0	0.93
	10.0	0.92
	12.0	0.915
	14.0	0.91
Soil, sand [10]	2.5	0.70
	3.0	0.71
	4.0	0.72
	5.0	0.80
	6.0	0.89
	7.0	0.92
	9.0	0.89
	11.0	0.88
	13.0	0.85
Sphalerite; zinc blende; zinc sulphide; ZnS [45]	20	0.95
	24	1.00
	28	0.16
	32	0.42
	36	0.59
	40	0.67
	44	0.74
	48	0.76
	52	0.87
	56	0.88
	60÷200	0.877
Stone, pebble [10]	2.5	0.77
	3.0	0.77
	4.0	0.78
	5.0	0.83
	6.0	0.86
	7.0	0.84
	8.0	0.80
	9.0	0.79
	10.0	0.76
	11.0	0.77
	12.0	0.79
	13.0	0.89

State of surface	λ [μm]	$\varepsilon_{\lambda,n}$
Sulphur dioxide; SO_2 temperature 300 K, optical path $5.16 \cdot 10^3$ Pa·m [21, 38, 55, 110]	3.83	0.00
	3.87	0.03
	3.92	0.20
	3.94	0.42
	3.98	0.43
	4.00	0.04
	4.02	0.03
	4.11	0.05
	4.19	0.12
	4.26	0.49
	4.35	0.10
	4.41	0.04
	4.50	0.00
temperature 300 K, optical path $19.85 \cdot 10^3$ Pa·m [21, 38, 55 110]	3.78	0.00
	3.87	0.04
	3.92	0.37
	3.93	0.80
	3.96	0.81
	4.05	0.70
	4.06	0.40
	4.07	0.08
	4.08	0.07
	4.11	0.08
	4.15	0.20
	4.23	0.74
	4.24	0.97
	4.29	0.99
	4.35	0.97
	4.38	0.50
	4.41	0.20
	4.45	0.10
	4.52	0.04
	4.64	0.00
mixture of SO_2+N_2, $p_{SO_2} = 10^4$ Pa, $p = 4 \cdot 10^5$ Pa, temperature 300 K [21, 38, 55, 110]	6.93	0.00
	6.96	0.04
	7.00	0.21
	7.10	0.40
	7.18	0.55
	7.23	0.80
	7.31	0.96
	7.41	0.99
	7.50	0.96
	7.54	0.83

Table 3.2.3. (*continued*)

State of surface	λ [μm]	$\varepsilon_{\lambda,n}$
	7.62	0.50
	7.63	0.20
	7.66	0.12
	7.83	0.00
mixture of SO_2+N_2; P_{CO_2} = 2.4·10^3 Pa, p = 3.2·10^4 Pa, temperature 300 K [21, 38, 55, 110]	7.00	0.00
	7.06	0.10
	7.22	0.54
	7.35	0.56
	7.47	0.53
	7.50	0.20
	7.62	0.03
	7.70	0.00
mixture of SO_2+N_2, P_{SO_2} = 69.9·10^3 Pa, p = 39.92·10^4 Pa, temperature 300 K [21, 38, 55, 110]	7.83	0.00
	7.97	0.10
	8.11	0.20
	8.52	0.80
	8.60	0.86
	8.75	0.88
	8.83	0.86
	8.88	0.80
	8.97	0.55
	9.26	0.30
	9.43	0.14
	9.64	0.01
	9.73	0.00
Tantalum carbide; TaC temperature of measurement 2000 K [131]	0.6	0.48
	0.8	0.39
	1.0	0.33
	2.0	0.26
	3.0	0.25
	4.0	0.24
	6.0	0.22
	8.0	0.21
	10.0	0.23
Tantalum nitride; TaN hot-moulded sample, grains of powder d ≅ 1 μm measurement in atmosphere of nitrogen, temperature 1990 K [131]	0.4	0.85
	0.6	0.81
	0.8	0.81
	1.0	0.79
	2.0	0.73
	3.0	0.70
	4.0	0.69
hot-moulded sample, grains of powder d ≅ 1 μm, measurement in vacuum [131]	0.4	0.56
	0.6	0.51
	0.8	0.48
	1.0	0.45
	2.0	0.38
	3.0	0.38
Tantalum oxide; Ta_2O_5 sintered at temperature 1673 K, density 6.51 g/cm^3 [131]	1.0	0.35
	2.0	0.09
	3.0	0.15
	4.0	0.20
	6.0	0.68
	8.0	0.82
	10.0	0.88
	15.0	0.78
Tantalum silicide; $TaSi_2$ temperature of measurement 1273 K, 95% $TaSi_2$, sintered at temperature 1773 K, density 4.78 g/cm^3 [131]	1.0	0.79
	2.0	0.77
	3.0	0.85
	4.0	0.85
	6.0	0.84
	8.0	0.85
	10.0	0.87
	15.0	0.87

Table 3.2.3. (*continued*)

State of surface	λ [μm]	$\varepsilon_{\lambda,n}$
Tellurium oxide; TeO_2 thickness of plate 3.05 mm, temperature of measurement 533÷643 K [131]	2.5	0.09
	3.0	0.14
	4.0	0.12
	5.0	0.17
	6.0	0.60
	8.0	0.98
	10.0	0.99
	15.0	0.89
thickness of plate 7.65 mm, temperature of measurement 533÷643 K [131]	2.5	0.12
	3.0	0.29
	4.0	0.20
	5.0	0.49
	6.0	0.86
Thallium chloride; TlCl smooth surface [45]	20	0.88
	30	0.94
	40	0.95
	50	0.99
	60	0.72
	80	0.38
	90	0.41
	100	0.37
	120	0.30
	140	0.36
	160	0.40
	200	0.41
Tile, concrete [40]	0.5	0.21
	1.0	0.17
	1.5	0.15
	2.0	0.18
	2.5	0.32
	3.0	0.69
	3.5	0.68
	4.0	0.73
	4.5	0.79
	5.0	0.91
	6.0	0.94
	8.0	0.94
	10.0	0.95
Titanium oxide; TiO_2 temperature of measurement 1223 K, density 3.87 g/cm^3, sintered at temperature 1673 K [131]	1.0	0.26
	2.0	0.14
	3.0	0.21
	4.0	0.30
	6.0	0.49
	8.0	0.76
	10.0	0.82
	15.0	0.89
Titanium silicide; $TiSi_2$ sintered at temperature 1573 K, density 2.82 g/cm^3, temperature of measurement 1223 K [131]	1.0	0.82
	2.0	0.86
	3.0	0.79
	4.0	0.76
	6.0	0.70
	8.0	0.75
	10.0	0.76
Uranium oxide UO_2 temperature of sample 2000 K [45, 83, 131]	0.458	0.822
	0.5154	0.831
	0.647	0.839
temperature of sample 2500 K [45, 83, 131]	0.458	0.826
	0.5145	0.835
	0.647	0.840
temperature of sample 3000 K [45, 83, 131]	0.458	0.833
	0.5145	0.844
	0.647	0.851
temperature of sample 3500 K [45, 83, 131]	0.458	0.855
	0.5145	0.866
	0.647	0.871
temperature of sample 3750 K [45, 83, 131]	0.458	0.900
	0.5145	0.918
	0.647	0.900

Table 3.2.3. (*continued*)

State of surface	λ [μm]	$\varepsilon_{\lambda,n}$
Water; H_2O		
thickness of layer 0.01 mm		
[24, 27, 40. 72]	0.5	0.00
	1.0	0.00
	1.5	0.01
	2.0	0.07
	2.5	0.04
	3.0	0.90
	3.5	0.09
	4.0	0.06
	4.5	0.31
	5.0	0.13
	5.5	0.18
	6.0	0.65
	6.5	0.85
	7.0	0.84
	8.0	0.73
	9.0	0.75
	10.0	0.81
	11.0	0.76
thickness of layer 0.2 mm		
[24, 27, 40, 72]	0.5	0.02
	1.0	0.10
	1.5	0.11
	2.0	0.21
	2.5	0.24
	3.0	0.98
	3.5	0.31
	4.0	0.26
	4.5	0.52
	5.0	0.40
	5.5	0.47
	6.0	0.95
	6.5	0.91
	7.0	0.96
	8.0	0.92
	9.0	0.94
	10.0	0.94
	11.0	0.84
thickness of layer 0.5 mm		
[24, 27, 40, 72]	0.5	0.03
	1.0	0.12
	1.5	0.38
	2.0	0.82
	2.5	0.99
	3.0	0.99
	3.5	0.94
	4.0	0.85
	4.5	0.99
	5.0	0.99
	5.5	0.98
	6.0	0.99
	6.5	0.98
	7.0	0.99
	8.0	0.99
	9.0	0.99
	10.0	1.00
	11.0	0.98
very large		
thickness of layer		
[10]	1.00	0.9804
	1.05	0.9805
	1.10	0.9806
	1.20	0.9807
	1.30	0.9809
	1.40	0.9809
	1.50	0.9812
	1.60	0.9814
	1.70	0.9815
	1.80	0.9818
	1.90	0.9821
	2.00	0.9826
	2.20	0.9837
	2.40	0.9853
	2.50	0.9863
	2.60	0.9875
	2.70	0.9904
	2.74	0.9925
	2.77	0.9910
	2.80	0.9858
	2.85	0.9800
	2.90	0.9753
	2.95	0.9714
	3.00	0.9660
	3.02	0.9651
	3.07	0.9561
	3.10	0.9571
	3.16	0.9588
	3.20	0.9601
	3.30	0.9635
	3.40	0.9663
	3.50	0.9695
	3.60	0.9720
	3.70	0.9744
	3.80	0.9764
	3.90	0.9775
	4.03	0.9781
	4.10	0.9782
	4.20	0.9784
	4.30	0.9785
	4.40	0.9786
	4.50	0.9787
	4.60	0.9789
	4.70	0.9790
	4.80	0.9792
	4.90	0.9795

Table 3.2.3. (*continued*)

State of surface	λ[μm]	$\varepsilon_{\lambda n}$	State of surface	λ[μm]	$\varepsilon_{\lambda,n}$
	5.00	0.9798		12.50	0.9785
	5.10	0.9802		13.00	0.9698
	5.20	0.9807		13.50	0.9645
	5.30	0.9814		14.00	0.9590
	5.40	0.9821		14.50	0.9528
	5.50	0.9827		15.00	0.9488
	5.60	0.9834			
	5.70	0.9843			
	5.80	0.9860			
	5.85	0.9875	Water vapour; H_2O pressure $97.4\cdot10^3$ Pa, thickness of layer 0.4 mm [19, 55, 110, 125]	2.3791	0.000
	5.90	0.9850		2.4372	0.060
	6.00	0.9798		2.4983	0.233
	6.04	0.9783		2.5625	0.693
	6.10	0.9772		2.6071	0.956
	6.20	0.9754		2.6300	0.940
	6.30	0.9766		2.6534	0.923
	6.40	0.9777		2.7013	0.960
	6.50	0.9788		2.7185	0.963
	6.60	0.9795		2.7412	0.925
	6.70	0.9800		2.7765	0.920
	6.80	0.9802		2.8560	0.500
	6.90	0.9803		2.9402	0.200
	7.00	0.9805		2.9786	0.135
	7.20	0.9813		3.0123	0.173
	7.40	0.9823		3.0295	0.156
	7.60	0.9827		3.0698	0.173
	7.80	0.9832		3.1244	0.128
	8.00	0.9834		3.2132	0.225
	8.20	0.9836		3.2254	0.206
	8.40	0.9838		3.2717	0.148
	8.60	0.9843		3.3332	0.111
	8.80	0.9849		3.4485	0.040
	9.00	0.9856		3.4875	0.000
	9.20	0.9865			
	9.40	0.9873			
	9.60	0.9882			
	9.80	0.9891	pressure $97.4\cdot10^3$ Pa, thickness of layer 0.9 mm [19, 55, 110, 125]	2.3510	0.000
	10.00	0.9901		2.3791	0.033
	10.20	0.9908		2.4372	0.124
	10.40	0.9915		2.4983	0.400
	10.60	0.9922		2.5625	0.827
	10.80	0.9927		2.6171	0.987
	10.90	0.9930		2.6300	0.973
	11.00	0.9927		2.7013	0.993
	11.10	0.9923		2.7185	0.963
	11.20	0.9916		2.7765	0.920
	11.30	0.9909		2.8560	0.620
	11.40	0.9903		2.9402	0.293
	11.50	0.9898		2.9916	0.218
	11.60	0.9891		3.0295	0.249
	11.70	0.9881		3.1244	0.180
	11.80	0.9870		3.2132	0.120
	12.00	0.9853		3.2254	0.286

Table 3.2.3. (*continued*)

State of surface	$\lambda[\mu m]$	$\varepsilon_{\lambda,n}$
	3.2920	0.200
	3.3332	0.186
	3.4485	0.060
	3.5105	0.000
pressure $97.4 \cdot 10^3$ Pa,		
thickness of layer 1.5 mm		
[19, 55, 110, 125]	2.3237	0.000
	2.3791	0.067
	2.4372	0.266
	2.4983	0.567
	2.5625	0.920
	2.5955	1.000
	2.7413	1.000
	2.7765	0.977
	2.8560	0.746
	2.9402	0.400
	3.0125	0.314
	3.0295	0.364
	3.0988	0.357
	3.1244	0.264
	3.1491	0.163
	3.2132	0.450
	3.2254	0.413
	3.2717	0.300
	3.3332	0.240
	3.4485	0.100
	3.5308	0.000
pressure $97.4 \cdot 10^5$ Pa,		
thickness of layer 3.4 mm		
[19, 55, 110, 125]	2.2827	0.000
	2.3237	0.107
	2.3791	0.111
	2.4372	0.293
	2.4983	0.716
	2.5625	0.951
	2.5825	1.000
	2.7765	1.000
	2.8560	0.847
	2.9402	0.533
	2.9659	0.507
	2.9974	0.447
	3.0295	0.508
	3.0522	0.520
	3.1244	0.400
	3.1441	0.271
	3.2132	0.546
	3.2254	0.514
	3.2717	0.425
	3.3332	0.367
	3.4485	0.140
	3.5720	0.000
pressure $97.4 \cdot 10^3$ Pa,		
thickness of layer 6.0 mm		
[19, 55, 110, 125]	2.2937	0.000
	2.3237	0.047
	2.3791	0.180
	2.4372	0.411
	2.4983	0.840
	2.5625	0.980
	2.5793	1.000
	2.8030	1.000
	2.8560	0.933
	2.9402	0.667
	2.9786	0.633
	3.0123	0.557
	3.0295	0.627
	3.0522	0.640
	3.1244	0.527
	3.1441	0.450
	3.1853	0.507
	3.2254	0.608
	3.2717	0.536
	3.3332	0.480
	3.3508	0.468
	3.4485	0.200
	3.5720	0.020
	3.6052	0.000
pressure $97.4 \cdot 10^3$ Pa,		
thickness of layer 10 mm		
[19, 55, 110, 125]	2.2606	0.000
	2.2707	0.011
	2.3237	0.105
	2.3791	0.310
	2.4372	0.680
	2.4983	0.933
	2.5625	1.000
	2.8295	1.000
	2.9402	0.767
	3.0123	0.687
	3.0295	0.740
	3.0622	0.760
	3.1244	0.627
	3.1441	0.550
	3.1907	0.790
	3.2254	0.733
	3.3332	0.660
	3.3508	0.633
	3.4485	0.380
	3.5720	0.100
	3.7047	0.000

Table 3.2.3. (*continued*)

State of surface	$\lambda[\mu m]$	$\varepsilon_{\lambda,n}$
pressure $97.4 \cdot 10^3$ Pa, thickness of layer 30 mm [19, 55, 110, 125]	2.2406	0.000
	2.2707	0.048
	2.3237	0.173
	2.3791	0.424
	2.4372	0.760
	2.4983	0.973
	2.5304	1.000
	2.8215	1.000
	2.8560	1.000
	2.9402	0.913
	2.9835	0.871
	3.0295	0.894
	3.0698	0.900
	3.1244	0.800
	3.1441	0.654
	3.2132	0.929
	3.2254	0.856
	3.2717	0.800
	3.3332	0.783
	3.3508	0.771
	3.4485	0.481
	3.5720	0.156
	3.7047	0.028
	3.7394	0.000
pressure $97.4 \cdot 10^3$ Pa, thickness of layer 100 mm [19, 55, 110, 125]	2.2060	0.000
	2.2202	0.033
	2.2707	0.149
	2.3237	0.321
	2.2791	0.673
	2.4372	0.900
	2.4850	1.000
	2.9192	1.000
	2.9402	0.986
	2.9835	0.970
	3.0295	0.980
	3.0698	0.985
	3.1244	0.900
	3.1441	0.851
	3.2132	0.987
	3.2254	0.980
	3.3332	0.924
	3.4485	0.667
	3.5720	0.280
	3.7047	0.100
	3.8476	0.000
pressure $97.4 \cdot 10^3$ Pa, thickness of layer 500 mm [19, 55, 110, 125]	2.1488	0.000
	2.1718	0.040
	2.2202	0.180
	2.2707	0.400
	2.3237	0.673
	2.3791	0.920
	2.4372	1.000
	3.3332	1.000
	3.4485	0.920
	3.5720	0.547
	3.7047	0.253
	3.8476	0.053
normal pressure, thickness of layer 109 cm, temperature 400 K [19, 55, 110, 125]	0.85	0.00
	0.95	0.77
	0.98	0.00
	1.07	0.00
	1.13	0.14
	1.20	0.00
	1.26	0.00
	1.36	0.75
	1.40	0.60
	1.48	0.11
	1.50	0.12
	1.53	0.00
	1.72	0.00
	1.80	0.30
	1.80	0.70
	1.83	0.83
	1.90	0.70
	1.91	0.72
	2.03	0.00
	2.27	0.00
	2.50	0.40
	2.52	0.70
	2.64	1.00
	2.77	1.00
	2.93	0.48
	2.95	0.50
	2.97	0.40
	3.00	0.44
	3.03	0.35
	3.05	0.36
	3.15	0.11
	3.19	0.39
	3.20	0.26
	3.25	0.24
	3.28	0.28
	3.37	0.20

Table 3.2.3. (*continued*)

State of surface	$\lambda[\mu m]$	$\varepsilon_{\lambda,n}$
	3.40	0.05
	3.46	0.00
	4.43	0.00
	5.00	0.50
	5.10	0.60
	5.14	0.75
	5.20	0.72
	5.23	0.83
	5.33	0.81
	5.50	0.99
	6.00	1.00
	6.26	0.67
	6.50	1.00
	7.00	1.00
	7.30	0.95
	7.30	0.90
	7.50	0.80
	7.50	0.93
	7.67	0.50
	7.75	0.70
	7.79	0.52
	7.83	0.72
	7.87	0.54
	8.00	0.71
	8.04	0.17
	8.20	0.30
	8.30	0.13
	8.54	0.10
	10.00	0.10
	10.20	0.11
	10.30	0.08
	10.80	0.16
	10.95	0.11
	11.00	0.17
	11.06	0.15
	11.40	0.16
	11.50	0.25
	11.56	0.19
	11.80	0.17
	12.25	0.33
	12.45	0.25
	12.70	0.27
	12.92	0.25
	13.29	9.43
	13.40	0.32
	13.66	0.30
	14.23	0.60
	14.50	0.45
	14.95	0.50
	15.00	0.64
	15.58	0.75
	15.70	0.73
	16.00	0.74
	16.18	0.68
	16.36	0.81
	17.20	0.90
	17.70	0.88
	19.50	1.00
Yttrium oxide; Y_2O_3 moulded and sintered sample at temperature 2023 K, density 4 g/cm^3 [131]	1.0	0.29
	2.0	0.28
	3.0	0.27
	4.0	0.25
	5.0	0.29
	6.0	0.40
	7.0	0.61
	8.0	0.69
	10.0	0.77
	15.0	0.78
Zirconium boride; ZrB_2 temperature 1604 K [131]	0.4	0.97
	0.5	0.91
	0.7	0.68
	0.9	0.59
	1.0	0.56
	2.0	0.45
	3.0	0.42
	4.0	0.40
temperature 1702 K [131]	0.55	0.86
	0.6	0.82
	0.7	0.75
	0.8	0.67
	0.9	0.59
	1.0	0.55
	2.0	0.45
	3.0	0.43
	4.0	0.42
	5.0	0.405
temperature 2200 K [131]	0.4	0.95
	0.5	0.89
	0.55	0.86
	0.7	0.73
	0.8	0.68
	0.9	0.65
	1.0	0.62
	2.0	0.52
	3.0	0.47
	4.0	0.44
	5.0	0.41

Table 3.2.3. (*continued*)

State of surface	λ[μm]	$\varepsilon_{\lambda,n}$
temperature 2330 K [131]	0.4	0.92
	0.5	0.89
	0.55	0.86
	0.6	0.83
	0.7	0.78
	0.8	0.74
	0.9	0.705
	1.0	0.64
	2.0	0.55
	3.0	0.52
	4.0	0.51
	5.0	0.50
temperature 2480 K [131]	0.3	0.90
	0.4	0.89
	0.5	0.87
	0.7	0.82
	1.0	0.73
	2.0	0.65
	3.0	0.62
	4.0	0.60
	5.0	0.59
Zirconium carbide; ZrC smooth surface, temperature 2100 K [131]	0.6	0.67
	0.8	0.64
	1.0	0.59
	2.0	0.44
	3.0	0.34
	4.0	0.27
smooth surface, temperature 2270 K [131]	0.6	0.62
	0.8	0.58
	1.0	0.56
	2.0	0.43
	3.0	0.37
	4.0	0.34
smooth surface, temperature 2470 K [131]	0.6	0.60
	0.8	0.57
	1.0	0.54
	2.0	0.43
	3.0	0.38
	4.0	0.36
smooth surface, temperature 2670 K [131]	0.6	0.535
	0.8	0.505
	1.0	0.49
	2.0	0.43
	3.0	0.405
	4.0	0.40
Zirconium nitride; ZrN temperature of measurement 1895 K [131]	0.4	0.87
	0.6	0.92
	0.8	0.83
	1.0	0.60
	2.0	0.52
	3.0	0.54
temperature of measurement 1968 K [131]	0.4	0.94
	0.6	0.88
	0.8	0.74
	1.0	0.66
	2.0	0.42
	3.0	0.45
temperature of measurement 2064 K [131]	0.4	0.87
	0.6	0.92
	0.8	0.83
	1.0	0.66
	2.0	0.59
	3.0	0.58
temperature of measurement 2287 K [131]	0.4	0.90
	0.6	0.80
	0.8	0.74
	1.0	0.73
	2.0	0.70
	3.0	0.70
Zirconium oxide; ZrO_2 moulded sample, temperature of measurement 900÷1900 K; $\Delta\varepsilon_{\lambda,n} \cong 0.2$ [131]	2.0	0.21
	3.0	0.24
	4.0	0.30
	5.0	0.49

State of surface	λ[μm]	$\varepsilon_{\lambda,n}$	
	6.0	0.68	
	7.0	0.82	
	8.0	0.89	
	9.0	0.93	
	10.0	0.95	
	15.0	0.93	
[45]	0.3	0.95	
	0.4	0.71	
	0.5	0.50	
	0.6	0.41	
	0.8	0.38	
	1.2	0.39	
	1.6	0.38	
	2.0	0.38	
	2.4	0.40	
	2.7	0.37	
temperature of moulding 3010 K, temperature of measurement 1940 K [131]	1.0	0.19	
	2.0	0.18	
	3.0	0.18	
	4.0	0.20	
	5.0	0.40	
	6.0	0.58	
	8.0	0.79	
	10.0	0.86	
	15.0	0.79	
Zircinium silicate; $ZrSiO_4$ 99% $ZrSiO_4$ [113]	1.0	0.28	
	2.0	0.21	
	3.0	0.24	
	4.0	0.32	
	5.0	0.63	
	6.0	0.88	
	7.0	0.91	
	8.0	0.94	
	9.0	0.96	
	10.0	0.91	
	11.0	0.85	
	12.0	0.89	
	13.0	0.91	
	14.0	0.92	
	15.0	0.92	

Table 3.2.4. Spectral (red) emissivity in normal direction $\varepsilon_{\lambda=0.65,n}$ of inorganic materials

State of surface	T[K]	$\varepsilon_{\lambda=0.65,n}$
Berylium oxide; BeO		
polished surface		
[128, 131]	1400	0.075
	1600	0.10
	1800	0.125
	2000	0.145
	2200	0.16
smooth surface		
[128, 131]	1200	0.21
	1400	0.22
	1600	0.23
	1800	0.24
Boron nitride; BN		
smooth surface		
[131]	1100	0.64
	1300	0.635
	1500	0.63
	1700	0.625
	1900	0.62
Carbon; coal; C		
polished surface		
[24, 56, 63, 72,		
133, 135]	1285	0.789
	1335	0.789
	1385	0.789
	1435	0.789
	1485	0.789
	1535	0.789
	1585	0.789
	1635	0.789
	1685	0.789
	1735	0.789
	1785	0.789
	1835	0.789
	1885	0.789
	1935	0.789
	1985	0.789
	2035	0.789
	2085	0.789
	2135	0.789
smooth surface,		
λ = 0.66 μm		
[56, 63, 72,		
133, 135]	1000	0.989
	1100	0.901
	1200	0.905
	1300	0.908
	1400	0.912
	1500	0.916
	1600	0.919
	1700	0.923
	1800	0.926
	1900	0.930
	2000	0.933
	2100	0.937
rough surface,		
λ = 0.66 μm		
[56, 63, 72,		
133, 135]	1000	0.934
	1100	0.938
	1200	0.942
	1300	0.947
	1400	0.951
	1500	0.955
	1600	0.959
	1700	0.964
	1800	0.968
	1900	0.972
	2000	0.976
	2100	0.980
Ceramics,		
fireproof 9606		
[45]	1100	0.38
	1200	0.415
	1300	0.455
	1400	0.485
	1500	0.525
	1600	0.565

Table 3.2.4. (*continued*)

State of surface	T[K]	$\varepsilon_{\lambda=0.65,n}$
Ceramics,		
fireproof 9608		
[45]	1100	0.47
	1200	0.475
	1300	0.48
	1400	0.484
	1500	0.49
	1600	0.50
Cerium oxide; CeO_2		
density 6.878 g/cm^3		
[131]	1400	0.24
	1600	0.23
	1700	0.225
Chromium nitride; CrN		
[170]	1400	0.59
	1600	0.54
	1800	0.50
	2000	0.46
	2200	0.40
	2300	0.38
Chromium nitride; Cr_2N		
smooth surface		
[131]	1100÷1900	0.69
Chromium silicide; CrSi		
[131]	1073÷1873	0.79
Chromium silicide; $CrSi_2$		
[131]	1073÷1873	0.79
Chromium silicide; Cr_3Si_2		
[131]	1073÷1873	0.79
Cobalt silicide; CoSi		
[131]	1073	0.67
	1173	0.71
	1273	0.75
	1373	0.79
	1473	0.83
	1573	0.86
Electrographite;		
Acheson graphite;		
carbon; C		
rough surface		
[45, 131]	1200÷2000	0.79
$\Delta\varepsilon_{\lambda=0.65,n} = 0.1$		
[131]	1200	0.91
	1600	0.89
	2000	0.87
	2400	0.86
	2800	0.85
	3200	0.84

State of surface	T[K]	$\varepsilon_{\lambda=0.65,n}$
Erbium oxide; Er_2O_3		
[131, 140]	500	0.56
	700	0.58
	900	0.72
	1100	0.76
	1300	0.74
	1500	0.68
	1700	0.63
	1900	0.55
Gadolinium oxide; Gd_2O_3		
polished surface		
[131]	500	0.01
	700	0.01
	900	0.02
	1100	0.03
	1300	0.22
	1500	0.35
	1700	0.47
	1900	0.53
Gadolinium silicide; $GdSi_2$		
[131]	1073	0.80
	1173	0.80
	1273	0.81
	1373	0.81
	1473	0.82
	1573	0.82
	1673	0.82
	1773	0.83
	1873	0.83
Gallium nitride; GaN		
[131]	1100	0.68
	1200	0.73
	1300	0.77
	1400	0.81
	1500	0.85
Graphite; C		
polished surface		
λ = 0.66 µm		
[41, 131]	1000	0.805
	1100	0.806
	1200	0.806
	1300	0.806
	1400	0.806
	1500	0.807
	1600	0.807
	1700	0.807
	1800	0.807
	1900	0.808
	2000	0.808
	2100	0.808

Table 3.2.4. (*continued*)

State of surface	T[K]	$\varepsilon_{\lambda=0.65,n}$
surface ground with		
abrasive paper nr 320,		
λ = 0.66 μm		
[41, 131]	1000	0.873
	1100	0.875
	1200	0.878
	1300	0.880
	1400	0.882
	1500	0.884
	1600	0.887
	1700	0.889
	1800	0.892
	1900	0.894
	2000	0.897
	2100	0.899
surface ground with		
abrasive paper nr 180,		
λ = 0.60 μm		
[41, 131]	1000	0.923
	1100	0.926
	1200	0.929
	1300	0.931
	1400	0.934
	1500	0.937
	1600	0.939
	1700	0.942
	1800	0.945
	1900	0.947
	2000	0.950
	2100	0.953
surface ground with		
abrasive paper nr 80,		
λ = 0.66 μm		
[41, 131]	1000	1.000
	1100	1.000
	1200	1.000
	1300	1.000
	1400	1.000
	1500	1.000
	1600	1.000
	1700	1.000
	1800	1.000
	1900	1.000
	2000	1.000
	2100	1.000
surface oxidized,		
λ = 0.66 μm		
[41, 131]	1000	1.000
	1100	1.000
	1200	1.000
	1300	1.000
	1400	1.000
	1500	1.000
	1600	1.000
	1700	1.000
	1800	1.000
	1900	1.000
	2000	1.000
	2100	1.000
Graphite polycrystalline;		
carbon; C		
polished surface		
[41, 131]	1285	0.769
	1335	0.770
	1385	0.771
	1435	0.772
	1485	0.773
	1535	0.774
	1585	0.775
	1635	0.776
	1685	0.777
	1735	0.778
	1785	0.779
	1835	0.779
	1885	0.780
	1935	0.781
	1985	0.782
	2035	0.783
	2085	0.784
	2135	0.785
surface film,		
rms = 5.08-7.62 μm		
[41, 131]	1000	0.778
	1100	0.778
	1200	0.779
	1300	0.779
	1400	0.779
	1500	0.779
	1600	0.780
	1700	0.780
	1800	0.780
	1900	0.780
	2000	0.780
	2100	0.781
ground surface		
rms = 0.762÷1.78 μm		
[41, 131]	1000	0.813
	1100	0.813
	1200	0.813
	1300	0.813
	1400	0.813
	1500	0.813
	1600	0.813
	1700	0.814

State of surface	T[K]	$\varepsilon_{\lambda=0.65,n}$
	1800	0.814
	1900	0.814
	2000	0.814
	2100	0.814
ground surface, rms = 1.02÷1.32 μm [41, 131, 140]	1000	0.884
	1100	0.884
	1200	0.884
	1300	0.884
	1400	0.884
	1500	0.885
	1600	0.885
	1700	0.885
	1800	0.885
	1900	0.885
	2000	0.885
	2100	0.885
rough surface [41, 131, 140]	1100	0.929
	1200	0.918
	1300	0.905
	1400	0.896
	1500	0.888
	1600	0.883
	1700	0.876
	1800	0.871
	1900	0.868
	2000	0.866
	2100	0.866
α - surface [131]	1000÷2500	0.73÷0.74
c - surface [131]	1400÷2500	0.9
Hafnium carbide; HfC smooth uniform surface [131]	1300	0.73
	1500	0.71
	1700	0.695
	1900	0.68
	2100	0.67
	2300	0.65
	2500	0.63
	2700	0.61
	2900	0.60
	3000	0.59
fine-grained sand [131]	1100÷2100	0.77
Hafnium oxide; HfO_2 [131]	2200	0.71
	2400	0.77
	2600	0.81
Lutetium oxide; Lu_2O_3 [131]	500	0.145
	600	0.145
	700	0.15
	800	0.27
	900	0.47
	1000	0.53
	1200	0.56
	1400	0.565
Magnesium silicide; Mg_2Si [131]	1073	0.67
	1173	0.68
	1273	0.69
Manganese silicide; $MnSi_2$ [131]	1073	0.70
	1173	0.73
	1273	0.76
	1373	0.80
	1473	0.83
Manganese silicide; Mn_2Si [131]	1073	0.68
	1173	0.71
	1273	0.75
	1373	0.78
Molybdenum silicide; $MoSi_2$ [131]	1073÷2273	0.75
Molybdenum silicide; Mo_3Si [131]	1073 1873	0.77
Molybdenum silicide; Mo_5Si_3 [131]	1073÷1973	0.75
Nickel silicide; $NiSi_2$ [131]	1073÷1173	0.67
	1173	0.71
	1273	0.75
	1373	0.78
	1473	0.82
Niobium carbide; NbC $\Delta\varepsilon_{\lambda=0.65,n} = 0.2$ [131]	1200	0.67
	1600	0.65
	2000	0.64
	2400	0.63
	2800	0.61
	3200	0.60

Table 3.2.4. (*continued*)

State of surface	T[K]	$\varepsilon_{\lambda=0.65,n}$
Niobium nitride; NbN smooth surface [10, 83]	1100÷2000	0.83
Niobium nitride; Nb_2N smooth surface [10, 83]	1100÷2000	0.82
Niobium silicide; $NbSi_2$ [131]	1073÷1973	0.80
Periclase; magnesium oxide; MgO moulded powder [131]	1000	0.16
	1200	0.20
	1400	0.24
	1600	0.32
	1800	0.43
Rhenium silicide; $ReSi_2$ [131]	1073	0.70
	1173	0.74
	1273	0.77
	1373	0.80
	1473	0.83
	1573	0.86
	1673	0.89
Samarium oxide; Sm_2O_3 [131]	600	0.24
	700	0.37
	800	0.49
	900	0.56
	1000	0.60
	1200	0.64
	1400	0.65
Scandium nitride; ScN [131]	1100	0.79
	1300	0.805
	1500	0.82
	1700	0.84
	1900	0.86
	2100	0.88
Silicon; Si ground surface [45]	1000	0.65
	1100	0.623
	1200	0.59
	1300	0.567
	1400	0.54
	1500	0.513
	1600	0.485
	1650	0.471
Silite; SiC $\Delta\varepsilon_{\lambda=0.65,n} = 0.2$ [66, 131, 132]	1300	0.95
	1500	0.94
	1700	0.93
	1900	0.92
	2100	0.91
	2300	0.90
	2500	0.88
Tantalum boride [10]	2093	0.7
Tantalum carbide; TaC [10]	2093	0.67
Tantalum silicide; $TaSi_2$ [131]	1073÷2073	0.74
Thorium oxide; ThO_2 [131]	1200	0.13
	1400	0.11
	1600	0.13
	1800	0.20
Titanium carbide; TiC $\Delta\varepsilon_{\lambda=0.65,n} = 0.15$ [131]	1200÷3000	0.9
Titanium nitride; TiN smooth surface [131]	1100÷1900	0.81÷0.79
Titanium silicide; $TiSi_2$ [131]	1073÷1873	0.82
Titanium silicide; Ti_5Si_3 sintered at temperature 1573 K, density 1.93 g/cm^3 [131]	1073÷1873	0.74
Tungsten silicide; W_3Si [131]	1073	0.79
	1273	0.81
	1473	0.82
	1673	0.83
	1873	0.84
	2073	0.85
	2173	0.86
Uranium carbide; UC_2 clean surface		$0.539-0.02T\cdot10^{-3}$
Uranium oxide; UO_2 polished surface [131]	2073÷2372	0.40

Table 3.2.4. (*continued*)

State of surface	T[K]	$\varepsilon_{\lambda=0.65,n}$	
powder [131]	2073÷2373	0.51	
polished surface [131]	3133	0.416	
Vanadium nitride; V_3N smooth surface	1100÷2100	0.82	
Yttrium nitride; YN smooth surface	1100÷2100	0.77	
Yttrium oxide; Y_2O_3 [131]	600÷1000	0.01	
	1200	0.16	
Zirconium boride; ZrB [10]	2093	0.70	
Zirconium boride; ZrB_2 moulded sample [131]	1300	0.86	
	1500	0.83	
	1700	0.80	
	1900	0.77	
	2100	0.74	
	2300	0.71	
Zirconium carbide; ZrC $\Delta\varepsilon_{\lambda=0.65,n} = 0.2$ [131]	1000÷3400	0.8÷0.6	
Zirconium nitride; ZrN [131]	1100÷2300	0.73÷0.76	
Zirconium silicide; $ZrSi_2$ [131]	1073÷2173	0.72	

Table 3.2.5. Bidirectional spectral reflectivity in normal direction $r_{\lambda,n}$ of inorganic materials

State of surface	λ [μm]	$r_{\lambda,n}$
Albite; sodium feldspar; $Na(AlSi_3O_8)$ smooth surface [24]	7.0	0.002
	8.0	0.01
	8.25	0.08
	8.5	0.37
	8.75	0.40
	9.0	0.205
	9.25	0.20
	9.5	0.27
	9.75	0.44
	10.0	0.40
	10.5	0.22
	11.0	0.14
	11.5	0.13
	12.0	0.12
Aluminium antimonide; AlSb synthetic monocrystal, polished surface [152]	20	0.24
	25	0.22
	26	0.205
	27	0.18
	28	0.15
	29	0.09
	30	0.74
	31	0.81
	32	0.62
	34	0.40
	36	0.35
	38	0.32
	40	0.30
	45	0.31
	50	0.30
Aluminium oxide LA 603; Al_2O_3 95% Al_2O_3 [140]	1.0	0.56
	3.0	0.55
	5.0	0.52
	7.0	0.33
	9.0	0.00
	11.0	0.39
	13.0	0.78
Alumino-silicate, potassium; $K(AlSi_3O_8)$ smooth surface [24]	7.0	0.005
	8.0	0.01
	8.25	0.04
	8.50	0.15
	8.75	0.35
	9.0	0.22
	9.5	0.40
	10.0	0.30
	10.5	0.16
	11.0	0.11
Andalusite; alumino-silicate; $Al_2(SiO_4)O$ smooth surface [76]	9.0	0.00
	10.0	0.13
	11.0	0.10
	12.0	0.06
	13.0	0.03
	14.0	0.18
	15.0	0.22
	16.0	0.17
	17.0	0.10
	18.0	0.10
	19.0	0.10
	20.0	0.09

Table 3.2.5. (*continued*)

State of surface	λ[μm]	$r_{\lambda,n}$
Anhydrite;		
calcium sulphate;		
$CaSO_4$		
[24]	1.0÷8.0	0.04
	9.0	0.59
	10.0	0.10
	12.0	0.08
	16.0	0.30
	20.0	0.08
Asphalt		
dry [70]	0.40	0.151
	0.42	0.154
	0.44	0.162
	0.46	0.177
	0.50	0.213
	0.54	0.247
	0.58	0.267
	0.62	0.272
	0.66	0.265
	0.68	0.251
	0.72	0.205
	0.76	0.149
	0.77	0.149
	0.78	0.151
	0.82	0.190
	0.84	0.201
Barite;		
barium sulphate;		
$BaSO_4$		
smooth surface		
[24]	6.0	0.05
	8.0	0.15
	9.0	0.47
	10.0	0.16
	12.0	0.06
	14.0	0.06
	16.0	0.23
	18.0	0.08
	20.0	0.06
Barium fluoride; BaF_2		
polished surface		
[152]	10.0	0.04
	20.0	0.005
	30.0	0.20
	40.0	0.91
	50.0	0.96
	60.0	0.46
	70.0	0 32
	80.0	0.29

State of surface	λ[μm]	$r_{\lambda,n}$
Barium titanite;		
$BaTiO_3$		
polished surface,		
synthetic monocrystal		
[152]	2.0	0.15
	6.0	0.13
	10.0	0.06
	12.0	0.03
	14.0	0.27
	18.0	0.78
	20.0	0.40
	21.0	0.24
	22.0	0.30
	24.0	0.77
	30.0	0.83
	50.0	0.898
	70.0	0.875
	90.0	0.89
	110.0÷170.0	0.90
Bentonite;		
alumino-silicate, hydrated;		
$Al_2O_3 \cdot 3Si_2 \cdot H_2O$		
[76]	8.0	0.02
	9.0	0.25
	9.5	0.50
	10.0	0.10
	10.5	0.62
	11.0	0.18
	12.0	0.07
	13.0	0.58
	14.0	0.18
	16.0	0.08
	18.0	0.01
	20.0	0.52
	21.0	0.35
	22.0	0.51
	24.0	0.16
	25.0	0.40
	26.0	0.72
Bismuth telluride;		
Bi_2Te_3		
synthetic monocrystal,		
smooth surface		
[152]	7.8	0.70
	8.0	0.685
	9.0	0.67
	10.0	0.651
	11.0	0.65
	12.0	0.649
	13.0	0.646
	14.0	0.643
	15.0	0.64

Table 3.2.5. (*continued*)

State of surface	λ[μm]	$r_{\lambda,n}$
Boron nitride; BN smooth surface [131]	0.3	0.45
	0.4	0.62
	0.5	0.74
	0.6	0.89
	0.7	0.75
	0.8	0.74
	0.9	0.76
	1.0	0.80
	1.5	0.82
	2.0	0.60
Boron silicide; B_4Si content of B_4Si-95%, temperature of sintering 1433 K, density 1.32 g/cm^3 [131]	0.2	0.13
	0.3	0.09
	0.4	0.07
	0.6	0.06
	0.8	0.065
	1.0	0.065
	2.0	0.09
	3.0	0.13
Brick, building red, room temperature [105]	0.5	0.36
	1.0	0.45
	1.5	0.72
	2.0	0.70
	3.0	0.22
	4.0	0.53
	4.5	0.40
	5.0	0.20
	6.0	0.14
	7.0	0.10
	8.0	0.09
	9.0	0.10
new [70]	0.40	0.100
	0.45	0.111
	0.50	0.121
	0.52	0.134
	0.54	0.167
	0.56	0.214
	0,58	0.267
	0.60	0.310
	0.62	0.342
	0.64	0.362
	0.65	0.370
Brick, chamotte, white [19, 26, 96]	0.5	0.90
	1.0	0.94
	2.0	0.90
	3.0	0.41
	4.0	0.63
	5.0	0.20
	6.0	0.15
	7.0	0.10
	8.0	0.17
	8.5	0.26
	9.0	0.41
Brick, chamotte, white, glazed [26, 96, 141]	0.5	0.82
	1.0	0.83
	2.0	0.83
	3.0	0.34
	4.0	0.29
	5.0	0.08
	7.0	0.08
	9.0	0.06
Cadmium telluride; CdTe synthetic monocrystal, polished surface [71, 131]	20.0	0.23
	30.0	0.21
	40.0	0.17
	50.0	0.11
	60.0	0.24
	70.0	0.34
	80.0	0.41
	90.0	0.35
	100.0	0.34
	150.0	0.33
	200.0	0.32
	300.0	0.31
	400.0	0.30
Calcite; calcspar; calcium carbonate; $CaCo_3$ monocrystal, polished surface [24, 152]	5.0	0.04
	6.0	0.60
	6.5	0.76
	7.0	0.23
	8.0	0.17
	10.0	0.12
	11.0	0.31
	12.0	0.10
	14.0	0.13

Table 3.2.5. (*continued*)

State of surface	λ[μm]	$r_{\lambda,n}$
	25.0	0.01
	30.0	0.46
	35.0	0.66
	40.0	0.63
	50.0	0.44
	60.0	0.33
Calcium titanate; $CaTiO_3$ synthetic monocrystal, polished surface [152]	40.0	0.64
	200.0	0.80
	400.0	0.68
	500.0	0.17
	600.0	0.83
	800.0	0.47
	1000.0	0.04
	1200.0	0.02
	1300.0	0.08
Cancrynite; $(Na,K)Al_6Si_6 \cdot O_{24} \cdot (CO_3) \cdot 2H_2O$ [76]	9.0	0.04
	10.0	0.25
	12.0	0.06
	14.0	0.08
	18.0	0.06
	22.0	0.09
Carbon; C smooth surface [10, 140]	0.66	0.067
rough surface [10, 140]	0.66	0.090
Carborundum monocrystal; SiC [133]	1.0	0.15
	3.0	0.18
	5.0	0.19
	7.0	0.19
	9.0	0.16
	10.0	0.07
	11.0	0.70
	12.0	0.97
	13.0	0.64
	14.0	0.55
Celestite; strontium sulphate; $SrSO_4$ [24]	4.0÷7.0	0.04
	8.0	0.13
	9.0	0.45
	10.0	0.15
	12.0	0.08
	14.0	0.08
	16.0	0.27
	18.0	0.08
Ceramics, fireproof 9606 [45]	0.3	0.96
	0.4	0.61
	0.5	0.41
	0.6	0.39
	0.7	0.36
	0.8	0.29
	1.0	0.18
	1.2	0.16
	1.6	0.08
	2.0	0.05
	2.4	0.08
	2.7	0.20
Ceramics, fireproof 9608 [45]	0.3	0.92
	0.4	0.16
	0.5	0.11
	0.6	0.12
	0.7	0.13
	0.8	0.12
	1.0	0.15
	1.2	0.16
	1.6	0.22
	2.0	0.34
	2.4	0.52
	2.7	0.73
Cerussite; lead carbonate; $PbCO_3$ [24]	0.6	0.87
	0.95	0.91
	4.4	0.29
	8.8	0.1
Cesium bromide; CsBr [152]	50.0	0.04
	60.0	0.03
	70.0	0.015
	77.0	0.0
	80.0	0.01
	85.0	0.05
	90.0	0.34
	95.0	0.50
	100.0	0.53
	110.0	0.78
	120.0	0.88
	130.0	0.87

Table 3.2.5. (*continued*)

State of surface	λ[μm]	$r_{\lambda,n}$
	140.0	0.58
	150.0	0.43
	160.0	0.37
	170.0	0.34
	180.0	0.32
	190.0	0.30
	200.0	0.29
	210.0	0.285
	220.0	0.28
Cesium iodide; CsJ polished surface [152]	50.0	0.07
	60.0	0.05
	70.0	0.04
	80.0	0.03
	90.0	0.02
	100.0	0.01
	150.0	0.84
	200.0	0.31
Chromium oxide; Cr_2O_3 sintered sample, room temperature of measurement [131]	0.2	0.1
	0.3	0.07
	0.4	0.065
	0.6	0.06
	0.8	0.058
	1.0	0.057
	2.0	0.06
	3.0	0.07
powder [10, 131]	0.54	0.24
	0.6	0.27
	0.95	0.45
	4.4	0.33
	8.0	0.50
Chromium silicide; Cr_3Si 99% Cr_3Si, sintering at temperature 1473 K, density 3.22 g/cm^3, temperature of measurement 298 K [131]	0.3	0.07
	0.4	0.09
	0.5	0.12
	0.6	0.13
	0.8	0.13
	1.0	0.13
	2.0	0.15
99% Cr_3Si, sintering at temperature 1373 K, density 3.18 g/cm^3, temperature of measurement 298 K [131]	0.3	0.12
	0.4	0.16
	0.5	0.20
	0.6	0.23
	0.8	0.24
	1.0	0.245
	2.0	0.32
Clay dry [70]	0.40	0.355
	0.41	0.378
	0.42	0.418
	0.43	0.464
	0.44	0.499
	0.45	0.528
	0.46	0.549
	0.47	0.567
	0.48	0.582
	0.49	0.599
	0.50	0.610
	0.52	0.633
	0.54	0.651
	0.56	0.669
	0.58	0.681
	0.60	0.700
	0.62	0.721
	0.64	0.743
	0.65	0.757
wet [70]	0.40	0.102
	0.42	0.135
	0.44	0.148
	0.46	0.156
	0.48	0.169
	0.50	0.172
	0.52	0.194
	0.54	0.218
	0.56	0.241
	0.58	0.267
	0.60	0.287
	0.62	0.314
	0.64	0.330
	0.65	0.337
	0.66	0.348
	0.68	0.350
	0.70	0.345
	0.72	0.340
	0.75	0.333
	0.76	0.335
	0.78	0.335

State of surface	λ [μm]	$r_{\lambda,n}$
	0.80	0.326
	0.82	0.311
	0.84	0.312
	0.86	0.320
Cobalt carbonate; $CoCO_3$ red [24]	0.6	0.03
	0.95	0.42
	4.4	0.14
	8.8	0.13
Concrete smooth surface [50, 76, 96, 130]	0.5	0.32
	1.0	0.40
	1.5	0.37
	2.0	0.31
	2.5	0.26
	3.0÷6.0	0.07
	6.5	0.23
	6.75	0.38
	7.0	0.33
	7.5	0.22
	8.0	0.17
	9.0	0.19
Conglomerate thick, small, dry [70]	0.40	0.124
	0.44	0.147
	0.48	0.170
	0.52	0.193
	0.54	0.210
	0.56	0.234
	0.58	0.270
	0.60	0.301
	0.62	0.320
	0.64	0.331
Copper oxide; CuO powder [24]	0.95	0.24
Corundum sintered; aluminium oxide; Al_2O_3 sintered sample at temperature of measurement 298 K, $\Delta r_{\lambda,n}$ = 0.2 [62, 131]	0.25	0.30
	0.3	0.40
	0.4	0.65
	0.5	0.78
	0.6	0.83
	0.8	0.85
	1.0	0.87
	2.0	0.90
wet [9]	0.5	0.85
	1.0	0.86
	2.0	0.66
	3.0	0.08
	4.0	0.47
	5.0	0.34
	6.0	0.09
	7.0	0.12
	8.0	0.06
dry [9]	0.5÷3.0	0.90
	4.0	0.81
	5.0	0.69
	6.0	0.48
Disthene; cyanite alumino-silicate; $Al_2(SiO_4)O$ [76]	9.0	0.13
	10.0	0.45
	11.0	0.15
	12.0	0.08
	13.0	0.04
	14.0	0.40
	15.0	0.38
	16.0	0.30
	17.0	0.29
	18.0	0.36
	19.0	0.30
	20.0	0.22
Dolomite; calcium-magnesium carbonate; $CaMg(CO_3)_2$ [24]	5.0	0.03
	6.0	0.19
	8.0	0.06
	10.0	0.04
	11.0	0.20
	12.0	0.06
	14.0	0.10
	16.0	0.05
Flagstone pavement dry [70]	0.40	0.204
	0.42	0.240
	0.44	0.259
	0.46	0.260
	0.48	0.262

Table 3.2.5. (*continued*)

State of surface	λ[μm]	$r_{\lambda,n}$
	0.50	0.290
	0.52	0.327
	0.54	0.353
	0.56	0.367
	0.58	0.397
	0.60	0.420
	0.62	0.429
	0.64	0.420
	0.65	0.422
Fluorite; calcium fluoride; CaF_2 polished surface [24, 152]	10.0	0.02
	20.0	0.11
	30.0	0.78
	35.0	0.93
	40.0	0.61
	60.0	0.27
	80.0	0.23
	300.0	0.20
Galena; galenite; lead glance; lead sulphide; PbS natural monocrystal [152]	20.0	0.34
	30.0	0.23
	40.0	0.12
	50.0	0.61
	60.0	0.80
	70.0	0.84
	80.0	0.81
	90.0	0.77
	100.0	0.77
	150.0	0.80
	200.0	0.76
Gallium antimonide; GaSb synthetic monocrystal, polished surface [152]	25.0	0.23
	30.0	0.20
	32.5	0.09
	35.0	0.46
	37.5	0.70
	40.0	0.43
	45.0	0.36
Gallium arsenide; GaAs synthetic monocrystal, polished surface [152]	20.0÷30.0	0.32

State of surface	λ[μm]	$r_{\lambda,}$
	40.0	0.23
	41.3	0.13
	42.5	0.64
	45.0	0.54
	47.5	0.42
Gallium phosphide; GaP synthetic monocrystal, polished surface [152]	12.0	0.24
	16.0	0.22
	20.0	0.18
	24.0	0.06
	25.0	0.66
	26.0	0.92
	28.0	0.54
	32.0	0.34
	36.0	0.31
	38.0	0.30
Germanium; Ge powder [24]	0.3÷0.4	0.4
	0.4÷0.8	0.53
	0.8÷2.6	0.6
	2.6÷5.0	0.63
synthetic monocrystal [152]	30.0	0.33
	40.0	0.34
	50.0	0.36
	60.0	0.37
	70.0	0.40
	80.0	0.41
	90.0	0.42
	100.0	0.42
	110.0	0.43
	120.0	0.44
	130.0	0.45
	140.0	0.46
	150.0	0.49
	160.0	0.50
	170.0	0.51
	180.0	0.52
	190.0	0.53
	200.0	0.54
	210.0	0.55
Glass, ampouled, aluminium, Corning 1723 [45]	0.32	0.04
	0.4	0.07
	0.8	0.06
	1.2	0.05
	1.6	0.07

Table 3.2.5. (*continued*)

State of surface	λ [μm]	$r_{\lambda,n}$
	2.0	0.07
	2.4	0.06
	2.6	0.04
Glass, arsenic, sulphide; As_2S_3 [152]	2.0	0.24
	5.0	0.25
	10.0	0.17
	15.0	0.15
	20.0	0.12
	25.0	0.09
	30.0	0.29
	32.5	0.38
	35.0	0.35
	40.0	0.28
	45.0	0.20
	50.0	0.005
Glass, borosilicate, Corning 7740 [45]	0.3	0.03
	0.5	0.07
	1.0	0.03
	1.5	0.04
	2.0	0.04
	2.5	0.03
Glass, borosilicate, Corning 7940 thickness of plate 12.7 mm [131]	0.4	0.07
	0.6	0.07
	0.8	0.055
	1.0	0.045
	1.2	0.05
	1.6	0.062
	2.0	0.054
	2.4	0.027
	2.6	0.013
Glass, borosilicate, PPG 3235 [45]	0.3	0.03
	0.4	0.06
	0.5	0.07
	1.0	0.03
	2.0	0.04
	3.0	0.08
Glass, borosilicate, Pyrex 774 [45]	2.8	0.12
	3.0	0.11
	3.2	0.10
	3.6	0.11
	4.0	0.11
	5.0	0.09
	6.0	0.06
	7.0	0.04
	8.0	0.05
	9.0	0.28
	9.1	0.30
	9.5	0.23
	10.0	0.14
	11.0	0.08
	12.0	0.07
Glass, borosilicate, Vicor [131]	0.4	0.07
	0.6	0.07
	1.0	0.045
	1.4	0.055
	1.8	0.055
	2.2	0.047
	2.6	0.013
Glass, bottle, dark green thickness of plate 1.59 mm [45]	2.4	0.03
	2.8	0.05
	3.0	0.04
	3.5	0.03
	4.0	0.03
	4.2	0.03
	4.4	0.03
	4.6	0.03
	5.0	0.03
	6.0	0.03
	8.0	0.04
	9.0	0.15
	9.5	0.22
	10.0	0.18
	11.0	0.13
	12.0	0.07
	12.5	0.07
Glass, bottle, light green [45]	2.4	0.03
	2.8	0.09
	3.0	0.07
	3.5	0.05
	4.0	0.05
	4.2	0.06
	4.4	0.07
	4.6	0.09
	5.0	0.08
	6.0	0.06

Table 3.2.5. (*continued*)

State of surface	λ[μm]	$r_{\lambda,n}$
	8.0	0.04
	9.0	0.15
	9.5	0.22
	10.0	0.18
	11.0	0.13
	12.0	0.07
	12.5	0.07
Glass, lead, oscilograph [45]	2.5	0.05
	3.0	0.10
	3.5	0.10
	4.0	0.10
	4.5	0.10
	5.0	0.10
	6.0	0.07
	7.0	0.06
	8.0	0.05
	9.0	0.19
	9.5	0.30
	10.0	0.26
	11.0	0.16
	12.0	0.10
Glass, metallized, electricity-conductive; Battelle 54726 [45]	0.28	0.04
	0.32	0.07
	0.36	0.10
	0.40	0.18
	0.50	0.10
	0.60	0.17
	0.80	0.08
	1.2	0.05
	1.6	0.068
	2.0	0.11
	2.4	0.21
	2.6	0.26
	2.7	0.23
Glass, metallized, electricity-conductive; LOF 81 E 19778 [45]	0.3	0.30
	0.4	0.78
	0.5	0.79
	0.6	0.79
	0.8	0.72
	1.0	0.65
	1.4	0.63
	1.8	0.57
	2.2	0.49
	2.6	0.24
	2.7	0.13
Glass, metallized, electricity-conductive; LOF PB 19195 [45]	0.3	0.13
	0.4	0.10
	0.5	0.07
	0.6	0.057
	0.8	0.05
	1.2	0.14
	1.6	0.25
	2.0	0.29
	2.4	0.34
	2.7	0.32
Granite - embankment dry [70]	0.40	0.210
	0.42	0.212
	0.44	0.220
	0.46	0.236
	0.50	0.279
	0.54	0.312
	0.58	0.339
	0.62	0.347
	0.66	0.335
	0.70	0.310
	0.74	0.265
	0.78	0.244
	0.80	0.250
	0.83	0.283
	0.85	0.309
Graphite; C polished surface	0.66	0.19
surface ground with abrasive paper No 320 [140]	0.66	0.178
surface ground with abrasive paper No 180 [140]	0.66	0.125
surface ground with abrasive paper No 180 [140]	0.66	0.104
oxidized surface [131, 140]	0.66	0.08
Graphite polycrystalline; carbon; C deposited layer [131, 140]	0.66	0.210
grinded surface [131, 140]	0.66	0.092÷0.160

State of surface	λ[μm]	$r_{\lambda,n}$
polished surface [131, 140]	0.66	0.080
Gypsum mortar [9, 96, 130]	0.5	0.76
	1.0	0.84
	2.0	0.52
	3.0	0.08
	4.0	0.36
	5.0	0.27
	6.0	0.12
	7.0	0.13
	8.0	0.10
	9.0	0.09
Haematite; bloodstone; iron oxide; Fe_2O_3 phase [131]	0.3	0.23
	0.4	0.32
	0.6	0.27
	0.8	0.24
	1.0	0.21
Indium antimonide; InSb synthetic monocrystal, polished surface, temperature of measurement 248 K [45, 152]	20.0	0.62
	30.0	0.66
	45.0	0.74
	48.0	0.76
	51.0	0.80
	54.0	0.11
	57.0	0.12
	60.0	0.34
	75.0	0.64
	90.0	0.70
	105.0	0.75
	120.0	0.79
	135.0	0.81
	150.0	0.83
	165.0	0.84
	180.0	0.82
	195.0	0.78
	200.0	0.77
synthetic monocrystal, polished surface, temperature of measurement 278 K [45, 152]	20.0	0.65
	30.0	0.68
	45.0	0.78
	48.0	0.80
	51.0	0.83
	54.0	0.17
	57.0	0.19
	60.0	0.38
	75.0	0.69
	90.0	0.75
	105.0	0.83
	120.0	0.86
	135.0	0.59
	150.0	0.50
	165.0	0.51
	180.0	0.54
	195.0	0.61
	200.0	0.64
synthetic monocrystal, polished surface, temperature of measurement 317 K [45, 152]	65.0	0.67
	75.0	0.77
	90.0	0.82
	105.0	0.85
	120.0	0.57
	135.0	0.39
	150.0	0.32
	165.0	0.34
	180.0	0.39
	195.0	0.46
	200.0	0.52
synthetic monocrystal, polished surface, temperature of measurement 372 K [45, 152]	20.0	0.69
	30.0	0.75
	45.0	0.85
	48.0	0.87
	51.0	0.91
	54.0	0.24
	57.0	0.44
	60.0	0.63
	75.0	0.85
	90.0	0.49
	105.0	0.31
	120.0	0.22
	135.0	0.20
	150.0	0.22
	165.0	0.24
	180.0	0.29
	195.0	0.34
	200.0	0.36

Table 3.2.5. (*continued*)

State of surface	λ[μm]	$r_{\lambda,n}$
synthetic monocrystal, polished surface, temperature of measurement 404 K [45, 152]	66.0	0.78
	69.0	0.85
	72.0	0.84
	75.0	0.66
	90.0	0.36
	105.0	0.25
	120.0	0.21
Indium arsenide; InAs synthetic monocrystal, polished surface [152]	24.0	0.28
	28.0	0.27
	30.0	0.27
	34.0	0.26
	38.0	0.20
	40.0	0.20
	42.0	0.31
	44.0	0.78
	46.0	0.85
	48.0	0.65
Indium phosphide; InP synthetic monocrystal, polished surface [152]	24.0	0.16
	26.0	0.11
	28.0	0.09
	30.0	0.71
	32.0	0.71
	34.0	0.48
	36.0	0.37
	38.0	0.28
	40.0	0.24
	45.0	0.23
Lead chloride; $PbCl_2$ polished surface [152]	90.0	0.58
	100.0	0.50
	120.0	0.53
	140.0	0.50
	160.0	0.47
	180.0	0.49
	200.0	0.52
	220.0	0.53
Lead chromate (chrome yellow); lead chrome; $PbCrO_4$ [24]	0.6	0.70
	4.4	0.41
	8.8	0.05
Lead oxide, yellow; PbO crystal powder [24]	0.6	0.52
	4.4	0.51
	8.8	0.26
Lead selenide; PbSe synthetic monocrystal [152]	30.0	0.23
	40.0	0.16
	50.0	0.19
	60.0	0.42
	70.0	0.61
	80.0	0.66
	90.0	0.63
	100.0	0.61
	120.0	0.62
	150.0	0.60
	200.0	0.53
Lime, burnt; quicklime; calcium oxide; CaO [24]	0.6	0.85
	4.4	0.22
	8.8	0.04
Limestone; calcium carbonate; $CaCO_3$ dry [70]	0.40	0.223
	0.42	0.269
	0.44	0.308
	0.46	0.342
	0.48	0.375
	0.50	0.410
	0.52	0.457
	0.54	0.500
	0.56	0.541
	0.58	0.571
	0.60	0.597
	0.62	0.615
	0.64	0.631
	0.65	0.639
Lithium fluoride; LiF polished surface [152]	13.0	0.03
	14.0	0.17
	15.0	0.63
	17.5	0.76
	20.0	0.72
	22.5	0.86
	25.0	0.92
	27.5	0.89
	30.0	0.83
	35.0	0.61

Table 3.2.5. (*continued*)

State of surface	λ[μm]	$r_{\lambda,n}$
	40.0	0.36
	50.0	0.28
Loam from the bottom of the drain dry [70]	0.40	0.200
	0.42	0.193
	0.44	0.193
	0.46	0.210
	0.48	0.224
	0.49	0.220
	0.50	0.210
	0.52	0.174
	0.54	0.174
	0.56	0.188
	0.58	0.188
	0.60	0.193
	0.62	0.198
	0.64	0.225
	0.65	0.250
	0.66	0.268
	0.67	0.295
	0.68	0.323
	0.69	0.350
	0.70	0.382
	0.71	0.415
	0.72	0.435
	0.73	0.470
	0.74	0.490
	0.75	0.510
	0.76	0.528
	0.78	0.538
	0.80	0.539
	0.82	0.538
	0.84	0.540
	0.85	0.544
Magnesite; magnesium carbonate; $MgCO_3$ powder [24]	0.6	0.85
	0.95	0.89
	4.4	0.11
	8.8	0.04
Magnesium fluoride; MgF_2 polished surface [152]	2.0	0.03
	5.0	0.02
	7.5	0.015
	10.0	0.00
	12.0	0.00
	15.0	0.01
	20.0	0.27
	22.5	0.46
	25.0	0.52
	27.5	0.27
	30.0	0.14
	32.5	0.00
	35.0	0.33
	37.5	0.47
	40.0	0.61
	42.5	0.36
	45.0	0.20
	50.0	0.14
Manganese carbonate; $MnCO_3$ [24]	6.0	0.04
	6.5	0.39
	7.0	0.28
	8.0	0.12
	10.0	0.09
	11.0	0.08
	11.5	0.18
	12.0	0.10
	14.0	0.10
	16.0	0.07
Manganese oxide; MnO moulded, granularity d = 53÷63 μm [131]	0.25	0.10
	0.3	0.10
	0.4	0.09
	0.6	0.14
	0.8	0.21
	1.0	0.27
	1.5	0.62
	2.0	0.70
Molybdenum oxide (molybdenum anhydride); molybdenum trioxide; MoO_3 sintered at temperature 1723 K [131]	0.25	0.13
	0.3	0.12
	0.35	0.11
	0.4	0.16
	0.5	0.58
	0.6	0.61
	0.7	0.48
	0.8	0.43
	0.9	0.43
	1.0	0.45

Table 3.2.5. (*continued*)

State of surface	λ[μm]	$r_{\lambda,n}$
	1.5	0.69
	2.0	0.90
	2.5	0.96
Molybdenum silicide; $MoSi_2$ [131]	0.25	0.23
	0.3	0.24
	0.4	0.24
	0.6	0.23
	0.8	0.22
	1.0	0.21
	1.5	0.19
	2.0	0.18
	2.5	0.15
Nephelite; nepheline; sodium aluminosilicate; $Na(AlSiO_4)$ [76]	8.0	0.01
	10.0	0.49
	12.0	0.10
	14.0	0.08
	16.0	0.09
	18.0	0.04
	20.0	0.09
	22.0	0.20
	24.0	0.21
Nickel oxide; NiO moulded powder, sintered at temperature 1673 K, density 5.28 g/cm^3 [131]	0.25	0.09
	0.3	0.10
	0.4	0.10
	0.6	0.12
	0.8	0.17
	1.0	0.15
	2.0	0.28
	3.0	0.39
Nickel sulphide; NiS [24]	4.0÷8.0	0.04
	9.0	0.22
	10.0	0.06
	12.0	0.07
	14.0	0.07
	16.0	0.14
	18.0	0.08
Olivine; magnesium-iron silicate; $(Mg,Fe)_2SiO_4$ [76]	8.0	0.01
	10.0	0.07
	11.0	0.20
	12.0	0.09
	14.0	0.03
	16.0	0.07
	18.0	0.12
	20.0	0.12
	22.0	0.22
	24.0	0.31
Pavement dry [70]	0.40	0.115
	0.44	0.136
	0.48	0.158
	0.52	0.177
	0.56	0.191
	0.60	0.201
	0.64	0.204
	0.68	0.195
	0.72	0.176
	0.76	0.168
	0.80	0.179
	0.83	0.196
	0.85	0.199
Periclase; magnesium oxide; MgO moulded plate, temperature of measurement 298 K [40, 41, 131]	0.3	0.95
	0.4	0.96
	0.6	0.97
	0.8	0.97
	1.0	0.96
	2.0	0.91
	3.0	0.70
	4.0	0.74
	6.0	0.43
	8.0	0.48
	10.0	0.22
	15.0	0.77
	20.0	1.00
	30.0	0.46

Table 3.2.5. (*continued*)

State of surface	λ[μm]	$r_{\lambda,n}$
polished surface monocrystal [152]	3.0	0.05
	4.0	0.04
	6.0	0.03
	8.0	0.02
	10.0	0.01
	12.0	0.00
	13.0	0.05
	14.0	0.64
	15.0	0.73
	16.0	0.85
	18.0	0.92
	20.0	0.94
	22.0	0.89
	23.0	0.81
	24.0	0.77
	25.0	0.48
Plaster smooth surface [19, 44, 96]	0.5	0.76
	1.0	0.89
	2.0	0.59
	3.0	0.08
	3.5	0.15
	4.0	0.20
	4.5	0.11
	5.0	0.18
	6.0	0.13
	7.0	0.18
	8.0	0.15
	8.5	0.98
	9.0	0.82
Platinum black electrolytic [24]	0.94	0.0085÷0.0122
	4.4	0.014÷0.017
	8.8	0.010÷0.036
chemically plated [24]	0.6	0.02
	0.95	0.03
Porcelain white [96, 130]	0.5	0.54
	1.0	0.34
	2.0	0.34
	3.0	0.22
	3.5	0.26
	4.0	0.23
	5.0	0.07
	6.0	0.07
	8.0	0.11
	9.0	0.52
Potassium bromide; KBr polished surface [24, 152]	50.0	0.00
	60.0	0.10
	70.0	0.59
	80.0	0.79
	90.0	0.44
	100.0	0.28
	120.0	0.20
Potassium chloride; KCl polished surface [24, 152]	40.0	0.02
	50.0	0.20
	60.0	0.87
	65.0	0.89
	70.0	0.76
	80.0	0.33
	90.0	0.24
	100.0	0.22
Potassium iodide; KJ polished surface [24, 152]	60.0	0.00
	70.0	0.07
	80.0	0.38
	90.0	0.76
	100.0	0.56
	120.0	0.23
	150.0	0.20
Quartz, crystalline; SiO_2 [152]	7.9	0.02
	8.0	0.11
	8.25	0.60
	8.50	0.77
	8.60	0.47
	8.70	0.78
	8.80	0.80
	9.0	0.86
	9.5	0.46
	10.0	0.18
	11.0	0.095
	12.0	0.04
	12.25	0.01
	12.50	0.40
	12.75	0.20
	13.0	0.13
	13.5	0.095
	14.0	0.07
	14.5	0.09

Table 3.2.5. (*continued*)

State of surface	λ[μm]	$r_{\lambda,n}$
	15.0	0.06
	15.5	0.045
	16.0	0.04
Quartz melted smooth surface, thickness of plate 1 mm [131]	2.0	0.04
	3.0	0.035
	4.0	0.03
	7.0	0.015
	8.0	0.52
	9.0	0.63
	9.5	0.72
	10.0	0.19
	15.0	0.08
	20.0	0.51
	21.0	0.92
	22.5	0.295
	25.0	0.53
	30.0	0.19
	40.0	0.14
	50.0	0.00
Road in winter soggy, yellowed after rain [70]	0.40	0.626
	0.42	0.651
	0.44	0.669
	0.46	0.679
	0.48	0.684
	0.50	0.697
	0.52	0.718
	0.54	0.749
	0.56	0.773
	0.58	0.781
	0.60	0.781
	0.62	0.774
	0.64	0.774
	0.65	0.790
Road, soil-surfaced dry, strongly rammed, on black earth [70]	0.40	0.061
	0.42	0.064
	0.44	0.068
	0.46	0.066
	0.48	0.063
	0.50	0.070
	0.52	0.082
	0.54	0.083
	0.56	0.081
	0.58	0.087
	0.60	0.088
	0.62	0.090
	0.65	0.093
dry, rammed, sandy, under cloudy sky [70]	0.40	0.056
	0.44	0.061
	0.48	0.071
	0.52	0.076
	0.56	0.089
	0.60	0.097
	0.62	0.102
	0.64	0.113
	0.66	0.126
	0.68	0.133
	0.72	0.140
	0.76	0.138
	0.80	0.145
	0.83	0.171
dry, rammed, sandy [70]	0.40	0.091
	0.44	0.100
	0.48	0.117
	0.52	0.129
	0.56	0.141
	0.60	0.153
	0.64	0.175
	0.68	0.185
	0.72	0.178
	0.76	0.178
	0.80	0.180
	0.83	0.191
dry, rammed on mugwort [70]	0.40	0.081
	0.44	0.076
	0.46	0.076
	0.50	0.087
	0.54	0.099
	0.58	0.108
	0.62	0.112
	0.65	0.113
	0.68	0.149
	0.70	0.161
	0.74	0.156
	0.75	0.167
	0.78	0.202
	0.80	0.223
	0.82	0.242
	0.84	0.257

Table 3.2.5. (*continued*)

State of surface	λ[μm]	$r_{\lambda,n}$
dry, rammed on lixiviated black earth [70]	0.40	0.072
	0.44	0.092
	0.48	0.093
	0.50	0.105
	0.52	0.119
	0.54	0.127
	0.58	0.131
	0.60	0.141
	0.64	0.150
	0.68	0.165
	0.72	0.190
	0.74	0.213
	0.76	0.249
	0.78	0.291
	0.80	0.320
	0.82	0.333
	0.84	0.340
dry, rammed on chestnut soil [70]	0.40	0.154
	0.44	0.183
	0.45	0.188
	0.46	0.188
	0.48	0.182
	0.50	0.187
	0.54	0.191
	0.56	0.190
	0.58	0.186
	0.60	0.178
	0.62	0.165
	0.65	0.141
dry, rammed on submugwort soil [70]	0.40	0.121
	0.44	0.140
	0.48	0.156
	0.50	0.161
	0.52	0.170
	0.54	0.188
	0.56	0.206
	0.58	0.217
	0.60	0.220
	0.62	0.218
	0.65	0.209
	0.66	0.220
	0.68	0.262
	0.70	0.279
	0.74	0.289
	0.78	0.302
	0.80	0.312
	0.82	0.330
	0.85	0.361
wet, muddy, on submugwort soil [70]	0.40	0.027
	0.44	0.028
	0.46	0.033
	0.48	0.036
	0.50	0.038
	0.52	0.040
	0.54	0.047
	0.56	0.055
	0.58	0.066
	0.60	0.073
	0.62	0.070
	0.65	0.049
dry, covered with a layer of loess dust [70]	0.40	0.136
	0.42	0.141
	0.44	0.142
	0.46	0.142
	0.48	0.147
	0.52	0.160
	0.56	0.168
	0.58	0.177
	0.60	0.185
	0.62	0.187
	0.64	0.182
	0.66	0.183
	0.67	0.208
	0.68	0.237
	0.69	0.273
	0.70	0.338
	0.71	0.400
	0.72	0.455
	0.73	0.505
	0.74	0.554
	0.75	0.595
	0.76	0.620
	0.77	0.625
	0.78	0.628
	0.80	0.627
	0.82	0.621
	0.85	0.616
Rock open to the view [70]	0.40	0.075
	0.42	0.089
	0.44	0.100
	0.46	0.100
	0.48	0.094
	0.50	0.096

Table 3.2.5. (*continued*)

State of surface	λ[μm]	$r_{\lambda,n}$
	0.54	0.108
	0.58	0.110
	0.60	0.111
	0.62	0.110
	0.65	0.088
Salt, domestic; table salt; common salt; sodium chloride; NaCl polished surface [24, 135]	35.0	0.05
	40.0	0.24
	50.0	0.88
	60.0	0.80
	70.0	0.36
	80.0	0.28
	90.0	0.24
	100.0	0.23
Sand [24]	1.0	0.40
	1.5	0.47
	2.0	0.50
	3.0	0.54
from a river, wet [70]	0.40	0.080
	0.42	0.095
	0.44	0.108
	0.46	0.118
	0.48	0.130
	0.50	0.145
	0.52	0.162
	0.54	0.179
	0.56	0.191
	0.58	0.202
	0.60	0.220
	0.62	0.240
	0.64	0.260
	0.65	0.280
	0.66	0.282
	0.68	0.317
	0.70	0.347
	0.71	0.371
	0.72	0.380
	0.74	0.394
	0.76	0.406
	0.80	0.434
	0.82	0.471
	0.84	0.535
	0.86	0.590
	0.87	0.608
from a river, dry [70]	0.40	0.148
	0.45	0.169
	0.50	0.190
	0.52	0.200
	0.54	0.215
	0.56	0.231
	0.58	0.241
	0.60	0.246
	0.62	0.249
	0.64	0.250
	0.65	0.250
	0.66	0.252
	0.68	0.276
	0.70	0.291
	0.72	0.322
	0.74	0.343
	0.76	0.359
	0.80	0.392
	0.84	0.440
from a river, dry [70]	0.40	0.105
	0.45	0.114
	0.50	0.132
	0.55	0.145
	0.60	0.160
	0.65	0.181
	0.66	0.181
	0.68	0.202
	0.70	0.235
	0.72	0.270
	0.74	0.285
	0.76	0.281
	0.77	0.284
	0.78	0.308
	0.80	0.348
	0.82	0.362
	0.84	0.410
from the Florida [24]	0.3÷0.4	0.15
	0.4÷0.8	0.40
	0.8÷2.6	0.50
	2.6÷5.0	0.30
from the Main [24]	0.3÷0.4	0.08
	0.4÷0.8	0.25
	0.8÷2.6	0.33
	2.6÷5.0	0.31
	5.0	0.48

Table 3.2.5. (*continued*)

State of surface	λ[μm]	$r_{\lambda,n}$
with stones, wet [70]	0.40	0.078
	0.45	0.084
	0.50	0.093
	0.55	0.113
	0.56	0.117
	0.58	0.121
	0.60	0.117
	0.62	0.109
	0.64	0.121
	0.65	0.128
	0.66	0.126
	0.68	0.118
	0.70	0.099
	0.71	0.098
	0.72	0.112
	0.74	0.157
	0.76	0.191
	0.78	0.220
	0.80	0.235
	0.82	0.244
	0.84	0.256
	0.85	0.258
dry [70]	0.40	0.179
	0.44	0.202
	0.48	0.224
	0.52	0.256
	0.54	0.273
	0.58	0.301
	0.62	0.327
	0.65	0.349
dry, light-grey [70]	0.40	0.190
	0.42	0.239
	0.44	0.262
	0.50	0.309
	0.55	0.352
	0.60	0.398
	0.64	0.428
	0.65	0.430
Sand on the dunes with clearly outlined undulations without shadows [70]	0.40	0.156
	0.44	0.159
	0.48	0.184
	0.50	0.211
	0.52	0.240
	0.54	0.256
	0.56	0.261
	0.58	0.259
	0.60	0.256
	0.62	0.257
	0.64	0.264
	0.66	0.278
	0.68	0.323
	0.70	0.364
	0.72	0.406
	0.74	0.448
	0.76	0.488
	0.78	0.527
	0.80	0.559
	0.82	0.591
	0.84	0.620
	0.86	0.648
	0.87	0.661
the shadows are transverse with respect to undulations [70]	0.40	0.125
	0.42	0.142
	0.44	0.160
	0.46	0.180
	0.48	0.203
	0.50	0.234
	0.52	0.265
	0.54	0.285
	0.56	0.300
	0.58	0.308
	0.60	0.311
	0.62	0.309
	0.64	0.302
	0.66	0.295
	0.68	0.287
	0.70	0.283
	0.72	0.281
	0.76	0.280
	0.80	0.280
	0.84	0.280
Sandstone brick-red [70]	0.40	0.117
	0.42	0.125
	0.44	0.130
	0.46	0.146
	0.48	0.156
	0.50	0.178
	0.52	0.185
	0.54	0.211
	0.56	0.250
	0.58	0.284
	0.60	0.280
	0.62	0.329
	0.64	0.371
	0.65	0.368

Table 3.2.5. (*continued*)

State of surface	λ[μm]	$r_{\lambda,n}$
light-grey [70]	0.40	0.400
	0.42	0.430
	0.44	0.470
	0.46	0.515
	0.48	0.557
	0.50	0.578
	0.52	0.588
	0.54	0.598
	0.56	0.610
	0.58	0.625
	0.60	0.640
	0.62	0.652
	0.64	0.655
	0.65	0.655
Sapphire; aluminium oxide; Al_2O_3 natural or synthetic monocrystal [152]	10.0	0.0
	11.0	0.63
	12.0	0.83
	14.0	0.91
	15.0	0.88
	16.0	0.62
	16.2	0.75
	17.0	0.58
	17.6	0.72
	18.0	0.57
	19.0	0.33
	20.0	0.18
	20.5	0.07
	21.0	0.87
	22.0	0.82
	22.5	0.65
	23.0	0.72
	24.0	0.49
	25.0	0.41
	26.0	0.33
	26.2	0.52
	27.0	0.40
	28.0	0.37
	29.0	0.35
Scheelite; calcium tungstate; $CaWO_4$ polished surface [152]	1.0÷5.0	0.08
	10.0	0.02
	12.5	0.91
	15.0	0.18
	20.0	0.07
	22.5	0.36
	25.0	0.09
	30.0	0.06
	35.0	0.78
	50.0	0.40
	90.0	0.30
	130.0	0.29
Selenium; Se crystalline [24]	1.0	0.31
	2.0	0.28
	4.0	0.28
	6.0	0.29
	8.0	0.29
	10.0	0.295
	12.0	0.295
	14.0	0.31
Siderite; ironstone; spathic iron; iron carbonate; $FeCO_3$ [24]	6.0	0.16
	6.5	0.34
	7.0	0.19
	8.0	0.12
	10.0	0.25
	11.0	0.25
	12.0	0.09
	14.0	0.09
	16.0	0.07
Silica polycrystalline; silicon dioxide; SiO_2 sintered [131]	0.2	0.15
	0.3	0.13
	0.4	0.73
	0.5	0.84
	0.6	0.94
	0.8	0.97
	1.0	0.99
Silicon; Si synthetic monocrystal polished surface, temperature 298 K [152]	20.0	0.28
	50.0	0.33
	80.0	0.29
	100.0	0.30
	150.0	0.39
	200.0	0.42

Table 3.2.5. (*continued*)

State of surface	$\lambda[\mu m]$	$r_{\lambda,n}$
Silicon nitride; Si_3N_4 smooth surface, density 1.82 g/cm^3 [131]	0.2	0.19
	0.3	0.28
	0.4	0.27
	0.5	0.33
	0.6	0.32
	0.8	0.29
	1.0	0.27
	1.5	0.23
	2.0	0.19
	3.0	0.18
Silite; SiC polished surface [152]	2.0	0.20
	4.0	0.18
	6.C	0.16
	8.0	0.12
	10.0	0.00
	11.0	0.96
	12.0	0.98
	14.0	0.55
	16.0	0.34
	18.0	0.31
	22.0	0.30
Smithsonite; zinc spar; zinc carbonate; $ZnCO_3$ [24]	5.0	0.02
	6.0	0.10
	6.5	0.33
	7.0	0.17
	8.0	0.11
	10.0	0.06
	11.0	0.04
	11.5	0.18
	12.0	0.05
	14.0	0.09
	16.0	0.07
Snow recent thick (obstructive) [24]	0.3÷0.4	0.35
	0.4÷0.8	0.40
	0.8÷2.6	0.15
	2.6÷5.0	0.18
	5.0	0.26
recent [70]	0.40	0.830
	0.45	0.820
	0.50	0.800
	0.55	0.780
	0.60	0.755
	0.65	0.735
	0.70	0.710
	0.75	0.685
	0.80	0.655
	0.85	0.620
	0.90	0.570
covered with ice [70]	0.40	0.720
	0.50	0.740
	0.60	0.755
	0.70	0.760
	0.80	0.760
	0.90	0.750
dry, slightly frozen [70]	0.40	0.520
	0.45	0.470
	0.50	0.470
	0.55	0.470
	0.60	0.480
	0.65	0.510
Sodalite; $Na_8(AlSiO_4)_6Cl$ [76]	9.0	0.08
	10.0	0.80
	12.0	0.09
	14.0	0.10
	16.0	0.07
	18.0	0.03
	20.0	0.15
	21.0	0.42
	22.0	0.28
	23.0	0.36
	24.0	0.22
Sodium fluoride; NaF polished surface [152]	20.0	0.04
	25.0	0.56
	26.0	0.47
	30.0	0.86
	35.0	0.93
	40.0	0.75
	45.0	0.37
	50.0	0.29

Table 3.2.5. (*continued*)

State of surface	λ[μm]	$r_{\lambda,n}$
Soil in a mountain area in the form of dry rubble [70]	0.40	0.079
	0.45	0.087
	0.50	0.098
	0.52	0.109
	0.54	0.130
	0.56	0.142
	0.58	0.146
	0.60	0.150
	0.62	0.160
	0.63	0.166
	0.64	0.160
	0.65	0.150
in the tundra, in the form of dry rubble on a mountain slope [70]	0.40	0.152
	0.42	0.156
	0.44	0.161
	0.46	0.170
	0.48	0.170
	0.50	0.178
	0.52	0.197
	0.54	0.205
	0.56	0.209
	0.58	0.218
	0.60	0.223
	0.62	0.221
	0.64	0.210
	0.65	0.200
in the tundra, partly shaded, in the form of rubble on a mountain slope [70]	0.40	0.051
	0.42	0.060
	0.45	0.064
	0.50	0.065
	0.55	0.073
	0.60	0.082
	0.62	0.081
	0.64	0.072
	0.65	0.065
on the top of a hill [70]	0.40	0.083
	0.44	0.088
	0.48	0.096
	0.52	0.119
	0.56	0.140
	0.60	0.154

State of surface	λ[μm]	$r_{\lambda,n}$
	0.64	0.167
	0.65	0.173
	0.68	0.195
	0.69	0.220
	0.70	0.242
	0.71	0.287
	0.72	0.322
	0.73	0.344
	0.74	0.353
	0.76	0.376
	0.77	0.442
	0.78	0.515
	0.79	0.544
	0.80	0.560
Soil, clay wet [70]	0.40	0.025
	0.42	0.032
	0.44	0.038
	0.46	0.039
	0.50	0.039
	0.55	0.039
	0.60	0.039
	0.65	0.039
	0.70	0.044
	0.75	0.048
	0.80	0.050
	0.81	0.050
Soil-lixivited black earth damp [70]	0.40	0.054
	0.42	0.058
	0.44	0.067
	0.46	0.073
	0.48	0.073
	0.50	0.079
	0.55	0.086
	0.60	0.091
	0.65	0.103
	0.70	0.119
	0.75	0.141
	0.80	0.162
	0.83	0.188
Soil, marshy very damp [70]	0.40	0.032
	0.45	0.041
	0.50	0.044
	0.52	0.050
	0.54	0.059
	0.56	0.063
	0.58	0.064
	0.60	0.061

Table 3.2.5. (*continued*)

State of surface	λ[μm]	$r_{\lambda,n}$
	0.62	0.064
	0.63	0.071
	0.64	0.081
	0.66	0.092
	0.68	0.097
	0.70	0.095
	0.72	0.085
	0.73	0.087
	0.74	0.095
	0.76	0.121
Soil, mugwort soggy [70]	0.40	0.050
	0.42	0.051
	0.44	0.056
	0.46	0.059
	0.48	0.060
	0.50	0.060
	0.52	0.054
	0.54	0.061
	0.56	0.076
	0.58	0.077
	0.60	0.083
	0.61	0.087
	0.62	0.085
	0.65	0.076
damp [70]	0.40	0.038
	0.44	0.047
	0.48	0.051
	0.52	0.051
	0.56	0.059
	0.60	0.069
	0.64	0.077
	0.68	0.091
	0.72	0.115
	0.76	0.132
	0.78	0.141
	0.80	0.157
	0.82	0.182
	0.84	0.224
	0.85	0.250
grey, dry [70]	0.40	0.037
	0.42	0.041
	0.44	0.041
	0.46	0.035
	0.48	0.038
	0.50	0.043
	0.52	0.049
	0.54	0.049
	0.56	0.050
	0.58	0.059
	0.60	0.060
	0.62	0.060
	0.64	0.060
	0.66	0.061
	0.68	0.069
	0.70	0.079
	0.72	0.089
	0.74	0.103
	0.76	0.155
	0.78	0.212
	0.80	0.228
	0.82	0.230
	0.84	0.225
Soil, sandy damp [70]	0.40	0.029
	0.42	0.030
	0.44	0.036
	0.46	0.041
	0.50	0.044
	0.54	0.050
	0.58	0.060
	0.62	0.069
	0.66	0.078
	0.70	0.074
	0.74	0.058
	0.78	0.060
	0.82	0.073
	0.85	0.077
dry [70]	0.40	0.094
	0.44	0.118
	0.48	0.141
	0.52	0.143
	0.56	0.150
	0.60	0.163
	0.65	0.189
Soil, strong black earth damp [70]	0.40	0.016
	0.44	0.017
	0.48	0.020
	0.52	0.020
	0.56	0.020
	0.60	0.021
	0.65	0.025
dry [70]	0.40	0.024
	0.44	0.023
	0.48	0.025
	0.52	0.026
	0.56	0.030
	0.60	0.030
	0.64	0.036
	0.68	0.042

Table 3.2.5. (*continued*)

State of surface	λ[μm]	$r_{\lambda,n}$
	0.72	0.048
	0.76	0.054
	0.80	0.059
	0.84	0.071
Sphalerite; zinc blende; zinc sulphide; ZnS polished surface [152]	2.0	0.12
	5.0	0.20
	10.0	0.21
	15.0	0.11
	20.0	0.11
	25.0	0.04
	27.5	0.01
	30.0	0.83
	35.0	0.31
	40.0	0.43
	45.0	0.32
	50.0	0.20
Stone dry [70]	0.40	0.164
	0.44	0.195
	0.48	0.217
	0.50	0.220
	0.54	0.220
	0.58	0.219
	0.60	0.227
	0.62	0.235
	0.65	0.234
soggy [70]	0.40	0.059
	0.45	0.061
	0.50	0.072
	0.52	0.081
	0.54	0.088
	0.56	0.091
	0.58	0.092
	0.60	0.090
	0.62	0.089
	0.65	0.074
dry road in the tundra [70]	0.40	0.106
	0.44	0.116
	0.46	0.125
	0.48	0.140
	0.50	0.160
	0.54	0.191
	0.58	0.220
	0.62	0.229
	0.65	0.232
	0.67	0.278
	0.69	0.315
	0.71	0.322
	0.72	0.325
	0.74	0.330
	0.76	0.330
	0.78	0.340
	0.80	0.346
	0.82	0.355
	0.84	0.352
	0.86	0.342
	0.88	0.355
	0.90	0.364
dry road in the northern forest zone, USSR [70]	0.40	0.172
	0.42	0.215
	0.44	0.241
	0.46	0.257
	0.48	0.270
	0.50	0.272
	0.52	0.288
	0.54	0.310
	0.56	0.321
	0.58	0.326
	0.60	0.341
	0.62	0.356
	0.64	0.376
	0.65	0.386
	0.66	0.395
	0.68	0.404
	0.70	0.401
	0.72	0.391
	0.74	0.381
	0.76	0.386
	0.78	0.410
	0.82	0.463
	0.85	0.501
Stone, lining white [96]	0.5	0.60
	1.0	0.57
	2.0	0.57
	3.0	0.40
	3.5	0.47
	4.0	0.39
	5.0	0.08
	6.0	0.09
	7.0	0.04
	7.5	0.03
	8.0	0.10
	9.0	0.40

State of surface	λ[μm]	$r_{\lambda,n}$
Stone, cobble; cobble stone; dry [70]	0.40	0.100
	0.42	0.123
	0.44	0.145
	0.46	0.161
	0.48	0.172
	0.50	0.182
	0.52	0.209
	0.53	0.211
	0.54	0.202
	0.56	0.174
	0.58	0.182
	0.60	0.200
	0.62	0.204
	0.64	0.217
	0.65	0.222
Strontium fluoride; SrF_2 polished surface [152]	10.0	0.02
	20.0	0.00
	30.0	0.76
	40.0	0.96
	50.0	0.47
	60.0	0.30
	70.0	0.25
	80.0	0.24
Strontium titanate; $SrTiO_3$ synthetic monocrystal, polished surface [71, 152]	2.0	0.14
	6.0	0.11
	10.0	0.03
	11.0	0.01
	12.0	0.03
	14.0	0.86
	16.0	0.87
	18.0	0.70
	20.0	0.14
	21.0	0.11
	22.0	0.39
	26.0	0.90
	30.0	0.93
	50.0	0.96
	55.0	0.79
	70.0	0.95
	90.0	0.945
Sulphur crystalline; S_8 polished surface [24, 152]	3.0	0.08
	6.0	0.10
	9.0	0.10
	12.0	0.07
	14.0	0.07
Tantalum oxide; Ta_2O_5 sintered at temperature 1673 K, density 6.51 g/cm^3 [131]	0.2	0.15
	0.3	0.19
	0.4	0.89
	0.5	0.99
Tantalum silicide; $TaSi_2$ temperature of measurement 298 K, 95% $TaSi_2$ sintered at temperature 1773 K, density 4.78 g/cm^3 [131]	0.3	0.10
	0.4	0.11
	0.6	0.17
	0.8	0.20
	1.0	0.205
	2.0	0.26
	3.0	0.40
Tellurium; Te [64]	0.6	0.49
	0.8	0.48
	1.0	0.50
	2.0	0.52
	4.0	0.57
	7.0	0.68
Thallium bromide; TlBr polished surface [152]	50.0	0.08
	60.0	0.05
	70.0	0.02
	80.0	0.07
	90.0	0.32
	100.0	0.61
	150.0	0.87
	200.0	0.87
	250.0	0.65
	300.0	0.59
	350.0	0.58
	400.0	0.57

Table 3.2.5. (*continued*)

State of surface	λ[μm]	$r_{\lambda,n}$	State of surface	λ[μm]	$r_{\lambda,n}$
Thallium bromide-thallium chloride mixture; KRS-6 polished surface [152]	50.0	0.02	Thorium oxide; ThO_2	0.6	0.86
	60.0	0.10		4.4	0.47
	70.0	0.37		8.8	0.07
	80.0	0.58	Tile, red [24]	0.5	0.37
	90.0	0.62		1.0	0.66
	100.0	0.80		2.0	0.71
	150.0	0.89		3.0	0.26
	200.0	0.75		4.0	0.53
	250.0	0.64		5.0	0.19
	300.0	0.605		6.0	0.15
	400.0	0.595		7.0	0.13
Thallium bromide-thallium iodide mixture; KRS-5 polished surface [152]	50.0	0.12		8.0	0.10
	60.0	0.11		9.0	0.33
	70.0	0.08	new [70]	0.40	0.075
	80.0	0.03		0.44	0.099
	90.0	0.06		0.48	0.130
	100.0	0.22		0.52	0.171
	150.0	0.84		0.56	0.214
	250.0	0.74		0.58	0.232
	300.0	0.68		0.60	0.243
	400.0	0.62		0.62	0.249
Thallium chloride; TlCl polished surface [152]	50.0	0.03		0.64	0.253
	60.0	0.35		0.65	0.255
	70.0	0.69		0.66	0.257
	80.0	0.80		0.67	0.275
	90.0	0.85		0.68	0.305
	100.0	0.87		0.69	0.420
	150.0	0.88		0.70	0.545
	200.0	0.70		0.71	0.631
	250.0	0.64		0.72	0.690
Thallium iodide; TlJ [76]	25.0÷50.0	0.17		0.74	0.731
	60.0	0.14		0.76	0.748
	80.0	0.07		0.78	0.750
	100.0	0.08		0.80	0.747
	150.0	0.68		0.82	0.741
	200.0	0.57		0.84	0.740
	300.0	0.515		0.86	0.748
				0.88	0.761
			Tile, white [105]	0.5	0.92
				1.0	0.93
				2.0	0.93
				2.5	0.65
				3.0	0.31
				4.0	0.24
				5.0	0.09
				6.0	0.07
				7.0	0.06
				8.0	0.05
				9.0	0.04

Table 3.2.5. (*continued*)

State of surface	λ[μm]	$r_{\lambda,n}$
Tile, white, glazed		
[96]	0.5	0.81
	1.0	0.84
	2.0	0.83
	2.5	0.56
	3.0	0.26
	4.0	0.27
	5.0	0.06
	6.0	0.05
	7.0	0.05
	8.0	0.06
	9.0	0.07
Tile, white, unglazed		
[96]	0.5	0.81
	1.0	0.84
	2.0	0.83
	2.5	0.56
	3.0	0.26
	4.0	0.48
	5.0	0.17
	6.0	0.12
	7.0	0.07
	8.0	0.09
	9.0	0.23
Titanium oxide; TiO_2 moulded, granularity d = 53÷63 μm, $\Delta r_{\lambda,n} = 0.2$		
[131]	0.25	0.10
	0.3	0.13
	0.4	0.27
	0.5	0.70
	0.6	0.68
	0.8	0.63
	1.0	0.57
	2.0	0.50
	4.0	0.16
	6.0	0.12
	8.0	0.08
	10.0	0.02
	15.0	0.81
	20.0	0.38
	30.0	0.72
	40.0	0.94
natural or synthetic monocrystal		
[152]	2.0	0.17
	4.0	0.15
	6.0	0.13
	8.0	0.09
	10.0	0.04
	11.0	0.00
	12.0	0.10
	13.0	0.85
	14.0	0.92
	16.0	0.95
	18.0	0.96
	20.0	0.92
	22.0	0.27
	24.0	0.91
	26.0	0.70
	27.0	0.36
	28.0	0.72
	29.0	0.87
	30.0	0.92
	40.0	0.96
	50.0	0.93
	60.0	0.81
	70.0	0.76
	90.0	0.70
	110.0	0.69
	130.0	0.68
	140.0	0.67
Titanium silicide; $TiSi_2$ temperature of measurement 298 K		
[131]	0.3	0.12
	0.4	0.20
	0.6	0.19
	0.8	0.20
	1.0	0.22
	2.0	0.31
Titanium silicide; Ti_5Si_3 temperature of measurement 298 K		
[131]	0.3	0.16
	0.4	0.16
	0.6	0.17
	0.8	0.19
	1.0	0.205
	2.0	0.27
Topaz; $(AlSiO_4)\cdot(F\cdot OH)_2$		
[76]	8.0	0.01
	10.0	0.36
	11.0	0.59
	12.0	0.17
	14.0	0.04

Table 3.2.5. (*continued*)

State of surface	$\lambda[\mu m]$	$r_{\lambda,n}$
	15.0	0.73
	16.0	0.60
	17.0	0.25
	18.0	0.50
	20.0	0.49
Tungsten silicide; WSi_2 temperature of measurement 298 K [131]	0.3	0.18
	0.4	0.18
	0.6	0.17
	0.8	0.15
	1.0	0.14
	2.0	0.10
	3.0	0.23
Uranium oxide; UO_2 temperature of sample 2000 K [10, 41, 131]	0.458	0.182
	0.5145	0.167
	0.647	0.163
temperature of sample 2500 K [10, 41, 131]	0.458	0.178
	0.5145	0.164
	0.647	0.160
temperature of sample 3000 K [10, 41, 131]	0.458	0.171
	0.5145	0.160
	0.647	0.156
temperature of sample 3500 K [10, 41, 131]	0.458	0.149
	0.5145	0.138
	0.647	0.136
temperature of sample 3750 K [10, 41, 131]	0.458	0.111
	0.5145	0.136
	0.647	0.100
Water, H_2O [10]	1.00	0.0196
	1.05	0.0195
	1.10	0.0194
	1.20	0.0193
	1.30	0.0191
	1.40	0.0191
	1.50	0.0188
	1.60	0.0186
	1.70	0.0185
	1.80	0.0182
	1.90	0.0179
	2.00	0.0174
	2.20	0.0163
	2.40	0.0147
	2.50	0.0137
	2.60	0.0125
	2.70	0.0096
	2.74	0.0075
	2.77	0.0090
	2.80	0.0142
	2.85	0.0200
	2.90	0.0248
	2.95	0.0287
	3.00	0.0340
	3.02	0.0349
	3.07	0.0439
	3.10	0.0429
	3.16	0.0412
	3.20	0.0399
	3.30	0.0366
	3.40	0.0337
	3.50	0.0305
	3.60	0.0280
	3.70	0.0256
	3.80	0.0236
	3.90	0.0225
	4.03	0.0219
	4.10	0.0218
	4.20	0.0216
	4.30	0.0215
	4.40	0.0214
	4.50	0.0213
	4.60	0.0211
	4.70	0.0210
	4.80	0.0208
	4.90	0.0206
	5.00	0.0202
	5.10	0.0198
	5.20	0.0193
	5.30	0.0186
	5.40	0.0180
	5.50	0.0173
	5.60	0.0167
	5.70	0.0157
	5.80	0.0140
	5.85	0.0125
	5.90	0.0150
	6.00	0.0202
	6.04	0.0217
	6.10	0.0228

Table 3.2.5. (*continued*)

State of surface	λ [μm]	$r_{\lambda,n}$
	6.20	0.0246
	6.30	0.0234
	6.40	0.0222
	6.50	0.0212
	6.60	0.0206
	6.70	0.0200
	6.80	0.0198
	6.90	0.0197
	7.00	0.0195
	7.20	0.0187
	7.40	0.0177
	7.60	0.0173
	7.80	0.0168
	8.00	0.0167
	8.20	0.0164
	8.40	0.0162
	8.60	0.0157
	8.80	0.0152
	9.00	0.0144
	9.20	0.0135
	9.40	0.0127
	9.60	0.0118
	9.80	0.0109
	10.00	0.0099
	10.20	0.0092
	10.40	0.0085
	10.60	0.0078
	10.80	0.0073
	10.90	0.0070
	11.00	0.0073
	11.10	0.0077
	11.20	0.0084
	11.30	0.0092
	11.40	0.0097
	11.50	0.0102
	11.60	0.0109
	11.70	0.0119
	11.80	0.0130
	12.00	0.0147
	12.50	0.0216
	13.00	0.0302
	13.50	0.0355
	14.00	0.0410
	14.50	0.0472
	15.00	0.0512
in the Kubani river, which is turbid [70]	0.40	0.153
	0.45	0.177
	0.50	0.193
	0.52	0.200
	0.54	0.204
	0.56	0.208
	0.58	0.204
	0.60	0.200
	0.65	0.180
	0.70	0.155
	0.75	0.122
	0.80	0.091
	0.84	0.066
in the mountain river Djamagat [70]	0.40	0.100
	0.45	0.119
	0.50	0.139
	0.55	0.161
	0.60	0.184
	0.65	0.201
in a drain, very turbid, of chocolate colour [70]	0.40	0.070
	0.45	0.099
	0.50	0.131
	0.55	0.169
	0.56	0.175
	0.58	0.184
	0.60	0.177
	0.62	0.157
	0.64	0.134
	0.65	0.125
	0.66	0.124
	0.68	0.135
	0.70	0.142
	0.72	0.151
	0.74	0.162
	0.76	0.174
	0.78	0.190
	0.80	0.212
	0.82	0.245
	0.84	0.280
	0.86	0.309
in a clean state with the reflection of a blue sky [70]	0.40	0.027
	0.42	0.025
	0.44	0.028
	0.46	0.033
	0.48	0.036
	0.50	0.038
	0.52	0.040
	0.54	0.047
	0.56	0.055
	0.58	0.066

Table 3.2.5. (*continued*)

State of surface	λ [μm]	$r_{\lambda,n}$
	0.60	0.073
	0.62	0.070
	0.64	0.057
	0.65	0.049
Zinc oxide; ZnO synthetic monocrystal [152]	2.5	0.112
	5.0	0.111
	10.0	0.081
	15.0	0.00
	16.0	0.021
	17.0	0.40
	18.0	0.80
	19.0	0.89
	20.0	0.90
	22.5	0.91
	25.0	0.65
	27.5	0.43
	30.0	0.36
	32.5	0.32
	35.0	0.31
	38.0	0.30
Zinc white; zinc oxide, ZnO temperature of measurement 300 K [131]	0.25	0.02
	0.3	0.02
	0.4	0.03
	0.5	1.00
	0.6	0.97
	0.8	0.965
	1.0	0.96
	1.5	0.96
	2.0	0.86
	2.5	0.57
	3.0	0.41
	4.0	0.18
	5.0	0.12
	6.0	0.11
	8.0	0.11
	10.0	0.09
	15.0	0.02
	20.0	0.90
	30.0	0.37
	40.0	0.31
powder [50, 130]	1.0	0.74
	2.0	0.41
	3.0	0.02
	4.0	0.04
	5.0	0.06
	6.0	0.04
	7.0	0.07
	8.0	0.10
	9.0	0.13
Zirconium oxide; ZrO_2 moulded samples [24]	0.25	0.11
	0.30	0.14
	0.40	0.25
	0.50	0.68
	0.60	0.78
	0.70	0.82
	0.80	0.83
	1.0	0.84
	2.0	0.96
Zirconium silicide; $ZrSi_2$ $\Delta r_{\lambda,n} = 0.05$ [131]	0.2÷3.0	0.15

Table 3.2.6. Spectral transmissivity in normal direction $\tau_{\lambda,n}$ of inorganic materials

State of surface and thickness of layer	λ [μm]	$\tau_{\lambda,n}$
Air thickness of layer 1.852 km, moisture content 9.19 g/m^3 H_2O [10, 22, 38, 72, 135]	0.61	0.600
	1.000	0.675
	1.025	0.710
	1.050	0.720
	1.075	0.680
	1.100	0.500
	1.125	0.370
	1.150	0.360
	1.175	0.500
	1.200	0.680
	1.225	0.720
	1.250	0.740
	1.275	0.735
	1.300	0.700
	1.325	0.585
	1.350	0.400
	1.375	0.200
	1.400	0.030
	1.450	0.086
	1.475	0.210
	1.500	0.480
	1.525	0.600
	1.550	0.760
	1.575	0.780
	1.600	0.785
	1.625	0.790
	1.650	0.795
	1.675	0.790
	1.700	0.780
	1.725	0.760
	1.750	0.730
	1.775	0.670
	1.800	0.510
	1.825	0.300
	1.850	0.900
	1.950	0.050
	1.975	0.270
	2.000	0.500
	2.025	0.490
	2.050	0.690
	2.075	0.700
	2.100	0.770
	2.125	0.790
	2.150	0.800
	2.175	0.790
	2.200	0.775
	2.225	0.780
	2.250	0.800
	2.275	0.800
	2.300	0.790
	2.325	0.770
	2.350	0.730
	2.375	0.680
	2.400	0.630
	2.425	0.600
	2.450	0.535
	2.475	0.500
	2.500	0.425
	2.525	0.070
	2.550	0.010
	2.925	0.030
	2.950	0.175
	2.975	0.195
	3.000	0.165
	3.025	0.270
	3.050	0.255
	3.075	0.160
	3.100	0.190
	3.125	0.140
	3.150	0.330
	3.175	0.320
	3.200	0.600
	3.225	0.110
	3.250	0.118
	3.275	0.330

Table 3.2.6. (*continued*)

State of surface and thickness of layer	λ [μm]	$\tau_{\lambda,n}$	State of surface and thickness of layer	λ [μm]	$\tau_{\lambda,n}$
	3.300	0.270		4.825	0.565
	3.325	0.450		4.850	0.435
	3.350	0.250		4.875	0.310
	3.375	0.435		4.900	0.580
	3.400	0.410		4.925	0.240
	3.425	0.465		4.950	0.580
	3.450	0.706		4.975	0.180
	3.475	0.670		5.000	0.320
	3.500	0.850		5.025	0.420
	3.525	0.830		5.050	0.120
	3.550	0.910		5.075	0.395
	3.575	0.900		5.100	0.150
	3.600	0.865		5.125	0.180
	3.625	0.855		5.150	0.220
	3.650	0.865		5.175	0.050
	3.675	0.900		5.200	0.175
	3.700	0.940		5.225	0.030
	3.725	0.790		5.275	0.060
	3.750	0.940		5.300	0.020
	3.775	0.920		7.55	0.090
	3.800	0.900		7.60	0.017
	3.825	0.880		7.65	0.013
	3.850	0.880		7.70	0.018
	3.875	0.900		7.75	0.021
	3.900	0.920		7.80	0.015
	3.925	0.910		7.85	0.130
	3.950	0.920		7.90	0.250
	3.975	0.910		7.95	0.330
	4.000	0.920		8.00	0.400
	4.025	0.925		8.05	0.580
	4.050	0.900		8.10	0.485
	4.075	0.860		8.20	0.660
	4.100	0.830		8.25	0.660
	4.125	0.800		8.30	0.660
	4.150	0.770		8.35	0.640
	4.175	0.700		8.40	0.660
	4.200	0.340		8.45	0.620
	4.225	0.050		8.50	0.620
	4.275	0.100		8.55	0.780
	4.500	0.350		8.60	0.790
	4.525	0.290		8.65	0.700
	4.550	0.400		8.70	0.765
	4.575	0.500		8.75	0.680
	4.600	0.690		8.80	0.790
	4.625	0.625		8.85	0.770
	4.650	0.660		8.90	0.780
	4.675	0.520		8.95	0.730
	4.700	0.700		9.00	0.780
	4.725	0.630		9.05	0.800
	4.750	0.570		9.10	0.790
	4.775	0.635		9.15	0.800
	4.800	0.470		9.20	0.780

Table 3.2.6. (*continued*)

State of surface and thickness of layer	$\lambda[\mu m]$	$\tau_{\lambda,n}$
	9.25	0.780
	9.30	0.750
	9.35	0.740
	9.40	0.770
	9.45	0.790
	9.50	0.795
	9.55	0.820
	9.60	0.815
	9.65	0.795
	9.70	0.780
	9.75	0.800
	9.80	0.780
	9.85	0.800
	9.90	0.805
	9.95	0.800
	10.00	0.810
	10.10	0.800
	10.15	0.790
	10.20	0.770
	10.25	0.740
	10.30	0.780
	10.35	0.785
	10.40	0.780
	10.45	0.740
	10.50	0.715
	10.55	0.720
	10.60	0.735
	10.65	0.760
	10.70	0.760
	10.75	0.750
	10.80	0.720
	10.85	0.700
	10.90	0.740
	10.95	0.730
	11.00	0.715
	11.05	0.695
	11.10	0.750
	11.15	0.765
	11.20	0.670
	11.25	0.670
	11.30	0.690
	11.35	0.695
	11.40	0.720
	11.45	0.680
	11.50	0.700
	11.55	0.690
	11.60	0.700
	11.65	0.720
	11.70	0.620
	11.75	0.510
	11.80	0.700
	11.85	0.650
	11.90	0.670
	11.95	0.700
	12.00	0.720
	12.05	0.715
	12.10	0.625
	12.15	0.700
	12.20	0.715
	12.25	0.675
	12.30	0.590
	12.35	0.675
	12.40	0.620
	12.45	0.540
	12.50	0.645
	12.55	0.320
	12.60	0.440
	12.65	0.620
	12.70	0.650
	12.75	0.470
	12.80	0.590
	12.85	0.580
	12.90	0.400
	12.95	0.515
	13.00	0.500
	13.05	0.445
	13.10	0.470
	13.15	0.520
	13.20	0.500
	13.25	0.330
	13.30	0.250
	13.35	0.290
	13.40	0.200
	13.45	0.140
	13.50	0.110
	13.55	0.190
	13.60	0.150
	13.65	0.095
	13.70	0.195
	13.75	0.210
	13.80	0.050
	13.85	0.010
thickness of layer 5.556 km, moisture content 9.19 g/m^3 H_2O [10, 22, 38, 72, 135]	0.61	0.216
	1.000	0.307
	1.025	0.360
	1.050	0.370
	1.075	0.315
	1.100	0.125
	1.125	0.051

Table 3.2.6. (*continued*)

State of surface and thickness of layer	λ [μm]	$\tau_{\lambda,n}$	State of surface and thickness of layer	λ [μm]	$\tau_{\lambda,n}$
	1.150	0.047		2.550	$0.1\cdot10^{-5}$
	1.175	0.125		2.925	$0.27\cdot10^{-4}$
	1.200	0.315		2.950	$0.54\cdot10^{-2}$
	1.225	0.370		2.975	$0.74\cdot10^{-2}$
	1.250	0.405		3.000	$0.45\cdot10^{-2}$
	1.275	0.396		3.025	0.0196
	1.300	0.343		3.050	0.0166
	1.325	0.200		3.075	0.0042
	1.350	0.064		3.100	0.0068
	1.375	0.008		3.125	0.0028
	1.400	$0.27\cdot10^{-4}$		3.150	0.0360
	1.450	$0.512\cdot10^{-3}$		3.175	0.0328
	1.475	0.009		3.200	0.2160
	1.500	0.110		3.225	0.0013
	1.525	0.216		3.250	0.0058
	1.550	0.440		3.275	0.0360
	1.575	0.475		3.300	0.0196
	1.600	0.485		3.325	0.0918
	1.625	0.495		3.350	0.0156
	1.650	0.500		3.375	0.0830
	1.675	0.495		3.400	0.069
	1.700	0.475		3.425	0.100
	1.725	0.440		3.450	0.343
	1.750	0.390		3.475	0.300
	1.775	0.300		3.500	0.615
	1.800	0.133		3.525	0.570
	1.825	0.027		3.550	0.755
	1.850	0.0007		3.575	0.730
	1.950	0.0001		3.600	0.650
	1.975	0.0196		3.625	0.625
	2.000	0.125		3.650	0.650
	2.025	0.118		3.675	0.730
	2.050	0.330		3.700	0.835
	2.075	0.343		3.725	0.495
	2.100	0.460		3.750	0.835
	2.125	0.495		3.775	0.780
	2.150	0.512		3.800	0.730
	2.175	0.495		3.825	0.680
	2.200	0.465		3.850	0.680
	2.225	0.495		3.875	0.730
	2.250	0.512		3.900	0.780
	2.275	0.512		3.925	0.755
	2.300	0.495		3.950	0.780
	2.325	0.460		3.975	0.755
	2.350	0.390		4.000	0.780
	2.375	0.315		4.025	0.790
	2.400	0.250		4.050	0.730
	2.425	0.216		4.075	0.635
	2.450	0.154		4.100	0.570
	2.475	0.125		4.125	0.512
	2.500	0.077		4.150	0.460
	2.525	$0.35\cdot10^{-3}$		4.175	0.343

Table 3.2.6. (*continued*)

State of surface and thickness of layer	$\lambda [\mu m]$	$\tau_{\lambda,n}$
	4.200	0.039
	4.225	$0.13 \cdot 10^{-3}$
	4.475	0.001
	4.500	0.043
	4.525	0.025
	4.550	0.064
	4.575	0.125
	4.600	0.330
	4.625	0.240
	4.650	0.286
	4.675	0.140
	4.700	0.343
	4.725	0.250
	4.750	0.185
	4.775	0.256
	4.800	0.104
	4.825	0.180
	4.850	0.080
	4.875	0.030
	4.900	0.195
	4.925	0.014
	4.950	0.195
	4.975	0.006
	5.000	0.033
	5.025	0.074
	5.050	0.002
	5.100	0.003
	5.125	0.006
	5.150	0.010
	5.175	$0.13 \cdot 10^{-3}$
	5.200	0.005
	5.225	$0.27 \cdot 10^{-4}$
	5.275	$0.22 \cdot 10^{-3}$
	5.300	$0.8 \cdot 10^{-5}$
	7.55	$0.73 \cdot 10^{-3}$
	7.60	$0.45 \cdot 10^{-2}$
	7.65	$0.22 \cdot 10^{-2}$
	7.70	$0.54 \cdot 10^{-2}$
	7.75	$0.93 \cdot 10^{-2}$
	7.80	$0.34 \cdot 10^{-2}$
	7.85	$0.22 \cdot 10^{-2}$
	7.90	0.016
	7.95	0.037
	8.00	0.064
	8.05	0.020
	8.10	0.140
	8.20	0.216
	8.25	0.288
	8.30	0.288
	8.35	0.262
	8.40	0.288
	8.45	0.238
	8.50	0.238
	8.55	0.475
	8.60	0.495
	8.65	0.343
	8.70	0.445
	8.75	0.315
	8.80	0.495
	8.85	0.460
	8.90	0.475
	8.95	0.390
	9.00	0.475
	9.05	0.512
	9.10	0.495
	9.15	0.512
	9.20	0.475
	9.25	0.475
	9.30	0.420
	9.35	0.405
	9.40	0.460
	9.45	0.495
	9.50	0.500
	9.55	0.550
	9.60	0.542
	9.65	0.500
	9.70	0.475
	9.75	0.512
	9.80	0.475
	9.85	0.512
	9.90	0.520
	9.95	0.512
	10.00	0.530
	10.05	0.530
	10.10	0.512
	10.15	0.495
	10.20	0.460
	10.25	0.420
	10.30	0.475
	10.35	0.485
	10.40	0.475
	10.45	0.405
	10.50	0.365
	10.55	0.370
	10.60	0.396
	10.65	0.440
	10.70	0.440
	10.75	0.420
	10.80	0.370
	10.85	0.343
	10.90	0.405
	10.95	0.390
	11.00	0.365
	11.05	0.335

Table 3.2.6. (*continued*)

State of surface and	λ[μm]	$\tau_{\lambda,n}$
	11.10	0.420
	11.15	0.445
	11.20	0.306
	11.25	0.300
	11.30	0.330
	11.35	0.335
	11.40	0.370
	11.45	0.316
	11.50	0.343
	11.55	0.330
	11.60	0.343
	11.65	0.370
	11.70	0.238
	11.75	0.133
	11.80	0.343
	11.85	0.275
	11.90	0.300
	11.95	0.343
	12.00	0.370
	12.05	0.365
	12.10	0.245
	12.15	0.343
	12.20	0.365
	12.25	0.307
	12.30	0.205
	12.35	0.307
	12.40	0.238
	12.45	0.158
	12.50	0.270
	12.55	0.033
	12.60	0.085
	12.65	0.238
	12.70	0.275
	12.75	0.104
	12.80	0.205
	12.85	0.195
	12.90	0.640
	12.95	0.136
	13.00	0.125
	13.05	0.088
	13.10	0.104
	13.15	0.140
	13.20	0.125
	13.25	0.036
	13.30	0.016
	13.35	0.025
	13.40	0.008
	13.45	$0.28\cdot10^{-2}$
	13.50	$0.13\cdot10^{-2}$
	13.55	$0.69\cdot10^{-2}$
	13.60	$0.34\cdot10^{-2}$
	13.65	$0.86\cdot10^{-3}$
	13.70	$0.74\cdot10^{-2}$
	13.75	$0.93\cdot10^{-2}$
	13.80	$0.13\cdot10^{-3}$
	13.85	$0.10\cdot10^{-5}$
thickness of layer 11.112 km, moisture content 9.19 g/m^3 H_2O [10, 22, 38, 72, 135]	0.610	0.047
	1.000	0.094
	1.025	0.130
	1.050	0.137
	1.075	0.100
	1.100	0.016
	1.125	0.026
	1.150	0.022
	1.175	0.016
	1.200	0.100
	1.225	0.137
	1.250	0.164
	1.275	0.158
	1.300	0.118
	1.325	0.040
	1.350	0.004
	1.375	$0.64\cdot10^{-4}$
	1.450	$0.26\cdot10^{-6}$
	1.475	$0.87\cdot10^{-4}$
	1.500	0.012
	1.525	0.046
	1.550	0.194
	1.575	0.225
	1.600	0.235
	1.625	0.245
	1.650	0.250
	1.675	0.245
	1.700	0.225
	1.725	0.194
	1.750	0.152
	1.775	0.900
	1.800	0.018
	1.825	$0.73\cdot10^{-3}$
	1.850	$0.53\cdot10^{-6}$
	1.950	$0.16\cdot10^{-7}$
	1.975	$0.38\cdot10^{-3}$
	2.000	0.016
	2.025	0.014
	2.050	0.109
	2.075	0.118
	2.100	0.212
	2.125	0.245

Table 3.2.6. (*continued*)

State of surface and thickness of layer	$\lambda[\mu m]$	$\tau_{\lambda,n}$
	2.150	0.262
	2.175	0.245
	2.200	0.215
	2.225	0.245
	2.250	0.262
	2.275	0.262
	2.300	0.245
	2.325	0.212
	2.350	0.152
	2.375	0.099
	2.400	0.063
	2.425	0.047
	2.450	0.024
	2.475	0.016
	2.500	0.006
	2.525	$0.12 \cdot 10^{-6}$
	2.950	$0.29 \cdot 10^{-7}$
	2.975	$0.55 \cdot 10^{-4}$
	3.000	$0.20 \cdot 10^{-4}$
	3.025	$0.39 \cdot 10^{-3}$
	3.050	$0.28 \cdot 10^{-3}$
	3.075	$0.18 \cdot 10^{-4}$
	3.100	$0.47 \cdot 10^{-4}$
	3.125	$0.76 \cdot 10^{-3}$
	3.150	$0.13 \cdot 10^{-2}$
	3.175	$0.11 \cdot 10^{-2}$
	3.200	0.047
	3.225	$0.17 \cdot 10^{-5}$
	3.250	$0.34 \cdot 10^{-4}$
	3.275	$0.13 \cdot 10^{-2}$
	3.300	$0.39 \cdot 10^{-3}$
	3.325	$0.84 \cdot 10^{-2}$
	3.350	$0.24 \cdot 10^{-3}$
	3.375	$0.69 \cdot 10^{-2}$
	3.400	$0.48 \cdot 10^{-2}$
	3.425	0.010
	3.450	0.118
	3.475	0.090
	3.500	0.378
	3.525	0.325
	3.550	0.570
	3.575	0.530
	3.600	0.420
	3.625	0.390
	3.650	0.420
	3.675	0.530
	3.700	0.700
	3.725	0.245
	3.750	0.700
	3.775	0.610
	3.800	0.530
	3.825	0.465
	3.850	0.460
	3.875	0.530
	3.900	0.610
	3.925	0.570
	3.950	0.640
	3.975	0.570
	4.000	0.610
	4.025	0.625
	4.050	0.530
	4.075	0.405
	4.100	0.325
	4.125	0.262
	4.150	0.212
	4.175	0.118
	4.200	$0.16 \cdot 10^{-2}$
	4.225	$0.16 \cdot 10^{-7}$
	4.475	$0.1 \cdot 10^{-5}$
	4.500	$0.18 \cdot 10^{-2}$
	4.525	$0.60 \cdot 10^{-3}$
	4.550	$0.41 \cdot 10^{-2}$
	4.575	0.016
	4.600	0.108
	4.625	0.058
	4.650	0.084
	4.675	0.020
	4.700	0.118
	4.725	0.063
	4.750	0.034
	4.775	0.066
	4.800	0.011
	4.825	0.033
	4.850	0.007
	4.875	$0.89 \cdot 10^{-3}$
	4.900	0.038
	4.925	$0.19 \cdot 10^{-3}$
	4.950	0.038
	4.975	$0.34 \cdot 10^{-4}$
	5.000	0.001
	5.025	0.006
	5.050	$0.30 \cdot 10^{-5}$
	5.075	$0.39 \cdot 10^{-2}$
	5.100	$0.12 \cdot 10^{-4}$
	5.125	$0.34 \cdot 10^{-4}$
	5.150	$0.11 \cdot 10^{-3}$
	5.175	$0.16 \cdot 10^{-5}$
	5.200	$0.29 \cdot 10^{-4}$
	5.275	$0.47 \cdot 10^{-7}$
	7.55	$0.53 \cdot 10^{-6}$
	7.60	$0.20 \cdot 10^{-4}$
	7.65	$0.49 \cdot 10^{-5}$
	7.70	$0.29 \cdot 10^{-4}$
	7.75	$0.86 \cdot 10^{-4}$

Table 3.2.6. (*continued*)

State of surface and thickness of layer	λ[μm]	$\tau_{\lambda,n}$	State of surface and thickness of layer	λ[μm]	$\tau_{\lambda,n}$
	7.80	$0.12 \cdot 10^{-4}$		10.45	0.162
	7.85	$0.49 \cdot 10^{-5}$		10.50	0.132
	7.90	$0.24 \cdot 10^{-3}$		10.55	0.134
	7.95	$0.13 \cdot 10^{-2}$		10.60	0.156
	8.00	$0.41 \cdot 10^{-2}$		10.65	0.194
	8.05	0.038		10.70	0.194
	8.10	0.020		10.75	0.178
	8.20	0.047		10.80	0.134
	8.25	0.082		10.85	0.118
	8.30	0.082		10.90	0.162
	8.35	0.069		10.95	0.152
	8.40	0.082		11.00	0.132
	8.45	0.057		11.05	0.112
	8.50	0.057		11.10	0.178
	8.55	0.225		11.15	0.198
	8.60	0.245		11.20	0.090
	8.65	0.118		11.25	0.090
	8.70	0.197		11.30	0.108
	8.75	0.100		11.35	0.112
	8.80	0.245		11.40	0.137
	8.85	0.212		11.45	0.099
	8.90	0.225		11.50	0.118
	8.95	0.152		11.55	0.108
	9.00	0.225		11.60	0.118
	9.05	0.262		11.65	0.137
	9.10	0.245		11.70	0.057
	9.15	0.262		11.75	0.018
	9.20	0.225		11.80	0.118
	9.25	0.225		11.85	0.075
	9.30	0.176		11.90	0.090
	9.35	0.164		11.95	0.118
	9.40	0.212		12.00	0.137
	9.45	0.245		12.05	0.132
	9.50	0.250		12.10	0.060
	9.55	0.304		12.15	0.118
	9.60	0.294		12.20	0.132
	9.65	0.250		12.25	0.094
	9.70	0.225		12.30	0.042
	9.75	0.262		12.35	0.094
	9.80	0.225		12.40	0.057
	9.85	0.262		12.45	0.025
	9.90	0.270		12.50	0.073
	9.95	0.262		12.55	$0.11 \cdot 10^{-2}$
	10.00	0.282		12.60	$0.72 \cdot 10^{-2}$
	10.05	0.282		12.65	0.057
	10.10	0.262		12.70	0.075
	10.15	0.245		12.75	0.011
	10.20	0.212		12.80	0.042
	10.25	0.178		12.85	0.038
	10.30	0.225		12.90	$0.41 \cdot 10^{-2}$
	10.35	0.235		12.95	0.019
	10.40	0.225		13.00	0.016

Table 3.2.6. (*continued*)

State of surface and thickness of layer	λ[μm]	$\tau_{\lambda,n}$
	13.05	$0.78\cdot10^{-2}$
	13.10	0.011
	13.15	0.020
	13.20	0.016
	13.25	0.013
	13.30	$0.24\cdot10^{-3}$
	13.35	$0.60\cdot10^{-3}$
	13.40	$0.64\cdot10^{-4}$
	13.45	$0.75\cdot10^{-5}$
	13.50	$0.17\cdot10^{-5}$
	13.55	$0.47\cdot10^{-4}$
	13.60	$0.12\cdot10^{-4}$
	13.65	$0.74\cdot10^{-6}$
	13.70	$0.55\cdot10^{-4}$
	13.75	$0.86\cdot10^{-4}$
	13.80	$0.16\cdot10^{-7}$
thickness of layer 22.224 km, moisture content 9.19 g/m^3 H_2O [10, 22, 38, 72, 135]	0.61	$0.22\cdot10^{-2}$
	1.000	$0.89\cdot10^{-2}$
	1.025	0.017
	1.050	0.019
	1.075	$0.99\cdot10^{-2}$
	1.100	$0.24\cdot10^{-3}$
	1.125	$0.65\cdot10^{-5}$
	1.150	$0.44\cdot10^{-5}$
	1.175	$0.24\cdot10^{-3}$
	1.200	$0.99\cdot10^{-2}$
	1.225	0.019
	1.250	0.027
	1.275	0.025
	1.300	0.014
	1.325	$0.16\cdot10^{-2}$
	1.350	$0.17\cdot10^{-4}$
	1.500	$0.15\cdot10^{-3}$
	1.525	$0.22\cdot10^{-2}$
	1.550	0.038
	1.575	0.050
	1.600	0.055
	1.625	0.066
	1.650	0.063
	1.675	0.060
	1.700	0.050
	1.725	0.038
	1.750	0.023
	1.775	$0.81\cdot10^{-2}$
	1.800	$0.31\cdot10^{-3}$
	1.825	$0.54\cdot10^{-6}$
	1.950	$0.24\cdot10^{-9}$
	1.975	$0.15\cdot10^{-6}$
	2.000	$0.24\cdot10^{-3}$
	2.025	$0.20\cdot10^{-3}$
	2.050	0.012
	2.075	0.014
	2.100	0.045
	2.125	0.060
	2.150	0.068
	2.175	0.060
	2.200	0.046
	2.225	0.060
	2.250	0.068
	2.275	0.068
	2.300	0.067
	2.325	0.045
	2.350	0.232
	2.375	$0.99\cdot10^{-2}$
	2.400	$0.39\cdot10^{-2}$
	2.425	$0.22\cdot10^{-2}$
	2.450	$0.56\cdot10^{-3}$
	2.475	$0.24\cdot10^{-3}$
	2.500	$0.35\cdot10^{-4}$
	3.025	$0.15\cdot10^{-6}$
	3.050	$0.76\cdot10^{-7}$
	3.150	$0.17\cdot10^{-5}$
	3.175	$0.12\cdot10^{-5}$
	3.200	$0.22\cdot10^{-2}$
	3.275	$0.17\cdot10^{-5}$
	3.300	$0.15\cdot10^{-6}$
	3.325	$0.71\cdot10^{-4}$
	3.350	$0.59\cdot10^{-7}$
	3.375	$0.48\cdot10^{-4}$
	3.400	$0.23\cdot10^{-4}$
	3.425	$0.10\cdot10^{-3}$
	3.450	0.014
	3.475	$0.81\cdot10^{-2}$
	3.500	0.143
	3.525	0.106
	3.550	0.325
	3.575	0.280
	3.600	0.176
	3.625	0.152
	3.650	0.176
	3.675	0.280
	3.700	0.490
	3.725	0.060
	3.750	0.490
	3.775	0.370
	3.800	0.280
	3.825	0.210
	3.850	0.210
	3.875	0.280
	3.900	0.370

Table 3.2.6. (*continued*)

State of surface and thickness of layer	λ[μm]	$\tau_{\lambda,n}$	State of surface and thickness of layer	λ[μm]	$\tau_{\lambda,n}$
	3.925	0.325		8.80	0.060
	3.950	0.370		8.85	0.045
	3.975	0.325		8.90	0.051
	4.000	0.370		8.95	0.023
	4.025	0.390		9.00	0.051
	4.050	0.280		9.05	0.069
	4.075	0.164		9.10	0.060
	4.100	0.106		9.15	0.069
	4.125	0.690		9.20	0.051
	4.150	0.450		9.25	0.051
	4.175	0.014		9.30	0.031
	4.200	$0.24\cdot10^{-5}$		9.35	0.027
	4.500	$0.34\cdot10^{-5}$		9.40	0.045
	4.525	$0.36\cdot10^{-6}$		9.45	0.060
	4.550	$0.17\cdot10^{-4}$		9.50	0.063
	4.575	$0.24\cdot10^{-3}$		9.55	0.093
	4.600	0.012		9.60	0.087
	4.625	$0.34\cdot10^{-2}$		9.65	0.063
	4.650	0.007		9.70	0.051
	4.675	$0.39\cdot10^{-3}$		9.75	0.068
	4.700	0.014		9.80	0.051
	4.725	$0.39\cdot10^{-2}$		9.85	0.069
	4.750	$0.12\cdot10^{-2}$		9.90	0.073
	4.775	$0.44\cdot10^{-2}$		9.95	0.069
	4.800	$0.12\cdot10^{-3}$		10.00	0.080
	4.825	$0.11\cdot10^{-2}$		10.05	0.080
	4.850	$0.48\cdot10^{-4}$		10.10	0.069
	4.875	$0.80\cdot10^{-6}$		10.15	0.060
	4.900	$0.14\cdot10^{-2}$		10.20	0.045
	4.925	$0.36\cdot10^{-5}$		10.25	0.032
	4.950	$0.14\cdot10^{-2}$		10.30	0.051
	5.000	$0.11\cdot10^{-5}$		10.35	0.051
	5.025	$0.30\cdot10^{-4}$		10.40	0.051
	5.075	$0.19\cdot10^{-4}$		10.45	0.026
	5.150	$0.13\cdot10^{-7}$		10.50	0.017
	7.90	$0.60\cdot10^{-7}$		10.55	0.019
	7.95	$0.17\cdot10^{-5}$		10.60	0.024
	8.00	$0.17\cdot10^{-4}$		10.65	0.038
	8.05	$0.15\cdot10^{-2}$		10.70	0.038
	8.10	$0.38\cdot10^{-3}$		10.75	0.032
	8.20	$0.22\cdot10^{-2}$		10.80	0.019
	8.25	$0.68\cdot10^{-2}$		10.85	0.014
	8.30	$0.68\cdot10^{-2}$		10.90	0.026
	8.35	$0.47\cdot10^{-2}$		10.95	0.023
	8.40	$0.68\cdot10^{-2}$		11.00	0.017
	8.45	$0.33\cdot10^{-2}$		11.05	0.013
	8.50	$0.33\cdot10^{-2}$		11.10	0.032
	8.55	0.051		11.15	0.039
	8.60	0.060		11.20	0.008
	8.65	0.014		11.25	0.008
	8.70	0.039		11.30	0.012
	8.75	0.010		11.35	0.013

State of surface and thickness of layer	λ [μm]	$\tau_{\lambda,n}$
	11.40	0.019
	11.45	0.010
	11.50	0.014
	11.55	0.012
	11.60	0.014
	11.65	0.019
	11.70	$0.33 \cdot 10^{-2}$
	11.75	$0.32 \cdot 10^{-3}$
	11.80	0.014
	11.85	$0.57 \cdot 10^{-2}$
	11.90	$0.81 \cdot 10^{-2}$
	11.95	0.014
	12.00	0.019
	12.05	0.017
	12.10	$0.36 \cdot 10^{-2}$
	12.15	0.014
	12.20	0.017
	12.25	$0.89 \cdot 10^{-2}$
	12.30	$0.18 \cdot 10^{-2}$
	12.35	$0.89 \cdot 10^{-2}$
	12.40	$0.33 \cdot 10^{-2}$
	12.45	$0.63 \cdot 10^{-3}$
	12.50	$0.54 \cdot 10^{-2}$
	12.55	$0.12 \cdot 10^{-5}$
	12.60	$0.52 \cdot 10^{-4}$
	12.65	$0.33 \cdot 10^{-2}$
	12.70	$0.57 \cdot 10^{-2}$
	12.75	$0.12 \cdot 10^{-3}$
	12.80	$0.18 \cdot 10^{-2}$
	12.85	$0.15 \cdot 10^{-2}$
	12.90	$0.17 \cdot 10^{-4}$
	12.95	$0.34 \cdot 10^{-3}$
	13.00	$0.25 \cdot 10^{-3}$
	13.05	$0.60 \cdot 10^{-4}$
	13.10	$0.12 \cdot 10^{-3}$
	13.15	$0.39 \cdot 10^{-3}$
	13.20	$0.25 \cdot 10^{-3}$
	13.25	$0.17 \cdot 10^{-5}$
	13.30	$0.59 \cdot 10^{-7}$
	13.35	$0.36 \cdot 10^{-6}$
thickness of layer 32.336 km, moisture content 9.19 g/m^3 H_2O [10, 22, 38, 72, 135]	0.61	$0.10 \cdot 10^{-3}$
	1.000	$0.83 \cdot 10^{-3}$
	1.025	$0.22 \cdot 10^{-2}$
	1.050	$0.26 \cdot 10^{-2}$
	1.075	$0.99 \cdot 10^{-3}$
	1.100	$0.38 \cdot 10^{-5}$
	1.175	$0.38 \cdot 10^{-5}$
	1.200	$0.99 \cdot 10^{3}$
	1.225	$0.26 \cdot 10^{-2}$
	1.250	$0.45 \cdot 10^{-2}$
	1.275	$0.40 \cdot 10^{-2}$
	1.300	$0.16 \cdot 10^{-4}$
	1.325	$0.64 \cdot 10^{-7}$
	1.500	$0.18 \cdot 10^{-5}$
	1.525	$0.10 \cdot 10^{-3}$
	1.550	$0.73 \cdot 10^{-2}$
	1.575	0.012
	1.600	0.013
	1.625	0.015
	1.650	0.016
	1.675	0.015
	1.700	0.012
	1.725	$0.73 \cdot 10^{-2}$
	1.750	$0.35 \cdot 10^{-2}$
	1.775	$0.73 \cdot 10^{-3}$
	1.800	$0.56 \cdot 10^{-5}$
	2.000	$0.38 \cdot 10^{-5}$
	2.025	$0.28 \cdot 10^{-5}$
	2.050	$0.13 \cdot 10^{-2}$
	2.075	$0.17 \cdot 10^{-2}$
	2.100	$0.96 \cdot 10^{-2}$
	2.125	0.015
	2.150	0.018
	2.175	0.015
	2.200	0.010
	2.225	0.015
	2.250	0.018
	2.275	0.018
	2.300	0.015
	2.325	0.010
	2.350	$0.35 \cdot 10^{-2}$
	2.375	$0.99 \cdot 10^{-3}$
	2.400	$0.25 \cdot 10^{-3}$
	2.425	$0.10 \cdot 10^{-3}$
	2.450	$0.13 \cdot 10^{-5}$
	2.475	$0.38 \cdot 10^{-5}$
	3.200	$0.16 \cdot 10^{-3}$
	3.425	$0.10 \cdot 10^{-5}$
	3.450	$0.17 \cdot 10^{-2}$
	3.475	$0.74 \cdot 10^{-3}$
	3.500	0.054
	3.525	0.034
	3.550	0.184
	3.575	0.148
	3.600	0.074
	3.625	0.059
	3.650	0.074
	3.675	0.148
	3.700	0.340
	3.725	0.015

Table 3.2.6. (*continued*)

State of surface and thickness of layer	λ[μm]	$\tau_{\lambda,n}$	State of surface and thickness of layer	λ[μm]	$\tau_{\lambda,n}$
	3.750	0.340		9.20	0.011
	3.775	0.325		9.25	0.011
	3.800	0.148		9.30	$0.55\cdot10^{-2}$
	3.825	0.097		9.35	$0.44\cdot10^{-2}$
	3.850	0.097		9.40	$0.95\cdot10^{-2}$
	3.875	0.148		9.45	0.019
	3.900	0.225		9.50	0.016
	3.925	0.184		9.55	0.028
	3.950	0.225		9.60	0.026
	3.975	0.184		9.65	0.016
	4.000	0.225		9.70	0.011
	4.025	0.245		9.75	0.018
	4.050	0.148		9.80	0.011
	4.075	0.066		9.85	0.018
	4.100	0.340		9.90	0.020
	4.125	0.018		9.95	0.018
	4.150	0.010		10.00	0.022
	4.175	$0.16\cdot10^{-2}$		10.05	0.022
	4.575	$0.38\cdot10^{-5}$		10.10	0.018
	4.600	$0.13\cdot10^{-2}$		10.15	0.015
	4.625	$0.20\cdot10^{-2}$		10.20	$0.95\cdot10^{-2}$
	4.650	$0.60\cdot10^{-3}$		10.25	$0.56\cdot10^{-2}$
	4.675	$0.76\cdot10^{-5}$		10.30	0.011
	4.700	$0.17\cdot10^{-2}$		10.35	0.013
	4.725	$0.25\cdot10^{-3}$		10.40	0.011
	4.750	$0.41\cdot10^{-4}$		10.45	$0.43\cdot10^{-2}$
	4.775	$0.29\cdot10^{-3}$		10.50	$0.23\cdot10^{-2}$
	4.800	$0.13\cdot10^{-2}$		10.55	$0.26\cdot10^{-2}$
	4.825	$0.35\cdot10^{-4}$		10.60	$0.38\cdot10^{-2}$
	4.900	$0.55\cdot10^{-4}$		10.65	$0.73\cdot10^{-2}$
	4.950	$0.55\cdot10^{-4}$		10.70	$0.73\cdot10^{-2}$
	8.05	$0.55\cdot10^{-4}$		10.75	$0.57\cdot10^{-2}$
	8.10	$0.75\cdot10^{-5}$		10.80	$0.26\cdot10^{-2}$
	8.20	$0.16\cdot10^{-3}$		10.85	$0.16\cdot10^{-2}$
	8.25	$0.67\cdot10^{-3}$		10.90	$0.43\cdot10^{-2}$
	8.30	$0.67\cdot10^{-3}$		10.95	$0.35\cdot10^{-2}$
	8.35	$0.32\cdot10^{-3}$		11.00	$0.23\cdot10^{-2}$
	8.40	$0.67\cdot10^{-3}$		11.05	$0.14\cdot10^{-2}$
	8.45	$0.16\cdot10^{-3}$		11.10	$0.56\cdot10^{-2}$
	8.55	0.011		11.15	$0.77\cdot10^{-2}$
	8.60	0.015		11.20	$0.73\cdot10^{-3}$
	8.65	$0.16\cdot10^{-2}$		11.25	$0.73\cdot10^{-3}$
	8.70	$0.76\cdot10^{-2}$		11.30	$0.13\cdot10^{-2}$
	8.75	$0.99\cdot10^{-3}$		11.35	$0.14\cdot10^{-2}$
	8.80	0.015		11.40	$0.26\cdot10^{-2}$
	8.85	0.010		11.45	$0.99\cdot10^{-3}$
	8.90	0.011		11.50	$0.15\cdot10^{-2}$
	8.95	$0.35\cdot10^{-2}$		11.55	$0.13\cdot10^{-2}$
	9.00	0.011		11.60	$0.15\cdot10^{-2}$
	9.05	0.018		11.70	$0.18\cdot10^{-3}$
	9.10	0.015		11.80	$0.15\cdot10^{-2}$
	9.15	0.018		11.85	$0.43\cdot10^{-3}$

Table 3.2.6. (*continued*)

State of surface and thickness of layer	$\lambda[\mu m]$	$\tau_{\lambda,n}$
	11.90	$0.73\cdot10^{-3}$
	11.95	$0.15\cdot10^{-2}$
	12.00	$0.26\cdot10^{-2}$
	12.05	$0.23\cdot10^{-2}$
	12.10	$0.22\cdot10^{-3}$
	12.15	$0.15\cdot10^{-2}$
	12.20	$0.23\cdot10^{-2}$
	12.25	$0.82\cdot10^{-3}$
	12.30	$0.74\cdot10^{-4}$
	12.35	$0.82\cdot10^{-3}$
	12.40	$0.18\cdot10^{-3}$
	12.45	$0.16\cdot10^{-4}$
	12.50	$0.38\cdot10^{-3}$
	12.65	$0.19\cdot10^{-3}$
	12.70	$0.43\cdot10^{-3}$
	12.75	$0.13\cdot10^{-5}$
	12.80	$0.75\cdot10^{-4}$
	12.85	$0.55\cdot10^{-4}$
	12.95	$0.63\cdot10^{-5}$
	13.00	$0.40\cdot10^{-5}$
	13.10	$0.13\cdot10^{-5}$
	13.15	$0.76\cdot10^{-5}$
	13.20	$0.40\cdot10^{-5}$
thickness of layer 44.448 km, moisture content 9.19 g/m^3 H_2O [10, 22, 38, 72, 135]	0.61	$0.47\cdot10^{-5}$
	1.000	$0.79\cdot10^{-4}$
	1.025	$0.29\cdot10^{-3}$
	1.050	$0.36\cdot10^{-3}$
	1.075	$0.99\cdot10^{-4}$
	1.200	$0.99\cdot10^{-4}$
	1.225	$0.36\cdot10^{-3}$
	1.250	$0.73\cdot10^{-3}$
	1.275	$0.63\cdot10^{-3}$
	1.300	$0.20\cdot10^{-3}$
	1.325	$0.26\cdot10^{-5}$
	1.525	$0.47\cdot10^{-5}$
	1.550	$0.14\cdot10^{-2}$
	1.575	$0.25\cdot10^{-2}$
	1.600	$0.31\cdot10^{-2}$
	1.625	$0.36\cdot10^{-2}$
	1.650	$0.39\cdot10^{-2}$
	1.675	$0.36\cdot10^{-2}$
	1.700	$0.25\cdot10^{-2}$
	1.725	$0.14\cdot10^{-2}$
	1.750	$0.54\cdot10^{-3}$
	1.775	$0.66\cdot10^{-4}$
	2.050	$0.14\cdot10^{-3}$
	2.075	$0.20\cdot10^{-3}$
	2.100	$0.20\cdot10^{-3}$
	2.125	$0.36\cdot10^{-2}$
	2.150	$0.46\cdot10^{-2}$
	2.175	$0.31\cdot10^{-2}$
	2.200	$0.22\cdot10^{-2}$
	2.225	$0.36\cdot10^{-2}$
	2.250	$0.46\cdot10^{-2}$
	2.275	$0.46\cdot10^{-2}$
	2.300	$0.36\cdot10^{-2}$
	2.325	$0.20\cdot10^{-2}$
	2.350	$0.54\cdot10^{-3}$
	2.375	$0.99\cdot10^{-4}$
	2.400	$0.15\cdot10^{-4}$
	2.425	$0.47\cdot10^{-5}$
	3.200	$0.47\cdot10^{-5}$
	3.450	$0.20\cdot10^{-2}$
	3.475	$0.66\cdot10^{-4}$
	3.500	0.021
	3.525	0.011
	3.550	0.106
	3.575	0.079
	3.600	0.031
	3.625	0.023
	3.650	0.031
	3.675	0.079
	3.700	0.240
	3.725	$0.36\cdot10^{-2}$
	3.750	0.240
	3.775	0.138
	3.800	0.079
	3.825	0.044
	3.850	0.044
	3.875	0.079
	3.900	0.138
	3.925	0.106
	3.950	0.138
	3.975	0.154
	4.000	0.079
	4.025	0.027
	4.050	0.079
	4.075	0.027
	4.100	0.011
	4.125	$0.48\cdot10^{-2}$
	4.150	$0.20\cdot10^{-2}$
	4.175	$0.20\cdot10^{-3}$
	4.600	$0.14\cdot10^{-3}$
	4.625	$0.11\cdot10^{-4}$
	4.650	$0.49\cdot10^{-4}$
	4.700	$0.19\cdot10^{-3}$
	4.725	$0.15\cdot10^{-3}$
	4.750	$0.14\cdot10^{-5}$
	4.775	$0.19\cdot10^{-4}$
	4.825	$0.11\cdot10^{-5}$

Table 3.2.6. (*continued*)

State of surface and thickness of layer	λ[μm]	$\tau_{\lambda,n}$
	8.05	$0.21\cdot10^{-5}$
	8.20	$0.47\cdot10^{-5}$
	8.25	$0.46\cdot10^{-4}$
	8.30	$0.46\cdot10^{-4}$
	8.35	$0.22\cdot10^{-4}$
	8.40	$0.46\cdot10^{-4}$
	8.45	$0.11\cdot10^{-4}$
	8.50	$0.11\cdot10^{-4}$
	8.55	$0.26\cdot10^{-2}$
	8.60	$0.37\cdot10^{-2}$
	8.65	$0.20\cdot10^{-3}$
	8.70	$0.15\cdot10^{-2}$
	8.75	$0.99\cdot10^{-4}$
	8.80	$0.36\cdot10^{-2}$
	8.85	$0.26\cdot10^{-2}$
	8.90	$0.26\cdot10^{-2}$
	8.95	$0.54\cdot10^{-3}$
	9.00	$0.26\cdot10^{-2}$
	9.05	$0.47\cdot10^{-2}$
	9.10	$0.36\cdot10^{-2}$
	9.15	$0.47\cdot10^{-2}$
	9.20	$0.26\cdot10^{-2}$
	9.25	$0.26\cdot10^{-2}$
	9.30	$0.96\cdot10^{-3}$
	9.35	$0.73\cdot10^{-3}$
	9.40	$0.20\cdot10^{-2}$
	9.45	$0.46\cdot10^{-2}$
	9.50	$0.39\cdot10^{-2}$
	9.55	$0.85\cdot10^{-2}$
	9.60	$0.75\cdot10^{-2}$
	9.65	$0.39\cdot10^{-2}$
	9.70	$0.26\cdot10^{-2}$
	9.75	$0.48\cdot10^{-2}$
	9.80	$0.26\cdot10^{-2}$
	9.85	$0.48\cdot10^{-2}$
	9.90	$0.53\cdot10^{-2}$
	9.95	$0.48\cdot10^{-2}$
	10.00	$0.64\cdot10^{-2}$
	10.05	$0.64\cdot10^{-2}$
	10.10	$0.48\cdot10^{-2}$
	10.15	$0.36\cdot10^{-2}$
	10.20	$0.20\cdot10^{-2}$
	10.25	$0.10\cdot10^{-2}$
	10.30	$0.26\cdot10^{-2}$
	10.35	$0.26\cdot10^{-2}$
	10.40	$0.26\cdot10^{-2}$
	10.45	$0.68\cdot10^{-3}$
	10.50	$0.30\cdot10^{-3}$
	10.55	$0.35\cdot10^{-3}$
	10.60	$0.59\cdot10^{-3}$
	10.65	$0.14\cdot10^{-2}$
	10.70	$0.14\cdot10^{-2}$
	10.75	$0.10\cdot10^{-2}$
	10.80	$0.35\cdot10^{-3}$
	10.85	$0.20\cdot10^{-3}$
	10.90	$0.68\cdot10^{-3}$
	10.95	$0.54\cdot10^{-3}$
	11.00	$0.36\cdot10^{-3}$
	11.05	$0.16\cdot10^{-3}$
	11.10	$0.10\cdot10^{-2}$
	11.15	$0.16\cdot10^{-2}$
	11.20	$0.66\cdot10^{-4}$
	11.25	$0.66\cdot10^{-4}$
	11.30	$0.14\cdot10^{-3}$
	11.35	$0.16\cdot10^{-3}$
	11.40	$0.35\cdot10^{-3}$
	11.45	$0.99\cdot10^{-4}$
	11.50	$0.20\cdot10^{-3}$
	11.55	$0.16\cdot10^{-3}$
	11.60	$0.20\cdot10^{-3}$
	11.65	$0.35\cdot10^{-3}$
	11.70	$0.11\cdot10^{-4}$
	11.80	$0.20\cdot10^{-3}$
	11.85	$0.33\cdot10^{-4}$
	11.90	$0.66\cdot10^{-4}$
	11.95	$0.20\cdot10^{-3}$
	12.00	$0.35\cdot10^{-3}$
	12.05	$0.30\cdot10^{-3}$
	12.10	$0.13\cdot10^{-4}$
	12.15	$0.20\cdot10^{-3}$
	12.20	$0.30\cdot10^{-3}$
	12.25	$0.79\cdot10^{-4}$
	12.30	$0.31\cdot10^{-5}$
	12.35	$0.79\cdot10^{-4}$
	12.40	$0.11\cdot10^{-4}$
	12.50	$0.29\cdot10^{-4}$
	12.65	$0.11\cdot10^{-4}$
	12.70	$0.33\cdot10^{-4}$
Aluminium antimonide; AlSb synthetic monocrystal thickness of plate 0.023 mm [152]	0.7	0.00
	0.8	0.01
	0.9	0.02
	1.0	0.03
	1.2	0.03
	1.4	0.025
	1.6	0.03
	1.8	0.035
	2.0	0.095
	2.2	0.010
	2.4	0.095
	2.6	0.08

Table 3.2.6. (*continued*)

State of surface and thickness of layer	λ [μm]	$\tau_{\lambda,n}$
	2.8	0.06
	3.0	0.045
	3.5	0.02
	4.0	0.007
	4.2	0.002
monocrystal, thickness of plate 3 mm, over λ = 4.0 μm $\tau_{\lambda,n}$ decreases when temperature increases [131]	0.8	0.76
	1.6	0.76
	2.4	0.76
	3.2	0.76
	4.0	0.76
	5.2	0.52
	5.6	0.37
	6.0	0.20
polycrystalline, thickness of plate 0.127 mm [131]	0.8	0.46
	1.6	0.47
	2.4	0.48
	3.2	0.50
	4.0	0.50
	4.8	0.49
	5.6	0.45
	6.0	0.33
	7.2	0.14
	8.0	0.04
polycrystalline, thickness of plate 0.254 mm [131]	0.8	0.33
	1.6	0.34
	2.4	0.36
	3.2	0.40
	4.0	0.43
	4.8	0.40
	5.6	0.31
	6.4	0.17
	7.2	0.06
	7.6	0.03
Ammonia; NH_3 mixture of argon and ammonia in gaseous state, 95% NH_3 thickness of layer 100 mm, pressure 10^5 Pa [125]	2.000	1.000
	2.122	0.985
	2.154	0.953
	2.188	0.800
	2.222	0.615
	2.239	0.612
	2.273	0.626
	2.300	0.700
	2.326	0.800
	2.362	0.926
	2.380	0.948
	2.436	0.978
	2.493	0.985
	2.522	0.985
	2.744	0.950
	2.818	0.550
	2.884	0.450
	2.951	0.500
	3.000	0.700
	3.073	0.715
	3.199	0.863
	3.311	0.885
	3.369	0.800
	3.467	0.900
	3.610	0.981
	3.715	0.996
	3.868	0.985
	4.000	0.989
	4.169	0.996
	5.000	0.993
	5.082	0.989
	5.105	0.971
	5.129	0.982
	5.164	0.975
	5.224	0.971
	5.284	0.964
	5.346	0.921
	5.395	0.932
	5.420	0.865
	5.559	0.579
	5.585	0.500
	5.623	0.385
	5.649	0.214
	5.689	0.286
	5.728	0.100
	5.768	0.352
	5.794	0.122
	5.834	0.232
	5.861	0.141
	5.888	0.271
	5.929	0.393
	5.951	0.339
	5.979	0.389
	6.000	0.239

Table 3.2.6. (*continued*)

State of surface and thickness of layer	λ[μm]	$\tau_{\lambda,n}$
	6.008	0.271
	6.043	0.085
	6.071	0.093
	6.131	0.011
	6.209	0.186
	6.255	0.128
	6.309	0.437
	6.346	0.286
	6.401	0.167
	6.475	0.118
	6.531	0.092
	6.557	0.178
	6.595	0.118
	6.626	0.200
	6.664	0.174
	6.702	0.381
	6.761	0.400
	6.788	0.392
	6.823	0.661
	6.859	0.559
	6.985	0.873
	6.938	0.752
	6.958	0.921
	7.000	0.835
	7.174	0.954
	7.215	0.941
	7.257	0.944
	7.443	0.752
	7.555	0.900
	7.638	0.507
	7.709	0.854
	7.763	0.811
	7.843	0.838
	7.889	0.823
	7.925	0.854
	7.966	0.800
	8.000	0.827
	8.035	0.789
	8.159	9.827
	8.349	0.489
	8.401	0.661
	8.511	0.325
	8.544	0.559
	8.643	0.185
	8.736	0.400
	8.810	0.085
	8.885	0.257
	9.000	0.041
	9.108	0.056
	9.241	0.029
	9.320	0.026
	9.394	0.044
	9.462	0.052
	9.531	0.074
	9.613	0.060
	9.708	0.278
	9.753	0.100
	9.824	0.154
	9.869	0.112
	10.000	0.170
mixture of argon and ammonia in gaseous state, 25% NH_3 thickness of layer 100 mm pressure 10^5 Pa [125]	2.122	1.000
	2.154	0.978
	2.188	0.955
	2.222	0.909
	2.239	0.850
	2.273	0.854
	2.326	0.873
	2.362	0.922
	2.380	0.974
	2.493	1.000
	2.522	0.996
	2.744	0.873
	2.818	0.800
	2.884	0.727
	2.951	0.750
	3.000	0.719
	3.073	0.888
	3.199	0.974
	3.369	0.985
	3.610	1.000
	5.164	1.000
	5.420	0.956
	5.585	0.800
	5.702	0.550
	5.728	0.636
	5.742	0.426
	5.768	0.622
	5.814	0.448
	5.834	0.519
	5.861	0.433
	5.951	0.689
	5.979	0.559
	6.008	0.585
	6.043	0.381
	6.131	0.078
	6.246	0.548
	6.255	0.485

State of surface and thickness of layer	λ[μm]	$\tau_{\lambda,n}$	State of surface and thickness of layer	λ[μm]	$\tau_{\lambda,n}$
	6.309	0.615	mixture of argon and ammonia in gaseous state, 2% NH_3, thickness of layer 100 mm pressure 10^5 Pa [125]		
	6.327	0.574		2.188	1.000
	6.456	0.519		2.273	0.986
	6.475	0.422		2.380	1.000
	6.489	0.481		2.744	1.000
	6.504	0.425		2,818	0.952
	6.531	0.500		2.884	0.888
	6.595	0.441		2.951	0.911
	6.626	0.515		3.000	0.877
	6.664	0.555		3.199	1.000
	6.702	0.529		5.420	1.000
	6.761	0.800		5.585	0.915
	6.788	0.674		5.649	0.823
	6.823	0.785		5.689	0.900
	6.985	0.650		5.728	0.746
	7.000	0.862		5.768	0.861
	7.174	0.974		5.834	0.777
	7.215	0.981		5.861	0.831
	7.421	0.800		5.888	0.660
	7.555	0.944		5.929	0.893
	7.709	0.858		5.951	0.878
	7.763	0.880		5.979	0.830
	7.840	0.835		6.000	0.754
	7.925	0.877		6.131	0.512
	7.966	0.831		6.209	0.864
	8.000	0.865		6.255	0.830
	8.035	0.819		6.309	0.800
	8.102	0.926		6.346	0.882
	8.208	0.815		6.401	0.792
	8.280	0.900		6.475	0.736
	8.349	0.727		6.531	0.742
	8.401	0.752		6.557	0.821
	8.544	0.674		6.626	0.752
	8.736	0.559		6.664	0.832
	8.762	0.707		6.702	0.787
	8.810	0.433		6.788	0.810
	8.885	0.515		6.823	0.792
	8.902	0.357		6.856	0.912
	9.000	0.414		6.980	1.000
	9.108	0.381		7.000	0.986
	9.241	0.271		7.079	1.000
	9.320	0.207		8.000	1.000
	9.394	0.271		8.092	0.988
	9.462	0.214		8.117	1.000
	9.531	0.475		8.159	0.974
	9.613	0.300		8.246	1.000
	9.708	0.500		8.349	0.970
	9.753	0.357		8.401	1.000
	10.000	0.529			

Table 3.2.6. (*continued*)

State of surface and thickness of layer	λ [μm]	$\tau_{\lambda,n}$
	8.511	0.900
	8.544	0.950
	8.643	0.819
	8.736	0.885
	9.000	0.782
	9.108	0.758
	9.241	0.712
	9.320	0.748
	9.394	0.850
	9.462	0.762
	9.531	0.872
	9.613	0.776
	9.708	0.915
	9.753	0.938
	9.824	0.900
	9.869	0.932
	10.000	0.940
Ammonium phosphate; (ammonium monobasic phosphate); $NH_4H_2PO_4$ synthetic monocrystal, thickness of plate 8 mm [152]	0.4	0.79
	0.6	0.86
	0.8	0.90
	1.0	0.92
	1.2	0.88
	1.4	0.56
	1.5	0.20
	1.6	0.00
Arsenic sulphide; As_2S_3 polished surface, thickness of plate 2 mm [63, 71]	2.0÷8.0	0.70
	9.0	0.68
	10.0	0.60
	11.0	0.60
	12.0	0.52
	13.0	0.22
	14.0	0.04
Barium fluoride; BaF_2 polished surface, thickness of plate 9 mm [71, 152]	1.0÷7.0	0.96
	9.0	0.94
	11.0	0.62
	13.0	0.07
	14.0	0.000
Barium oxide; BaO synthetic monocrystal, thickness of plate 0.7 mm [131, 152]	0.31	0.04
	0.41	0.44
	0.62	0.74
	1.24	0.85
	2.5	0.87
Barium titanite; $BaTiO_3$ synthetic monocrystal, thickness of plate 0.094 mm [75, 152]	2.0	0.68
	4.0	0.70
	6.0	0.70
	7.0	0.67
	8.0	0.45
	9.0	0.39
	10.0	0.25
	12.0	0.01
Beryllium oxide; BeO smooth surface [131]	2.0÷7.0	0.1÷0.8
	7.0÷20.0	0.0
	20.0÷200.0	0.4÷0.9
Cadmium fluoride; CdF_2 polished surface, thickness of plate 5.0 mm [152]	0.25	0.13
	0.5	0.58
	1.0	0.90
	2.0	0.90
	4.0	0.91
	6.0	0.92
	8.0	0.91
	10.0	0.60
	12.0	0.03
Cadmium sulphide; CdS polished surface, thickness of plate 3.02 mm [152]	1.0	0.66
	2.0	0.72
	4.0	0.73
	6.0	0.73
	8.0	0.74
	10.0	0.76
	12.0	0.73
	14.0	0.72
	16.0	0.36
	17.0	0.05

Table 3.2.6. (*continued*)

State of surface and thickness of layer	λ [μm]	$\tau_{\lambda,n}$
	18.0	0.06
	20.0	0.08
	22.0	0.16
Cadmium telluride; CdTe synthetic monocrystal [71, 152]	1.0÷10.0	0.37
Calcite; Calc spar; calcium carbonate; $CaCO_3$ polished surface, thickness of plate 2.4 mm [152]	1.0	0.84
	2.0	0.80
	2.25	0.40
	2.38	0.76
	2.5	0.29
	2.63	0.77
	2.75	0.22
	2.88	0.76
	3.0	0.67
	3.25	0.26
	3.5	0.00
	3.75	0.21
	4.0	0.01
	4.25	0.21
	4.5	0.07
	5.0	0.33
	5.25	0.16
	5.5	0.00
Carbon; C thin layer on plate of melted quartz [131]	200	0.048
	250	0.008
	300	0.029
	400	0.132
	500	0.194
	600	0.243
	700	0.288
	800	0.299
	900	0.314
	1000	0.316
	1100	0.326
	1200	0.331
	1300	0.332
	1400	0.320
	1500	0.306

State of surface and thickness of layer	λ [μm]	$\tau_{\lambda,n}$
Cassiterite; tin oxide; SnO_2 natural monocrystal, polished surface, thickness of plate 0.5 mm [152]	0.5	0.04
	1.0	0.16
	2.0	0.46
	3.0	0.45
	4.0	0.67
	5.0	0.70
	6.0	0.66
	7.0	0.33
	7.5	0.04
	8.0	0.00
Cerargyrite; horn silver; silver chloride; AgCl polished surface, thickness of plate 5 mm [71, 152]	2.5	0.76
	5.0	0.79
	7.5÷17.5	0.80
	20.0	0.77
	22.5	0.64
	25.0	0.35
Cesium bromide; CsBr thickness of plate 5 mm [152]	16	0.90
	24	0.90
	28	0.88
	32	0.86
	36	0.73
	40	0.54
	44	0.32
	48	0.15
	50	0.10
thickness of plate 6 mm [71, 152]	15÷24	0.90
	28	0.89
	32	0.86
	36	0.74
	40	0.54
	44	0.33
	48	0.17
	52	0.08

Table 3.2.6. (*continued*)

State of surface and thickness of layer	$\lambda[\mu m]$	$\tau_{\lambda,n}$
Cesium iodide; CsJ thickness of plate 5 mm [71, 152]	1÷40	0.92
	45	0.88
	50	0.72
	55	0.51
	60	0.41
	65	0.16
	70	0.08
	75	0.02
Chile saltpetre; Chile nitrate; sodium nitrate; $NaNO_3$ natural or synthetic monocrystal, polished surface, thickness of plate 1.75 mm [152]	1.0	0.63
	2.0	0.70
	3.0	0.68
	4.0	0.22
	5.0	0.73
	6.0	0.02
	8.0	0.02
	9.0	0.07
	10.0	0.16
	11.0	0.05
	12.0	0.02
	13.0	0.59
	14.0	0.59
	15.0	0.50
Copper bromide; CuBr polished surface, thickness of plate 2.92 mm [152]	0.5	0.55
	1.0	0.66
	2.0	0.71
	4.0	0.777
	8.0÷16.0	0.78
	20.0	0.77
	24.0	0.67
	28.0	0.18
	32.0	0.01
Copper chloride; CuCl polished surface, thickness of plate 3.92 mm [152]	0.5	0.52
	1.0	0.68
	2.0	0.77
	3.0÷4.0	0.77
	12.0	0.78
	14.0	0.77
	16.0	0.75
	18.0	0.66
	20.0	0.24
	22.0	0.05
Corundum, sintered; aluminium oxide; Al_2O_3 monocrystal, thickness 1 mm, $\tau_{\lambda,n}$ decreases with increase of temperature for $\lambda \geq 4.0$ m [131]	0.8	0.90
	1.6	0.92
	2.4	0.93
	3.2	0.92
	4.0	0.90
	5.2	0.74
	5.6	0.63
	6.0	0.50
	6.4	0.33
	6.8	0.11
monocrystal, thickness of plate 1.97 mm, temperature 555 K [131]	0.8	0.835
	1.6	0.84
	2.4	0.845
	3.2	0.85
	4.0	0.85
	4.4	0.87
	4.8	0.87
	5.2	0.83
	5.6	0.72
	6.0	0.53
monocrystal, thickness of plate 1.97 mm, temperature 830 K [131]	0.8	0.835
	1.6	0.84
	2.4	0.845
	3.2	0.85
	4.0	0.85
	4.4	0.85
	4.8	0.83
	5.2	0.775
	5.6	0.66
	6.0	0.43

Table 3.2.6. (*continued*)

State of surface and thickness of layer	λ[μm]	$\tau_{\lambda,n}$
monocrystal, thickness of plate 1.97 mm, temperature 1300 K [131]	0.8	0.83
	1.6	0.84
	2.4	0.845
	3.2	0.85
	4.0	0.85
	4.4	0.835
	4.8	0.79
	5.2	0.70
	5.6	0.54
	6.0	0.25
monocrystal, thickness of plate 1.97 mm, temperature 1730 K [131]	0.8	0.835
	1.6	0.84
	2.4	0.845
	3.2	0.85
	4.0	0.84
	4.4	0.805
	4.8	0.70
	5.2	0.565
	5.6	0.37
	6.0	0.13
monocrystal, thickness of plate 1.97 mm, temperature 2030 K [131]	0.8	0.835
	1.6	0.84
	2.4	0.845
	3.2	0.85
	4.0	0.83
	4.4	0.77
	4.8	0.65
	5.2	0.51
	5.6	0.32
	6.0	0.08
monocrystal, thickness of plate 3 mm [131]	0.8	0.76
	1.6	0.76
	2.4	0.76
	3.2	0.76
	4.0	0.755
	4.4	0.73
	4.8	0.66
	5.2	0.53
	5.6	0.38
	6.0	0.20
polycrystal, 96÷99% Al_2O_3, thickness of plate 0.127 mm [131]	0.8	0.46
	1.6	0.47
	2.4	0.48
	3.2	0.50
	4.0	0.50
	4.8	0.49
	5.6	0.45
	6.4	0.32
	7.2	0.14
	8.0	0.04
polycrastal, 96÷99% Al_2O_3, thickness of plate 0.254 mm [131]	0.8	0.335
	1.6	0.34
	2.4	0.36
	3.2	0.40
	4.0	0.425
	4.8	0.40
	5.6	0.31
	6.4	0.17
	7.2	0.05
	8.0	0.00
Diamond; C thickness of plate 1.8 mm [152]	2.0	0.67
	3.0	0.52
	4.0	0.31
	5.0	0.10
	6.0	0.54
	7.0	0.61
	8.0÷24.0	0.62
Fluorite; calcium fluoride; CaF_2 polished surface, thickness of plate 1 mm [71, 152]	1.0÷6.0	0.95
	7.0	0.93
	8.0	0.92
	9.0	0.86
	10.0	0.75
	11.0	0.57
	12.0	0.31
	13.0	0.11

Table 3.2.6. (*continued*)

State of surface and thickness of layer	λ[μm]	$\tau_{\lambda,n}$
polished surface, thickness of plate 10 mm [152]	1.0÷5.0	0.92
	6.0	0.90
	7.0	0.76
	8.0	0.46
	9.0	0.11
Galena; galenite; lead glance; lead sulphide; PbS natural monocrystal [152]	2.0	0.01
	3.0	0.02
	4.0	0.154
	5.0	0.154
	6.0	0.135
	7.0	0.104
	8.0	0.063
	9.0	0.031
	10.0	0.018
	11.0	0.010
	12.0	0.007
Gallium antimonide; GaSb synthetic monocrystal, thickness of plate 0.504 mm [152]	1.5	0.00
	1.75	0.19
	2.0	0.32
	2.5	0.31
	3.0	0.20
	3.5	0.11
	4.0	0.05
	4.5	0.03
Gallium arsenide; GaAs synthetic monocrystal, thickness of plate 0.89 mm [131]	1.0	0.15
	1.25	0.47
	2.5	0.54
	5.0÷15.0	0.55
Gallium phosphide; GaP polished surface, thickness of plate 0.386 mm, room temperature [131]	6.0	0.44
	8.0	0.37
	10.0	0.34
	12.0	0.32
	12.5	0.26
	13.0	0.27
polished surface, thickness of plate 0.18 mm, room temperature [131]	13.0	0.38
	13.5	0.10
	14.0	0.13
	15.0	0.37
	16.0	0.34
	18.0	0.16
	20.0	0.21
	22.0	0.10
	24.0	0.04
Germanium; Ge synthetic monocrystal, thickness of plate 2 mm at temperature 298 K [35, 63, 71]	2.0	0.24
	2.5	0.41
	5.0	0.51
	10.0	0.50
	15.0	0.42
	20.0	0.23
	22.5	0.10
	25.0	0.17
	30.0	0.00
	35.0	0.08
	40.0	0.17
	50.0	0.20
synthetic monocrystal, thickness of plate 2 mm, resistivity 30 cm, temperature 473 K [135, 152]	2.0	0.01
	3.0	0.11
	4.0	0.09
	5.0	0.06
	6.0	0.03
	7.0	0.00

Table 3.2.6. (*continued*)

State of surface and thickness of layer	λ [μm]	$\tau_{\lambda,n}$
Glass, arsenic sulphide; As_2S_3 thickness of plate 6.05 mm [152]	2.0	0.70
	3.0	0.57
	4.0	0.72
	6.0	0.72
	7.0	0.70
	8.0	0.64
	8.5	0.35
	9.0	0.43
	10.0	0.31
	11.0	0.33
	12.0	0.08
	13.0	0.02
Glass BS-37A thickness of plate 2.5 mm [152]	1.0	0.84
	2.0	0.85
	3.0	0.86
	4.0	0.86
	5.0	0.43
	5.5	0.03
Glass IR-11 thickness of plate 2.03 mm [152]	2.0	0.88
	2.5	0.87
	2.6	0.87
	2.7	0.87
	2.8	0.83
	2.9	9.60
	3.0	0.40
	3.2	0.53
	3.4	0.63
	3.6	0.72
	3.8	0.78
	4.0	0.80
	4.2	0.79
	4.4	0.76
	4.6	0.72
	4.8	0.67
	5.0	0.48
	5.2	0.27
	5.4	0.14
	5.6	0.09
	5.8	0.05
	6.0	0.02
thickness of plate 8 mm [152]	2.0	0.88
	2.5	0.84
	2.6	0.83
	2.7	0.83
	2.8	0.62
	2.9	0.27
	3.0	0.05
	3.2	0.16
	3.4	0.31
	3.6	0.43
	3.8	0.52
	4.0	0.61
	4.2	0.59
	4.4	0.52
	4.6	0.45
	4.8	0.36
	5.0	0.13
	5.2	0.01
Glass selenium arsenide (SeAs) polished surface, thickness of plate 2.0 mm [152]	1.0	0.25
	2.0	0.675
	4.0	0.705
	6.0	0.71
	8.0	0.71
	10.0	0.70
	12.0	0.61
	12.5	0.18
	12.75	0.04
	13.0	0.54
	14.0	0.57
	15.0	0.59
	16.0	0.55
Gypsum; hydrated calcium sulphate; $CaSO_4 \cdot 2H_2O$ thickness of plate 0.648 mm [24]	1.5	0.78
	2.0	0.74
	2.5	0.25
	3.0	0.04
	3.5	0.18
	4.0	0.36
	4.5	0.00
	5.0	0.10
	5.5	0.11
	6.0	0.00

Table 3.2.6. (*continued*)

State of surface and thickness of layer	λ[μm]	$\tau_{\lambda,n}$
Indium antimonide; InSb synthetic, monocrystal, polished surface, thickness of plate 0.25 mm [71]	7.0	0.03
	8.0	0.07
	9.0	0.12
	10.0	0.14
	12.0	0.16
	14.0	0.15
	16.0	0.15
synthetic, monocrystal, thickness of plate 0.15 mm, temperature 78 K [152]	5.0	0.41
	6.0	0.43
	7.0	0.43
	8.0	0.43
	9.0	0.43
	10.0	0.425
	15.0	0.43
	20.0	0.43
	25.0	0.43
	27.5	0.28
	30.0	0.35
	-	-
	55.0	0.07
	75.0	0.32
	95.0	0.26
	115.0	0.34
	135.0	0.335
synthetic monocrystal, thickness of plate 0.15 mm, temperature 297 K [152]	7.0	0.01
	8.0	0.25
	9.0	0.30
	10.0	0.305
	15.0	0.32
	20.0	0.32
	25.0	0.284
	27.7	0.17
	30.0	0.18
	-	-
	55.0	0.015
	75.0	0.035
	95.0	0.035
	115.0	0.01
	135.0	0.00
Lead chloride; $PbCl_2$ polished surface, thickness of plate 2.4 mm [152]	10.0÷14.0	1.00
	15.0	0.98
	20.0	0.57
	23.0	0.11
Lead fluoride; PbF_2 polished surface, thickness of plate 10 mm [152]	6.0÷8.0	1.00
	9.0	0.95
	10.0	0.84
	11.0	0.54
	12.0	0.22
	13.0	0.03
Lithium fluoride; LiF polished surface, thickness of plate 0.1 mm [71, 135, 152]	4.0	0.97
	6.0	0.97
	8.0	0.91
	10.0	0.70
	12.0	0.23
	14.0	0.02
polished surface, thickness of plate 1 mm [71, 135, 152]	5.0	0.96
	6.0	0.92
	8.0	0.48
	10.0	0.04
	12.0	0.00
	100.0	0.00
	200.0	0.14
	300.0	0.36
	400.0	0.56
	500.0	0.60
polished surface, thickness of plate 10 mm [71, 135, 152]	4.0	0.91
	6.0	0.63
	7.0	0.09
	8.0	0.00
Magnesium fluoride; MgF_2 polished surface, thickness of plate 1 mm [152]	1.0	0.68
	2.0	0.90
	2.65	0.77
	3.0	0.93

Table 3.2.6. (*continued*)

State of surface and thickness of layer	λ [μm]	$\tau_{\lambda,n}$
	4.0	0.94
	5.0	0.91
	6.0	0.94
	7.0	0.88
	8.0	0.43
	9.0	0.20
polished surface, thickness of plate 1.75 mm, temperature 300 K [71, 152]	1.0	0.30
	2.0	0.75
	2.65	0.64
	3.0	0.83
	4.0	0.88
	5.0	0.89
	6.0	0.87
	7.0	0.80
	8.0	0.60
	9.0	0.21
	10.0	0.03
polished surface, thickness of plate 1.75 mm, temperature 1073 K [71]	1.0	0.30
	2.0	0.75
	2.65	0.64
	3.0	0.83
	4.0	0.88
	5.0	0.89
	6.0	0.83
	7.0	0.60
	8.0	0.27
	9.0	0.00
monocrystal, polished surface, thickness of plate 3 mm [152]	2	0.96
	4	0.96
	6	0.95
	7	0.83
	8	0.51
	9	0.16
	10	0.06
Mica synthetic monocrystal, thickness of plate 0.03 mm [41, 131, 152]	2.5	0.97
	2.6	0.95
	2.7	0.03
	2.8	0.96
	2.9	0.96
	3.0	0.94
	3.5	0.97
	4.0	0.99
	4.5	0.96
	5.0	0.93
	5.5	0.63
	6.0	0.72
	6.5	0.78
	7.0	0.51
	7.5	0.57
	8.0	0.38
	8.5	0.08
	8.9	0.00
	9.0	0.00
synthetic monocrystal, thickness of plate 0.08 mm [41, 152]	1.0	0.98
	1.4	0.91
	1.5	0.97
	1.6	0.98
	2.0	0.98
	2.1	0.98
	2.2	0.82
	2.3	0.91
	2.4	0.95
	2.5	0.92
	2.6	0.30
	2.7	0.03
	2.8	0.88
	2.9	0.91
	3.0	0.97
	3.5	0.96
	4.0	0.95
	4.5	0.90
	5.0	0.57
	5.5	0.27
	6.0	0.38
	6.5	0.48
	7.0	0.32
	7.5	0.23
	8.0	0.11
	8.5	0.02
Periclase; magnesium oxide; MgO polished surface, monocrystal, thickness of plate 0.468 mm [152]	2.0	0.87
	4.0	0.87
	6.0	0.87

Table 3.2.6. (*continued*)

State of surface and thickness of layer	λ[μm]	$\tau_{\lambda,n}$
	8.0	0.76
	9.0	0.65
	10.0	0.09
	10.5	0.13
	10.75	0.05
polished surface, monocrystal, thickness of plate 0.076 mm [152]	9.0	0.86
	10.0	0.59
	10.5	0.67
	11.0	0.40
	11.5	0.18
	12.0	0.20
	12.5	0.27
	13.0	0.07
Potassium aluminate; IR-11 thickness of plate 2.03 mm [152]	2.0	0.88
	2.5	0.87
	3.0	0.39
	3.5	0.69
	4.0	0.88
	4.5	0.73
	5.0	0.49
	5.5	0.12
	6.0	0.03
thickness of plate 8.00 mm [152]	2.0	0.87
	2.5	0.85
	3.0	0.04
	3.5	0.44
	4.0	0.59
	4.5	0.45
	5.0	0.12
Potassium bromide; KBr polished surface, thickness of plate 0.1 mm [152]	20÷34	0.90
	42	0.72
	50	0.17
	54	0.03
polished surface, thickness of plate 1.0 mm [152]	20÷24	0.87
	28	0.82
	34	0.63
	42	0.10
	46	0.01

State of surface and thickness of layer	λ[μm]	$\tau_{\lambda,n}$
polished surface, thickness of plate 10 mm [152]	20	0.84
	24	0.71
	28	0.34
	34	0.05
polished surface, thickness of plate 0.41 mm [152]	100	0.03
	200	0.14
	300	0.34
	350	0.48
	400	0.54
	500	0.60
	600	0.60
Potassium chloride; KCL polished surface, thickness of plate 0.1 mm [152]	14÷24	0.93
	28	0.89
	34	0.65
	42	0.06
polished surface, thickness of plate 1.0 mm [152]	14÷16	0.94
	20	0.89
	24	0.82
	28	0.47
	34	0.03
polished surface, thickness of plate 10 mm [152]	14	0.91
	16	0.87
	20	0.56
	24	0.10
	28	0.00
Potassium hydrogen phosphate; KH_2PO_4 synthetic monocrystal, polished surface, thickness of plate 10 mm [152]	0.4	0.75
	0.6	0.82
	0.8	0.87
	1.0	0.87
	1.2	0.81
	1.4	0.48
	1.5	0.12

State of surface and thickness of layer	λ [μm]	$\tau_{\lambda,n}$
Potassium iodide; KJ polished surface, thickness of plate 0.83 mm [152]	3.0	0.95
	10.0	0.94
	20.0	0.91
	30.0	0.82
	40.0	0.44
	50.0	0.12
Quartz crystalline; SiO_2 thickness of plate 4.55 mm [152]	1.0	0.92
	2.0	0.91
	2.25	0.89
	2.5	0.71
	2.75	0.355
	2.9	0.19
	3.0	0.30
	3.25	0.665
	3.50	0.54
	3.75	0.215
	4.00	0.10
	4.25	0.03
thickness of plate 10 mm [152]	100	0.515
	200	0.68
	300	0.75
	400	0.80
	500	0.80
	600	0.80
thickness of plate 40 mm [152]	100	0.06
	200	0.27
	300	0.495
	400	0.66
	500	0.73
	600	0.75
Quartz, melted; SiO_2 thickness of plate 0.064 mm [152]	2.0	0.96
	3.0	0.96
	4.0	0.95
	4.25	0.945
	4.50	0.945
	4.75	0.90
	5.00	0.58
	5.25	0.50
	5.40	0.38
	5.75	0.60
	6.00	0.50
	6.25	0.47
	6.50	0.55
	6.75	0.47
	7.00	0.44
	7.25	0.36
	7.50	0.20
	8.0÷10.0	0.00
	10.5	0.14
	11.0	0.27
	11.50	0.18
thickness of plate 0.10 mm [152]	2.0	0.95
	3.0	0.945
	4.0	0.95
	4.25	0.945
	4.50	0.93
	4.75	0.80
	5.00	0.33
	5.25	0.28
	5.40	0.25
	5.75	0.36
	6.00	0.24
	6.25	0.20
	6.50	0.29
	6.75	0.21
	7.00	0.17
	7.25	0.12
	7.50	0.08
	8.0÷10.0	0.00
	10.5	0.02
	11.25	0.09
	11.50	0.03
thickness of plate 0.18 mm [152]	2.0	0.95
	3.0	0.945
	4.0	0.94
	4.25	0.925
	4.50	0.84
	4.75	0.69
	5.0	0.11
	5.25	0.09
	5.40	0.01
	5.75	0.12
	6.0	0.06
	6.25	0.04
	6.5	0.09
	7.0	0.03
	7.5	0.00
thickness of plate 0.51 mm [152]	2.0	0.95
	3.0	0.945

Table 3.2.6. (*continued*)

State of surface and thickness of layer	λ [μm]	$\tau_{\lambda,n}$
	3.5	0.945
	4.0	0.91
	4.25	0.83
	4.50	0.64
	4.75	0.44
	5.0	0.00
thickness of plate 0.55 mm		
[152]	100	0.37
	200	0.58
	300	0.68
	400	0.715
	500	0.75
	600	0.78
thickness of plate 0.7 mm, temperature 293 K		
[131]	0.225	0.34
	0.25	0.59
	0.275	0.67
	0.3	0.71
	0.325	0.76
	0.35	0.835
	0.375	0.92
	0.4	0.94
thickness of plate 0.7 mm, temperature 1173 K		
[131]	0.225	0.00
	0.25	0.28
	0.275	0.56
	0.3	0.64
	0.325	0.74
	0.35	0.835
	0.375	0.92
	0.4	0.94
thickness of plate 0.76 mm		
[152]	2.0	0.95
	3.0	0.945
	3.5	0.945
	4.0	0.87
	4.25	0.78
	4.50	0.55
	4.75	0.20
	5.0	0.00
thickness of plate 0.8 mm, temperature 298 K		
[131]	2.0÷3.0	1.00
	3.5	0.97
	4.0	0.88
	4.5	0.59
	4.75	0.44
	5.0	0.01
thickness of plate 0.8 mm, temperature 1873 K		
[131]	2.0÷3.0	1.00
	3.5	0.90
	4.0	0.53
	4.5	0.06
	4.75	0.05
	5.0	0.00
thickness of plate 1.9 mm, temperature 293 K		
[132]	0.225	0.66
	0.25	0.87
	0.275	0.92
	0.3	0.92
	0.325	0.92
	0.35	0.92
	0.375	0.91
	0.4	0.91
thickness of plate 1.9 mm, temperature 1173 K		
[131]	0.225	0.45
	0.25	0.74
	0.275	0.83
	0.3	0.90
	0.325	0.90
	0.35	0.91
	0.375	0.91
	0.4	0.91
thickness of plate 8.30 mm		
[152]	100	0.02
	200	0.06
	300	0.15
	400	0.25
	500	0.37
	600	0.46
thickness of plate 29.85 mm, temperature 288 K		
[131]	2.0	0.87
	2.5	0.84
	2.75	0.37
	3.0	0.57
	3.5	0.42
	3.75	0.01

State of surface and thickness of layer	λ [μm]	$\tau_{\lambda,n}$
Salt, domestic; table salt; common salt; sodium chloride; NaCl polished surface, thickness of plate 0.1 mm [71, 152]	5.0÷16.0	0.95
	20.0	0.94
	24.0	0.84
	28.0	0.60
	34.0	0.06
thickness of plate 0.32 mm [152]	100	0.03
	150	0.14
	200	0.23
	250	0.32
	300	0.39
	350	0.44
	400	0.52
	450	0.57
	500	0.62
	600	0.61
polished surface, thickness of plate 1.0 mm [71, 152]	5.0÷10.0	0.94
	15.0	0.905
	16.0	0.90
	20.0	0.67
	24.0	0.24
	28.0	0.02
polished surface, thickness of plate 10 mm [71, 152]	5.0÷12.0	0.93
	16.0	0 66
	20.0	0.04
Sapphire; aluminium oxide; Al_2O_3 natural or synthetic monocrystal, thickness of plate 1.00 mm [152]	1.0	0.87
	2.0	0.91
	3.0	0.92
	4.0	0.88
	5.0	0.78
	6.0	0.52
	6.5	0.26
	7.0	0.04
Scheelite; calcium tungstate; $CaWO_4$ polished surface, thickness of plate 2,54 mm [146]	0.5	0.74
	1.0	0.81
	2.0	0.80
	3.0	0.80
	4.0	0.69
	4.5	0.85
	5.0	0.70
	6.0	0.07
	6.5	0.03
Selenium; Se crystal, thickness of plate 0.198 mm [152]	1.0	0.50
	2.0	0.51
	2.5	0.52
	2.7	0.52
	2.8	0.56
	2.9	0.49
	3.0	0.495
	3.2	0.52
	3.3	0.51
	3.4	0.43
	3.5	0.47
	4.0	0.55
crystal, thickness of plate 0.35 mm [152]	1.6	0.22
	2.0	0.23
	2.5	0.245
	2.7	0.255
	2.8	0.32
	2.9	0.27
	3.0	0.275
	3.5	0.29
	4.0	0.325
polycrystal, thickness of plate 0.155 mm [152]	10.0	0.33
	15.0	0.325
	20.0	0.33
	25.0	0.28
	30.0	0.19
	40.0	0.17
	50.0	0.14
	100.0	0.18

Table 3.2.6. (*continued*)

State of surface and thickness of layer	λ[μm]	$\tau_{\lambda,n}$
amorphous, thickness of plate 0.54 mm [152]	15.0	0.71
	17.0	0.70
	18.0	0.69
	20.0	0.46
	21.0	0.47
	22.0	0.63
	23.0	0.66
	24.0	0.63
	25.0	0.57
amorphous, thickness of plate 2,06 mm [152]	0.8	0.40
	1.0	0.69
	1.2	0.71
	2.0	0.71
	5.0	0.71
	10.0	0.71
	12.0	0.70
	13.0	0.57
	15.0	0.68
	20.0	0.09
	22.0	0.37
	23.0	0.41
	24.0	0.36
amorphous, thickness of plate 5.62 mm [152]	0.8	0.29
	1.0	0.65
	1.2	0.67
	2.0	0.71
	5.0	0.71
	10.0	0.70
	12.0	0.68
	13.0	0.42
	15.0	0.67
	20.0	0.01
	22.0	0.26
	23.0	0.31
	24.0	0.19
Silicon carbide; SiC synthetic, monocrystal, thickness of plate 0.27 mm [152]	1.0	0.62
	3.0	0.66
	3.5	0.65
	4.0	0.65
	4.2	0.50
	4.4	0.63
	4.6	0.60
	4.8	0.57
	5.0	0.58
	6.0	0.01
	6.8	0.49
	7.0	0.10
	7.5	0.01
	8.0	0.04
	8.5	0.02
	9.0	0.02
	10.0	0.02
	11.0	0.00
	13.0	0.00
	14.0	0.01
	15.0	0.08
	16.0	0.17
	17.0	0.26
	19.0	0.33
	21.0	0.36
	23.0	0.37
	25.0	0.38
Silicone [35]	1.0	0.25
	2.0	0.96
	3.0	0.77
	4.0	0.67
	6.0	0.61
	7.0	0.57
	8.0	0.55
Silicon; Si synthetic, monocrystal, polished surface, resistivity 150 Ω•cm, thickness of plate 0.5 mm [152]	1.0	0.25
	2.0	0.50
	4.0	0.54
	6.0	0.544
	8.0	0.535
	9.0	0.47
	10.0	0.535
	12.0	0.50
	13.0	0.498
	13.5	0.445
	14.0	0.48
	15.0	0.525
	16.0	0.40
	16.5	0.33
	17.0	0.44
	18.0	0.45
	20.0	0.47
	22.0	0.48

Table 3.2.6. (*continued*)

State of surface and thickness of layer	λ[μm]	$\tau_{\lambda,n}$
synthetic monocrystal, polished surface, temperature 298 K, thickness of plate 5 mm [71, 152]	1.25	0.55
	1.5	0.58
	2.0	0.58
	3.0	0.58
	4.0	0.58
	5.0	0.58
	6.0	0.58
	7.0	0.48
	8.0	0.49
synthetic monocrystal, polished surface, temperature 673 K, thickness of plate 5 mm [71, 152]	1.25	0.16
	1.5	0.49
	2.0	0.53
	6.0	0.41
	7.0	0.39
	8.0	0.36
Silver, bromide-silver chloride mixture; KRS-13 polished surface, thickness of plate 9.5 mm [152]	1.0	0.19
	2.0	0.53
	3.0	0.66
	4.0	0.69
	5.0÷10.0	0.90
	15.0	0.28
	20.0	0.57
	25.0	0.29
Sodium fluoride; NaF polished surface, thickness of plate 0.1 mm [152]	2.0÷12.0	0.98
	14.0	0.91
	16.0	0.80
	18.0	0.53
	20.0	0.14
	22.0	0.01
polished surface, thickness of plate 1.0 mm [152]	2.0÷8.0	0.97
	10.0	0.92
	12.0	0.77
	14.0	0.45
	16.0	0.11
	17.0	0.02
polished surface, thickness of plate 10 mm [152]	2.0÷8.0	0.92
	10.0	0.51
	12.0	0.07
Sphalerite; zinc blende; zinc sulphide; ZnS thickness of plate 1.78 mm [152]	2.0÷8.0	0.69
	10.0	0.67
	12.0	0.62
	14.0	0.40
	14.5	0.22
	15.0	0.00
	50.0	0.20
polycrystalline, thickness of plate 3.8 mm [152]	1.2	0.20
	1.4	0.38
	1.6	0.43
	1.8	0.53
	2.0	0.57
	2.2	0.59
	2.4	0.64
	2.6	0.66
	2.8	0.67
	3.0	0.68
	4.0	0.72
	5.0	0.73
	7.0	0.73
	9.0	0.73
	11.0	0.62
	12.0	0.60
	13.0	0.48
	14.0	0.22
	15.0	0.00
Spinel; magnesium-aluminate; $MgO \cdot Al_2O_3$ synthetic monocrystal, thickness of plate 5.44 mm [152]	0.4	0.70
	0.5	0.79
	0.6	0.83
	0.7	0.85
	0.8	0.87

Table 3.2.6. (*continued*)

State of surface and thickness of layer	$\lambda[\mu m]$ [1]	$\tau_{\lambda,n}$
	0.9	0.88
	1.0	0.89
	2.0	0.80
	2.5	0.80
	3.0	0.28
	3.5	0.87
	3.7	0.88
	4.0	0.85
	4.5	0.70
	5.0	0.50
	5.5	0.00
Stibine, antimony hydride; SbH_3 thickness of plate o.45 mm [76]	1.0	0.40
	2.0	0.48
	4.0	0.45
	6.0	0.49
	8.0	0.54
	10.0	0.55
	12.0	0.49
	14.0	0.20
thickness of plate 4.9 mm [76]	1.0	0.11
	2.0	0.43
	4.0	0.41
	6.0	0.43
	8.0	0.48
	10.0	0.45
	12.0	0.20
	14.0	0.02
Strontium fluoride; SrF_2 polished surface, thickness of plate 10 mm [152]	6.0	1.0
	7.0	1.0
	8.0	0.98
	9.0	0.90
	10.0	0.72
	11.0	0.37
Strontium titanate; $SrTiO_3$ synthetic monocrystal, polished surface, thickness of plate 1.0 mm [152]	1.0	0.74
	2.0	0.75
	2.85	0.69
	3.0	0.76
	4.0	0.77
	4.66	0.73
	5.0	0.74
	6.0	0.58
	7.0	0.12
synthetic monocrystal, polished surface, thickness of plate 10 mm [152]	1.0	0.74
	2.0	0.75
	2.9	0.70
	3.0	0.76
	4.0	0.75
	4.66	0.39
	5.0	0.41
	6.0	0.03
	7.0	0.00
Sulphur, crystalline; S_8 monocrystalline, thickness of plate 0.4 mm [152]	1.0	0.61
	2.0	0.88
	3.0	0.71
	4.0	0.75
	5.0	0.77
	6.0	0.80
	7.0	0.75
	8.0	0.79
	9.0	0.79
	10.0	0.78
	11.0	0.67
	11.5	0.37
	11.75	0.11
	12.0	0.33
	14.0	0.59
	16.0	0.59
	18.0	0.59
	20.0	0.50
	22.0	0.015
	24.0	0.52
	28.0	0.71
	30.0	0.62
	32.0	0.68
monocrystalline, thickness of plate 0.90 mm [152]	20.0	0.33
	30.0	0.45
	40.0	0.13
	50.0	0.09
	60.0	0.46
	65.0	0.43
	70.0	0.42
	75.0	0.60

Table 3.2.6. (*continued*)

State of surface and thickness of layer	λ[μm]	$\tau_{\lambda,n}$
	80.0	0.56
	90.0	0.62
	95.0	0.50
	100.0	0.60
	105.0	0.65
	110.0	0.59
	120.0	0.63
	130.0	0.37
monocrystalline, thickness of plate 2.75 mm [152]	20.0	0.16
	30.0	0.41
	40.0	0.05
	50.0	0.00
	60.0	0.14
	65.0	0.11
	70.0	0.15
	75.0	0.24
	80.0	0.28
	90.0	0.37
	95.0	0.30
	100.0	0.36
	105.0	0.47
	110.0	0.42
	120.0	0.33
	130.0	0.18
Tellurium; Te thickness of plate 0.23 mm, temperature 100 K [131]	35.0	0.21
	40.0	0.26
thickness of plate 1.23 mm [71, 131]	4.0÷22.5	0.30
	25.0	0.295
	30.0	0.27
	35.0	0.01
	40.0	0.08
thickness of plate 0.09 mm, temperature 100 K [131]	100.0	0.08
thickness of plate 0.09 mm, temperature 300 K [131]	45.0	0.04
	50.0	0.08
	75.0	0.12
Tellurium oxide; TeO_2 thickness of plate 3.05 mm, temperature of measurement 533÷643 K [131]	0.5	0.73
	0.6	0.84
	0.7	0.85
	0.8	0.85
	1.0	0.85
	1.5	0.85
	2.0	0.85
	3.0	0.79
	4.0	0.85
	5.0	0.81
	6.0	0.73
thickness of plate 7.65 mm, temperature of measurement 533÷643 K [131]	0.6	0.80
	0.7	0.81
	0.8	0.82
	1.0÷2.0	0.83
	3.0	0.51
	4.0	0.67
	5.0	0.60
	6.0	0.20
Thallium, bromide-thalium chloride mixture; KRS-6 polished surface, thickness of plate 3.5 mm [152]	1.0	0.695
	2.0	0.73
	4.0	0.74
	6.0	0.743
	8.0	0.748
	10.0	0.75
	20.0	0.76
	30.0	0.40
polished surface, thickness of plate 6.0 mm [71]	4.0	0.61
	7.0	0.80
	10.0	0.81
	13.0	0.80
	16.0	0.77
	19.0	0.66
	22.0	0.53
	25.0	0.39
	28.0	0.15

Table 3.2.6. (*continued*)

State of surface and thickness of layer	λ[μm]	$\tau_{\lambda,n}$
	31.0	0.08
	34.0	0.06
Thallium bromide-thallium iodide mixture; KRS-5 polished surface, thickness of plate 2.4 mm [71, 152]	0.5	0.00
	0.6	0.47
	0.7	0.56
	0.8	0.59
	0.9	0.61
	1.0	0.62
	2.0	0.66
	4.0	0.67
	6.0	0.68
	8.0	0.685
	10.0	0.69
	20.0	0.695
	30.0	0.70
	40.0	0.59
	50.0	0.25
	60.0	0.06
Thallium bromide; TlBr thickness of plate 6 mm [152]	22.0	0.605
	24.0	0.605
	26.0	0.595
	28.0	0.55
	30.0	0.495
	32.0	0.495
	34.0	0.34
	36.0	0.27
Thallium chloride; TlCl polished surface, thickness of plate 6.0 mm [152]	23.0	0.45
	24.0	0.34
	25.0	0.25
	26.0	0.16
	27.0	0.09
	28.0	0.05
Titanium oxide; TiO_2 natural or synthetic monocrystal, thickness of plate 6 mm [131, 152]	1.0÷4.0	0.70
	5.0	0.38
	6.0	0.02

State of surface and thickness of layer	λ[μm]	$\tau_{\lambda,n}$
Water; H_2O thickness of layer 3 μm [10, 49, 76, 126, 139]	2.0	0.98
	2.5	0.91
	3.0	0.00
	3.5	0.84
	4.0	0.94
	4.5	0.74
	5.0	0.80
	5.5	0.85
	6.0	0.29
	6.5	0.57
	7.0	0.63
	8.0	0.65
	9.0	0.67
	10.0	0.61
thickness of layer 9 μm [10, 49, 76, 126, 139]	2.0	0.95
	2.5	0.90
	3.0	0.00
	3.5	0.60
	4.0	0.88
	4.5	0.60
	5.0	0.70
	5.5	0.75
	6.0	0.17
	7.0	0.46
	8.0	0.50
	9.0	0.53
	10.0	0.46
thickness of layer 18 μm [10, 49, 76, 126]	2.0	0.89
	2.5	0.59
	3.0	0.00
	3.5	0.35
	4.0	0.70
	4.5	0.35
	5.0	0.37
	5.5	0.50
	6.0	0.01
	7.0	0.13
	8.0	0.17
	9.0	0.22
	10.0	0.16
thickness of layer 0.6 mm [10, 49, 76, 126, 139]	0.6	1.0
	0.8	0.99
	1.0	0.98

State of surface and thickness of layer	λ [μm]	$\tau_{\lambda,n}$
	1.2	0.93
	1.4	0.48
	1.6	0.29
	1.7	0.52
	1.8	0.28
	2.0	0.01
	2.2	0.18
	2.4	0.01
thickness of layer 4 mm [10, 49, 76, 126, 139]	0.6	1.0
	0.8	0.99
	1.0	0.84
	1.1	0.92
	1.2	0.84
	1.3	0.82
	1.4	0.08
thickness of layer 8 mm [10, 49, 76, 126, 139]	0.6	1.0
	0.8	0.98
	0.9	0.97
	1.0	0.73
	1.1	0.85
	1.2	0.41
	1.3	0.39
	1.4	0.04
thickness of layer 22 mm [10, 49, 76, 126, 139]	0.6	1.0
	0.8	0.94
	0.9	0.88
	1.0	0.40
	1.1	0.64
	1.2	0.09
	1.3	0.08
	1.4	0.02
Ytterbium oxide; Yb_2O_3 [41, 131]	2.5	0.71
	3.0	0.63
	3.5	0.65
	4.0	0.82
	5.0	0.83
	6.0	0.73
	8.0	0.21
	9.0	0.03
	10.0	0.00
Yttrium oxide; Y_2O_3 temperature of measurement 298 K [131]	2.5	0.81
	3.0	0.82
	4.0	0.84
	5.0	0.85
	6.0	0.86
	7.0	0.74
	8.0	0.33
	9.0	0.03
	10.0	0.00
Zinc oxide; ZnO synthetic monocrystal, thickness of plate 0.06 mm temperature 78 K [152]	6.5	0.44
	7.0	0.43
	7.5	0.39
	8.0	0.35
	8.5	0.29
	9.0	0.24
synthetic monocrystal, thickness of plate 0.06 mm, temperature 300 K [152]	6.5	0.515
	7.0	0.501
	7.5	0.475
	8.0	0.445
	8.5	0.415
	9.0	0.340

3.3. TABLES OF RADIATIVE PROPERTIES OF ORGANIC AND BIOLOGICAL MATERIALS

Table 3.3.1. Hemispherical total emissivity ε_T of organic and biological materials

State of surface	T[K]	ε_T
Carbon black from acetylene [76]	123	0.97
	323÷2973	0.99
from tristearin [76]	323	0.95
	673	0.95
from camphor [76]	123	0.94
	323	0.98
	673	0.99
Ebonite; hard rubber [12]	293	0.94
Enamel, cellulose various colours [12]	290÷350	0.58÷0.79
Enamel, chlorinated [12]	290÷350	0.61÷0.70
Enamel, epoxy [12]	290÷350	0.6÷0.75
Enamel, oleoresinous water proof various colours [12]	290÷350	0.62÷0.78
Enamel, phthalic various colours [12]	290÷350	0.56÷0.79
Enamel, phthalic, oil proof various colours [12]	290÷350	0.6÷0.8
Enamel, phthalic, resisting in tropic various colours [12]	290÷350	0.61÷0.71
Enamel, polyurethan [12]	290÷350	0.55÷0.8
Enamel polyvinyle [12]	290÷350	0.63÷0.75
Grease on the polished iron, thickness of layer 0.003 mm [12]	293	0.35
on the polished iron, thickness of layer 0.08 mm [12]	293	0.44
on the polished iron, thickness of layer 0.12 mm [12]	293	0.75
on the polished iron, thick layer [12] (over 0.12 mm)	293	0.82
Lacquer white [12]	298÷368	0.8÷0.95
mat, black [12]	303÷368	0.96÷0.98
silvery [12]	373	0.27÷0.68
Lacquer, aluminium pigmented shiny [12]	293÷363	0.26÷0.39
mat [12]	313÷363	0.5
Lacquer, nitrocellulose [12]	290÷350	0.64÷0.75
Oil, machine [12]	293	0.07
	308	0.16
	323	0.5
	333	0.6
Paint, aluminium pigmented [12]	293÷373	0.26÷0.67

Table 3.3.1. (*continued*)

State of surface	T[K]	ε_T
Paint, black mat [12]	313÷373	0.96÷0.98
Paint, oil various colours [12]	363÷373	0.92÷0.96
on the steel, thickness of layer 0.05 mm [12]		0.44
on the steel, thickness of layer 0.1 mm [12]		0.65
on the steel, thickness of layer 0.2 mm [12]		0.80
Paint, oil primer white [12]	290÷350	0.59÷0.71
grey [12]	290÷350	0.66÷0.74
Paint, phthalic, chromate primer [12]	290÷350	0.69÷0.74
Paint, phthalic primer [12]	290÷350	0.73÷0.80
Paint, white on a copper base [12]	363	0.95
Paper various colours [12]	368	0.89
Paper, roofing [12]	294	0.91
Rubber soft grey [12]	293	0.89
hard black [12]	293÷313	0.94
Shellac mat, black [12]	348	0.91
Varnish, oil on the polished steel, thickness of layer less than 0.02 mm [12]	313	0.06
on the polished steel, thickness of layer 0.02 mm [12]	313	0.22
on the polished steel, thickness of layer 0.1 mm [12]	313	0.45
on the polished steel, thickness of layer 0.2 mm [12]	313	0.65
on the polished steel, thickness of layer more than 0.2 mm [12]	313	0.81
Varnish, resin black [12]	313	0.95
Varnish, spirit black [12]	295	0.77÷0.87
colourless [12]	313	0.83
Velvet, black [76]	313	0.97
Wood [12]	293÷313	0.9÷0.94

Table 3.3.2. Total emissivity in normal direction $\varepsilon_{T,n}$ of organic and biological materials

State of surface	T[K]	$\varepsilon_{T,n}$
Carbon black [19]	323÷1223	0.96
lamp black [19, 67, 83, 84]	293÷673	0.95÷0.97
with water glass [19, 44, 67]	293÷473	0.96
	673÷758	0.96÷0.95
Cloth		
woolen [19]	293	0.75
woolen, black [19]	293	0.98
Cotton [19]		0.76
Dye cresol red; $C_{21}H_{18}O_5S$ powder [1, 9]	373	0.4
Dye methylene blue; $C_{16}H_{18}N_3SCL$ powder [1, 9]	373	0.8
Ebonite; hard rubber [10, 19]		0.89
Enamel, cellulose various colours [2, 9, 10]		0.5÷0.8
Enamel for radiators [19]	373	0.925
Enamel, oleoresinous, waterproof various colours [2, 9, 10]		0.62÷0.76
Enamel, phthalic, for common purpose various colours [2, 9, 10]		0.63÷0.78
Enamel, phthalic, oilproof various colours [2, 9, 10]		0.6÷0.8
Enamel, phthalic, tropical, oilproof		
various colours [2, 9, 10]		0.6÷0.75
Enamel, white		
glazed surface [1, 2, 19, 44]	293	0.90
	313÷370	0.80÷0.95
on the iron [44]	293	0.90
Enamel, aluminium pigmented		
various sorts [19, 44, 55]	373	0.27÷0.67
after heating at temperature 400 K [19]	423÷588	0.35
glazed surface [1, 44]	293	0.4
Lacquer		
white [12]	298÷368	0.8÷0.95
black, bright [1, 12, 19]	298	0.88
black, mat [12, 19, 44, 55]	303÷368	0.96÷0.98
Lacquer, aluminium pigmented		
bright [12]	293÷363	0.2÷0.4
mat [12]	313÷363	0.5
Lacquer, phenol-formaldehyde, plastic filled with wood flour [19, 26]	353	0.935
Lacquer, stoving [26]	373	0.925
Leather [19]		0.75÷0.80
Linoleum [19]	293	0.855
Oil, vegetable [19]		0.82

Table 3.3.2. (*continued*)

State of surface	T[K]	$\varepsilon_{r,n}$
Paint, aluminium pigmented [12]	293÷373	0.26÷0.67
Paint, aluminium pigmented - 10% Al 10% Al and 22% shellac [76, 78]	363	0.52
Paint, aluminium pigmented - 26% Al 26% Al and 27% shellac [76, 78]	363	0.30
Paint, minium [19]	293÷373	0.93
Paint, oil various colours [19, 26, 44]	273÷473	0.88÷0.96
Paint, phthalic, primer [12]	290÷350	0.93
Paper [19, 26]	293÷368	0.7÷0.92
white [19, 44, 76]	293	0.7÷0.9
	313	0.95
black [19, 44]		0.90
mat black [19]		0.94
red [19]		0.76
blue [19]		0.84
green [19]		0.85
yellow [19]		0.72
Paper, roofing [1, 19, 26, 44, 55]	293	0.91÷0.93
Polyethylene thickness of foil 0.2 mm [24, 45, 61, 131, 132]	473	0.29
	623	0.33
	773	0.37
thickness of foil 0.4 mm [24, 45, 61, 131, 132]	473	0.46
	623	0.51
	773	0.56
thickness of foil 0.6 mm [24, 45, 61, 131, 132]	473	0.53
	623	0.59
	773	0.63
Resin, epoxy; resin epoxide [45]	75÷625	0.92÷0.91
Resin, phenolic, CTL - 9L - LD [45]	75÷625	0.89÷0.88
Resin, phenolic, Plyophen 5023 [45]	75÷625	0.825÷0.93
Resin, polyester, Polylite [45]	300	0.725
Resin, polyester, Vibrin X-1068 [45]	50÷600	0.89÷0.88
Resin, polyester, Vibrin 135 [45]	50÷600	0.89÷0.88
Rubber grey, soft [1, 19, 44, 55]	300	0.86
hard, bright [76, 78]	313	0.94
hard, black rough [1, 19, 44, 55]	300	0.95
Shellac bright, black [1, 44]	293	0.82
mat, black [1, 44]	348÷423	0.91
Silk [19]	293	0.77
Tar [19]		0.79÷0.84
Teflon polytetrafluoroethylene, $CF_2 = CF_2$ thickness of plate 1 mm [44, 45]	473	0.30
	623	0.33
	773	0.35
thickness of plate 5 mm [44, 45]	473	0.45
	623	0.49
	773	0.52
Varnish on the aluminium alloy, after first heating [45]	200	0.982
	300	0.985
	400	0.965
	450	0.950

Table 3.3.2. (*continued*)

State of surface	T [K]	$\varepsilon_{r,n}$
Varnish, alkyd on the aluminium alloy [45]	100÷300	0.90
	350	0.895
	400	0.88
	450	0.855
after second heating on the aluminium alloy [45]	100	0.92
	150	0.93
	200	0.94
	250	0.95
	300	0.96
	350	0.96
	400	0.945
	450	0.92
	500	0.89
after third heating on the aluminium alloy [45]	75÷475	0.87÷0.88
after first heating, on the steel [45]	100	0.726
	150	0.728
	200	0.732
	250	0.748
	300	0.783
	350	0.786
	400	0.824
	475	0.828
after second heating, on the steel [45]	100	0.776
	200	0.780
	300	0.794
	475	0.823
Varnish, oil [45]	75÷225	0.98
	300	0.975
	350	0.955
	400	0.915
	450	0.82
after first heating, on the steel [45]	100	0.922
	150	0.969
	200	0.992
	250	0.996
	300	0.992
	350	0.980
	400	0.949
	450	0.830
after second heating, on the steel [45]	100÷450	0.89÷0.91
on polished surface of steel, thickness of layer more than 0.02 mm [12]	313	0.22
on polished surface of steel, thickness of layer more than 0.2 mm [12]	313	0.81
Varnish, resin black [76, 78]	313	0.95
Varnish, shellac black bright on the steel [1, 19]	293	0.82
mat surface [1, 19]	350÷463	0.91
Varnish, spirit [76, 78]	313	0.83
black [12]	295	0.82
black, bright [19]	300	0.82
Wood raw white surface [10, 19, 44]	293	0.7÷0.8
shaved surface [19, 44]	293÷313	0.8÷0.9
ground surface [19]		0.5 0.7
Wood, beech shaved surface [19, 26, 76]	313÷343	0.935
Wood, oak shaved surface [19, 55]	294÷313	0.895÷0.9
Wood, spruce ground surface [109]	311	0.82
Wood, walnut smooth surface [76, 78]	313	0.83

Table 3.3.3. Spectral emissivity in normal direction $\varepsilon_{\lambda,n}$ of organic and biological materials

State of surface	λ [μm]	$\varepsilon_{\lambda,n}$
Beef frozen, temperature 325 K [83]	4.0	0.81
	8.0	0.83
	12.0	0.83
	16.0	0.85
	20.0	0.86
	24.0	0.87
Glass, organic; methacrylate, polymethyl polished surface [10]	2.5÷13	0.6÷0.8
mat surface [10]	2.5	0.62
	3.5	0.59
	5.0	0.72
	7.0	0.78
	9.0	0.83
	11	0.77
	13	0.70
Resin, epoxy; resin, epoxyde [45]	0.3	0.93
	0.6	0.84
	0.8	0.76
	1.0	0.79
	1.4	0.83
	1.8	0.88
	2.2	0.93
	2.6	0.96
Resin, phenolic, CTL - 9L - LD [45]	0.3	0.95
	0.4	0.93
	0.6	0.93
	0.8	0.77
	1.0	0.65
	1.4	0.66
	1.8	0.70
	2.2	0.83
	2.6	0.89
Resin, phenolic, Plyophen 5023 [45]	0.3	0.96
	0.4	0.94
	0.6	0.81
	0.8	0.57
	1.0	0.58
	1.4	0.63
	1.8	0.66
	2.2	0.78
	2.6	0.85
Resin, polyester, Polylite [45]	0.4	0.53
	0.6	0.35
	1.0	0.26
	1.4	0.18
	1.6	0.15
	1.8	0.26
	2.0	0.25
Resin, polyester, Vibrin 135 [45]	0.3	0.95
	0.4	0.70
	0.6	0.38
	0.8	0.53
	1.0	0.60
	1.4	0.69
	1.8	0.78
	2.2	0.87
	2.6	0.86

Table 3.3.4. Spectral (red) emissivity in normal direction $\varepsilon_{\lambda=0.65,n}$ of organic and biological materials

State of surface	T[K]	$\varepsilon_{\lambda=0.65,n}$
Paint, alkyd green, Alclad 2024 [45]	350	0.725
	400	0.79
	450	0.83
Paint, alkyd, with zinc chromate, Alclad 2024 [45]	350	0.51
	400	0.565
	450	0.57

Table 3.3.5. Bidirectional spectral reflectivity in normal direction $r_{\lambda,n}$ of organic and biological materials

State of surface	λ [μm]	$r_{\lambda,n}$
Alder young stand of trees with young foliage [70]	0.40	0.030
	0.50	0.037
	0.52	0.058
	0.54	0.089
	0.55	0.096
	0.56	0.087
	0.58	0.071
	0.60	0.067
	0.62	0.058
	0.64	0.051
	0.66	0.062
	0.68	0.078
	0.70	0.142
	0.72	0.262
	0.73	0.374
	0.74	0.446
	0.75	0.469
	0.76	0.476
	0.77	0.476
	0.78	0.472
	0.80	0.461
	0.82	0.447
	0.84	0.481
	0.86	0.414
Aspen a mature stand of trees with young foliage [70]	0.40	0.041
	0.44	0.043
	0.48	0.042
	0.50	0.052
	0.52	0.080
	0.54	0.117
	0.55	0.125
	0.56	0.128
	0.58	0.117
	0.60	0.095
	0.62	0.080
	0.64	0.060
	0.66	0.060
	0.68	0.075
	0.69	0.092
	0.70	0.124
	0.71	0.175
	0.72	0.268
	0.73	0.335
	0.74	0.364
	0.75	0.380
a mature stand of trees with a full foliage [70]	0.40	0.020
	0.42	0.020
	0.44	0.022
	0.46	0.027
	0.48	0.031
	0.50	0.030
	0.52	0.042
	0.54	0.065
	0.56	0.073
	0.58	0.065
	0.60	0.058
	0.62	0.059
	0.64	0.058
	0.66	0.060
	0.68	0.088
	0.69	0.128
	0.70	0.170
	0.71	0.280
	0.72	0.327
	0.73	0.387
	0.74	0.415
	0.75	0.432

Table 3.3.5. (*continued*)

State of surface	λ [μm]	$r_{\lambda,n}$
a mature stand of trees a late verdure [70]	0.40	0.030
	0.44	0.031
	0.46	0.034
	0.48	0.037
	0.50	0.044
	0.52	0.065
	0.54	0.088
	0.56	0.098
	0.58	0.080
	0.60	0.065
	0.64	0.063
	0.66	0.070
	0.68	0.095
	0.69	0.135
	0.70	0.175
	0.71	0.280
	0.72	0.360
	0.73	0.436
	0.74	0.495
	0.75	0.535
a mature stand of trees, autumn colouring [70]	0.40	0.047
	0.42	0.048
	0.44	0.050
	0.46	0.054
	0.48	0.068
	0.50	0.081
	0.52	0.112
	0.54	0.161
	0.56	0.188
	0.58	0.210
	0.60	0.220
	0.62	0.208
	0.64	0.191
	0.65	0.186
young stand of trees in winter [70]	0.40	0.056
	0.50	0.072
	0.60	0.090
	0.64	0.091
young stand of trees with young foliage [70]	0.40	0.034
	0.44	0.043
	0.48	0.048
	0.50	0.054
	0.52	0.082
	0.53	0.106
	0.54	0.109
	0.55	0.124
	0.56	0.122
	0.58	0.095
	0.60	0.084
	0.62	0.077
	0.64	0.073
	0.66	0.077
	0.68	0.090
	0.70	0.142
	0.71	0.236
	0.72	0.402
	0.73	0.452
	0.74	0.484
	0.75	0.512
	0.76	0.531
	0.77	0.550
	0.78	0.560
	0.80	0.575
	0.82	0.583
	0.84	0.588
	0.86	0.590
	0.88	0.590
young stand of trees with a full foliage [70]	0.40	0.040
	0.44	0.048
	0.48	0.051
	0.50	0.061
	0.52	0.102
	0.53	0.118
	0.54	0.123
	0.55	0.129
	0.56	0.128
	0.58	0.103
	0.60	0.090
	0.62	0.082
	0.64	0.079
	0.66	0.075
	0.68	0.108
	0.69	0.138
	0.70	0.191
	0.71	0.400
	0.72	0.440
	0.73	0.518
	0.74	0.562
	0.75	0.589
	0.76	0.604
	0.77	0.618
	0.78	0.674
	0.80	0.680

Table 3.3.5. (*continued*)

State of surface	λ[μm]	$r_{\lambda,n}$
Bark		
on a growing birch		
[70]	0.40	0.202
	0.50	0.215
	0.60	0.231
	0.65	0.269
	0.68	0.285
	0.70	0.298
	0.72	0.300
	0.74	0.301
	0.80	0.291
	0.85	0.284
pine [90]	0.8	0.21
	1.2	0.26
	1.6	0.29
	2.0	0.28
	2.4	0.25
	2.6	0.22
Barley		
before developing ears		
[70]	0.40	0.027
	0.42	0.028
	0.44	0.029
	0.46	0.034
	0.48	0.035
	0.50	0.037
	0.51	0.046
	0.52	0.072
	0.54	0.110
	0.55	0.113
	0.56	0.103
	0.58	0.081
	0.60	0.065
	0.62	0.052
	0.64	0.042
	0.65	0.040
after developing ears		
[70]	0.40	0.028
	0.42	0.038
	0.44	0.038
	0.46	0.042
	0.48	0.046
	0.50	0.047
	0.52	0.094
	0.54	0.131
	0.55	0.132
	0.56	0.137
	0.58	0.120
	0.59	0.110
	0.60	0.118
	0.61	0.122
	0.62	0.117
	0.64	0.100
	0.66	0.092
	0.68	0.108
	0.69	0.135
	0.70	0.191
	0.71	0.281
	0.72	0.394
	0.73	0.456
	0.74	0.495
	0.75	0.525
	0.76	0.552
	0.78	0.586
	0.80	0.607
	0.82	0.623
	0.84	0.638
	0.85	0.642
ripe [70]	0.40	0.071
	0.44	0.122
	0.48	0.185
	0.52	0.238
	0.56	0.272
	0.60	0.290
	0.65	0.290
Birch		
mature stand of trees		
in winter [70]	0.4	0.072
	0.5	0.071
	0.6	0.100
	0.7	0.100
	0.8	0.103
	0.9	0.110
young foliage,		
mature stand of trees		
[70]	0.40	0.044
	0.50	0.060
	0.52	0.085
	0.54	0.121
	0.56	0.123
	0.58	0.109
	0.60	0.09
	0.62	0.082
	0.64	0.078
a full foliage,		
mature stand of trees		
[70]	0.40	0.033
	0.42	0.037
	0.44	0.044
	0.46	0.044
	0.48	0.044
	0.50	0.045
	0.52	0.065

Table 3.3.5. (*continued*)

State of surface	λ [μm]	$r_{\lambda,n}$
	0.54	0.106
	0.56	0.110
	0.58	0.088
	0.60	0.083
	0.62	0.068
	0.64	0.069
a late verdure, mature stand of trees [70]	0.40	0.084
	0.50	0.144
	0.52	0.185
	0.54	0.233
	0.56	0.249
	0.58	0.241
	0.60	0.229
	0.62	0.220
	0.64	0.205
	0.66	0.208
	0.68	0.220
	0.70	0.225
	0.72	0.295
	0.74	0.332
	0.76	0.368
	0.78	0.390
	0.80	0.405
	0.82	0.418
	0.84	0.421
	0.86	0.422
young stand of trees in winter [70]	0.4	0.058
	0.5	0.044
	0.6	0.050
young stand of trees young foliage [70]	0.4	0.047
	0.5	0.053
	0.52	0.090
	0.54	0.138
	0.56	0.142
	0.58	0.117
	0.6	0.160
young stand of trees a full foliage [70]	0.4	0.026
	0.5	0.040
	0.52	0.052
	0.54	0.085
	0.56	0.088
	0.58	0.071
	0.6	0.068
young stand of trees a late verdure [70]	0.4	0.059
	0.44	0.061
	0.48	0.075
	0.5	0.095
	0.52	0.120
	0.54	0.169
	0.56	0.178
	0.6	0.160
	0.64	0.143
	0.66	0.160
	0.68	0.180
	0.70	0.245
	0.72	0.305
	0.74	0.370
	0.76	0.450
	0.78	0.517
	0.80	0.529
	0.82	0.517
	0.84	0.500
	0.86	0.535
young stand of trees, a full foliage [70]	0.40	0.051
	0.42	0.052
	0.44	0.062
	0.46	0.063
	0.48	0.065
	0.50	0.066
	0.52	0.096
	0.54	0.138
	0.55	0.148
	0.56	0.136
	0.58	0.121
	0.60	0.113
	0.62	0.106
	0.64	0.104
	0.65	0.094
dwarf, a full foliage [70]	0.40	0.020
	0.50	0.029
	0.52	0.040
	0.54	0.066
	0.56	0.067
	0.58	0.048
	0.60	0.050
	0.62	0.053
	0.64	0.042
	0.65	0.040
	0.66	0.048
	0.68	0.075
	0.70	0.120
	0.76	0.320

Table 3.3.5. (*continued*)

State of surface	λ[μm]	$r_{\lambda,n}$
	0.80	0.385
	0.82	0.396
	0.84	0.392
	0.86	0.380
	0.88	0.365
	0.90	0.363
Bread wheaten dough [24, 70, 93]	0.4	0.26
	0.5	0.46
	0.7	0.71
	0.9	0.70
	1.0	0.60
	1.1	0.65
	1.2	0.42
	1.3	0.44
	1.5	0.23
white crust of slack-baked bread [24, 70, 93]	0.4	0.15
	0.5	0.36
	0.7	0.67
	0.8	0.74
	0.9	0.73
	1.0	0.66
	1.1	0.69
	1.2	0.52
	1.3	0.53
	1.5	0.34
brown crust of baked bread [24, 70, 93]	0.4	0.06
	0.5	0.06
	0.7	0.14
	0.9	0.38
	1.1	0.56
	1.3	0.57
	1.5	0.40
Buckwheat before full blossoming [70]	0.40	0.045
	0.42	0.064
	0.44	0.069
	0.46	0.068
	0.48	0.070
	0.50	0.078
	0.52	0.112
	0.53	0.135
	0.54	0.151
	0.55	0.156
	0.56	0.127
	0.57	0.116
	0.58	0.120
	0.60	0.139
	0.61	0.140
	0.62	0.142
	0.64	0.137
	0.65	0.136
Bulrusch light green in the lake inshore [70]	0.40	0.054
	0.42	0.049
	0.44	0.060
	0.46	0.060
	0.48	0.060
	0.50	0.070
	0.52	0.096
	0.54	0.119
	0.55	0.120
	0.56	0.112
	0.58	0.112
	0.60	0.088
	0.62	0.078
	0.63	0.078
	0.64	0.074
	0.65	0.082
	0.66	0.100
	0.68	0.138
	0.69	0.155
	0.70	0.175
	0.71	0.230
	0.72	0.259
	0.73	0.342
	0.74	0.467
	0.75	0.542
	0.76	0.571
	0.78	0.605
	0.80	0.622
	0.82	0.629
	0.84	0.625
Cabbage during growing [70]	0.40	0.046
	0.41	0.056
	0.42	0.072
	0.43	0.082
	0.44	0.095
	0.45	0.096
	0.46	0.096
	0.47	0.098
	0.48	0.101
	0.49	0.100
	0.50	0.104

Table 3.3.5. (*continued*)

State of surface	λ[μm]	$r_{\lambda,n}$
	0.51	0.110
	0.52	0.129
	0.53	0.154
	0.54	0.163
	0.55	0.176
	0.56	0.144
	0.57	0.136
	0.58	0.118
	0.60	0.109
	0.62	0.105
	0.63	0.102
	0.64	0.104
	0.65	0.105
	0.66	0.091
	0.68	0.073
	0.69	0.072
	0.70	0.077
	0.71	0.164
	0.72	0.258
	0.73	0.438
	0.74	0.562
	0.75	0.629
	0.76	0.657
	0.77	0.666
	0.78	0.669
	0.80	0.666
	0.82	0.650
	0.84	0.623
Carbon, black paint [24]	0.6	0.031
	0.95	0.034
from acetylene [24]	0.95	$42\text{-}82\cdot10^{-4}$
	4.4	$71\text{-}92\cdot10^{-4}$
	8.8	$112\cdot10^{-4}$
from tar [24]	0.95	$126\cdot10^{-4}$
from tristearin [24]	0.95	$91\text{-}125\cdot10^{-4}$
	4.4	$67\cdot10^{-4}$
	8.8	$144\cdot10^{-4}$
from camphor [24]	0.95	$130\text{-}136\cdot10^{-4}$
	4.4	$95\cdot10^{-4}$
	8.8	$162\cdot10^{-4}$
from paraffin [24]	0.95	0.0097

State of surface	λ[μm]	$r_{\lambda,n}$
Clover white, during blossoming [70]	0.40	0.050
	0.42	0.068
	0.44	0.091
	0.46	0.078
	0.48	0.054
	0.50	0.048
	0.52	0.074
	0.54	0.148
	0.56	0.180
	0.58	0.162
	0.60	0.137
	0.61	0.130
	0.62	0.144
	0.64	0.118
	0.65	0.110
red, during blossoming [70]	0.40	0.078
	0.42	0.079
	0.44	0.089
	0.46	0.079
	0.48	0.059
	0.50	0.054
	0.52	0.078
	0.54	0.132
	0.56	0.139
	0.58	0.159
	0.60	0.155
	0.62	0.167
	0.64	0.156
	0.65	0.152
red with young grass [70]	0.40	0.059
	0.42	0.065
	0.44	0.069
	0.46	0.068
	0.48	0.062
	0.50	0.066
	0.51	0.090
	0.52	0.131
	0.53	0.141
	0.54	0.151
	0.55	0.152
	0.56	0.142
	0.58	0.109
	0.60	0.099
	0.62	0.082
	0.63	0.084
	0.65	0.063

Table 3.3.5. (*continued*)

State of surface	λ[μm]	$r_{\lambda,n}$
Corn with green crops [70]	0.40	0.055
	0.42	0.064
	0.44	0.069
	0.45	0.073
	0.46	0.070
	0.48	0.064
	0.52	0.055
	0.56	0.050
	0.60	0.055
	0.64	0.062
	0.68	0.078
	0.70	0.100
	0.71	0.110
	0.72	0.124
	0.74	0.169
	0.76	0.231
	0.78	0.319
	0.80	0.442
	0.82	0.591
	0.84	0.626
	0.86	0.633
Cotton before blossoming [70]	0.40	0.058
	0.44	0.071
	0.48	0.077
	0.52	0.094
	0.54	0.133
	0.55	0.141
	0.56	0.137
	0.58	0.115
	0.60	0.106
	0.61	0.102
	0.62	0.090
	0.64	0.076
	0.65	0.078
	0.66	0.095
	0.67	0.118
	0.68	0.180
	0.69	0.261
	0.70	0.339
	0.71	0.392
	0.72	0.450
	0.73	0.493
	0.74	0.534
	0.75	0.550
	0.76	0.570
	0.77	0.584
	0.78	0.596
	0.79	0.610
	0.80	0.621
	0.82	0.646
	0.84	0.659
	0.86	0.666
blossoming [70]	0.40	0.039
	0.42	0.047
	0.44	0.054
	0.48	0.061
	0.50	0.070
	0.51	0.081
	0.52	0.103
	0.53	0.118
	0.54	0.108
	0.56	0.095
	0.58	0.083
	0.60	0.052
	0.61	0.051
	0.62	0.045
	0.63	0.035
	0.65	0.031
Duckweed [70]	0.40	0.046
	0.42	0.044
	0.44	0.042
	0.46	0.045
	0.48	0.046
	0.50	0.048
	0.52	0.060
	0.54	0.070
	0.56	0.070
	0.58	0.069
	0.60	0.064
	0.62	0.064
	0.65	0.068
	0.66	0.076
	0.68	0.117
	0.70	0.165
	0.71	0.190
	0.72	0.206
	0.73	0.217
	0.74	0.221
	0.75	0.227
	0.76	0.229
	0.78	0.221
	0.80	0.214
	0.82	0.221
	0.84	0.235
	0.86	0.251
	0.87	0.258

Table 3.3.5. (*continued*)

State of surface	$\lambda[\mu m]$	$r_{\lambda,n}$
Elm a mature stand of trees, young foliage [70]	0.40	0.026
	0.48	0.036
	0.50	0.043
	0.52	0.072
	0.54	0.109
	0.56	0.117
	0.58	0.092
	0.60	0.073
	0.62	0.060
	0.64	0.061
a mature stand of trees, a full foliage [70]	0.40	0.033
	0.48	0.039
	0.50	0.039
	0.52	0.058
	0.54	0.096
	0.56	0.105
	0.58	0.084
	0.60	0.066
	0.62	0.052
	0.64	0.050
a mature stand of trees, a late, stongly-dusty verdure [70]	0.75	0.522
	0.76	0.543
	0.78	0.567
	0.80	0.578
	0.82	0.587
	0.84	0.601
	0.86	0.622
Enamel white [96]	0.5	0.70
	1.0	0.65
	2.0	0.36
	3.0	0.12
	4.0	0.07
	5.0	0.03
	6.0	0.03
	8.0	0.05
Enamel, epoxy white on the aluminium [22]	0.5	0.89
	1.0	o.91
	3.0	0.07
	5.0	0.09
	7.0	0.07
	11.0	0.07
	15.0	0.09
	19.0	0.14
	23.0	0.21
Enamel, silicone resin whit powder of aluminium [22]	0.5	0.74
	1.0	0.72
	3.0	0.78
	5.0	0.79
	7.0	0.785
	11.0	0.78
	15.0	0.80
	19.0	0.83
	23.0	0.80
Fallow green, flowery [70]	0.40	0.021
	0.41	0.014
	0.42	0.019
	0.44	0.025
	0.46	0.028
	0.48	0.031
	0.50	0.036
	0.52	0.050
	0.54	0.070
	0.56	0.070
	0.58	0.066
	0.60	0.062
	0.62	0.061
	0.64	0.060
	0.66	0.072
	0.68	0.123
	0.69	0.148
	0.70	0.152
	0.72	0.157
	0.74	0.156
	0.76	0.156
Grass dusty [70]	0.40	0.031
	0.42	0.036
	0.43	0.045
	0.44	0.045
	0.46	0.043
	0.48	0.043
	0.50	0.049
	0.52	0.058
	0.54	0.083
	0.56	0.081
	0.58	0.082
	0.60	0.071
	0.62	0.074

Table 3.3.5. (*continued*)

State of surface	λ[μm]	$r_{\lambda,n}$
	0.64	0.069
	0.66	0.070
	0.67	0.076
	0.68	0.090
	0.69	0.118
	0.70	0.150
	0.71	0.218
	0.72	0.320
	0.73	0.360
	0.74	0.371
	0.75	0.380
	0.76	0.385
	0.78	0.391
	0.80	0.392
	0.82	0.391
	0.84	0.390
	0.86	0.376
	0.88	0.355
half-dry in autumn on the sand ground [70]	0.40	0.035
	0.42	0.048
	0.44	0.061
	0.46	0.070
	0.48	0.071
	0.50	0.071
	0.52	0.080
	0.54	0.096
	0.56	0.103
	0.58	0.110
	0.60	0.113
	0.62	0.113
	0.64	0.109
Hay dry [70]	0.40	0.050
	0.41	0.052
	0.42	0.053
	0.43	0.057
	0.44	0.063
	0.45	0.061
	0.46	0.063
	0.47	0.072
	0.48	0.081
	0.49	0.080
	0.50	0.088
	0.51	0.094
	0.52	0.100
	0.53	0.116
	0.54	0.115
	0.55	0.119
	0.56	0.113
	0.57	0.127

State of surface	λ[μm]	$r_{\lambda,n}$
	0.58	0.136
	0.59	0.137
	0.60	0.134
	0.61	0.129
	0.62	0.133
	0.63	0.143
	0.64	0.147
	0.65	0.156
	0.66	0.165
	0.68	0.178
	0.70	0.185
	0.72	0.193
	0.74	0.208
	0.76	0.222
	0.78	0.244
	0.80	0.281
	0.81	0.300
	0.82	0.314
	0.83	0.335
	0.84	0.370
	0.85	0.397
	0.86	0.421
	0.87	0.449
	0.88	0.471
	0.89	0.489
	0.90	0.494
Heather thick brush woods before full-blossoming [70]	0.40	0.018
	0.50	0.027
	0.52	0.029
	0.54	0.041
	0.56	0.044
	0.58	0.040
	0.60	0.040
	0.62	0.054
	0.64	0.050
	0.66	0.052
	0.68	0.070
	0.70	0.107
	0.72	0.115
	0.74	0.190
	0.75	0.200
	0.76	0.202
	0.78	0.193
	0.80	0.180
	0.81	0.177
	0.82	0.188
	0.84	0.193
	0.86	0.202
	0.90	0.209

Table 3.3.5. (*continued*)

State of surface	λ [μm]	$r_{\lambda,n}$
Juniper a full foliage [70]	0.40	0.028
	0.42	0.024
	0.44	0.033
	0.46	0.039
	0.48	0.035
	0.50	0.043
	0.52	0.067
	0.54	0.096
	0.55	0.105
	0.56	0.100
	0.58	0.079
	0.60	0.070
	0.64	0.069
	0.66	0.078
	0.68	0.080
	0.70	0.118
	0.72	0.160
	0.74	0.240
	0.76	0.294
	0.78	0.309
	0.80	0.319
	0.85	0.340
	0.90	0.352
Larch young stand of trees in winter [70]	0.40	0.034
	0.42	0.038
	0.44	0.040
	0.46	0.042
	0.48	0.045
	0.50	0.045
	0.60	0.054
	0.62	0.054
	0.64	0.054
young stand of trees with young foliage [70]	0.40	0.024
	0.42	0.024
	0.44	0.031
	0.46	0.035
	0.48	0.036
	0.50	0.044
	0.52	0.072
	0.54	0.095
	0.55	0.099
	0.56	0.096
	0.58	0.075
	0.60	0.064
	0.64	0.059
young stand of trees with a full foliage [70]	0.40	0.041
	0.42	0.041
	0.44	0.044
	0.46	0.048
	0.48	0.042
	0.50	0.039
	0.52	0.064
	0.54	0.089
	0.56	0.088
	0.58	0.068
	0.60	0.056
	0.62	0.050
	0.64	0.049
Lichen on a turf path [70]	0.40	0.027
	0.45	0.027
	0.50	0.040
	0.55	0.052
	0.60	0.063
	0.65	0.074
	0.70	0.091
	0.75	0.115
	0.80	0.128
	0.84	0.139
	0.86	0.150
	0.88	0.169
	0.90	0.190
Lime-tree a mature stand of trees in winter [70]	0.40	0.068
	0.42	0.062
	0.44	0.064
	0.46	0.064
	0.48	0.066
	0.50	0.072
	0.60	0.086
	0.65	0.089
a full foliage [70]	0.40	0.034
	0.42	0.032
	0.44	0.042
	0.46	0.046
	0.48	0.047
	0.50	0.047
	0.52	0.081
	0.54	0.111
	0.56	0.105
	0.58	0.082

Table 3.3.5. (*continued*)

State of surface	λ[μm]	$r_{\lambda,n}$
	0.59	0.068
	0.60	0.078
	0.62	0.064
	0.64	0.055
with autumn colouring [70]	0.40	0.034
	0.42	0.032
	0.44	0.030
	0.46	0.030
	0.48	0.055
	0.50	0.045
	0.52	0.034
	0.54	0.051
	0.56	0.080
	0.57	0.098
	0.58	0.088
	0.60	0.082
	0.62	0.081
	0.63	0.081
	0.65	0.062
Linoleum [105]	1.0	0.15
	3.0	0.05
	7.0÷8.0	0.17
Maize during ripening [70]	0.40	0.032
	0.42	0.034
	0.44	0.034
	0.45	0.042
	0.46	0.044
	0.48	0.037
	0.49	0.036
	0.50	0.042
	0.52	0.068
	0.54	0.085
	0.55	0.089
	0.56	0.086
	0.58	0.078
	0.60	0.080
	0.61	0.082
	0.62	0.078
	0.64	0.075
Meadow poor, mountainous with drying grass [70]	0.40	0.029
	0.45	0.041
	0.50	0.055
	0.52	0.068
	0.54	0.081
	0.56	0.087
	0.58	0.084
	0,60	0.085
	0.62	0.093
	0.65	0.112
	0.66	0.120
	0.68	0.135
	0.70	0.149
	0.72	0.170
	0.73	0.215
	0.74	0.258
	0.75	0.277
	0.76	0.290
	0.78	0.305
	0.80	0.311
	0.82	0.321
	0.84	0.339
mowed [70]	0.40	0.044
	0.45	0.059
	0.50	0.081
	0.52	0.096
	0.54	0.109
	0.56	0.116
	0.58	0.117
	0.60	0.113
	0.62	0.125
	0.64	0.140
	0.66	0.150
	0.68	0.162
	0.70	0.188
	0.72	0.210
	0.74	0.250
	0.76	0.272
	0.78	0.289
	0.80	0.297
	0.82	0.303
	0.84	0.323
grown over with sparse low grass [70]	0.40	0.034
	0.42	0.040
	0.44	0.042
	0.45	0.042
	0.46	0.044
	0.48	0.052
	0.49	0.049
	0.50	0.056
	0.52	0.067
	0.54	0.078
	0.56	0.082
	0.58	0.082
	0.60	0.084

Table 3.3.5. (*continued*)

State of surface	λ[μm]	$r_{\lambda,n}$
	0.62	0.093
	0.64	0.099
	0.66	0.118
	0.68	0.120
	0.70	0.137
	0.72	0.159
	0.74	0.201
	0.75	0.236
	0.76	0.264
	0.77	0.280
	0.78	0.288
	0.80	0.299
	0.82	0.315
	0.84	0.350
	0.86	0.377
	0.88	0.392
	0.90	0.402
before mowing [70]	0.40	0.019
	0.42	0.027
	0.43	0.034
	0.44	0.034
	0.46	0.040
	0.48	0.036
	0.50	0.044
	0.52	0.062
	0.54	0.089
	0.56	0.095
	0.58	0.087
	0.60	0.080
	0.62	0.084
	0.64	0.079
	0.65	0.074
grown over with sparse low grass on upland ground [70]	0.40	0.024
	0.42	0.040
	0.46	0.056
	0.50	0.060
	0.54	0.082
	0.56	0.085
	0.58	0.096
	0.59	0.086
	0.60	0.098
	0.61	0.102
	0.63	0.099
	0.65	0.090
with sun altitude 25° [70]	0.40	0.022
	0.44	0.026
	0.48	0.032
	0.50	0.040
	0.51	0.049
	0.52	0.066
	0.53	0.086
	0.54	0.102
	0.55	0.115
	0.56	0.110
	0.58	0.087
	0.60	0.070
	0.62	0.061
	0.64	0.054
	0.66	0.068
	0.67	0.085
	0.68	0.111
	0.69	0.180
	0.70	0.345
	0.71	0.450
	0.72	0.515
	0.73	0.530
	0.74	0.540
	0.75	0.551
	0.76	0.565
	0.77	0.578
	0.78	0.593
	0.79	0.608
	0.80	0.622
	0.81	0.638
	0.82	0.654
	0.83	0.672
	0.84	0.690
	0.85	0.710
with sun altitude 45° [70]	0.40	0.027
	0.44	0.035
	0.48	0.040
	0.50	0.041
	0.51	0.054
	0.52	0.089
	0.54	0.126
	0.55	0.131
	0.56	0.121
	0.58	0.093
	0.60	0.073
	0.62	0.077
	0.64	0.070
	0.66	0.073
	0.68	0.081
	0.69	0.118
	0.70	0.167
	0.71	0.228
	0.72	0.332
	0.73	0.404
	0.74	0.495
	0.75	0.558

Table 3.3.5. (*continued*)

State of surface	λ [μm]	$r_{\lambda,n}$
	0.76	0.592
	0.77	0.616
	0.78	0.635
	0.80	0.664
	0.82	0.686
	0.84	0.700
with blossoming clover and timothy [70]	0.40	0.030
	0.42	0.042
	0.44	0.050
	0.46	0.052
	0.48	0.052
	0.50	0.055
	0.52	0.075
	0.54	0.107
	0.55	0.112
	0.56	0.113
	0.58	0.110
	0.60	0.105
	0.62	0.096
	0.64	0.087
	0.66	0.085
	0.68	0.105
	0.70	0.138
	0.71	0.164
	0.72	0.182
	0.73	0.215
	0.74	0.251
	0.75	0.269
	0.76	0.299
	0.77	0.322
	0.78	0.342
	0.79	0.359
	0.80	0.377
	0.81	0.399
	0.82	0.426
	0.83	0.481
	0.84	0.528
	0.85	0.543
with clover and timothy, mown and soggy under cloudy sky [70]	0.40	0.037
	0.44	0.039
	0.48	0.041
	0.50	0.050
	0.52	0.078
	0.54	0.122
	0.55	0.132
	0.56	0.130
	0.58	0.107
	0.60	0.102
	0.62	0.089
	0.64	0.080
	0.65	0.075
with abundantly blossoming buttercup [70]	0.40	0.026
	0.44	0.033
	0.48	0.045
	0.50	0.049
	0.52	0.054
	0.53	0.081
	0.54	0.100
	0.55	0.125
	0.56	0.142
	0.58	0.123
	0.59	0.126
	0.60	0.132
	0.61	0.132
	0.62	0.127
	0.64	0.110
	0.66	0.112
	0.67	0.122
	0.68	0.143
	0.69	0.173
	0.70	0.198
	0.71	0.238
	0.72	0.259
	0.73	0.284
	0.74	0.351
	0.75	0.412
	0.76	0.427
	0.77	0.440
	0.78	0.450
	0.79	0.470
	0.80	0.515
	0.81	0.610
	0.82	0.700
	0.83	0.759
	0.84	0.788
	0.85	0.805
richness [70]	0.40	0.015
	0.42	0.018
	0.44	0.022
	0.46	0.027
	0.48	0.027
	0.50	0.029
	0.52	0.050
	0.54	0.093
	0.55	0.104
	0.56	0.097
	0.58	0.073
	0.60	0.061
	0.62	0.059

Table 3.3.5. (*continued*)

State of surface	λ[μm]	$r_{\lambda,n}$
	0.64	0.047
	0.66	0.045
	0.68	0.068
	0.70	0.100
	0.71	0.123
	0.72	0.150
	0.73	0.172
	0.74	0.205
	0.75	0.249
	0.76	0.292
	0.78	0.342
	0.80	0.368
	0.82	0.389
	0.84	0.411
	0.86	0.437
	0.88	0.463
	0.90	0.489
in nadir, with thick low growing grass [70]	0.40	0.054
	0.42	0.050
	0.44	0.050
	0.46	0.047
	0.48	0.046
	0.50	0.051
	0.52	0.067
	0.54	0.085
	0.56	0.094
	0.58	0.083
	0.60	0.087
	0.62	0.080
	0.64	0.075
	0.65	0.075
Millet ripening [70]	0.40	0.020
	0.42	0.020
	0.44	0.020
	0.46	0.022
	0.48	0.024
	0.50	0.026
	0.52	0.050
	0.53	0.075
	0.54	0.083
	0.56	0.078
	0.58	0.049
	0.60	0.030
	0.62	0.058
	0.64	0.041
	0.66	0.040
	0.68	0.076
	0.69	0.115
	0.70	0.140
	0.71	0.195
	0.72	0.213
	0.74	0.291
	0.76	0.360
	0.78	0.432
	0.80	0.480
	0.82	0.513
	0.84	0.521
Moss on the rocks [70]	0.40	0.016
	0.42	0.016
	0.44	0.018
	0.46	0.020
	0.48	0.022
	0.50	0.027
	0.52	0.039
	0.54	0.048
	0.56	0.054
	0.58	0.051
	0.60	0.050
	0.62	0.050
	0.64	0.045
	0.65	0.040
on bare subsoil [70]	0.40	0.020
	0.42	0.016
	0.44	0.013
	0.46	0.014
	0.48	0.018
	0.50	0.020
	0.52	0.029
	0.54	0.033
	0.56	0.037
	0.58	0.040
	0.60	0.043
	0.62	0.049
	0.65	0.056
	0.66	0.068
	0.68	0.076
	0.70	0.083
	0.72	0.099
	0.74	0.118
	0.76	0.132
	0.78	0.155
	0.80	0.178
	0.82	0.211
	0.84	0.251
	0.86	0.288
	0.88	0.310
	0.90	0.329

Table 3.3.5. (*continued*)

State of surface	λ [μm]	$r_{\lambda,n}$
reindeer, withered [70]	0.40	0.018
	0.42	0.030
	0.44	0.055
	0.46	0.080
	0.48	0.085
	0.50	0.078
	0.52	0.082
	0.54	0.092
	0.56	0.092
	0.58	0.087
	0.60	0.085
	0.62	0.090
	0.64	0.090
	0.66	0.105
	0.68	0.123
	0.70	0.148
	0.72	0.183
	0.74	0.227
	0.76	0.270
	0.78	0.298
	0.80	0.318
	0.82	0.323
	0.84	0.323
	0.86	0.320
	0.88	0.310
	0.90	0.300
Oak young stand of trees in winter [70]	0.40	0.050
	0.50	0.050
	0.60	0.066
	0.65	0.065
a mature stand of trees with a full foliage [70]	0.40	0.044
	0.50	0.045
	0.52	0.084
	0.54	0.129
	0.55	0.150
	0.56	0.140
	0.58	0.089
	0.60	0.085
	0.62	0.060
	0.64	0.053
a mature stand of trees with autumn colouring [70]	0.40	0.071
	0.44	0.102
	0.48	0.106
	0.50	0.122
	0.51	0.145
	0.52	0.190
	0.53	0.220
	0.54	0.240
	0.55	0.255
	0.56	0.267
	0.58	0.281
	0.60	0.282
	0.62	0.289
	0.64	0.315
Oats during earning [70]	0.40	0.044
	0.42	0.040
	0.44	0.041
	0.46	0.045
	0.48	0.036
	0.50	0.041
	0.52	0.077
	0.54	0.111
	0.55	0.113
	0.56	0.102
	0.58	0.078
	0.60	0.061
	0.62	0.051
	0.64	0.040
	0.65	0.032
	0.66	0.030
	0.68	0.045
	0.69	0.063
	0.70	0.152
	0.71	0.455
	0.72	0.620
	0.73	0.690
	0.74	0.738
	0.75	0.770
	0.76	0.794
	0.78	0.824
	0.80	0.835
	0.82	0.842
	0.84	0.851
	0.86	0.862
	0.88	0.872
	0.90	0.880
Paint black, mat, thickness of layer 0.05 mm [90]	0.8	0.05
	1.2	0.06
	1.6	0.09
	2.0	0.12
	2.4	0.15

Table 3.3.5. (*continued*)

State of surface	λ[μm]	$r_{\lambda,n}$
Paint BS 381C Nr 224 dark green, thickness of layer 0.05 mm [90]	0.8÷2.6	0.1÷0.08
Paint BS 381 Nr 261 creamy, thickness of layer 0.075 mm [90]	0.8	0.48
	1.2	0.39
	1.6	0.29
	2.0	0.24
	2.4	0.17
	2.6	0.14
Paint BS 381 Nr 298 green [90]	0.8÷2.6	0.11÷0.09
Paint "Delux" red, one layer of thickness 0.0025 mm [90]	0.8	0.55
	1.2	0.68
	1.6	0.73
	2.0	0.70
	2.4	0.65
	2.6	0.69
red, some layers, sum of thickness 0.1 mm [90]	0.8	0.79
	1.2	0.72
	1.6	0.73
	2.0	0.74
	2.4	0.65
	2.6	0.70
Paint with lithophone [90]	0.5	0.76
	1.0	0.76
	2.0	0.59
	3.0	0.13
	4.0	0.24
	5.0	0.20
	6.0	0.15
	7.0	0.17
	8.0	0.18
	9.0	0.17
Paint zinc white pigmented [96]	0.5	0.83
	1.0	0.78
	2.0	0.41
	3.0	0.02
	4.0	0.05
	5.0	0.06
	6.0	0.06
	7.0	0.06
	8.0	0.10
	9.0	0.12
Paper white [24]	0.3÷0.4	0.08
	0.4÷0.8	0.30
	0.8÷2.6	0.30
	2.6÷5.0	0.15
Paper, roofing [105]	0.5÷8.0	0.12
	9.0	0.55
Pasture [70]	0.40	0.019
	0.42	0.027
	0.44	0.029
	0.46	0.028
	0.48	0.028
	0.50	0.040
	0.52	0.052
	0.54	0.063
	0.56	0.070
	0.58	0.064
	0.60	0.060
	0.62	0.053
	0.64	0.054
	0.66	0.065
	0.68	0.079
	0.70	0.102
	0.72	0.123
	0.74	0.148
	0.76	0.180
	0.78	0.220
	0.80	0.255
	0.82	0.262
	0.84	0.249
	0.85	0.250
in the mountains area [70]	0.40	0.034
	0.44	0.042
	0.48	0.049
	0.50	0.056
	0.52	0.075
	0.54	0.101
	0.56	0.107
	0.58	0.092
	0.60	0.088
	0.62	0.089
	0.64	0.089

Table 3.3.5. (*continued*)

State of surface	λ[μm]	$r_{\lambda,n}$	State of surface	λ[μm]	$r_{\lambda,n}$
with in mould area				0.76	0.444
[70]	0.40	0.030		0.77	0.456
	0.42	0.039		0.78	0.461
	0.44	0.041	Pea		
	0.46	0.043	during ripening		
	0.48	0.045	[70]	0.40	0.115
	0.50	0.049		0.41	0.117
	0.51	0.059		0.42	0.128
	0.52	0.075		0.43	0.137
	0.54	0.094		0.44	0.136
	0.55	0.098		0.45	0.121
	0.56	0.096		0.46	0.118
	0.58	0.081		0.47	0.125
	0.60	0.085		0.48	0.142
	0.62	0.079		0.49	0.157
	0.65	0.075		0.50	0.170
	0.66	0.090		0.51	0.192
	0.67	0.126		0.52	0.230
	0.68	0.162		0.53	0.278
	0.69	0.208		0.54	0.296
	0.70	0.250		0.55	0.308
	0.71	0.279		0.56	0.311
	0.72	0.291		0.57	0.313
	0.74	0.306		0.58	0.315
	0.76	0.303		0.59	0.330
	0.78	0.311		0.60	0.351
	0.80	0.314		0.61	0.386
	0.82	0.318		0.62	0.412
	0.85	0.339		0.63	0.350
in the steppe area				0.64	0.304
[70]	0.40	0.026		0.65	0.292
	0.42	0.030	Peat		
	0.44	0.039	peat-bog [70]	0.40	0.021
	0.46	0.043		0.42	0.023
	0.48	0.045		0.44	0.028
	0.50	0.058		0.46	0.030
	0.52	0.075		0.48	0.032
	0.54	0.135		0.50	0.033
	0.55	0.151		0.52	0.033
	0.56	0.145		0.54	0.038
	0.58	0.121		0.56	0.037
	0.60	0.110		0.58	0.038
	0.62	0.100		0.60	0.043
	0.65	0.098		0.62	0.045
	0.66	0.100		0.64	0.046
	0.67	0.122		0.66	0.047
	0.68	0.145		0.68	0.057
	0.69	0.181		0.70	0.069
	0.70	0.222		0.72	0.090
	0.71	0.265		0.74	0.102
	0.72	0.283		0.76	0.108
	0.74	0.385		0.80	0.113
	0.75	0.424			

Table 3.3.5. (*continued*)

State of surface	λ[μm]	$r_{\lambda,n}$
	0.82	0.124
	0.84	0.138
	0.86	0.144
	0.90	0.153
exposed, dry holm [70]	0.40	0.045
	0.42	0.047
	0.44	0.049
	0.46	0.051
	0.48	0.057
	0.50	0.063
	0.52	0.070
	0.54	0.077
	0.56	0.079
	0.58	0.077
	0.60	0.077
	0.62	0.082
	0.64	0.086
	0.68	0.093
	0.72	0.110
	0.74	0.118
	0.78	0.142
	0.80	0.149
	0.84	0.173
	0.86	0.184
	0.88	0.171
	0.90	0.175
Pine in winter a mature stand of trees [70]	0.40	0.018
	0.50	0.016
	0.55	0.021
	0.60	0.019
	0.65	0.017
a mature stand of trees with young foliage [70]	0.40	0.021
	0.50	0.026
	0.52	0.034
	0.54	0.046
	0.56	0.048
	0.58	0.040
	0.60	0.041
	0.62	0.042
	0.64	0.038
	0.66	0.045
	0.68	0.063
	0.70	0.095
	0.71	0.114
	0.72	0.150
	0.73	0.210
	0.74	0.250
	0.75	0.269
	0.76	0.281
	0.77	0.290
	0.78	0.299
	0.80	0.309
	0.82	0.310
	0.84	0.310
	0.86	0.310
a mature stand of trees with young foliage [70]	0.40	0.032
	0.42	0.030
	0.43	0.033
	0.44	0.033
	0.45	0.032
	0.46	0.034
	0.48	0.035
	0.49	0.033
	0.50	0.032
	0.52	0.041
	0.54	0.051
	0.55	0.054
	0.56	0.053
	0.58	0.046
	0.60	0.045
	0.62	0.053
	0.64	0.049
young stand of trees with young foliage [70]	0.40	0.051
	0.42	0.059
	0.44	0.066
	0.46	0.070
	0.48	0.074
	0.50	0.084
	0.52	0.096
	0.54	0.121
	0.56	0.130
	0.58	0.129
	0.60	0.128
	0.62	0.130
	0.64	0.131
	0.66	0.128
	0.68	0.129
	0.70	0.142
	0.72	0.215
	0.73	0.226
	0.74	0.276
	0.75	0.302

Table 3.3.5. (*continued*)

State of surface	λ[μm]	$r_{\lambda,n}$
young stand of trees with a full foliage [70]	0.40	0.052
	0.42	0.052
	0.44	0.048
	0.46	0.052
	0.48	0.053
	0.50	0.060
	0.52	0.075
	0.54	0.099
	0.55	0.104
	0.56	0.107
	0.58	0.096
	0.60	0.092
	0.62	0.092
	0.64	0.089
	0.66	0.095
	0.68	0.108
	0.70	0.130
	0.72	0.170
	0.73	0.190
	0.74	0.210
	0.76	0.238
	0.78	0.270
	0.80	0.303
	0.84	0.376
Plantain upper part of the leaf [70]	0.40	0.060
	0.45	0.058
	0.50	0.054
	0.51	0.060
	0.52	0.096
	0.54	0.136
	0.55	0.142
	0.56	0.140
	0.58	0.105
	0.60	0.087
	0.64	0.064
	0.68	0.056
	0.70	0.132
	0.71	0.300
	0.72	0.344
	0.73	0.400
	0.74	0.424
	0.75	0.452
	0.76	0.464
	0.77	0.472
	0.78	0.478
	0.80	0.485
Polyethylene [24, 45]	1	0.660
	2	0.583
	3	0.686
	4	0.608
	5	0.520
	6	0.575
	7	0.150
	8	0.245
	9	0.316
	10÷11	0.340
	12	0.345
	13	0.542
	14	0.940
	15	0.500
	16	0.435
	17	0.570
	18	0.250
	19	0.350
	20	0.260
	21	0.265
	22÷23	0.415
	24	0.580
Potato plant out of blossoming [70]	0.40	0.047
	0.41	0.047
	0.42	0.051
	0.43	0.050
	0.44	0.050
	0.46	0.053
	0.48	0.055
	0.50	0.060
	0.52	0.068
	0.54	0.099
	0.55	0.108
	0.56	0.109
	0.58	0.090
	0.60	0.070
	0.62	0.072
	0.64	0.060
	0.66	0.042
	0.68	0.050
	0.70	0.093
	0.71	0.132
	0.72	0.175
	0.73	0.246
	0.74	0.336
	0.75	0.382
	0.76	0.404
	0.78	0.417
	0.80	0.428

Table 3.3.5. (*continued*)

State of surface	λ[μm]	$r_{\lambda,n}$
	0.82	0.440
	0.84	0.434
	0.85	0.400
Resin, polyester, Polylite		
[45]	0.4	0.91
	0.6	0.90
	1.0	0.83
	1.4	0.82
	1.8	0.85
	2.0	0.83
Resin, polyester, Vibrin X-1068		
[45]	0.3	0.97
	0.4	0.93
	0.6	0.69
	0.8	0.64
	1.0	0.69
	1.2	0.72
	1.6	0.77
	1.8	0.86
	2.2	0.91
	2.6	0.95
Rubber		
[105]	0.5	0.37
	1.0	0.33
	3.0	0.20
	5.0	0.24
	7.0	0.24
	9.0	0.25
Rye		
ering winter rye		
[70]	0.40	0.072
	0.42	0.091
	0.44	0.108
	0.46	0.120
	0.48	0.120
	0.50	0.131
	0.52	0.151
	0.54	0.191
	0.55	0.214
	0.56	0.216
	0.58	0.180
	0.59	0.160
	0.60	0.164
	0.62	0.169
	0.64	0.160
	0.66	0.155
	0.68	0.168
	0.70	0.262
	0.71	0.370
	0.72	0.400
	0.73	0.493
	0.74	0.555
	0.75	0.600
	0.76	0.630
	0.77	0.658
	0.78	0.678
	0.79	0.697
	0.80	0.713
	0.81	0.732
	0.82	0.753
	0.83	0.771
	0.84	0.780
	0.85	0.791
out of bloom winter rye		
[70]	0.40	0.027
	0.44	0.039
	0.48	0.041
	0.50	0.050
	0.52	0.074
	0.54	0.101
	0.55	0.108
	0.56	0.107
	0.58	0.082
	0.60	0.077
	0.63	0.073
	0.65	0.070
earing spring rye		
[70]	0.40	0.032
	0.42	0.040
	0.44	0.052
	0.46	0.069
	0.48	0.072
	0.50	0.079
	0.52	0.093
	0.54	0.129
	0.55	0.157
	0.56	0.169
	0.58	0.140
	0.60	0.105
	0.62	0.107
	0.64	0.092
	0.65	0.085
Sedge		
[70]	0.40	0.045
	0.42	0.049
	0.46	0.056
	0.50	0.065
	0.51	0.078
	0.52	0.099
	0.53	0.134
	0.54	0.154
	0.55	0.164

Table 3.3.5. (*continued*)

State of surface	λ [μm]	$r_{\lambda,n}$
	0.56	0.157
	0.58	0.126
	0.60	0.116
	0.62	0.090
	0.64	0.085
	0.65	0.100
	0.66	0.126
	0.67	0.151
	0.68	0.182
	0.69	0.248
	0.70	0.308
	0.71	0.361
	0.72	0.385
	0.73	0.465
	0.74	0.620
	0.75	0.740
	0.76	0.770
	0.77	0.790
	0.78	0.808
	0.80	0.832
	0.82	0.845
Skin [9, 24]	0.3	0.03
	0.4	0.17
	0.5	0.34
	0.6	0.31
	0.7	0.46
	0.8	0.23
	0.9	0.17
	1.0	0.18
	2.0	0.07
	4.0	0.05
	6.0÷14	0.04
Spruce in winter a mature stand of trees [70]	0.40	0.023
	0.42	0.025
	0.44	0.028
	0.46	0.027
	0.48	0.026
	0.50	0.022
	0.52	0.036
	0.54	0.051
	0.56	0.048
	0.58	0.044
	0.60	0.044
	0.62	0.040
	0.64	0.043
	0.66	0.030
	0.68	0.028
	0.70	0.040
	0.72	0.062
	0.74	0.075
	0.80	0.085
	0.88	0.087
a mature stand of trees with young foliage [70]	0.40	0.034
	0.42	0.038
	0.44	0.044
	0.46	0.048
	0.48	0.051
	0.50	0.054
	0.52	0.094
	0.54	0.147
	0.55	0.157
	0.56	0.154
	0.58	0.121
	0.60	0.114
	0.62	0.111
	0.64	0.100
	0.66	0.097
	0.68	0.082
	0.70	0.080
	0.72	0.088
	0.74	0.165
	0.75	0.174
	0.78	0.176
	0.80	0.173
	0.85	0.182
a mature stand of trees with a full foliage [70]	0.40	0.027
	0.46	0.026
	0.50	0.033
	0.52	0.045
	0.54	0.063
	0.56	0.062
	0.58	0.051
	0.60	0.050
	0.62	0.050
	0.64	0.048
a mature stand of trees, a late verdure [70]	0.40	0.049
	0.50	0.069
	0.52	0.085
	0.54	0.105
	0.56	0.111
	0.58	0.100
	0.60	0.093
	0.62	0.096
	0.64	0.090

Table 3.3.5. (*continued*)

State of surface	λ [µm]	$r_{\lambda,n}$
young stand of trees, in winter [70]	0.40	0.010
	0.45	0.011
	0.50	0.010
	0.55	0.024
	0.60	0.020
	0.65	0.015
young stand of trees with a young foliage [70]	0.40	0.031
	0.50	0.048
	0.52	0.065
	0.54	0.090
	0.56	0.100
	0.58	0.088
	0.60	0.085
	0.64	0.078
	0.66	0.078
	0.68	0.080
	0.70	0.010
	0.71	0.126
	0.72	0.178
	0.73	0.194
	0.74	0.229
	0.75	0.240
young stand of trees with a full foliage [70]	0.40	0.020
	0.50	0.040
	0.52	0.064
	0.54	0.097
	0.56	0.104
	0.60	0.074
	0.62	0.066
	0.64	0.065
	0.66	0.066
	0.68	0.075
	0.70	0.097
	0.71	0.120
	0.72	0.138
	0.73	0.185
	0.74	0.250
	0.75	0.277
	0.76	0.293
	0.77	0.313
	0.78	0.334
	0.79	0.360
	0.80	0.377
	0.82	0.406
young stand of trees a late verdure [70]	0.52	0.037
	0.54	0.049
	0.56	0.049
	0.58	0.037
	0.60	0.034
	0.62	0.033
	0.64	0.030
Steppe [70]	0.40	0.018
	0.42	0.022
	0.44	0.024
	0.46	0.025
	0.48	0.028
	0.50	0.030
	0.52	0.038
	0.54	0.041
	0.56	0.041
	0.58	0.042
	0.60	0.042
	0.62	0.042
	0.64	
	0.66	0.081
	0.67	0.099
	0.68	0.122
	0.69	0.145
	0.70	0.163
	0.71	0.185
	0.72	0.218
	0.73	0.240
	0.74	0.270
	0.75	0.291
	0.76	0.311
	0.77	0.331
	0.78	0.344
	0.79	0.362
	0.80	0.384
	0.81	0.428
	0.82	0.462
	0.83	0.488
	0.84	0.506
	0.85	0.522
with sun-burnt grass [70]	0.40	0.025
	0.42	0.034
	0.44	0.036
	0.46	0.048
	0.48	0.057
	0.50	0.057
	0.52	0.071
	0.54	0.084

Table 3.3.5. (*continued*)

State of surface	λ[μm]	$r_{\lambda,n}$
	0.56	0.089
	0.58	0.088
	0.60	0.086
	0.62	0.083
	0.65	0.100
	0.66	0.112
	0.68	0.133
	0.70	0.151
	0.72	0.168
	0.74	0.211
	0.76	0.252
	0.78	0.261
	0.80	0.264
	0.82	0.280
	0.83	0.290
Straw in sheaves of oat [70]	0.40	0.052
	0.42	0.075
	0.44	0.094
	0.46	0.121
	0.48	0.149
	0.50	0.173
	0.52	0.195
	0.54	0.215
	0.56	0.223
	0.58	0.220
	0.60	0.217
	0.62	0.215
	0.64	0.215
	0.65	0.215
	0.66	0.227
	0.68	0.238
	0.70	0.274
	0.72	0.312
	0.74	0.373
	0.76	0.410
	0.78	0.441
	0.80	0.495
	0.82	0.550
of wheat [70]	0.40	0.044
	0.42	0.076
	0.44	0.095
	0.46	0.125
	0.48	0.150
	0.50	0.173
	0.52	0.187
	0.54	0.200
	0.56	0.217
	0.58	0.234
	0.60	0.251
	0.62	0.265
	0.64	0.277
	0.65	0.282
	0.66	0.285
	0.68	0.280
	0.70	0.277
	0.72	0.286
	0.74	0.344
	0.76	0.378
	0.78	0.418
	0.80	0.475
	0.82	0.535
of rye [70]	0.40	0.041
	0.42	0.062
	0.44	0.084
	0.46	0.123
	0.48	0.140
	0.50	0.167
	0.52	0.193
	0.54	0.219
	0.56	0.233
	0.58	0.248
	0.60	0.258
	0.62	0.266
	0.64	0.269
	0.65	0.272
	0.66	0.271
	0.68	0.270
	0.70	0.290
	0.72	0.250
	0.74	0.291
	0.76	0.345
	0.78	0.358
	0.80	0.397
	0.82	0.459
of lentil [70]	0.40	0.037
	0.42	0.050
	0.44	0.048
	0.46	0.052
	0.48	0.066
	0.50	0.072
	0.52	0.075
	0.54	0.080
	0.56	0.085
	0.58	0.085
	0.60	0.082
	0.62	0.085
	0.64	0.094
	0.65	0.096
	0.66	0.115
	0.67	0.127
	0.68	0.152
	0.69	0.198

Table 3.3.5. (*continued*)

State of surface	λ[μm]	$r_{\lambda,n}$
	0.70	0.246
	0.71	0.290
	0.72	0.311
	0.74	0.312
	0.76	0.315
	0.78	0.301
	0.80	0.300
	0.82	0.319
Stubbly-field after oat [70]	0.40	0.043
	0.45	0.053
	0.50	0.068
	0.52	0.088
	0.53	0.097
	0.55	0.101
	0.60	0.105
	0.65	0.111
	0.66	0.127
	0.68	0.148
	0.70	0.197
	0.72	0.239
	0.74	0.272
	0.76	0.301
	0.78	0.330
	0.80	0.361
	0.82	0.400
	0.84	0.448
	0.85	0.460
after lentil [70]	0.40	0.035
	0.45	0.041
	0.50	0.052
	0.55	0.063
	0.60	0.071
	0.65	0.077
	0.66	0.081
	0.68	0.098
	0.70	0.119
	0.72	0.147
	0.74	0.172
	0.76	0.183
	0.80	0.192
	0.85	0.190
after barley [70]	0.40	0.050
	0.42	0.061
	0.44	0.072
	0.46	0.079
	0.48	0.082
	0.50	0.100
	0.52	0.109
	0.54	0.133
	0.56	0.151
	0.58	0.158
	0.60	0.161
	0.62	0.173
	0.65	0.210
Sunflower blossoming field [70]	0.40	0.056
	0.45	0.053
	0.50	0.063
	0.52	0.078
	0.54	0.102
	0.55	0.110
	0.56	0.099
	0.58	0.081
	0.60	0.076
	0.65	0.076
	0.66	0.077
	0.68	0.085
	0.70	0.102
	0.71	0.138
	0.72	0.215
	0.73	0.328
	0.74	0.415
	0.75	0.450
	0.76	0.493
	0.77	0.512
	0.78	0.528
	0.79	0.538
	0.80	0.533
	0.81	0.511
	0.82	0.487
	0.83	0.468
	0.84	0.453
	0.85	0.445
Teflon; polytetrafluoroethylene; $CF_2 = CF_2$ [45, 131]	1.0	0.800
	2.0	0.245
	3.0	0.317
	4.0	0.335
	5.0	0.135
	6.0	0.025
	7.0	0.260
	9.0	0.200
	10	0.090
	11	0.060
	12	0.260
	13	0.844
	14	0.062

Table 3.3.5. (*continued*)

State of surface	λ[μm]	$r_{\lambda,n}$
	15÷17	0.000
	18	0.150
	19	0.130
	20	0.940
	21÷22	0.479
	22÷24	0.000
Tomato plantation, in the late summer [70]	0.40	0.022
	0.42	0.030
	0.44	0.030
	0.46	0.030
	0.48	0.030
	0.50	0.037
	0.52	0.051
	0.54	0.063
	0.56	0.059
	0.58	0.048
	0.60	0.040
	0.62	0.038
	0.64	0.036
	0.65	0.036
	0.66	0.052
	0.67	0.078
	0.68	0.115
	0.69	0.208
	0.70	0.277
	0.71	0.323
	0.72	0.339
	0.74	0.343
	0.76	0.355
	0.78	0.387
	0.80	0.386
	0.82	0.345
	0.84	0.328
Vetch before blossoming [70]	0.40	0.60
	0.42	0.059
	0.44	0.067
	0.45	0.066
	0.46	0.072
	0.47	0.079
	0.48	0.076
	0.49	0.076
	0.50	0.082
	0.52	0.104
	0.54	0.120
	0.56	0.119
	0.58	0.115
	0.60	0.110
	0.62	0.118
	0.64	0.106
Weeds thick brushwood (withering, acquiring a brown tint), with a cloudy sky [70]	0.40	0.050
	0.42	0.050
	0.44	0.054
	0.46	0.060
	0.50	0.076
	0.54	0.038
	0.58	0.090
	0.62	0.112
	0.65	0.117
Wheat during blossoming [70]	0.40	0.019
	0.44	0.030
	0.48	0.033
	0.50	0.038
	0.51	0.040
	0.52	0.050
	0.54	0.086
	0.56	0.103
	0.58	0.072
	0.60	0.059
	0.62	0.052
	0.64	0.042
	0.66	0.028
	0.68	0.040
	0.70	0.073
	0.71	0.100
	0.72	0.145
	0.73	0.187
	0.74	0.232
	0.75	0.265
	0.76	0.290
	0.77	0.315
	0.78	0.324
	0.79	0.351
	0.80	0.363
	0.81	0.380
	0.82	0.410
	0.83	0.448
	0.84	0.478
	0.85	0.498

Table 3.3.5. (*continued*)

State of surface	λ [μm]	$r_{\lambda,n}$
before developing ears		
[70]	0.40	0.031
	0.42	0.041
	0.44	0.049
	0.46	0.052
	0.48	0.051
	0.50	0.046
	0.52	0.060
	0.54	0.100
	0.55	0.119
	0.56	0.120
	0.58	0.097
	0.59	0.096
	0.60	0.091
	0.62	0.071
	0.64	0.064
	0.65	0.070
Willow coppice a late verdure		
[70]	0.40	0.105
	0.48	0.090
	0.50	0.095
	0.52	0.105
	0.54	0.114
	0.56	0.112
	0.58	0.103
	0.65	0.080
	0.66	0.080
	0.68	0.078
	0.70	0.113
	0.71	0.200
	0.72	0.280
	0.73	0.328
	0.74	0.369
	0.76	0.418
	0.78	0.452
	0.80	0.483
	0.82	0.510
	0.84	0.530
Willow herb thick, brushwood in time of blossoming		
[70]	0.40	0.074
	0.42	0.090
	0.43	0.109
	0.44	0.105
	0.46	0.102
	0.48	0.101
	0.50	0.098
	0.51	0.099
	0.52	0.110
	0.54	0.126
	0.55	0.122
	0.56	0.119
	0.58	0.114
	0.59	0.111
	0.60	0.127
	0.62	0.146
	0.64	0.157
	0.65	0.166
	0.66	0.170
	0.67	0.185
	0.68	0.218
	0.69	0.250
	0.70	0.345
	0.71	0.420
	0.72	0.500
	0.73	0.580
	0.74	0.638
	0.75	0.685
	0.76	0.715
	0.78	0.780
	0.80	0.778
	0.82	0.772
	0.84	0.765
	0.86	0.757
	0.88	0.750
	0.90	0.750
Wood		
[104, 105]	1.0	0.82
	2.0	0.54
	3.0	0.13
	4.0	0.25
	5.0	0.32
	6.0	0.32
	7.0	0.30
	8.0	0.33
on a darkened old bridge		
[70]	0.40	0.182
	0.42	0.188
	0.44	0.197
	0.46	0.207
	0.48	0.225
	0.50	0.250
	0.52	0.270
	0.54	0.284
	0.56	0.290
	0.58	0.282
	0.60	0.272
	0.62	0.266
	0.65	0.262
	0.66	0.264
	0.68	0.298
	0.70	0.350

Table 3.3.5. (*continued*)

State of surface	$\lambda[\mu m]$	$r_{\lambda,n}$
	0.71	0.372
	0.72	0.382
	0.76	0.414
	0.80	0.447
	0.84	0.477
darkened wall of a house [70]	0.40	0.142
	0.42	0.153
	0.44	0.161
	0.46	0.168
	0.48	0.180
	0.50	0.196
	0.52	0.204
	0.54	0.207
	0.56	0.203
	0.58	0.219
	0.60	0.232
	0.62	0.242
	0.64	0.245
	0.65	0.243
on a darkened old roof [70]	0.40	0.095
	0.45	0.105
	0.50	0.118
	0.52	0.132
	0.54	0.143
	0.55	0.146
	0.60	0.159
	0.65	0.172
	0.66	0.175
	0.68	0.180
	0.70	0.202
	0.72	0.237
	0.74	0.251
	0.76	0.259
	0.78	0.266
	0.80	0.278
	0.82	0.311
	0.84	0.349
	0.85	0.371
Wood, pine [90]	0.8	0.19
	1.2	0.32
	1.6	0.34
	2.0	0.31
	2.4	0.21
	2.6	0.27
Wormwood blossoming [70]	0.40	0.058
	0.42	0.054
	0.46	0.056
	0.50	0.067
	0.54	0.072
	0.58	0.081
	0.62	0.081
	0.65	0.091
	0.66	0.100
	0.68	0.139
	0.70	0.174
	0.72	0.188
	0.74	0.195
	0.76	0.200
	0.78	0.201
	0.80	0.215
	0.82	0.227

Table 3.3.6. Spectral transmissivity in normal direction $\tau_{\lambda,n}$ of organic and biological materials

State of surface and thickness of layer	λ [μm]	$\tau_{\lambda,n}$
Alcohol, butyl; butanol; $CH_3 \cdot CH_2 \cdot CH_2 \cdot CH_2 \cdot OH$ thickness of layer 5 cm [76]	0.86	0.92
	0.89	0.80
	0.93	0.42
	1.04	0.44
	1.12	0.63
	1.15	0.02
Alcohol, isobutyl; $(CH_3)_2CHCH_2OH$ thickness of layer 5 cm [76]	0.86	0.53
	0.89	0.50
	0.93	0.47
	1.04	0.36
	1.12	0.03
Aminobenzene (aniline); $C_6H_5NH_2$ thickness of layer 5 cm [76]	0.86	0.74
	0.88	0.34
	0.89	0.72
	0.93	0.80
	1.04	0.13
	1.08	0.60
	1.12	0.16
Benzene; benzol; C_6H_6 thickness of layer 5 cm [76]	0.86	0.95
	0.88	0.30
	0.93	1.00
	0.97	0.98
	1.04	0.93
	1.08	0.93

State of surface and thickness of layer	λ [μm]	$\tau_{\lambda,n}$
	1.11	0.30
	1.15	0.00
Bread crumb of slack-baked bread, thickness of layer 11 mm [24, 76, 105]	0.4	0.03
	0.5	0.065
	0.7	0.18
	0.9	0.215
	1.0	0.16
	1.1	0.19
	1.2	0.08
	1.3	0.09
	1.5	0.035
crumb baked bread, thickness of layer 11 mm [24, 76, 105]	0.4	0.04
	0.5	0.10
	0.7	0.21
	0.9	0.23
	1.0	0.196
	1.1	0.22
	1.2	0.13
	1.3	0.14
	1.5	0.06
white crust of slack-baked bread, thickness of layer 6 mm [24, 76, 105]	0.4	0.04
	0.5	0.08
	0.7	0.19
	0.9	0.22
	1.0	0.19
	1.1	0.215
	1.2	0.12
	1.3	0.14
	1.5	0.07

Table 3.3.6. (*continued*)

State of surface and thickness of layer	λ [μm]	$\tau_{\lambda,n}$
brown crust of baked bread, thickness of layer 6 mm [24, 76, 105]	0.4	0.02
	0.5	0.03
	0.7	0.09
	0.9	0.18
	1.0	0.19
	1.1	0.215
	1.2	0.14
	1.3	0.15
	1.5	0.105
Butyl chloride; $CH_3 \cdot CH_2 \cdot CH_2 \cdot CH_2Cl$ thickness of layer 5 cm [76]	0.86	0.89
	0.9	0.46
	0.96	0.92
	1.0	0.66
	1.09	0.83
	1.2	0.01
Carbon tetrachloride; tetrachloromethane; CCl_4 thickness of layer 5 cm [76]	0.86	0.84
	0.93	0.94
	1.00	0.94
	1.08	0.91
	1.15	0.81
Carboxyhaemoglobin thickness of layer 10 mm [76]	0.7	0.17
	0.8	0.18
	0.9	0.11
	1.0	0.07
	1.1	0.23
	1.2	0.16
	1.3	0.15
Cellophane thickness of layer 0.3 mm [24]	1.0	0.90
	2.0	0.89
	3.0	0.02
	3.5	0.09
	4.0	0.64
	5.0	0.68
	5.5	0.77
	6.0	0.37
	7.0	0.09
	9.0	0.06
	10	0.06
	11	0.20
	12	0.26
	13	0.20
Cellulose, acetate; acetylcellulose thickness of layer 0.5 mm [24, 45, 160]	1.0	0.88
	2.0	0.91
	4.0	0.97
	5.0	0.98
	6.0	0.87
	7.0	0.97
	7.5	0.93
	8.0	0.96
	10	0.93
	11	0.98
	12	0.87
Chlorophyll thin layer [24]	0.8	0.13
	1.0	0.75
	2.0	0.80
	3.0	0.70
	3.5	0.49
	4.0	0.79
	5.0	0.80
	6.0	0.50
	8.0	0.42
	10	0.52
	12	0.55
	14	0.52
Glass, organic; metacrylate polymethyl thickness of plate 0.02 mm [152]	1.25	0.98
	1.50	0.98
	1.75	0.98
	2.0	1.00
	2.4	0.97
	3.0	0.96
	3.1	0.97
	3.2	0.85
	3.3	0.56
	3.4	0.48
	3.5	0.78
	3.6	0.90
	3.7	0.94
	3.8	0.97

Table 3.3.6. (*continued*)

State of surface and thickness of layer	λ[μm]	$\tau_{\lambda,n}$
	3.9	0.98
	4.0	0.98
	4.5	0.98
	5.0	0.98
thickness of plate 0.5 mm [63]	1.0	0.62
	1.5	0.66
	2.0	0.86
	2.25	0.17
	2.5	0.57
	3.0	0.48
	3.5	0.00
	4.0	0.31
	4.5	0.49
	5.0	0.47
	5.5	0.17
	6.0	0.00
thickness of plate 3 mm [24]	0.8	0.87
	1.0	0.96
	1.2	0.81
	1.4	0.65
	1.6	0.30
thickness of plate 5 mm [24]	0.8	0.85
	1.0	0.88
	1.2	0.75
	1.4	0.09
	1.5	0.14
	1.6	0.07
Lacquer, vinyl thickness of layer 0.02 mm [24, 62]	1.0	0.99
	2.0	0.92
	3.0	0.88
	3.5	0.79
	4.0	0.95
	5.0	0.88
	6.0	0.79
	7.0	0.44
	8.0	0.21
	9.0	0.15
	10	0.30
	12	0.64
	14	0.92
	15	0.83
Lard thickness of layer 5 mm [24]	0.5	0.18
	0.6	0.20
	0.7	0.27
	0.8	0.35
	0.9	0.50
	1.0	0.70
	1.1	0.89
	1.2	0.98
	1.3	0.90
	1.4	0.43
	1.5	0.26
Oxyhaemoglobin thickness of layer 1.013 mm [9, 76]	0.6	0.04
	0.7	0.59
	0.8	0.37
	0.9	0.20
	1.0	0.24
	1.1	0.42
	1.2	0.63
	1.3	0.71
	1.4	0.51
thickness of layer 10 mm [76]	1.1	0.00
	1.2	0.05
	1.3	0.11
Paraffin thickness of layer 0.6 mm, temperature of melting 341÷345 K [76]	10	0.16
	12	0.12
	15	0.21
thickness of layer 1.2 mm, temperature of melting 341÷345 K [76]	20	0.55
	25	0.67
	30	0.71
	40	0.75
	50	0.79
thickness of layer 2.0 mm, temperature of melting 341÷345 K [76]	20	0.44
	25	0.51
	30	0.54
	40	0.57
	50	0.61

Table 3.3.6. (*continued*)

State of surface and thickness of layer	λ [μm]	$\tau_{\lambda,n}$
	60	0.63
	80	0.67
	100	0.68
	150	0.77
	200	0.82
thickness of layer 13 mm, temperature of melting 341÷345 K [76]	50	0.08
	60	0.09
	80	0.10
	100	0.10
	150	0.13
	200	0.18
	250	0.31
thickness of layer 2.05 mm, temperature of melting 315÷317 K [76]	20	0.20
	25	0.30
	30	0.39
	40	0.47
	50	0.53
	60	0.58
	80	0.64
	100	0.74
thickness of layer 3.7 mm, temperature of melting 315÷317 K [76]	30	0.23
	40	0.29
	50	0.35
	60	0.42
	80	0.54
	100	0.68
Polychlorotrifluoroethylen; $(CF_2 - CFCl)_n$ thickness of plate 3.0 mm [152]	0.8	0.77
	1.0	0.82
	1.5	0.84
	2.0	0.86
	2.5	0.87
	2.75	0.85
	3.0	0.57
	3.2	0.73
	3.4	0.68
	3.6	0.72
	3.8	0.60
	4.0	0.11
	4.5	0.0
	4.6	0.04

State of surface and thickness of layer	λ [μm]	$\tau_{\lambda,n}$
	4.7	0.16
	4.8	0.52
	4.9	0.44
	10.0	0.26
Polyethylene thickness of plate 0.025 mm [62]	1.0	0.24
	2.0	0.57
	3.0	0.73
	3.5	0.41
	4.0	0.70
	5.0	0.84
	6.0	0.85
	6.5	0.74
	7.0	0.82
	9.0	0.85
	11.	0.84
	13	0.78
thickness of plate 0.24 mm [152]	24	0.84
	26	0.77
	28	0.77
	30	0.73
	32	0.76
	34	0.80
	36	0.69
	38	0.53
thickness of plate 3.21 mm [152]	100	0.64
	200	0.72
	300	0.79
	400	0.84
	500	0.87
	600	0.90
Polyvinyl, chloride thickness of plate 0.3 mm [9]	1.0	0.67
	2.0	0.58
	3.0	0.03
	4.0	0.09
	5.0	0.08
	6.0	0.00
thickness of plate 3.2 mm [9]	0.5	0.40
	1.0	0.37
	2.0	0.21
	3.0	0.01

Table 3.3.6. (*continued*)

State of surface and thickness of layer	λ[μm]	$\tau_{\lambda,n}$
Resin, glyptal thickness of layer 0.3 mm [24]	0.4	0.60
	0.5	0.57
	0.6	0.63
	0.8	0.68
	1.0	0.63
	1.2	0.50
	1.4	0.39
	1.6	0.38
	1.8	0.29
	2.0	0.24
Resin, glyptal with lin seed oil thickness of layer 0.26 mm [24]	0.4	0.74
	0.5	0.80
	0.6	0.75
	0.8	0.82
	1.0	0.78
	1.2	0.63
	1.4	0.48
	1.6	0.50
	1.8	0.34
	2.0	0.24
Resin, maleic thickness of layer 0.43 mm [24]	0.4	0.55
	0.5	0.42
	0.6	0.46
	0.8	0.51
	1.0	0.53
	1.2	0.39
	1.4	0.30
	1.6	0.34
	1.8	0.11
	2.0	0.10
Resin, phenolformaldehyde; PF resins nonplasticized, thickness of layer 0.40 mm [24]	0.5	0.95
	0.6	1.00
	0.8	0.97
	1.0	0.88
	1.2	0.64
	1.4	0.45
	1.6	0.45
	1.8	0.31
	2.0	0.18
plasticized, thickness of layer 0.31 mm [24]	0.5	0.29
	0.6	0.30
	0.8	0.42
	1.0	0.42
	1.2	0.28
	1.4	0.16
	1.6	0.19
	1.8	0.11
	2.0	0.05
Rubber, natural, hydrochlorided; rubber hydrochloride; $(C_5H_9Cl)_n$ thickness of layer 0.045 mm [45, 152]	2.0	0.86
	3.0	0.34
	3.5	0.09
	4.0	0.62
	5.0	0.81
	6.0	0.54
	7.0	0.41
	8.0	0.42
	9.0	0.09
	10	0.58
	11	0.71
	12	0.54
	13	0.65
Silk thickness of layer 1.013 mm [76]	0.7	0.71
	0.8	0.80
	0.9	0.77
	1.0	0.74
	1.1	0.76
	1.2	0.73
	1.3	0.70
Skin thickness of layer 0.43 mm [9, 24]	0.8	0.54
	1.0	0.59
	1.2	0.59
	1.4	0.20
	1.6	0.43
	1.8	0.36
	1.9	0.03
	2.0	0.17

Table 3.3.6. (*continued*)

State of surface and thickness of layer	λ [μm]	$\tau_{\lambda,n}$
	2.2	0.25
	2.4	0.09
thickness of layer 1.60 mm [9, 24]	0.8	0.15
	1.0	0.17
	1.2	0.17
	1.4	0.02
	1.6	0.01
	1.8	0.05
	2.0	0.01
	2.2	0.005
Teflon; polytetrafluoroethylene; $CF_2 = CF_2$ thickness of plate 0.025 mm [152]	2.0	0.93
	3.0	0.94
	3.5	0.94
	4.0	0.88
	4.2	0.82
	4.4	0.93
	4.6	0.94
	4.8	0.94
	5.0	0.94
	5.5	0.89
	6.0	0.92
	6.5	0.90
	7.0	0.87
	7.2	0.89
	7.4	0.82
	7.6	0.70
	7.8	0.22
	8.0	0.02
	8.5	0.02
	8.8	0.02
	9.0	0.55
	9.5	0.81
	10.0	0.85
	10.2	0.88
	10.4	0.86
	10.6	0.81
	10.8	0.78
	11.0	0.79
	11.5	0.82
	12.0	0.88
	12.5	0.72
	13.0	0.52
	13.2	0.68
	13.4	0.60
	13.8	0.42
	14.0	0.44
	14.2	0.48
	14.4	0.55
	14.6	0.57
	14.8	0.57
	15.0	0.48
thickness of plate 0.13 mm [152]	2.0	0.58
	2.5	0.68
	3.0	0.74
	3.5	0.76
	3.7	0.77
	3.8	0.65
	3.9	0.62
	4.0	0.52
	4.1	0.48
	4.2	0.33
	4.3	0.45
	4.4	0.70
	4.5	0.76
	4.6	0.78
	4.7	0.80
	4.8	0.81
	4.9	0.80
	5.0	0.75
	5.2	0.66
	5.3	0.60
	5.4	0.60
	5.5	0.55
	5.7	0.64
	5.9	0.64
	6.0	0.67
	6.2	0.63
	6.4	0.47
	6.6	0.60
	6.8	0.46
	6.9	0.41
	7.1	0.43
	7.2	0.54
	7.4	0.34
	7.5	0.26
	7.6	0.15
	8.0	0.02
	9.0	0.02
	9.2	0.13
	9.4	0.28
	9.6	0.35
	9.8	0.44
	10.0	0.48
	10.4	0.50

Table 3.3.6. (*continued*)

State of surface and thickness of layer	λ [μm]	$\tau_{\lambda,n}$
	10.7	0.43
	11.0	0.47
	11.2	0.47
	11.6	0.42
	11.8	0.35
	12.2	0.35
	12.4	0.34
	12.5	0.31
	12.7	0.14
	13.0	0.08
	13.2	0.17
	13.6	0.04
	14.0	0.03
	14.2	0.04
	14.3	0.03
	14.7	0.10
	15.0	0.03
thickness of plate 0.2 mm [62]	0.5	0.85
	1.0	0.89
	2.0	0.94
	3.0	0.96
	4.0	0.85
	4.5	0.44
	5.0	0.84
	6.0	0.76
	7.0	0.45
	8.0	0.02
thickness of plate 0.25 mm [152]	2.0	0.33
	2.5	0.44
	2.8	0.48
	3.0	0.52
	3.5	0.58
	3.7	0.60
	3.8	0.43
	4.0	0.30
	4.2	0.12
	4.3	0.28
	4.4	0.51
	4.5	0.60
	4.6	0.61
	4.7	0.63
	4.8	0.65
	4.9	0.63
	5.0	0.55
	5.2	0.42
	5.3	0.35
	5.4	0.35
	5.5	0.32
	5.6	0.29
	5.9	0.41
	6.0	0.46
	6.2	0.46
	6.4	0.23
	6.6	0.33
	6.8	0.20
	6.9	0.17
	7.0	0.17
	7.2	0.28
	7.4	0.11
	7.6	0.02
	8.0	0.02
	9.0	0.02
	9.2	0.02
	9.5	0.10
	10.0	0.22
	10.3	0.24
	10.7	0.19
	11.0	0.22
	11.2	0.22
	11.5	0.18
	11.7	0.15
	12.0	0.12
	12.3	0.11
	12.5	0.10
	12.7	0.03
	13.0	0.02
	13.2	0.03
	13.5	0.02
	14.0	0.01
	14.5	0.01
	14.6	0.02
	14.8	0.01
	15.0	0.01
thickness of plate 0.76 mm [152]	2.0	0.09
	3.0	0.28
	3.5	0.34
	4.0	0.05
	4.2	0.01
	4.4	0.20
	4.6	0.33
	4.8	0.38
	5.0	0.23
	5.5	0.03
	6.0	0.13
	6.5	0.01
	7.0	0.01
	8.0	0.01

Table 3.3.6. (*continued*)

State of surface and thickness of layer	λ[μm]	$\tau_{\lambda,n}$	State of surface and thickness of layer	λ[μm]	$\tau_{\lambda,n}$
thickness of plate 1.81 mm [152]	100	0.65			
	200	0.75			
	300	0.82			
	400	0.85			
	500	0.89			
	600	0.91			
thickness of plate 3.0 mm [152]	1.0	0.82			
	1.5	0.83			
	2.0	0.87			
	2.5	0.87			
	3.0	0.58			
	3.5	0.71			
thickness of plate 4 mm [62]	1.0	0.74			
	2.0	0.83			
	3.0	0.61			
	4.0	0.15			
	4.5	0.03			
	5.0	0.32			
	6.0	0.08			
	6.5	0.01			

REFERENCES

1 Altgauzen A.P., Gutman M.B., Malyshev S.A., Svenchanskii A.D. and Smolenskii L.A., Low-temperature electroheating (in Russian), Izd. Energiya, Moskva 1968.

2 Altgauzen A.P., Smelyanskii M.Ya. and Shevtsov M.S., Electroheating equipment (in Russian), Izd. Energiya, Moskva 1967, pp. 27-30.

3 Bennett H.E., Measurement of specular reflectance at normal incidence. In: H.Blau and H.Fischer (eds.), Radiative transfer from solids materials, The MacMillan Co., New York 1962, pp. 166-180.

4 Bergquan J.B. and Seban R.A., Spectral radiation from alumina powder on a metallic substrate, J. of Heat Transfer, 1 (1971), p. 36-40.

5 Birkebak R.C. and Eckert E.R.G., Effects of roughness of metal surface on angular distribution of monochromatic reflected radiation, J. of Heat Transfer, 1 (1965), pp. 85-94.

6. Birkebak R.C., Sparrow E.M., Eckert E.R.G. and Ramsey J.W., Effect of surface roughness of metallic surfaces, J. of Heat Transfer, 2 (1964), pp. 193-199.

7 Blickensderfer R., Deardorf D.K. and Lincoln R.L., Normal total emittance at 400-850K and normal spectral reflectance at room temperature of Be, Hf, Nb, Ta, Ti, V and Zr, J. of the Less-Common Metals, 51 (1971), pp. 13-23.

8 BN-69/2780-1, Infrared heating-general terms (in Polish), Zjednoczenie Urządzeń Technologicznych TECHMA, Warszawa 1969.

9 Borkhert R. and Yubitz V., Technology of infrared heating (in Russian), Gosudarstvennoe Energeticheskoe Izd. Moskva-Leningrad 1969.

10 Bramson M.A., Infrared radiation of hot bodies (in Russian), Izd. Nauka, Moskva 1964.

11 Brügel W., Physik und Technik der Ultrarotstrahlung, Curt R. Vincentz

Verlag, Hannover 1951.

12 Bauchman K and Jungnickel G., Wärmeübertragung and Be- und Verarbeitungsmaschinen, 3, Politechnika Wrocławska, Wrocław 1978, pp. 212-260.

13 Buraczewski C., Thermal radiation heat transfer in non-enclosures with emitting and absorbing gases (in Polish), Instytut Maszyn Przepływowych, Polska Akademia Nauk, Poznań 1965.

14 Burakowski T., The influence of chromium upon emissivity of chromium stainless steels used in furnaces for heat treatments (in Polish), Doctoral Dissertation, Politechnika Warszawska, Warszawa 1972.

15 Burakowski T., Radiative properties of chromium stainless steel (in Polish), Metaloznawstwo i Obróbka Cieplna, 3 (1973)

16 Burakowski T., Emissivity of resistance alloys (in Polish), Instytut Mechaniki Precyzyjnej, Warszawa 1976.

17 Burakowski T., The measurements of emissivity of resistance alloys (in Polish), Metaloznawstwo i Obróbka Cieplna, 21 (1976) pp. 24-28.

18 Burakowski T., Giziński J. and Sala A., Infrared and its application (in Polish), Wyd. Ministerstwa Obrony Narodowej, Warszawa 1963.

19 Burakowski T., Giziński J. and Sala A., Infrared radiaters (in Polish), Wydawnictwa Naukowo-Techniczne, Warszawa 1970.

20 Caren R.P. and Liu C.K., Thermal radiation from microscopically roughened dielectric surface, J. of Heat Transfer, 1 (1972), pp.73-79.

21 Chan S.H. and Tien C.L., Infrared radiation properties of sulfur dioxide, J. of Heat Transfer, 2 (1971), pp. 172-178.

22 Chapman A.J., Heat transfer, MacMillan Publishing Co., New York 1974.

23 Davisson C. and Weeks J.R.Jr., The relation between the total thermal emissive power of a metal and its electrical resistivity, J. of the Optical Society of America and Review of Scientific Instruments, 5 (1924), pp. 581-605.

24 Deribere M., Les applications pratiques des rayons infrarouges, Dunod, Paris 1954.

25 Detkov S.P. and Beregovoi A.N., Extrapolation of emissivity of gases (in Russian), Inzheneriino-Fizicheskii Zhurnal, 5 (1974), pp. 833-839.

26 Eckert E.R.G., Einführung in den Wärme- und Stoffaustauch, Springer--Verlag, Berlin 1959.

27 Eckert E.R.G. and Drake R.M.Jr., Analysis of heat and mass transfer, McGraw-Hill Book Co., New York 1972.

28 Eckert E.R.G., Hartnett J.P. and Irvine T.F.Jr., Measurement of total emissivity of porous materials in use for transpiration cooling, Jet

Propulsion, 4 (1956), pp. 280-282.

29 Edwards D.K., Absorption by infrared bands of carbon dioxide gas elevated pressures and temperatures, J. of the Optical Society of America, 6 (1960), pp. 617-626.

30 Edwards D.K., Radiation interchange in a nongray enclosure containing an isothermal carbon dioxide-nitrogen gas mixture, J. of Heat Transfer, 1 (1963), pp. 1-11.

31 Edwards D.K., Radiation. In: Heat Transfer 1970, Fourth International Heat Transfer Conference, Paris - Versailles 1970, Elsevier Scientific Publishing Co., Amsterdam 1971, pp. 199-208.

32 Eryon N.D. and Gliksman L.R., An experimental an analytical study of radiative and conductive heat transfer in molten glass, J. of Heat Transfer, 2 (1973), pp. 224-230.

33 Farsberg C.H. and Domoto G.A., Thermal-radiation properties of thin metallic films on dielectric, J. of Heat Transfer, 4 (1972), pp.467-472.

34 Flugge S. and Trendelenburg F., Ergebnisse der Exakten Naturwissenschaften, Springer-Verlag, Berlin 1959.

35 Forsyte W.E. and Adams E.G., Radiating characteristics of tungsten and tungsten lamps J. of the Optical Society of America, 2 (1945), pp. 108-113.

36 Gardon R., The emission of radiation by transparent materials, The MacMillan Co., New York 1962.

37 Gauffé A., Transmission de la chaleur par rayonnement, Eyrolles Gauthier Villars, Paris 1968.

38 Gebhart B., Heat transfer, McGraw-Hill Book Co., New York 1971.

39 Geeraert B., Les pyrometres a radiation, Elektrowärme International, 5 (1968), pp. 155-158.

40 Giedt W.H., Principles of engineering heat transfer, D. Van Nostrand Co., New York 1957.

41 Gier J.T., Dunkle R.V. and Beyans J.T., Measurement of absolute spectral reflectivity from 1,0 to 15 microns, J. of the Optical Society of America, 7 (1974), pp. 558-562.

42 Giziński J., Influence of thin layer oxides upon total emissivity of metals (in Polish), Biuletyn Instytutu Mechaniki Precyzyjnej, 4 (1969). pp. 37-41.

43 Goel T.C. and Unvala B.A., Spectral emittance of titanium at high temperatures, Rev. International Hautes Temperatures and Refractory, 7 (1970), pp.197-201.

44 Gogół W., Heat transfer - tables and diagrams (in Polish), Politechnika Warszawska, Warszawa 1976.

45 Goldsmith A., Waterman T.E. and Hirschborn H.J., Handbook of solid materials, The MacMillan Co., New York 1961.

46 Gordon A.R. and Muchnik G.F., Determination of total emissivity as a function of surface roughness (in Russian), Teplofizika Vysokikh Temperatur, 2 (1964), pp. 292-294.

47 Gordov A.N., The pirometry (in Russian), Izd. Metallurgiya, Moskva 1964.

48 Gray W.A. and Muller R., Engineering calculations in radiative heat transfer, Pergamon Press, Oxford 1974.

49 Grigorev B.A., Impulse radiation heating. Part I, Character of impulse radiation and radiative heat transfer (in Russian), Izd. Nauka, Moskva 1974.

50 Grigull U., Die Grundgesetze der Warmeubertragung, Springer-Verlag, Berlin 1963.

51 Gruzin V.G., Measurement of temperature of liquid iron alloys (in Russian), Metallurgizdat, Moskva 1955.

52 Hackford H.L., Infrared radiation (in Polish), Wydawnictwa Naukowo-Techniczne, Warszawa 1963.

53 Harrison T.R., Radiation pirometry and its underlying principles of radiant heat transfer, John Wiley Sons, New York 1960.

54 Hitchcock J.E., Formulation of an irradiation factor method for surface radiation problems, J. of Heat Transfer, 3 (1970), pp. 412-417.

55 Hobler T., Heat transfer and heaters (in Polish), Wydawnictwa Naukowo-Techniczne, Warszawa 1968.

56 Holter M.R., Nudelman S., Suits G.H., Wolfe W.L. and Zissis G.J., Fundamentals of infrared technology, The MacMillan Co., New York 1962.

57 Houchens A.F. and Hering R.G., Surface roughness effects on equilibrium temperature. In: AIAA 10th Aerospace Sciences Meeting, January 17-19, San Diego 1972.

58 Houghton J.T. and Smith S.D., Infrared physics, Oxford at the Clarendon Press, Oxford 1966.

59 Howard J.N., Burch D.E. and Williams D., Infrared transmission of synthetic atmosphere. IV. Application of theoretical band models, J. of the Optical Society of America, 5 (1956), pp. 334-338.

60 Hsieh T.C., Hashemi and Grief R., Shock tube measurement of the emission

of carbon dioxide in the 2,7 micron region, J. of Heat Transfer, 3 (1975), pp. 397-399.

61 Irvine T.F.Jr., Thermal radiation properties of solids. In: W. Ibele (ed), Modern developments in heat transfer, Academic Press, New York 1963, pp. 213-224.

62 Ivanov Yu.A. and Tyapkin B.V., Infrared technology in military equipment (in Russian), Izd. Sovetskoe Radio, Moskva 1963.

63 Jamieson J.A., McFee R.H., Plass G.N., Grube R.H. and Richards R.G., Infrared physics and engineering, McGraw-Hill Book Co., New York 1963.

64 Jeżewski M. and Kalisz J., Tables of physical quantities (in Polish), Państwowe Wydawnictwo Naukowe, Warszawa 1957.

65 Jody B.J., Jain P.C. and Saxena S.C., Determination of thermal properties from steady-state heat transfer measurements on a heated tungsten wire in vacuum and helium gas, of Heat Transfer, 4 (1975), pp. 605-609.

66 Kartashova V.I. and Filipov O.K., Emissivity of globar in infrared spectrum (in Russian), Optiko-Mekhanicheskaya Promyshlennost, 3 (1975), p. 70.

67 Klyuchnikov A.D. and Ivantsov G.P., Thermal radiation heat transfer in furnaces (in Russian), Izd. Energiya, Moskva 1970.

68 Knop H.W.Jr., The emissivity of iron-tungsten and iron-cobalt alloys, Physical Rev. 10 (1948), pp. 1413-1416.

69 Kreith F., Principles of heat transfer, International Textbook Co., Scranton 1959.

70 Krinov E.L., Spectral reflectivity of natural materials (in Russian), Izd. Akademiya Nauk SSSR, Moskva 1947.

71 Krizhanowskii R.E. and Shtern Z.Yu., Heat properties of non-metals (oxides) (in Russian), Izd. Energiya, Leningrad 1973.

72 Kruze P.W., Macglauchin L.D., and Macquistan R.B., Elements of infrared technology generation, transmission and defection, John Wiley Sons, New York 1962.

73 Larrabee R.D., Spectral emissivity of tungsten, J. of the Optical Society of America, 6 (1959), pp. 619-625.

74 La Toison M., Infrarouge et applications thermiques, Centrex, Eindhoven 1964.

75 Laws W.R., Heating to 1250°C with an infra-red radiation furnace, Steel Times, 26 (1965), pp. 302-324.

76 Lecompte J., Le rayonnement infrarouge, Gauthier-Villars Editeur, Paris 1949.

77 Lin J.C. and Greif R., Theoretical determination of absorption with an emphasis on high temperatures and a specific application to carbon monoxide, J. of Heat Transfer, 4 (1973), pp. 535-538.

78 Madejski J., Theory of heat transfer (in Polish), Państwowe Wydawnictwo Naukowe, Warszawa 1963.

79 Marple D.T.P., Spectral emissivity of rhenium, J. of the Optical Society of America, 7 (1956), pp. 490-494.

80 Mazur M., Infrared heating (in Polish), Państwowe Wydawnictwa Techniczne, Warszawa 1953.

81 Mazur M., Industrial electroheating equipment, Wydawnictwa Naukowo-Techniczne, Warszawa 1964.

82 McAdams W.H., Heat transmission, McGraw-Hill Book Co. Inc, New York 1942.

83 McCulloch J.W. and Sunderland J.E., Measurement of monochromatic emittance of nonconductors at moderate temperatures, J. of Heat Treatment, 2 (1970), pp. 231-236.

84 Middleton W.E.K. and Wyszecki G., Colors produced by reflection at grazing incidence from rough surface, J. of the Optical Society of America, 11 (1957), pp. 1020-1023.

85 Mukhin V.M., Investigation methods of total emissivity of metals with the increasing oxidation (in Russian). In: Trudy VNIIETO, Investigations of industrial infrared electroheating, Izd. Energiya, Moskva 1970, pp. 64-72.

86 Mukhin V.M. and Evadis E.I., Investigation of changes of total emissivity of metals as a function of temperature and heating time (in Russian), Elektrotermiya, Moskva, pp. 34-35.

87 Nagata Ken-ichi and Nishiwaki Jien, Reflection of light from filmed rough surface - determinations of film thickness and rms roughness, Japanese J. of Applied Physics, 2 (1967), pp. 251-257.

88 Nakamoto K., Infrared spectra of inorganic and coordination compounds, John Willey Sons, New York 1963.

89 Neuer G., Verfahren zum Messen das emissionsgrads fester Stoffe bei hohen Temperaturen, VDI-2, 1 (1972), pp. 38-43.

90 Nicolle R.L., Irvine J. and Bowden F.G., Near-infrared diffuse reflectivities of natural and made materials, British J. Applied Physics,

2 (1969), pp. 201-204.

91 Novitskii L.A., Petrechenko B.I. and Varakina L.P., Investigation of total emissivity of solid materials and coatings at low temperature (in Russian), Izmeritelnaya Tekhnika, 12 (1966), pp. 28-30.

92 Padalka V.G. and Shklyarevskii I.N., Investigation of microcharacteristic of copper for optical constants and infrared and resistivity at temperature 82 and 295 K (in Russian), Optika i Spektroskopiya, 2 (1962), pp. 291-296.

93 Pasaglia E., Measurement of physical properties. In: R.F. Bunshah (ed.), Techniques of Metals Research, John Willey Sons, New York 1972, pp. 2-90.

94 Pattison J.R., The total emissivity of some refractory materials above 900°C, Transactions of the British Ceramic Society, 11 (1955), pp. 698-705.

95 Penner S.S., Quantitative moleculary spectroscopy and gas emissivities, Pergamon Press, London 1959.

96 Pepperhoff W., Temperaturstrahlung, Verlag von dr Dietrich Steinkopff, Darmstadt 1956.

97 Petrov B.A., Total emissivity of materials at high temperatures (in Russian), Izd. Nauka, Moskva 1969.

98 Plaksin I.N. and Solnyshkin N.I., Infrared spectroscopy of film reactants on the minerals (in Russian). Izd. Nauka, Moskva 1969.

99 Polyanskii W.K. and Kovalevskii L.V., Scattering of coherent radiation on the rough surface (in Russian), Optika i Spektroskopiya, 5 (1971), pp. 764-787.

100 Price P.H., Radiation-interaction with cuduction and convection. In: Heat Transfer 1970. Fourth International Heat Transfer Conference, Paris-Versailles 1970, vol. 9, Elsevier Publishing Company, Amsterdam 1971, pp. 209-214.

101 Price D.J. and Lowery H., The emissivity characteristics of hot metals, with special reference to the infra-red, Paper of the Steel Casting Research Committee, 7, pp. 523-546.

102 Quinn T.J., Emissivity and temperature measurement, Rev. International Temperature and Refractory, 7 (1970), pp. 180-191.

103 Ramanathan K.G., Yen S.H. and Estalote E.A., Total hemispherical emissivities of copper, aluminium and silver, Applied Optics, 11 (1977), pp. 2810-2817.

104 Razujevic K., Tables of heating with diagram (in Polish), Wydawnictwa Naukowo-Techniczne, Warszawa 1966.

105 Reinders H., Warmeanstausch durch Strahlung Einfuhrung in die technische Anvendung, VDI-Verlag, Düsseldorf 1961.

106 Reprinceva S.M. and Federovich N.W., Radiation heat transfer in dispersion medium (in Russian), Izd. Nauka i Tekhnika, Minsk 1968.

107 Rezaikina N.N. and Ryzhov L.N., Optical characteristics of materials investigated with the methods of infrared spectroscopy. In: Trudy VNIIETO, Investigation of industial infrared electroheating, Izd. Energiya, Moskva 1970, pp. 112-115.

108 Roger C.R., Yen S.H. and Ramanathen K.G., Temperature variation of total hemispherical emissivity of stainless steel AISI 304, J. of Optical Society of America, 10 (1979), pp. 1384-1390.

109 Rohsenow W.M. and Hartnett J.P. (eds.), Handbook of heat transfer. E.R.G. Eckert, Relations and properties, McGraw-Hill Book Co., New York 1973, p. 15-1, 15-28.

110 Rohsenow W.M. and Hartnett J.P. (eds.), Handbook of heat transfer. S.S. Penner, Equilibrium radiation properties of gases, McGraw-Hill Book Co., New York 1973, pp. 15-72, 15-93.

111 Rudolph H.H., Methoden zur Bestimmung des Emissionsgrades von technischen Oberflachen bei hoheren temperaturen, Elektrowärme, 5 (1960), pp. 151-157.

112 Rutgers G.A.W., Temperature radiation of solids in: S. Flugge (ed.), Handbuch der Physik, Licht und Materie II, Springer-Verlag, Berlin 1958, pp. 143-170.

113 Sacadura J.F.O., Determination des facteurs d'emission normaux de materiaux dielectriques entre 1000 et 1500°C, Rev. International Hautes Températures et Réfractaires, 8 (1971), pp. 101-110.

114 Sala A., Investigation of relationship between the roughness of metal surface and its total emissivity in the normal direction $\varepsilon_{T,n}$, Acta IMEKO 1967, PO-132.

115 Sala A., Influence of metal surface roughness upon total emissivity for various oxidation (in Polish), Doctoral Dissertation, Politechnika Warszawska, Wydział Mechaniki Precyzyjnej, Warszawa 1968.

116 Sala A., Investigation of relationship between roughness and total emissivity $\varepsilon_{T,n}$ of metal surface (in Polish), Biuletyn Instytutu Mechaniki Precyzyjnej, 4 (1969), pp. 42-53.

117 Sala A., Total emissivity ε_T as a function of changes of oxides layer on steel 1H18N9T, Prace Instytutu Mechaniki Precyzyjnej (in Polish), 69/3-A (1970), pp. 24-36.

118 Sala A., Adaptation and experimental verification of theory of cavity for the definition of function of total emissivity and surface roughness (in Polish), Prace Instytutu Mechaniki Precyzyjnej, 71/1-A (1971), pp. 11-20.

119 Sala A., Diffusion of radiation on oxidized metalic surface (in Polish) Prace Instytutu Mechaniki Precyzyjnej, 73/3-A (1971), pp. 37-46.

120 Sala A., Spectral emissivity of steel 1H18N9T and NC6 as a function of structural changes (in Polish), Prace Instytutu Mechaniki Precyzyjnej, 74/A-4 (1971), pp. 26-35.

121 Sala A., Experimental-analitycal function of total emissivity for grooved surface (in Polish), Metaloznawstwo i Obróbka Cieplna, 6 (1973), pp. 70-75.

122 Sala A., Emissivity of metals and alloys as a function of surface quality (in Polish), Habilitation Dissertation, Instytut Mechaniki Precyzyjnej, Warszawa 1973.

123 Sala A., Geometrical distribution of emissivity and reflectivity of flat metal surface for various oxidation and roughness (in Polish), Metaloznawstwo i Obróbka Cieplna, 11 (1974), pp. 25-35.

124 Sala A., Influence of roughness changes of oxides layer upon the total emissivity ε_T (in Polish), Metaloznawstwo i Obróbka Cieplna, 23 (1976) pp. 33-38.

125 Sala A., Optical properties of protective atmospheres (in Polish), Metaloznawstwo i Obróbka Cieplna, 24 (1976), pp. 34-43.

126 Sala A. and Sobusiak T., Investigation of absorptivity radiation in thin layer of water from protective atmospheres (in Polish), Metaloznawstwo i Obróbka Cieplna, 5 (1973), pp. 37-41.

127 Seban R.A., The emissivity of transition metals in the infrared, J. of Heat Transfer, 2 (1965), pp. 173-176.

128 Seifert R.L., The spectral emissivity and total emissivity of beryllium oxide, Physical Rev., 10 (1948), pp. 1181-1187.

129 Sennett R.S. and Scott G.D., The structure of evaporated metal films and their optical properties, J. of the Optical Society of America, 4 (1950), pp. 203-211.

130 Severinets G.N., Application of gas infrared heaters to drying, stoving and heating processes (in Russian), Izd. Nedra, Leningrad

1970.

131 Sheindlin A.E., Radiative properties of solid materials (in Russian), Izd. Energiya, Moskva 1974.

132 Siegiel R. and Howell J.R., Thermal radiation heat transfer, McGraw-Hill Book Co., New York 1972.

133 Silverman S., The emissivity of globar, J. of the Optical Society of America, 11 (1948), p. 989.

134 Simon I., Infrared radiation (in Polish), Państwowe Wydawnictwo Naukowe, Warszawa 1978.

135 Smith R.A., Jones F.E. and Chasmar R.P., The defection and measurement of infra-red radiation, Oxford at the Clarendon Press, Oxford 1957.

136 Sparrow E.M., Heinisch R.P. and Tien K.K., Effect of film thickness on the infrared reflectance of very thin metallic films, J. of Heat Transfer 4 (1973), pp. 534-535.

137 Stokes A.R., The theory of the optical properties of inhomogenneous materials, E and F.N. Spon Limited, London 1963.

138 Sully A.H., Brandes F.A. and Waterhause R.B., Some measurement of the total emissivity of metals and pure refractory oxides and the variation of emissivity with temperature, British J. of Applied Physics, 3 (1952).

139 Summer W., Ultra-violet and infra-red engineering, Sir Isaac Pitman and Sons, Ltd., London 1962.

140 Svet D.Ya., Thermal radiation of metals and some materials (in Russian), Izd. Metallurgiya, Moskva 1964.

141 Tewfik O.E., Yang Ji-Wu, Emissivity measurement of porous materials, J. of Heat Transfer, 1 (1963), pp. 79-80.

142 Thomas L.K., Normal spectral emissivity of Ta-W and Nb-Mo alloys, J. of Applied Physics, 8 (1968), pp. 3737-3742.

143 Thomas L.K., Thermal radiation from rough tungsten surfaces in normal and off-normal directions, J. of Applied Physics, 10 (1968), pp. 4681-4687.

144 Tien C.L., Heat transfer by laminar flow from a rotating cone, J. of Heat Transfer, 3 (1960), pp. 252-255.

145 Tien C.L., Chan C.K. and Cunnigton G.R., Infrared radiation of thin plastic films, J. of Heat Transfer, 1 (1972), pp. 41-45.

146 Tishchenko W.G., Pyrometry of liquid metals (in Russian), Izd. Nauka Dumka, Kiev 1964.

147 Toor J.S. and Viskanta R., Experiment and analysis of directional effects on radiant heat transfer, J. of Heat Transfer, 4 (1972), pp. 459-466.

148 Torrance K.E. and Sparrow E.M., Biangular reflectance of an electric nonconductor as a function of wavelength and surface roughness, J. of Heat Transfer, 2, pp. 283-292.

149 Turner A.F., Reflectance properties of thin and multilayers. In: H. Blau and H. Fischer (eds.) Radiative transfer from solid materials, The MacMillan Co., New York 1962, pp. 24-58.

150 Unvala B.A. and Goel T.C., Thermal conductivity, electrical resistivity and total emittance of titanium at high temperatures, Rev. Int. Hautes Temper. et Refract., 7, pp. 341-345.

151 Verret D.P. and Ramanathan K.G., Total hemispherical emissivity of tungsten, J. of Optical Society of America, 9 (1978), pp. 1167-1172.

152 Voronkowa E.M., Grechushnikov B.N. and Petrov I.P., Optical materials for infrared technology (in Russian), Izd. Nauka, Moskva 1955.

153 Wahlin H.B. and Knop H.W.Jr., The spectral emissivity of iron and cobalt, Physical Rev., 6 (1948), pp. 687-689.

154 Ward L., The temperature coefficient of reflectivity of nickel, Proc. Phys. Soc., 9-B, pp. 862-865.

155 Ward L., The variation with temperature of spectral emissivites of iron, nickel, cobalt, Proc. Phys. Soc., 3 (1956), pp. 339-343.

156 Wójcik L.A. and Sievers A.J., Total hemispherical emissivity of W(100), J. of Optical Society of America, 70, 4 (1980), pp. 443-450.

157 Wong H.Y. and Aggarwall S.R., Total hemispherical emittance of stainless steel AISI 321, J. of the Iron and Steel Instytute, 8, 209 (1971), pp. 635-637.

158 Wood W.D., Doem H.W. and Lucks C.F., Thermal radiative properties, Planum Press, New York 1964.

159 Worthing A.G., Temperature radiation emissivities and emittances, Optical and Radiation Pyrometry, pp. 1164-1187.

160 Zhbankov R.G., Infrared spectrums of cellulose and its derivatives (in Russian), Akademiya Nauk BSSR, Institut Fiziki, Minsk 1964.

INDEX OF MATERIALS

	ε_T	$\varepsilon_{T,n}$	$\varepsilon_{\lambda,n}$	$\varepsilon_{\lambda=0.65n}$	$r_{\lambda,n}$	$\tau_{\lambda,n}$
Alcohol isobutyl; $(CH_3)_2CHCH_2OH$	-	-	-	-	-	412
AlCu2MG2Ni1Si ⟶ aluminium-copper-magnesium alloy; PA29N*; H18 (duralumin)	-	185	-	-	-	-
AlCu4MgA ⟶ aluminium-copper-magnesium alloy; PA25* (duralumin)	-	185	-	-	-	-
AlCu4Mg1 ⟶ aluminium-copper-magnesium alloy; PA7N*; 2024 (duralumin)	-	185	-	-	-	-
Alder	-	-	-	-	385	-
$Al(F,OH)_2SiO_4$ ⟶ topaz	-	-	-	-	337	-
Alloy 260 ⟶ copper-zinc-alloy, CuZn30; M70* (brass without alloing agent)	-	189	-	-	-	-
Alloy 606 ⟶ copper-aluminium alloy; BA5* (aluminium bronze)	-	188	216	240	-	-
AlMg3* ⟶ aluminium-magnesium alloy; PA11N*	-	-	214	-	-	-
Al_2O_3 ⟶ aluminium oxide; LA603	-	263	-	-	312	-
Al_2O_3 ⟶ corundum sintered; aluminium oxide	-	270	285	-	317	359
Al_2O_3 ⟶ sapphire; aluminium oxide	-	-	-	-	330	368
$Al(OH)_3$ ⟶ aluminium hydroxide; alumina trihydrate	-	263	-	-	-	-
$Al_2O_3 \cdot 3SiO_2 \cdot H_2O$ ⟶ alumino silicate hydrated; bentonite	-	-	-	-	313	-
AlSb ⟶ aluminium antominide	-	-	-	-	312	353
$Al_2(SiO_4)O$ ⟶ andalusite; alumino silicate	-	-	-	-	312	-
$Al_2(SiO_4)O$ ⟶ cyanite; alumino silicate; disthene	-	-	-	-	317	-
Alumina trihydrate ⟶ aluminium hydroxide; $Al(OH)_3$	-	263	-	-	-	-
Aluminium; Al	175	184	214	-	246	-

	ε_T	$\varepsilon_{T,n}$	$\varepsilon_{\lambda,n}$	$\varepsilon_{\lambda=0.65n}$	$r_{\lambda,n}$	$\tau_{\lambda,n}$
AM-350 ⟶ steel, chromium-nickel-molybdenum stainless; OOH17N14M*	-	206	226	243	-	-
Aminobenzeze (aniline); $C_6H_5NH_2$	-	-	-	-	-	412
Ammonia; NH_3	-	263	-	-	-	354
Ammonium monobasic phosphate ⟶ ammonium phosphate; $NH_4H_2PO_4$	-	-	-	-	-	357
Ammonium phosphate (ammonium monobasic phosphate); $NH_4H_2PO_4$	-	-	-	-	-	357
Andalusite; alumino-silicate; $Al_2(SiO_4)O$	-	-	-	-	312	-
Anhydrite; calcium sulphate; $CaSO_4$	-	-	-	-	313	-
Aniline ⟶ aminobenzen; $C_6H_5NH_2$	-	-	-	-	-	412
Antimony; Sb	175	186	215	-	247	-
Arsenic sulphide; As_2S_3	-	-	-	-	-	357
Asbestos	259	264	-	-	-	-
Asbestos, cement	259	-	-	-	-	-
Aspen	-	-	-	-	385	-
Asphalt	-	-	-	-	313	-
As_2S_3 ⟶ arsenic sulphide	-	-	-	-	-	357
As_2S_3 ⟶ glass, arsenic sulphide	-	-	-	-	319	361
Au ⟶ gold	176	189	217	240	249	-
B ⟶ boron	-	-	215	-	-	-
Ba ⟶ barium	-	-	-	-	247	-
BA5* ⟶ copper-aluminium alloy; Alloy 606; (aluminium bronze)	-	188	216	240	-	-
$BaCl_2$ ⟶ barium chloride	-	264	-	-	-	-
BaF_2 ⟶ barium fluoride	-	-	-	-	313	357
Baidonal ⟶ steel chromium-aluminium heat resistant; H17J5*	-	204	-	-	-	-
BaO ⟶ barium oxide	-	-	-	-	-	357
Barite; barium sulphate; $BaSO_4$	-	-	-	-	313	-
Barium; Ba	-	-	-	-	247	-
Barium chloride; $BaCl_2$	-	264	-	-	-	-
Barium fluoride; BaF_2	-	-	-	-	313	357

	ε_T	$\varepsilon_{T,n}$	$\varepsilon_{\lambda,n}$	$\varepsilon_{\lambda=0.65n}$	$r_{\lambda,n}$	$\tau_{\lambda,n}$
Barium oxide; BaO	-	-	-	-	-	357
Barium sulphate → barite; $BaSO_4$	-	-	-	-	313	-
Barium sulphide; BaS	-	264	-	-	-	-
Barium titanite; $BaTiO_3$	-	-	-	-	313	357
Bark	-	-	-	-	387	
Barley	-	-	-	-	387	
BaS → barium sulphide	-	264	-	-	-	-
$BaSO_4$ → barite; barium sulphate	-	-	-	-	313	-
$BaTiO_3$ → barium titanite	-	-	-	-	313	357
Battele 54726 → glass, metallized electricity conductive, Battele 54726	-	271	289	-	320	-
B_4C → boron carbide	-	264	-	-	-	-
Be → berylium	-	186	-	240	247	-
Beef	-	-	383	-	-	-
Bentonite; alumino-silicate, hydrated; $Al_2O_3 \cdot 3Si_2 \cdot H_2O$	-	-	-	-	313	-
Benzene; benzol; C_6H_6	-	-	-	-	-	412
Benzol → benzene; C_6H_6	-	-	-	-	-	412
BeO → beryllium oxide	-	264	281	306	-	357
Beryllium; Be	-	186	-	240	247	-
Beryllium oxide; BeO	-	264	281	306	-	357
Beryllium sulphate; $BeSO_4$	-	264	-	-	-	-
$BeSO_4$ → beryllium sulphate	-	264	-	-	-	-
Bi → bismuth	-	186	-	-	-	-
Birch	-	-	-	-	387	-
Bismuth; Bi	-	186	-	-	-	-
Bismuth telluride; Bi_2Te_3	-	-	-	-	313	-
Bi_2Te_3 → bismuth telluride	-	-	-	-	313	-
Bloodstone → haematite; iron oxide; Fe_2O_3	260	271	289	-	321	-
BN → boron nitride	-	265	281	306	314	-
Boron; B	-	-	215	-	-	-
Boron carbide; B_4C	-	264	-	-	-	-
Boron nitride - BN	-	265	281	306	314	-
Boron silicide - B_4S_i	-	-	281	-	314	-

	ε_T	$\varepsilon_{T,n}$	$\varepsilon_{\lambda,n}$	$\varepsilon_{\lambda=0.65n}$	$r_{\lambda,n}$	$\tau_{\lambda,n}$
Brass → copper-zinc-alloy	176	188	217	-	-	-
Brass without alloing agent → copper-zinc-alloy; CuZn30; Alloy 260; M70*	-	189	-	-	-	-
Bread	-	-	-	-	389	412
Brick, building, red	259	-	282	-	314	-
Brick, burnt, red	-	265	-	-	-	-
Brick, calcium-silicate	-	265	282	-	-	-
Brick, chamotte	-	265	282	-	-	-
Brick, chamotte, porous	-	-	282	-	-	-
Brick, chamotte, white	-	-	-	-	314	-
Brick, chamotte, white, glazed	-	-	-	-	314	-
Brick, chrome-magnesite	-	265	-	-	-	-
Brick, chrome-magnesite, refractory	-	265	-	-	-	-
Brick, corundum	-	265	-	-	-	-
Brick, Dinas	-	265	282	-	-	-
Brick, dinas, refractory	-	265	-	-	-	-
Brick, magnesite	-	265	-	-	-	-
Brick, magnesite-silicate with iron oxide	-	265	-	-	-	-
Brick, sillimanite; $Al(AlSiO_5)$	-	265	-	-	-	-
BS-37A → glass, BS-37A	-	-	-	-	-	362
BS 381C Nr 224 → paint, BS 381C Nr 224	-	-	-	-	400	-
BS 381C Nr 261 → paint, BS 381C Nr 261	-	-	-	-	400	-
BS 381 C Nr 298 → paint, BS 381C Nr 298	-	-	-	-	400	-
B_4Si → boron silicide	-	-	281	-	314	-
Buckwheat	-	-	-	-	389	-
Bulrusch	-	-	-	-	389	-
Butanol → alcohol butyl; $CH_3 \cdot CH_2 \cdot CH_2 \cdot CH_2OH$	-	-	-	-	-	412
Butyl chloride; $CH_3 \cdot CH_2 \cdot CH_2 \cdot CH_2 \cdot Cl$	-	-	-	-	-	413

	ε_T	$\varepsilon_{T,n}$	$\varepsilon_{\lambda,n}$	$\varepsilon_{\lambda=0.65n}$	$r_{\lambda,n}$	$\tau_{\lambda,n}$
C ⟶ carbon	-	-	-	-	315	358
C ⟶ carbon; coal	-	-	-	306	-	-
C ⟶ diamond	-	-	-	-	-	360
C ⟶ electrographite; Acheson graphite; carbon	259	270	286	307	-	-
C ⟶ graphite	-	-	-	307	320	-
C ⟶ graphite policrystalline; carbon	260	-	289	308	320	-
C 1010 ⟶ steel, carbon 10*	-	204	-	-	-	-
C 1015 ⟶ steel, carbon 15*	-	204	-	-	-	-
C 1020 ⟶ steel, carbon 20*	-	204	-	-	-	-
C 1050 ⟶ steel, carbon 50*	-	204	-	-	-	-
Cabbage	-	-	-	-	389	-
$CaCO_3$ ⟶ calcite; calcium carbonate; calcspar	-	265	-	-	314	358
$CaCO_3$ ⟶ limestone	-	273	292	-	-	-
$CaCO_3$ ⟶ limestone; $CaCO_3$	-	-	-	-	322	-
Cadmium; Cd	-	-	-	-	247	-
Cadmium fluoride; CdF_2	-	-	-	-	-	357
Cadmium sulphide; CdS	-	-	-	-	-	357
Cadmium telluride; CdTe	-	-	-	-	314	358
CaF_2 ⟶ calcium fluoride; fluorite	-	270	-	-	318	360
Calcined soda ⟶ soda ash; sodium carbonate; $NaCO_3$	-	275	-	-	-	-
Calcite; calcspar; $CaCO_3$; calcium carbonate	-	265	-	-	314	358
Calcium carbonate ⟶ calcite; calcspar; $CaCO_3$	-	265	-	-	314	358
Calcium carbonate ⟶ limestone; $CaCO_3$	-	-	-	-	322	-
Calcium fluoride ⟶ fluorite; CaF_2	-	270	-	-	318	360
Calcium hydrate ⟶ calcium hydroxide; $Ga(OH)_2$; lime	-	272	-	-	-	-

	ε_T	$\varepsilon_{T,n}$	$\varepsilon_{\lambda,n}$	$\varepsilon_{\lambda=0.65n}$	$r_{\lambda,n}$	$\tau_{\lambda,n}$
Calcium hydroxide ⟶ lime; calcium hydrate; $(CaOH)_2$	-	272	-	-	-	-
Calcium magnesium-carbonate ⟶ dolomite; $CaMg(CO_3)_2$	-	-	-	-	317	-
Calcium nitrate; $Ca(NO_3)_2$	-	266	-	-	-	-
Calcium oxide ⟶ lime burnt; quicklime; CaO	-	272	-	-	322	-
Calcium sulphate ⟶ anhydrite; $CaSO_4$	-	-	-	-	313	-
Calcium titanate; $CaTiO_3$	-	-	-	-	315	-
Calcium tungstate ⟶ scheelite; $CaWO_4$	-	-	-	-	330	368
Calcspar ⟶ calcite; calcium carbonate; $CaCO_3$	-	265	-	-	314	358
$CaMg(CO_3)_2$ ⟶ dolomite; calcium magnesium-carbonate	-	-	-	-	317	-
Cancrynite; $(Na,K(Al_6Si_6O_{24})CO_3)$ $2H_2O$	-	-	-	-	315	-
$Ca(NO_3)_2$ ⟶ calcium nitrate	-	266	-	-	-	-
CaO ⟶ calcium oxide; lime burnt; quicklime	-	272	-	-	322	-
$Ca(OH)_2$ ⟶ calcium hydrate; calcium hydroxide; lime	-	272	-	-	-	-
Carbolithe, fr. ⟶ silite	-	-	295	-	-	-
Carbon ⟶ C; coal	-	-	-	306	-	-
Carbon; C	-	-	-	-	315	358
Carbon; coal; C	-	-	-	306	-	-
Carbon ⟶ electrographite; Acheson graphite; C	259	270	285	307	-	-
Carbon ⟶ graphite policrystal-line; C	260	-	289	308	320	-
Carbon, black	377	379	-	-	390	-
Carbon dioxide; CO_2	-	266	282	-	-	-
Carbon oxide; CO	-	268	284	-	-	-

	ε_T	$\varepsilon_{T,n}$	$\varepsilon_{\lambda,n}$	$\varepsilon_{\lambda \doteq 0.65n}$	$r_{\lambda,n}$	$\tau_{\lambda,n}$
Carbon silicide	-	-	284	-	-	-
Carbon tetrachloride; tetrachloromethane; CCl_4	-	-	-	-	-	413
Carborundum monocrystal; SiC	-	-	-	-	315	-
Carboxyhaemoglobin	-	-	-	-	-	413
$CaSO_4$ ⟶ anhydrite; calcium sulphate	-	-	-	-	313	-
$CaSO_4 \cdot H_2O$ ⟶ gypsum; hydrated calcium sulphate	260	271	-	-	-	362
Cassiterite; tin oxide; SnO_2	-	269	-	-	-	358
Cast iron, carbon	175	186	-	240	-	-
Cast iron, silicon	-	-	-	240	-	-
$CaTiO_3$ ⟶ calcium titanate	-	-	-	-	315	-
$CaWO_4$ ⟶ calcium tungstate; scheelite	-	-	-	-	330	368
CCl_4 ⟶ carbon tetrachloride; tetrachloromethane	-	-	-	-	-	413
Cd ⟶ cadmium	-	-	-	-	247	-
CdF_2 ⟶ cadmium fluoride	-	-	-	-	-	357
CdS ⟶ cadmium sulphide	-	-	-	-	-	357
CdTe ⟶ cadmium telluride	-	-	-	-	314	358
Celestite; $SrSO_4$; strontium sulphate	-	-	-	-	315	-
Cellophane	-	-	-	-	-	413
Cellulose, acetate; acetylcellulose	-	-	-	-	-	413
Cement	-	269	-	-	-	-
Cement, fireproof	-	269	-	-	-	-
Cement, Portland	-	-	284	-	-	-
CeO_2 ⟶ cerium oxide	259	269	-	307	-	-
Ceramics, fireproof 9606	259	-	-	306	315	-
Ceramics, fireproof 9608	259	-	-	307	315	-
Ceramics, mullite, heat insulating	-	269	-	-	-	-
Ceramics, sillimanite with 25% pottery clay	-	269	-	-	-	-

	ε_T	$\varepsilon_{T,n}$	$\varepsilon_{\lambda,n}$	$\varepsilon_{\lambda=0.65n}$	$r_{\lambda,n}$	$\tau_{\lambda,n}$
Cerargyrite; horn silver; silver chloride; AgCl	-	-	-	-	-	358
Cerium oxide; CeO_2	259	269	-	307	-	-
Cerusite; lead carbonate; $PbCO_3$	-	-	-	-	315	-
Cesium bromide; CsBr	-	-	-	-	315	358
Cesium iodide; CsJ	-	-	-	-	316	359
$CF_2{=}CF_2$ ⟶ polytetrafluoro-ethylene; teflon	-	380	-	-	408	417
$(CF_2{-}CFCl)_n$ ⟶ polychloro-trifluoroethylen	-	-	-	-	-	415
C_6H_6 ⟶ benzene; benzol	-	-	-	-	-	412
$CH_3\cdot CH_2\cdot CH_2\cdot CH_2Cl$ ⟶ butyl chloride	-	-	-	-	-	413
$(CH_3)_2CH_2\cdot CH_2\cdot OH$ ⟶ alcohol isobutyl	-	-	-	-	-	412
$CH_3\cdot CH_2\cdot CH_2\cdot CH_2\cdot OH$ ⟶ alcohol butyl; butanol	-	-	-	-	-	412
$(C_5H_9Cl)_n$ ⟶ rubber natural hydrochlorided; rubber hydrochloride	-	-	-	-	-	416
Chile nitrate ⟶ chile saltpetre; sodium nitrate; $NaNO_3$	-	269	-	-	-	359
Chile saltpetre; Chile nitrate; sodium nitrate; $NaNO_3$	-	269	-	-	-	359
China clay ⟶ porcelain clay; kaolin	-	272	-	-	-	-
Chlorophyll	-	-	-	-	-	413
Chrome yellow ⟶ lead chromate; lead chrome; $PbCrO_4$	261	-	-	-	322	-
$C_6H_5NH_2$ ⟶ aminobenzen; aniline	-	-	-	-	-	412
$C_{16}H_{18}N_3SCl$ ⟶ dye methylene blue	-	379	-	-	-	-
$C_{21}H_{18}O_5S$ ⟶ dye cresol red	-	379	-	-	-	-
Chromium; Cr	175	186	215	-	-	-
Chromium-nickel-zinc alloy; CrNi25Zn20	-	187	-	-	-	-

	ε_T	$\varepsilon_{T,n}$	$\varepsilon_{\lambda,n}$	$\varepsilon_{\lambda=0.65n}$	$r_{\lambda,n}$	$\tau_{\lambda,n}$
Copper trioxide; Cu_2O_3	259	-	-	-	-	-
Copper oxide ⟶ cuprite; red copper ore; ruby copper; Cu_2O	-	270	-	-	-	-
Copper sulphate; cupric sulphate; $CuSO_4$	-	270	-	-	-	-
Copper-tin alloy (red bronze)	176	188	-	-	248	-
Copper-zinc alloy (brass)	176	188	217	-	-	-
Copper zinc alloy; CuZn30, M70* Alloy 260 (brass without alloing agent)	-	189	-	-	-	-
Coralum, fr. ⟶ corundum, sintered	-	-	286	-	-	-
Cork	-	270	-	-	-	-
Corn	-	-	-	-	391	-
Corning 1723 ⟶ glass, ampouled, aluminium	-	-	287	-	318	-
Corning 7740 ⟶ glass, borosilicate, Corning 7740	-	-	287	-	319	-
Corning 7900 ⟶ glass, borosilicate, Corning 7900	-	271	-	-	-	-
Corning 7940 ⟶ glass, borosilicate, Corning 7940	-	271	-	-	319	-
Corundum; Al_2O_3	259	-	-	-	-	-
Corundum, sintered; aluminium oxide; Al_2O_3	-	270	285	-	317	359
Corundum sintered; fr. Coralum	-	-	286	-	-	-
CoSi ⟶ cobalt silicid	-	-	-	307	-	-
Cotton	-	379	-	-	391	-
Cr ⟶ chromium	175	186	215	-	-	-
CrN ⟶ chromium nitride	-	-	-	307	-	-
Cr_2N ⟶ chromium nitride	-	-	-	307	-	-
CrNi25Zn20 ⟶ chromium-nickel-zinc alloy	-	187	-	-	-	-
Cr_2O_3 ⟶ chromium oxide	-	269	285	-	316	-

	ε_T	$\varepsilon_{T,n}$	$\varepsilon_{\lambda,n}$	$\varepsilon_{\lambda=0.65n}$	$r_{\lambda,n}$	$\tau_{\lambda,n}$
CrSi → chromium silicide	-	-	-	307	-	-
Cr_3Si → chromium silicide	-	-	285	-	316	-
$CrSi_2$ → chromium silicide	-	-	-	307	-	-
Cr_3Si_2 → chromium silicide	-	-	-	307	-	-
CsBr → caesium bromide	-	-	-	-	315	358
CsJ → caesium iodide	-	-	-	-	316	359
CTL-9L-LD → resin phenolic, CTL-9L-LD	-	380	383	-	-	-
Cu → copper	176	187	216	-	248	-
CuBr → copper bromide	-	-	-	-	-	359
$Cu(C_2H_3O_2)_2$ → cuprite acetate	-	270	-	-	-	-
CuCl → copper chloride	-	-	-	-	-	359
CuO → copper oxide	259	270	-	-	317	-
Cu_2O → cuprite; copper oxide; red copper ore; ruby copper	-	270	-	-	-	-
Cu_2O_3 → copper trioxide	259	-	-	-	-	-
Cupric sulphate → copper sulphate; $CuSO_4$	-	270	-	-	-	-
Cuprite acetate; $Cu(C_2H_3O_2)_2$	-	270	-	-	-	-
Cuprite; copper oxide; red copper ore; ruby copper; Cu_2O	-	270	-	-	-	-
$CuSO_4$ → copper sulphate; cupric sulphate	-	270	-	-	-	-
CuZn30 → copper-zinc alloy; M70* Alloy 260 (brass without alloing agent)	-	189	-	-	-	-
Cyanite aluminosilicate → disthene; $Al_2(SiO_4)O$	-	-	-	-	317	-
Delux → paint "Delux"	-	-	-	-	400	-
Diamond; C	-	-	-	-	-	360
Diatomaceous earth powder	-	270	-	-	-	-
Disthene; cyanite aluminosilicate; $Al_2(SiO_4)O$	-	-	-	-	317	-

	ε_T	$\varepsilon_{T,n}$	$\varepsilon_{\lambda,n}$	$\varepsilon_{\lambda=0.65n}$	$r_{\lambda,n}$	$\tau_{\lambda,n}$
Dolomite; calcium-magnesium carbonate; $CaMg(CO_3)_2$	-	-	-	-	317	-
Duckweed	-	-	-	-	391	-
Duralumin ⟶ aluminium-copper-magnesium alloy; AlCu2MgNi1Si; PA29N*; H18	-	185	-	-	-	-
Duralumin ⟶ aluminium-copper-magnesium alloy; AlCu4MgA; PA25*	-	185	-	-	-	-
Duralumin ⟶ aluminium-copper-magnesium alloy; AlCu4Mg1; PA7N*; 2024	-	185	-	-	-	-
Dye cresol red; $C_{21}H_{18}O_5S$	-	379	-	-	-	-
Dye methylene blue; $C_{16}H_{18}N_3SCl$	-	379	-	-	-	-
Ebonite; hard rubber	377	379	-	-	-	-
ЭИ435** ⟶ nickel-chromium-iron alloy; nimonic 75	-	200	-	-	-	-
ЭИ481** ⟶ nickel-chromium-iron alloy; nimonic	-	201	-	-	-	-
ЭИ607** ⟶ nickel-chromium-iron alloy; nimonic 95	-	201	-	242	-	-
Electrographite; Acheson graphite; carbon; C	259	270	286	307	-	-
Elm	-	-	-	-	392	-
En 60 ⟶ steel, high chromium, stainless, H17*	-	210	-	-	-	-
En 56A ⟶ steel chromium, stainless; 1H13*	-	209	-	-	-	-
Enamel	-	-	-	-	392	-
Enamel, cellulose	377	379	-	-	-	-
Enamel, chlorinated	377	-	-	-	-	-
Enamel, epoxy	377	-	-	-	392	-
Enamel, for radiators	-	379	-	-	-	-
Enamel, oleoresinous, waterproof	377	379	-	-	-	-
Enamel, pigmented aluminium	-	379	-	-	-	-

	ε_T	$\varepsilon_{T,n}$	$\varepsilon_{\lambda,n}$	$\varepsilon_{\lambda=0.65n}$	$r_{\lambda,n}$	$\tau_{\lambda,n}$
Enamel, phthalic, for common purpose	377	379	-	-	-	-
Enamel, phthalic, oilproof	377	379	-	-	-	-
Enamel, phthalic, resisting in tropic	377	-	-	-	-	-
Enamel, phthalated, tropical oilproof	-	379	-	-	-	-
Enamel, polyurethan	377	-	-	-	-	-
Enamel, polyvinyle	377	-	-	-	-	-
Enamel, silicone resin	-	-	-	-	392	-
Enamel, white	-	379	-	-	-	-
Er ⟶ erbium	-	-	-	240	-	-
Erbium; Er	-	-	-	240	-	-
Erbium oxide; Er_2O_3	-	-	-	307	-	-
Er_2O_3 ⟶ erbium oxide	-	-	-	307	-	-
Eu ⟶ europium	-	-	-	-	248	-
Europium; Eu	-	-	-	-	248	-
Fallow	-	-	-	-	392	-
Fe ⟶ iron	176	189	219	240	250	-
$FeCO_3$ ⟶ iron carbonate; ironstone; siderite; spathic iron	-	-	-	-	330	-
Fe_2O_3 ⟶ bloodstone; haematite; iron oxide	260	271	289	-	321	-
$FeSO_4$ ⟶ iron sulphate	-	272	-	-	-	-
Flagstone povement	-	-	-	-	317	-
Fluorite; calcium fluoride; CaF_2	-	270	-	-	318	360
GaAs ⟶ gallium arsenide	-	-	-	-	318	361
Gadolinium; Gd	-	-	-	-	248	-
Gadolinium boride; GdB	-	270	-	-	-	-
Gadolinium oxide; Gd_2O_3	-	-	-	307	-	-
Gadolinium silicide; $GdSi_2$	-	-	-	307	-	-
Galena; galenite; lead glance; lead sulphide; PbS	-	-	286	-	318	-
Galenite ⟶ galena; lead glance; lead sulphide; PbS	-	-	-	-	318	-

	ε_T	$\varepsilon_{T,n}$	$\varepsilon_{\lambda,n}$	$\varepsilon_{\lambda=0.65n}$	$r_{\lambda,n}$	$\tau_{\lambda,n}$
Gallium antimonide; GaSb	-	-	-	-	318	361
Gallium arsenide; GaAs	-	-	-	-	318	361
Gallium nitride; GaN	-	-	-	307	-	-
Gallium phosphide; GaP	-	-	-	-	318	361
GaN ⟶ gallium nitride	-	-	-	307	-	-
GaP ⟶ gallium phosphide	-	-	-	-	318	361
GaSb ⟶ gallium antimonide	-	-	-	-	318	361
Gd ⟶ gadolinium	-	-	-	-	248	-
GdB ⟶ gadolinium boride	-	270	-	-	-	-
Gd_2O_3 ⟶ gadolinium oxide	-	-	-	307	-	-
$GdSi_2$ ⟶ gadolinium silicide	-	-	-	307	-	-
Ge ⟶ germanium	259	-	286	-	318	361
Germanium; Ge	259	-	286	-	318	361
Glass	260	270	287	-	-	-
Glass, ampouled, aluminium	-	270	-	-	-	-
Glass, ampouled, aluminium-Corning 1723	-	-	287	-	318	-
Glass, arsenic, sulphide; As_2S_3	-	-	-	-	319	361
Glass, borosilicate-Corning 7740	-	-	287	-	319	-
Glass, borosilicate-Corning 7900	-	271	-	-	-	-
Glass, borosilicate-Corning 7940	-	271	-	-	319	-
Glass, borosilicate-Pitsburgh 3255	-	271	-	-	-	-
Glass, borosilicate-PPG 3235	-	-	287	-	319	-
Glass, borosilicate-Pyrex	-	271	-	-	-	-
Glass, borosilicate-Pyrex 774	-	-	287	-	319	-
Glass, borosilicate-Vicor	-	-	287	-	319	-
Glass, bottle, dark green	-	-	288	-	319	-
Glass, bottle, light green	-	-	288	-	319	-
Glass, BS-37A	-	-	-	-	-	362
Glass, IR-11	-	-	-	-	-	362
Glass, lead, 21% PbO	-	-	288	-	-	-
Glass, lead, 30% PbO	-	-	288	-	-	-
Glass, lead, 35% PbO	-	-	288	-	-	-
Glass, lead, 50% PbO	-	-	288	-	-	-
Glass, lead, 67% PbO	-	-	289	-	-	-

	ε_T	$\varepsilon_{T,n}$	$\varepsilon_{\lambda,n}$	$\varepsilon_{\lambda=0.65n}$	$r_{\lambda,n}$	$\tau_{\lambda,n}$
Glass, lead, oscilograph	-	-	288	-	320	-
Glass, metallized electricity conductive; Battele 54726	-	271	289	-	320	-
Glass, metallized electricity conductive; LOF 81E 19778	-	271	289	-	320	-
Glass, metallized electricity conductive; LOF PB 19195	-	271	289	-	320	-
Glass, organic; methacrylate polymethyl	-	-	383	-	-	414
Glass, selenium arsenide; (SeAs)	-	-	-	-	-	362
Glass, soda, lime	-	271	-	-	-	-
Globar, am. ⟶ silite, globar, am.	-	-	295	-	-	-
Gold; Au	176	189	217	240	249	-
Granite	-	271	-	-	-	-
Granite-embankment	-	-	-	-	320	-
Graphite; C	-	-	-	307	320	-
Graphite, polycrystalline; carbon; C	260	-	289	308	320	-
Grass	-	-	-	-	392	-
Grease	377	-	-	-	-	-
Gypsum; hydrated calcium sulphate; $CaSO_4 2H_2O$	260	271	-	-	-	362
Gypsum mortar	-	-	-	-	321	-
H 15* ⟶ steel, chromium; stainless; 51430	-	210	-	-	-	-
H 17 ⟶ steel, high chromium, stainless	-	210	-	-	-	-
H 18* ⟶ aluminium-copper-magnesium alloy; AlCu2Mg2Ni1Si; duralumin	-	185	-	-	-	-
H 21 ⟶ steel, chromium-tungsten-nickel-vanadium; WWN 1*	-	210	-	-	-	-
H 28 ⟶ steel, high chromium, stainless	181	211	228	-	-	-

	ε_T	$\varepsilon_{T,n}$	$\varepsilon_{\lambda,n}$	$\varepsilon_{\lambda=0.65n}$	$r_{\lambda,n}$	$\tau_{\lambda,n}$
Haematite; bloodstone; iron oxide; Fe_2O_3	260	271	289	-	321	-
Hafnium; Hf	176	189	218	240	249	-
Hafnium bromide; HfB	-	271	-	-	-	-
Hafnium carbide; HfC	-	272	-	309	-	-
Hafnium nitride; HfN	-	272	290	-	-	-
Hafnium oxide; HfO_2	261	272	-	309	-	-
Hard rubber ⟶ ebonite	377	379	-	-	-	-
Hastelloy B ⟶ nickel-molybdenum-iron alloy	-	202	223	242	-	-
Hastelloy C ⟶ nickel-molybdenum-chromium alloy	-	202	223	242	-	-
Hay	-	-	-	-	393	-
HCl ⟶ hydrochloric acid; muriatic acid	-	-	290	-	-	-
Heather	-	-	-	-	393	-
Hf ⟶ hafnium	176	189	218	240	249	-
HF ⟶ hydrogen fluoride	-	-	290	-	-	-
HfB ⟶ hafnium bromide	-	271	-	-	-	-
HfC ⟶ hafnium carbide	-	272	-	309	-	-
HfN ⟶ hafnium nitride	-	272	290	-	-	-
HfO_2 ⟶ hafnium oxide	261	272	-	309	-	-
Hg ⟶ mercury	177	190	-	-	250	-
H 5M* ⟶ steel, chromium-molybdenum, heat-resistant; 51501	-	206	-	-	-	-
H 17J5* ⟶ steel, chromium-aluminium, heat-resistant; baidonal	-	204	-	-	-	-
H 17N5M3* ⟶ steel, chromium-nickel-molybdenum, stainless	180	-	-	243	-	-
H 17N9J** ⟶ steel, chromium-nickel-aluminium, stainless	179	-	226	-	-	-
H 18N9* ⟶ steel, chromium-nickel, stainless; 302; 30302	-	206	226	-	-	-

	ε_T	$\varepsilon_{T,n}$	$\varepsilon_{\lambda,n}$	$\varepsilon_{\lambda=0.65n}$	$r_{\lambda,n}$	$\tau_{\lambda,n}$
H 25N20* ⟶ steel, chromium-nickel, stainless; 310; 30310	180	-	-	-	-	-
H_2O ⟶ water	261	277	299	-	338	373
H_2O ⟶ water vapour	-	277	300	-	-	-
Horn silver ⟶ cerargyrite; silver chloride; AgCl	-	-	-	-	-	358
H 25T* ⟶ steel, high chromium, stainless	-	210	-	-	-	-
Hydrated calcium sulphate ⟶ gypsum; $CaSO_4 \cdot 2H_2O$	260	271	-	-	-	362
Hydrochloric acid; muriatic acid; HCl	-	-	290	-	-	-
Hydrogen fluoride; HF	-	-	290	-	-	-
Hydrogen oxide ⟶ water; H_2O	261	277	-	-	-	-
Hydrogen oxide ⟶ water vapour; H_2O	-	277	-	-	-	-
Ice	-	272	-	-	-	-
InAs ⟶ indium arsenide	-	-	-	-	322	-
Inconel ⟶ nickel-chromium-iron alloy	-	193	-	-	-	-
Inconel B ⟶ nickel-chromium-iron alloy	-	193	-	-	-	-
Inconel X ⟶ nickel-chromium-iron alloy	-	193	222	242	-	-
Indium antimonide; InSb	-	-	291	-	321	362
Indium arsenide; InAs	-	-	-	-	322	-
Indium phosphide; InP	-	-	-	-	322	-
InP ⟶ indium phosphide	-	-	-	-	322	-
InSb ⟶ indium antimonide	-	-	291	-	321	362
Ir ⟶ iridium	176	-	218	240	250	-
IR-11 ⟶ glass, IR-11	-	-	-	-	-	362
IR-11 ⟶ potassium aluminate, IR-11	-	-	-	-	-	365
Iridium; Ir	176	-	218	240	250	-
Iron; Fe	176	189	219	240	250	-

	ε_T	$\varepsilon_{T,n}$	$\varepsilon_{\lambda,n}$	$\varepsilon_{\lambda=0.65n}$	$r_{\lambda,n}$	$\tau_{\lambda,n}$
Iron carbonate ⟶ $FeCO_3$; ironstone; siderite; spathic iron	-	-	-	-	330	-
Iron-cobalt alloy	-	-	-	240	-	-
Iron oxide ⟶ haematite; Fe_2O_3; bloodstone	260	271	289	-	321	-
Ironstone ⟶ iron carbonate; siderite; spathic iron; Fe_2O_3	-	-	-	-	330	-
Iron sulphate; $FeSO_4$	-	272	-	-	-	-
Iron-tungsten alloy	177	-	-	241	-	-
Juniper	-	-	-	-	394	-
$K(AlSi_3O_8)$ ⟶ alumino-silicate potassium	-	-	-	-	312	-
Kanthal Al ⟶ steel, chromium-aluminium, heat-resistant	-	205	-	-	-	-
Kanthal DSD ⟶ steel, chromium-aluminium, heat-resistant	-	205	-	-	-	-
Kaolin; porcelain clay; China clay	-	272	-	-	-	-
KBr ⟶ potassium bromide	-	-	-	-	325	365
KCl ⟶ potassium chloride	-	-	-	-	325	365
K_2CO_3 ⟶ potassium carbonate	-	274	-	-	-	-
KH_2PO_4 ⟶ potassium hydrogen phosphate	-	-	-	-	-	365
KJ ⟶ potassium iodide	-	-	-	-	325	365
KRS-6 ⟶ thallium bromide-thallium chloride mixture	-	-	-	-	336	372
KRS-5 ⟶ thallium bromide-thallium iodide mixture	-	-	-	-	336	372
KRS-13 ⟶ silver bromide-silver chloride mixture	-	-	-	-	-	370

	ε_T	$\varepsilon_{T,n}$	$\varepsilon_{\lambda,n}$	$\varepsilon_{\lambda=0.65n}$	$r_{\lambda,n}$	$\tau_{\lambda,n}$
Lacquer	377	379	-	-	-	-
Lacquer, aluminium pigmented	377	379	-	-	-	-
Lacquer, nitrocellulose	377	-	-	-	-	-
Lacquer, phenol-formaldehyde, plastic filled with wood flour	-	380	-	-	-	-
Lacquer, stoving	-	380	-	-	-	-
Lacquer vinyl	-	-	-	-	-	414
Lantanum bromide	-	272	-	-	-	-
Larch	-	-	-	-	394	-
Lard	-	-	-	-	-	414
Lead, Pb	177	190	-	-	-	-
Lead acetate; $Pb(C_2H_5O_2)_2$	-	272	-	-	-	-
Lead carbonate ⟶ cerussite; $PbCO_3$	-	-	-	-	315	-
Lead chloride; $PbCl_2$	-	-	-	-	321	363
Lead chromate (chrome yellow); lead chrome; $PbCrO_4$	261	-	-	-	322	-
Lead chrome ⟶ lead chromate (chrome yellow); $PbCrO_4$	261	-	-	-	322	-
Lead fluoride; PbF_2	-	-	-	-	-	363
Lead glance ⟶ galena; lead sulphide; galenite; PbS	-	-	286	-	318	-
Lead oxide ⟶ minium; red lead; Pb_3O_4	-	273	-	-	-	-
Lead oxide, yellow; PbO	-	272	-	-	322	-
Lead selenide; PbSe	-	-	292	-	322	-
Lead sulphide ⟶ galena; lead glance; PbS	-	-	286	-	318	-
Lead sulphide; $PbSO_4$	-	272	286	-	318	-
Leather	-	380	-	-	-	-
Lichen	-	-	-	-	394	-
Li_2CO_3 ⟶ Lithium carbonate	-	273	-	-	-	-
LiF ⟶ lithium fluoride	-	-	-	-	322	363
Lime-tree	-	-	-	-	394	-
Lime; calcium hydrate; calcium hydroxide; $Ca(OH)_2$	-	272	-	-	-	-

	ε_T	$\varepsilon_{T,n}$	$\varepsilon_{\lambda,n}$	$\varepsilon_{\lambda=0.65n}$	$r_{\lambda,n}$	$\tau_{\lambda,n}$
Lime burnt; quicklime; calcium oxide; CaO	-	272	-	-	322	-
Lime mortar	261	273	-	-	-	-
Lime plaster	261	273	-	-	-	-
Limestone; $CaCO_3$	-	273	292	-	-	-
Linoleum	-	380	-	-	395	-
Lithium carbonate; Li_2CO_3	-	273	-	-	-	-
Lithium fluoride; LiF	-	-	-	-	322	363
Loam from the bottom of the drain	-	-	-	-	323	-
LOF 81E 19778 → glass metallized electricity conductive	-	271	289	-	320	-
LOF PB 19195 → glass metallized electricity conductive	-	271	289	-	320	-
Lu → lutetium	-	-	-	-	250	-
Lu_2O_3 → lutetium oxide	-	-	-	309	-	-
Lutetium, Lu	-	-	-	-	250	-
Lutetium oxide, Lu_2O_3	-	-	-	309	-	-
M 70* → copper-zinc alloy; CuZn30; Alloy 260 (brass without alloing agent)	-	189	-	-	-	-
Magnesite; magnesium carbonate; $MgCO_3$	261	273	292	-	323	-
Magnesium; Mg	-	190	-	241	250	-
Magnesium carbonate → magnesite; $MgCO_3$	261	273	292	-	323	-
Magnesium cyanogen; magnesium dicyan	-	-	292	-	-	-
Magnesium dicyan → magnesium cyanogen	-	-	292	-	-	-
Magnesium fluoride; MgF_2	-	-	-	-	323	363
Magnesium-iron silicate → olivine; $(Mg, Fe)_2SiO_4$	-	-	-	-	324	-

	ε_T	$\varepsilon_{T,n}$	$\varepsilon_{\lambda,n}$	$\varepsilon_{\lambda=0.65n}$	$r_{\lambda,n}$	$\tau_{\lambda,n}$
Magnesium oxide → MgO; periclase	-	273	294	310	324	364
Magnesium silicide; Mg_2Si	-	-	-	309	-	-
Maize	-	-	-	-	395	-
Manganese carbonate; $MnCO_3$	-	273	-	-	323	-
Manganese oxide; MnO	-	-	-	-	323	-
Manganese silicide; $MnSi_2$	-	-	-	309	-	-
Manganese silicide; Mn_2Si	-	-	-	309	-	-
Manganin → copper-manganese-nickel alloy	-	188	217	240	-	-
Marble, grey	-	273	-	-	-	-
Marble, white	261	-	-	-	-	-
Meadow	-	-	-	-	395	-
Mercury; Hg	177	190	-	-	250	-
Methacrylate polymethyl → glass, organic	-	-	383	-	-	414
Mg → magnesium	-	190	-	241	250	-
MgAl30 → aluminium-magnesium alloy	-	-	-	-	247	-
$MgAl_2O_4$ → spinel, magnesium-aluminate	-	-	-	-	-	370
$MgCO_3$ → magnesite; magnesium carbonate	261	273	292	-	323	-
MgF_2 → magnesium fluoride	-	-	-	-	323	363
$(MgFe)_2SiO_4$ → magnesium-iron silicate, olivine	-	-	-	-	324	-
MgO → magnesium oxide; periclase	-	273	294	310	324	364
$MgOAl_2O_3$ → spinel	-	-	-	-	-	
Mg_2Si → magnesium silicide	-	-	-	309	-	-
$Mg_3(Si_4O_{10})(OH)_2$ → talc	-	277	-	-	-	-
Mica	-	-	-	-	-	364
Millet	-	-	-	-	398	-
Minium; lead oxide; red lead; Pb_3O_4	-	273	-	-	-	
$MnCO_3$ → manganese carbonate	-	273	-	-	322	-
MnO → manganese oxide	-	-	-	-	322	-

	ε_T	$\varepsilon_{T,n}$	$\varepsilon_{\lambda,n}$	$\varepsilon_{\lambda=0.65n}$	$r_{\lambda,n}$	$\tau_{\lambda,n}$
$MnSi_2$ ⟶ manganese silicide, $MnSi_2$	-	-	-	309	-	-
Mn_2Si ⟶ manganese silicide, Mn_2Si	-	-	-	309	-	-
Mo ⟶ molybdenum	177	190	219	241	250	-
Molybdenum; Mo	177	190	219	241	250	-
Molybdenum anhydride ⟶ molybdenum oxide; molybdenum trioxide; MoO_3	-	-	-	-	323	-
Molybdenum oxide; molybdenum trioxide (molybdenum anhydride); MoO_3	-	-	-	-	323	-
Molybdenum-rhenium alloy; MoRe 8	177	-	-	-	-	-
Molybdenum-rhenium alloy; MoRe 20	177	-	-	-	-	-
Molybdenum-rhenium alloy; MoRe 47	177	-	-	-	-	-
Molybdenum silicide; $MoSi_2$	-	273	293	309	324	-
Molybdenum silicide; Mo_3Si	-	-	-	309	-	-
Molybdenum silicide; Mo_5Si_2	-	-	-	309	-	-
Molybdenum trioxide ⟶ molybdenum oxide; molybdenum anhydride; MoO_3	-	-	-	-	323	-
Molybdenum-tungsten alloy; MoW20	177	-	-	241	-	-
Molybdenum-tungsten alloy; MoW25	-	-	-	241	-	-
Molybdenum-tungsten alloy; MoW30	177	-	-	241	-	-
Molybdenum-tungsten alloy; MoW37	-	-	-	241	-	-
Molybdenum-tungsten alloy; MoW87; WMo13	-	-	-	241	-	-
Molybdenum-tungsten alloy; WMo13	-	-	-	241	-	-
MoNb10 ⟶ niobium-molybdenum alloy; MoNb10	-	-	223	-	-	-
MoNb20 ⟶ niobium-molybdenum alloy; MoNb20	-	-	223	-	-	-
MoNb30 ⟶ niobium-molybdenum alloy; MoNb30	-	-	223	-	-	-
MoNb40 ⟶ niobium-molybdenum alloy; MoNb40	-	-	223	-	-	-

	ε_T	$\varepsilon_{T,n}$	$\varepsilon_{\lambda,n}$	$\varepsilon_{\lambda=0.65n}$	$r_{\lambda,n}$	$\tau_{\lambda,n}$
MoNb50 → niobium-molybdenum alloy; MoNb50	-	-	223	-	-	-
Monel → nickel-copper-iron alloy	-	201	-	-	-	-
Monel K → nickel-copper-aluminium alloy	-	202	202	-	-	-
MoO_3 → molybdenum anhydride, molybdenum oxide; molybdenum trioxide	-	-	-	-	323	-
MoRe8 → molybdenum-rhenium alloy; MoRe8	177	-	-	-	-	-
MoRe20 → molybdenum-rhenium alloy; MoRe20	177	-	-	-	-	-
MoRe47 → molybdenum-rhenium alloy; MoRe47	177	-	-	-	-	-
$MoSi_2$ → molybdenum silicide, $MoSi_2$	-	273	293	309	324	-
Mo_3Si → molybdenum silicide, Mo_3Si	-	-	-	309	-	-
Mo_5Si_3 → molybdenum silicide, Mo_5Si_3	-	-	-	309	-	-
Moss	-	-	-	-	398	-
Mullite; $3Al_2O_3 \cdot 2SiO_2$	-	273	293	-	-	-
Muriatic acid → hydrochloric acid; HCl	-	-	290	-	-	-
MoW 20 → molybdenum-tungsten alloy; MoW 20	177	-	-	241	-	-
MoW 25 → molybdenum-tungsten alloy; MoW 25	-	-	-	241	-	-
MoW 30 → molybdenum-tungsten alloy, MoW 30	177	-	-	241	-	-
MoW 37 → molybdenum-tungsten alloy, MoW 37	-	-	-	241	-	-
MoW 87 → molybdenum-tungsten alloy, MoW 87	-	-	-	241	-	-

	ε_T	$\varepsilon_{T,n}$	$\varepsilon_{\lambda,n}$	$\varepsilon_{\lambda=0.65n}$	$r_{\lambda,n}$	$\tau_{\lambda,n}$
N-155 ⟶ steel stainless chromium-nickel, N-155	-	207	226	243	-	-
$Na(AlSiO_4)$ ⟶ nepheline, nephelite; sodium aluminosilicate	-	-	-	-	324	-
$Na(AlSi_3O_8)$ ⟶ albite; sodium feldspar	-	-	-	-	312	-
$Na_8(AlSiO_4)_6Cl$ ⟶ sodalite	-	-	-	-	331	-
Na_2C_2 ⟶ sodium carbide	-	275	-	-	-	-
NaCl ⟶ salt domestic; salt table; salt common; sodium chloride	-	-	-	-	328	367
$NaCO_3$ ⟶ soda ash; calcined soda; sodium carbonate	-	275	-	-	-	-
NaF ⟶ sodium fluoride	-	-	-	-	331	370
$(Na,K(Al_6Si_6\cdot O_{24})CO_3)2H_2O$ ⟶ cancrynite	-	-	-	-	315	-
$NaNO_3$ ⟶ chile saltpetre; chile nitrate; sodium nitrate	-	269	-	-	-	359
Nb ⟶ niobium	178	202	223	242	250	-
NbC ⟶ niobium carbide	261	-	293	309	-	-
NbMo10 ⟶ niobium-molybdenum alloy, NbMo10	-	-	223	-	-	-
NbMo20 ⟶ niobium-molybdenum alloy, NbMo20	-	-	223	-	-	-
NbMo30 ⟶ niobium-molybdenum alloy, NbMo30	-	-	223	-	-	-
NbMo40 ⟶ niobium-molybdenum alloy, NbMo40	-	-	223	-	-	-
NbMo50 ⟶ niobium-molybdenum alloy, NbMo50	-	-	223	-	-	-
NbMo60 ⟶ niobium-molybdenum alloy, NbMo60	-	-	223	-	-	-

	ε_T	$\varepsilon_{T,n}$	$\varepsilon_{\lambda,n}$	$\varepsilon_{\lambda=0.65n}$	$r_{\lambda,n}$	$\tau_{\lambda,n}$
Nickel; Ni	177	190	221	241	250	-
Nickel-chromium alloy; N80H20*; nichrome	-	191	-	-	-	-
Nickel-chromium-iron alloy; inconel	-	193	-	-	-	-
Nickel-chromium-iron alloy; inconel B	-	193	-	-	-	-
Nickel-chromium-iron alloy; inconel X	-	193	222	242	-	-
Nickel-chromium-iron alloy; nikrothal 20***	-	193	-	-	-	-
Nickel-chromium-iron alloy; nikrothal 30***	-	195	-	-	-	-
Nickel-chromium-iron alloy; nikrothal 40***	-	195	-	-	-	-
Nickel-chromium-iron alloy; nikrothal 60***	-	196	-	-	-	-
Nickel-chromium-iron alloy; nikrothal 70***	-	198	-	-	-	-
Nickel-chromium-iron alloy; nikrothal 80***	-	199	-	-	-	-
Nickel-chromium-iron alloy; nimonic 75; ЭИ 435 **	-	200	-	-	-	-
Nickel-chromium-iron alloy; nimonic; ЭИ481**	-	201	-	-	-	-
Nickel-chromium-iron alloy; nimonic 95; ЭИ607**	178	201	-	242	-	-
Nickel-copper-aluminium alloy; monel K	-	202	222	-	-	-
Nickel-copper-iron alloy; monel	-	201	-	-	-	-
Nickel-molybdenum-chromium alloy; hastelloy C	-	202	223	242	-	-
Nickel-molybdenum-iron alloy; hastelloy B	-	202	223	242	-	-

	ε_T	$\varepsilon_{T,n}$	$\varepsilon_{\lambda,n}$	$\varepsilon_{\lambda=0.65n}$	$r_{\lambda,n}$	$\tau_{\lambda,n}$
Nickel oxide; NiO	261	273	293	-	324	-
Nickel-rhenium alloy; NiRe8	178	-	-	-	-	-
Nickel-rhenium alloy; NiRe11	178	-	-	-	-	-
Nickel-rhenium alloy; NiRe16	178	-	-	-	-	-
Nickel-rhenium alloy; NiRe24	178	-	-	-	-	-
Nickel-chromium alloy; NiRe30	178	-	-	-	-	-
Nickel silicide; $NiSi_2$	-	-	-	309	-	-
Nickel sulphide; NiS	-	-	-	-	324	-
Nikrothal 20*** ⟶ nickel-chromium-iron alloy	-	193	-	-	-	-
Nikrothal 30*** ⟶ nickel-chromium-iron alloy	-	195	-	-	-	-
Nikrothal 40*** ⟶ nickel-chromium-iron alloy	-	195	-	-	-	-
Nikrothal 60*** ⟶ nickel-chromium-iron alloy	-	196	-	-	-	-
Nikrothal 70*** ⟶ nickel-chromium-iron alloy	-	198	-	-	-	-
Nikrothal 80*** ⟶ nickel-chromium-iron alloy	-	199	-	-	-	-
Nimonic ⟶ nickel-chromium-iron alloy; ЭИ 481**	-	201	-	-	-	-
Nimonic 75 ⟶ nickel-chromium-iron alloy; ЭИ 435**	-	200	-	-	-	-
Nimonic 95 ⟶ nickel-chromium-iron alloy; ЭИ 607**	178	201	-	242	-	-
NiO ⟶ nickel oxide	261	273	293	-	324	-
Niobium; Nb	178	202	223	242	250	-
Niobium carbide; NbC	261	-	293	309	-	-
Niobium-molybdenum alloy; NbMo10	-	-	223	-	-	-
Niobium-molybdenum alloy; NbMo20	-	-	223	-	-	-
Niobium-molybdenum alloy; NbMo30	-	-	223	-	-	-
Niobium-molybdenum alloy; NbMo40	-	-	223	-	-	-
Niobium-molybdenum alloy; NbMo50	-	-	223	-	-	-
Niobium-molybdenum alloy; NbMo60	-	-	223	-	-	-

	ε_T	$\varepsilon_{T,n}$	$\varepsilon_{\lambda,n}$	$\varepsilon_{\lambda=0.65n}$	$r_{\lambda,n}$	$\tau_{\lambda,n}$
Niobium-molybdenum alloy; NbMo70	-	-	224	-	-	-
Niobium-molybdenum alloy; NbMo80	-	-	224	-	-	-
Niobium-molybdenum alloy; NbMo90	-	-	224	-	-	-
Niobium nitride; NbN	-	-	-	310	-	-
Niobium nitride; Nb_2N	-	-	-	310	-	-
Niobium silicide; $NbSi_2$	-	-	-	310	-	-
Niobium-tungsten alloy; NbW10	178	-	-	242	-	-
Niobium-tungsten alloy; NbW15	179	-	-	-	-	-
NiRe8 ⟶ nickel-rhenium alloy; NiRe8	178	-	-	-	-	-
NiRe11 ⟶ nickel-rhenium alloy; NiRe11	178	-	-	-	-	-
NiRe16 ⟶ nickel-rhenium alloy, NiRe16	178	-	-	-	-	-
NiRe24 ⟶ nickel-rhenium alloy, NiRe24	178	-	-	-	-	-
NiRe30 ⟶ nickel-rhenium alloy, NiRe30	178	-	-	-	-	-
NiS ⟶ nickel sulphide	-	-	-	-	324	-
NiS_2 ⟶ nickel silicide	-	-	-	309	-	-
Nitrogen oxide; NO	-	-	293	-	-	-
NO ⟶ nitrogen oxide	-	-	293	-	-	-
Oak	-	-	-	-	399	-
Oats	-	-	-	-	399	-
Oil, machine	377	-	-	-	-	-
Oil, vegetable	-	380	-	-	-	-
Olivine; magnesium-iron silicate; $(Mg,Fe)_2SiO_4$	-	-	-	-	324	-
Os ⟶ osmium; Os	-	-	-	242	-	-
Osmium; Os	-	-	-	242	-	-
Oxyhaemoglobin	-	-	-	-	-	414

	ε_T	$\varepsilon_{T,n}$	$\varepsilon_{\lambda,n}$	$\varepsilon_{\lambda=0.65n}$	$r_{\lambda,n}$	$\tau_{\lambda,n}$
PA 25* ⟶ aluminium-copper magnesium alloy; AlCu4MgA (duralumin)	-	185	-	-	-	-
Paint	-	-	-	-	399	-
Paint, alkyd green; Alclad 2024	-	-	-	384	-	-
Paint, alkyd, with zinc chromate; Alclad 2024	-	-	-	384	-	-
Paint, aluminium pigmented	377	380	-	-	-	-
Paint aluminium pigmented 10% Al	-	380	-	-	-	-
Paint aluminium pigmented 26% Al	-	380	-	-	-	-
Paint, black	378	-	-	-	-	-
Paint, BS 381 C Nr 224	-	-	-	-	400	-
Paint, BS 381 C Nr 261	-	-	-	-	400	-
Paint, BS 381 Nr 298	-	-	-	-	400	-
Paint, Delux	-	-	-	-	400	-
Paint, minium	-	380	-	-	-	-
Paint, oil	378	380	-	-	-	-
Paint, oil primer	378	-	-	-	-	-
Paint, phthalic, chromate primer	378	-	-	-	-	-
Paint, phthalic primer	378	380	-	-	-	-
Paint, white	378	-	-	-	-	-
Paint with lithophone	-	-	-	-	400	-
Paint, zinc white pigmented	-	-	-	-	400	-
Palladium; Pd	179	202	224	242	251	-
PA7N* ⟶ aluminium-copper alloy; AlCu4Mg1; 2024 (duralumin)	-	185	-	-	-	-
PA11N* ⟶ aluminium-magnesium alloy; AlMg3*	-	-	214	-	-	-
PA25* ⟶ aluminium-copper-magnesium alloy; AlCu4MgA	-	185	-	-	-	-
PA29N* ⟶ aluminium-copper-magnesium alloy; AlCu2Mg2Ni1Si (duralumin)	-	185	-	-	-	-
Paper	378	380	-	-	400	-
Paper, roofing	378	380	-	-	400	-

	ε_T	$\varepsilon_{T,n}$	$\varepsilon_{\lambda,n}$	$\varepsilon_{\lambda=0.65n}$	$r_{\lambda,n}$	$\tau_{\lambda,n}$
Paraffin	-	-	-	-	-	415
Pasture	-	-	-	-	400	-
Pavement	-	-	-	-	324	-
Pb ⟶ lead	177	190	-	-	-	-
$Pb(C_2H_5O_2)_2$ ⟶ lead acetate	-	272	-	-	-	-
$PbCl_2$ ⟶ lead chloride	-	-	-	-	322	363
$PbCO_3$ ⟶ cerussite; lead carbonate	-	-	-	-	315	-
$PbCrO_4$ ⟶ chrome yellow; lead chromate; lead chrome	261	-	-	-	322	-
PbF_2 ⟶ lead fluoride	-	-	-	-	-	363
PbO ⟶ lead oxide, yellow	-	272	-	-	322	-
Pb_3O_4 ⟶ lead oxide; minium; red lead	-	273	-	-	-	-
PbS ⟶ galena; lead glance; lead sulphide	-	-	286	-	318	-
PbSe ⟶ lead selenide	-	-	292	-	322	-
$PbSO_4$ ⟶ lead sulphide	-	272	-	-	-	-
Pd ⟶ palladium	179	202	224	242	251	-
Pea	-	-	-	-	401	-
Peat	-	-	-	-	401	-
Periclase; magnesium oxide; MgO	-	273	294	310	324	364
PF resins ⟶ resin phenolformaldehyde	-	-	-	-	-	416
PH15-7Mo ⟶ steel, chromium-nickel, stainless; PH15-7Mo	-	207	227	243	-	-
PH17-7 ⟶ steel, chromium-nickel, stainless, PH17-7	-	-	227	243	-	-
Pine	-	-	-	-	402	-
Pitsburgh 3255 ⟶ glass, borosilicate, Pitsburgh 3255	-	271	-	-	-	-
Plantain	-	-	-	-	403	-
Plaster	-	-	-	-	325	-
Platinum; Pt	179	202	224	-	251	-

	ε_T	$\varepsilon_{T,n}$	$\varepsilon_{\lambda,n}$	$\varepsilon_{\lambda=0.65n}$	$r_{\lambda,n}$	$\tau_{\lambda,n}$
Platinum black	-	274	-	-	325	-
Platinum-rhodium alloy; PtRh10	179	203	-	-	-	-
Polychlorotrifluoroethylene; $(CF_2\ CF^{Cl})_n$	-	-	-	-	-	415
Polyethylene	-	380	-	-	403	415
Polytetrafluoroethylene ⟶ teflon $CF_2=CF_2$	-	380	-	-	408	417
Polyvinyl, chloride	-	-	-	-	-	416
Porcelain	-	274	-	-	325	-
Porcelain clay ⟶ kaolin; China clay	-	272	-	-	-	-
Potassium aluminate; IR-11	-	-	-	-	-	365
Potassium bromide; KBr	-	-	-	-	325	365
Potassium carbonate; K_2CO_3	-	274	-	-	-	-
Potassium chloride; KCl	-	-	-	-	325	365
Potassium hydrogen phosphate; KH_2PO_4	-	-	-	-	-	365
Potassium iodide; KJ	-	-	-	-	325	365
Potato plant	-	-	-	-	403	-
PPG 3235 ⟶ glass borosilicate PPG 3235	-	-	287	-	319	-
Pt ⟶ platinum	179	202	224	-	251	-
PtRh 10 ⟶ platinum-rhodium alloy, PtRh 10	179	203	-	-	-	-
Pyrex ⟶ glass borosilicate, Pyrex	-	271	-	-	-	-
Pyrex 774 ⟶ glass borosilicate, Pyrex 774	-	-	287	-	319	-
Quartz crystalline; SiO_2	-	-	-	-	325	365
Quartz, melted	261	274	-	-	326	366
Quick lime ⟶ calcium oxide; CaO; lime burnt	-	272	-	-	322	-

	ε_T	$\varepsilon_{T,n}$	$\varepsilon_{\lambda,n}$	$\varepsilon_{\lambda=0.65n}$	$r_{\lambda,n}$	$\tau_{\lambda,n}$
RbCl ⟶ rubidium chloride	-	274	-	-	-	-
Re ⟶ rhenium	-	203	224	242	-	-
Red bronze ⟶ copper-tin alloy	-	188	-	-	248	-
Red copper ore ⟶ cuprite; copper oxide; Cu_2O; ruby copper	-	270	-	-	-	-
Red lead ⟶ minium; lead oxide; Pb_3O_4	-	273	-	-	-	-
$ReSi_2$ ⟶ rhenium silicide	-	-	-	310	-	-
Resin, epoxide ⟶ resin epoxy	-	380	383	-	-	-
Resin, epoxy; resin epoxide	-	380	383	-	-	-
Resin, glyptal	-	-	-	-	-	416
Resin, glyptal with lin seed oil	-	-	-	-	-	416
Resin, maleic	-	-	-	-	-	416
Resin, phenolformaldehyde; PF resin	-	-	-	-	-	416
Resin, phenolic CTL-9L-LD	-	380	383	-	-	-
Resin, phenolic, plypphen 5023	-	380	383	-	-	-
Resin, polyester, Polylite	-	380	383	-	404	-
Resin, polyester, Vibrin 135	-	380	383	-	-	-
Resin, polyester, Vibrin X-1068	-	380	-	-	404	-
Rh - rhodium	179	203	225	243	251	-
Rhenium; Re	-	203	224	242	-	-
Rhenium silicide; $ReSi_2$	-	-	-	310	-	-
Rhodium; Rh	179	203	225	243	251	-
Road in winter	-	-	-	-	326	-
Road, soil-surfaced	-	-	-	-	326	-
Rock	-	-	-	-	327	-
Ru ⟶ ruthenium	-	-	-	243	-	-
Rubber	378	380	-	-	404	-
Rubber hydrochloride ⟶ rubber natural hydrochlorided; $(C_5H_9Cl)_n$	-	-	-	-	-	416

	ε_T	$\varepsilon_{T,n}$	$\varepsilon_{\lambda,n}$	$\varepsilon_{\lambda=0.65n}$	$r_{\lambda,n}$	$\tau_{\lambda,n}$
Rubber, natural, hydrochlorided; rubber hydrochloride; $(C_5H_9Cl)_n$	-	-	-	-	-	416
Rubidium chloride; RbCl	-	274	-	-	-	-
Ruby copper ⟶ copper oxide; Cu_2O; cuprite; red copper ore	-	270	-	-	-	-
Ruthenium; Ru	-	-	-	243	-	-
Rye	-	-	-	-	404	-
S_8 ⟶ sulphur crystalline	-	-	-	-	335	371
Salt common ⟶ salt, domestic, salt table; sodium chloride; NaCl	-	-	-	-	328	367
Salt domestic; salt common; salt table; sodium chloride; NaCl	-	-	-	-	328	367
Salt table ⟶ salt common, salt domestic, sodium chloride, NaCl	-	-	-	-	328	367
Samarium boride	-	274	-	-	-	-
Samarium oxide; Sm_2O_3	-	274	-	310	-	-
Sand	-	274	-	-	328	-
Sand on the dunes	-	-	-	-	329	-
Sandstone	261	-	-	-	329	-
Sapphire; aluminium oxide; Al_2O_3	-	-	-	-	330	368
Sb ⟶ antimony	175	186	215	-	247	-
SnH_3 ⟶ stibine antimony hydride	-	-	-	-	-	370
Scandium nitride; ScN	-	-	-	310	-	-
Scheelite; calcium tungstate; $CaWO_4$	-	-	-	-	330	368
ScN ⟶ scandium nitride	-	-	-	310	-	-
Se ⟶ selenium	-	-	-	-	330	368
SeAs ⟶ glass selenium arsenide	-	--	-	-	-	362
Sedge	-	-	-	-	404	-
Selenium; Se	-	-	-	-	330	368

	ε_T	$\varepsilon_{T,n}$	$\varepsilon_{\lambda,n}$	$\varepsilon_{\lambda=0.65n}$	$r_{\lambda,n}$	$\tau_{\lambda,n}$
Shellac	378	380	-	-	-	-
Si ⟶ silicon	261	274	295	310	-	369
SiC ⟶ carborundum monocrystal	-	-	-	-	315	-
SiC ⟶ silicon carbide	-	-	-	-	-	369
SiC ⟶ silite	261	-	-	310	331	-
Siderite; iron carbonate; ironstone; spathic iron; $FeCO_3$	-	-	-	-	330	-
Silica monocrystalline; silicon dioxide; SiO_2	-	274	-	-	-	-
Silica, polycrystalline; silicon dioxide; SiO_2	-	274	294	-	330	-
Silicon; Si	261	274	295	310	330	369
Silicon carbide; SiC	-	-	-	-	-	369
Silicon dioxide ⟶ silica monocrystalline; SiO_2	-	274	-	-	-	-
Silicon dioxide ⟶ silica polycrystalline; SiO_2	-	274	294	-	330	-
Silicone	-	274	295	-	-	369
Silicon nitride; Si_3N_4	-	274	295	-	331	-
Silicon oxide sintered; SiO	-	-	295	-	-	-
Sillimanite	-	275	-	-	-	-
Silite	-	-	295	-	-	-
Silite; SiC	261	-	-	310	331	-
Silite; Carbolithe, fr.	-	-	295	-	-	-
Silite, globar, am.	-	-	295	-	-	-
Silk	-	380	-	-	-	417
Silver; Ag	179	203	225	243	251	256
Silver-aluminium alloy, AgAl 1	-	-	225	-	-	-
Silver-aluminium alloy, AgAl 1.7	-	-	225	-	-	-
Silver-aluminium alloy, AgAl 3.5	-	-	225	-	-	-
Silver-aluminium alloy, AgAl 12	-	-	225	-	-	-
Silver bromide-silver chloride mixture; KRS-13	-	-	-	-	-	370

	ε_T	$\varepsilon_{T,n}$	$\varepsilon_{\lambda,n}$	$\varepsilon_{\lambda=0.65n}$	$r_{\lambda,n}$	$\tau_{\lambda,n}$
Silver chloride ⟶ cerargyrite; AgCl; horn silver	-	-	-	-	-	358
Silver-zinc alloy, AgZn 7	-	-	225	-	-	-
Silver-zinc alloy, AgZn 37	-	-	225	-	-	-
Si_3N_4 ⟶ silicon nitride	-	274	295	-	331	-
SiO ⟶ silicon oxide, sintered	-	-	295	-	-	-
SiO_2 ⟶ silica monocrystalline, silicon dioxide	-	274	-	-	-	-
SiO_2 ⟶ silica polycrystalline, silicon dioxide	-	274	294	-	330	-
SiO_2 ⟶ quartz crystalline	-	-	-	-	325	365
SiO_2 ⟶ quartz, melted	-	-	-	-	-	366
Skin	-	-	-	-	405	417
Slag, blast furnace	-	275	-	-	-	-
Slag, boiler	-	275	-	-	-	-
Smithsonite; zinc carbonate; zinc spar; $ZnCO_3$	-	275	-	-	331	-
Sm_2O_3 ⟶ samarium oxide	-	274	-	310	-	-
Sn ⟶ tin	181	212	-	-	254	-
SnO_2 ⟶ cassiterite; tin oxide	-	269	-	-	-	358
Snow	-	275	-	-	331	-
SO_2 ⟶ sulphur dioxide	-	275	296	-	-	-
Soda ash; calcined soda; sodium carbonate; $NaCO_3$	-	275	-	-	-	-
Sodalite; $Na(AlSiO_4)_6Cl$	-	-	-	-	331	-
Sodium aluminosilicate ⟶ nepheline; nephelite; $Na(AlSiO_4)$	-	-	-	-	324	-
Sodium carbide; Na_2C_2	-	275	-	-	-	-
Sodium carbonate ⟶ calcined soda; soda ash; $NaCO_3$	-	275	-	-	-	-
Sodium chloride ⟶ salt domestic; salt common; salt table; NaCl	-	-	-	-	328	367
Sodium feldspar ⟶ albite; $Na(AlSi_3O_8)$	-	-	-	-	312	

	ε_T	$\varepsilon_{T,n}$	$\varepsilon_{\lambda,n}$	$\varepsilon_{\lambda=0.65n}$	$r_{\lambda,n}$	$\tau_{\lambda,n}$
Sodium fluoride; NaF	-	-	-	-	331	370
Sodium nitrate ⟶ chile nitrate; chile saltpetre; $NaNO_3$	-	269	-	-	-	359
Soil	-	-	296	-	332	-
Soil, clay	-	-	-	-	332	-
Soil lixivited - black earth	-	-	-	-	332	-
Soil marshy	-	-	-	-	332	-
Soil mugwort	-	-	-	-	333	-
Soil, sandy	-	-	296	-	333	-
Soil, strong black earth	-	-	-	-	333	-
Sphatic iron ⟶ iron carbonate; ironstone; siderite; $FeCO_3$	-	-	-	-	330	-
Sphalerite; zinc blende; zinc sulphide; ZnS	-	-	296	-	334	370
Spinel, magnesium-aluminate; $MgAl_2O_4$	-	-	-	-	-	370
Spruce	-	-	-	-	405	-
Sr ⟶ strontium	-	-	-	-	254	-
SrF_2 ⟶ strontium fluoride	-	-	-	-	335	371
$Sr(OH)_2$ ⟶ strontium hydrate; strontium hydroxide	-	275	-	-	-	-
$SrSO_4$ ⟶ celestite; strontium sulphate	-	-	-	-	315	-
$SrTiO_3$ ⟶ strontium titanate	-	-	-	-	335	371
Steel carbon	-	203	-	-	253	-
Steel carbon 10*; C1010	-	204	-	-	-	-
Steel carbon 15*; C1015	-	204	-	-	-	-
Steel carbon 20*; C1020	-	204	-	-	-	-
Steel carbon 50*; C1050	-	204	-	-	-	-
Steel, cast	-	204	-	-	-	-
Steel, chromium, stainless, H15*; 51430	-	210	-	-	-	-
Steel, chromium, stainless, 1H13*; En56A	-	209	-	-	-	-

	ε_T	$\varepsilon_{T,n}$	$\varepsilon_{\lambda,n}$	$\varepsilon_{\lambda=0.65n}$	$r_{\lambda,n}$	$\tau_{\lambda,n}$
Steel, chromium-aluminium, heat-resistant, H17J5; baidonal	-	204	-	-	-	-
Steel, chromium-aluminium, heat-resistant (kanthal AL)	-	205	-	-	-	-
Steel, chromium-aluminium, heat-resistant (kanthal DSD)	-	205	-	-	-	-
Steel, chromium-molybdenum, heat-resistant, H5M*; 51501	-	206	-	-	-	-
Steel, chromium-nickel, stainless, H18N9*; 302; 30302	-	206	226	-	-	-
Steel, chromium-nickel, stainless, H25N20*; 310; 30310	180	-	-	-	-	-
Steel, chromium-nickel, stainless, N155	-	207	226	243	-	-
Steel, chromium-nickel, stainless, PH15-7Mo	-	207	227	243	-	-
Steel, chromium-nickel, stainless, PH17-7Mo	-	-	227	243	-	-
Steel, chromium-nickel, stainless, OH18N10*, 801 Grade C	180	207	-	-	-	-
Steel, chromium-nickel, stainless, OH18N10T*; 821 Grade Ti	180	207	227	244	-	-
Steel, chromium-nickel, stainless, 1H18N9T*; 321; 30321	-	208	228	-	253	-
Steel, chromium-nickel, stainless, 303	-	209	-	-	-	-
Steel, chromium-nickel, stainless, 304	180	-	-	-	-	-
Steel, chromium-nickel-aluminium, stainless, H17N9J*	179	-	226	-	-	-
Steel, chromium-nickel-molybdenum, stainless, OOH17N14M*; AM-350	-	206	226	243	-	-
Steel, chromium-nickel-molybdenum, stainless, OH17N14M2*; 316LC	-	206	226	243	-	-

	ε_T	$\varepsilon_{T,n}$	$\varepsilon_{\lambda,n}$	$\varepsilon_{\lambda=0.65n}$	$r_{\lambda,n}$	$\tau_{\lambda,n}$
Steel, chromium-nickel-molybdenum, stainless, H17N5M3*	180	-	-	243	-	-
Steel, chromium-nickel-niobium, stainless, OH18N12Nb*; 821 Grade Nb	180	206	-	-	-	-
Steel, chromium-tungsten-nickel-vanadium WWN1*; H21	-	210	-	-	-	-
Steel, high carbon, cast	-	210	-	-	-	-
Steel, high chromium stainless, H17*; En60	-	210	-	-	-	-
Steel, high chromium, stainless, H25T*	-	210	-	-	-	-
Steel, high chromium, stainless, H28*	181	211	228	-	-	-
Steel, low carbon	-	211	-	-	-	-
Steppe	-	-	-	-	406	-
Stibine, antimony hydride; SbH_3	-	-	-	-	-	370
Stone	-	-	-	-	334	-
Stone, pebble	-	-	296	-	-	-
Stone lining white	-	-	-	-	334	-
Stone, cobble; cobble stone	-	-	-	-	335	-
Straw	-	-	-	-	407	-
Strontium; Sr	-	-	-	-	254	-
Strontium fluoride; SrF_2	-	-	-	-	335	371
Strontium hydrate ⟶ strontium hydroxide; $Sr(OH)_2$	-	275	-	-	-	-
Strontium hydroxide; strontium hydrate; $Sr(OH)_2$	-	275	-	-	-	-
Strontium sulphate ⟶ celestite; $SrSO_4$	-	-	-	-	315	-
Strontium titanate; $SrTiO_3$	-	-	-	-	335	371
Stubbly - field	-	-	-	-	408	-
Sulphur crystalline; S_8	-	-	-	-	335	371
Sulphur dioxide; SO_2	-	275	206	-	-	-
Sunflower	-	-	-	-	408	-

	ε_T	$\varepsilon_{T,n}$	$\varepsilon_{\lambda,n}$	$\varepsilon_{\lambda=0.65}$	$r_{\lambda,n}$	$\tau_{\lambda,n}$
Ta ⟶ tantalum	181	211	228	244	254	-
TaC ⟶ tantalum carbide	261	-	297	310	-	-
Talc; $Mg_3(Si_4O_{10})(OH)_2$	-	277	-	-	-	-
TaN ⟶ tantalum nitride	-	277	297	-	-	-
TaNb1 ⟶ tantalum-niobium alloy, TaNb1	-	-	-	244	-	-
TaNb10 ⟶ tantalum-niobium alloy, TaNb10	-	-	-	244	-	-
TaNb20 ⟶ tantalum-niobium alloy, TaNb20	-	-	-	244	-	-
TaNb30 ⟶ tantalum-niobium alloy, TaNb30	-	-	-	244	-	-
TaNb40 ⟶ tantalum-niobium alloy, TaNb40	-	-	-	244	-	-
TaNb50 ⟶ tantalum-niobium alloy, TaNb50	-	-	-	244	-	-
TaNb60 ⟶ tantalum-niobium alloy, TaNb60	-	-	-	244	-	-
TaNb95 ⟶ tantalum-niobium alloy, TaNb95	-	-	-	244	-	-
TaNb99 ⟶ tantalum-niobium alloy, TaNb99	-	-	-	244	-	-
Tantalum; Ta	181	211	228	244	254	-
Tantalum boride	-	-	-	310	-	-
Tantalum carbide; TaC	261	-	297	310	-	-
Tantalum-niobium alloy, NbTa1	-	-	-	244	-	-
Tantalum-niobium alloy, NbTa5	-	-	-	244	-	-
Tantalum-niobium alloy, NbTa40	-	-	-	244	-	-
Tantalum-niobium alloy, NbTa50	-	-	-	244	-	-
Tantalum-niobium alloy, TaNb 1	-	-	-	244	-	-
Tantalum-niobium alloy, TaNb10	-	-	-	244	-	-
Tantalum-niobium alloy, TaNb20	-	-	-	244	-	-
Tantalum-niobium alloy, TaNb30	-	-	-	244	-	-
Tantalum-niobium alloy, TaNb40	-	-	-	244	-	-
Tantalum-niobium alloy, TaNb50	-	-	-	244	-	-
Tantalum-niobium alloy, TaNb60	-	-	-	244	-	-

	ε_T	$\varepsilon_{T,n}$	$\varepsilon_{\lambda,n}$	$\varepsilon_{\lambda=0.65n}$	$r_{\lambda,n}$	$\tau_{\lambda,n}$
Tantalum-niobium alloy, TaNb95	-	-	-	244	-	-
Tantalum-niobium alloy, TaNb99	-	-	-	244	-	-
Tantalum nitride; TaN	-	277	297	-	-	-
Tantalum oxide; Ta_2O_5	261	-	297	-	335	-
Tantalum silicide; $TaSi_2$	-	-	297	310	335	-
Tantalum-tungsten alloy, TaW10	-	211	232	245	-	-
Tantalum-tungsten alloy, TaW30	-	212	232	245	-	-
Tantalum-tungsten alloy, TaW50	-	-	232	-	-	-
Tantalum-tungsten alloy, TaW90	-	-	232	-	-	-
Tantalum-tungsten alloy, WTa10	-	-	232	-	-	-
Tantalum-tungsten alloy, WTa50	-	-	232	-	-	-
Ta_2O_5 ⟶ tantalum oxide	261	-	297	-	335	-
Tar	-	380	-	-	-	-
$TaSi_2$ ⟶ tantalum silicide	-	-	297	310	335	-
TaW10 ⟶ tantalum-tungsten alloy, TaW10	-	211	232	245	-	-
TaW30 ⟶ tantalum-tungsten alloy, TaW30	-	212	232	245	-	-
TaW50 ⟶ tantalum-tungsten alloy, TaW50	-	-	232	-	-	-
TaW90 ⟶ tantalum-tungsten alloy, TaW90	-	-	232	-	-	-
Te ⟶ tellurium	-	-	-	-	335	372
Teflon; polytetrafluoroethylene; $CF_2=CF_2$	-	380	-	-	408	417
Tellurium; Te	-	-	-	-	335	372
Tellurium oxide; TeO_2	-	-	298	-	-	372
TeO_2 ⟶ tellurium oxide	-	-	298	-	-	372
Tetrachloromethane ⟶ carbon tetrachloride; CCl_4	-	-	-	-	-	413
Th ⟶ thorium	-	212	-	245	-	-
Thallium bromide; TlBr	-	-	-	-	335	373
Thallium, bromide-thallium chloride mixture; KRS-6	-	-	-	-	336	372

	ε_T	$\varepsilon_{T,n}$	$\varepsilon_{\lambda,n}$	$\varepsilon_{\lambda=0.65n}$	$r_{\lambda,n}$	$\tau_{\lambda,n}$
Thallium bromide-thallium iodide mixture; KRS-5	-	-	-	-	336	372
Thallium carbonate; Tl_2CO_3	-	277	-	-	-	-
Thallium chloride; TlCl	-	-	298	-	336	373
Thallium iodide; TlJ	-	-	-	-	336	-
$Th(NO_3)_4$ ⟶ thorium nitrate	-	277	-	-	-	-
ThO_2 ⟶ thorium oxide	-	277	-	310	336	-
Thorium; Th	-	212	-	245	-	-
Thorium nitrate; $Th(NO_3)_4$	-	277	-	-	-	-
Thorium oxide; ThO_2	-	277	-	310	336	-
Ti ⟶ titanum	181	212	232	245	254	-
TiC ⟶ titanicum carbide	261	-	-	310	-	-
Tile, concrete	-	277	298	-	-	-
Tile, red	-	277	-	-	336	-
Tile, silicate	-	277	-	-	-	-
Tile, white	-	-	-	-	336	-
Tile, white,glazed	-	-	-	-	336	-
Tile, white, unglazed	-	-	-	-	336	-
Tin; Sn	181	212	-	-	254	-
TiN ⟶ titanium nitride	-	277	-	310	-	-
Tin oxide ⟶ SnO_2 cassiterite	-	269	-	-	-	358
TiO_2 ⟶ titanium oxide	-	277	298	-	337	373
$TiSi_2$ ⟶ titanium silicide, $TiSi_2$	-	-	298	310	337	-
Ti_5Si_3 ⟶ titanium silicide, $TiSi_3$	-	-	-	310	337	-
Titanium; Ti	181	212	232	245	254	-
Titanium-aluminium alloy	181	-	-	245	-	-
Titanium-aluminium-vanadium alloy	-	-	233	-	-	-
Titanium carbide; TiC	261	-	-	310	-	-
Titanium nitride; TiN	-	277	-	310	-	-
Titanium oxide; TiO_2	-	277	298	-	337	373
Titanium silicide; $TiSi_2$	-	-	298	310	337	-
Titanium silicide; Ti_5Si_3	-	-	-	310	337	-
TlBr ⟶ thallium bromide	-	-	-	-	335	373
TlCl ⟶ thallium chloride	-	-	298	-	336	373
$TlCO_3$ ⟶ thallium carbonate	-	277	-	-	-	-

	ε_T	$\varepsilon_{T,n}$	$\varepsilon_{\lambda,n}$	$\varepsilon_{\lambda=0.65}$	$r_{\lambda,n}$	$\tau_{\lambda,n}$
TlJ ⟶ thallium iodide	-	-	-	-	336	-
Tomato	-	-	-	-	409	-
Topaz; $[Al(F,OH)_2SiO_4]$	-	-	-	-	337	-
Tungsten; W	181	212	233	-	255	-
Tungsten oxide; WO_3	-	277	-	-	-	-
Tungsten-rhenium alloy, WRe27	182	-	-	245	-	-
Tungsten silicide; W_3Si	-	-	-	310	-	-
Tungsten silicide; WSi_2	-	-	-	-	338	-
U ⟶ uranium	-	-	-	245	-	-
UC_2 ⟶ uranium carbide	261	-	-	310	-	-
UO_2 ⟶ uranium oxide	-	-	298	310	338	-
Uranium; U	-	-	-	245	-	-
Uranium carbide; UC_2	261	-	-	310	-	-
Uranium oxide; UO_2	-	-	298	310	338	-
V ⟶ vanadium	182	212	238	245	255	-
Vanavium; V	182	212	238	245	255	-
Vanadium nitride; V_3N	-	-	-	311	-	-
Varnish	-	381	-	-	-	-
Varnish, alkyd	-	381	-	-	-	-
Varnish, oil	378	381	-	-	-	-
Varnish, resin	378	381	-	-	-	-
Varnish, shellac	-	381	-	-	-	-
Varnish, spirit	378	381	-	-	-	-
Velvet black	378	-	-	-	-	-
Vetch	-	-	-	-	409	-
Vibrin 135 ⟶ resin polyester, Vibrin 135	-	380	383	-	-	-
Vibrin X-1068 ⟶ resin polyester, Vibrin X-1068	-	380	-	-	404	-
Vicor ⟶ glass, borosilicate, Vicor	-	-	287	-	319	-
V_3N ⟶ vanadium nitride	-	-	-	311	-	-

	ε_T	$\varepsilon_{T,n}$	$\varepsilon_{\lambda,n}$	$\varepsilon_{\lambda=0.65}$	$r_{\lambda,n}$	$\tau_{\lambda,n}$
W ⟶ tungsten	181	212	233	-	255	-
Water; hydrogen oxide; H_2O	261	277	299	-	338	373
Water vapour; hydrogen oxide; H_2O	-	277	300	-	-	-
Weeds	-	-	-	-	409	-
Wheat	-	-	-	-	409	-
Willow	-	-	-	-	410	-
Willow herb	-	-	-	-	410	-
WMo13 ⟶ molybdenum-tungsten alloy, WMo13	-	-	-	241	-	-
WO_3 ⟶ tungsten oxide	-	277	-	-	-	-
Wood	378	381	-	-	410	-
Wood, beech	-	382	-	-	-	-
Wood, oak	-	382	-	-	-	-
Wood, pine	-	-	-	-	411	-
Wood, spruce	-	382	-	-	-	-
Wood, walnut	-	382	-	-	-	-
Warmwood	-	-	-	-	411	-
WRe27 ⟶ tungsten-rhenium alloy, WRe27	182	-	-	245	-	-
WSi_2 ⟶ tungsten silicide, WSi_2	-	-	-	-	338	-
W_3Si ⟶ tungsten silicide, W_3Si	-	-	-	310	-	-
WTa10 ⟶ tantalum-tungsten alloy, WTa10	-	-	232	-	-	-
WTa50 ⟶ tantalum-tungsten alloy, WTa50	-	-	232	-	-	-
WWN 1* ⟶ steel chromium-tungsten-nickel-vanadium, H21	-	210	-	-	-	-
Y ⟶ yttrium	-	-	-	245	-	-
Yb ⟶ ytterbium	-	-	239	-	-	-
YB ⟶ yttrium boride	-	280	-	-	-	-
Yb_2O_3 ⟶ ytterbium oxide	-	-	-	-	-	374
YN ⟶ yttium nitride	-	-	-	311	-	-
Y_2O_3 ⟶ yttrium oxide	-	-	303	311	-	374

	ε_T	$\varepsilon_{T,n}$	$\varepsilon_{\lambda,n}$	$\varepsilon_{\lambda=0.65n}$	$r_{\lambda,n}$	$\tau_{\lambda,n}$
ZnS ⟶ sphalerite; zinc blende; zinc sulphide	-	-	296	-	334	370
Zr ⟶ zirconium	183	213	239	245	255	-
ZrB ⟶ zirconium boride, ZrB	-	280	-	311	-	-
ZrB_2 ⟶ zirconium boride, ZrB_2	262	-	303	311	-	-
ZrC ⟶ zirconium carbide	262	-	304	311	-	-
ZrN ⟶ zirconium nitride	-	280	304	311	-	-
ZrO_2 ⟶ zirconium oxide	-	280	304	-	340	-
$ZrSi_2$ ⟶ zirconium silicide	-	280	-	311	340	-
$ZrSiO_4$ ⟶ zirconium silicate	-	-	305	-	-	-
00H17N14M* ⟶ steel, chromium-nickel-molybdenum, stainless, AM-350	-	206	226	243	-	-
0H18N10* ⟶ steel, chromium-nickel, stainless; 801 Grade C	180	207	-	-	-	-
0H18N10T* ⟶ steel, chromium-nickel, stainless; 821 Grade Ti	180	207	227	244	-	-
0H17N14M2* ⟶ steel, chromium-nickel-molybdenum, stainless; 316L3	-	206	226	243	-	-
0H18N12Nb* ⟶ steel, chromium-nickel-niobium; stainless, 821 Grade Nb	180	206	-	-	-	-
1H13* ⟶ steel, chromium, stainless; EN56A	-	209	-	-	-	-
1H18N9T* ⟶ steel, chromium-nickel, stainless, 321; 30321	-	208	228	-	253	-
2Cu2Pa9* ⟶ aluminium-zinc alloy; 7075	-	186	-	-	-	-
$3Al_2O_3\ 2SiO_2$ ⟶ mullite	-	273	293	-		-
10* ⟶ steel, carbon	-	204	-	-	-	-
15* ⟶ steel, carbon	-	204	-	-	-	-
20* ⟶ steel, carbon	-	204	-	-	-	-

	ε_T	$\varepsilon_{T,n}$	$\varepsilon_{\lambda,n}$	$\varepsilon_{\lambda=0.65n}$	$r_{\lambda,n}$	$\tau_{\lambda,n}$
50* ⟶ steel, carbon	-	204	-	-	-	-
302 ⟶ steel, chromium-nickel, stainless, H18N9*; 30302	-	206	226	-	-	-
303 ⟶ steel, chromium-nickel, stainless	-	209	-	-	-	-
304 ⟶ steel, chromium-nickel, stainless	180	-	-	-	-	-
310 ⟶ steel, chromium-nickel, stainless; H25N20*; 30310	180	-	-	-	-	-
316LC ⟶ steel, chromium-nickel; molybdenum, stainless, OH17N14M2*	-	206	226	243	-	-
321 ⟶ steel, chromium-nickel, stainless, 1H18N9T*; 303021	-	208	228	-	253	-
801 Grade C ⟶ steel, chromium-nickel, stainless, OH18N10*	180	207	-	-	-	-
821 Grade Nb ⟶ steel, chromium-nickel-niobium, stainless; OH18N12Nb*	180	206	-	-	-	-
821 Grade Ti ⟶ steel, chromium-nickel, stainless; OH18N10T*	180	207	227	244	-	-
2024 ⟶ aluminium-copper-magnesium alloy; AlCu3Mg1; PA7N* (duralumin)	-	185	-	-	-	-
5023 ⟶ resin, phenolic, plyophen 5023	-	380	383	-	-	-
7075 ⟶ aluminium-zinc alloy; 2Cu2Pa9*	-	186	-	-	-	-
9606 ⟶ ceramics, fireproof*, 9606	259	-	-	306	315	-
9608 ⟶ ceramics, fireproof*, 9608	259	-	-	307	315	-
30302 ⟶ steel, chromium-nickel, stainless, H18N9*; 302	-	206	226	-	-	-